WATER REUSE

WATER REUSE

Edited by
E. JOE MIDDLEBROOKS

ANN ARBOR SCIENCE
PUBLISHERS INC /THE BUTTERWORTH GROUP

PREFACE

The need for wastewater reuse has been recognized for many years, but until the past decade, little effort was devoted to implementing direct reuse systems. Indirect reuse and agricultural application of waste waters in arid areas were the principal reuse programs employed in the past. Little effort was made to study these processes and the impact that reuse would have on the environment and public health. The recent edict by the U.S. Environmental Protection Agency, requiring that land application of waste water be considered as a feasible alternative in any system supported by federal funds, has regenerated interest in the old practice of wastewater farming. This resurgence in wastewater farming has generated numerous studies on performance, economics, toxics and the impact on ground water and health. The advent of direct recycle into potable water supplies has caused considerable controversy and has produced many useful studies. Of particular concern is the persistence of trace organic materials involved in indirect and direct reuse of treated wastewaters.

This book opens with a chapter that summarizes state-of-the-art in wastewater reuse; chapters presenting details of each major area of activity follow. Readers interested in an overview only will find the introductory chapter an excellent primer on wastewater reuse. Engineering design, plant performance, detailed data on trace organics, and the results of viral and microbial analyses are presented, along with the legal and institutional constraints associated with wastewater reuse. The entire field of wastewater reuse is discussed, including potable reuse. Social and esthetic aspects of wastewater reuse are considered, and detail is devoted to health and safety considerations in wastewater reuse. An analysis of risks associated with potable reuse of waste water is also presented.

The latest activities in all aspects of wastewater reuse are presented in this text by leading authorities in the field. The information is useful to consulting engineers, environmental scientists, administrators, attorneys, social scientists, regulatory personnel, politicians and academic institutions. Design data, case histories, performance data, monitoring information, health information, social implications, legal and organizational structures, and background information needed to analyze the desirability of water reuse are presented. The book is an excellent resource for any level of interest.

E. Joe Middlebrooks

E. Joe Middlebrooks occupies the Newman Chair of Natural Resources Engineering, Department of Agricultural Engineering, Clemson University, Clemson, South Carolina. He is former Dean of the College of Engineering, Utah State University at Logan. Holder of BCE and MSE degrees from the University of Florida at Gainesville, and a PhD from Mississippi State University, Joe Middlebrooks was subsequently a Special Postdoctoral Research Fellow sponsored by FWPCA at the University of California, Berkeley, where he advanced his earlier work on algal growth and eutrophication.

At Utah State University and the Utah Water Research Laboratory from 1970 to 1982, his activities centered around temperature and toxicity relationships, water and wastewater treatment process design, evaluation and modification, and algal growth and assessment of eutrophication control. Previously he taught at the University of Florida, University of Arizona in Tucson and Mississippi State University at State College. He was named Outstanding Educator of America in 1973.

His professional experience is extensive, including industry and government as well as the educational field. A registered Professional Engineer, he is a consultant to both engineering and industrial firms, as well as to the nation of Iraq, and is a member of the Editorial Advisory Board of Ann Arbor Science Publishers.

Past President of the Association of Environmental Engineering Professors and active in numerous other scientific and professional societies, Dr. Middlebrooks is author or co-author of more than 150 publications, including nine books, five of which were published by Ann Arbor Science.

CONTENTS

Section 2
Aquaculture and Wetlands

Section 3
Municipal and Industrial Reuse

Section 4
Viral and Bacterial Removal and Monitoring

Section 5
Health Effects and Persistent Compounds

CHAPTER 1

INTRODUCTION AND OVERVIEW

E. J. Middlebrooks
Department of Agricultural Engineering
Clemson University
Clemson, South Carolina

M. J. Humenick
College of Engineering
University of Texas
Austin, Texas

WASTEWATER RECLAMATION IN RETROSPECT

The history of wastewater reclamation for human consumption is very limited, and information has been accumulated for only a few years. Direct reuse of wastewater for human consumption in the United States first occurred at Chanute, Kansas, in 1956 and 1957 [1]. This experiment was brought on by a drought period that necessitated the introduction of sewage treatment plant effluent into the raw drinking water impoundment. The water that was produced by the conventional water treatment process was of poor quality and probably not used extensively for drinking [1].

In 1968, the city of Windhoek, Southwest Africa, installed a treatment facility that directly reroutes reclaimed wastewater in with the new supply [2]. The reclaimed water comprises 28.6% of the daily water supplied to the city, and the population of 50,000 people has been drinking reclaimed water since October 1968 [2]. All indications are that the water is of a very high quality.

Many other applications of reclaimed wastewater have been practiced in the United States since the early part of the twentieth century. Examples of this are the use of septic tank and Imhoff tank effluents being used to irrigate crops in California in 1906 [3]. However, the California State Board of Health did not adopt its initial regulations governing the use of sewage for irrigation purposes until 1918. These regulations prohibited the use of raw sewage, septic or Imhoff tank effluents or similar sewages (or water polluted by such sewages) for the irrigation of most garden crops to be eaten raw by humans. Crops that were to be eaten cooked could be irrigated, but only if done so at least 30 days before harvesting. Fruits, nut trees and melons that did not come in contact with the sewage were excluded from the 1918 regulations. These regulations were modified in 1933, then were eliminated by the passage of the Dickey Water Pollution Act of 1949, whereby it became necessary to gain permission from the State Regional Water Pollution Control Boards to reuse reclaimed wastewater. A 1967 revision of the Water Quality Control Act by the State Legislature added a chapter on wastewater reclamation and reuse to the Water Code. However, the State Board of Health has not defined the standards that would be required for the direct use of wastewater in the State of California. Essentially the same situation exists in most states.

It is well established that people have been drinking reclaimed sewage for years along the Ohio and Mississippi river system. For the past 100 years or more the water discharged upstream after dilution in the river has been used by the cities downstream, and there is little, if any, epidemiology data that indicate a significant health hazard has resulted from this indirect reuse of wastewater effluents.

Except in isolated cases, the acceptance of the direct reuse of reclaimed sewage effluents does not appear to be in the immediate plans of State Boards of Health and other public health officials. Their logic in refusing to approve the direct reuse of reclaimed wastewaters is based on the fact that many viruses had been found in wastewaters, including the polioviruses, Coxsackie viruses, ECHO viruses, reoviruses and adenoviruses. A summary of the types of viruses that have been isolated from wastewaters is shown in Table I. Although many diseases can be caused by these viruses the same diseases can be caused by many other organisms, making it impossible to incriminate specific viruses. Also, because these viruses are present in a community whether they occur in sewage or not, and are spread throughout the area by immune carriers, the disease can eventually erupt anyway.

Table I. Viruses Isolated from Wastewater [4-9]

Group	Type
Adenovirus	1, 2, 3, 5, 7, 9
Coxsackie virus	A-1, 2, 4, 5, 6, 9, 10, 13,18
	B-1, 2, 3, 4, 5, 6
Echo virus	1, 3, 4, 6, 7, 8, 9, 11, 12, 14, 19, 20, 21
Poliovirus	1, 2, 3
Reovirus	1, 3

Finding a pathogenic virus in wastewater is relatively easy; however, proving that wastewater has caused disease because of a particular virus is very difficult. Mosley [10] reviewed the literature in 1967 and concluded that polio may have been transmitted by sewage contamination, but that the only epidemics of a supposed viral disease that definitely can be linked to such contamination are those of infectious hepatitis. A similar conclusion was reached by Bancroft et al. [11] and Berger [12]. However, several investigators have shown that sewage is quite capable of spreading infectious hepatitis, and many outbreaks of gastroenteritis have been traced to sewage contamination without a definite etiologic agent, either bacterial or viral, being proven [13-16].

Although it is very difficult to prove that viral diseases have been spread by wastewater, the possibility remains that diseases can be spread in this manner. Most of the difficulty associated with trying to identify the cause of a disease results from an inability to measure effectively the viral organisms in the laboratory. There is evidence that wastewater has caused outbreaks of disease other than infectious hepatitis, and the fact that numerous pathogenic viruses are found in sewage certainly indicates that the possibility of contamination in reclaimed water does exist.

Acceptance of reclaimed water for drinking purposes will be determined by three factors: the scarcity of water; the economics of providing water from directly recycled, reclaimed sewage compared with other means; and the degree of safety that can be ascertained. The first point, scarcity of water, is certainly a major concern at many locations, and in such situations the economics do not become an issue at all. In areas where fresh water can be obtained, the alternatives become much more important. The safety of reclaimed water is something of a uncertainty if all the possibilities of contamination are considered; however, the evidence obtained in South Africa, and experiences at the Santee Recreation Project in California and the Indian Creek Reservoir near Lake Tahoe indicate that there is very little danger associated with a well-treated wastewater effluent [17,18]. The technology certainly exists to provide a quality of water that is acceptable for human consumption and meets all known public health standards. The only criterion that is not met by modern technology is the one of ultimate assurance of safety. This is a factor that is not built into our present water treatment systems, and there is no more certainty associated with the conventional water systems than can be assured with a well-engineered wastewater reclamation plant.

Use of reclaimed water for all purposes can easily be instituted at controlled and isolated locations such as military installations; however, in marginal cases where other sources of supply are available, it may be wise to consider uses other than direct human consumption. Possible uses for reclaimed water in the United States other than direct human consumption are listed in Table II.

Table II. Applications of Reclaimed Wastewater [19]

A. Irrigation
   1. Agricultural
   2. Landscape
      a. Parks
      b. Golf Courses
      c. Green Belts
      d. Lawns
B. Industrial
   1. Cooling Water
   2. Wash Water
   3. Boiler Feed
C. Recreational Impoundments
   1. Santee, California
   2. Indian Creek Reservoir
D. Ground-water Recharge
   1. Surface Spreading
   2. Direct Injection
E. Groundwater Protection
   1. Injection to Prevent Salt Water Intrusion
F. Oil Field Repressurization
G. Others
   1. Fire Protection
   2. Refuse Compaction
   3. Soil Compaction

The application of reclaimed water for irrigation has a long history, originating in England in the nineteenth century where sewage farms and sewage collection systems were first developed. In the United States, the practice of using raw sewage for irrigation was abandoned in the early part of this century, and by the 1930s a minimum of primary treatment was required before water could be used for agriculture. The use of reclaimed water in California for irrigating crops was discussed earlier, and this history is very similar to that in other areas of the United States.

Many areas have used reclaimed wastewater for landscape irrigation. The first in the state of California was at the McQueen Plant operated by the city of San Francisco. Secondary effluent was used to fill the Golden Gate Park ornamental lakes and to irrigate portions of the shrubbery and grass. Many cities in the United States currently use effluent from secondary treatment processes for this purpose.

Although little reclaimed water was used in industrial applications in the past, significant changes have taken place recently, as evidenced by the large increase in cooling-water reuse in many industries. As our water sources continue to diminish and effluent discharge standards become more stringent, industry will be forced to reuse its waste streams to meet its water requirements.

Many areas are contemplating the development of recreational impoundments similar to the one developed by the Santee Recreation Project in California in 1961 [17]. Water produced from reclaimed sewage at Santee has been used to fill a series of lakes for recreational purposes which include boating, picnicking and fishing. In 1965 these facilities were expanded to include swimming. The Santee Project has been one of the most significant in demonstrating the feasibility and social acceptability of using reclaimed water.

The replenishment of groundwater supplies has been used extensively and has been very successful in controlling salt-water intrusion along the coast of Southern California. Drinking water aquifers are recharged by either surface spreading or direct injection of secondary effluent. It appears that the quality of secondary treated effluent is adequate for the recharge of drinking water aquifers; however, the injection of high concentrations of salts and pollutants have caused taste and odor problems in some recharge areas. This indicates that a high degree of treatment is necessary to ensure the the existing water is not contaminated and made unfit to use. At present there is an ongoing project conducted by the Los Angeles Sanitation District to determine health effects associated with groundwater recharge in the area. Another use of reclaimed water has been the injection of reclaimed wastewaters for the purpose of oil field repressurization.

The quality requirements for a reclaimed water are related to its intended use. Thus, in isolated areas, it probably will be necessary to construct a single system that will produce a quality of effluent acceptable for human consumption. In certain instances it may be desirable to employ some of the less sophisticated treatment techniques to provide water for irrigation, recharge or other uses. An overview of many of the current techniques in wastewater reclamation follows. Details are given in other chapters of this book and in the *Proceedings of the 1979 Water Reuse Symposium* [20].

WASTEWATER RECLAMATION PROCESSES

Although some form of conventional wastewater treatment usually precedes advanced treatment processes, discussion of the conventional processes is omitted here because of the readily available descriptions and discussions in textbooks. However, it should be kept in mind that a partly or completely conventional process usually must precede if most of the advanced techniques are to be successful. The unit operations and processes being used for wastewater reclamation are considered individually, and a summary of various proposed processes is given in Table III.

Table III. Treatment Efficiencies and Costs for Various Treatment Processes

Process	SS (mg/l)	COD (mg/l)	BOD5 (mg/l)	Turb (JTU)	P (mg/l)	N (mg/l)	Cost ($/mgd)		Reference
							1	10	
Chemical Treatment (Solids Contact)	0-7	17	2.9-9.5	0.2-2.9	0.08-0.9	5-10.7	60-130	36-90	21-24
Granular or Mixed Media Filtration with Chemical	0-5	13-17	3.1-5.8	0.2-10	0.05	5	50	24-40[a]	25-29
Intermittent Sand Filt	0-3	–	3-5	–	–	–	30[b]	–	30
Sand Filtration with Chem Coag	2-5	26	2-3	0.8-3.5	0.15-1.5	–	–	–	31
Extended Aeration	35	–	20	–	–	–	–	200	28
Total Contaminant	–	–	–	–	–	–	40-55	–	32
Activated Carbon	0.6	10-12	1.0	1.2	0.3	6.6	160	27[c]-179[d]	21, 25, 27, 33-35
Reverse Osmosis	–	0-1.0	–	0.27	0.2[e]	3.5[f]	–	300-600	26, 35-37
Electrodialysis	–	8.0	–	–	7.8[e]	8.2[f]	200	70-500	21, 35, 36, 38
Ion Exchange	–	3.7	–	0.0	8.8[e]	4.2[f]	–	170-220[g]	35, 36, 39
Dissolved Air Flotation	6	3	–	–	–	–	110	40-70 60[h]	29
Microstraining	5	6	3	–	–	–	18	15	40-42
Ultrafiltration	0	20	<1	<0.1	–	–	–	–	43

[a] For 7.5 mgd plant.
[b] Amortization $90,000; 20 yr at 5%.
[c] For 7.5 mgd plant.
[d] Amortization $3,359,000; 20 yr at 5%.
[e] PO_4 only.
[f] NH_3-NO_3 only.
[g] Amortization $1,370,000; 20 yr at 5%.
[h] For 20 mgd plant.

Modification of Biological Processes

The removal of nitrogen and phosphorus by aerobic biological process modification has resulted in significant reductions in nutrients. The alternatives in biological treatment are summarized in Figure 1.

The quantity of nitrogen and phosphorus removed is directly related to the carbon available to the organisms. Assuming that biological mass can be expressed by the empirical formula $C_5H_7NO_2$, approximately 0.12 lb of nitrogen will be required per pound of biological mass produced. The phosphorus requirement is approximately one-fifth of the nitrogen. If it were possible to adjust the wastewater characteristics to provide a balanced substrate, all of the nutrients could be removed by biological growth. However, because of the large imbalance between carbon, nitrogen and phosphorus, only 30-40% of the nitrogen and 20-40% of the phosphorus is removed from domestic sewage by biological processes.

Studies have shown that by adjusting the characteristics of the raw waste it is possible to remove essentially all of the nitrogen from wastewaters [45]. But the imbalance with phosphorus is extremely difficult to overcome and soluble phosphorus remains in the effluent.

Anaerobic denitrification has been successful in removing nitrates from wastewaters and may serve as a means of reducing the load of nutrients applied to more sophisticated processes [46,47]. The process has been most successful with irrigation return flows, and a reasonable degree of success has been obtained with sewage [48]. Figure 2 shows three systems that incorporate nitrogen removal involving nitrification followed by denitrification [44].

Algae harvesting has been successful in removing nutrients when the substrate can be adjusted to provide optimum conditions for algal growth [52].

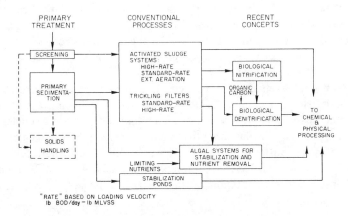

Figure 1. Alternatives in biological treatment [44].

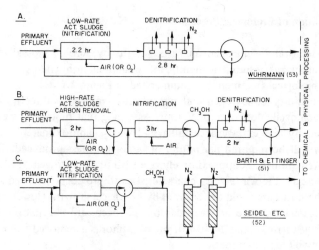

Figure 2. Biological systems incorporating nitrogen removal [44].

The same principles apply to algal growth as discussed above for aerobic processes. Biological processes have been criticized as contributing to the nutrient problems; however, the application of certain modifications can be beneficial.

Nitrogen Removal

Ammonia removal by air stripping is a modification of the process used in water treatment to remove undesirable gases. Ammonia stripping is easily controlled and, when it follows phosphorus removal by lime treatment at high pH values, it is economical. However, effective use of the process is limited to air temperatures above 32°F. If the additional cost of housing the process becomes necessary, the process has been found to be uneconomical [53].

Selective ion exchange is a very reliable ammonia removal process, but the process is relatively complex and the costs high. Breakpoint chlorination is simple and easy to control but costs are prohibitive except as a supplemental polishing process following other treatment processes. Breakpoint chlorination can destroy the buffering capacity of a water, which can be harmful in many natural environments. Extensive information is available in an EPA Technology Transfer Series publication on the design of nitrogen removal facilities [54].

Solids-Contact Treatment and Phosphorus Removal

Solids-contact chemical treatment is an effective way to precipitate inorganics and remove suspended solids. The basic idea of this method is to contact raw wastewater and fresh chemicals with previously formed solids. These solids provide a large surface on which chemical precipitation and/or flocculation can occur. The circulated solids also promote flocculation by offering a large quantity of floc to contact the newly-formed precipitate.

Removals of suspended matter, heavy metals and phosphorus have been achieved with this unit operation to meet present and anticipated water quality standards. Process design for phosphorus removal is available in an EPA Technology Transfer Series publication [55]. The system is often based on lime, although any coagulant such as alum or iron salts with or without polymer addition may be used with equal success. It may include a complete solids handling circuit for lime reuse (Figure 3). In the process, the wastewater stream enters a central reaction zone where fresh chemicals and previously precipitated solids are brought together. This combination is mixed as it is pumped up through the reaction zone by a recirculating turbine. Ample time is allowed in the larger outer reaction zone for coagulation and flocculation. There is a continuous flow of slurry from the reaction well to the clarification zone where liquid-solids separation occurs. Although the liquid is only in the unit from one to two hours, the solids detention time varies from six hours to as long as two weeks. Solids that settle to the bottom are mechanically raked to the central area. They are then reused by being pumped up through the recirculation drum, and the excess solids are removed to be thickened, dewatered and incinerated. For lime treatment, the incinerator also may act as a calciner. Normal sizing range is 0.25-1.0 gm per sq ft of clarification area. Effluent quality and costs for solids-contact treatment are summarized in Table III.

Granular Media Filtration

Granular media filtration is probably the most overlooked unit operation for upgrading wastewater quality. The filters are generally used for liquid-solids separation. The simple design and operation of this process make it applicable to wastewater streams containing up to 200 mg/ℓ suspended solids. Automation based on easily measured parameters gives minimum operation and maintenance costs (Figure 4).

Filtration rates can range from 25-50 gpm/ft^2 for coarse solids to 2-5 gpm/ft^2 for colloidal suspensions. The versatility of filter bed designs (media size and depths) is such that nearly any effluent quality can be achieved. Granular

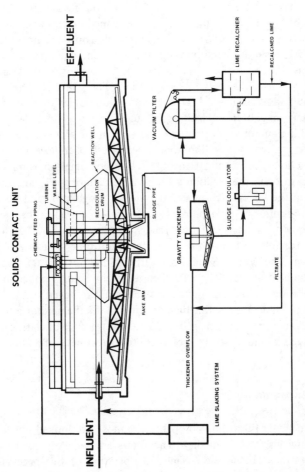

Figure 3. Complete lime chemical treatment facility (courtesy of Envirotech Corp.).

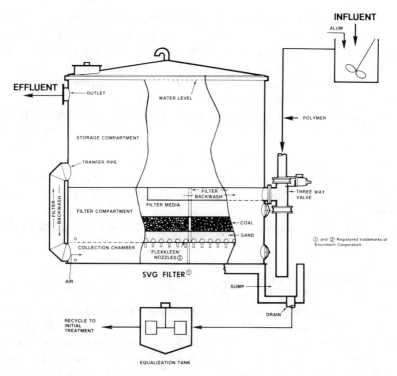

Figure 4. Typical flowsheet for SS and P removal by granular media filtration (courtesy of Envirotech Corp.).

media filters also can be used as chemical treatment units by adding inorganic coagulants (e.g., alum) and/or polymers to achieve both a high degree of liquid-solids separation and precipitation of other pollutants such as phosphorus. Process design information and methodology for municipal wastewater filtration is readily available [56,57].

Oil removal is another application for granular media filters. They can handle free oil and, to a limited extent, emulsified oil. Where both suspended matter and oil must be removed, this process has been very successful. Generally, some form of chemical addition is needed for good performance when treating water containing emulsified oil and solids [58].

The wastewater is passed through one or more layers of coal, sand, garnet, etc. As it flows through the granular material, suspended solids are removed by physical screening, sedimentation and interparticle action. The liquid headloss increases until the filter reaches its removal capacity, at which point the wastewater flow is stopped and the filter must be cleaned.

In wastewater applications, cleaning the media is difficult because of slime growths throughout the filter media. These growths and the collected solids can be removed by a combination of air-water backwash. Air is first added to expand the bed and scrub the media by particle collision. This is followed by hydraulic flushing to remove the waste materials and restratify the media according to specific gravity of the media particles.

Precoat filters, either semicontinuous pressure designs or continuous vacuum types, provide a similar "bed" for solids removal. Because the precoat material is wasted together with the filtered solids, the economics of this operation must be reviewed carefully to determine if they are more favorable than those of a corresponding granular filter. One distinct advantage of the precoat unit is that filtrate clarity is almost always less than 1 mg/ℓ of suspended solids.

Effluent quality and costs for granular media filtration are summarized in Table III.

Activated Carbon Adsorption

Soluble organic materials generally can be removed from wastewater by activated carbon adsorption; however, certain organic materials do not adsorb on activated carbon, or do so very slowly. The primary design criteria for the process are contact time and adsorption capacity. Laboratory or pilot tests will determine the contact time required to achieve the desired effluent quality. Such testing should be done carefully since an adsorption unit involves substantial capital investment and operating costs (Figure 5).

Use of granular activated carbon (GAC) is similar to that of granular media filters. Only the carbon regeneration requires special skills. In a typical system the wastewater stream is passed by gravity through a packed bed. The carbon depth, surface area and contactor geometry can be varied over a wide range to obtain particular process and operational results. For higher quality effluent, two or more countercurrent stages are used. Spent carbon is regenerated in a multiple hearth furnace, which volatilizes adsorbed organics.

Powdered activated carbon (PAC) is used as an alternative to GAC when organic removal is required intermittently or the carbon dose is very low. Application of PAC is achieved by mixing the carbon with the water and then removing the carbon by sedimentation. A typical application is the addition of PAC to the flocculation unit in a water treatment plant and removal of the carbon along with other flocculated materials in the following sedimentation basin. PAC generally is not reused because of problems with its regeneration. Effluent quality and costs for activated carbon adsorption are summarized in Table III.

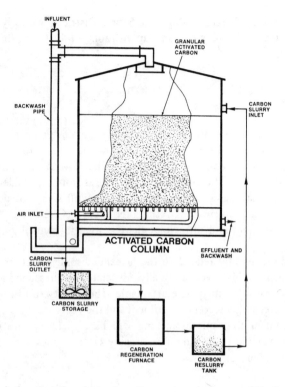

Figure 5. Typical activated carbon adsorption system (courtesy of Envirotech Corp.).

Biological Activated Carbon

A relatively new process which shows much promise for organics removal is biological activated carbon (BAC), a process applied to wastewater and potable water treatment. The basic system consists of ozonation followed by granular activated carbon adsorption. The activated carbon attains a biofilm which aids in the destruction of organic compounds.

Although the system is not entirely understood at this time, its success is explained [59,60] as follows: Preozonation is used to render refractory organics more amenable to adsorption and biological degradation along with supplying oxygen for the biofilm attached to the granular carbon. The role of the carbon is to adsorb organics, and the role of the biofilm is to oxidize both sorbed and dissolved organics as the water passes through the columns. The net result of the system is to remove organics, especially those of a refractory nature, and extend the life of the carbon. One U.S. application of this system is at the Cleveland Regional Sewer District's Westerly Wastewater Treatment Plant. There are many applications of BAC for drinking water treatment in Europe.

Microscreening

A microscreen is composed of a rotating drum covered with a screen type mat which acts as a simple filter to remove particles from waste streams (Figure 6). The units employ continuous gravity flow through the submerged portion of the drum. The cleared effluent passes from the inside of the drum to a clear-water storage chamber. Backwashing with sprays of product water is used to clean the screen. Backwashing requirements are usually less than 5% of the feed volume, and ultraviolet light is used to control growths on the screen.

Flow rate is directly related to the suspended solids concentration. At the low suspended solids levels normally associated with secondary effluents, loading rates of 4-7 gpm/ft^2 are attainable. The quality of effluent obtained from microscreening a secondary effluent and the costs are shown in Table III.

It is doubtful that microscreening will have many applications, except where it is desirable to provide a reasonably clear secondary effluent for irrigation of shrubbery or lawns.

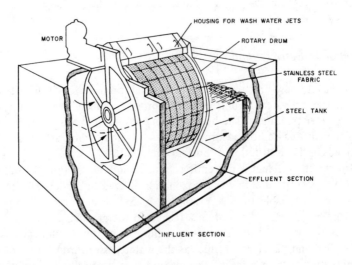

Figure 6. View of a microstrainer [61].

Membrane Processes

Electrodialysis and reverse osmosis are both membrane processes that have been utilized to remove nutrients and salts from wastewaters. Electrodialysis uses an induced electric current to separate the cationic and anionic components of a wastewater with membranes which forces ions to pass through from the dilute solution into the concentrated side. Ions are removed in direct proportion to their concentration, and approximately 30% of the ions are removed with one pass.

Reverse osmosis requires that the wastewater be pumped at pressures up to 350 psi through cellulose acetate membranes against the natural osmotic pressure. Sieving, surface tension and hydrogen bonding are some of the mechanisms thought to control ion removal in the cellulose acetate membrane.

Ultrafiltration employs a relatively coarse membrane at relatively low pressures (25-50 psi). Ultrafiltration is most successful in separating suspended solids and large molecule colloidal solids ranging in size from 0.002-10 microns. The removal process is achieved by physical screening through molecular sized openings rather than by molecular diffusion employed in reverse osmosis.

Comparison of Membrane Processes and Ion Exchange

Dryden [35] conducted a series of studies on the effectiveness and economics of the membrane processes and ion exchange. The characteristics of the influent and the effluent for all three processes are shown in Table IV. Electrodialysis appeared unable to remove soluble organic compounds as shown by the small percentage reduction in the COD. Nutrient and TDS removal were also lowest in the electrodialysis tests.

An economic analysis of the three processes showed the ion exchange to be most economical at all practical influent TDS concentrations. A comparison of the total costs and the influent TDS for the three processes is shown in Figure 7. The costs include pretreatment by carbon filtration but do not include brine disposal; therefore, if the plants were not located so that brine disposal could be handled by disposal to the ocean, the total costs for all of the processes would be much higher.

Although the reverse osmosis process was the most expensive, it was pointed out that this unit was far from being perfected. Developments since Dryden's study have shown that reverse osmosis is competitive. One or a combination of these three processes are employed in most water reclamation systems.

Table IV. Water Quality Characteristics of Membrane Processes and Ion Exchange Resins [35]

Characteristic	Reverse Osmosis[a]			Electrodialysis[b]			Ion Exchange[c]		
	Influent (mg/l)	Effluent (mg/l)	% Removal	Influent (mg/l)	Effluent (mg/l)	% Removal	Influent (mg/l)	Effluent (mg/l)	% Removal
COD	8.7	1.0	88.5	9.4	8.0	14.9	10.0	3.7	63.0
NH_3-N	10.1	1.1	89.2	9.0	5.1	43.4	15.6	3.8	75.6
NO_3-N	4.9	2.4	51.0	6.2	3.1	50.0	2.9	0.4	86.2
PO_4-P	10.9	0.2	98.4	10.1	7.8	22.8	8.8	0.1	98.8
TDS	750.0	59.0	92.1	705.0	465.0	34.0	611.0	81.0	86.7

[a] 15,000 gpd pilot plant; influent carbon treated; average based on 40 grab samples; average water recovery: 75%.
[b] 15,000 gpd pilot plant; influent carbon treated; averaged based on ten 24-hr composite samples; average water recovery: 83%.
[c] 5,000 gpd pilot plant; influent carbon treated; average based on twenty 16-hr composite samples; average water recovery: 87%.

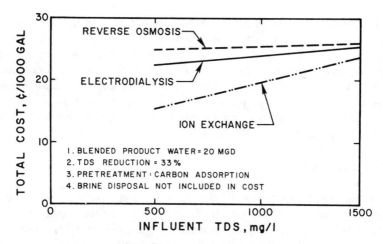

Figure 7. Comparison of demineralization costs [35].

Distillation

Distillation is accomplished by a variety of processes including flash distillation, differential distillation, and steam distillation. Because of the high costs of distillation, it is unlikely that this process will be competitive except in isolated cases. However, when it appears to be feasible or expeditious to incorporate distillation, it must be considered.

Foam Separation

Experimental results show that foam separation can be used to remove colloidal organic matter and dissolved portions by adsorption on the bubbles. However, it is unlikely that adequate surface-active agents will be found in many wastewaters to make this process effective. It is uneconomical and unwise to add surface-active agents to the wastewater to take advantage of this process. Also, these added ions must be removed later by another process.

Chemical Oxidation and Disinfection

There are potential applications for chemical treatment in water reuse systems. Historically, chemicals such as chlorine and its derivatives, permanganate and oxygen, have been used in water and wastewater treatment for disinfection and the removal of various inorganic and organic materials such as iron, manganese, sulfide, cyanide, taste, odor, and color-producing compounds [62].

The above applications are still valid for consideration in reuse systems; however, recent work on chemical oxidation of contaminants appears promising. Examples of these processes are breakpoint chlorination for nitrogen removal, ozonation, ozonation plus ultraviolet radiation, ozonation plus ultrasound for removal of organics and/or disinfection, and the use of chlorine dioxide in disinfection to prevent trihalomethane production in drinking water [63]. These processes are essentially in a developmental stage in the United States but the technology is improving.

A Comparison of Processes

In the preceding overview of the many processes available to produce a reusable product from wastewater, Table III presented costs and efficiency data. Further comparisons of reclamation processes have been made by Eliassen and Tchobanoglous [19] and these are shown in Table V.

An important component of treatment that is frequently omitted when reporting costs is the ultimate disposal of the contaminants removed from wastewater. Eliassen and Tchobanoglous [19] also prepared Table VI which summarizes some of the more important methods applicable to ultimate disposal of contaminants.

A true economic picture can be obtained only when all components of the system are considered. All candidate systems must be analyzed in detail to ensure reasonable cost and efficiency data.

WASTEWATER RECLAMATION SYSTEMS

There are many processes that can be combined to produce an end product from wastewater that would be acceptable for uses ranging from irrigation to human consumption. Many combinations have been tried on an experimental basis, and several systems have been constructed to produce water for human contact, such as the Santee Project, the South Lake Tahoe Project, and Windhoek Project. These systems and many others have been discussed by Kaufman and Humenick [44], Eliassen and Tchoganoglous [19] and in the Proceedings of the Water Reuse Symposium [20].

It was recognized as early as the 1930s that the most efficient biological systems did not produce an effluent acceptable for high quality reuse. Several systems were developed in an attempt to produce better effluents. Two examples of these attempts are shown in Figure 8. Another early attempt to improve effluents was the Guggenheim process shown in Figure 9.

Since these early developments, many modifications in existing systems and new systems have been developed. Equipment manufacturers have developed or purchased such proprietary devices and processes as those shown

Table V. Comparison of Reclamation Processes [19]

Process Description	Treatment Provided (substance removed)	Removal Efficiency (%)	Type of Waste to be Disposed of	Estimated Removal Costs		
				($/acre foot)	(¢/1000 gal)	(¢/100 m³)
Secondary Treatment (Raw Sewage)	Suspended Solids Organic Matter	90 SS 90 BOD	Sludge	10-40	3-12	80-320
Anaerobic Denitrification (Irrigation Wastewater)	Nitrate-Nitrogen	80-95	None	6-10	2-3	52-80
Algae Harvesting (Irrigation Wastewater)	Nitrate-Nitrogen	50-90	Algae	6-12	2-4	52-106
Ammonia Stripping (Raw Sewage)	Ammonia-Nitrogen	80-95	—	3-10	1-3	26-80
Ion Exchange (Filter Effluent)	Nitrogen Phosphorus	80-98	Liquid	55-82	17-25	450-660
Electrodialysis (Carbon Filter Effluent)	Dissolved Solids	10-40	Liquid	82-245	25-75	660-2000
Chemical Precipitation (A/S Effluent)	Phosphorus	88-95	Sludge	12-25	4-8	106-210
Carbon Adsorption (A/S or Sand Filter Effluent)	Organic Material	90-98	Liquid	12-25	4-8	106-210
Sorption (Sand Filter Effluent)	Phosphorus	90-98	Liquid	12-25	4-8	106-210
Filtration (A/S Effluent)	Suspended Solids	50-90	Liquid and Sludge	3-10	1-3	26-80
Reverse Osmosis (Carbon Filter Effluent)	Dissolved Solids	65-95	Liquid	75-125	25-40	220-1040
Distillation (Carbon Filter Effluent)	Dissolved Solids	90-98	Liquid	120-325	40-100	1040-4000
Foam Separation (A/S Effluent)	ABS	—	Liquid	3-6	1-2	26-52

Table VI. Ultimate Contaminant Disposal Methods and Costs [19]

Type of Waste	Method of Disposal	Cost[a]	Remarks
Liquid	Soil Spreading	$0.001-0.3/1000 gal	Provisions must be made to prevent ground-water contamination.
	Shallow Well Injection		
	Deep Well Injection	$13-27/1000 gal	Disposal site should be available (porous strata, natural or artificial cavities, etc.)
			Used as a wetting agent.
	Landfill	–	Provisions must be made to prevent ground-water contamination.
	Evaporation Pond	$8-64/1000 gal	Truck, rail hauling or pipeline needed to transport the wastes.
	Ocean Outfall	[b]	–
Sludge	Lagooning	–	Used as a wetting agent.
	Landfill	–	Depends on waste characteristics.
	Recovery of Products	$4-80/gpd[c]	Heat value may be recovered for use.
	Wet Combustion	$1.4-35/1000 gal[d]	
	Incineration	$10-100/gpd[c]	Concentration of sludge is needed. Ash must be disposed of.
		$5-55/1000 gal[d]	
Ash	Landfill	$0.75-1.0/ton	Depends on waste characteristics.
	Soil Conditioner	–	–

[a]Costs vary with plant capacity and land costs.
[b]$25/ft for 8 in. diam/pipe.
[c]Construction cost.
[d]Operation and maintenance cost.

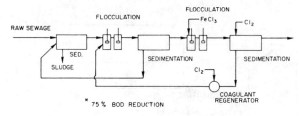

A - STEVENSON (CHLORINATED SLUDGE) PROCESS, 1933 (54)

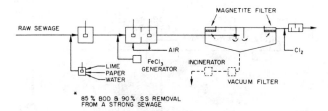

B - LAUGHLIN PROCESS, 1933 (55)

Figure 8. Wastewater reclamation systems (Stevenson Process and Laughlin Process) [44].

in Figures 10-12. The well-publicized Windhoek, South Lake Tahoe and Water Factory 21 systems are illustrated in Figures 13-15. A system developed and operated by the Environmental Protection Agency near the District of Columbia is shown in Figure 16.

These are only a few of the systems that have been tried, but they illustrate the many alternatives available for application to the wide range of problems that will be encountered in wastewater reclamation.

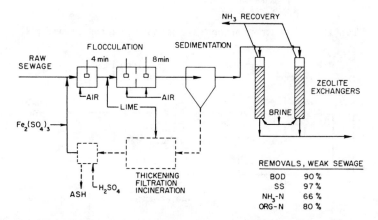

Figure 9. Wastewater reclamation systems (the Guggenheim process) [44].

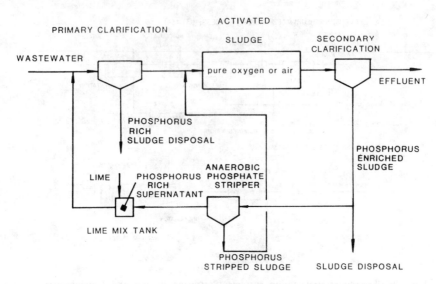

Figure 10. Water reclamation systems (Phostrip process, Union Carbide).

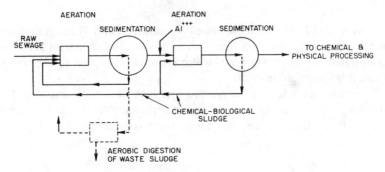

Figure 11. Wastewater reclamation systems (system Attisholz-Sweden) [44].

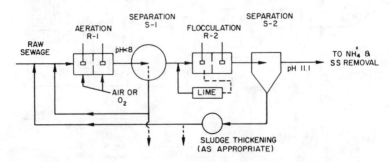

Figure 12. Wastewater reclamation systems (integrated biological-chemical process)[44].

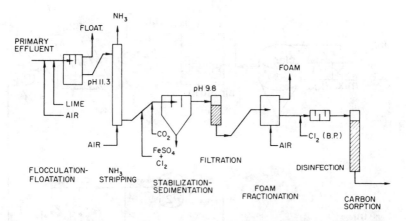

Figure 13. Wastewater reclamation systems (Windhoek, South Africa, process) [44].

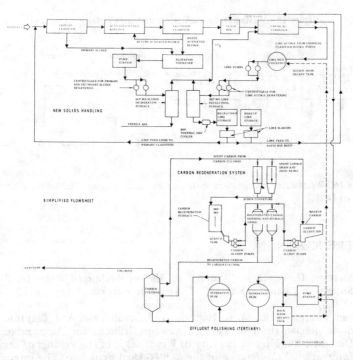

Figure 14. Wastewater reclamation systems (South Lake Tahoe Public Utility District).

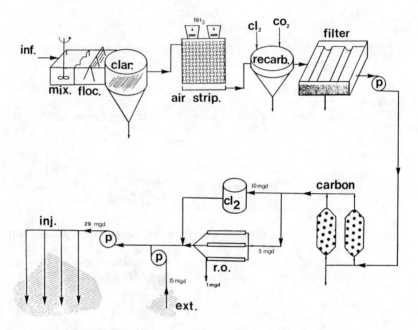

Figure 15. Wastewater reclamation systems (Water Factory 21).

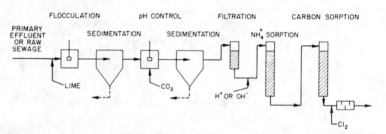

Figure 16. Wastewater reclamation systems (chemical systems: "EPA-Blue Plains" process) [44].

REFERENCES

1. Metzler, D. F., et al. "Emergency Use of Reclaimed Water for Potable Supply at Chanute, Kansas," *J. Am. Water Works Assoc.* 50:1021-1057 (1958).
2. van Vuuren, R. J., M. R. Henzen, G. J. Stander and A. J. Clayton. "Full Scale Reclamation of Purified Sewage Effluent for Augmentation of Domestic Supplies of the City of Windhoek," in *Proceedings of the Fifth International Conference of the IAWPC* (San Francisco: The International Association on Water Pollution Research, 1970).

3. Ward, P. C., and H. J. Ongerth. "Reclaimed Wastewater Quality—The Public Health Viewpoint," in *Proceedings of Wastewater Reclamation and Reuse Workshop* (Berkeley: The University of California, 1970).
4. Chin, T. D. Y., W. H. Mosley, S. Robinson and C. R. Gravelle. "Detection of Enteric Viruses in Sewage and Water. Relative Sensitivity of the Method," in *Transmission of Viruses by the Water Route,* Gerald Berg, Ed. (New York: Interscience Publishers, Div. of John Wiley and Sons, 1967).
5. Englund, B., R. E. Leach, B. Adame and R. Shiosaki. "Virologic Assessment of Sewage Treatment at Santee, California," in *Transmission of Viruses by the Water Route,* Gerald Berg, Ed. (New York: Interscience Publishers, Div. of John Wiley and Sons, 1967).
6. Foliguet, J. M., L. Schwartzbrod and O. G. Gaudin. "La pollution Virale des Eaux Usees, de Surface et d'Alimentation," *Bull. World Health Org.* 35:737-749 (1966).
7. Kalter, S. "Picornaviruses in Water," in *Transmission of Viruses by the Water Route,* Gerald Berg, Ed. (New York: Interscience Publishers, Div. of John Wiley and Sons, 1967).
8. Lund, E., and C. E. Hedstrom. "Recovery of Viruses from a Sewage Treatment Plant," in *Transmission of Viruses by the Water Route,* Gerald Berg, Ed. (New York: Interscience Publishers, Div. of John Wiley and Sons, 1967).
9. Metcalf, T. G., and W. C. Stiles. "Enteroviruses Within an Estuarine Environment," *Am. J. Epidemiol.* 88:379-391 (1968).
10. Mosley, J. W. "Transmission of Viral Diseases by Drinking Water," in *Transmission of Viruses by the Water Route,* Gerald Berg, Ed. (New York: Interscience Publishers, Div. of John Wiley and Sons, 1967).
11. Bancroft, P. M., W. E. Engelhard and C. A. Evans. "Poliomyelitis in Huskerville (Lincoln), Nebraska," *J. Am. Med. Assoc.* 164:836-847 (1957).
12. Berger, B. B., et al. "Engineering Evaluation of Virus Hazard in Water," *J. San. Eng. Div., Proc. Am. Soc. Civil Eng.* 96:111-150 (1970).
13. Green, D. M., S. S. Scott, D. A. E. Mowat, E. J. M. Shearer and J. M. Thomson. "Water-borne Outbreak of Viral Gastroenteritis and Sonne Dysentary," *J. Hyg., Cambridge* 66:383-392 (1968).
14. Lobel, H. O., A. L. Bisno, M. Goldfield and J. E. Prier. "A Waterborne Epidemic of Gastroenteritis with Secondary Person-to-Person Spread," *Am. J. Epidemiol.* 89:384-392 (1969).
15. Werner, S. B., P. H. Jones, W. M. McCormack, E. A. Ager and P. T. Holm. "Gastroenteritis Following Ingestion of Sewage Polluted Water: An Outbreak at a Logging Camp on the Olympic Peninsula," *Am. J. Epidemiol.* 89:277-285 (1969).
16. Whatley, T. R., G. W. Comstock, H. J. Garber and F. S. Sanchez. "A Waterborne Outbreak of Infectious Hepatitis in a Small Maryland Town," *Am. J. Epidemiol.* 87:138-147 (1968).
17. Merrill, J. C., Jr., W. F. Jopling, R. F. Bott, A. Katko and H. E. Pintler. "The Santee Recreation Project Santee, California, Final Report," U.S. Dept. of the Interior, Federal Water Pollution Control Adm., Cincinnati, Ohio (1967).
18. Porcella, D. B., P. H. McGauhey and G. L. Dugan. "Response to Tertiary Effluent in Indian Creek Reservoir," *J. Water Pollution Control Fed.* 44(11):2148-2161 (1972).

19. Eliassen, R., and G. Tchobanoglous. "Reclamation of Wastewater in the United States," paper presented at the International Symposium on Water for Tomorrow, Milan, Italy, 1968.

20. AWWA Research Foundation. *Proceedings of the Water Reuse Symposium* (Washington, D.C., March 25-30, 1979).

21. Smith, C. V., and D. DiGregorio. "Advanced Wastewater Treatment—An Overall Survey," *Chem. Eng.* (April 27, 1970), pp. 71-74.

22. Goodman, B. L., and K. A. Mikkelson. "Advanced Wastewater Treatment —Removing Phosphorus and Suspended Solids," *Chem. Eng.* (April 27, 1970), pp. 75-83.

23. Convey, J. J. "Treatment Techniques for Removing Phosphorus from Municipal Wastewaters," Environmental Protection Agency, Cincinnati, OH, Water Pollution Contract Research Series 17010 (January 29, 1970).

24. Kreissel, J. F. "Phosphorus Removal Practice," paper presented at the Sanitary Engineering Institute, University of Wisconsin, Madison, WI, March 9-10, 1971.

25. Culp, R. L. "Design for Media Filtration and Granular Carbon Treatment," paper presented at the Design Seminar for Wastewater Treatment Facilities, Environmental Protection Agency, Dallas, TX, July 27-28, 1971.

26. Kavanaugh, M. "Discussion—Granular Media Filtration in Wastewater Reuse," in *Proceedings of the Wastewater Reclamation and Reuse Workshop* (Lake Tahoe, CA, 1970).

27. Shell, G. L., and D. G. Burns. "Process Design to Meet New Water Quality Standards," paper presented at the Pennsylvania WPCA 42nd Annual Conference, State College, PA, August 5, 1970.

28. Shell, G. L. Personal communication (September 20, 1971).

29. Bare, R. B. "Algae Removal from Waste Stabilization Lagoon Effluents Utilizing Dissolved Air Flotation," Unpublished M.S. Thesis, Utah State University, Logan, UT (1970).

30. Imhoff, K., and G. M. Fair. *Sewage Treatment* (New York: John Wiley & Sons, Inc., 1940).

31. Hannah, S. A. "Chemical Precipitation," paper presented at the Advanced Waste Treatment and Water Reuse Symposium, FWQA, Dallas, TX, January 12-14, 1971.

32. Michel, R. L. "Construction Costs of Municipal Wastewater Treatment Plants (1967-69)," Office of Water Programs, Environmental Protection Agency, Washington, DC (May 1970).

33. Masse, A. N. "Use of Activated Carbon for Wastewater Treatment," paper presented at the Design Seminar for Wastewater Treatment, Dallas, TX, July 27-28, 1971.

34. English, J. N., A. N. Masse, C. W. Carry and J. B. Pitkin. "Removal of Organics from Wastewater by Activated Carbon," *Chem. Eng. Prog.* 67(107):147-153 (1970).

35. Dryden, F. D. "Mineral Removal by Ion Exchange, Reverse Osmosis and Electrodialysis," in *Proceeding of the Wastewater Reclamation and Reuse Workshop* (Lake Tahoe, CA, 1970).

36. Cohen, J. M. "Demineralization of Wastewaters," paper presented at the Advanced Waste Treatment and Water Reuse Symposium, FWQA, Dallas, TX, January 12-14, 1971.

37. Nusbaum, J., J. H. Sleigh, Jr. and S. S. Kremen. "Study and Experiments in Waste Water Reclamation by Reverse Osmosis," FWQA, Department

of the Interior, Washington, D.C., Water Pollution Control Series 17040 (May 1970).

38. Smith, D. J., and J. L. Eisenman. "Electrodialysis in Advanced Waste Treatment," FWPCA, Department of the Interior, Washington, DC, Publication WP-20, AWTR-18.

39. Abrams, I. M. "Absorbent Resins for Color and General Organic Removal," in *Proceedings of the Wastewater Reclamation and Reuse Workshop* (Berkeley: University of California, 1970).

40. Cohen, J. M. "Solids Removal Process," paper presented at the Advanced Waste Treatment Seminar, San Francisco, CA, October 28-29, 1970.

41. Advanced Waste Treatment Research Laboratory. "Current Status of Advanced Waste Treatment Processes," in *Suspended and Colloidal Solids Removal,* Report PPB 1703, Cincinnati, OH (July 1, 1970).

42. "Microstraining and Disinfection of Combined Sewer Overflows," Cochrane Division, Cane Co., Federal Water Quality Administration, U.S. Department of Interior, Water Pollution Control Research Series 11023 EVO 06/70.

43. Kugelman, I. J., W. A. Schwartz and J. M. Cohen. "Advanced Waste Treatment Plants for Treatment of Small Waste Flows," paper presented at the Advanced Waste Treatment and Water Reuse Symposium, FWQA, Dallas, TX, January 12-14, 1971.

44. Kaufman, W. J., and M. J. Humenick. "Candidate Process Systems—An Overview," in *Proceedings of the Wastewater Reclamation and Reuse Workshop* (Berkeley: University of California, June 25-27, 1970).

45. Metcalf and Eddy Engineers, "Report to the Spring Creek Committee Upon Disposal of Sewage and Industrial Wastes on the Spring Creek Watershed, Centre County, Pennsylvania," Boston, MA (1961).

46. McCarty, P. L. "Feasibility of the Denitrification Process for Removal of Nitrate Nitrogen from Agricultural Drainage Waters," California State Department of Water Resources, Report to the San Joaquin District (1966).

47. "Removal of Nitrogen from Tile Drainage—A Summary Report," Environmental Protection Agency and California Department of Water Resources (1971).

48. Barth, E. F. "Total Treatment Using Chemical and Physical Processes," in *Proceedings of Wastewater Reclamation and Reuse Workshop* (Berkeley: University of California, 1970).

49. Wuhrmann, K. "Nitrogen Removal in Sewage Treatment Processes," *Verh. Internat. Verein. Limnol.* 15:580-596 (1964).

50. Barth, E. F., and M. B. Ettinger. "Managing Continuous Flow Biological Denitrification," paper presented at the Seventh Industrial Water and Waste Conference, Texas Water Pollution Control Association, University of Texas, Austin, June 1-2, 1967.

51. Seidel, D. F., and R. W. Crites. "Evaluation of Denitrification Processes," *J. San. Eng. Div. Proc. of Am. Soc. Civil Eng.* 96(SA2):267-277 (1970).

52. Oswald, W. J., D. G. Crosby and C. G. Goluche. "Removal of Pesticides and Algal Growth Potential from San Joaquin Valley Drainage Waters—A Feasibility Study," California State Department of Water Resources, Report to the San Joaquin District (1964).

53. Culp, R. L., and G. L. Culp. *Advanced Wastewater Treatment* (New York: Van Nostrand Reinhold Company, 1971).

54. U.S. Environmental Protection Agency Technology Transfer. "Process Design Manual for Nitrogen Removal (October 1975).

55. U.S. Environmental Protection Agency Technology Transfer. "Process Design Manual for Phosphorus Removal," Publication EPA 625/1-76-001a (April 1976).

56. Cleasby, J. L., and E. R. Baumann. "Wastewater Filtration Design Considerations," Environmental Protection Agency Technology Transfer, EPA-625/4-74-007a (July 1974).

57. Fitzpatrick, J. A., and C. L. Swanson. "Evaluation of Full-Scale Tertiary Wastewater Filters," U.S. Environmental Protection Agency Report, EPA-600/2-80-005 (May 1980).

58. Humenick, M. J., and B. J. Davis. "High Rate Filtration of Refinery Oily Wastewater Emulsions," *J. Water Poll. Control Fed.* 50(8):1953-1965 (1978).

59. Rice, R. G., G. W. Miller, C. M. Robson and W. Kuhn. "Biological Activated Carbon," in *Carbon Adsorption* (Ann Arbor: Ann Arbor Science Publishers, Inc., 1978).

60. Weber, W. J., and W. E. Wing. "Integrated Biological and Physicochemical Treatment for Reclamation of Wastewater," *Prog. Water Technol.* 10(1-2/G.B.):217-233 (1978).

61. Truesdale, G. A., A. E. Birkbeck and D. Shaw. "A Critical Examination of Some Methods of Further Treatment of Effluents from Percolating Filters," paper presented at the Institute of Sewage Purification Conference, Paper No. 4, July 9-12, 1963.

62. Weber, W. J., Jr. "Oxidation Processes," in *Physicochemical Processes for Water Quality Control* (New York: Wiley-Interscience, 1972).

63. Ohio River Valley Water Sanitation Commission, "Water Treatment Process Modifications for Trihalomethane Control and Organic Substances in the Ohio River," U.S. Environmental Protection Agency Report, EPA-600/2-80-028 (March 1980).

64. Stevenson, R. A. "Clarification of Sewage with Ferric Chloride," *Water Works and Sewerage* 80:89-91 (1933).

65. Zack, S. I. "The Laughlin Process of Sewage Treatment," *Sew. Works J.* 5(3):458-481 (1933).

SECTION 1

FACTORS AFFECTING SYSTEM DESIGN

CHAPTER 2

LEGAL AND INSTITUTIONAL CONSIDERATIONS IN WATER REUSE

Loretta C. Lohman
  Industrial Economics Division
  University of Denver Research Institute
  Denver, Colorado 80208

INTRODUCTION

This chapter focuses on the legal and institutional considerations that affect a planned reuse of water. Planned reuse is defined as a deliberate second or repetitive use of water by the same or another user, withholding the water from a public water course and delaying its availability or reallocation. Unplanned reuse is a regular occurrence throughout the world as water is withdrawn, used, and then returned to a stream system and thus remains available for further use downstream.

The distinction between planned and unplanned reuse influences such factors as effects upon water quality, water quantity and the character of a stream. Planned reuse may encompass a deliberate alteration of the existing hydrologic cycle at the reuse site. Therefore, to envisage properly the legal and institutional factors that shape any deliberate reuse of water in the United States, it must be understood that the hydrologic connection of water within a river basin can involve states, regions and even the nation as the water flow crosses state or national boundaries. The flow of water, whether determined naturally or manipulated artificially, is subject to our particular federal system of government, and the power to manage water is divided or shared throughout the nation, among the states, within regions and in special entities such as Indian Tribes.

This diversity of governmental responsibilities and authorities makes it difficult to distinguish the legal from the institutional since they are intertwined in a complex set of interrelationships. Indeed, most legal-institutional studies of water management do not clearly distinguish the two, and generally settle into a primarily legal exposition. Further, the ordinary assumption that the term institution signifies a corporate body—a university, a government, a particular agency—makes understanding of the two elements even more difficult. The distinction should be made that it is the legal responsibilities and obligations of the various strata and agencies of government—federal, state, regional, local, and special entities—that provide the framework in which institutional activities take place.

The institutional is a less tangible but nevertheless clearly distinguishable process. It is the actual operating activities, as shaped by administrative policy and precedent, of a person or agency charged with application of the legally defined procedures germane to water reuse. This definition directs the focus onto the agencies and actors most concerned with the process of the allocation and uses of water and, consequently, onto those factors that can affect planned reuse of water. The following sections examine first the pervasive structure imposed by law, and then the diverse elements making up institutional process.

LEGAL CONSIDERATIONS

The legal factors applicable to water reuse generally flow downward from the federal level to state, region or town. This is true for most areas of water use. Even though water law, particularly that concerning water rights, displays complex interrelationships within and between each statutory level, it is important to recognize federal preeminence as it is exercised through policy guidelines, especially throughout the western states served by the Water and Power Resources Service (formerly the Bureau of Reclamation). Federal dominance has been assured through the gathering of financial resources and technical expertise into the national government and by the constitutional subsumption of state interests to the national welfare [1]. Although national in scope, the following overview of legal factors affecting water reuse will cite examples specific to the semi-arid Colorado River Basin where reuse projects are not only underway but have also encountered serious obstacles in existing law.

Elements of Federal Control

Planned water reuse may be subject to the rules, regulations and policies (either mandated or interpreted) of a number of federal statutes and the responsible agencies. Guiding any agency activity related to water is the National Water Policy represented for the President by the Water Resources Council. Major federal departments and agencies which may be involved in planned reuse and are responsible for implementing this policy while providing water project funding or administering programs affecting or affected by water supplies or water quality are: the U.S. Environmental Protection Agency (EPA); Water and Power Resources Service; the U.S. Geological Survey (USGS); the U.S. Soil Conservation Service (SCS); The Bureau of Land Management (BLM); the Department of Interior (USDI); the Economic Development Administration (EDA); the Council on Environmental Quality (CEQ); the Department of Housing and Urban Development (HUD); the Farmers Home Administration (FmHA); the Community Services Administration (CSA); the U.S. Army Corps of Engineers; the Food and Drug Administration (FDA). In short, matters concerning water supplies or water quality can be of concern to numerous and varied federal agencies and programs.

Even an abbreviated survey of legislation that responds to the national water policy and can affect water reuse demonstrates a range almost as broad as the agencies involved. The major acts include:

- the National Water Pollution Control Act (NWPCA) and its 1972 and 1977 Amendments,
- the National Pollution Discharge Elimination System (NPDES) permits,
- the National Environmental Policy Act (NEPA),
- the Reclamation Act of 1902 and subsequent legislation relating to reclamation,
- the Safe Drinking Water Act,
- the Fish and Wildlife Coordination Act,
- the National Historic Preservation Act, and
- the Endangered and Threatened Species Preservation Act.

Although the above is not an inclusive listing, these legislative acts contain many of the elements which could affect water reuse and reclamation.

The activities of the agencies in implementing legislatively defined duties are directly responsive to national water policy goals as enunciated for or by the President. Current national policy requires that best practicable efforts at conservation and prevention of water waste be exercised by affected water users before any attempts can be made to justify a new water project or expand an existing one. This single requirement sweeps across the broad spectrum of water uses and supplies. Further, the policy calls for increased cost sharing on the part of the states for water projects, full compliance with environmental mitigation procedures, renegotiation of project water prices, and recalculation of project operating, maintenance and repair costs. The National Water Policy guides the administrative implementation and regulations of agencies constructing water works and administering water quality (EPA, EDA, HUD, CSA) and it will have a special impact on the major construction programs of the Water and Power Resources Service and the Corps of Engineers. The new policy may in turn directly affect state and regional efforts to optimize water use, satisfy river compact commitments, and engage in long-range planning. For example, the efforts of many states to maintain an agricultural base can be hampered if water is reallocated from existing supplies rather than supplied by newly constructed storage facilities. As the manifesto from which discretionary water management receives guidance, the national water policy bears a direct influence, in both favorable and negative ways, on water reuse projects. The following brief overview of specific legal elements further illustrates this policy impact.

National Water Pollution Control Act

The NWPCA and its 1972 Amendments (Public Law 92-500) constitute the most direct controls over wastewater discharges and hence on water reuse [2]. Various sections of the act establish stream standards, effluent limitations which are tied to best practicable technology, toxic effluent standards and thermal discharge standards. Each of these sections can determine the type of treatment required before discharge and can thus act as an economic or practical incentive for water reuse in place of treatment and discharge. Water discharges are directly controlled by the NPDES permits required by the NWPCA and issued by either the EPA or an EPA-designated state agency. The permits address all point-source discharges and are written, ideally, to require any discharge to meet all federal and state standards. Permits can be written in such a manner that the allowable discharge or the constraints on proper retention could provide an economic incentive for water reuse (e.g., power plant cooling water could be reused rather than expensively desalted or disposed of by evaporation in a lined pond). Finally, NWPCA provisions for the Section 208 area wide waste treatment plans have essentially required that each state or portion thereof devise a plan to control or reduce all current and future sources of water pollution, including water used in energy production and accounting for energy consumed in water treatment. Many 208 plans do (or can) provide incentives for water reuse, particularly in water-short areas such as Clark County, Nevada.

In light of the established impetus of NWPCA, it is interesting to note that its 1977 Amendments (Public Law 95-217) may be interpreted as having a negative impact on water reuse. This conundrum was discovered by the State of California during its efforts to develop multipurpose water reclamation projects. It seems that EPA, in interpreting the provision that priority funding of wastewater treatment plants eligible for aid under the Clean Water Construction Grant Program go to those projects that will result in compliance with the "enforceable requirements of this Act" (i.e., pollution control), has drawn administrative regulations which could preclude assistance for multipurpose wastewater reclamation projects unless pollution control is clearly the primary purpose [3]. Since treatment designed to facilitate water reuse, or water reuse included as part of the treatment, might not meet "best practicable" criteria, or be the most cost effective, or be primarily for pollution control, federal funding assistance could be jeopardized by efforts to create innovative reuse projects. On the other hand, that instance in which water reuse is the best means of pollution control ought to receive priority funding under this EPA criterion.

National Environmental Policy Act

The NEPA requires that a detailed environmental impact statement (EIS) be included in every recommendation, report, proposal, legislation or any major federal action (such as funding a water reuse project) that significantly affects the quality of the human environment. Many states require a similar statement in areas where the federal requirement does not apply. A reuse project could be subject to state or federal EIS for a number of reasons, including:

- groundwater recharge with reclaimed water could significantly affect potable water supplies;
- withholding a discharge for reuse could affect the character or quantity of instream waters impacting on wildlife habitat, fish, or downstream water users;
- reusing water for ornamental or crop irrigation could adversely impact neighbors or the food chain; and
- the esthetics of a reuse project could affect an entire area.

Any type of reuse project could be subject to an EIS for reasons similar to those listed. We should also note that the nonspecific environmental acts for fish, wildlife, endangered species, and historic sites are most frequently brought into play under NEPA procedures. Even the 1970 Clean Air Act can be applied during an EIS procedure to block the siting of an industry at the source of a reusable water supply if air quality might be endangered. Clearly factors such as these could delay or impede any water project, no matter how desirable the project may seem as regards optimal utilization of the water resource.

Because the time, expense and potential legal and social controversy surrounding approval of an EIS can make an innovative project much less attractive, the Council on Environmental Quality has promulgated regulations streamlining the procedure [4]. Even so, the very nature of the assessment brings diverse elements—legal, administrative, special interest—together in a program designed to resolve conflicts the procedure itself has, in part, abetted.

Safe Drinking Water Act

Since 1962 the U.S. Public Health Service has maintained national drinking water standards that were codified by the 1977 passage of Public Law 93-523. The act set, with a procedure for updating, maximum allowable standards for contaminants in a drinking supply watercourse and in water injected underground. Any discharge of reused water or mingling of reused water with potential drinking water that fails to meet safety standards is subject to cease and desist orders brought by a damaged or concerned party. Thus the standards could preclude reuse if disposal of effluent by other than discharge is prohibitively expensive or practically impossible.

Food and Drug Administration

The FDA, through its power to police foods within or contributory to the human food chain, is an example of a tangential agency's ability to restrict certain types of water reuse. For example, land application of previously used water could fall under FDA control if the crop grown is indiscriminately used as human food or insufficiently removed from the direct food chain. Water reused in food processing would also have to meet stringent FDA standards. Although they stand in indirect relationship to water reuse, the duties of the FDA should not be discounted.

Federal-State Relations

The impact on the states of general federal policy, as well as federal water policy, has produced several areas of peculiar sensitivity and responsibility. In terms of water reuse projects, two areas of mutual federal-state activity are worth delineating: the Winters Doctrine as it affects present use of water and future development of water supplies; and interstate and international water compacts or treaties as they constrain consumption- and define quality parameters.

Winters Doctrine

Also known as the reservation doctrine, the Winters Doctrine reserved water on Indian Reservations for future tribal use. It was expanded by the U.S. Supreme Court in its *Arizona* v. *California* (1963) decision to include waters claimed for future use on federal lands generally [1]. Since those waters now reserved for future use currently flow off of the federal and tribal lands, they have already been appropriated for other uses outside of the nonpublic lands. Not only is much of this water already in use, but the actual quantity of water so reserved has never been determined and may well exceed the amount of unappropriated water available. In any case, if reserved water were to be quantified and put into use, it would likely reduce the available amount of water already appropriated by downstream or junior water users. The paradox of the Winters Doctrine is simply that the exercising of rights to reserve waters could not only impede existing water uses but it also could effectively preclude planned reuses due to the uncertain water supply and the need for costly investment in facilities.

Although all existing water appropriations would not be threatened if reserved rights were put to use, there are several instances where some such rights could be impaired. For example, water appropriated privately after the date of federal reservation has no protection in time as the reserved rights are not subject to state water laws. On Indian tribal lands the reserved water, generally agreed to be the amount necessary to irrigate all arable lands within the Indian reservation, has priority over all appropriations granted after the date of water reservation. Thus, these lingering uncertainties over ownership of water rights constrain planned water reuse in many areas of the country.

Interstate Compacts and Treaties

Although properly a matter between states or nations, compacts and treaties concerning streams that cross legal boundaries can affect several governmental levels. The Colorado River Compact of 1922 and the Mexican Treaty and Protocol of 1944 concerning Colorado River waters are perhaps the clearest examples of interrelated federal-state impact. Meeting the requirements of Compact and Treaty has involved each level and every branch of government—judicial, executive, and legislative—and all aspects of water itself—quantity, quality, flow, storage and uses.

The waters of the Colorado River, as they flow through seven states and into Mexico, become more saline as they travel downstream and are subject to great uncertainty as to yearly replenishment. The Compact limits each state in its consumption of water (defined by the U.S. Supreme Court as diversions from a stream less return flows) a matter still in dispute because of uncertain flows, while the Treaty and its subsequent Minute 242 obligate delivery of a specific quantity of water of a specific quality to Mexico. Thus, the benefits of or incentives for a planned water reuse depend on the amount of water available for each state's consumption as modified by previous allocations, quality considerations, policy guidelines and other compact or treaty obligations.

Although they represent an extreme example, the difficulties in allocating and using Colorado River water clearly illustrate the type of constraints a compact or treaty can exercise upon water reuse, at least until the point just prior to ocean discharge. Compacts or treaties can control the amount of water available, can impose penalties for quality degradation, can be a source of dispute particularly where water is in short supply, and can affect international relations.

Addressed comprehensively in an entire issue of *Natural Resources Journal* [5], the uncertain effects of existing compacts and treaties on water reuse exemplify the complexity and mutual needs that have grown around the development of water resources which do not follow governmental boundaries. This problem presents an encapsulated picture of the interrelationship of law and policy as they meld to form the administrative process.

State Law

State law specifically addressing water is generally divided into quantitative and qualitative matters. Water quality law in particular derives from and generally depends upon the guidance provided by federal legislation, especially that setting drinking water standards, stream standards, and permissible limits for effluent discharge. However, states can and sometimes do enact standards which are more comprehensive or rigid than federal criteria. More important to water reuse, the states exercise, without strong federal guidance, the right to appropriate and use water, to determine the proper or beneficial use of water, and to protect water users within state boundaries and state guidelines. In addition, many states oversee all aspects of wastewater treatment, reclamation, and permissible reuses as part of their protective responsibilities.

Water Rights Law

By far the most critical aspect of state law affecting water reuse is the manner by which it determines water allocation for its use, a determination which is supreme to all but a few reserved rights or federal projects. In the United States a right to use water is obtained either through the riparian system or under the appropriation doctrine. Although each state has its own particular statutory system, the systems retain enough similarities to allow general descriptions of the two water allocation methods:

1. The riparian system, the oldest method of determining water rights, gives the right to use water to the owner of land abutting a natural watercourse. Found primarily in the more humid Eastern states and based in English common law, riparian rights have, in recent years, been modified by most states to include principles of reasonable use and pollution control through institution of various types of permitting procedures controlling the type of use and setting an absolute limit on duration of a water right.

2. The appropriation doctrine, unrestricted by property ownership, is a priority system of "first in time, first in right," that establishes the right to use water on the basis of the date of appropriation. The absolute priority date is that on which the water was first put to beneficial use as determined by the responsible state official under definitions and standards set by law or policy.

Under both the riparian and appropriation systems of water allocation, the states are developing greater controls in reaction to population growth, limitations on usable water supplies, and the damages caused by increasing amounts and types of pollution. In either system it is likely that a state has or can obtain enough control over water to protect present water users from diminishing water supplies or water quality in a normal year while at the same time providing for optimal use of the resources as further development takes place.

Reuse of water might damage the holders of water rights downstream by either reducing the quantity of water available or by changing the quality to which the user has become accustomed. For example, land application of wastewater could deprive downstream users of an amount of water—effluent —which made up a portion of their water right. There have even been instances when downstream users have demanded their right to "dirty" water because they had come to rely on the nutrients it contained. Users affected in these or similar ways could sue for damages both real and future or could obtain cease and desist order against a reuse project. Further, in any state with antiparalleling statutes, the sale of previously used water could be judged to be unfair competition with fresh water purveyors and thus be restrained or prohibited.

Determination of water ownership, of points of diversion, and the concurrent importance of historic patterns of water use and flow, use preferences, and definitions of beneficial use are extemely complex and vary somewhat from state to state. However, any innovative use or reuse of water could affect or run counter to such determinations. For example, a proposed exchange in which irrigation water is to be used by a city before it is applied to the land can be constrained by the short duration of the irrigation season. Such an exchange would reduce historic river flow in those months when irrigation does not take place since withdrawals do not occur without irrigation. Such questions must be addressed when considering projects that change the character or pattern of existing water flow.

Water Quality Law

As noted earlier, state water quality legislation is basically derivative of federal law and must meet at least those standards defined by national legislation or policy. The states generally classify their waters into various groups, such as: recreation water, agricultural water, water supporting specific aquatic life, domestic water supply and existing (and to be maintained) high-quality water. These classifications in turn determine the appropriate uses for the water and the permissible discharges into the stream. In addition, many states can require that a minimum stream flow be maintained to preserve aquatic or wildlife habitat, or recreational and aesthetic values.

Also noted previously, one of the major federal and state tools for protecting or enhancing water quality is the NPDES permit. Because of the flexibility possible in developing the provisions of a single permit or a series of permits, states can use them as a positive enhancement for water reuse. Items that a discharge permit might control include:

- identification of any diseases or parasites that could affect public health or aquatic life;
- limitations on the amounts of total dissolved solids (salt), heavy metals or other nonorganic constituents;
- limitations on concentrations of carcinogens, mutagens or any number of pollutants;
- determination of reliability of any pretreatment system;
- identification of emergency procedures; and
- delineation of disposal facilities to prevent undesirable discharge from overflow or seepage.

Permits might also be issued for a stream reach in order to achieve a balanced waste load through dilution rather than through enforcement of the same requirements for each discharge. The permit may disallow any discharge, and, instead, require and sometimes define proper disposal. Use of the NPDES permit to enhance the attractiveness of planned reuse can have the benefits of optimizing utilization of a resource, cleansing waters and protecting public health.

On the other hand, protection of public health can present serious barriers to water reuse. Although such barriers are primarily the result of administrative activities, they are also generated by broad grants of legal authority to certain agencies under which specific rules and regulations can be drawn to address any type of reuse. Such rules most often regulate reuses that might involve direct human contact or human ingestion of products derived from reused water, and include definition of the mix of reclaimed and fresh water used in groundwater replenishment [6] and control of the timing between reuse irrigation and grazing on a pasture [7]. Such public health rules exemplify the need to discuss or allude to the legal structure as part of the institutional setting from which some rules derive and to which other rules contribute.

INSTITUTIONAL CONSIDERATIONS

The statutory definitions and mandates defining obligations and regulations in the administration of water—allocation, use, quality, health and safety—provide a framework, a forum, in which institutional activity takes place. Not only is it technically difficult to distinguish legal from institutional, but the ubiquitous nature of the legal, personal, political and social factors influencing institutional activity precludes precise, empirical analysis. Drawing a narrow, focused definition of institution that closely examines the agencies and the behavior of the actors that constitute it is but one step. It is further necessary to delineate the various factors which come into play, be they policy initiatives, methods of decision making or direct influences on the actor/agency, none of which is completely separate from the other or from the framework provided by the legal structure. In a sense it is a circular discussion reflecting a multitude of components similar to any activity of civilized society.

Each agency involved in water matters, in addition to having a specific legal character, is part of a larger governmental entity that has broader and more varied purposes and goals. Further, the personalities who shape and are shaped by an agency perspective bring all their prejudices and experiences to bear on agency management. Thus it is that an agency may, from its inception, contain impulses contrary to other agencies at any governmental level. Adding to this potentially volatile situation is the interplay of various actor's personalities within an agency which adds further complexity to any administrative procedure, creating a pyramidal system which moves from broad mandate through specific obligation to accomplishment as interpreted by personal perspective.

To distinguish "legal" from "institutional" and to assist in outlining as clearly as possible their respective influences on each other and on water resource management, a brief vocabulary is introduced: (1) *structure* is that which is written—an observable, quantifiable law or bodily creation of law; (2) *process* is the amorphous gathering of all those diverse elements that meet in an actor or agency designated to act for the structure; and (3) *meld* represents the interaction of the static structure and the fluid forces of society and personality that make up process. Meld can be neutral, positive or negative. Where structure is law and process is institutional activity, meld—the area in which structure and process merge or conflict—is the institution.

Patterns of Influence Derived from Structure

Obviously, no management activity takes place in a vacuum. The agency or actor who makes a decision that can influence water reuse is subject not only to his particular organizational structure but to the myriad factors of the process which affect the manner in which he reaches and carries out his decision. He may represent the structure, but he is part of the process; metaphorically, his is the process of interpreting all these factors. Thus, while the influence of structure over process may seem dominant, an examination of the forces derivative of the legally constructed framework illustrates that structure is but one, albeit major, component of the picture.

For convenience we shall arbitrarily divide the elemental patterns of influence drawn from the structural framework into categories similar to those of the previously discussed legal components: national interest, federal-state concerns, and state policies. And we shall attempt to illustrate the structure behind each institutional pattern described.

National Interest

The emerging national interest as it affects water management reflects contemporary changes in the legal framework and the transfer of policy activities from the legislative to the executive. Historically, national activities concerning water were dependent upon a legislative agenda responsive to the requirements or desires of local beneficiaries (states, irrigators, cities, etc.), requirements that had been hammered out in bargaining sessions between localities and presented to the legislature as a unit. The advent of environmental concerns, of values questioning development for narrow benefits, and of recognition of the finiteness of water resources, has redefined the scope of national interest. Elements of national interest that can institutionally affect water management now include:

- national water policy emphasizing water conservation and prevention of waste;
- interpretation of policy by funding agencies that favor or hinder a particular approach to water treatment or management;
- requirements to meet quality standards that are not fully understood, defined or practicable;
- redefinition of costs and benefits to count aesthetic losses or to devalue benefits to a narrow group;
- broadening of EIS criteria to require more complete assessments;
- rigid standards on permissible effluent discharges; and
- lack of interagency coordination on goals and projects.

These broad, environmentally oriented concerns affect, either as law or policy, structural water programs and create conflict, even about the appropriate uses of water. Peripheral environmental concerns—air quality, threatened and endangered species, etc.—impede projects that would reuse and consume low quality waters. All of these factors mesh to create an atmosphere of conflicting regulations and demands. A complex managerial system, in which federal regulations aim at requiring state or local entities to take specific actions in order to receive funding or approval for a project, has emerged as a major cause of institutional disarray.

The historic rules through which local coalitions presented water development packages to a responsive Congress no longer enjoy universal application. The development of groups with national constituencies (Sierra Club, Friends of the Earth, etc.) and the passage of environmental legislation have altered the legislative-constituent balance, which has, in turn, altered the traditional perspective that focused water management and development decisions. The traditional water resource development agencies—SCS, Corps of Engineers, and Water and Power Resources Service—have adapted and expanded their functions to include multiobjective planning, broader cost-benefit analyses, and mitigation procedures for environmental changes, all of which affect the method of action, the speed with which it occurs and even whether it does occur. While these agencies can still be extremely responsive to traditional coalitions, the requirements for obtaining legislative approvals now demand greater and more flexible capability to deal with broader uncertainties.

The creation of new agencies or expansion of existing responsibilities to include aspects of water resource management has also increased the complexity of the process and brought about uncertain conditions for positive meld. These agencies, operating without a traditional pattern for dealing with water management, are more responsive to the new national constituencies and to the changing perception of the resource. EPA, HUD and the like not only have important funding responsibilities but exercise regulatory powers that affect all types of resource development and management but at the same time exhibit little comprehension of the effort and costs involved in the traditional patterns. To these new agencies the process is emerging based on different rules and peopled by a different cast of characters.

Thus, even internally, the national interest incorporates structure into process in a pattern of institutional conflict, of new versus old, of youth against experience. The conflict is not only a matter of law but also of method, of a complex pattern of constituencies, funding priorities, knowledge and politics. The conflict can be as simple as a proposed reuse favored over new development.

Federal-State Concerns

Further down the pyramid, we find the varied forces in complex convergence on any given area of water management, as is amply illustrated in those areas potentially affected by reserved water rights or by river basin compacts and treaties.

The basic problem with reserved water is that it is not yet clear who is claiming how much water at what locations, when, and for what purposes.

Until the waters are quantified, the appropriateness of the use and the allocation's effect on existing water users and uses cannot be determined. Nor can the reasonableness of a reserved right be challenged until the amount of water and intended use is known. Thus, the implications of the reservation doctrine add complexity and uncertainty to water resource management and to any entity wishing to make use of the resource.

The patterns of tension and influence created by river basin compacts and treaties, however, are known. The Colorado River Compact and the Treaty with Mexico provide vivid examples of those nuances faced by water managers at all governmental levels. The seven states of the Colorado River Basin (Wyoming, Colorado, Utah, New Mexico, Nevada, Arizona and California) offer a classic illustration of cooperation for mutual goals reached only after sharp internal debates through the mechanism of the Colorado River Basin Salinity Control Forum as spurred by provisions of the Treaty with Mexico.

The Forum was created by the seven states in response to several questions generated by passage of Public Law 92-500, and through debate and discussion, it has provided a tool of coalition to urge desired federal actions for development of water resources and construction of salinity control projects along the Colorado River [8]. Its goal is to make the responsibility of salinity control and other large water projects a national burden while preserving each state's right to develop its own water resources. The consensus format developed by the Forum, and aided by the seven states' implicit agreement as to the course of Colorado River development, has created a unique pattern of interstate cooperation in response to the national demands imposed by the Treaty. The region's traditional pattern of cooperation to obtain Colorado River developments, resulting after years of effort, is the forerunner of the methods established by the Forum.

Although threatened by the exigencies of a newly conservative national water policy, the demands of environmental concerns, and the differing levels of intrastate development and types of demands for water, a water resource consensus has ameliorated these isolating factors by means of Forum appointments as well as by Forum creation. Both water allocation and water quality interests represent each state on the Forum. And in working together under the constraints of consensus proceedings, these interests compromise on both intra and interstate policy differences in order to present a single view to the federal agencies they address. Not only does meeting the broad range of issues imposed by salinity control imperatives and resource development needs remain a collective Basin effort [9], but also the development of interpersonal professional linkages establishes a degree of regional comity and a basis for positive meld.

The formal and informal personal ties developed through the Forum and other water-related regional organizations do not always carry over to intrastate activities. Nor does regional comity based upon Colorado River problems necessarily extend to a consensus on the broad range of water management and related issues facing each state. Varying state policies as well as differing agency agendas can, in themselves, become institutional obstacles to any area of water development including planned reuse. Regional unity to obtain specific benefits does not necessarily indicate a comity extending any further than the instance, particularly when each unit is subject to the demands of a wide variety of needs, regulations and policies.

On the other side, the need to satisfy the international agreements concerning the Colorado River places one aspect of the national interest in conflict with another. To preserve amity with Mexico, the water of the Colorado River as it crosses the border must not exceed the internationally negotiated level of salinity. The salinity levels can be maintained if the states stop and possibly undo some upstream developments and irrigation, or if the more financially able federal government constructs a series of expensive and inefficient desalting plants and other structural projects. The states are unwilling to shoulder what they consider to be a national burden, and thus they place the federal government in the paradoxical position of "robbing Peter to pay Paul." In order to fulfill its commitment to Mexico the government must violate some of the basic tenets of its water policy, particularly its cost-benefit ratios and conservation guidelines.

State Policies

As the other prime branch of our federal system of government, the state exercises vital control over many aspects of water resources: allocation of water, determination of appropriate uses, and control over quality and health factors. The governor and legislature can exercise policy influence similar to that of the federal branch. This state function directly impacts on the process that most directly affects the course of resource management on an ordinary, day-to-day basis.

State policies, as defined by the executive or legislature, can have particular impact on water resource management and can determine a broad range of attitudes toward water reuse. California, for example, has received both gubernatorial and legislative directives making the research and development of possible water reuses a major state goal. On the other hand, Colorado's policy is to preserve irrigated agriculture and develop fully all available water resources. In terms of institutional analysis, what such policies state is less important than their existence and the control they exert on the process of resource administration.

Several methods of exercising policy control are noteworthy in their relationship to both state and federal agencies and programs. One innovative method results from policymakers' interpretation of the public interest to allow guidance on resource use from outside the traditional procedures. Another stems from the establishment of agreements with federal funding agencies which codify states' rights in project planning and development. A third vehicle is the use of traditional politicking to garner public support for a particular position.

Any one of these methods can have immediate impact on the process of water allocation, water uses, water quality decisions and permissible discharges since they are localized and immediate to the actors and agencies who make and implement the resource decisions. For example, if it is state policy to achieve development before seeking optimization, the administrative interest will not be in water reuse.

Depending on the legal, constitutional and traditional support framing an agency/administrator, state policy can affect water resource management in varying degrees. Matters pertaining to water quality, because of its recent prominence and broad constituency, are apt to be most subject to policy influence. Matters of water allocation, supported as they are by legal and traditional antecedents, will more likely follow a predictable, historic pattern.

It is at the state level that the interrelationship of legal and policy factors is most noticeable in the resulting administrative action. As many state laws are derived from federal legislation so are state policies influenced by federal policy. State policy may directly parallel federal policy as a state maneuvers to obtain favorable position for federal benefits, or it may articulate itself contrarily to serve as a focal point in creating opposition to or impetus for change of a particlar federal policy. The influence of federal policy can blend with the very natural desire of each state to develop fully its own resources to create a complex institutional meld. Furthermore, each state's attempt to reach full internal development in a manner consistent with its own enunciated policy can threaten the interests of neighboring states and severely strain the coalitions necessary to obtain federal support or to modify federal policy.

The Nature of Resource Decision-Making

The manner in which a decision is reached is another key institutional element. Students of public administration, political science and sociology have all been developing languages appropriate to describing clearly the institutional processes of water resource management. These efforts include the terminologies of distributive and regulatory politics, incrementalism, synchronic and diachronic decisions and explanations of domain. Each of these vocabularies can shed light on the kinds of relationships and factors that make up institutional process affecting water reuse. A brief examination of elements of this complex language illustrates the increasingly sophisticated tools of policy analysis and clarifies various types of resource decision making, and emphasizes the usefulness of our use of the relatively simple terms of structure, process and meld as analytic tools.

Distributive-Regulatory Politics

First described by Lowi [10], regulatory and distributive politics represent two of three modes of decision making pertinent to water resource management. The third component, redistributive politics, represents an ideological and well-defined program of management which bears little resemblance to the institutional pattern found in water resources.

Analysis or description by way of regulatory politics most adequately reflects the contemporary situation of traditional water use and development as it conflicts with environmental, aesthetic, and conservation values. Although not the traditional means of approaching resource development, regulatory bargaining has supplanted tradition in seeking and making trade-offs and establishing broad sets of rules. Much of the regulatory effect can be the result of negotiation during the NEPA environmental assessment process as it merges with cost-benefit analysis or other planning tools.

A classic example of water resource distributive politics has been described by Ingram as she traced the passage of the 1968 Colorado River Basin Act [11]. In this traditional distributive process the seven states of the Colorado River Basin formed a coalition in which the various state officials, federal legislators, etc., bargained to reach a consensus position to be presented to the federal legislative and executive branches. Basin-wide support for the Act was obtained by promises of future support for other development projects, by inclusion of certain unrelated items in the Act, and by the long-reaching influence a cohesive position was likely to effect. Most decision making in the Colorado River Basin has been and remains distributive through the activities of the Forum [12].

Incrementalism

Popularly known as "the science of muddling through," incrementalism [13] can be broken into categories roughly congruent to those of distributive-regulatory-redistributive politics. From a basic point of view, simple incrementalism, as a means of decision analysis, provides the means for making only slight variations on the status quo. It could be considered a mode of making small additions to the traditional pattern. In a more complex manner, disjointed incrementalism takes a variety of factors into account and produces a decision similar to that obtained by regulatory politics. Finally, strategic analysis presumes that there is a "best" choice and attempts to identify it in much the way redistributive politics attempts to impose the correct program.

Of these three broad types of incrementalism, of ways of reaching decisions, disjointed incrementalism best represents muddling through in a scientific manner. It represents the broad range of available knowledge as it is tempered by social needs, regional variations and political realities. In institutional terms, incrementalism represents the ability to adapt, to change in response to a national water policy, to account for broad rather than narrow benefits and to explore a variety of resource management tools, and it does so in an evolutionary manner.

Synchronic-Diachronic Policymaking

In a logical extension of incrementalism, Williams [14] postulates that structure and process can have distinctly different institutional effects as they meld. Diachronic policymaking creates significant modification of a structure —such as the creation of EPA or passage of the 1972 Amendments to the Water Pollution Control Act. Synchronic policymaking represents, on the other hand, an evolutionary, adaptive approach in which the process gradually modifies the structure, an example being the Water and Power Resources Service's recent consideration of low quality water use rather than facilities development as a water source for energy industries.

Federal activity in water policy combines diachronic and synchronic policymaking, although it has of late deemphasized drastically traditional modes of water resource decisions in favor of creating and implementing new modes and new means of implementation. On the other hand, state policymaking continues along a synchronic path, casting new structure into the position of "pretender" when thrown against historic agencies and traditional methods of resource management.

Domain

The final factor in our categorization of resource decision making is the concept of domain. Mileti et al. [15] describe the concept in sociological investigations of disaster response as that response which fits within the fixed and accustomed pattern of normal and proper agency activity without impinging upon another's territory or opening the primary agency to outside infringement. This sense of turf is a strong inducement to behave in accepted, traditional ways. Only when the need is urgently clear does multiorganizational communication or networking take place—the ensuing process and meld are broadened accordingly.

Water resource management contains many examples of domain mentality, particularly as agencies pursue their goals without consideration of all impacts (e.g., EPA or a state water quality agency mandating a specific quality discharge which causes withdrawal of water supplies from downstream users). However, the activities of the Forum are the result of the development of a multiorganizational network responding to a clearly defined need surpassing the boundaries of any one state or agency.

Role of Agency/Administrator

The foregoing discussion of patterns of influence and the nature of decision making provides the basic blocks for constructing the finishing pieces of our institutional diorama. The primary element, without which we would have management by automation, is the actors on the line, those who people the water management agencies and take the daily actions necessary to administer the varied elements of the resource. At all levels of government these actors and their agencies can be roughly categorized into the interests of water quality and water allocation, with some peripheral interest indicated by nonwater environmental and health-related agencies and actors. These actors/agencies in turn serve as the locus for a variety of specific clientele (irrigators, power plants, anglers, etc.) who are seeking to preserve or advance their particular interests.

As discussed earlier, the differing structures of agencies and the varied traditional processes involved in administration create inherent conflicts which are exacerbated by the demands of the agency clientele. Delineating the structure and process generally applicable to each interest category illustrates the occasion for tension and conflict:

- Allocation agencies attempt to achieve maximum beneficial use of a water resource. These agencies function within well-developed legal and administrative histories which have often created an elite corps of actors who share common professional histories, mastery of difficult and arcane technicalities concerning water use, and similar training and occupational biases. Leadership within allocation agencies, although ostensibly subject to political control, tends to follow generational patterns, thereby creating a strong sense and record of tradition. Allocation clientele respond with their own traditional expectations, invoking the market system as the primary source for change.
- Water quality agencies attempt to maintain or achieve a given standard of quality even if it precludes full utilization of the resource. These agencies are diachronic in character, representing a relatively new area of management in which the basic research (as to what is a pollutant or what are desirable parameters) is ongoing. Quality agencies are required to protect and enhance water quality, using changing standards based on an incomplete empirical foundation. Personnel, due to the new and generally unsophisticated skill pool, are frequently drawn from agency clientele.

These differing roles do not, in themselves, create institutional problems. However, the distinct backgrounds and experiences of the roles and their interpreters provide fertile soil for the quick growth of adversary postures.

It seems obvious that the solution to avoiding the tensions and conflicts arising from these separate agencies would be the merging of all water resource functions into a single umbrella agency. However, this solution encounters the same barriers as creation of a new agency without the quick political payoff desired from the immediately obvious action of creating a new agency. Merging agencies bring established policies and clientele into conflict with developing agendas and widely diverse clientele. Add to this the political difficulty of undertaking and explaining a complex restructuring which violates each agency's sense of domain and drive for self-perpetuation, and one can construct a vivid scenario for a conflict situation.

Finally, the adversarial climate created by the differing structures and clientele demands affects the role perceptions of the primary actors who actually shape the legal structure into the administrative process. This role perception is determined by the agency framework, by the clientele served, by the accepted decision-making mode, and by the personal background, training and beliefs of the actor. Role perception guides as well as responds to policy development and both role perception and policy in turn guide implementation. Finally, role identification can affect the types of relationships established between agencies and levels of government.

Allocation Actor

Thus the role behavior of each water resource actor represents the culmination of his perception of all the structure and process as filtered through his personal screen into the meld. In this scenario the allocator, because of history and circumstance, is the elite and is, of course, influenced by his elite position. The incremental nature of the allocation process allows a more discretionary approach to resource management and can critically affect innovation such as water reuse. For example, in fulfilling his obligations to ensure beneficial use, prevent waste, and protect prior water users, the allocator can manipulate requests for new water or changes in use of water which could encourage or even mandate reuse. These discretionary tools may include:

- limiting an allocation by defining reuse as part of the appropriated beneficial use;
- factoring the amount of historic consumption of water in a restrictive manner;
- defining a site-specific reuse as the "public interest";
- making reuse part of the permission for a change of use or change of diversion point; and
- bringing the power of office to bear in negotiating a reuse.

All of these potential administrative tools could remain passive, neither impeding nor assisting a reuse proposal. On the other hand, an allocator could hamper reuse by rigidly following legal and traditional patterns and fulfilling full appropriations or requests for change whenever water was available and no other users damaged.

By reason of his established position in his domain, an allocation actor can, if he chooses, remain relatively unaffected by the demands or requirements of a water quality actor. Legal precedent and traditional process are on his side should he choose an adversarial role.

Quality Actor

The advent of regulatory politics as represented by water quality actors has caused a number of breaches in the formerly static or marginally incremental arena of water resource management. The very nature of quality agencies/actors leads to a process too often based on seemingly immutable national goals that can preclude the flexibility or administrative discretion needed to manage best a resource as ephemeral as water. Furthermore, the strength of the quality agency-clientele relationship with its combined ideological and scientific roots contributes to a hardening of positions just where more elastic concepts, such as "reasonable degradation" and "waste-load allocation," would lead to a focus for compromise with allocation interests and to quicker implementation of any rules. Within his narrow, discretionary limits a quality actor could affect reuse by:

* writing NPDES permits to include design standards for evaporation disposal which could be costly;
* allowing discharge of waters offering no impairment of in-stream quality;
* allowing discharges of waters where sufficient dilution capability exists, even if reused; or
* writing NPDES permits for entire stream reaches incorporating one user's reuse with another's inability to reuse with commensurate incentives.

The quality actor's efforts and some discretionary processes can be hampered by peripheral agency action such as health agency requirements that a noncontact, nonpotable reuse be severely limited or heavily diluted (e.g., mixing standards for groundwater replenishment, or landscape irrigation restriction to nonpublic areas), or air quality agency insistence that a power plant site which could reuse water is unacceptable to existing air ambience.

Conflict and Compromise

By once again turning to the Colorado River Basin we can trace the emergence of interagency communication and cooperation. The creation of the Forum, initially in response to what the states perceived as unreasonable federal regulatory pressures, was also a recognition of the need to mesh allocation and quality concerns to achieve optimization of a limited water resource. Although interstate differences persist, the states, particularly through the Forum, have responded to their mutual interests in fully developing compact allocations and solving the salinity problem. Intrastate activity has also begun to create, through various means, multiorganization linkages.

Two states have linked their water managers by creation of state policy promoting water reuse. Two others have taken advantage of the absence in the Upper Basin of severe pollution problems and have coupled that with specific state development policies to achieve a degree of interagency cooperation. In another, the allocation actor, by use of his tenure and his active membership on the body setting water quality policy, has attempted effectively to guide the quality agenda, thus himself playing the role of the interagency link. In only two of the Colorado River Basin states has a mutuality of interest failed to develop clearly except in matters confronting the entire seven state unit of the Colorado River Basin.

By no means do the policies, together or individually, of water management in the seven states act as positive forces for reuse. Some administrators consider water reuse to be of minimal value from either a supply optimization or quality enhancing viewpoint, and thus force the management process into a passive or negative attitude toward reuse. A belief in the value of some reuse can, on the other hand, turn the process toward a pursuit of reuse opportunities. To date, studies indicate the predominance of the negative or passive attitude among water resource managers [16].

In the end, shaped by various forces, guided by the structure of his agency and by state or federal policies, the individual actor with his sense of domain and his established method of decision making determines the scope and dimension of the institution and its effect on water reuse. If the actor has himself become an "institution" in terms of longevity and the accumulation of power, his ability to control the institutional process is enhanced. If the milieu is not fully at the actor's command, confusion may be the resulting process. In either case, the actor's role perception determines whether or not the institutional process will be a negative or positive consideration in water reuse. And his role perception is affected by his guiding structure and all the influencing factors of policy, personality, methodology and communication.

A FINAL NOTE

This chapter has traced the intertwined motifs of legal and institutional considerations applicable to water reuse and water resource management in general. Even though we have separated and considered the two as structure and process it is essential to recognize that all of the legal discussion is but one factor of the institutional. It is this separation which enabled us to account for all the varied elements that make up the complex meld which forms the institution responsible for carrying out the law (structure) of water resource management. From this analysis of institutional complexities (law, personality, policy, etc.) as they affect water reuse, several themes emerge which are worthy of summation.

First, the task of defining and describing the institutions themselves is an exercise in complexity. Every level of government—in its laws and policies, as well as the social pressure groups to which it is subject (irrigators, environmentalists, etc.)—impacts upon the institutional process we have described as it addresses a water resource. Since the definition itself is so complex, there is no simple way to describe impacts on water reuse, either for analytical purposes or for practical goals.

Second, the demands of regulatory politics—all the pressures for environmental protection, conservation, preservation of historic modes versus development—have disrupted the distributive process traditional to water resource management. To some degree these conflicting forces have negated successful coalitions while sometimes creating new ones. Again, analysis discovers no common tool except, perhaps, clearly identified urgent need, for bringing widely contradictory interests into an arena of compromise.

Third, in the end it is the actors and their activities that are at the heart of the entire issue. They are the institutions wherein structure and process meld. They are the focus of all the complex and contradictory forces which affect or impact on water resources. And they are why it is not possible to do more than generalize as to the institutional considerations which must be accounted for in planning a water reuse project until a site-specific proposal is presented.

REFERENCES

1. Jamail, M. H., J. R. McCain and S. J. Ullery. *Federal-State Water Use Relations in the American West: An Evolutionary Guide to Future Equilibrium* (Tucson, AZ: The University of Arizona, Office of Arid Lands Studies, 1978), pp. 3-5.
2. MacArthur, G. *Some Ecological Implications of the 1972 Amendments to the Federal Water Pollution Control Act* (Denver, CO: Rocky Mountain Center on Environment, September 17, 1973).

3. "Background Information Regarding Funding of Water Reclamation by U.S. Environmental Protection Agency" (Sacramento, CA: California State Water Resources Control Board, Office of Water Recycling, April 17, 1979).

4. "Implication of Procedural Provisions; Final Regulations," Council on Environmental Quality—National Environmental Policy Act. 43:230 *Federal Register* IV:55977-56007 (November 29, 1978).

5. Darling, W. J., Ed. "International Symposium on the Salinity of the Colorado River," *Natural Resources* J. 15(1):1-239 (January 1975).

6. Bookman-Edmonston Engineering, Inc. "Annual Survey Report on Groundwater Replenishment," Glendale, AZ (1978).

7. "Supplemental Information and Comments Regarding Proposed Changes to Parts I & II of Wastewater Disposal Regulation," Utah Division of Health, Salt Lake City, UT (April 12, 1978) and personal communication with Lynn Thatcher, Director of Bureau of Water Quality.

8. "Water Quality Standards for Salinity Including Numeric Criteria and Plan of Implementation for Salinity Control," Colorado River Basin Salinity Control Forum, June 1975; and, "Proposed 1978 Revision" (August 1978).

9. Mann, D. E. "Water Policy and Decision-Making in the Colorado River Basin," Lake Powell Research Project Bulletin No. 24 (Washington, DC: National Science Foundation, July 1976), pp. 1-23.

10. Lowi, T. J. "American Business, Public Policy, Case Studies and Political Theory," *World Politics* 16(4):677-745 (1964).

11. Ingram, H. M. *Patterns of Politics in Water Resources Development: A Case Study of New Mexico's Role in the Colorado River Basin Bill* (Albuquerque, NM: University of New Mexico, Institute for Social Research and Development, December 1969).

12. Peterson, D. and A. B. Crawford. *Value and Choices in the Development of the Colorado River Basin* (Tucson, AZ: University of Arizona Press, 1978).

13. Lindblom, C. E. "Still Muddling, Not Yet Through," *Public Adm. Rev.* 39(6):517-526 (1979).

14. Williams, B. A. "Beyond 'Incrementalism,' Organizational Theory and Public Policy." *Policy Studies J.* 7(4):683-689 (1979).

15. Mileti, D. S., T. E. Drabek and J. E. Haas. *Human Systems in Extreme Environments: A Sociological Perspective.* Programs on Technology, Environment and Man, Monograph #21 (Boulder, CO: The University of Colorado, Institute of Behavioral Science, 1975), pp. 35-36.

16. Ralph Stone and Company, Inc. *Wastewater Reclamation: Socio-Economics, Technology, and Public Acceptance* (Washington, DC: Office of Water Resources Research, 1974).

CHAPTER 3

INFLUENCE OF SOCIAL FACTORS ON PUBLIC
ACCEPTANCE OF RENOVATED WASTEWATER

Betty H. Olson
  University of California
  Program in Social Ecology
  Environmental Analysis
  Irvine, California

William Bruvold
  Division of Social Behavior
  School of Public Health
  University of California
  Berkeley, California

INTRODUCTION

Incentives

The importance of public opinion in determining and implementing policy for the use of renovated wastewater is an extremely timely and interesting issue. Water as a resource has increasingly gained attention in water-short areas of the Southwestern and Northeastern United States. The pressures of burgeoning populations and increased agriculture have placed a strong need for alternative water supplies. Legislation in the 1970s beginning with PL 92-500 enhanced the feasibility of widespread use of reclaimed wastewater to augment current supplies through requiring higher levels of treatment and providing cost incentives [1]. Because the success or failure of many of these projects will be determined to a large degree by public acceptance, the motivating factors behind public approval or rejection of renovated wastewater must be understood. What are the overriding issues for public acceptance: economic concerns, emotional and psychological inputs, or direct benefit? These are some of the questions which this chapter explores in determining the importance of social factors in public acceptance of various uses of reclaimed water.

Intentional Reuse

There are only two examples to date for direct reuse for drinking water: Chanute, Kansas [2] and Windhoek, South Africa [3,4].

During the drought period of 1952-1957 the town of Chanute, Kansas, comprised of 1200 individuals, made direct use of effluent from its sewage treatment plant which received domestic and industrial waste. The normal source of water, the Neosho River, had ceased to flow in the summer of 1956. Other water sources either were not immediately available or were not within practical consideration. The sewage effluent had a lower concentration of fecal coliforms than the usual river water, due to upstream sewage discharges. The river was dammed and a pond constructed in the river bed (with a detention time of approximately 17 days) so that sewage plant effluent entered 1 mile above the waterworks influent. The system was used for a total of five months. Initially the only special treatment was recirculation of effluent in the sewage treatment plant. During the last three months both the sewage effluent and the raw water intake were chlorinated. It was estimated that there were approximately 10 cycles in the five-month period with some supplementation by rainfall [2,5].

The water treatment plant did not function properly when effluent was used as the water source. Break point in chlorination was difficult to achieve and foaming and rapid clogging of the filters occurred. During this time period, the tap water never failed to meet drinking-water standards, though it was pale yellow in color, had a musty unpleasant taste, and foamed in the glass due to its high organic content. The water was high in fluoride, sodium, total solids and organics. Coliform organisms were found on three different occasions during this time period but were within mandated standards [2,5].

Public acceptance of this quality of water was poor; the sales of bottled water flourished and 75 new wells were drilled in the area. The highly mineralized nature of the well water made it unsuitable for consumption. The railroad company voluntarily brought 150,000 gallons of drinking water to the town. In this situation the use of reclaimed water was a highly publicized occurrence and the quality of the product was extremely poor. A 1969 telephone survey of 39 Chanute residents conducted by Ackerman [4] revealed that only 61% of these individuals eventually accepted drinking the treated wastewater. Another study indicated that water managers in the area believed the public had been less accepting of drinking reclaimed water than the 1969 survey showed [2,6].

In the 1960s because of the severe arid conditions of southwest South Africa, plans were initiated to expand the water supply to the growing, centrally located city of Windhoek. Water officials realized that the socio-economomic development of the city and its environs depended on efficient use and reuse of water resources. Investigations examining the feasibility of importing water revealed problems of transport over long distances, high financial burdens and complications of international negotiations. The Kunene and Okavango Rivers located to the north within Angola were unreliable water sources due to political issues [3].

The solution was determined to be the construction of a full-scale wastewater treatment plant to produce effluent of drinking water quality. The Windhoek plant was designed to treat 220 m^3 of water per hour and had a total capacity of 5000 m^3/day. In September 1968 a critical situation in water supply and demand arose which caused the speedy commissioning of the full-scale recycling plant and the integration of this source of supply into Windhoek's domestic water supply [3]. The plant was officially opened on January 31, 1969.

At this time, the city was still experiencing drought conditions and 15% of the total water supply came from reclaimed wastewater. In 1976 6000 m^3 could be supplied as reclaimed water or approximately 19% of the demand. In 1977 25% of the total demand for consumable water could be met through utilization of 40% of the tertiary treated wastewater.

The Windhoek reclamation facility has operated intermittently, depending on rainfall. It was active from November 1968 to June 1969. Winter rain decreased the need for water so supplementation of supplies with reclaimed water ceased until October 1969. Renovated wastewater was again added to supplies obtained by conventional methods during the periods of October 1969 to March 1970, November 1970 to February 1971, and April to May 1972. Abundant rains and new impounded water resources decreased the need for water supplementation after 1972. During 1975 and 1976 the Windhoek plant was modified, incorporating the experience of the Stander Plant in Pretoria. Domestic supplies in Windhoek have been periodically supplemented with reclaimed water from January to May 1977.

At Windhoek, South Africa, there have been no reports of adverse public opinion to the direct use of renovated wastewater. Mr. Odendall, of the National Institute of Water Research, South Africa, in June 1969 reported that public acceptance of the reclaimed water was 100%. The use of reclaimed water in this instance was initiated without determining public opinion. Also, it is unclear from various reports the degree of choice that the population had in selecting its water supply [5].

Unintentional Reuse

If, as some authors have pointed out, all water is simply recycled water, then the question of social factors involved in public acceptance of reclaimed water is often obviated. Several examples of unintentional reuse are available on an international basis.

In South Africa, the Hartebeestpoort Dam receives a flow of 400,000 m³/day from the Crocodile River, of which approximately 4700 m³/day is composed of treated sewage. During the dry season the amount of treated effluent can account for 50% of total water volume of the dam. Hartebeestpoort Dam is used for aquatic sports and as a municipal water supply. Occasional public complaints concerning the taste and odor of the domestic water supply have been reported, even though the water is treated prior to distribution.

The Bon Accord Dam in South Africa receives water from the Apies River, which had been heavily polluted along its course by treated sewage, agricultural run off and industrial effluents (45,000 m³/day). During the dry season, the upper regions are essentially fed by treated sewage from the Daspoort Sewage Works (19,000 m³/day) and effluent from a power station and steel foundry (± 10,000 m³/day). The average retention time of the dam is 4-5 months and marked signs of eutrophication had been observed at the time of the study. During the dry season approximately 3500 m³/day of water are released from the dam, and 15,000-20,000 m³/day of treated sewage enter the river from the Rooiwal Sewage Works approximately 5-km downstream. These two sources serve as the water supply for a town of 18,000 inhabitants located 20-km downstream [5].

In a study conducted by Gavis [7] it was found that approximately one-third of the U.S. population currently drinks water from streams in which 3.3% of the water has been used upstream and returned after use to the stream. In extreme cases it is estimated that this return flow may comprise 25% of the total river volume. In certain areas of the world where less pure sources are available, as much as 100% of the flow during dry periods may be from industrial or municipal wastewater [8]. It is estimated that the proportion of the Thames River which is comprised of sewage effluent at the intake for the city of London is approximately 13% [9].

Thus, it appears that the public has to a greater or lesser degree been consuming blended water (sewage effluents and natural waters) for centuries. In the first two examples no direct information was collected on how the public felt about the water, although in the case of the Hartbeestpoort Dam consumer complaints did occur. The public may have embarked on a number of alternatives to avoid using the water, such as purchasing water elsewhere for drinking or cooking. However, without any knowledge of public reactions, it is unclear if the delivery of poor quality water resulted in acceptance or rejection of these water supplies.

The example of the Thames River most likely reflects the norm in surface water supplies where some contamination of the source water is present but comprises a small percentage of the total volume. In such cases consumers are probably unaware of the presence of sewage effluents in the water supply. A 1978 University of California study found public knowledge of water resources to be extremely poor. This study of 241 individuals living in Anaheim and Irvine, California, reported that 67% of the respondents could not identify the source of their drinking water. Formal education supplied only 17% of the information on water resources; the majority of information was obtained from newspapers and magazines. Only 18.3% of the 241 people queried in this study believed that their education on water resources was adequate.

Pratte and Litsky's study [10] into the feasibility of industrial water reuse in Massachusetts showed that even many industrial personnel lack knowledge concerning the basic tenets of water reuse and lack faith in reclamation technology.

ATTITUDES TOWARD REUSE

Determining what factors influence public acceptance of various uses of reclaimed water is difficult because the public generally is uninformed about water resource issues or water reclamation processes. To compound the difficulty, the task of acquiring useful and concrete information is also dependent on the wording and framework of questions used to elicit responses.

Variable Findings

A study conducted at Clark University [6] during 1970-1971 showed that public acceptance of reclaimed water was significantly decreased when the feeling was introduced that authorities were not sure of the health hazards involved. The 1972 Carley study [11] of 447 people in Denver, Colorado, asked questions in terms of protecting mountain environments. Depending on how the question was phrased, the acceptance rates for drinking reclaimed water varied from 37.8% to 84.1%. The Gallup Poll [12] survey of 2927 individuals asked, "Would you be willing to drink recycled sewage?" The low positive response rate of 38.2% in this case is not surprising.

The type of questionnaire and the setting of the questionnaire is extremely important, as has been pointed out by Bruvold [13]. For example, when Johnson [14] interviewed only residents in single family dwellings who knew the source of their community water supply, the results described the attitudes of a narrow segment of the population. The Gallup Poll, however, used a complex procedure to draw a probability sample of the entire United States. The 1972 Bruvold study [15] had interviewers go to the home; other surveys such as the Stone study [16] have used the telephone, and Olson and Pratte [17] and Olson et al. [18] used mail questionnaires. The framework of how questions are asked is extremely important.

The majority of early studies to assess public attitudes toward reclaimed water were fócused on areas where reclamation practices were not currently in use. There are several notable exceptions such as the Bruvold and Ongreth study [19] which compared communities throughout the State of California using reclaimed water for various purposes with those without reclamation projects of similar size, geographic location, and socioeconomic status.

Stable Findings

The different survey techniques and the lack of standardization of questionnaires appear upon an initial perusal of the literature to produce inconsistent results. But Bruvold and Ward [20] who examined public acceptance of reclaimed water by separating various uses into categories found a distinct ordering pattern in public acceptance. Their study of two California communities that were using reclaimed water reported that the most important factor in acceptance of reclaimed water is the degree of bodily contact involved.

They divided the uses of reclaimed water into three categories: ingestion, bodily contact but no ingestion, and noncontact uses. Several interesting factors became apparent through these divisions. Though 48%-55% of the population were opposed to drinking reclaimed water, a much lower percentage, 32%-22%, of the population was opposed to bodily contact and when noncontact use was viewed, opposition fell even lower to anywhere from 20% to 0%. As less body contact was involved, the expressed opposition decreased dramatically.

These results were later extended to a larger sample of ten communities in California [21]. This later study confirmed that as the personal contact with reclaimed water decreased, public acceptance increased. These findings have been substantiated by Olson and Pratte [17] and Olson et al. [18] using the questionnaire format developed by Bruvold [15]. At Clark University a pilot study [6] of 220 respondents from Glouchester, Massachusetts; Wilmington, Delaware; and Indianapolis and Kokomo, Indiana during the summer of 1970 divided the uses of reclaimed water in a similar fashion and their findings substantiated those of Bruvold and Ward [20].

A telephone survey conducted by Stone [16] for the U.S. Office of Water Resources of 1000 respondents in ten communities in California again found that water reuse with less intimate body contact was widely accepted.

By rank ordering the data from these studies by degree of acceptance with 1 = highest acceptance and 13 = lowest acceptance (Table I), it is apparent

Table I. Rank Order of Acceptance of Uses of Reclaimed Water Categorized by Degree of Contact (1 = highest acceptance; 13 = lowest acceptance)

	Stone [16]	Bruvold and Ward [20]	Olson and Pratte [17]	Olson et al. [18]
Noncontact				
Park/golf course	2	1	1	1
Garden irrigation	2	2	2	2
Scenic lakes	6.5	3	3	3.5
Factory/cooling processes	2	5	4	6.5
Boating/fishing	4.5	6	5	3.5
Toilet	4.5	4	6	5
Farm irrigation	6.5	7	7	6.5
Body Contact				
Laundry	9	8	8	8
Beaches	8	9	9	9
Bathing	10	10	10	10
Human Consumption				
Cooking	11.5	11	12.5	11
Food canning	11.5	12	13	13
Drinking	13	13	12.5	12

that regardless of varying percentages of acceptance by these investigations, higher order uses are always the most strongly opposed. As the degree of contact decreases, opposition by the public rapidly decreases. It is the supposition of Bruvold et al. [28] that an opposition rate $< 20\%$ (ranks of $\pm$ 7) by the public will not affect implementation of a proposed project.

Sociodemographic

Education

Education is a factor which has repeatedly been shown to be important in the acceptance of reclaimed water. The effect of education may be of greatest importance in areas where a high proportion of the population is not college educated.

In the Clark University study [6] only 26% of the respondents with a grade school education approved of drinking reclaimed water compared to 63% acceptance by those with some college. Johnson [14] and Carley [11] also demonstrated that education has an effect on the acceptance of reclaimed water. Johnson found that individuals with eight or fewer years of formal education were negatively disposed to the acceptance of reclaimed water (63.6%), those with some high school or the completion of high school were less negatively predisposed, but were still significantly more negative than individuals who had attended or completed college. Johnson's finding in relation to education showed that over 90% of individuals with some college were willing to drink reclaimed water.

The extensive survey conducted by the Gallup Poll in 1973 for the American Water Works Association [12] examined 2927 individuals and found that only 38.2% of the population responded positively to drinking recycled sewage. In this poll, 63% of those with eight or fewer years of education indicated they would object to drinking recycled sewage, whereas 41% of those with a college education were similarly inclined.

The Olson et al. study [18] surveyed a highly educated sample from Irvine and Anaheim, California. Three percent of the people had attended high school, 15% were high school graduates and 82% had attended or graduated from college. High school graduates were less accepting of the use of reclaimed water for swimming, drinking and laundry than college graduates. The effect of education appears to be attenuated after the educational level of some college is attained.

Nontraditional education also seems to have a bearing on the acceptance of reclaimed water. Johnson further reported that people with knowledge about reclaimed water were more highly predisposed (86.3%) to drinking it than were those with little or no knowledge of renovated wastewater [14]. Exposure to water resources or natural sciences in an occupation has been shown to be important. The study by Olson et al. [18] showed that individuals who had learned about natural resources in their occupations were more in favor of reclaimed water for drinking and swimming than those who had not. This study also revealed that there was no difference between acceptance of reclaimed water for laundry purposes between these two groups.

All the factors involved in education are not clearly understood as a comparison of the following two studies illustrate. A telephone survey of 291 households in Baltimore County, Maryland [22] found that reclaimed water would be most widely accepted in high status communities that are populated by young educated adults with relatively high incomes. These findings are extremely interesting in light of the Olson et al. study [18] conducted in 1978. The affluent, young, well-educated community of Ivrine, California, which is using reclamation practices, showed no difference in acceptance patterns compared to the community of Anaheim, California, which has no reclamation projects, a mean income approximately 50% lower and a 50% less educated public. Further, respondents from Irvine were willing to pay 25% more on their current water bill to avoid drinking reclaimed water. These differences may be explained by several hypotheses. One is that the reality of use decreases its acceptability. This community being highly educated may envision more potential dangers or may be willing or able to incur higher costs because of higher incomes. Regardless of the exact reason behind these differences, the need to understand the interaction among various aspects of regional, socioeconomic and educational factors is apparent.

Gender

The influence of gender on the acceptance of reclaimed water is interesting and has been shown to be an important factor in the acceptance of reclaimed water. The Carley study [11] in 1972 and the 1973 Gallup Poll [12] for the American Water Works Association found that women were less accepting of reclaimed water than men. Two studies made in California in 1976 and in 1978 also showed that regardless of education level—except for men and women who have achieved doctorate degrees—women were less accepting than men of reclaimed water for drinking, swimming and laundry [17,18]. The reason why women, regardless of almost all educational levels, were more negative must still be elucidated, but several hypotheses suggest the woman's role as caretaker of the family or perhaps her more intimate contact with substances entering domestic sewage could account for this difference.

Age

Several studies have demonstrated that age is correlated negatively with the acceptance of reclaimed water [11,12,23]. The Gallup Poll [12] was the first extensive survey to suggest a relationship between age and acceptance of renovated wastewater. In his most recent study Bruvold [23] used sophisticated sampling techniques to select 1394 individuals from ten communities in California. The results indicated an inverse relationship between age and acceptance of reclaimed water. Older individuals and those who had resided longest at their current address were less favorable toward uses of reclaimed water.

Beliefs

Knowledge and Experience

Early investigations of attitudes toward drinking reclaimed water often included such factors as geographic location and past experiences with or knowledge of reclamation projects.

In 1965 Baumann and Kasperson [24] conducted a telephone survey of 36 communities and 722 individuals in Texas, Kansas, Illinois and Massachusetts. The major focus of this survey was respondents' willingness to drink reclaimed water. The authors felt that this preliminary investigation indicated regional variations in consumer acceptance. Overall the results showed that 62% of the respondents were willing to drink reclaimed water. Acceptance ranged from a low of 45% in Massachusetts to a high of 74% in Illinois. Firm positions were held by 92% of the respondents from these four states on drinking reclaimed water with only 8% of those surveyed undecided on the issue.

Clark University conducted 408 interviews [6] in five cities. In Lubbock, Texas, a city with an adequate water supply, 66% of those surveyed were against drinking reclaimed water. In San Angelo, Texas, which had faced a severe water shortage, only 38% were against such water reuse. Of the others interviewed, opposition came from 61% in Kokomo, Indiana, 49% in Colorado Springs, and only 21% in Santee, California.

The authors reasoned that these differences were based on previous positive or negative experiences with reclaimed water or on prior experience with a water crisis situation. The reclamation effort in Santee demonstrated a mixed, public response. Santee used reclaimed water for recreational lakes and golf-course irrigation. The lakes were completed in 1961 and the public was allowed to use this area for picnicking, boating and scenic beauty, but primary contact was prohibited. Later, fishing was inaugurated but a capture-release program was maintained until June 1964. In the water-short areas of California and Colorado the creation of such recreational lakes can be seen by the public as a positive benefit. However, it should be pointed out that at Santee primary contact with reclaimed water was never allowed by the California State Department of Public Health. Eventually, a recall election against water managers resulted in their resignation and the golf course is no longer irrigated with reclaimed water.

More recent research suggests the degree of past knowledge or experience has perhaps been overestimated. Bruvold and Ongreth [19] found that individuals from California cities that had reclamation projects were only slightly more favorable to drinking reclaimed water than those from cities that did not have such projects. This finding was substantiated by Olson et al. [18] in their 1978 study comparing the cities of Irvine (reclamation projects) and Anaheim, California (no reclamation projects). The difference in acceptance between the two cities was less than 3%. In the Irvine studies conducted by Olson and Pratte [17] and Olson et al. [18] there was no difference in acceptance of intimate contact uses in light of ongoing reclamation practices in the community even after quite severe drought conditions had been experienced in the state. These communities would be expected to be highly favorable based on sociodemographic factors and past experience with reclamation projects.

The Johnson study [14] found that individuals who felt their water supply was inadequate to meet future demands favored the use of reclaimed water. The University of California study conducted in 1978 in Irvine and Anaheim was unable to demonstrate that the lack of water supplies to meet future needs had a statistically significant effect on acceptance of reclaimed water for swimming, laundry or drinking purposes. The Clark University study [6] reported that water supply crises in Glouchester, Massachusetts, and San Antonio, Texas, had engendered a more positive feeling in reclaiming wastewater in terms of drinking than would normally be expected.

Present Water Quality

The effect of the quality of the present water supply on attitudes toward reclaimed water has been addressed by several researchers. However, their conclusions are divergent. Johnson [14] reported that those who perceived their water quality as poor would be willing to drink reclaimed water. Stone [16] maintained that individuals who perceived a high quality of water would be more favorable toward drinking reclaimed water due to greater faith in water treatment practices. The 1978 University of California study was unable to demonstrate any statistically significant relationship between perceived quality and more favorable attitudes toward using reclaimed water for laundry, swimming or drinking purposes.

The Johnson study [14] surveyed citizens from five areas of the United States to compare areas where consumers had high or low perceived quality of present water supplies and where consumers would perceive an extreme scarcity or abundance of water supplies. Residents of Portland, Oregon and Tucson, Arizona felt that their water was moderately turbid, and those from Philadelphia and Cincinnati felt that their water had a heavy to moderate odor.

These individuals were also asked how much more they would be willing to pay to avoid the use of reclaimed water. Of those who viewed their source of water as being polluted, 56% were not willing to pay any more for the use of reclaimed water, 59% who were not sure of the quality of their water were also not willing to pay any additional costs for the use of reclaimed water. Of the individuals who perceived their present water supply as nonpolluted, 74% were willing to pay more to keep their present supply. Further questioning revealed that of those who perceived their water quality as polluted, only 17.4% were willing to incur a cost increase of 50% to keep their present water source compared to 18.2% for those who felt their present water was unpolluted. Increases of greater than 51% of current water costs to maintain the present supply were approved of by 15-18% of individuals who were not sure of their water quality or felt that it was unpolluted. Only 7% of the individuals who thought their water supply was polluted were willing to accept an increase of more than 51% to keep the water source.

Johnson concluded that an environment where water supply sources are most polluted would provide the setting for greater consumer acceptance of renovated wastewater. These findings indicate that the Northeast, the Ohio Valley, the Midwest and the Lower Plains would be more favorable toward using renovated wastewater than would the Pacific Northwest, New England, and the Southeast. The Great Lakes, the Missouri Basin, the Lower Mississippi Basin and the Southwest would be moderately predisposed to the use of recycled water.

Economics

Early studies indicated that the price or the cost involved in water supply appeared to have little impact on the acceptance of reclaimed water. The Clark University [16] and the Johns Hopkins [22] studies found that there was no effect between cost increases in the water bill and acceptance of reclaimed water. However, when wastewater reclamation is viewed to be economically beneficial, individuals in Orange County, California, were more in favor of its uses for drinking, swimming and laundry [22]. This study further showed that only 8.3% of the respondents from Irvine, California, a city utilizing reclaimed water for irrigation purposes, would be willing to drink reclaimed water rather than pay a higher price for water compared to 30.5% of the respondents from the City of Anaheim, a nonreclamation city. Interestingly, the majority of respondents from both cities were willing to pay approximately 25% more on their water bill to avoid using reclaimed water for drinking. These findings differ from those of the Clark University and the Johns Hopkins studies. Drinking reclaimed water is more than a far-off supposition in California. These data support a hypothesis set forward by Bruvold [25] that as the reality of the situation comes closer, the response becomes more negative. Thus, even in a highly educated, affluent area such as Orange County individuals would be willing to incur significant increased cost in order to avoid the higher uses of reclaimed water.

Psychological Variables

The importance of psychological variables and their influence on the acceptance of reclaimed water was first addressed by Bruvold in the generalized terms of psychological repugnance which correlated negatively with acceptance of reclaimed water.

Baumann and Sims [26] demonstrated a negative but insignificant correlation between psychological variables and acceptance of reclaimed water. Olson and Pratte [17] extended these earlier efforts by defining six psychological variables. Several of these (aversion to the unclean, overconcern with health, and aversion to human waste) were found to negatively correlate with the acceptance of reclaimed water. Faith in science and technology, aversion to change and ecological concern had no effect on acceptance of reclaimed water for drinking, swimming or laundry purposes. A later study [18] found that experiences in childhood regarding attitudes toward cleanliness were important in predicting a person's negative response toward the use of reclaimed water for swimming, laundry and drinking purposes.

Using a chi square analysis Johnson [14] demonstrated that individuals who believed that technology could solve almost any problem had a higher acceptance of reclaimed water and that individuals who felt that it was safe to drink the product of polluted sources were also more willing to drink renovated wastewater. Respondents were also queried concerning how they felt other individuals in the community would accept reclaimed water. Approximately 70% felt that other individuals in the community would either not be sure or would reject the use of reclaimed water. Only 29% of those who approved the use of reclaimed water thought that other local residents would approve, whereas nearly 77% of those who disapproved thought that the community would reflect their attitude.

From these studies it would appear that approximately 10% of the population is extremely negative toward all uses of reclaimed water. It is highly unlikely that this group will be positively influenced on water reclamation issues by educational or other types of media. However, using Bruvold's [21] supposition that if 20% or less of the population is negative there would be no difficulty in the initiation of water reclamation practices, then this segment of the population becomes unimportant in the implementation of reclamation projects.

Governmental Factors

Public Officials

An extensive study made in 1973 of 158 federal, state and local water resources officials [27] found that public officials often underestimate the public acceptance of reclaimed water. A survey made after the Chanute reuse experience found that water managers believed the public to be far more negative than was actually the case. A 1969 survey of Chanute residents [6] showed the public acceptance rate for drinking renovated wastewater as 61%, but public officials felt only 1.4% of the public would be predisposed to drinking reclaimed water. Further, ten officials from the Philadelphia Water Treatment Plant felt that consumers would not approve of renovated wastewater [14].

All studies conducted on public acceptance of drinking reclaimed water have indicated a higher acceptance on the part of the respondents than water management officials would have predicted. The more negative attitude of public officials is not too surprising, for they are most likely to interact with individuals who openly express their negative attitudes toward such issues. Public health officials and engineers are charged with protecting health. Therefore, they are likely to be conservative in their outlook toward the safety of using renovated wastewater with the viewpoint that it is better to error on the side of safety. This conservative attitude is reflected in how officials and engineers envision the public's attitude as a whole. Though officials have often been criticized for holding such negative views, perhaps in reality a proper balance is struck between a more accepting and less knowledgeable public, and the better informed, more conservative water managers.

Local Government

The Johnson study [14] found no relationship between respondents' attitudes and the acceptance of reclaimed water. The 1978 University of California study found similar results. This latter study reasoned that low interest in government affairs accounted for the lack of relationship between acceptance of reclaimed water and attitude toward government. The same study showed that only 11.3% of the population from both Irvine and Anaheim reported an active involvement in community affairs and that 64% of the population had never attended a local community meeting.

Attitudes of Industrial Managers Toward Water Reuse

Pratte and Litsky [10] examined the potential for wastewater reuse by industry as a means of water conservation in Boston. This city has been named as one of the five areas facing drought conditions during the next decade unless strong conservation measures are undertaken.

Industrial officials of plants which used a minimum of 3.8×10^5 liters of municipal water per day were surveyed. Only 4 of the 26 firms used just the minimum, the majority used at least 9.6×10^5 liter/day of fresh water.

The results of the study indicated that 42% of the industries were willing to use reclaimed water if it met industrial quality requirements, 23% agreed to use reclaimed water under certain conditions and 35% indicated that perhaps they would agree to use reclaimed water. But none of the industries interviewed indicated that they would refuse to use reclaimed water that met their quality standards.

Only 35% of the industries interviewed agreed to accept reclaimed water that met drinking water standards at the cost of their present water. Overall, industrial officials agreed that the use of reclaimed water could conserve fresh water, but 89% of those surveyed attached additional needs such as: economic feasibility, proximity to sewage treatment plants, and superiority of recycling as a conservation technique.

Additional interviews concerning the feasibility of industrial reuse were conducted with public officials from the Metropolitan District Commission, Department of Environmental Quality Engineering, Department of Environmental Management, Water Resources Commission, New England River Basins Commission, Environmental Protection Agency (Region 1) and the Massachusetts Association of Conservation Commissions.

The opinion of public officials toward water reuse in industry revealed divergent views on the issue. Of the 12 officials interviewed from the agencies, 6 believed that water reuse by industry had potential in Massachusetts, and 6 believed that it did not. When public officials were asked to consider reuse as a means of state water conservation only one felt that it would enhance conservation, three were undecided and eight believed that it would not contribute to the conservation effort. Again, as has been shown in attitude studies of the general public, this study demonstrated that industry officials, like the general public, were more accepting of the use of reclaimed water than were public officials.

CONCLUSIONS

Public Opinion Data

The results of several major public studies of public opinion regarding reclaimed water use in the United States have been reviewed and analyzed in this chapter. It is a fair recapitulation to say that although percentage results on specific uses vary from study to study, the ordering of results by favorability of opinion toward specific uses has been remarkably stable (Table I) [25]. The use of reclaimed water for drinking and other close contact or ingestive purposes is always least favored. Uses of reclaimed water for medium contact purposes such as laundry or dairy pasture irrigation are next least favored. Use of reclaimed water for low contact purposes such as golf course irrigation is quite highly favored, and uses involving essentially no human contact such as green belt irrigation are most favored. Thus it may be said that public opinion toward specific uses of reclaimed water becomes more favorable as the degree of human contact decreases or, conversely, that public opinion toward specific uses of reclaimed water becomes less favorable as the degree of human contact increases.

This generalization holds over several major studies and is not reversed or reordered by such factors as gender, age, education, belief and knowledge of respondents. The conclusion seems clear: use reclaimed water for lower contact purposes first, then increase the degree of contact as local acceptance improves. Such advice is sound and likely generally correct, but is it borne out by actual experience? Unfortunately, there are not good answers to this important question, as actual adoptions of reclaimed water for higher contact uses have been rare and temporary, such as in Chanute, Kansas or Santee, California.

What will happen as communities begin widespread visible use of reclaimed water? Will the acceptance follow the "ladder" of public opinion as indicated by the several studies done on this matter in the 1970s? Unfortunately what people say they will do in response to a survey and what they actually will do in a real situation may not be the same. Research studies of actual adoptions of reclaimed water by various communities are needed to determine if the advice that emanates from the public opinion research is sound.

Option Rating Data

In an attempt to move closer to actual studies of the adoption of reclaimed water, Bruvold [23] researched ten California communities that were seriously considering specific uses of reclaimed water in the aftermath of the 1975-1976 West Coast drought. This research is not yet adoption research, but at least here specific options for reuse were in the proposal stage and respondents had to seriously think about which, if any, they could accept and support with their own tax monies. This research, which was several steps closer to actual behavior than were earlier public opinion surveys, yielded results that only partially confirmed the survey findings. The option rating research showed that people, in general, were very much opposed to ingestive use of reclaimed water. When people knew the options were real and they soon could be receiving reclaimed water in their own homes their responses to such proposals became strongly negative—even more negative than the public opinion data might indicate. However, as degree of contact involved lessened, support for reclamation options became quite positive, with some receiving 70-30 percentage splits in favor of the reclamation option.

These results confirm the public opinion data which indicate that attitude will change from unfavorable to favorable as the degree of contact in reuse decreases. But there was a new result in the option rating data not seen in the public opinion data, and this was that opposition began to rise for the lowest contact and disposal options, a result not seen earlier. When thinking of their own community, people opposed the use of reclaimed water for drinking and also opposed its use for a very low contact purpose and especially for disposal. Nevertheless, they approved its use and the additional cost of treatment necessary for medium contact purposes such as park and agricultural irrigation because they felt that such uses would best conserve scarce water resources, best protect public health and best protect the environment. The generalized rating option results compared to the generalized public opinion results are schematically presented in Figure 1.

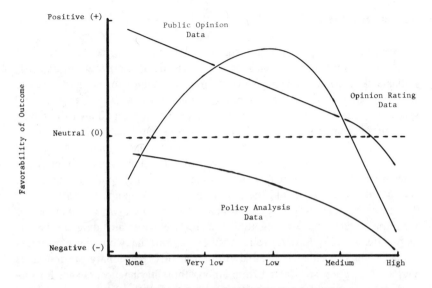

Figure 1. Generalized results of public opinion, option rating and policy analyses findings.

Policy Analysis Data

Bruvold et al. [28] have recently completed a policy analysis assessing the relative disbenefits of several uses of reclaimed water spanning the range from very low contact to ingestive use. The policy analysis systematically considered health effects, environmental effects, treatment costs, distribution system costs and public opinion data for each of 11 specific uses of reclaimed water. The relative disbenefits calculated were groundwater recharge by spreading (-5), groundwater recharge by injection (-8), industrial use where human contact was likely (-6), industrial use where human contact was unlikely (-1), irrigation of fodder and fiber crops (-2), irrigation of food crops (-5), irrigation of parks and playgrounds (-4), restricted recreational lakes (0-1), nonrestricted recreational lakes (-5), direct potable reuse (-10), and residential lawn irrigation (-5). The general trend revealed by this analysis is illustrated in Figure 1, which shows that the policy analysis data roughly parallel the public opinion data except that the policy findings are consistently more negative, showing some increasing fall off as ingestive use is approached.

Recommendations

What do all the presently available findings indicate for public acceptance of various possible uses of reclaimed water? The data, summarized in Figure 1, indicate that options involving low to medium degrees of contact have the best chances of being realized. Options involving no contact or very low degrees of contact face the risk of public opposition even though the public opinion and policy analysis data appear favorable. Options involving ingestion or very high levels of intimate contact appear very unsuitable because public opinion, option rating and policy analysis results all are extremely negative. Low to medium range contact options receive favorable public opinion and option rating responses, and the policy analysis data are not too unfavorable. Therefore, the best way to proceed may be to propose two or three low- to medium-contact uses to community decision makers, including the voters, as practicable [29] in order to identify one option that would have the widespread support needed for actual fruition in this recessionary tax-conscious period. An option so selected from among those having most chance for success ought to produce more actual adoptions of advanced uses of reclaimed water. Subsequent studies will then deal with public attitudes toward existing uses, not potential or proposed uses.

REFERENCES

1. U.S. Public Law 92-500, The National Water Pollution Control Act Amendments of 1972.
2. Tefler, J. G. In *Water Reuse,* C. K. Lawrence, Ed. (New York: American Institute of Chemical Engineers, 1967), pp. 101-105.
3. Pratte, J. *Trends of Wastewater Reclamation: South Africa–1977,* Final Report for the State Water Resources Board of California, September 1977, pp. 1-40.
4. Ackerman, N. "Water Reuse in the U.S.," Unpublished M.A. Thesis, Department of Geography, Southern Illinois University, Carbondale, IL (1971).
5. Hart, O., and L. R. J. Van Vuuren, "Water Reuse in South Africa," in *Water Renovation and Reuse,* H. Shuval, Ed. (New York: Hillel Academic Press, 1977), pp. 355-395.
6. Kasperson, R. E., and J. X. Kasperson, *Water Reuse and the Cities* (Hanover, NH: Clark University Press, 1977), pp. 121-141.
7. Gavis, J. *Wastewater Reuse,* U.S. National Water Commission Reports, NWC-EES-71-003, July 1971 (Washington, D.C., 1971).
8. World Health Organization, *Reuse of Effluents: Methods of Wastewater Treatment and Health Safeguards,* WHO Technical Reports, 517 Geneva (1973).
9. Eden, G. E., D. A. Bailey and K. Jones, "Water Reuse in the United Kingdom," in *Water Renovation and Reuse,* H. Shuval, Ed. (New York: Academic Press, 1977), pp. 397-428.

10. Pratte, J., and W. Litsky, "The Potential for Municipal Wastewater Reuse by Industry in Massachusetts," in *Water Reuse Symposium,* R. Heaton, Ed. 2:1235-1246 (1979).
11. Carley, R. L. "Wastewater Reuse and Public Opinion," Unpublished M.S. Thesis, Department of Civil and Environmental Engineering, University of Colorado, Boulder, CO (1972).
12. American Water Works Association. "On the Use of Reclaimed Water as a Public Water Supply Source," AWWA Policy Statement, *J. Am. Water Works Assoc.* 63:609 (1971).
13. Bruvold, W. H. "Affective Response Towards Uses of Reclaimed Water," Experimental Publications System, Issue 3 Dec., 1971 Manuscript 1170, pp. 1-12.
14. Johnson, J. F. "Renovated Wastewater: An Alternative Source of Municipal Supply in the U.S., (Chicago, IL: University of Chicago, Department of Geography Research Papers, 135, 1971).
15. Bruvold, W. H. "Using Reclaimed Wastewater—Public Opinion," *J. Water Poll. Control Fed.* 94:1690-1696 (1972).
16. Ralph Stone and Company, Inc. *Wastewater Reclamation: Social Economics and Technical Public Acceptance* (Springfield, VA: National Technical Information Service, 1974).
17. Olson, B. H., and J. Pratte. "Psychological Factors Affecting Public Acceptance of Water Reuse," *J. Water Tech. Prog.* 19 1/2:319-321 (1978).
18. Olson, B. H., J. A. Henning, R. A. Marschack and M. G. Rigby. "Educational and Social Factors Effecting Public Acceptance of Reclaimed Water," in *Water Reuse Symposium,* R. Heaton, Ed. 2:1219-1230 (1979).
19. Bruvold, W. H., and H. J. Ongreth. "Public Use and Evaluation of Reclaimed Water," *J. Am. Water Works Assoc.* 66:294-297 (1974).
20. Bruvold, W. H., and P. C. Ward. "Public Attitudes Toward Uses of Reclaimed Wastewater," *Water Sew. Works* 117:120-122 (1970).
21. Bruvold, W. H. "Public Opinion and Knowledge Concerning New Water Sources in California," *Water Resources Res.* 8:1145-1150 (1972).
22. Baummer, J. C., and P. G. Robertson, Eds. *Reclamation of Wastewater for Municipal Supply* (Baltimore, MD: Department of Geography, Johns Hopkins University, 1973), Ch. 4.
23. Bruvold, W. H. *Public Attitudes Toward Community Wastewater Reclamation and Reuse Options.* (Davis, CA: University of California Water Resources Center, 1979).
24. Baumann, D. D., and R. E. Kasperson. "Public Acceptance of Renovative Wastewater: Myth and Reality," *Water Resources Res.* 10:667-674 (1974).
25. Bruvold, W. H. "Human Perception and Evaluation of Water Quality," *Critical Reviews in Environmental Control* 5:153-231 (1975).
26. Baumann, D. D., and J. Sims. "Renovative Wastewater for Drinking: The Question of Public Acceptance," *Water Resources Res.* 10:675-682 (1974).
27. Dworkin, D., and D. Baumann. *An Evaluation of Water Reuse for Municipal Supply* (Alexandria, VA: Institute for Water Resources, U.S. Army Corps of Engineers, 1973), pp. 513-519.
28. Bruvold, W. H., B. H. Olson and M. Rigby. "Public Policy for the Use of Reclaimed Water," *Environ. Manag.* (in press).
29. Bruvold, W. H., and J. Crook. "Water Reuse and Recycling: Gaining Public Acceptance," *Consulting Eng.* 53:126-129 (1979).

CHAPTER 4

FEDERAL INCENTIVES AND REQUIREMENTS INFLUENCING WASTEWATER REUSE

Richard E. Thomas
U.S. Environmental Protection Agency
Office of Water Program Operations
Washington, DC

INTRODUCTION

This chapter presents an overview of guidelines applicable to reuse of municipal wastewaters as mandated by recent federal laws and the U.S. Environmental Protection Agency (EPA) extension of this mandate into federal guidance for management of the EPA Construction Grants Program. Reuse of municipal wastewater and recycle of the nutrients it contains through the innovative/alternative technology thrust of the EPA Construction Grants Program is only one aspect of conserving water supplies through reuse. However, the magnitude of the program (with over 12,000 projects and eventual expenditures likely to exceed 50 billion U.S. dollars) will produce an obvious impact on the technology for reusing and recycling municipal wastewater throughout the world.

The major elements which provide the framework for criteria applied to reuse and recycle systems funded by the federally assisted Construction Grants Program are outlined here. More specific information is contained in the documents and publications highlighted in this chapter.

FEDERAL LEGISLATION

Early federal legislation for water pollution control in the United States had no provisions to encourage reuse or recycling of wastewaters as a conservation practice. In fact, the Federal Water Pollution Control Act of 1956 instituted a grants program containing prohibitions and omissions that discouraged development and use of many recycling or reuse alternatives.

The Federal Water Pollution Control Act Amendments of 1972 was the first federal legislation to contain provisions that encouraged recycle and reuse. The encouragement offered in that Act was reiterated in the Clean Water Act of 1977 and was given the added impetus of many financial incentives that encompass wastewater reuse. These financial incentives are integral factors in the innovative and alternative (I/A) technology program spelled out in the Clean Water Act of 1977.

The projected influence of these incentives and requirements on the future of wastewater reuse planning and implementation is considered here. The factors to be covered include: (1) financial incentives to stimulate more use, (2) requirements to spend a specific part of funds appropriated in fiscal years 1979, 1980 and 1981, (3) cost-effectiveness allowances, and (4) new areas of eligibility. These factors are covered from the perspective of the new law, the EPA promulgation of construction grant regulations, and the EPA issuance of program guidance for implementing the law and the new revisions to construction grant regulations.

Early Federal Legislation

Sullivan [1] has summarized federal legislation for water pollution control from 1899 until passage of the Federal Water Pollution Control Act Amendments of 1972. The 1899 Rivers and Harbors Act prohibited discharge of refuse into navigable waters; the Public Health Service Act of 1912 provided for surveys and studies of the effect of water pollution on human life; and the Oil Pollution Act of 1924 prohibited discharge of oil into coastal waters. But the legislation passed from 1899 to 1948 was not enforced vigorously.

The Water Pollution Control Act of 1948 was the first legislation with broad coverage for abatement of water pollution. It included support for research and technical assistance to the states while maintaining the policy that the states were to have the primary role in abating pollution. The Federal Water Pollution Control Act of 1956 signalled a change in the federal policy by establishing a federal grants program to assist states and municipalities in planning and building facilities for treatment of wastewaters. This first venture of federal funding for abating water pollution was actually structured to discourage wastewater reuse. It did not make any form of recycling or recla-mation of wastewater eligible for grant assistance. All of the five billion dollars provided by this act was directed to construction of facilities using conven-tional treatment technologies. However, the impetus for use of recycling and reuse technologies to become a focal point of later federal involvement came from the increased support for research and development in the 1956 Act.

The 1972 Act (PL 92-500)

The Federal Water Pollution Control Act Amendments of 1972 mandated a sweeping federal/state/local government program to reduce, prevent and eliminate water pollution [2]. This Act contains several provisions directly encouraging reuse concepts, including:

1. a mandate for the Administrator of EPA to encourage waste mangement alter-natives that would produce revenues from recycling sewage pollutants for production of agriculture, silviculture or aquaculture products;
2. a provision that grants made from funds authorized for any fiscal year begin-ning after June 30, 1974, include consideration of alternative technologies and that the works proposed for grant assistance provide for best technology adaptable to recycling or elimination of pollutant discharge; and
3. a definition that land which will be an integral part of the treatment process is an allowable cost but land for construction of conventional plants is a non-allowable cost.

The message of this act was straightforward: Planners should give serious con-sideration to alternatives that combine effective wastewater management with reuse of the wastewater and nutrients. This encouragement of reuse alterna-tives stimulated intense discussion between proponents and opponents but had only a moderate influence on the actual adoption of reuse and recycling technologies. The moderate influence on actual adoption was a major con-sideration for the Congress as they formulated the Clean Water Act of 1977.

Land application of wastewater has been the dominant process for reuse of wastwater throughout history. Tracking the use of land application is a good indicator of government and public attitudes toward reuse and reclamation. Thomas and Reed [3] have summarized the use of land application in the United States from 1940 through 1976. They note that although there was a dramatic increase in use following passage of the 1972 Act, the increase was comparable to the rate of increase in the 1940s without federal encourage-ments or funds. It is of interest to observe that passage of the 1956 Act with its disincentives for land treatment was followed by a period of decreasing re-use. Their tabular display suggests that wastewater reuse will experience in-creasing popularity and use under present federal policies and funding support.

The 1977 Act (PL 95-217)

The Clean Water Act of 1977 [4] requires grant applicants to analyze innovative and alternative technologies in wastewater treatment after September 30, 1978. This requirement and the legislative history of the act make it clear that the intention is to achieve increased wastewater reuse and recycling of nutrients. In the words of the Congressional Conference Committee report on the Clean Water Act, the EPA ". . . has been provided all of the legislative tools to require the utilization of such innovative and alternative wastewater treatment processes and techniques." These I/A technologies emphasizing nutrient recycling, wastewater reuse in agriculture and industry, and energy recovery are to be given preference over conventional wastewater technologies in the EPA program to abate water pollution.

The Clean Water Act was enacted into law on December 27, 1977, and it does more than carry the straightforward message of the 1972 Act that water reuse and nutrient recycling deserve serious consideration as viable waste management alternatives. It provides many financial incentives for I/A approaches to waste management and specifically designates many water reuse and nutrient recycle approaches as I/A technology. Among these financial incentives for I/A projects are the following:

1. The Federal share of a construction grant may be increased from 75% to 85%;
2. Two percent of allocated funds must be spent on increasing the federal share from 75% to 85% in fiscal years 1979 and 1980 (in 1981 this set-aside increases to 3%);
3. The federal government may participate with full construction grant funding in projects which are up to 15% more costly than the most cost-effective of the conventional alternatives; and
4. Projects which fail to meet design criteria may be eligible for 100% federal grants for modification or replacement.

These financial incentives underscore the intent of the Congress to force reuse as opposed to treatment and discharge. Water reuse technologies which were designated as eligible for the financial incentives include land treatment, agricultural reuse, direct reuse (nonpotable) and aquifer recharge.

IMPLEMENTING THE FEDERAL LAWS

Most of the identifiable pieces which collectively can be considered as federal guidelines for the reuse of wastewater are linked in some way to the EPA Construction Grants Program [5]. Section 201 of the Law, which authorizes EPA to make grants to municipalities for the construction of wastewater treatment facilities, also require that grant applicants evaluate "alternative waste management techniques . . . (which) provide for the application of best practicable waste treatment technology . . ." Section 201 further specifies that best practicable waste treatment technology (BPWIT) includes "reclaiming and recycling of water and confined disposal of pollutants so they will not migrate to cause water or other environmental pollution . . ."

Pursuant to Section 304 of the 1972 Act, the EPA published "Alternative Waste Management Techniques for Best Practicable Waste Treatment" [6]. Contained in this document are criteria for systems in three categories: treatment and discharge to navigable waters, land application and utilization practices, and wastewater reuse.

In a discussion of EPA guidance, Hais [7] noted that construction grant requirements are directly applicable only to the planning, design and construction of facilities which receive grant funding. Furthermore, the coverage of the rules and regulations under which the Construction Grants Program operates does not extend to the long-term operation of completed facilities, whether these facilities were assisted with grant funds or not. However, because of the magnitude of the Construction Grants Program (eventual expenditures likely to exceed $50 billion and more than 12,000 projects), its impact on municipal wastewater treatment practices in this country is obviously considerable. Consequently, even though the Construction Grants Program elements listed above are not regulatory in the formal sense of the word, their impact is often comparable to that of regulations. Where state regulations or guidelines do not exist to control a particular practice, applicable construction grant guidelines have in some instances been adapted to fill this void. In other cases, existing state guidelines have been revised as necessary and appropriate in recognition of federal requirements.

The effect of the federal guidelines was great enough that by 1978 Rhett stated: "There is a strengthening sense . . . that we are on the threshold of a revolution in the concept of wastewater treatment in the United States. There is growing optimism that alternatives which emphasize recycling will indeed become the standard in the near future" [8].

Elements Related to the 1972 Act

The Federal Water Pollution Control Act Amendments were enacted in October 1972. This Act detailed a comprehensive federal role in control of water pollution within a sweeping federal-state-local government campaign to clean up the nation's waters. The Administrator of EPA was instructed to encourage projects which recycle sewage pollutants through production of agriculture, silviculture or aquaculture products. All projects receiving grant funds after June 30, 1974 were required to provide for immediate or later application of reclamation or recycle technologies to the extent that such was practicable. The EPA has issued major program guidance documents on reuse concepts to supplement the Law and the EPA Regulations.

Best Practicable Waste Treatment Technology

The BPWTT criteria published by the EPA [6] in October 1975 contain requirements which for all intents and purposes have the impact of federal regulations for projects receiving support from the multibillion dollar federal grants program.

Municipal wastewater treatment plants are required by law (PL 92-500) to provide for the application of BPWTT by July 1, 1983. The principal reuse and recycle systems available for implementation include land treatment technologies and industrial reuse. A major objective of BPWTT criteria for all reuse and recycle systems is to protect surface and groundwaters for other uses. The criteria for protection of groundwater for systems which reuse wastewater while recycling nutrients describe three cases as follows:

Case 1: The groundwater can potentially be used for drinking water supply. In this case the groundwater resulting from the land application of wastewater, including the affected native groundwater, must meet the maximum contaminant levels for inorganic and organic chemicals specified in the National Interim Primary Drinking Water Regulations.

Case 2: The groundwater is presently being used for drinking water supply. In this case, the groundwater must meet the maximum microbiological contaminant levels specified in the National Interim Primary Drinking Water Regulations as well as the levels for chemicals specified in Case 1.

Case 3: The groundwater has uses other than drinking water supply. In this case groundwater criteria are to be developed by the EPA Regional Administrator based on the present or potential use of the groundwater.

The National Interim Primary Drinking Water Standards for inorganic and organic chemicals and microbiological contaminants presently include limits on arsenic, barium, cadmium, chromium, lead, mercury, nitrate, selenium, silver, fluoride, endrin, lindane, methoxychlor, toxaphene, chlorophenoxys (2,4-D and 2,4,5-TP Silvex) and coliform bacteria. Where a land treatment system results in a surface discharge (either effluent collected through underdrains or runoff from an overland flow system), the wastewater must meet the BPWTT standards for treatment and discharge. The minimum level of treatment required for municipal systems which treat and discharge to surface water is secondary treatment as defined by EPA in terms of five-day BOD, suspended solids and pH. It is important to note that in any case, the point at which the wastewater is measured for compliance with the federal BPWTT criteria is when it leaves the land application system (i.e., becomes part of the permanent groundwater or is surface discharged) and not at the point of application to the land.

EPA Policy Statement on Land Treatment
of Municipal Wastewater

On October 3, 1977, the Administrator of the EPA issused a policy statement [9] on land treatment of municipal wastewater. The policy statement highlighted the agency's intent to "press vigorously for publicly owned treatment works to utilize land treatment processes to reclaim and recycle municipal wastewater." It also required that construction grant applicants who do not select methods that encourage water conservation, wastewater reclamation and reuse must provide a complete justification for the rejection of land treatment.

With respect to "guidelines" for land treatment systems, the October 3, 1977 statement emphasized the importance of ensuring that federal, state, and local requirements and regulations are imposed at proper points in the treatment system. To this end, the statement indicate that "Whenever states insist upon placing unnecessarily stringent preapplication treatment requirements upon land treatment, such as requiring EPA secondary effluent quality in all cases prior to application on the land, the unnecessary wastewater treatment facilities will not be funded by EPA." This is not to say that secondary treatment or an even higher level of treatment may not be necessary for some land application projects. The intent is to cause arbitrary preapplication treatment requirements, such as a minimum of secondary treatment or better in all cases, to be revised to reflect a level of preapplication treatment appropriate for the given situation. Just as secondary treatment may be appropriate in some instances, the equivalent of primary treatment or even raw wastewater may be sufficient in others.

Elements Related to the 1977 Act

The Clean Water Act of 1977 continued congressional emphasis on water reuse as an important factor in the innovative and alternative technology program. In an article on the future of the Construction Grants Program, Senator Stafford [10] noted that the act still exhorts the EPA to recycle nutrients while reusing wastewaters. The EPA responded to the mandates of the 1977 Act with new and specific guidance on use of technologies which conserve energy and recycle nutrients while reclaiming or reusing wastewaters.

The 1977 Act contains specific sections that refer to innovative and alternative technology. The EPA, in fulfilling its mandate under the CWA of 1977, has published final regulations pertaining to I/A technology. Extensive reviews of the legislative history of the act along with consideration of the voluminous public comment and results of public hearings have led to the formulation of the I/A technology guidelines identified as Appendix E of the final regulations [11].

Innovative and Alternative Technology

Experience with wastewater management planning in the U.S. has indicated a strong tendency for contemporary designers to consider a narrow range of alternatives, both with respect to liquid processing technology and for residual and disposal. Innovative design concepts should include a broad range of reuse, beneficial recycling, and energy conservation and recovery opportunities, as well as the common treatment and discharge technologies. There are many recycling, reclamation and energy recovery opportunities that might be incorporated into an innovative concept of wastewater management. These are described and discussed in the EPA manual on I/A technology assessment [12].

Conceptually, innovative designs may embody a number of recycle/reclamation or reuse opportunities depending on the particular site variables and design objectives. Maximum consideration should be given to the identification of all potential conceptual approaches early in the planning process, especially those system designs incorporating alternative technologies that exhibit significant recycle, reclamation, energy recovery or revenue-generating potential.

The underlying concept of the I/A guidelines is the provision of a basic monetary incentive, i.e., a grant increase from 75% to 85% for the design and implementation of municipal treatment technology that represents an advancement of the current state-of-the-technology. Emphasis is on the use of technologies directed to meeting the specific national goals of: (1) greater recycling and reuse of water, nutrients and natural resources; (2) increased energy recovery and conservation; (3) improved cost-effectiveness in meeting specific water quality goals; and (4) improved toxics management.

The legislation, guidelines, and EPA policy have been structured to provide additional incentives to both the public and private segments of the municipal construction industry most directly responsible for implementing improved wastewater management systems. Specific efforts have been made to encourage the use of innovative concepts in the planning and design of municipal treatment facilities by providing indemnification of risk through the provision of 100% grants for modification of facilities which fail.

A major emphasis of innovative systems under the Clean Water Act of 1977 is the greater attention placed on multi-objective planning, intermedia impact considerations, and total systems design. Satisfaction of these objectives require a more systematic screening and evaluation of alternatives than has been generally employed in the past. More important, however, is the much greater effort needed in concept development and the formulation of innovative alternatives.

Land treatment of municipal wastewaters is a leading example of an alternative technology that emphasizes wastewater reuse coupled with nutrients recycle for beneficial crop production. The EPA has issued specific policy and construction grants guidance for this recycle and reuse technology.

Program Requirements Memorandum for Evaluation
of Land Treatment Alternatives

On November 15, 1978, the EPA issued Program Requirements Memorandum (PRM) 79-3, titled "Revision of Agency Guidance for Evaluation of Land Treatment Alternatives" [5]. This particular PRM was developed to provide guidance in support of the October 1977 policy statement issued by the Administrator of EPA.

Again, while a guidance document such as this PRM is specifically developed for implementing the EPA Construction Grants Program, it does constitute EPA policy. In this way it can be interpreted as having the effect of federal guidelines, particularly because these memoranda and similar issuances are widely circulated for review and comment prior to publication in final form.

PRM 79-3 establishes the EPA "Process Design Manual for Land Treatment of Municipal Wastewater" [13] as the principal reference for technical information upon which review of construction grant projects involving land treatment will be based. The PRM also cites three planning and design factors which have historically limited the use of land treatment in this country. These factors are: (1) overly conservative and, consequently, costly design of slow-rate (water reuse and nutrient recycle) systems; (2) failure to consider rapid infiltration as a proven water reclamation alternative; and (3) requirements of a substantially higher and more costly level of preapplication treatment than is needed to protect health and ensure design performance.

The PRM provides more specific guidance on four items related to the design of land treatment systems: site selection, loading rates and land area, preapplication treatment, and environmental effect. Much of the information pertinent to these factors is included in the PRM by reference to other documents such as the Design Manual and the BPWTT publication.

The Agency recognizes that no single value or even one set of values can be realistically applied to all locations considering the variability across the country in climate, geology, treatment needs and other factors affecting the design of land treatment systems to recycle nutrients or reuse wastewaters. For this reason the EPA guidelines are varied to suit a number of possible situations and include ranges of values wherever possible. For example, the range of application rates indicated in the PRM varies from 0.6-6 M/yr for slow-rate systems to 6-170 M/yr for rapid infiltration systems.

This concept of flexibility is particularly important with respect to EPA's position on preapplication treatment requirements. Some have misinterpreted

EPA statements as indicating that the Agency will not support any project that requires secondary treatment prior to application to the land. This is not the case. The Agency requires that the level of preapplication treatment be suited to the particular situation. In some cases primary treated or even raw sewage may be acceptable for application to the land, and in other instances treatment beyond that provided by secondary may be necessary. Clearly an arbitrary requirement that all wastewater must receive secondary treatment prior to land application is not consistent with the Agency's approach. The guidance on preapplication treament included in the PRM ranges from simple screening or comminution for overland flow in isolated areas with no public access to extensive BOD and suspended solids control with disinfection for reuse systems in public access areas such as parks and golf courses.

SUMMARY

Federal legislation to control water pollution was initiated with the Rivers and Harbors Act of 1899 but this act and subsequent legislation in the first half of the twentieth century did nothing to encourage wastewater reuse. The Federal Water Pollution Control Act of 1956 was a turning point. It included a modest construction grants program continuing support for use of conventional technologies, and it also included support for research and development of new technologies. The results of this research were coupled with a growing public concern for the quality of our environment to form a foundation for the Federal Water Pollution Control Act of 1972. This Act mandated an ambitious program to clean up the surface waters of our nation. The recycling of nutrients and reuse of wastewaters were central themes of this goal for cleaning waters that had been seriously polluted by past management of wastewaters. The recently formed Environmental Protection Agency was responsible for implementing the program, which was built around a three-year appropriation of 18 billion dollars in federal funding support. A congressional review of progress in 1977 revealed that the first five years of the federal funding program had not achieved the shift to reuse and recycle technology anticipated with passage of the 1972 Act. Congress took this into account as they formulated the Clean Water Act of 1977. In an attempt to stimulate greater use of reuse and recycle technologies, the 1977 Act provided financial incentives for these technologies. Innovative and alternative technologies including many reuse technologies receive an 85% federal share instead of the 75% share received by users of conventional technologies.

EPA implementation of the federal construction grant program for publicly owned treatment facilities has accomplished some substantial gains for reuse technologies as a result of the Federal Water Pollution Control Act of 1972. The knowledge gained from research and demonstration efforts started in the early 1960s has been incorporated into Agency policy and guidance documents throughout the 1970s. Reuse of partially treated wastewaters in agriculture and silviculture is gaining renewed popularity as a management option. Selected industrial use of treated wastewaters reduces demand on limited fresh water sources. Research and demonstration projects are rapidly advancing the knowledge for reuse of wastewater in aquaculture projects. EPA has issued a series of guidance documents to encourage proven and developing wastewater management technologies which include reuse. These guidance documents have wide impacts because the program which they guide may exceed 50 billion dollars and over 12,000 projects. In the wastewater treatment profession, there is a growing sense that the provisions of the Clean Water Act of 1977 may make reuse and recycle the standard in the near future.

REFERENCES

1. Sullivan, R. H. "Federal and State Legislative History and Provisions for Land Treatment of Municipal Wastewater Effluents and Sludges," in *Recycling Municipal Sludges and Effluents.*
2. U.S. Congress. "Federal Water Pollution Control Act Amendments of 1972" (Washington, DC: U.S. Government Printing Office, 1972).
3. Thomas, R. E., and Reed, S. C. "EPA Policy on Land Treatment and the Clean Water Act of 1977," *J. Water Poll. Control Fed.* 52(3):452 (1980).
4. U.S. Congress. "The Clean Water Act" (Washington, DC: U.S. Government Printing Office, 1977).
5. "Construction Grants Program Requirements Memorandum 79-3," U.S. Environmental Protection Agency (Washington, DC: Office of Water Programs Operations, 1978).
6. "Alternative Waste Management Techniques for Best Practicable Waste Treatment," U.S. Environmental Protection Agency, *Federal Register* 41(26):6190 (February 11, 1976).
7. Hais, A. B. "Federal Guidelines for Use of Land Treatment of Wastewater in the United States," in *State of Knowledge in Land Treatment of Wastewater.* H. L. McKim, Ed. (Hanover, NH: U.S. Army Corps of Engineers, 1978), p. 1.
8. Rhett, J. T. "Land Treatment: Achieving Water Quality through Effective Recycling of Wastewater," in *State of Knowledge in Land Treatment of Wastewater.* H. L. McKim, Ed. (Hanover, NII: U.S. Army Corps of Engineers, 1978), p. vii.
9. "EPA Policy on Land Treatment of Municipal Wastewater," U.S. Environmental Protection Agency (Washington, DC' Office of the Administrator, 1977).
10. Stafford, Sen. R. T. "The Future of the Construction Grants Program," *EPA J.* 4(7):5, 36-37 (1978).
11. "Municipal Wastewater Treatment Works–Construction Grants Program," U.S. Environmental Protection Agency, Federal Register 42(188):44022 (September 27, 1978).

12. "Innovative and Alternative Technology Assessment Manual," U.S. Environmental Protection Agency," Report No. EPA 430/9-78-009 (1978).
13. "Process Design Manual for Land Treatment of Municipal Wastewater," U.S. Environmental Protection Agency, Report No. EPA 625/1-77-008 (1977).

CHAPTER 5

WASTEWATER REUSE–AN ASSESSMENT
OF THE POTENTIAL AND TECHNOLOGY

Robert B. Williams
  Culp/Wesner/Culp
  Consulting Engineers
  Cameron Park, California

INTRODUCTION

Wastewater reuse has been practiced in one form or another for countless years in the United States, and probably in every other country in the world. It normally takes the form of effluent discharge to streams that subsequently are used as the water supply for a variety of beneficial uses such as agricultural irrigation or domestic water supplies. This method of disposal has become a generally accepted practice and is frequently called indirect reuse. Conversely, planned direct reuse has only been practiced on a much smaller scale, primarily for agricultural and industrial purposes.

Wastewater reuse can be divided into direct and indirect (or inadvertent) reuse, controlled or uncontrolled reuse and recycling. The term reuse is applied to wastewaters that are discharged and then withdrawn by a user other than the discharger. The terms direct and indirect are differentiated by inclusion of an intermediate step, popularly called "loss of identity", through discharge into a naturally occurring water regime. When a discharge effluent loses accountability, the downstream use is termed indirect. The terms controlled and uncontrolled refer to the extent of prior reuse planning or design. A system is controlled when the particular reuse was planned and intentional and the wastewater reclamation system is specifically designed for the particular project. Uncontrolled reuse is not a part of a planned water management strategy. The classical example of uncontrolled, indirect reuse was described in the opening paragraph, and, in fact, many communities, such as New Orleans, ingest water that has already been used as much as five times. Finally, the term recycling refers to the internal use of water by the original user prior to discharge to a treatment system or other point of disposal.

This chapter presents information on the need and potential for wastewater reuse and then focuses on the treatment technology for controlled, direct reuse of reclaimed wastewaters for a variety of beneficial uses. The information is based on recent reports prepared for the Department of the Interior Office of Water Research and Technology (OWRT) [1,2].

The term "beneficial use" has been defined as: "the many purposes that water serves in promoting the economic good and well-being of mankind" [3]. The beneficial uses of water considered to have the greatest potential for reuse are those projected to require either the greatest quantities of water, or those that are impacted because of projected water shortages. Data for water withdrawals were developed by the U.S. Water Resources Council (WRC) which divided the country into 21 regions as shown in Figure 1.

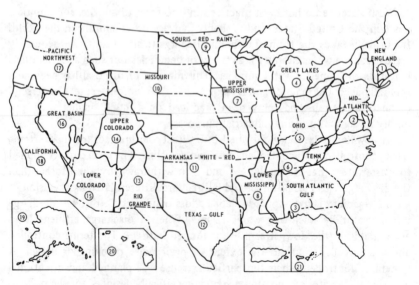

Figure 1. Water Resources Council regions.

EXISTING AND PROJECTED WATER REQUIREMENTS

Freshwater withdrawals and wastewater discharges are presented in a report [2] for the OWRT for agriculture, steam electric, manufacturing, municipal, minerals, public lands and fish hatcheries. The estimated amounts are shown in Figure 2, for the years 1975, 1985 and 2000. Agricultural irrigation and steam electric estimates are based on dry-year conditions. Figure 2 shows a nominal decrease in withdrawals for agriculture and steam electric uses, and a significant decrease (over 60%) projected for manufacturing requirements by the year 2000. Conversely, withdrawals for municipal, mineral, public lands and fish hatcheries are all projected to increase by the year 2000. Similar results are projected for the wastewater discharges from these categories.

Water withdrawals are presented differently in Figure 3, which includes estimated historic values. Figure 3 clearly shows the substantial industrial development between the mid-1950s and 1970, then a sudden drop in withdrawals between 1970 and 1975. This drop is the result of recycling and water conservation practices instituted within the industrial sector.

Freshwater withdrawals for 1975 are shown in Figure 4. It is estimated that agriculture and steam electric plants will continue to use over 75% of all fresh water withdrawn in the United States. To provide an overview of the current major water users in the nation, Figure 5 was developed. It shows that agricultural use dominates the area west of the Mississippi River, and steam electric and industrial uses dominate east of the Mississippi River. Approximately 99% of agricultural water use is for irrigation, the largest water user in the nation. Most of the demand centers in the Pacific Northwest, California and Missouri regions, as shown in Figure 6.

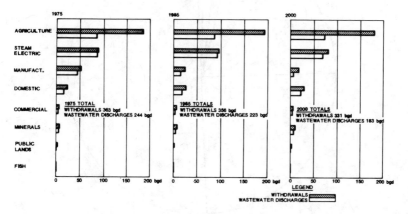

Figure 2. National freshwater withdrawals and discharges.

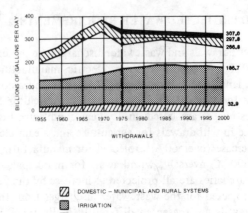

Figure 3. Historical and projected water requirement.

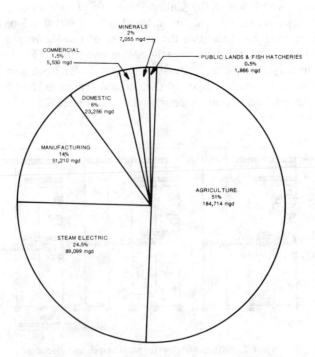

Figure 4. National freshwater withdrawals in 1975.

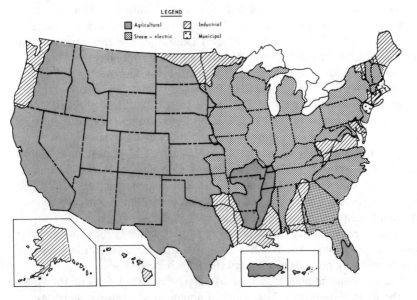

Figure 5. Dominant functional water use in each subregion.

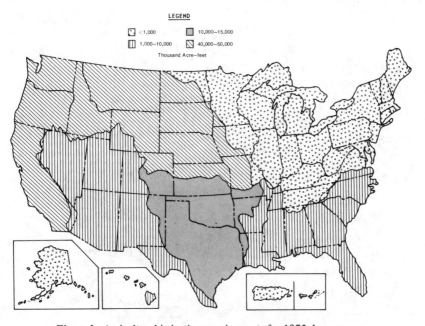

Figure 6. Agricultural irrigation requirements for 1975 dry-year use.

Cooling water for steam electric generating plants represents the next largest requirement for fresh water in the nation. Most of the demand is in the eastern United States, with the Great Lakes and Ohio regions accounting for over 50% of the total. The withdrawals are shown in Figure 7.

Industrial withdrawals have been estimated for the manufacturing and minerals industries listed in Table I. The primary metals, chemicals and paper industries accounted for about 77% of the total freshwater requirements for manufacturing industries in 1975. It is projected that these three industries will withdraw 72% of the industrial water in the year 2000. Most of the water use for manufacturing is in the Great Lakes, Ohio, Mid-Atlantic, South Atlantic-Gulf, Lower Mississippi and Texas-Gulf regions, as shown in Figure 8. The Great Lakes region accounts for 20% of all industrial use. The three highest industrial water users for each region are shown in Figure 9 to indicate the principal water-using industries.

In the manufacturing industries water is used primarily for cooling, boiler feed and processing. About 70% of all industrial use is for cooling, an area that represents a potential for wastewater reuse and recycling.

The nonmetal mining industries withdraw more than 60% of the water for the mineral industries. The principal nonmetal water-using industries are chemical and fertilizer mining and sand and gravel operations. Reclaimed wastewater could be used for several purposes, including processing in some mineral industries.

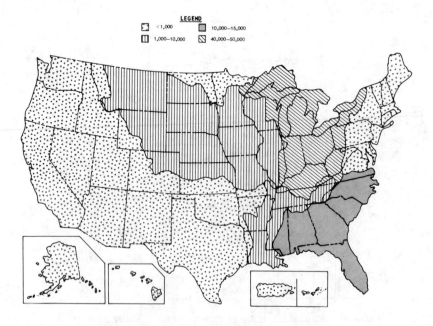

Figure 7. Steam electric water requirements in 1975.

Table I. Manufacturing and Mineral Industries

Manufacturing Industries	SIC No.[a]	Minerals Industries	SIC No.[a]
Primary Metals	33	Metal Mining	10
Chemicals and Allied Products	28	Non-Metal Mining	14
Paper and Allied Products	26	Fuels:	
Petroleum and Coal Products	29	Anthracite Mining	11
Food and Kindred Products	20	Bituminous Coal and	
Transportation Equipment	37	Lignite Mining	12
Textile Mills Products	22	Oil and Gas Extraction	13
Other Manufacturing	—		

[a]Standard Industrial Classification number.

Two other uses are included, although their water withdrawals are substantially lower: municipal requirements and landscape irrigation. The former represents the highest level of use in terms of procurement rights by the states and is the most familiar use. Withdrawals for municipal use, presented in Figure 10, are greatest in the Mid-Atlantic and California regions.

Landscape irrigation is included because reclaimed wastewater could be used to supply a significant part of the requirement. The estimated 1975 water withdrawals amount to 1.2 million ac-ft, and no single region used more than 14% of the total. The water withdrawals are shown in Figure 11, which shows the greatest use in the California, Great Lakes, Mid-Atlantic and South Atlantic Gulf regions.

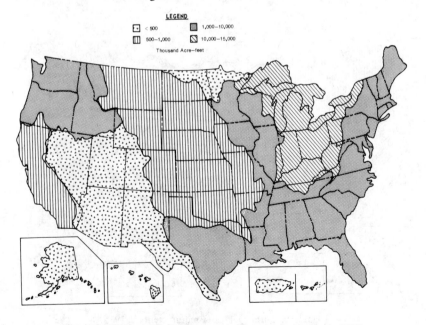

Figure 8. Water requirements in manufacturing industries in 1975.

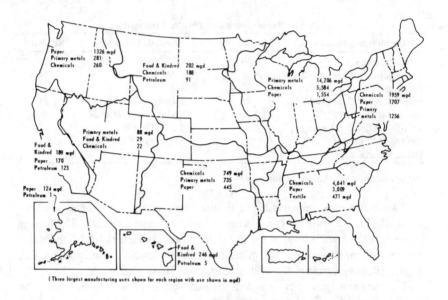

Figure 9. Major manufacturing freshwater withdrawals in 1975.

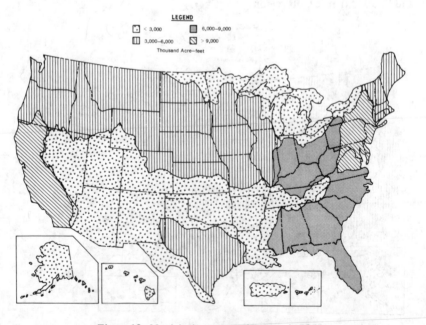

Figure 10. Municipal water requirements in 1975.

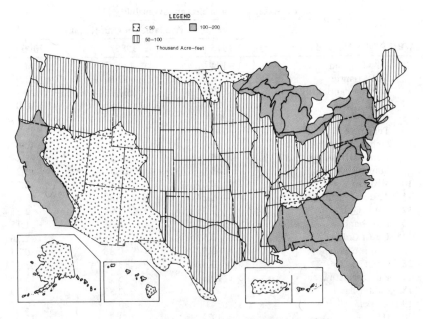

Figure 11. Water use for landscape irrigation in 1978.

Water Availability

The Water Resources Council concluded in the Second National Water Assessment that, over-all, the nation's water supplies are sufficient to meet the requirements for all beneficial uses. However, in some subregions and local areas, water requirements exceed available water sources. Competitive offstream uses of available water for the functional uses already described, coupled with environmental and in-stream requirements, are projected to result in severe shortages in many areas. The severity of the water shortages in some of the regions is demonstrated by the ratios of water availability to water requirements given in Table II. A ratio with a value of 5 or less for a particular region would indicate that local areas within that region would have severe shortages. The lower the ratio, the more severe is the projected water shortage. Areas that may be expected to experience water shortages are shown in Figure 12.

The supply problems include the following:

- shortages resulting from inadequate distribution systems,
- groundwater overdrafts,
- quality degradation of both surface and groundwater supplies,
- institutional constraints and
- competition between various uses.

Table II. Water Requirements and Availability [1]

WRC Region No. and Name	Unit Available per Unit Required [a]		
	1975	1985	2000
1 New England	164	122	74.2
2 Mid-Atlantic	44.2	32.8	22.9
3 South Atlantic-Gulf	47.8	34.3	23.1
4 Great Lakes	29.0	22.8	16.0
5 Ohio	1,000	71.1	41.3
6 Tennessee	131	63.8	37.3
7 Upper Mississippi	99.0	69.0	40.9
8 Lower Mississippi	106	92.7	75.6
9 Souris-Red-rainy	48.0	27.9	13.2
10 Missouri	3.04	2.52	2.37
11 Arkansas-White-Red	6.34	5.80	5.65
12 Texas Gulf	2.75	2.95	2.85
13 Rio Grande	1.12	1.08	1.15
14 Upper Colorado	3.53	2.95	2.76
15 Lower Colorado	0.92	0.86	0.84
16 Great Basin	3.42	3.45	3.25
17 Pacific Northwest	19.3	16.1	15.6
18 California	2.68	2.55	2.39
19 Alaska	15,604	4,393	1,980
20 Hawaii	12.2	11.5	11.0
21 Caribbean	15.1	13.8	17.3
Regions 1 through 18	11.1	9.92	8.93

[a]Ratio of available to required water, e.g., New England region in 1975: 164 gallons available for each gallon required.

Groundwater overdrafts represent a serious problem because they indicate that use exceeds natural recharge capabilities. Ground waters have vast storage capabilities which are being exhausted in many parts of the United States. Evidence of this manifests itself in declining springs and stream flows, all water intrusion problems and land subsidence. Groundwater mining—the term used to denote permanent loss of water from storage—will lead to severe dropping of the water tables and possibly even total depletion. Areas having excess groundwater withdrawals are shown in Figure 13.

Wastewater Reuse and Recycling

The total wastewater reuse in 1975 was estimated to be approximately 680 mgd or 760,000 ac-ft/yr. Reuse was distributed among three principal categories as shown in Table III.

The estimated annual reuse quantity was divided among about 536 locations in the United States, as shown in Figure 14, with the largest volume usage in Arizona, California and Texas. A breakdown of the usage and project locations is given in Table IV.

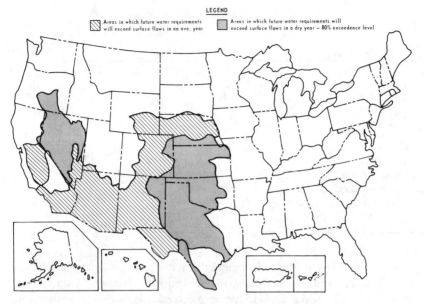

Figure 12. Projected water-shortage areas.

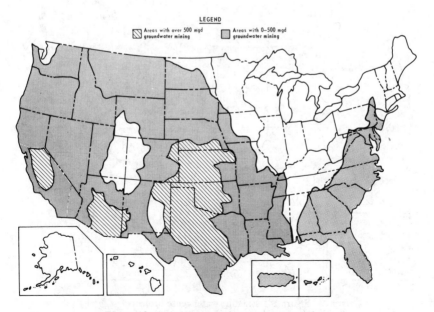

Figure 13. Existing rates of groundwater mining.

Table III. Existing Wastewater Reuse

Beneficial Use	Wastewater Reuse (1000 ac-ft/yr)	
	Total	Subtotals
Irrigation	471	
Agricultural		223
Landscape		37
Not Specified		211
Industrial	240	
Process		74
Cooling		159
Boiler Feed		7
Groundwater Recharge		
Other (recreation, fish and wildlife, etc.)	11	
	760	760

Water recycling is currently practiced in steam electric plants and in the manufacturing and mining industries. A significant increase in water recycling is projected which leads to the large decreases in freshwater withdrawals and wastewater discharges presented in Figures 2 and 3. Recycling in manufacturing industries and steam electric plants is projected to increase the most,

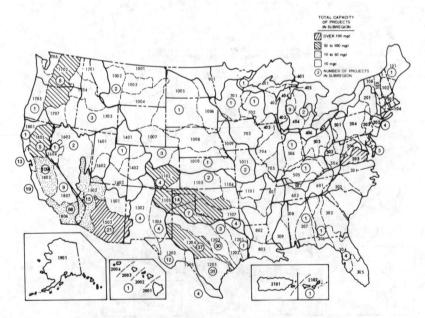

Figure 14. Existing water reuse projects.

Table IV. Volume and Location of Wastewater Reuse

WRC Region No. and Name	Number of Projects	Wastewater Reuse (1000 ac-ft/yr)
15 Lower Colorado	36	204.4
18 California	283	192.3
2 Mid Atlantic	7	130.5
11 Arkansas-White-Red	31	84.1
12 Texas-Gulf	102	57.4
17 Pacific Northwest-Columbia	14	29.0
13 Rio Grande	24	22.0
4 Great Lakes	13	10.8
3 South Atlantic-Gulf	6	7.4
10 Missouri Basin	9	6.8
16 Great Basin	3	5.6
7 Upper Mississippi	2	3.0
1 New England	1	3.0
14 Upper Colorado	1	1.2
5 Ohio	2	1.1
20 Hawaii	1	1.0
21 Caribbean	1	0.4
Total	536	760

as shown in Table V. The values shown in Table V indicate the number of times the supplied fresh water will be used prior to discharging and show a fourfold increase between 1975 and 2000. For example, by the year 2000 the manufacturing industries are projected to use the water over 17 times before discharging. This recycling results in a 62% decrease in withdrawals (Figure 2) by 2000 and decreases the requirement for manufacturing from 14% of the national total in 1975 to only 6% in 2000.

Table V. Existing and Projected Water Recycling

Category	Fresh Water Recycled (bgd)		Recycle Rate	
	1975	2000	1975	2000
Steam Electric	57	517	1.6	7.5
Manufacturing Industries	61	316	2.2	17.1
Minerals Industries	21	32	4.0	4.0
Totals	139	865	1.9	8.8

Potential For Wastewater Reuse

Wastewater discharges available for reuse are divided into treated and untreated quality categories. Treated wastewater includes:

* secondary effluent from municipal wastewater treatment plants (some uses require higher quality),
* industrial discharges treated by best available technology,
* discharges from steam electric plants and
* discharges from fish hatcheries.

Users requiring treatment greater than secondary quality consume small quantities when compared to those users that can use secondary effluent. Therefore, these users, which include domestic, industrial, fisheries and other uses, have not been quantified here. However, their treatment quality requirements and costs are presented later in this chapter.

Untreated wastewaters are considered to be irrigation return flows that can be collected and reused. Most agricultural return flows reach streams through groundwater flow or surface runoff, and are not available for direct, planned reuse. However, these discharges become a part of the stream flow and are frequently withdrawn at downstream locations for reuse. Reuse of this type is not planned, and has not been quantified. Wastewaters from livestock use, rural domestic use and public lands have not been quantified either because they represent relatively small quantities in generally remote locations.

Users considered capable of substituting wastewater for freshwater withdrawals are listed in Table VI. Comparing these users with those in Figure 2, which shows the annual consumptions, provides an indication of the potential for reuse. Treatment qualities, treatment systems to meet these qualities and the costs and energy requirements to build and operate these systems are all presented later in this chapter.

Agricultural irrigation represents the single, largest user of water currently, and is projected to remain the largest user through the year 2000. As such, agricultural irrigation presents opportunities for significant water conservation and reuse.

Table VI. General Water Quality Requirements

Use Category	Wastewater Quality Required
Agricultural Irrigation	Treated
Landscape Irrigation	Treated
Steam Electric	Treated or Untreated
Industrial Cooling	Treated or Untreated
Industrial, Other	Treated

Landscape irrigation requirements are small when compared to other major users, but are included because reclaimed wastewater can be used to supply a significant portion of the requirement. The 1975 landscape irrigation use is shown in Table VII.

Cooling water for steam electric generating plants is the second largest water user, with most of the demand in the eastern United States. Although recycling is projected to increase markedly, reclaimed wastewater could be used for a substantial part of the requirement for makeup water.

Industrial requirements for cooling and process applications are another good potential for reuse of wastewaters. However, quality requirements may be more stringent than secondary levels. Reuse of wastewater for many industrial applications is currently widely accepted.

Although groundwater recharge is not considered a use, per se, it represents a use for treated wastewater effluents that is beneficial, especially when the groundwater mining presently being experienced in many areas is considered (see Figure 13). It is being used successfully in several California projects where saltwater intrusion is a major problem.

Table VIII summarizes estimates for wastewater discharges available for reuse, water withdrawals that are capable of using wastewater instead of fresh water, actual and estimated potential water recycling and actual and estimated potential reuse. Reuse of available wastewater is projected to increase from 0.4% in 1975 to an estimated 4% in 2000. Recycling in the industrial sector is projected to increase substantially by the year 2000.

The quantity of wastewater that will actually be reused by the year 2000 will be determined by several factors, including: (1) the geographical location of discharges and potential users, (2) the timing of wastewater discharges and the requirements of potential users and (3) the availability and cost of alternative supplies. There are large regional differences in wastewater availability and reuse potential. In the eastern states and Alaska the quantities of available wastewater discharges and the withdrawals greatly exceed the available wastewater. This difference is primarily due to agricultural irrigation requirements.

Table VII. Landscape Irrigation Usage (Three Largest Users in 1975)

WRC Region No. and Name	1,000 ac-ft/yr	Percentage of Total Annual Water Use Within Region	Percentage of Total U.S. Landscape Irrigation
18 California	174	0.4	14
3 South Atlantic Gulf	133	0.5	11
4 Great Lakes	127	0.3	10

Table VIII. Summary of Recycle and Reuse Potential[a]

WRC Region No. and Name	Water Withdrawals Capable of Wastewater Reuse		Wastewater Discharges Available for Reuse				Wastewater Recycle		Wastewater Reuse	
			Treated		Untreated					
	1975	2000	1975	2000	1975	2000	Actual 1975	Estimated Potential 2000	Actual 1975	Estimated Potential 2000
1 New England	3,615	1,424	4,563	2,108	0	0	2,134	16,284	3	13
2 Mid-Atlantic	13,757	8,009	16,261	10,036	0	0	8,680	58,131	116	224
3 South Atlantic-Gulf	22,645	25,311	18,799	17,180	653	833	16,071	127,414	7	192
4 Great Lakes	38,564	20,410	40,059	20,737	0	0	9,589	97,417	10	290
5 Ohio	32,514	13,913	33,018	12,443	0	0	19,032	112,417	1	118
6 Tennessee	7,030	5,475	7,076	4,880	0	0	2,732	29,199	0	41
7 Upper Mississippi	10,311	5,359	11,151	5,093	0	0	8,155	66,139	3	48
8 Lower Mississippi	14,249	24,405	9,000	18,132	722	561	10,453	52,982	0	108
9 Souris-Red-Rainy	256	594	211	54	3	28	123	308	0	1
10 Missouri	40,385	47,016	5,126	5,791	1,767	1,901	5,476	37,795	6	1,139
11 Arkansas-White-Red	13,489	13,542	1,848	1,773	390	346	12,530	42,479	75	147
12 Texas Gulf	16,704	14,355	3,452	3,659	283	171	24,448	136,241	52	430
13 Rio Grande	6,588	5,771	278	308	240	174	2,469	4,124	20	370
14 Upper Colorado	7,564	8,251	223	353	587	559	2,650	7,598	1	61
15 Lower Colorado	8,710	7,272	354	522	545	362	2,766	8,815	182	733
16 Great Basin	8,280	7,190	470	591	640	454	1,586	11,511	4	62
17 Pacific Northwest	45,322	40,172	3,528	1,902	4,563	3,438	3,677	28,288	26	67
18 California	38,644	39,412	2,647	3,089	3,147	2,590	5,807	23,827	172	696
19 Alaska	196	592	234	295	0	0	438	3,704	0	1
20 Hawaii	1,813	1,173	300	205	178	88	197	808	1	7
21 Caribbean	619	616	318	452	0	0	29	44	1	6
Totals	331,255	290,262	158,916	109,603	13,718	11,505	139,042	865,525	680	4,754

[a] All figures are million gallons per day.

Beneficial Uses of Water

The beneficial uses of water are usually ranked in order of preferred use by each state. Typically, the generalized ranking would be domestic use, agricultural uses, industrial uses and then other uses. Within these categories several different specific beneficial uses require varying degrees of water quality and account for significantly different consumptive uses. Beneficial uses that are typically uncontrolled such as stream augmentation, navigation, aesthetics and hydroelectric power generation, are not included.

Nineteen specific beneficial uses identified as potential candidates for controlled reuse schemes are listed in Table IX, together with general descriptions of the basis for establishing the water quality criteria. Although the beneficial uses are presented in general order of procurement rights, many of the lower order uses may, in fact, require more stringent water qualities than those of some higher qualities. Procurement rights refer to the beneficial use that would take precedence if there were more than two competing uses for the same volume of water. The specific water quality criteria for each beneficial use are presented later.

Water Quality Criteria

The beneficial uses of water each require a specific water quality. If this quality is not naturally available or attainable by treatment, then the beneficial use is lost, although the reduced water quality may be suitable for another use. Water quality criteria for each of the beneficial uses are established in order to match up potential wastewater treatment systems.

The water qualities for each beneficial use are presented in terms of criteria. One definition of criteria is, "A criterion designates a means by which anything is tried in forming a correct judgment respecting it" [1]. It does not carry a connotation of authority or imply an ideal condition. To be useful, a criterion should be capable of quantitative evaluation by acceptable analytical procedures, and should be capable of definitive resolution.

The water quality criteria are summarized in Table X, and do not include the recommended data from the various references. Narratives describing the beneficial uses are not included, as these can be found in other reports [2,3]. There are reported limitations regarding the criteria and these generally refer to the differences in establishing water quality data. These differences include items such as factors of safety, lack of scientific data base, reasons other than public health and others. The drinking water quality recently established by the U.S. Environmental Protection Agency (EPA) is presented separately in Table X because it represents standards and not criteria.

Table IX. Beneficial Uses of Water

Beneficial Water Use	Description	Basis for Establishing Water Quality Criteria
1. Direct Potable	Water suitable for human ingestion and domestic and food processing uses directly.	Water qualities should meet the interim drinking water standards promulgated by the EPA.
2. Public Water Supply— Surface Water	Raw water supplies that are suitable for human ingestion and domestic, municipal, or food processing uses *after* conventional water treatment.	Surface waters that will meet potable standards *after* conventional water treatment of coagulation, clarification, filtration and disinfection. Contaminants that interfere with the treatment process should be limited. Human water consumption is assumed to average 2 liters/day and an average weight per individual of 70 kg.
3. Public Water Supply— Ground Water	Raw water supplies that are suitable for human ingestion and domestic, municipal or food processing uses *after* disinfection.	Ground waters that will meet potable standards *after* disinfection only. Human water consumption is assumed to average 2 liters/day, and an average weight per individual of 70 kg.
4. Agricultural Irrigation— Forage Crops	Waters used for the irrigation of forage crops or grasses only. Household uses or livestock and wildlife watering are not included.	Water supplies must be suitable for use with *no* water treatment. Irrigation waters are assumed to be used continuously through the seasons on all soil types. Soil conditions requiring special or different water qualities are not considered.
5. Agricultural Irrigation— Truck Crops	Waters used for the irrigation of crops used for human consumption. Household uses or livestock and wildlife watering are not included.	Water supplies must be suitable for use with *no* water treatment. Irrigation waters are assumed to be used continuously through the seasons on all soil types. Soil conditions requiring special or different water qualities are not considered.
6. Urban or Landscape Irrigation	Waters used for irrigation of grasses and shrubs within the urban community such as golf courses or parks.	Water supplies should be free of suspended solids and low in turbidity to ensure complete bacterial and viral kill. Potential aerosols should be bacterially safe.

7. Livestock and Wildlife Watering — Water used for consumption by livestock and wildlife (nondomestic vertebrates) excluding fish.

Water qualities suitable to assure both short- and long-term survival of nondomestic vertebrates, excluding fish. Contaminants that will taint the taste of the meat or milk must also be limited.

8. Power Plant and Industrial Cooling—Once Through — Water used by power plants and certain industries for the removal of heat on a once-through basis.

The once-through use of water for the condensation of steam requires large volumes of water and only the removal of contaminants that could clog or settle in the system and requires *no* prior treatment. Other water used as the site will be treated as necessary.

9. Power Plant and Industrial Cooling—Recirculation — Water used by power plants and certain industries for the removal of heat for recirculating water through cooling facilities.

Water qualities of the raw water supply should be suitably low in contaminants to avoid clogging, scaling, and slime growth accumulation. Contaminant build-up is assumed to be controlled by regular wastage or blowdown. The water should also be noncorrosive. Other water used at the site will be treated as necessary.

10. Industrial Boiler Make-Up Water — Water used for the generation of steam for power or processing needs.

Water qualities depend upon the operating pressures of the boiler system and raw water supplies receive extensive onsite treatment. If the steam is used for food processing, water qualities must also meet the requirements for potable use.

11. Industrial Water Supply—Food and Kindred Products — Water used for washing, rinsing, conveying or for the preparation of food products.

Water qualities in general must meet potable use standards. Water used to generate steam must also meet the requirements for boiler make-up water. Water is assumed to be bacteriologically safe. Certain situations which require even more stringent water qualities are not included.

12. Industrial Water Supply—Paper and Allied Products — Water used for cooking and grinding wood chips, washing of pulp and the transport of paper fibers

Contaminants that could clog equipment or cause slime growths, as well as affect the color or tex-

Table IX, continued

Beneficial Water Use	Description	Basis for Establishing Water Quality Criteria
	through the various processes.	ture and uniformity of the pulp must be limited. Hardness which can cause scale on equipment or contaminants causing corrosion or other problems are also assumed to be limited.
13. Industrial Water Supply—Chemicals and Allied Products	Water used for transportation, washing, and mixing the products as well as a chemical reaction media.	Water qualities are assumed to be suitable when the water does not cause unfavorable chemical reactions or when the chemical reactions are not delayed or do not require excessive quantities of chemicals. Water used for cooling or boiler make-up is not included.
14. Industrial Water Supply—Petroleum and Coal Products	Water used in the processes such as refining, desalting and fractionation and also for transportation of the products.	Surface or groundwaters low in dissolved salts and chlorides and iron. Cooling and boiler make-up waters are not included.
15. Industrial Water Supply—Primary Metals	Water used for the processing of ferrous and nonferrous metals.	Surface waters that will not corrode or plug equipment or create scale problems. Waters used for cooling or boiler make-up are covered elsewhere.
16. Primary Contact Recreation	Water in which the human body may be submerged for prolonged periods. Such activities include swimming or water skiing.	The water must be aesthetically enjoyable, virtually free of objectionable substances such as oil, scum, debris and aquatic growth, and free from pathogens and toxic substances that could cause skin or eye irritations. The water should be clear enough for some depth to reveal submerged objects. Ingestion of limited amounts of water should not cause illness.
17. Secondary Contact Recreation	Water in which the human body comes in contact for limited periods of time. Such	The water must be aesthetically acceptable, virtually free of objectionable substances such as oil,

	activities include wading, boating, and hiking.	debris, scum and massive growths of aquatic plants that would result in matting and odor. Ingestion of small amounts of water or total body immersion should not be harmful.
18. Warm Water Fishery	Water that will support and allow for the propagation of a warm water fishery.	Water quality must be suitable for supporting warm water fish. Substances that could taint the taste of fish are also to be limited. The European Inland Fisheries Advisory Commission indicated that toxic concentrations lower than those adversely affecting fish would not be toxic to other warm water organisms.
19. Cold Water Fishery	Water that will support and allow for the propagation of a cold water fishery.	Water quality must be suitable for supporting cold water fish. Substances that could taint the taste of fish are also to be limited. The European Inland Fisheries Advisory Commission indicated that toxic concentrations lower than those adversely affecting fish would not be toxic to other cold water organisms.

Table X. Water Quality Criteria

	Drinking Water Supply	Raw Water Supply	Agricultural Irrigation	Landscape Irrigation	Livestock and Wildlife	Once-Through Cooling	Recirculation Cooling	Boiler Feed Water	Food and Kindred Products	Paper and Allied Products	Chemical and Allied Products	Petroleum and Allied Products	Primary Metal Industries	Primary Contact Recreation	Cold Water Fisheries	Warm Water Fisheries
Acidity	—	—	—	—	—	—	—	—	0	—	—	—	—	—	—	—
Alkalinity	—	—	—	—	30–130	500	350	100	200	75	500	300	2	—	Max. Δ = 25%	—
Aluminum	—	5.0	5.0	5.0	5.0	1	0.1	0.1	—	—	—	—	—	—	—	—
Aquatic Growth	—	—	—	—	—	—	—	—	—	—	—	—	—	V.F.[a]	—	—
Arsenic	0.05	0.05	0.1	0.1	0.2	—	—	—	0.05	—	—	—	—	—	0.05	0.05
Bacteria	—	—	—	—	—	—	—	—	—	—	—	—	—	—	—	—
Fecal Coliforms	—	2,000	1,000	—	1,000	—	—	—	—	—	—	—	—	200 MPN/100 ml	—	—
Total Coliforms	—	20,000	—	—	5,000	—	—	—	—	—	—	—	—	200 MPN/100 ml	—	—
Barium	1	1	—	—	—	—	—	—	—	—	—	—	—	—	5.0	5.0
Beryllium	—	0.1	0.1	0.2	—	—	—	—	—	—	—	—	—	—	0.011–1.100	0.01–1.100
Bicarbonate	—	—	—	—	—	600	24	120	—	—	600	480	50	—	—	—
BOD	—	—	—	20	—	—	—	—	2.2	—	—	—	—	—	—	—
Boron	0.75	—	—	—	5.0	—	—	—	1.0	—	—	—	—	—	—	—
Cadmium	0.01	0.01	0.01	0.01	0.05	—	0.4	—	0.01	—	—	—	—	—	0.0004–0.0150	0.0004–0.0150
Calcium	—	—	—	—	—	200	50	—	100	50	250	100	—	—	—	—
Chloride	250	250	—	100	—	600	500	—	200	75	500	300	100	—	—	—
Chromium	0.05	0.05	0.1	0.05	1.0	—	—	—	0.05	—	—	—	—	—	0.1	0.1
Cobalt	—	—	0.2	0.05	1.0	75	—	—	—	—	—	—	—	—	—	—
COD	—	—	—	—	—	75	75	5.0	—	—	500	1,000	—	4	—	—
Color	15	75	—	—	—	1	1	—	5	—	—	25	—	30	Max. Δ = 10%	Max. = 10%
Copper	1	1.0	0.2	0.2	0.5	1	1	0.05	1.0	—	—	—	—	—	0.01–0.04	0.01–0.04
Cyanide	0.2	0.2	—	—	—	—	—	—	0.01	—	—	—	—	—	0.005	0.005
Fluoride	1.4–2.4	1.4–2.4	2.0	2.0	—	—	—	—	1.0	—	—	1.2	—	—	1.5	1.5
Hardness	—	—	—	—	—	850	650	0.25	200	100	1,000	900	100	—	—	—
Iron	0.3	0.3	5.0	5.0	0.3	1	0.5	1.0	0.2	0.1	10	0.3	—	—	0.5	0.5
Lead	0.05	0.05	5.0	5.0	0.1	0.5	0.05	0.3	0.05	0.05	—	—	—	—	0.004–0.150	0.004–0.150
Lithium	—	—	2.5	2.5	—	—	—	—	—	5.0	5.0	—	—	—	—	—
Magnesium	—	—	—	—	—	—	—	—	50	—	100	80	—	—	—	—
Manganese	0.05	0.05	0.2	0.2	0.05	0.5	0.5	0.1	0.1	0.05	2	0.3	—	—	1.0	1.0
MBAS	0.5	0.5	—	—	—	1	1	1	—	—	—	—	—	—	—	—
Mercury	0.002	0.002	—	—	0.001	—	—	—	—	—	—	—	—	—	0.00005	0.00005

Parameter												
Molybdenum	10	—	—	—	—	—	—	—	—	—	0.050-0.400	0.050-0.400
Nickel	—	0.1	0.01	—	—	—	—	—	—	—	—	—
Nitrogen	—	0.2	0.2	—	—	—	—	—	—	—	—	—
Ammonia	—	0.05	—	1	No Limit	—	—	10	—	—	0.02 Un-ionized	0.02 Un-ionized
Nitrate (as N)	10	10.0	—	10	10	—	—	10	10	—	—	—
Nitrite (as N)	—	1.0	—	—	1	—	—	0	0	—	—	—
Odor	3	V.F.[a]	FM	No Visible Floating	No Floating	—	—	—	V.F.[a]	V.F.[a]	No Visible Floating	No Visible Floating
Oil	—	V.F.	—	No Floating	—	—	—	—	1.0	—	No Visible Floating	No Visible Floating
Organics												
CCE	—	0.3	—	Aerobic Present 2.0	—	—	—	0.2	Aerobic	Aerobic	—	—
PCB	—	0.001	—	—	—	—	—	—	—	—	—	0.000001
Oxygen	—	—	—	Aerobic Present 2.0	—	—	—	—	5	30	—	—
Pesticides	—	—	—	1	0.007	—	—	—	0.007	—	—	—
pH	—	1.0-9.0	6.0-9.0	7.0-9.0	5.0-8.3	8.2-10.0	5.9-9.0	7.0-8.5	6.0-9.0	6.5-8.3	6.5-9.0	6.5-9.0
Phenols	—	0.01	—	1,000	1,000	—	—	0.001	0.007	—	—	—
Radionuclides	—	—	—	2	2	—	—	—	—	—	—	—
Selenium	0.01	0.01	0.02	0.05	0.05	50	20	0.01	20	—	0.0001-0.00025	0.05
Silica	—	—	—	0.05	—	—	—	50	—	—	—	—
Silver	0.05	0.05	—	50	50	20	—	0.05	—	—	0.0001-0.00025	0.001-0.00025
Solids												
Settleable	—	—	15	Minimized	5	100	—	10	5	Free From 5	25	25
Suspended	—	—	50	5,000	1,000	2	5	500	10,000	—	Max. of 2000	Max. of 2000
Total Dissolved	250	250	<1,200	3,000	1,000	200	500	500	2,500	1,000	30	—
Sulfate	—	—	600	680	250	1	—	250	850	300	75	—
Temperature	—	—	—	Avoid Changes in Freezing Dates	1	—	—	—	—	—	18.2	29.9
Turbidity	1	0.1	0.1	0.1	—	—	—	5	—	5	Max. Δ = 10%	Max. = 10%
Vanadium	—	0.1	—	—	—	—	—	5	—	—	0.050-0.600	0.050-0.600
Zinc	5	5.0	2.0	25.0	1	1	—	5	—	—	—	—
Radium, pCi/l	5	—	—	—	—	—	—	—	—	—	—	—
Endrin	0.002	—	—	—	—	—	—	—	—	—	—	—
Lindane	0.004	—	—	—	—	—	—	—	—	—	—	—
Methoxychlor	0.1	—	—	—	—	—	—	—	—	—	—	—
Toxaphene	0.005	—	—	—	—	—	—	—	—	—	—	—
2,4-D	0.1	—	—	—	—	—	—	—	—	—	—	—

[a] V.F. = virtually free.
[b] Avoid changes in freezing dates.

Influent Wastewater Characterization

Influent data from several pilot and full-scale wastewater treatment plants were obtained and analyzed, then the average influent concentrations for all available constituents were determined (see Table XI). These influent concentrations are intended to be representative of "typical" wastewaters, although they should not be used as a substitute for actual data, if available. However, it is anticipated that most waste waters should fall within the range of concentrations shown in Table XI.

These data have been used to develop predicted effluent qualities from selected wastewater treatment facilities, discussed later in this chapter. They are used also to demonstrate the use of statistical data computed from daily operating data from these wastewater treatment facilities. Finally, because trace constituents, in particular, are not usually measured on a regular basis, the data in Table XI may prove useful for comparative or other purposes.

WASTEWATER TREATMENT PROCESS EVALUATION

This section focuses on the suitability of unit processes for removing contaminants at the concentrations typically found in municipal wastewaters. Only a few selected processes are included here to demonstrate the efficiency of the processes, and how to make use of the data.

Based on published information and data obtained from the pilot and full-scale plants, Table XII has been developed to show the removal efficiencies of the treatment processes. This table can be used to identify the treatment processes that best remove the contaminants of concern, and thereby aid in selecting the combination of treatment processes to build up the treatment system.

The daily operating data from the 17 full-scale and in-depth pilot plant systems were analyzed in detail. Probability analyses were performed for all available contaminants, and the percentage removals across individual treatment units were computed. This analysis contrasts the traditional methods of analysis which only report on the average removal across the process. The principal purpose in dealing with the percentage removals is the ease with which they can be combined to synthetically develop effluent concentrations for a treatment system, including the reliability or probability of occurrence. If two or more processes are arranged in series, the over-all performance of the combined processes can be determined from the values for the individual processes. The individual unit treatment processes presented in Table XII were each statistically evaluated and the 10, 50 and 90 percentile probabilities of percent removal were computed. The method for

Table XI. Influent Wastewater Characterization–Average Concentrations[a]

| Constituent | Concentrations | | |
| | Range | | |
	Low	High	Average
BOD	82	297	181
COD	204	871	417
TSS	64[c]	406	192
NH_3-N	13.8	35	20
NO_3-N	0.09	1.52	0.61
Org-N	7.0	15.3	13
Ortho-P	3.0	13	6.8
Total-P	5.96	132	9.4
Alkalinity	91	426	211
Oil and Grease	35	104	61
pH[b]	7.0	7.6	7.2
Arsenic	0.002	0.020[c]	0.007
Barium	0.041	0.700[c]	0.235
Cadmium	0.002	0.02	0.008
Chromium	0.015	0.560	0.167
Copper	0.058	0.217	0.117
Fluoride	0.6	1.36	0.864
Iron	1.24	3.54	2.25
Lead	0.017	0.206	0.148[c]
Manganese	0.044	0.195	0.117
Mercury	0.66[c]	0.001[c]	0.112
Selenium	0.001	0.014	0.006
Silver	0.004	0.044	0.022
Zinc	0.125	1.195	0.419
Endrin[d]	0.003	0.007	0.005
Lindane[d]	0.7	0.021	0.260
Methoxychlor[d]	0.027	0.20	0.102
Toxaphene[d]	0.001	0.002	0.001
2,4-D[d]	0.017	0.4	0.152
2,4,5-TP Silvex[d]	0.07	0.14	0.110
TOC	63	171	102
Hardness	84	672	362
Color P-C Units	39	55	67
Turbidity, FTU	78	152	2.16
Foaming Agents	1.53	5259	107
TDS	554		2577

[a] All values in mg/l except as noted.
[b] pH units.
[c] Average from two data sources.
[d] ppb.

Table XII. Constituent Removal by Unit Processes

Constituent	Primary Treatment	Activated Sludge	Nitrification	Denitrification	Trickling Filters	RBC's	Coag.-Floc.-Sed.	Filtration after A.S.	Carbon Adsorption	Ammonia Stripping	Selective Ion Exch.	Breakpoint Chlorin.	Reverse Osmosis	Overland Flow	Irrigation	Infilt.-Perc.	Chlorination	Ozone
BOD	x	+	+	0	+	+	+	x	+		x		+	+	+	+		0
COD	x	+	+	0	+		+	x	x	0	x		+	+	+	+		+
TSS	+	+	+	0	+	+	+	+	+			+	+	+	+	+		
NH₃-N	0	+	+	x		+	0	x	x	+	+	+	+	+	+	+		
NO₃-N				+				x	0					x				
Phosphorus	0	x	+	+			+	+	+				+	+	+	+		
Alkalinity		x					x	+								x		
Oil and Grease	+	+	+				x		x					+	+	+		
Total Coliform		+	+		0		+		+			+		+	+	+	+	+
TDS													+					
Arsenic	x	x	x				x	+	0									
Barium		x	0				x	0										
Cadium	x	+	+		0	x	+	x	0							0		
Chromium	x	+	+		0	+	+	x	x									
Copper	x	+	+			+	+	+	0	x						+		
Fluoride							x		0							x		
Iron	x	+	+		x	+	+	+	+									
Lead	+	+	+		x	+	+	0	x							x		
Manganese	0	x	x		0		x	+	x				+					
Mercury	0	0	0		0		+	0	x	0								
Selenium	0	0	0				0	+	0									
Silver	+	+	+		x		+		x									
Zinc	x	x	+		+	+	+		+							+		
Color	0	x	x		0		+	x	+				+	+	+	+		+
Foaming Agents	x	+	+		+		x		+				+	+	+	+		0
Turbidity	x	+	+	0	x		+	+	+				+	+	+	+		
TOC	x	+	+	0	x		+	x	+	0	0		+	+	+	+		+

Symbols: 0 = 25% removal of influent concentration; x = 25-50%; + = >50%; + = >50%; Blank denotes no data, inconclusive results, or an increase.

combining the percentage removals for the individual unit processes that make up a preselected treatment system is as follows:

$$(C_1 + C_2 + \cdots\cdots + C_n) - (C_1 \times C_2 \times \cdots\cdots \times C_n)$$

where C_1, C_2, C_n = decimal equivalents of the percent removals for a particular contaminant through processes 1, 2 and n. Process n represents the last process in a sequence of processes that together make up the treatment system.

The values developed in this study were compared against the operational data obtained from two of the best operated advanced wastewater treatment (AWT) plants: the South Lake Tahoe plant and the Upper Occoquan plant. The results are shown in Table XIII. Table XIII shows reasonably good correlations between field data and computed data. The correlation for the 90 percentile value is not as good for the South Lake Tahoe plant, probably due to less efficient plant operations or lower carbon contact times. Influent wastewater data were not available for an extended period of time, and therefore the percent removals were selected as providing more reliable results.

Process reliability, in terms of the 10, 50 and 90 percentile removals, provides valuable information for the reuse systems to determine the consistency of treatment, and therefore the dependability of the supply. If a low reliability is computed for the removal of a constituent that is critical for a specific beneficial use, then some other means of controlling that parameter must be provided. In general, performance data from wastewater treatment processes do not follow normal statistical distribution; many external factors and process control practices can upset what might otherwise be a natural distribution of data. Using log-normal statistics, the data fall along two lines, which is characteristic of the treatment process data used in this study. The break normally occurs between the 40 and 70 percentile values, and could be caused by any number of items.

To demonstrate the available data for a few of the more frequently used treatment processes Tables XIV-XIX show the percentage removal values for the following processes:

- primary treatment,
- activated sludge,
- trickling filters,
- filtration (following biological secondary treatment),
- coagulation-sedimentation (lime addition) and
- activated carbon adsorption.

Included in the tables are the average parameter removals that have been reported in the literature. These values can be compared to the average values computed from data from the 17 pilot and full-scale plants, and any major differences can be noted when using the data. The data are not always representative of an exactly comparative process, such as upflow/downflow carbon columns, upflow columns and differing detention times, but they do present average field conditions.

Wastewater Treatment Systems

Table XX shows 19 beneficial uses of water presented earlier matched to wastewater treatment systems that produce effluent of a suitable quality to ensure the use. Three additional subcategories of beneficial uses are included.

Table XIII. Comparison of Measured to Calculated Data

| Parameter | South Tahoe Plant[a] | | | | Upper Occoquan[b] | | | |
| | Measured[c] | | Calculated[d] | | Measured[c] | | Calculated[d] | |
	50%	90%	50%	90%	50%	90%	50%	90%
BOD	1.2	3.0	0.1	16.2	100	99	99	90
COD	10	19	8.7	45.8	99	97	97	81
PO$_4$	0.2	0.5	0.02	0.09	100	100	100	99
MBAS	0.2	0.4	0.45	–	–	–	–	–
SS	–	–	–	–	100	99	100	89

[a]Data presented in terms of effluent parameter concentrations using average influent concentrations measured at the plant.

[b]Data presented in terms of percentage removals.

[c]Obtained from plant operating data.

[d]Computed by combining percent removal data developed in the study for the individual processer making up the treatment system.

Table XIV. Contaminant Removals for Primary Treatment[a]

| Constituent | Published Average[b] | Primary Treatment | | | |
| | | | Probability | | |
		Average	10%	50%	90%
BOD	34-47	42	66	42	15
COD		38	65	37	18
TSS	53-64	53	78	56	24
NH$_3$-N		18	36	15	2
Phosphorus		27	43	15	5
Alkalinity		+	+	+	+
Oil and Grease	61	65	91	65	30
Arsenic		34	–	40	–
Cadmium	30	38	75	37	5
Chromium	32-36	44	81	49	6
Copper	36-40	49	78	51	16
Fluoride		x	x	x	x
Iron	57	43	89	37	0
Lead	54	52	85	51	17
Manganese	27	20	52	17	0
Mercury		11	–	8	–
Selenium		0	0	0	0
Silver	46	55	82	51	21
Zinc	36-50	36	66	35	7
Color		15	43	12	0
Foaming Agents		27	–	13	–
Turbidity		31	55	30	6
TOC		34	–	33	–

[a]Values are given in terms of percent removal.

[b]Typical removals found in published literature.

Symbols: x = data inconclusive; – = insufficient data; 0 = no significant removal; + = increase.

Table XV. Contaminant Removals for Activated Sludge[a]

Constituent	Published Average [b]	Activated Sludge			
		Average	Probability		
			10%	50%	90%
BOD	80-96	89	97	93	74
COD	90	72	87	67	47
TSS	75-93	81	87	71	43
NH_3-N	52-87	63	80	60	35
NO_3-N		+	+	+	+
Phosphorus	67	45	58	41	18
Alkalinity		38	69	61	47
Oil and Grease	98	82	99	86	56
Arsenic	24	28	70	35	0
Barium		31	62	34	0
Cadmium	30-64	54	86	53	16
Chromium	36-96	74	89	75	57
Copper	59-85	76	93	77	55
Fluoride		x	x	x	x
Iron	48	72	89	75	50
Lead	43-63	69	86	70	45
Manganese	22	33	62	29	6
Mercury	68-76	13	31	11	0
Selenium		7	26	12	0
Silver	37-71	79	95	86	37
Zinc	44-75	49	71	49	17
Color		48	71	50	20
Foaming Agents		71	—	63	—
Turbidity		86	94	89	73
TOC		83	99	81	62

[a]Values are given terms of percent removal.
[b]Typical removals found in published literature.
Symbols: x = data inconclusive; − = insufficient data; 0 = no significant removal; + = increase.

All of the treatment processes include a biological secondary treatment system and represent typical, but not the only, treatment systems suitable for attaining various beneficial uses. These systems are briefly described in this section, including energy use, treatment costs and treatment efficiencies.

Each of the treatment systems includes a typical sludge handling system in order to obtain more realistic costs and energy requirements. The sludge handling systems were not optimized, but do represent typical systems in common use and are capable of treating the types and quantities of sludge that are generated in each treatment system.

The costs presented have been updated to January 1980 using the EPA index. These costs are also related to the Bureau of Labor Statistics (BLS) indices for individual components of construction, such as concrete, steel,

Table XVI. Contaminant Removals for Trickling Filters

Constituent	Published Average [a]	Trickling Filters			
		Average	Probability		
			10%	50%	90%
BOD	70-90	69	81	71	65
COD		58	71	59	45
TSS	69-92	63	84	66	32
Cadmium	5	0	0	0	0
Chromium	19	0	0	0	0
Copper	46	56	79	63	10
Lead	36	44	83	34	5
Manganese	16	–	–	–	–
Mercury	–	0	0	0	0
Silver	48	–	–	–	–
Zinc	56	57	85	62	14

[a]Typical removals found in published literature.
Symbols: – = insufficient data; 0 = no significant removal.

Table XVII. Contaminant Removals for the Filtration Process (Following Biological Secondary Treatment)

Constituent	Published Average [a]	Effluent Filtration			
		Average	Probability		
			10%	50%	90%
BOD	60-93	39	69	46	15
COD		34	58	33	14
TSS	37-95	73	96	81	41
NH_3-N		33	68	34	0
NO_3-N		56	99	52	5
Phosphorus		57	98	87	34
Alkalinity		83	93	85	70
Arsenic		67	100	64	30
Barium		+	+	+	+
Cadmium		32	68	22	0
Chromium		53	82	50	20
Copper		+	+	+	+
Iron		56	76	55	30
Lead		16	50	10	3
Manganese		80	89	84	56
Mercury		33	98	17	0
Selenium		90	100	100	43
Zinc		+	+	+	+
Color		31	49	29	14
Turbidity		71	89	76	47
TOC		33	47	28	15

[a]Typical removals found in published literature.
Symbols: 0 = no significant removal; + = increase.

Table XVIII. Contaminant Removals for the Coagulation-Sedimentation Process (Lime Addition)

Constituent	Published Average[a]	Lime Treatment			
		Average	Probability		
			10%	50%	90%
BOD	80-81	53	74	51	33
COD	81	52	70	51	35
TSS	88-90	65	88	70	34
NH_3-N	30	22	39	22	12
Phosphorus	95-98	91	97	94	84
Alkalinity		x	+	+	+
Oil and Grease		40	95	31	0
Arsenic	30	6	31	2	0
Barium	37	61	70	62	50
Cadmium	95	30	33	30	27
Chromium	11-97	56	80	55	26
Copper	84-97	55	55	56	25
Fluoride	3	50	62	55	32
Iron	40	87	95	91	74
Lead	97	44	75	39	16
Manganese	87-96	93	99	96	76
Mercury		0	0	0	0
Selenium	16	0	0	0	0
Silver	92-97	49	57	42	20
Zinc	90-98	78	97	78	45
Color		46	85	48	7
Foaming Agents		39	90	63	0
Turbidity		67	83	66	46
TOC	78	73	61	44	27

[a]Typical removals found in published literature.
Symbols: x = data inconclusive; 0 = no significant removal; + = increase.

buildings, electrical, mechanical, labor, timber and many others. Operation and maintenance costs were based on the following costs: labor at $10.00/hr, electricity at $0.04/kWh, fuel at $3.00/million btu, chlorine at $250.00/ton and lime at $65.00/ton. Construction costs and annual operation and maintenance costs and requirements for each of the treatment systems considered were generated by computer using a software package developed by CWC.

Secondary Treatment

The activated sludge and trickling filter processes were selected as being representative of secondary treatment systems. Flow schematics for these two systems are shown in Figures 15 and 16, respectively. The activated sludge process was sized using slow-speed, mechanical aerators.

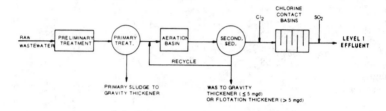

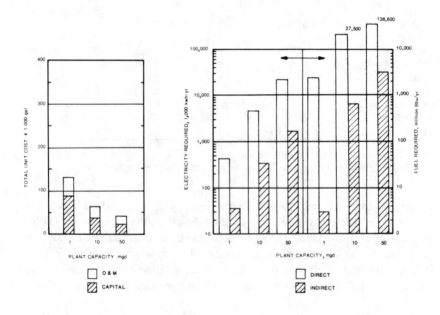

Figure 15. Activated sludge.

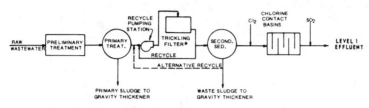

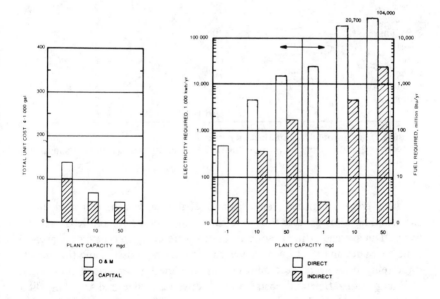

Figure 16. Trickling filter.

Table XIX. Contaminant Removals for the Activated Carbon Adsorption Process

Constituent	Published Average[a]	Activated Carbon Adsorption			
		Average	Probability		
			10%	50%	90%
BOD	85	53	80	60	22
COD	60	47	74	46	22
TSS	50-86	64	84	65	34
NH_3-N		39	58	26	10
NO_3-N		55	–	–	–
Phosphorus		88	99	97	89
Alkalinity		+	+	+	+
Oil and Grease		47	79	57	12
Arsenic		0	0	0	0
Barium		x	x	x	x
Cadmium	99	0	0	0	0
Chromium	97	48	85	53	15
Copper		49	62	43	20
Fluoride		13	x	x	x
Iron		73	91	79	32
Lead		32	54	25	4
Manganese		32	64	24	5
Mercury		0	0	0	0
Selenium	37	0	0	0	0
Silver	97	27	37	22	1
Zinc		66	91	66	32
Color	90	70	94	78	31
Foaming Agents		64	59	51	0
Turbidity		73	88	79	48
TOC		64	80	62	42

[a]Typical removals found in published literature.
Symbols: x = data inconclusive; – = insufficient data; 0 = no significant removal; + = increase.

The average constituent removals and reliabilities for the activated sludge and trickling filter process trains are shown in Tables XXI and XXII, respectively. The average effluent concentrations were computed using the average removal values and the typical raw wastewater characteristics, and the process train reliabilities were determined using the method described previously.

Energy utilization and costs for the activated sludge and trickling filter systems are presented also in graphical form in Figures 15 and 16. Both the direct and indirect energy requirements are presented. Direct energy refers to the use of fuel and electricity on the plant site; indirect energy refers to the fuel and electrical energy consumed in the manufacture and transport of chemicals for use at some point in the wastewater treatment system.

Secondary effluents are suitable for agricultural irrigation, livestock and wildlife watering, once-through cooling water and for primary metals. They

Table XX. Required Treatment Process to Achieve Various Beneficial Uses

Beneficial Use	Wastewater Treatment System
Agricultural Irrigation–Forage Crops	Secondary Treatment
Agricultural Irrigation–Truck Crops	Secondary Treatment
Urban Irrigation–Landscape	Filtration of Secondary Effluent
Livestock and Wildlife Watering	Secondary Treatment
Power Plant and Industrial Cooling	
Once-through	Secondary Treatment
Recirculation	Tertiary Lime Treatment
Industrial–Boiler Make-up	
Low Pressure	Ion Exchange–Tertiary Lime
Intermediate Pressure	Carbon Adsorption, Ion Exchange, Tertiary Lime
Industrial Water Supply	
Petroleum and Coal Products	Filtration of Secondary Effluent
Primary Metals	Secondary Treatment
Paper and Allied Products	Carbon Adsorption/Tertiary Lime
Chemicals and Allied Products	Carbon Adsorption/Tertiary Lime
Food and Kindred Products	Carbon Adsorption/Tertiary Lime
Fisheries	
Warm Water	Ion Exchange–Tertiary Lime
Cold Water	Ion Exchange–Tertiary Lime
Recreation	
Secondary Contact	Filtration of Secondary Effluent
Primary Contact	Ion Exchange–Tertiary Lime
Public Water Supply	
Groundwater Spreading	Infiltration–Percolation Basins
Groundwater Injection	Reverse Osmosis of Biolog/Phys-Chem
Surface Water	Carbon Adsorption–Ion Exchange–Tertiary Lime
Direct Potable	Reverse Osmosis of Biolog/Phys-Chem

are also suitable for groundwater recharge by surface spreading (infiltration/ percolation basins or rapid infiltration basins). Secondary treatment of wastewaters is the minimum level of treatment required for discharge into water bodies in the United States.

Filtration of Secondary Effluent

Effluent reuse for landscape irrigation is becoming more prevalent in recent years, especially in water-short areas. This irrigation includes parks, greenbelts, medians, golf courses, schools and other landscaped areas, and is a major factor behind the dual water system that has received considerable attention lately. However, these systems typically are small, requiring only a few thousand acre-feet per year.

The flow schematic for this treatment system is shown in Figure 17. The process includes the activated sludge process, flow equalization, filter supply

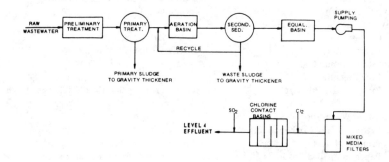

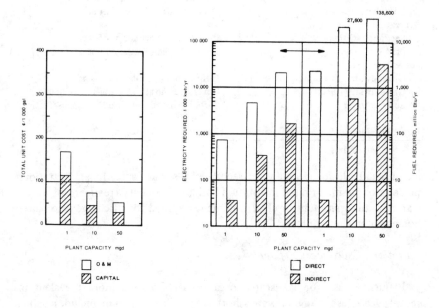

Figure 17. Activated sludge plus filtration.

Table XXI. Level 1a: Secondary Treatment with Activated Sludge–Average Process Train Performance[a]

Constituent	Average Removal (%)	Average Reliability			Average Effluent Concentration (mg/l)
		10%	50%	90%	
BOD	94	99	96	78	11
COD	83	95	79	57	71
TSS	91	97	92	57	17
NH_3-N	70	92	77	47	6.0
NO_3-N	+	+	+	+	+
Phosphorus	60	76	60	22	3.8
Alkalinity	38	69	61	47	131
Oil and Grease	94	100	95	69	3.7
Arsenic	52	70	61	0	0.003
Barium	31	62	34	0	0.162
Cadmium	71	96	70	20	0.002
Chromium	85	98	87	60	0.025
Copper	88	98	89	62	.014
Fluoride	x	x	x	x	x
Iron	84	99	84	50	0.360
Lead	85	98	85	54	0.022
Manganese	46	82	41	6	0.063
Mercury	23	31	18	0	0.086
Selenium	7	26	12	0	0.006
Silver	91	99	93	50	0.002
Zinc	67	90	67	23	0.138
TOC	89	99	87	62	11
Turbidity	90	97	92	75	11[a]
Color	56	83	56	20	29[b]
Foaming Agents	79	–	68	–	0.45

[a]TU.
[b]P-C units.
Symbols: x = data inconclusive; − = insufficient data; 0 = no significant removal; + = increase.

pumping and mixed media filters, followed by effluent disinfection. Filter supply pumping was included to cover all eventualities, although it may be optional in some instances where plant hydraulics favor gravity filtration. Flow equalization is provided to minimize filter size and improve filter efficiency.

The estimated average effluent concentrations of various constituents are shown in Table XXIII which also includes the computed average constituent removals and reliabilities for the 10, and 50 and 90 percentile points. The estimated unit costs as well as the primary and secondary energy requirements are shown graphically in Figure 17.

Table XXII. Level 1b: Secondary Treatment with Trickling Filters–Average Process Train Performance

Constituent	Average Removal (%)	Average Reliability			Average Effluent Concentration (mg/l)
		10%	50%	90%	
BOD	82	94	83	70	33
COD	74	90	74	55	108
TSS	83	96	85	48	33
Cadmium	38	75	37	5	0.005
Chromium	44	81	49	6	0.094
Copper	78	95	82	24	0.026
Lead	73	97	68	21	0.040
Mercury	11	–	8	–	0.100
Zinc	72	95	75	20	0.117

Symbol: – = insufficient data.

Table XXIII. Level 4: Filtration of Secondary Effluent–Average Process Train Performance

Constituent	Average Removal (%)	Average Reliability			Average Effluent Concentration (mg/l)
		10%	50%	90%	
BOD	96	100	99	84	7
COD	88	99	89	63	50
TSS	99	100	100	88	2
NH_3-N	80	96	78	36	4.0
NO_3-N	56	99	52	5	8.8
Phosphorus	83	100	94	49	1.6
Alkalinity	89	98	94	84	23
Oil and Grease	94	100	95	69	4
Arsenic	52	70	61	0	0.003
Barium	31	62	34	0	0.162
Cadmium	71	96	70	20	0.002
Chromium	85	98	87	60	0.025
Copper	88	98	89	62	0.014
Fluoride	x	x	x	x	x
Iron	84	99	84	54	0.360
Lead	85	98	85	54	0.022
Manganese	46	82	42	6	0.063
Mercury	23	31	18	0	0.086
Selenium	7	26	12	0	0.006
Silver	91	99	93	50	0.002
Zinc	67	90	67	23	0.138
TOC	90	99	87	62	10
Turbidity	97	97	92	75	3
Color	70	83	56	20	20
Foaming Agents	79	–	68	–	0.45

Symbols: x = data inconclusive; – = insufficient data; 0 = no significant removal.

Activated Sludge plus Chemical Coagulation

Chemical coagulation is used to reduce the phosphorus concentration of the waste water. Chemicals can be added at virtually any point in the treatment system, although some locations yield better results than others, depending on the chemical and the raw influent characteristic. The more frequently used systems include alum addition to the aeration basin, ferric chloride addition to the primary sedimentation tanks and tertiary lime treatment. Tertiary lime treatment provides the most reliable method for phosphorus removal, and therefore is the process included here. Other processes for the removal of phosphorus, notably the proprietary system Phostrip, A/O and Bardenpho, were not modeled because data were not available at the time the study was completed.

The water quality required for recycled cooling systems includes, as a minimum, phosphorus and suspended solids removal. Selection of tertiary lime treatment also removes constituents such as silica, which are harmful to the cooling process, but also increases the concentration of calcium. The number of concentrating cycles of water use, therefore, may be limited. In some instances nitrogen control is also necessary, which would require a different treatment scheme.

The treatment scheme for tertiary lime treatment is shown on Figure 18. Lime is added to the liquid stream from the secondary sedimentation basins, and the mixture is flocculated and clarified. The phosphorus and contaminant-laden sludge is removed for separate handling and the pH of the clarified wastewater is adjusted in a two-stage process. Clarified effluent is then filtered and disinfected. pH adjustment can be achieved by using carbon dioxide (flue gas), sulphuric acid or submerged burners. Typically, 400 mg/l of lime is needed to overcome wastewater alkalinity and raise the pH above 11.2.

The estimated average effluent concentrations of various constituents are shown in Table XXIV. This table also includes the computed average constituent removals and reliabilities for the 10, 50 and 90 percentile points. The estimated unit costs and the primary and secondary energy requirements are shown graphically on Figure 18.

Activated Sludge plus Chemical Coagulation and Ion Exchange

The process consisting of activated sludge plus chemical coagulation and ion exchange could be used as feed water for boilers. However, the water quality for boiler feed water is directly dependent upon the pressure at which the boiler is operated. The water quality from this treatment scheme is suitable for low-pressure boilers.

Secondary effluent is treated with lime and then clarified. The phosphorus-rich sludge is removed for processing (such as by recalcination) or disposal,

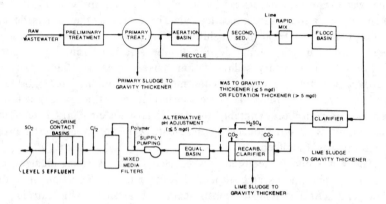

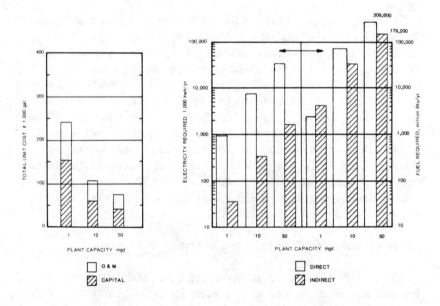

Figure 18. Activated sludge/tertiary lime/filtration.

Table XXIV. Level 5c: Tertiary Lime Treatment of Activated Sludge Effluent–Average Process Train Performance

Constituent	Average Removal (%)	Average Reliability			Average Effluent Concentration (mg/l)
		10%	50%	90%	
BOD	98	100	99	86	4
COD	95	99	94	80	21
TSS	97	100	98	81	6
NH$_3$-N	76	92	74	42	4.8
Phosphorus	97	100	98	88	0.28
Oil and Grease	94	100	95	69	4
Arsenic	61	98	63	0	0.003
Barium	79	95	79	52	0.049
Cadmium	98	100	98	87	0.002
Chromium	100	100	96	81	0
Copper	97	100	98	86	0.004
Fluoride	x	x	x	x	x
Iron	98	100	99	91	0.045
Lead	98	100	98	77	0.003
Manganese	97	100	98	85	0.004
Mercury	23	31	18	0	0.086
Selenium	7	26	12	0	0.006
Silver	76	100	99	80	0.005
Zinc	88	100	86	38	0.050
TOC	100	100	94	70	0
Turbidity	98	100	99	90	2
Color	75	97	74	36	17
Foaming Agents	79	–	68	–	0.454

Symbols: x = data inconclusive; – = insufficient data; 0 = no significant removal.

which is typical for small plants. The pH of the clarified wastewater is adjusted, as described before, and then filtered prior to passing through the selective ion exchange process. This process, using a naturally occurring zeolite (clinotilolite) which has an affinity for ammonia, removes the nitrogen in the wastewater stream. Finally, chlorine disinfection is provided prior to reuse.

The treatment scheme for this alternative is shown in Figure 19. Also included are the unit costs and the primary and secondary energy requirements. The computed constituent removals for the 10, 50 and 90 percentile points and the estimated average effluent concentrations of a number of constituents are shown in Table XXV.

Carbon Adsorption Following Tertiary Lime and Ion Exchange

This treatment scheme would be used for the production of high-quality water which is low in phosphorus, nitrogen, COD, suspended solids and numerous other constituents. Examples of beneficial uses include intermediate pressure boiler make-up water, and surface water supplies that are subse-

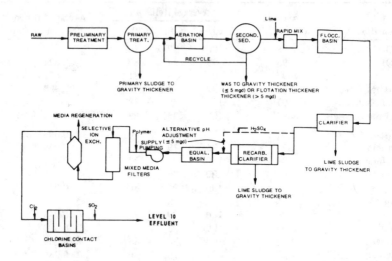

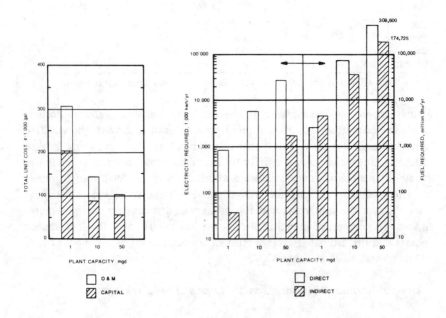

Figure 19. Activated sludge/tertiary lime/filtration/selective ion exhange.

Table XXV. Level 6b: Lime Tertiary Treatment Followed by Ion Exchange–Average Process Train Performance

Constituent	Average Removal (%)	Average Reliability			Average Effluent Concentration (mg/l)
		10%	50%	90%	
BOD	99	100	100	86	2
COD	96	100	95	80	17
TSS	99	100	99	81	2
NH$_3$-N	100	100	97	42	0
Phosphorus	97	100	98	88	0.4
Oil and Grease	94	100	95	69	4
Arsenic	61	98	63	0	0.003
Barium	79	95	79	52	0.049
Cadmium	98	100	98	87	0.002
Chromium	100	100	96	81	0
Copper	97	100	98	86	0.004
Fluoride	x	x	x	x	x
Iron	98	100	99	91	0.045
Lead	98	100	98	77	0.003
Manganese	97	100	98	85	0.004
Mercury	23	31	18	0	0.086
Selenium	7	26	12	0	0.006
Silver	76	100	99	80	0.005
Zinc	88	100	86	38	0.050
TOC	100	100	94	70	0
Turbidity	98	100	99	90	2
Color	75	97	74	36	17
Foaming Agents	79	–	68	–	0.454

Symbols: x = data inconclusive; – = insufficient data; 0 = no significant removal.

quently treated for potable use. An example of the latter is the Upper Occoquan Sewage Authority plant. The water quality requirements for cooling water uses were presented earlier.

The treatment scheme for this sophisticated process system is shown in Figure 20. Lime is added to secondary effluent and the mixture is flocculated and clarified. The phosphorus rich sludge is removed for separate handling, either by direct disposal or recalcination for reuse in the process. The high pH, which is effective for inactivating pathogenic organisms, is adjusted to neutral conditions in a two-stage process, in which either sulphuric acid or recarbonation (flue gas) was used. In some instances, submerged gas burners are more economical. Again, a lime dose of 400 mg/l was assumed to be necessary to raise the pH above 11.2

The estimated average effluent concentrations of several constituents are shown in Table XXVI. This table also identifies the computed average constituent removals and the removal reliabilities in terms of the 10, 50 and 90 percentile points. The estimated unit costs and the primary and secondary energy requirements are shown graphically on Figure 20.

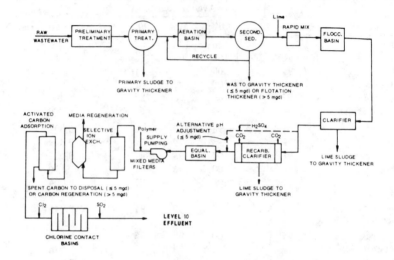

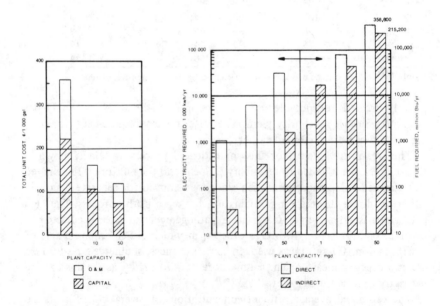

Figure 20. Activated sludge/lime/filtration/ion exchange/granular activated carbon.

Table XXVI. Level 10: Carbon Adsorption of Tertiary Lime—Ion Exchange Effluent—
Average Process Train Performance

Constituent	Average Removal (%)	Average Reliability			Average Effluent Concentration (mg/l)
		10%	50%	90%	
BOD	100	100	100	89	0
COD	98	100	97	84	8
TSS	100	100	100	92	0
NH_3-N	100	100	98	86	0
Phosphorus	100	100	100	99	0
Oil and Grease	97	100	98	73	2
Arsenic	61	98	63	0	0.003
Barium	79	95	79	52	0.049
Cadmium	98	100	98	87	0.0002
Chromium	100	100	98	84	0
Copper	98	100	99	89	0.002
Fluoride	13	x	x	x	0.752
Iron	99	100	100	94	0.023
Lead	99	100	98	78	0.001
Manganese	98	100	98	86	0.002
Mercury	23	31	18	0	0.086
Selenium	7	26	12	0	0.006
Silver	82	100	99	80	0.022
Zinc	96	100	90	57	0.017
TOC	100	100	98	83	0
Turbidity	100	100	100	95	0
Color	93	100	94	56	5
Foaming Agents	92	–	84	–	0.173

Symbols: x = data inconclusive; – = insufficient data; 0 = no significant removal.

Groundwater Recharge by Infiltration-Percolation Basins

Spreading basins (infiltration-percolation basins) have been used successfully to recharge ground water and control seawater intrusions in California.
Secondary effluent is conveyed to spreading basins located in areas having
coarse-textured, well-drained sands and gravels. As the water passes down
through the soil profile, biological, physical and chemical reactions rid the
water of most undesirable contaminations. Successful die-off of virus and
pathogenic organisms has occurred in these systems if the travel time to
groundwater table is sufficiently long.

These systems have been used as the treatment process in some locations
with good results. The percolate is collected in an underdrain system and
pumped out to the receiving stream. Used in this mode, the pretreatment
requirements are less stringent, only requiring primary treatment.

The treatment process used for this system includes activated sludge secondary treatment in order to approximate the costs of treatment as practiced in

California. The system would operate equally well using a lagoon system for pretreatment. The treatment system is shown in Figure 21, together with estimated unit costs and primary and secondary energy requirements.

Table XXVII presents the average effluent (percolate) concentrations of certain constituents. The computed, average constituent removals are shown in Table XXVII. For this treatment alternative there are few data, and removal reliabilities were not developed. However, based on a recent study [4], the percentage removals are included.

Nutrient Removal and Reverse Osmosis

The treatment process of nutrient removal and reverse osmosis was based on the proposed Denver Water Board's potable reuse program. The quality of the water from this system is projected to be suitable for direct reuse to the water distribution system. The Denver Water Board is currently designing a 1 mgd demonstration of the proposed treatment system (see Figure 22).

The system is suitable for achieving water qualities for direct potable reuse and for direct groundwater recharge by injection. The latter system is being practiced at the Orange County Water Factory 21 in Fountain Valley, California. The treatment processes are similar to those described for the carbon adsorption following tertiary lime and ion exchange, with the addition of ozone disinfection and reverse osmosis (RO). The RO process is extremely efficient at removing a majority of contaminants normally associated with wastewaters. It also serves as a back-up for several preceding processes in the

Table XXVII. Spreading Basins—Average Process Train Performance

Constituent	Average Removal (%)	Average Reliability			Average Effluent Concentration (mg/l)
		10%	50%	90%	
BOD	99	–	–	–	1.8
COD	87	–	–	–	54
TSS	99	–	–	–	1.9
NH_3-N	90	–	–	–	2
Phosphorus-Total	96	–	–	–	1.1
Phosphate	88	80	55	26	0.8
TOC	72	87	73	42	28.6
Fluoride	30	50	29	7	0.6
Cadmium	12	–	–	–	0.007
Chloride	+	–	–	–	–
Copper	72	–	–	–	0.03
Lead	23	–	–	–	0.11
Zinc	71	0	0	0	0.12

Symbols: – = insufficient data; + = increase in concentration.

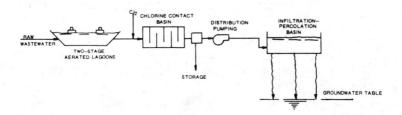

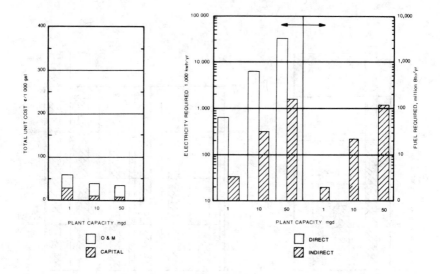

Figure 21. Groundwater recharge by I/P basins.

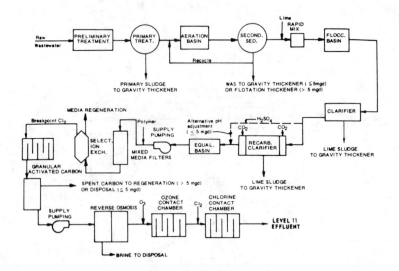

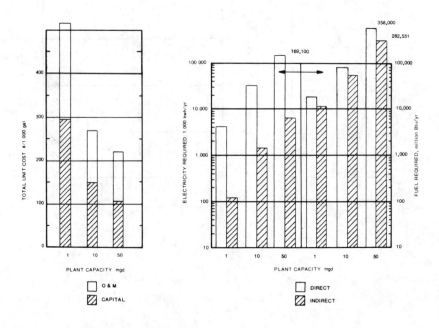

Figure 22. Nutrient removal and reverse osmosis.

event of their failure. This has been tested at Water Factory 21 by taking various processes from service and measuring differences in effluent quality. Any differences have been too small to detect or to make an impact on receiving water qualities.

The estimated average effluent concentrations are presented in Table XXVIII, based on information obtained from Water Factory 21. The reliability of constituent removal is also shown on Table XXVIII. Estimated unit costs and the primary and secondary energy requirements are shown graphically on Figure 22.

Table XXVIII. Process Train Performance, Nutrient Removal, Chemical Oxidation and Reverse Osmosis

Constituent	Average Removal (%)	Average Reliability			Average Effluent Concentration (mg/l)
		10%	50% +	90%	
		(% Removals)			
BOD	100	100	100	89	0
COD	100	100	100	97	<1
TSS	100	100	99	87	0
NH_3-N	100	100	91	68	0
Phosphorus	100	100	100	99	0
Oil and Grease	100	100	100	90	0
Arsenic	61	93	63	0	0.003
Barium	79	95	79	52	0.092
Cadmium	98	100	98	87	0.0002
Chromium	100	100	98	84	0
Copper	98	100	99	98	0.002
Iron	99	100	100	94	0.023
Lead	99	100	98	78	0.001
Manganese	98	100	98	86	0.002
Mercury	23	31	18	0	0.028
Selenium	7	26	12	0	0.006
Silver	82	100	99	80	0.004
Zinc	98	100	95	58	0.008
TOC	100	100	98	83	<1
Turbidity	100	100	100	95	<0.1[a]
Color	93	100	94	56	5[b]
Foaming Agents	92	—	84	—	0.17
TDS	95	—	—	—	129

[a]TU.
[b]P-C units.
Symbols: — = insufficient data.

REFERENCES

1. Culp/Wesner/Culp and Mark V. Hunter. "Water Reuse and Recycling, Volume 1, Evaluation of Needs and Potential," USDI Office of Water Research and Technology, OWRT/RU-79/1 (April 1979).
2. Culp/Wesner/Culp. "Water Reuse and Recycling, Volume 2, Evaluation of Treatment Technology," USDI Office of Water Research and Technology, OWRT/RU-79-2 (April 1979).
3. McKee, Jack E., and Harold W. Wolf. "Water Quality Criteria," The Resources Agency of California, State Water Resources Control Board, Publication No. 3-A (June 1976).
4. Culp/Wesner/Culp. "Long-Term Environmental Effects of Land Treatment, Rapid Infiltration," U.S. EPA Office of Water Program Operations, Final Draft, 1980.

CHAPTER 6

WASTEWATER REUSE IN ARID AREAS

Herman Bouwer
USDA Agricultural Research Service
U.S. Water Conservation Laboratory
Phoenix, Arizona

INTRODUCTION

Reuse of wastewater is the use of such water by entities or for purposes other than those that produced the wastewater. Recycling of wastewater is the use of the water by the same entity that produced it. In wastewater reuse, the wastewater is normally discharged (usually after some treatment) and transported to the place of use. In recycling, the wastewater is used on the premises of the entity producing it.

Both reuse and recycling of wastewater are expected to increase dramatically in the future as new water resources become increasingly scarce and expensive to develop, and as ground water is being depleted. Culp et al. [1] (see also the summaries [2,3]) predict that wastewater reuse in the United States will increase from 0.2% of freshwater withdrawals capable of using wastewater in 1975 to 1.7% in 2000, and that reuse of available wastewater will increase from 0.4% in 1975 to 4% in 2000. Since the volume of wastewater in 2000 will be much larger than that in 1975, this represents more than a ten-fold increase in wastewater reuse. Recycling of wastewater will probably increase so much that the total freshwater withdrawals in 2000 will actually be less than those in 1975.

This chapter considers only the deliberate or direct reuse of wastewater, which according to Culp et al. [1] presently amounts to 940 million m³ (760,000 ac-ft)/yr in the United States alone. Of this, 580 million m³

137

(471,000 ac-ft) are for irrigation, 300 million m^3 (240,000 ac-ft) for industrial use (mostly as cooling and process water), 47 million m^3 (38,000 ac-ft) for groundwater recharge, and 14 million m^3 (11,000 ac-ft) for recreation, fish and wildlife, and other purposes. To this must be added the incidental or indirect reuse of wastewater as from streams or lakes that receive sewage effluent or other wastewater, or ground water that is recharged by septic-tank outflows, leakage from sewers, seepage from sewage lagoons or from streams or lakes with sewage-contaminated water, etc. Since all wastewater after discharge becomes part of the hydrologic cycle again, indirectly all wastewater is eventually reused.

Direct reuse of wastewater is practiced most in arid areas and other regions where population growth and industrial activity are limited by available water resources.

The main sources of wastewater are sewage from cities and towns (municipal and industrial wastewaters), and tail water and deep percolation from irrigated agriculture. Tail water is the irrigation runoff from the lower end of the fields. Its quality is about the same as that of the original irrigation water and it normally is reused on lower fields or is pumped back to the same field. Deep percolation is the difference between the amount of irrigation water applied to the soil and that used by the crop for evapotranspiration. Deep percolation thus is the excess irrigation water that seeps through the root zone and eventually down to the underlying groundwater. Some deep percolation flow is necessary to leach salts out of the root zone brought there with the irrigation water. Deep percolation water is much saltier than the original irrigation water and also contains nitrate and other agricultural chemicals such as pesticides.

The amount of deep percolation water from irrigated fields is on the order of 0.1-1 m (0.3-3 ft) per year, depending on the irrigation efficiency and on the amount of irrigation water applied [4]. Taking a conservative figure of 0.15 m (0.5 ft) per year, the total deep percolation flow from the roughly 20 million irrigated ha (50 million acres) in the western United States is 30,000 million m^3 (25 million ac-feet) per year. This volume exceeds the volume of domestic sewage effluent produced by the entire population of the United States, assuming a rate of 378 liters (100 gal)/person/day. Thus, deep percolation or underground return flow from irrigated agriculture is a very significant source of wastewater.

Deep-percolation water eventually joins underlying ground water, where it is both a source of recharge and of quality deterioration, and from where it is reused again as wells deliver a mixture of native ground water and deep-percolation water. Where water tables are so high that they could adversely affect crop yield, underground drains are installed to intercept the deep-percolation water and discharge it into surface water. Deep-percolation water can also drain naturally into surface water, as in irrigated river valleys. A more detailed discussion of deep percolation from irrigated fields and its reuse potentials is given in the proceedings of the Deep Percolation Symposium held in Scottsdale, Arizona, in May 1980 [4]. The rest of this chapter is devoted to reuse of sewage effluent or municipal wastewater.

IRRIGATION

When sewage effluent is to be used for irrigation, both public health and agronomic aspects must be considered. Public health considerations are centered around pathogenic organisms (viruses, bacteria, protozoa, helminth eggs) that are or could be present in the effluent in great numbers and variety [5], and that could produce disease in farm workers who irrigate and handle the crops, in people who consume the crops, and in people who inhale effluent aerosols from spray-irrigated fields. Agronomic aspects deal with the effect of the sewage effluent on the yield of the crop (quantity, quality and time of harvest), the structure of the soil, and the accumulation of metals and other toxic substances in the soil.

Public Health Aspects

Several states have formulated quality criteria and/or permit systems for crop irrigation with sewage effluent. Such criteria always must tread a fine line between the quality or treatment that is theoretically desirable for safe irrigation with wastewater and what is practically achievable. Quality criteria often are developed more on the basis of what could be in the effluent and what might happen to the people than on documented evidence of disease caused by irrigation with effluent. Thus, although a good secondary effluent additionally disinfected so that the fecal coliform concentration is less than 1000/100 ml may be suitable for unrestricted irrigation, including sprinkler irrigation of fruit and vegetable crops consumed raw by people [6], state criteria normally are much more stringent. For unrestricted irrigation, for example, the State of California [7] requires that the effluent be adequately disinfected (7-day median coliform concentration not in excess of 2.2/100 ml and 30-day maximum coliform concentration not in excess of 23/100 ml), well-oxidized (organic matter stabilized), coagulated (colloidal and finely divided suspended matter removed), clarified (clarification of oxidized, coagulated effluent by further settling), and filtered (clarified wastewater which has passed through soils or filter media). The same standards apply to irrigation of parks and playgrounds. For irrigation of fodder, fiber, and seed crops and of orchards and vineyards, primary treatment is sufficient. Irrigation of pasture for milking animals requires an oxidized and disinfected effluent with a coliform count of less than 23/100 ml. This effluent then is also suitable for landscape irrigation (golf courses, cemeteries, highway plantings, etc.) [7].

Arizona standards [8] allow the use of undisinfected secondary effluent for irrigation of fiber or forage crops not intended for human consumption, and for orchard crops if the effluent does not contact the fruit (overhead sprinklers cannot be used). If the secondary effluent is chlorinated or otherwise disinfected to produce a total coliform concentration of less than 5000/100 ml and a fecal coliform concentration of less than 1000/100 ml, it can be used to irrigate crops for human consumption, provided that the produce is cooked or otherwise processed to destroy pathogenic organisms. Such effluent can also be used for irrigation of orchard crops where the effluent does contact the fruit, and of golf courses and cemeteries. If the secondary effluent has received additional treatment (for example, tertiary treatment, lagooning, or groundwater recharge) to reduce the biochemical oxygen demand (BOD) and suspended solids content both to less than 10 mg/l and the fecal coliform concentration to less than 200/100 ml, the effluent can be used for unrestricted irrigation. Such effluent can also be used to irrigate parks, lawns, school grounds, private yards and sport fields, or to provide a substantial portion of water in lakes used for primary-contact recreation.

The Arizona standards are currently being revised to make them more stringent. Unrestricted irrigation, for example, would require a further lowering of the fecal coliform concentration to a geometric mean of less than 2.2/100 ml with no single sample to exceed a count of 25/100 ml. Virus concentrations should not exceed 1 PFU (plaque forming unit) per 40 liters, roundworm (*Ascaris lumbricoides*) and tapeworm (*Taeniarhynchus saginatus*) eggs and *Entamoeba histolytica* must be nondetectable, turbidity must be below 1 TU (Jackson turbidity unit), and the pH must be between 4.5 and 9. Irrigation of fiber and seed crops would be permitted only if the average fecal coliform concentration were not to exceed a geometric mean of 5000/100 ml and not more than 10,000/100 ml for a single sample. Thus, even irrigation of nonedible crops could require secondary treatment followed by lagooning or chlorination. The basic philosophy behind the revised standards is increased protection of farm workers and others exposed to the effluent, and reduced risk from abuses or violations (for example, when secondary effluent officially intended for irrigation of nonedible crops is inadvertently or deliberately used to irrigate a vegetable patch). If, however, a municipality has complete control over the irrigation system and its operators, and it restricts public access to the fields and other facilities, the standards are less stringent and would allow, for example, irrigation of fiber and seed crops with effluent that has had primary treatment only.

Utah allows use of secondary effluent for irrigation of forage crops and, on a case by case basis, of surface-irrigated crops for human consumption if the plants are tall enough (cereals, corn, etc.) so that the edible part is a safe distance above the ground. If the secondary effluent has received adequate polishing treatment, it can also be used to irrigate pasture for dairy animals. Irrigation of parks, golf courses, lawns and other public places requires suitable advanced treatment of the effluent. Quality criteria for each treatment level are shown in Table I.

Irrigation with sewage effluent in the State of Washington is governed by the standards shown in Table II. Irrigation with sewage effluent in Colorado is controlled by limitations on effluent discharges in irrigation canals. Water in irrigation canals then must meet the standards shown in Table III. There are also limitations on metals and pesticides. Texas has comprehensive regulations on the design and operation of effluent irrigation systems, including a requirement that effluent applications be based on the nitrogen needs of the crop. This is to protect the underlying groundwater against nitrate pollution.

Some regulations also specify public-access control (fencing, time between irrigation and resumption of public access for golf courses, etc.). Where the effluent is applied with sprinkler or spray irrigation techniques, a buffer zone is usually required between the field and the nearest areas of public access to keep people from inhaling effluent-spray aerosols. Aerosols drift downwind from the irrigated field and can contain pathogenic organisms. Fecal bacteria have been observed to travel more than 300 m (1000 ft) through the air this

Table I. Wastewater Treatment Standards in Utah [9]

Parameter	Secondary Treatment	Polished Secondary Treatment	Advanced Treatment
BOD (mg/l) avg	25	15	10
7-day avg.	35	20	
maximum % removal	85		
Suspended Solids (mg/l)			
30-day avg	25	10	5
7-day avg	35	12	
maximum % removal	85		
Total Coliform (organisms/100 ml)			
30-day geometric mean	2000	200	
7-day geometric mean	25,000	250	
maximum			3
Fecal Coliform (organisms/100 ml)			
30-day geometric mean	25		
10-day geometric mean	250		
pH		6.5-9	

Table II. Sewage Effluent Standards for Irrigation in Washington [9]

Irrigation Use	Treatment Required	Total Coliforms	Other
Fodder, fiber, seed, forest	Primary, disinfection	230/100 ml	Settleable solids 5 ml liter/hr storage for 1 wk
Pasture for dairy animals	Secondary, disinfection	23/100 ml	
Golf courses, cemeteries, lawns, playgrounds	Secondary, disinfection	23/100 ml	
Orchards, vineyards, (surface irrigation only)	Secondary, disinfection	23/100 ml	
Food crops (surface irrigation only)	Secondary, disinfection	2.2/100 ml	
Food crops (spray irrigation)	Secondary, filtration, disinfection	2.2/100 ml no sample exceeds 23/100 ml in 30-day period	Turbidity 2 TU

The standard for primary treatment is reduction of BOD by 35% and of suspended solids by 55%.

way, depending on windspeed, relative humidity, sunlight and temperature [10]. Required buffer zones are normally in the range of 100-300 m (300-1000 ft). The best protection against aerosol drifts is use of surface irrigation methods (furrow or basin systems) wherever possible. Where the topography permits only sprinkler irrigation, low-pressure systems should be used to minimize formation of fine drops or mists, and sprinkling during wind or darkness should be avoided.

It should come as no surprise that the different states have different standards governing the use of sewage effluent for irrigation. If nothing else, this reflects the difficulty in accurate assessment of the risk factor. Sewage effluent is an important water resource that is quite suitable for crop irrigation. Regulations should be strict enough to permit this use without health risk, but not so strict that they virtually rule out irrigation with sewage effluent. When considering public health aspects, one is not only concerned about the consumer of the crop, but also about the farm workers and others who handle

Table III. Water Quality Standards for Irrigation Canals in Colorado [9]

	Maximum Limits (mg/l)	
	7-day average	30-day average
BOD$_5$	45	30
Suspended Solids		
Conventional treatment plant	45	30
Aerated stabilization pond	110	75
Nonaerated stabilization pond	160	105
Residual Chlorine	<0.5	
Oil and Grease	10	No visible sheen

or could be exposed to the effluent (including aerosols). Possible abuse of the wastewater reuse permit and the growing of crops in violation of the regulations must also be considered. For these reasons, most state regulations governing the use of sewage effluent for irrigation tend to be conservative. While this is desirable from the public health standpoint, regulations should also be sufficiently flexible and should be combined with a permit system on a case-by-case basis to avoid saddling communities with expensive treatment plants under circumstances where such treatment obviously is not necessary. If, for example, a community wants to use its sewage effluent for irrigation of nonedible crops or plants in a remote area with no public access and with a closed gravity irrigation system that is operated by well-trained personnel, there is no reason to require more than primary treatment or lagooning for such use.

Agronomic Aspects

When sewage effluent is to be used for irrigation, agronomic aspects must also be considered. These consist of crop and soil effects, and problems of water management that are created by the difference between the steady supply of sewage effluent and the seasonal variation in water requirements of the crop. Often, irrigation must continue, even when the crops don't need water or when the fields are idle. Alternatives are to use storage facilities (surface reservoirs or underground storage via groundwater recharge) or to dispose of the excess effluent in a different manner, such as discharge into surface water, irrigation of additional land (including forests or other nonagricultural land) and groundwater recharge. Where ground water below irrigated fields is to be protected against nitrate contamination, the amount of sewage effluent that can be applied is restricted by the nitrogen uptake of the crop. If the sewage effluent is used for irrigation of relatively large agricultural areas with existing farms and established cropping patterns, it is preferable to treat the effluent to the degree required by the crops traditionally grown. A change to crops especially adapted to a particular type sewage effluent may be possible for a small, individual farm or "sewage farm," but difficult for a number of established farms.

The main agronomic concerns about using sewage effluent for irrigation relate to the chemical constituents of the effluent. Typical increases in concentrations of major elements from tapwater to secondary effluent by one cycle of domestic water use in the United States are presented in Table IV, which also shows average concentrations. These concentrations apply to the United States, where the inhouse water use is relatively high (about 380 liters or 100 gal/person/day in Phoenix with its warm, dry climate and unrestricted supply of low-cost water). Higher concentrations can be expected where per capita water use is less. Sewage effluent also contains metals and other minor elements [12,13]. The concentrations of these elements in urban or domestic sewage effluent normally are too low to be of concern for irrigation (see for example, chemical composition of secondary effluent from Phoenix [14]). Excessive concentrations of heavy metals or other trace elements in municipal wastewater can be expected where the sewage contains appreciable industrial discharges. Source control then is the most effective solution to reduce the concentration of these and other undesirable chemicals in the effluent.

Table IV. Average Increases in Concentration from Tap Water to Secondary Effluent, and Average Concentrations in Secondary Effluent [11]

	Average Increase (mg/l)	Average Concentration (mg/l)
Organics	52	55
BOD	25	25
Sodium	70	135
Potassium	10	15
Ammonium-N	16	16
Calcium	15	60
Magnesium	7	25
Chloride	75	130
Nitrate-N	2	3
Nitrite-N	0.3	0.3
Bicarbonate (HCO_3)	100	300
Carbonate (CO_3)	0	0
Sulfate (SO_4)	30	100
Silica (SiO_3)	15	50
Phosphate (PO_4)	8	8
Hardness (as $CaCO_3$)	70	270
Alkalinity (as $CaCO_3$)	85	250
Total dissolved solids	320	730

The agronomic suitability of a certain sewage effluent can be assessed by comparing its chemical composition to guidelines developed for irrigation water quality, such as those in Tables V and VI. The salinity ranges in Table V were obtained by multiplying Ayers' [15] electrical conductivity values by 640. Salinity problems are related to the salt tolerance of the crop, which is shown in Table VII for field, vegetable and forage crops. Table V lists boron (see also Table VI), sodium and chloride as the main toxic ions. Plants differ in their tolerance to boron. A listing of crops in order of decreasing tolerance to boron is shown in Table VIII.

Table V. Guidelines for Interpretation of Water Quality for Irrigation [15]

Problems and Quality Parameters	No Problems	Increasing Problems	Severe Problems
Salinity effects on crop yield:			
Total dissolved-solids concentration (mg/l)	<480	480-1920	>1920
Deflocculation of clay and reduction in infiltration rate:			
Total dissolved-solids concentration (mg/l)	>320	<320	<128
Adjusted sodium adsorption ratio (SAR)	<6	6-9	>9
Specific ion toxicity:			
Boron (mg/l)	<0.5	0.5-2	2-10
Sodium (as adjusted SAR) if water is absorbed by roots only	<3	3-9	>9
Sodium (mg/l) if water is also absorbed by leaves	<69	>69	
Chloride (mg/l) if water is absorbed by roots only	<142	142-355	>355
Chloride (mg/l) if water is also absorbed by leaves	<106	>106	
Quality effects:			
Nitrogen in mg/l (excess N may delay harvest time and adversely affect yield or quality of sugar beets, grapes, citrus, avocados, apricots, etc.)	<5	5-30	>30
Bicarbonate as HCO_3 in mg/l (when water is applied with sprinklers, bicarbonate may cause white carbonate deposits on fruits and leaves)	<90	90-250	>520

Table VI. Recommended Maximum Limits in Milligrams per Liter for Trace Elements in Irrigation Water [6]

	Permanent Irrigation of All Soils	Up to 20 yr Irrigation of Fine-Textured Neutral to Alkaline Soils (pH 6-8.5)
Aluminum	5	20
Arsenic	0.1	2
Beryllium	0.1	0.5
Boron-sensitive crops[a]	0.75	2
semitolerant crops	1	
tolerant crops	2	
Cadmium	0.01	0.05
Chromium	0.1	1
Cobalt	0.05	5
Copper	0.2	5
Fluoride	1	15
Iron	5	20
Lead	5	10
Lithium: citrus	0.075	0.075
other crops	2.5	2.5
Manganese	0.2	10
Molybdenum	0.01	0.05 [b]
Nickel	0.2	2
Selenium	0.02	0.02
Vanadium	0.1	1
Zinc	2	10

[a] See Table VIII for boron sensitivity of crops.
[b] For acid soils only.

Table VII. Salt Tolerance of Crops: Electrical Conductivity (in mmho/cm at 25°) of Saturation Extract of Salinized Plots Which Produces 50% Reduction in Yield [16]

Field Crops		Vegetable Crops		Forage Crops	
Barley	17.6	Beets	11.6	Bermuda grass	18
Sugarbeets	16	Spinach	8	Tall wheatgrass	18
Cotton	16	Tomato	8	Crested wheatgrass	18
Safflower	14	Broccoli	8	Tall fescue	14.7
Wheat	14	Cabbage	7	Barleyhay	13.5
Sorghum	12	Potato	6	Perennial rye	13
Soybean	9	Corn	6	Harding grass	13
Sesbania	9	Sweet potato	6	Beardless wildrye	11
Paddy rice	8	Lettuce	5	Birdsfoot trefoil	10
Corn	7	Bell pepper	5	Alfalfa	8
Broadbean	6.5	Onion	4	Orchard grass	8
Flax	6.5	Carrot	4	Meadow foxtail	6.5
Beans	3.5	Beans	3	Clovers, alsike & red	4

Table VIII. Relative Tolerance of Plants to Boron, Listed in Decreasing Order of Tolerance Within Each Group [17]

Tolerant	Semitolerant	Sensitive
Athel (*Tamarix asphylla*)	Sunflower (native)	Pecan
Asparagus	Potato	Black walnut
Palm (*Phoenix canariensis*)	Acala cotton	Persian (English) walnut
Date Palm (*P. dactylifera*)	Pima cotton	Jerusalen artichoke
Sugar beet	Tomato	Navy bean
Mangel	Sweetpea	American elm
Garden beet	Radish	Plum
Alfalfa	Field pea	Pear
Gladiolus	Ragged Robin rose	Apple
Broadbean	Olive	Grape (sultanina & Malaga)
Onion	Barley	Kadota fig
Turnip	Wheat	Persimmon
Cabbage	Corn	Cherry
Lettuce	Milo	Peach
Carrot	Oat	Apricot
	Zinnia	Thornless blackberry
	Pumpkin	Orange
	Bell pepper	Avocado
	Sweet potato	Grapefruit
	Lima bean	Lemon

One of the main constituents of sewage effluent that can cause problems in crops is nitrogen, whose concentration in secondary sewage effluent generally is in the upper part of the 5-30 mg/l range where there are increasing problems. The N-concentration could even be above 30 mg/l, causing severe problems (Table V). A concentration of 30 mg/l corresponds to 300 kg per ha·m or about 80 lb/ac-ft. Thus, with irrigation applications of 1-2 m (3-7 ft)/yr, the amount of nitrogen applied to the land could readily exceed the nitrogen requirement of the crop which generally is in the range of 50-700 kg/ha (about 50-600 lb/acre)/yr (based on nitrogen removed in harvested crop). Ranges of nitrogen and other nutrient uptake for crops are shown in Table IX. The ranges are quite wide, primarily due to yield variations. Local data on crop uptake of nutrients can best be obtained from local offices of the Agricultural Extension Service or from State Agricultural Experiment Stations.

Too much nitrogen fertilization of crops can produce too much vegetative growth and not enough fruit or not enough fruits of adequate size; it can cause crop lodging, delay the maturity or harvestability of the crop, lower its sugar or starch content, or adversely affect texture, flavor or color of fruit and vegetable crops. For example, too much nitrogen on cotton delays maturity, causes rank growth, encourages diseases, increases the risk of boll rot and

Table IX. Nutrient Uptake Rates for Various Crops [10]

| | Uptake (lb/acre · yr)[a] | | |
	Nitrogen	Phosphorus	Potassium
Forage crops			
Alfalfa[b]	200-480	20-30	155-200
Bromegrass	116-200	35-50	220
Coastal Bermuda grass	350-600	30-40	200
Kentucky bluegrass	180-240	40	180
Quackgrass	210-250	27-41	245
Reed canary grass	300-400	36-50	280
Ryegrass	180-250	55-75	240-290
Sweet clover[b]	158	16	90
Tall fescue	135-290	26	267
Field crops			
Barley	63	15	20
Corn	155-172	17-25	96
Cotton	66-100	12	34
Milomaize	81	14	64
Potatoes	205	20	220-288
Soybeans[b]	94-128	11-18	29-48
Wheat	50-81	15	18-42
Forest crops			
Young deciduous	100		
Young evergreen	60		
Medium and mature deciduous	30-50		
Medium and mature evergreen	20-30		

[a]To convert to kg/ha, multiply values by 1.12.
[b]Legumes will also take nitrogen from the atmosphere and will not withstand wet conditions.

reduced lint quality, and could increase shedding of small squares at critical periods [18]. If sugarbeets receive more than 200 kg/ha or lb/acre of nitrogen, their yield and quality can be adversely affected [19]. The sugarbeets are bigger but their sugar content is lower. Since farmers are paid on the basis of both sugar content and yield, a lower sugar content means lower income, even if the total sugar yield is not affected by the excessive nitrogen. The general practice is to fertilize sugarbeets with nitrogen early in the growing season and to "starve" the plants for nitrogen as they mature. Sugarcane also shows a decrease in sugar content if it gets too much nitrogen late in the growing season [20].

Potatoes can have a lower starch content and fewer and smaller tubers when they receive more nitrogen than 200 kg/ha or lb/acre [19]. Grain crops tend to lodge under too much nitrogen (short-straw varieties and Mexican wheat may be more resistant to lodging [19]). The yield of Dixon Cling peaches in California is reduced when the nitrogen application is above or below the optimum range of 0.35-0.7 kg (0.75-1.5 lb)/fruit tree/yr [19]. The reduction is due to smaller fruit sizes. Most stonefruits show a delay in harvestability when they receive too much nitrogen. This can be very important, since a 1- to 2-week delay can mean the difference between sale for table consumption or sale to lower paying outlets such as the canning or juice markets. Some apple varieties have color development problems when overfertilized with nitrogen (Red Delicious, for example). Crops also tend to accumulate more nitrate when soil nitrogen levels are high. High nitrate concentrations in vegetable crops would be a problem when consumed by infants because of the danger of infant cyanosis or methemoglobenemia. The same problem could occur with forage crops for cattle.

Navel and valencia oranges on certain soils produce grainy, pulpy oranges with lower juice content when fertilized with more than 150 kg/ha or lb/acre of nitrogen during the summer. Too much nitrogen on valencias can also cause regreening of the oranges when ripe. The Fuerte avocado variety showed reduced fruit quality when fertilized with more than 150 kg/ha or lb/acre of nitrogen per year [19].

Where nitrogen concentrations in the effluent are too high for general crop irrigation, they can be reduced by blending with other water or by further treatment of the effluent to reduce the nitrogen concentration. The nitrogen concentration can be reduced by tertiary treatment (nitrification followed by denitrification, ammonia volitalization), by extended lagooning, by special treatment technology (Carrousel system [21]), or by using the effluent first for groundwater recharge with rapid-infiltration basins that are managed to stimulate denitrification in the underlying soil (a process that is discussed later in this chapter).

Phosphorus, which is the other main nutrient for plants, does not seem to have direct adverse effects on crops. Phosphate, however, could immobilize iron, copper or zinc in the soil and thus produce deficiency symptoms in the crops. Potassium in sewage effluent is not considered a problem.

Recommended maximum limits for metals and other trace substances in sewage effluent are shown in Table VI. The limits are lower for permanent irrigation of all soils than for temporary irrigation of neutral to alkaline fine-textured soils. A high soil pH, occurring naturally or as the result of added lime, helps immobilize metals and, hence, allows the use of irrigation water with higher metal concentrations than does acid soil.

Chlorine, which is used to reduce the bacterial and virus contents of the effluent, is toxic to plants, causing yellowing of leaves at high concentrations. However, such high concentrations will rarely be encountered in effluent irrigation because the chlorine dosages normally are relatively low and residual chlorine concentrations continue to decrease in the presence of the organic material in the effluent. No upper limits have been developed for residual chlorine concentrations, but a value of 0.5 mg/l has sometimes been suggested.

No information is available on the fate of chlorinated hydrocarbons, trihalomethanes and other confirmed or suspected carcinogenic or toxic organic compounds in the effluent when it is used for irrigation. The main concern about these compounds, which are also formed by reactions between chlorine and organic material and which resist biodegradation, is in the potable use of the water (see the discussions of Pre- and Posttreatment, Trace Organics, Injection Systems, and Direct Potable Reuse of Sewage Effluent). Little is known about possible uptake of chlorinated hydrocarbons and other trace organics by plants and about the effect that this may have on the consumer of those plants.

Irrigation water can adversely affect soil structure if it contains too much sodium in relation to calcium and magnesium and if the total dissolved salts (TDS) content is relatively low. The ratio of sodium to calcium and magnesium is expressed as the sodium adsorption ratio (SAR), defined as $Na[(Ca + Mg)/2]^{1/2}$ with the concentrations expressed in milliequivalents per liter. The SAR-values in Table V are the so-called adjusted values that take into account the effects of the carbonate and bicarbonate ions in the water on precipitation and dissolution of calcium. The adjusted SAR is obtained by multiplying SAR by a factor that is a function of the concentrations of carbonate and bicarbonate ions in the water [6,15]. Deflocculation is only a potential problem for soils with appreciable amounts of clay. Since the TDS content of sewage effluent will seldom be below the value of 320 mg/l where problems could start (Table V), deflocculation of clay will almost never be a problem in irrigation with sewage water.

Irrigation Systems

Irrigation is a very logical use of sewage effluent, since the water quality requirements are not as critical as for most other uses. Also, nitrogen and phosphorus, which are pollutants when discharged into surface water, have fertilizer value when the effluent is used on crops. High BOD loadings, which can be detrimental when effluent is discharged into surface water, are no problem on soil. Direct irrigation with sewage effluent thus is becoming an increasingly popular method of sewage disposal and utilization. In the sewage treatment business, the technique is often referred to as "slow-rate land treatment."

Land treatment systems are classified in the innovative and alternative (I/A) systems category and are eligible for greater financial support from the U.S. Environmental Protection Agency than conventional treatment and disposal systems. This includes 85% construction grants (versus 75% for conventional systems); 15% credit on cost effectiveness; 100% funding of cost of system evaluation, training persons, and disseminating technical information; 100% grants for modification and replacement if the I/A project fails; and higher priority on state project lists (Clean Water Act of 1977). Municipalities are now required to consider land treatment in any application for sewage treatment plant construction grants and they must give the reasons for rejection in case the land treatment option was not selected.

Many successful sewage-irrigation systems exist today and some are quite old (cf. Iskander [22] and other papers in the same proceedings; Jewell and Seabrook [23]). Early publications [24,25] about the technique often sound quite similar to what is written today. In California, 62 sewage irrigation systems existed in 1935. In 1978, almost 185 million m^3 (150,000 ac-ft) of effluent were used for crop irrigation and 35 million m^3 (28,000 ac-ft) for landscape irrigation [26].

A complete wastewater reuse system was installed in the mid-1970s by the Irvine Ranch Water District southeast of Los Angeles [27,28]. The District collects and treats about 18,000 m^3/day (4.8 mgd) of sewage effluent, mostly from domestic sources (about 50,000 people). Most of the water supply for the area is from the Colorado River, which already has a relatively high TDS content. To minimize the increase in TDS due to domestic use, the use of self-regenerating water softeners in the area is prohibited since these softeners discharge a high-salinity, high-sodium water into the sewer system. The sewage effluent goes through a shredder and aerated grid chamber and then receives primary treatment, aeration, aerobic digestion, secondary clarification, in-line coagulation, dual-media filtration, and chlorination. The final effluent meets the California requirements for unrestricted irrigation. The boron content of the water is relatively high (0.8-0.9 mg/l), which could be objectionable for certain ornamental plants.

The reclaimed water is distributed by a 50-km (30-mi) pipeline to serve every area of the District developed after 1975. People in older areas are also interested in the reclaimed sewage water, but the cost of retrofitting these areas with a separate distribution system is prohibitive. The effluent is used to irrigate agricultural crops (mainly citrus), greenbelts, recreational areas, golf courses (one course takes 3,800 m^3/day or about 1 mgd), lawns of private homes, and the campus of the University of California at Irvine (which also uses about 3,800 m^3/day or 1 mgd). All the irrigation piping is clearly identified on both sides with the words RECLAIMED WATER in green stenciled letters in at least three places per 13-ft section of pipe. All valves are housed in a valve box with green locking covers that are triangular or otherwise specially marked. Reclaimed wastewater is sold to the users for $0.056/m^3 ($0.159/100 ft^3), which is about half the domestic water price of $0.12/m^3 ($0.33/100 ft^3). With the expected population increases, the capacities of the sewage treatment plants and the reuse system are expected to increase to 130,000 m^3/day (35 mgd) in the next two decades.

In Arizona, the Buckeye Irrigation Company has a contract with the City of Phoenix to use 49 million m^3 (40,000 ac-ft) of secondary sewage effluent per year. The effluent has a TDS content of about 900 mg/l, and it is used to "dilute" pumped groundwater having a much higher salt content that furnishes the rest of the district's water requirements. In Texas, effluent from 135 towns and cities was used for irrigation in 1965 [29]. Large systems are near San Antonio, where 1600 ha (4000 acres) of grassland are irrigated, and near Lubbock, where more than 800 ha (2000 acres) of cropland (grain sorghum, wheat, cotton) are irrigated [30].

These are just a few examples of the many sewage irrigation systems in existence. The literature on sewage irrigation is about as extensive as the practice itself (see, for example, the annotated bibliography by Law [31]).

GROUNDWATER RECHARGE

Incidental groundwater recharge with sewage effluent regularly takes place from septic-tank drain fields and cesspools, and as seepage from losing streams [32] that receive sewage effluent. Some sewage treatment plants discharge their effluent in dry washes or ephemeral streams (Tucson, AZ, for example), from which the entire effluent flow eventually disappears as seepage. Deliberate recharge includes both spreading and well-injection techniques. Spreading is done with infiltration basins, which are called rapid-infiltration (RI) systems or infiltration-percolation (IP) systems. These systems also fall in the category of innovative or alternative systems and qualify for the same increased EPA grants and other benefits as the slow-rate systems discussed in the previous section. With proper design and management, RI systems significantly increase the quality of the sewage effluent as it percolates through the soils below the basin and becomes renovated water. Thus, RI systems can utilize partially treated sewage (primary or secondary effluent). The soil and aquifer system below the infiltration basin then serves as a natural advanced-treatment plant.

Where the aquifers are confined or suitable land for infiltration basins is not available, well-injection techniques can be used. The effluent then enters the aquifer directly from the wells. Since the pores in aquifers generally are much larger than those in surface soils or vadose zones and since anaerobic conditions prevail, the quality of the effluent water is improved less as it moves laterally through the aquifer than as it moves downward through soils below infiltration basins. Thus, the effluent may have to be treated to meet the requirements of its anticipated reuse before it is injected into the aquifer. Both types of recharge systems are discussed.

Rapid-Infiltration Systems

Flooding Cycles, Hydraulic Loading, and Nitrogen Transformations

Rapid-infiltration systems work best on soils that are coarse enough to yield high infiltration rates but fine enough to sufficiently "treat" the sewage water as it percolates downward. This combination is usually achieved with soils in the sandy loam to loamy sand range. Regular drying of the basins is necessary to restore infiltration rates which tend to decrease during inundation due to accumulation of suspended solids and to bacterial activity on the soil surface. Drying causes these solids to shrink and partially decompose, so that the infiltration rate is back to normal when the basin is flooded again. Periodic cleaning of the bottom (removal of accumulated materials or cultivation to break up surface layers) may also be required. In addition, regular drying is necessary to bring oxygen into the soil for oxidation of organic material and for nitrification of adsorbed ammonium.

The long-term average infiltration rate (which includes drying periods) of a RI basin or system is called the hydraulic loading rate, usually expressed in meters per year. Thus, if the infiltration rate averages 0.6 m/day (2 ft/day) during flooding and the basins are operated on a schedule of 1 week flooding and 1 week drying, the hydraulic loading rate would be 110 m/yr (365 ft/yr). The hydraulic capacity of the system would then be 1.1 million m^3/ha (365 ac-ft/acre)/yr, or 3,000 m^3/day/ha (0.326 mgd/acre) of infiltration basin. Hydraulic loading rates of existing RI systems typically are in the range of 15-100 m/yr (50-300 ft/yr), depending on soil, climate and wastewater characteristics.

The optimum sequence of flooding and drying periods depends on several factors; for a given system it can best be determined experimentally. The flooding and drying cycles also affect nitrogen transformations in the soil and nitrogen removal from the effluent water as it percolates downward. The effluent normally used for rapid infiltration systems has had secondary treatment, so that nitrogen is predominantly in the ammonium form. When the flooding periods are relatively short (for example, one day of flooding followed by four days of drying), the soil beneath the infiltration basins will be sufficiently aerobic to convert essentially all nitrogen in the sewage effluent to nitrate in the soil. Thus, if the sewage effluent contains 25 mg/l total nitrogen, the renovated water can be expected to contain about 25 mg/l of nitrate nitrogen. This is well above the upper limit of 10 mg/l for drinking water and it is close to the range where serious problems can be expected when the renovated water is used for general irrigation. When flooding periods are long (for example, a sequence of one month flooding then one month drying), anaerobic conditions will prevail in the soil during flooding so that ammonium is not converted to nitrate. Initially, the ammonium will be adsorbed by the soil's cation exchange complex. When this complex has become saturated with adsorbed ammonium, the rest of the ammonium will pass through the system and remain in the renovated water. In that case, most of the total nitrogen in the effluent can be expected to show up as ammonium nitrogen in the renovated water.

To maximize nitrogen removal, flooding should be stopped before the cation exchange complex becomes saturated with ammonium. The adsorbed ammonium can then be nitrified during drying. The length of the flooding period should then be limited so that the oxygen demand for nitrification of adsorbed ammonium does not exceed the amount of oxygen entering the soil during drying. Part of the nitrate formed during drying will be denitrified in the micro-anaerobic zones of the soil during the first few days of a drying period. The remaining nitrate will be leached out when flooding is resumed. Most of the nitrogen in the renovated water will then be in the nitrate form, but the total nitrogen concentration will be considerably lower than that of the effluent that went into the ground. In field studies, nitrogen removal percentages of 60-70% have been obtained at inundation schedules of 9 days flooding and 12 days drying, and hydraulic loading rates of 76 m/yr (250 ft/yr) [33]. Corresponding nitrogen loading rates were about 19,000 kg/ha (17,000 lb/acre)/yr. At this nitrogen load, crop uptake of nitrogen is insignificant. Thus, nitrogen removal in rapid infiltration systems is essentially all achieved by denitrification. A nitrogen removal of 60-70% would bring the nitrate concentration in the renovated water down to below the upper limit for drinking water and the total nitrogen concentration in the low end of the 5-30 mg/l range where increasing problems could be expected when the water is used for general irrigation.

The removal of nitrogen from sewage water in rapid-infiltration systems is proportional to the flow velocity of the water in the soil if the amount of oxygen that entered the soil during drying was adequate for complete nitrification of adsorbed ammonium. Laboratory studies indicated a potential for up to 80% nitrogen removal by adjusting flooding and drying cycles and regulating soil intake rates (by water-depth control or other means). For more detailed information on nitrogen transformations below rapid-infiltration basins and the management of such basins for maximum denitrification, reference is made to [33-36] and the literature cited therein.

Flushing Meadows Project

Since the performance of a rapid-infiltration system depends very much on the local conditions of soil, climate, hydrogeology, and wastewater characteristics, local experimentation often is needed to develop the best management schedule (flooding and drying sequence, hydraulic loading rate as affected by water depth in basins, and frequency of cultivating or cleaning basin bottoms). An example of such an experimental project is the Flushing Meadows Project near Phoenix, Arizona. This project was installed in 1967 in the Salt River bed about 2.4 km (1.5 mi) downstream from the 91st Avenue Sewage Treatment Plant. It consists of six parallel infiltration basis 6.1 by 213 m (20 by 700 ft) each and spaced 6.1 m (20 ft) apart. The soil consists of 0.9 m (3 ft) of fine, loamy sand underlain by sand and gravel strata to a depth of 73 m (240 ft) where a clay layer begins. The water table is at a depth of about 3 m (10 ft). In 12 years of research [14,33,34,37] this project has shown that:

1. The largest hydraulic capacity of the system was obtained with flooding periods of 2-3 weeks alternated with drying periods of 10-20 days. At this schedule and using a water depth of 0.3 m (1 ft) in the basins, the hydraulic loading rate was 122 m/yr (400 ft/yr). Vegetation or gravel layers in the basins did not offer particular advantages, and the best bottom condition was bare soil. Volunteer vegetation did not seem to have any undesirable effects, so weed control was not necessary. Sludge-like accumulations on the basin bottoms were periodically removed. After 10 years of operation and a total infiltration of 754 m (2473 ft), there were not signs of reductions in hydraulic loading or in hydraulic conductivity of the underlying aquifer.

2. Nitrogen removal from the effluent water as it seeped through the ground to become renovated water was about 30% at maximum hydraulic loading, but 65% if the loading rate was reduced to about 60 m (200 ft)/yr by using 9-day flooding and 12-day drying cycles, and reducing the water depth in the basins to 0.15 m (0.5 ft). The form and concentration of nitrogen in the renovated water sampled from the aquifer below the basins were slow to respond to a reduction in hydraulic loading [33]. In the tenth year of operation (1977), the renovated water contained 2.8 mg/l NH_4-N, 6.25 mg/l NO_3-N, and 0.58 mg/l organic N, for a total N-content of 9.6 mg/l. This is 65% less than the total N-content of the secondary sewage effluent, which averaged 27.4 mg/l (mostly as NH_4-N) in that year.

3. Phosphate removal increased with increasing distance of underground movement of the sewage water. After 9 m (30 ft) of downward movement to and in the underlying aquifer, removal was about 40% at high hydraulic loading and 80% at reduced hydraulic loading. Additional lateral movement of 61 m (200 ft) through the aquifer increased the removal to 94% in 1977 (i.e., a concentration of 0.51 mg/l PO_4-P in the renovated water versus 7.9 mg/l in the effluent). After ten years of operation and a total infiltration of 754 m (2473 ft), there were no signs of a decrease in phosphate removal.

4. Fluoride removal paralleled phosphate removal. Fluoride concentrations in 1977 were 2.08 mg/l for the effluent, 1.66 mg/l for renovated water sampled from wells between the basins, and 0.95 mg/l renovated water sampled from a well 30 m (100 ft) away from the basins.

5. Boron was not removed and was present at concentrations of 0.5-0.7 mg/l (0.59 mg/l in 1977) in both effluent and renovated water.

6. Concentrations of zinc, copper, cadmium and mercury were below maximum limits for drinking and permanent irrigation. The concentration of lead in the renovated water was 0.066 mg/l, which exceeded the 0.05 mg/l limit for drinking water but was well below the 5 mg/l limit for permanent irrigation (Table VI).

7. BOD_5 values were reduced from a range of 10-20 mg/l for the effluent to essentially zero for the renovated water. Corresponding COD-ranges were 30-60 and 10-20 mg/l, respectively. Not all organic carbon was removed from the effluent water as it seeped through the soil and aquifer. The total organic carbon (TOC) contents of the renovated water averaged 5.2 mg/l for the wells between the basins and 3.3 mg/l for the wells 30 m (100 ft) away from the basins. The average TOC content of the secondary sewage effluent was 19.2 mg/l. Trace organics in renovated wastewater presently are the main concern in potable use of such water.

8. The secondary sewage effluent contained 21 virus units (PFUs) per liter, but viruses could not be detected in renovated water sampled below the basins [38]. Fecal coliforms were present at concentrations of up to several hundred per 100 ml in the renovated water sampled below the basins, but could not be detected in renovated water sampled 91 m (300 ft) away from the basins. Hence, a lateral movement of about 100 m through the aquifer appears adequate to produce renovated water free from fecal coliforms. Laboratory studies showed that distilled water (rain) could remobilize previously adsorbed viruses in the soil only during the first 24 hr of a drying period, but that addition of $CaCl_2$ prevented most of this desorption [39]. This finding is important for managing rapid-infiltration systems in humid climates.

Aquifer Protection, System Design, and Management

The results of the Flushing Meadows Project indicate that RI systems can yield renovated water of sufficient quality for unrestricted irrigation and recreation. At the same time, however, the renovated water may not be as good as the native ground water. Thus, movement of renovated water into the aquifer should be controlled and the renovated water should be taken out of the aquifer again at some point to protect high quality indigenous ground water. This can be done with systems of the type shown in Figure 1. The top system represents the situation where the RI basins are located on high ground and the renovated water drains naturally into a stream or other surface water from where it can be used again. These systems are also used where reduction of stream or lake pollution is the main objective of the system. Sewage effluent, instead of being directly discharged into surface water, is

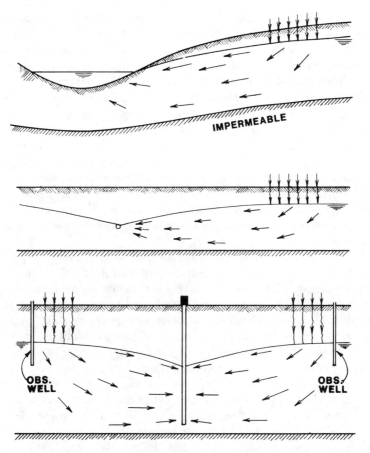

Figure 1. Schematic of effluent infiltration systems with removal of renovated water from aquifer by natural drainage into surface water (top) and interception with drains (center) or pumped wells (bottom).

then put through a RI system first so that it eventually drains into the surface water as renovated water with greatly reduced concentrations of undesirable substances. In the center system, there is no natural drainage to surface water, but the groundwater table is sufficiently high so that the renovated water can be collected with underground tile drains. Where the water table is too deep to be reached with underground drains, the renovated water can be collected from a series of wells (Figure 1, bottom). The infiltration basins can then be placed in two parallel strips, with a line of wells midway between the strips.

The desired distance between the infiltration basins and the point where the renovated water leaves the aquifer depends on the quality of the sewage effluent going into the ground, on the soil and aquifer materials, and on the quality requirement for the renovated water. In general, the underground flow distance should always be as large as possible. A reasonable distance may be about 100 m (300 ft). The underground detention time of the water in the soil and aquifer system should preferably be at least 1 month.

These systems should be designed so that the water table below the infiltration basins during flooding does not rise higher than about 0.3 m (1 ft) below the bottom of the basin. A rise above this level could "back up" the infiltrating water and reduce infiltration rates (see Refs. 40 and 41 for design procedures and Ref. 32 for methods for calculating the rise of groundwater mounds). When a basin is dried, the water table should fall relatively rapidly to a depth of about 1.2 m (4 ft) in sand and 1.8 m (6 ft) in finer soils to permit adequate entry of atmospheric oxygen into the soil for nitrification of adsorbed ammonium, decomposition of organic material, and other aerobic processes.

Rapid-infiltration systems with a number of small basins are preferable to those with a few large basins, because they offer greater flexibility in selecting optimum lengths of flooding and drying periods. Having a few reserve basins is also advisable. These basins can be used when other basins are undergoing repairs or when infiltration rates are below average (for example, due to long periods of rainy weather and resulting lack of drying and infiltration recovery in the basins, or due to breakdowns in the treatment plant resulting in discharges of sewage effluent having a high suspended solids content). Each basin should also have its own inflow and outflow control facility. The outflow drain should be located and the basin be graded so that all water in the basin can flow out by gravity when the inflow is stopped to start a drying period. There should be no low places in the basin where water can continue to stand and algae can develop. Infiltration rates will then become very low and it could take days before the last few centimeters of water disappear. As a result, infiltration recovery is poor unless, of course, the entire basin is dried for an unduly long period of time. Thus, the net effect of low, nondraining areas in infiltration basins is that they reduce the hydraulic loading rate of the basin, either by poor infiltration recovery in the low spots or by requiring extra long drying periods.

Spread of renovated water in the aquifer outside the system of infiltration basins and wells (Figure 1, bottom) can be avoided by managing infiltration rates and pumping the wells so that the water table below the periphery of the infiltration-renovation system never rises above the water table in the adjacent aquifer. This can be ascertained by monitoring water levels in observation wells at the outer edge of the infiltration basins (Figure 1, bottom).

The portion of the aquifer between the infiltration facility and the location where the renovated water leaves the aquifer again is essentially dedicated to advanced wastewater treatment. This principle can be extended to entire aquifers or major portions thereof, as, for example, below irrigated valleys or desert basins where urban developments are concentrated around the irrigated land and the sewage effluent from these developments is used as a water source for the irrigated land. The effluent could then be used for groundwater recharge with infiltration basins along the periphery of the valley or basin [42]. The renovated water would then move through the aquifer system to the lower, agricultural areas where it could be pumped from wells for irrigation.

In addition to protection of native ground water, managing the underground movement of renovated water in systems like those of Figure 1 has the advantage that the institution responsible for the rapid infiltration and subsequent collection of renovated water can maintain control over this water as long as it owns the land above the entire system. This is important in areas where groundwater rights are based on ownership of the overlying land (English and American rules; see Ref. 32 and references therein).

Pre- and Posttreatment, Trace Organics

A third advantage of collecting the renovated water with systems such as those in Figure 1 is that it is possible to use the most economical combination of sewage treatment before infiltration, quality improvement in the soil-aquifer system, and treatment of the renovated water after it has been collected. For the top system in Figure 1, this would be possible only if the renovated water drained into a very small stream or formed seeps and springs in an area without other sources of surface water.

Collection of renovated water and opportunity for posttreatment are especially important where the renovated water will eventually be used for drinking. The main source of concern about potable use of renovated sewage water is toxicity (including possible carcinogenicity) of the nondegradable organic carbon left in the water. At the Flushing Meadows Project [33], for example, such residual trace organics occurred at concentrations of 3-5 mg/l, expressed as TOC. Identification with chromatography and mass spectrophotometry of these organics in renovated water from laboratory columns filled with Flushing Meadows sand and flooded with secondary effluent from the City of Phoenix indicated presence of chlorinated hydrocarbons, trihalomethanes, and other potentially toxic materials [43]. Since chlorinated hydrocarbons are also formed by reactions between chlorine and organic compounds (precursors) normally present in sewage effluent, the effluent should not be chlorinated if it is to be used for groundwater recharge and if the soil-aquifer system can remove all microorganisms from the sewage water anyway. Most systems, except those with coarse sands and gravels and high water tables, can produce renovated water that is free from pathogens.

Methods for removing organic compounds from sewage effluent or renovated water include activated carbon adsorption, adsorption resins and reverse osmosis. Such removal will be much cheaper as posttreatment of the
renovated water where the TOC concentration is on the order of a few mg/l
than as pretreatment before infiltration where TOC concentrations are closer
to 20 mg/l (for secondary effluent). However, where the renovated water is
allowed to move through the aquifer in an uncontrolled manner and where
wells in the aquifer would pick up the renovated water for potable use, posttreatment of the water from each well may not be feasible and the much
more expensive pretreatment option may be needed.

An advantage of RI systems is that they can use primary effluent, thus
reducing the pretreatment costs to about one-fourth to one-half the cost of
combined primary-secondary treatment. Primary effluent will produce lower
hydraulic loading rates than secondary effluent (possibly half as much), so
that more land will be required. However, the higher BOD of the primary effluent is an additional carbon source for denitrifying bacteria. This should
enhance the removal of nitrogen from the sewage water by denitrification
in the soil below the basins. Also, the increased concentration of biodegradable organic carbon in the primary effluent may stimulate secondary utilization, which is the ability of microorganisms to break down normally
nondegradable organic compounds if other substrate is in plentiful supply
[44-46]. Thus, primary effluent may well produce renovated water with a
lower concentration of trace organics than would secondary effluent. This
is something to think about for those who advocate as much pretreatment
of sewage effluent as possible when it is used for groundwater recharge with
infiltration basins.

Effect of Algae on Infiltration in Phoenix Project

Rapid-infiltration basins generally are used only for infiltration. Crops are
not grown and weeds will generally not survive under the extremes of alternate flooding and drying. The water depth in the basins normally is 0.3-0.6 m
(1-2 ft). While large depths normally increase infiltration rates (especially
when infiltration is restricted at the surface of the soil because of accumulated solids), they could also lead to increased algae growth. Suspended algae
such as *Carteria* sp., can then reach bloom concentrations, which clog the bottom soil not only by formation of a fine algae "filter cake," but also by precipitation of $CaCO_3$ due to the CO_2-uptake of the algae from the water and
resulting increase in the pH. This clogging can produce severe additional reductions in infiltration rates and it can also reduce the recovery of infiltration during drying. On the other hand, if the water in the basin is shallow, infiltration causes a high turnover rate of the water and suspended algae do not

have time to develop. Because of this, small water depths may actually produce higher infiltration rates than large water depths during summer months.
In winter, algae are much less active and infiltration rates can be expected to
increase with increasing water depth.

The effect of algae on infiltration and hydraulic loading was dramatically
demonstrated at the RI project below the 23rd Avenue Sewage Treatment
Plant in Phoenix. This project, which consists of four 4-ha (10-acre) infiltration basins, first received secondary sewage effluent that had passed through
a 32-ha (80-acre) lagoon in which the average detention time was about three
days. This was enough to cause severe algae blooms in the summer. The main
bloom organism was *Carteria klebsii* Dillworth, which occurred in suspended
form. The sewage effluent thus entered the infiltration basins as a green
liquid with 50-100 mg/l suspended solids as algae. The resulting soil clogging
by algae and $CaCO_3$ precipitation reduced infiltration rates and slowed infiltration recovery during drying so that the hydraulic loading rate was only 22
m/yr (72 ft/yr). This rate was later increased to 76 m/yr or 250 ft/yr (and
possibly to close to 90 m/yr or 300 ft/yr by using larger water depths in the
winter) after construction of a by-pass channel in the 32-ha (80-acre) pond
(Figure 2) to convey secondary effluent directly to the infiltration basins.

The renovated water from the Phoenix project will be pumped from a
depth of 30-120 m or 100-400 ft (depth to groundwater table varies from 9-
21 m or 30-70 ft) from three wells which will be spaced symmetrically on the
center dike of the project. The renovated water will then be piped to an irri-
gation-district canal for unrestricted use. Observation wells at the periphery
of the basin area will be used to monitor groundwater levels so that the
project can be managed to minimize encroachment of renovated water into
native groundwater (similar to the bottom system of Figure 1).

Injection Wells

Recharge wells are basically pumped wells in reverse [32]. Because the
water enters the aquifer over a small area (the circumference of the screened
or open section of the well), they are very susceptible to clogging by suspended
solids in the injection water. Thus, suspended solids must be removed as
much as possible before injection. Even then, periodic pumping or redevelopment of the well may be necessary to maintain adequate injection rates. Clogging is not a problem when the receiving formation consists of cavernous or
fractured rock with large fissures or solution channels. In Florida, for example, secondary sewage effluent is injected directly into a saline aquifer of
fractured limestone that is separated from overlying fresh water aquifers by
impermeable material.

Figure 2. Rapid-infiltration system of Phoenix, AZ, with the four infiltration basins initally receiving secondary effluent from a lagoon (top) and later from a bypass channel around the lagoon (bottom left) to reduce algal growth in the effluent.

Injection Systems

Groundwater recharge by well injection generally is much more expensive than by infiltration basins, especially when wastewater is used, because of the additional pretreatment required. The economics of well injection are enhanced if injection produces more benefits than groundwater recharge alone, such as creation of a groundwater ridge to stop intrusion of seawater into a coastal aquifer [32]. An example of such a project is the Coastal Barrier Project in Orange County south of Los Angeles [47]. This project contains 23 injection wells spaced about 180 m (600 ft) apart on a line about 5.6 km (3.5 mi) inland from the coast. Each well is a multiple unit of four cased wells that recharge four confined aquifers (Figure 3) and have a combined capacity of 28 l/sec (450 gpm) per unit. In addition, the project has 31 monitoring wells on both sides of the injection line, and 5 producing wells drilled to a deeper aquifer free from saltwater intrusion to supply supplemental water for the injection wells. There are also 7 extraction wells about 3.2 km (2 mi) from the coast to assure a seaward gradient from the injection groundwater ridge and to regulate the amount of water that will eventually seep into the ocean [47].

The main water source for the injection wells is tertiary sewage effluent from Water Factory 21, which is an advanced treatment plant that receives conventional secondary (activated sludge) effluent from another plant. The capacity of Water Factory 21 is 56,700 m³/day (15 mgd). The advanced treatment processes include lime clarification, ammonia stripping, recarbonation, mixed media filtration, activated carbon adsorption, reverse osmosis and chlorination. Because the aquifers receiving the injection water furnish water for domestic, agricultural and industrial purposes, the State of California has imposed very stringent quality standards for the injection water (Table X). The TDS requirement can only be met by blending the tertiary effluent with demineralized water from the reverse osmosis (RO) process. The standards require at least 50% blending with RO-effluent or with pumped water from any of the five deep wells in the project. The injection water is monitored for viruses but so far viruses have not been detected in the tertiary effluent. The product water from the treatment process is pumped to the recharge wells which inject the water into the aquifers under pressure (maximum pressure is 1.4 kg/cm² or 20 psi). The optimum total injection flow is 45,360-75,600 m³/day (12-20 mgd), depending on inland groundwater levels.

The total cost of the injected water, which also includes some deep-well water, was $0.186/m³ ($700/million gal or about $230/ac-ft). Individual total costs per 3,785 m³ or million gallons were $494 for the advanced treatment, $620 for the reverse osmosis, and $52 for the injection system. Total cost of the blended water from the AWT and RO plants was $0.22/m³ or $836/million gal [47].

Use of tertiary effluent for groundwater recharge is also planned by El Paso, Texas [48]. Ten injection wells are expected to be necessary to handle the design flow of 38,000 m³/day (10 mgd). The wells would be located north of the city in an area where there are a number of city wells. Thus, the reclaimed water after groundwater recharge will be reused for municipal purposes. Expansion of the system is envisioned so that reuse would provide more than 25% of El Paso's water needs during the next 70 years.

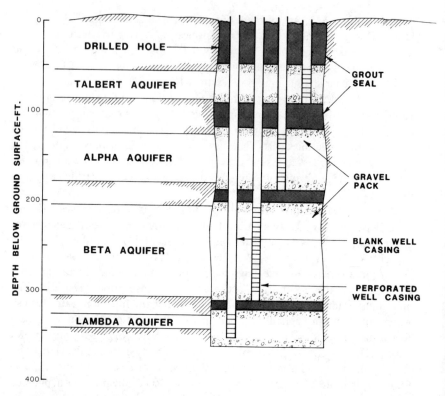

Figure 3. Schematic of multiple injection well of Coastal Barrier Project, Orange County, CA [47].

Movement of Contaminants

Some treatment can be expected when injected sewage water moves through an aquifer. If the effluent still contains biodegradable organic compounds, bacterial populations will establish themselves and decompose the organic materials. Under the anaerobic conditions of the aquifer, however, the decomposition will be slower and less complete than under aerobic conditions. On the other hand, some organic compounds are degraded better under anaerobic than under aerobic conditions. Anaerobic conditions will also stimulate denitrification of nitrate if suitable organic compounds are available as energy source for the denitrifying bacteria. Nitrification will not occur but ammonium can be adsorbed by the finer aquifer materials. When the capacity for ammonium adsorption has been reached, however, ammonium will advance farther and farther into the aquifer. Heavy-metal ions and nonbiodegradable organic compounds can also be adsorbed or otherwise immobilized. Phosphates can be adsorbed or precipitated, especially if the pH

Table X. Regulatory Agency Requirements for Injection Water Orange County Water District Wastewater Reclamation and Injection Project [47]

Constituent	Maximum Concentration (mg/l)
Ammonium	1.0
Sodium	110.0
Total Hardness (CaCO$_3$)	220.0
Sulfate	125.0
Chloride	120.0
Total Nitrogen	10.0
Fluoride	0.8
Boron	0.5
MBAS	0.5
Hexavalent Chromium	0.05
Cadmium	0.01
Selenium	0.01
Phenol	0.001
Copper	1.0
Lead	0.05
Mercury	0.005
Arsenic	0.05
Iron	0.3
Manganese	0.05
Barium	1.0
Silver	0.05
Cyanide	0.02
Electrical Conductivity	900 μmhos/cm
pH	6.5-8.0
Taste	none
Odor	none
Foam	none
Color	none
Filter Effluent Turbidity	1.0 JTU
Carbon Adsorption Effluent COD	30 mg/l
Chlorine Contact Basin Effluent	free chlorine residual

Reclaimed wastewater must be blended at least 50% with demineralized (reverse osmosis effluent) or deep well water. Reclaimed wastewater must be tested for virus.

is above 7 and enough calcium is present to form complex calcium phosphate compounds like apatite [33]. Bacteria and viruses in the water eventually die, but they can travel considerable distances [49].

Travel of contaminants through aquifers and their attenuation through dispersion, adsorption, precipitation, decomposition, volatilization, decay or die-off is governed by transport theory and, in the simpler cases, can be modeled with finite-difference or finite-element models using dispersion theory and various sink terms (see Refs. 32 and 50 and the references therein). Typical breakthrough curves of a contaminant at a certain distance downgradient from an injection well or other contaminant source are shown in Figure 4 [51]. The curves assume that the contaminant is injected at a constant rate and concentration. The piston or plug flow breakthrough curve is the idealized version that would be obtained if all contaminant molecules traveled in a straight line and at the same velocity. In reality, streamlines are tortuous and velocities are different, so that some pollutant molecules arrive sooner and others later than indicated by the average velocity of the water in the aquifer. This produces the typical sigmoid breakthrough curve when dispersion is the only attenuation mechanism (Figure 4). Eventually, the concentration of the contaminants in the water a distance away from the injection well becomes the same as that of the injection water itself (relative concentration becomes one). If the contaminant does not react with the aquifer (nonreactive or conservative contaminant or tracer such as nitrate in the absence of organic carbon, or chloride), the resulting sigmoid breakthrough curve theoretically intersects the piston flow curve at a relative concentration of 0.5.

If the contaminant is adsorbed in the aquifer material, the breakthrough is delayed, but eventually the concentration of the contaminant at a certain distance will be the same as that in the injection water when the final adsorption capacity of the aquifer for the contaminant in question is reached (Figure 4). If there is also precipitation of the contaminant or other removal or sink (volatilization, radioactive decay, biological decomposition, dieoff for microorganisms, etc.) the concentration of the contaminant at a distance from the

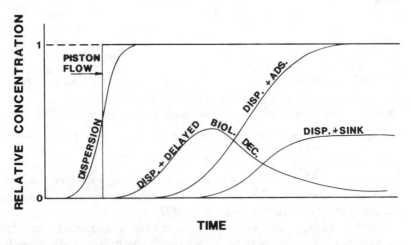

Figure 4. Hypothetical breakthrough curves of contaminants for piston flow, dispersion, dispersion plus adsorption, dispersion plus precipitation or other sink, and dispersion plus delayed biological decay.

injection well will never reach that of the concentration in the injection water, as indicated by the dispersion-and-sink curve in Figure 4. Biological decomposition sometimes takes time to develop, as bacterial populations need time to grow or bacteria need to develop a taste for a particular contaminant. This produces the typical breakthrough curve in Figure 4 where the relative concentration first increases but then decreases with time [51].

Prediction of travel of pollutants and other compounds in aquifers with transport theory requires considerable analysis and input information, particularly if the pollutants are also adsorbed or react otherwise with the aquifer material. A simpler approach was developed by Roberts et al. [51], who evaluated retardance and breakthrough relations by monitoring water quality parameters in a test well located a relatively small distance (8 m or 26 ft in their study) from an injection well receiving tertiary effluent from the Palo Alto sewage treatment plant. The results can then be used to predict arrival times at other points in the aquifer.

The technique consists of determining breakthrough curves in the test well for a conservative or nonreactive tracer like chloride, and for the reactive compound(s) in question. The fractional breakthroughs or relative concentrations (concentration of substance in water of test well divided by concentration in injected water) are then plotted against the cumulative volume injected, yielding curves as in Figure 5. If a compound is adsorbed by the aquifer material, its arrival at the test well will be delayed as indicated by the position of the breakthrough curve on the right side of the curve for the nonreactive substance. The amount of reactive compound stored in the aquifer is calculated as

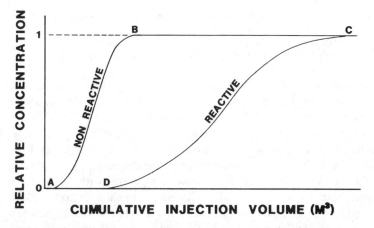

Figure 5. Breakthrough curves for a nonreactive and a reactive chemical in water moving through an aquifer.

$$M = C_o \times ABCD \tag{1}$$

where M =   amount of reactive compound stored in aquifer (grams)
         C_0 =   concentration of reactive compound in injection water (mg/l)
         ABCD =   area between breakthrough curves for nonreactive tracer and reactive compound (injection volume expressed in m^3).

If the aquifer is isotropic and there is no groundwater movement other than that caused by the injection system, M will be stored in a cylindrical volume around the injection well with a radius equal to the distance between the injection well and the test well. If the isochrone through the test well is not circular but distorted by anisotropy or other nonuniformities, the aquifer volume in which the reactive compound is stored is calculated as

$$V_{aq} = \frac{01BA}{n} \tag{2}$$

where V_{aq} =   aquifer volume in which reactive compound is stored
         01BA =   area above nonreactive breakthrough curve (injection volume expressed in m^3)
         n =   porosity of aquifer (volume fraction)

The amount of reactive compound stored per m^3 of aquifer is then calculated as M/V_{aq}.
The retardance of the reactive compound is calculated as

$$t_r = \frac{01CD}{01BA} \tag{3}$$

where t_r =   ratio between arrival times of reactive compound and nonreactive substance
         01CD =   area above breakthrough curve of reactive compound
         01BA =   area above breakthrough curve of nonreactive tracer.

If, for example, t_r is 100, the reactive compound moves 100 times slower in the aquifer than the water itself. Thus, if the rate of water flow in the aquifer is known, arrival times of the reactive compound at other points in the aquifer can be computed (assuming of course that conditions between the injection well and test well can be extrapolated to the rest of the aquifer). Reported values of t_r include 40 for cadmium and 5 for chloroform [52]. Much larger t_r-values may be expected for organic compounds with larger molecular weight. Other t_r values reported [51] were 36 for chlorobenzene, 80 for styrene, and more than 200 for di-chlorobenzene isomers. These t_r values apply to a confined aquifer consisting of fine and coarse sand with some gravel and clay deposits. The aquifer is at a depth of 15 m, is 1.5 m thick, and is located below the Santa Clara Valley wastewater treatment plant at Palo Alto, California. It could store 32 mg chlorobenzene/m^3 of aquifer, as calculated with Equations 1 and 2 [51]. Since cadmium seems to be one of the more mobile heavy metal ions in the underground environment, and chloroform one of the more mobile halogenated hydrocarbons, these two substances may well be useful indicators of the movement of metal and trace-organics in groundwater, particularly for monitoring spread of contaminants.

An interesting aspect of adsorption is that variations in the concentrations of the reactive compounds in injection water or other input water are greatly damped out when this water moves through an aquifer. Thus, where the concentration of chlorobenzene in the injected water ranged from about 700 to more than 40,000 ng/l, the concentration in the water from the test well 8 m (26 ft) away ranged from about 2500 to 5000 ng/l [51]. The ability of aquifers to damp out fluctuations of contaminant concentrations is utilized in bank filtration where wells are located at relatively small distance from a polluted river. The wells then pump river water that has first been filtered through the aquifer. Such a system gives the consumer more protection against accidental or slug discharges of contaminants than direct extraction of the water from the river.

General Aspects

Groundwater recharge with sewage effluent will be most effective if the pretreatment is specifically designed for recharge and for the subsequent reuse of the water. Considering escalating costs of energy and tertiary treatment, new schemes should be carefully scrutinized for energy and cost effectiveness and innovative or alternative technology should be considered. For example, anaerobic and aerobic lagoons in series and lime addition to precipitate phosphorus, volatilize ammonia, and kill microorganisms, are used as pretreatment of effluent for rapid infiltration in Israel [53]. The Carrousel system [21], which is based on the oxidation ditch principle, could offer an effective pretreatment because it has low energy requirements and it can remove nitrogen by denitrification. The Bardenpho process also removes nitrogen and phosphorus by bacterial action [54]. The effluent from such systems can be used for rapid-infiltration systems with recovery of renovated water (Figure 1). This water can then be reused as such for irrigation, recreation or industrial purposes, or it can be further treated for special industrial applications or for potable use (disinfection, removal of trace organics, blending) or for injection into aquifers with recharge wells and further general reuse. What is best for a given situation depends entirely on the local wastewater situation, on the water reuse needs, on the local soils and hydrogeology, and on the cost and availability of land.

INDUSTRIAL USE OF SEWAGE EFFLUENT

Cooling in steam-electric power plants represents the largest industrial reuse potential for municipal wastewater. Quality requirements for cooling water as formulated by the National Academy of Sciences and the National Academy of Engineering [6] are summarized in Table XI. Lime treatment, filtration and disinfection of the effluent are generally needed before these requirements are met. Sometimes, nitrification (trickling filter) or demineralization are also required.

Table XI. Quality Requirements of Water at Point of Use
for Cooling in Heat Exchangers [6]

	Once Through		Makeup for Re-circulation	
	Fresh	Brackish[a]	Fresh	Brackish[a]
Silica (SiO$_2$)	50	25	50	25
Aluminum (Al)	(b)	(b)	0.1	0.1
Iron (Fe)	(b)	(b)	0.5	0.5
Manganese (Mn)	(b)	(b)	0.5	0.02
Calcium (Ca)	200	420	50	420
Magnesium (Mg)	(b)	(b)	(b)	(b)
Ammonium (NH$_4$)	(b)	(b)	(b)	(b)
Bicarbonate (HCO$_3$)	600	140	24	140
Sulfate (SO$_4$)	680	2700	200	2700
Chloride (Cl)	600	19000	500	19000
Dissolved solids	1000	35000	500	35000
Copper (Cu)	(b)	(b)	(b)	(b)
Zinc (Zn)	(b)	(b)	(b)	(b)
Hardness (CaCO$_3$)	850	6250	650	6250
Alkalinity (CaCO$_3$)	500	115	350	115
pH (units)	5.0-8.3	6.0-8.3	(b)	(b)
Organics:				
Methylene blue active substances	(b)	(b)	1	1
Carbon tetrachloride extract	(c)	(c)	1	2
Chemical oxygen demand (COD)	75	75	75	75
Hydrogen sulfide (H$_2$S)	--	--	(b)	(b)
Dissolved oxygen (O$_2$)	Present	Present	(b)	(b)
Temperature	(b)	(b)	(b)	(b)
Suspended solids	5000	2500	100	100

[a]Brackish water—dissolved solids more than 1000 mg/l by definition 1963 Census of Manufacturers.
[b]Accepted as received (if meeting other limiting values); has never been a problem at concentrations encountered.
[c]No floating oil.

Quality parameters of sewage effluent directly used for cooling are shown in Table XII. Examples of users that give additional treatment to the effluent before using it for cooling are El Paso Products Co., Odessa, Texas; Champlin Refinery, Enid, Oklahoma; DOW Chemical Co., Midland, Michigan; and Texaco, Amarillo, Texas [1].

Use of sewage effluent for cooling water in a nuclear power plant is planned in Arizona, where the Palo Verde Nuclear Power Plant now under construction 72 km (45 mi) west of Phoenix will use most of Phoenix's sewage effluent in the next two decades [55]. The plant has a contract for 173 million m^3 (140,000 ac-ft) of effluent per year, which will be piped from the sewage treatment plant(s) to the project site. The first three reactor units, which have a capacity of 1270 MW each and are scheduled to go on line in 1982, 1984 and 1986, will use about 123 million m^3 (100,000 ac-ft) per year. Two other units were in the original plans but were cancelled later. A comparison of Phoenix secondary effluent characteristics with the desired quality for cooling water in the power plant is shown in Table XIII. To meet the requirements for cooling water, the effluent will be treated on-site to nitrify NH_4 and to reduce Ca, Mg, PO_4, HCO_3 and SiO_2. The treatment process consists of a trickling filter, two-stage lime softening, recarbonation, and gravity filtration. The reclaimed water will flow into a 32-ha (80-ac) storage reservoir with a capacity of 2.4 million m^3 (635 mg)—enough to cool three generator units operating at peak capacity for eight days. The cooling system will operate at 15 cycles of concentration, after which the brine will be disposed of in evaporation ponds.

The nuclear power project pays the city of Phoenix $0.81/1,000 m^3 ($1/ ac-ft) of effluent before using it for cooling to preserve its option to the wastewater. This will be raised to $1.62/1,000 m^3 ($2/ac-ft) when the official permit is issued. Once the power plant uses the water for cooling, the price will be equal to 40% of the cost of municipal and industrial water from the Central Arizona Project (an aquaduct from the Colorado River to central Arizona), but not less than $16.2/1,000 m^3 ($20/ac-ft) and not more than $24.3/1,000 m^3 ($30/ac-ft).

Municipal wastewater is also used in the steel industry, the chemical industry, pulp and paper plants, cement factories and mining. Typical applications include cooling, processing and slurry transport. Sewage effluent could also be used to transport heat from dry, hot rock formations to the surface for utilization of geothermal energy, in the same way as proposed for irrigation return flow [56].

Table XII. Municipal Wastewater Effluent Quality Used for Power Plant Cooling[a] [1]

Parameter	Nevada Power Co. Sunrise Station Las Vegas, NV	Clark County Sanitation District, Las Vegas, NV	City of Denton, TX	Southwestern Public Service Co., Lubbock, TX	City of Burbank, CA	City of Colorado Springs, CO
BOD	21	30	10	15	2	8
Suspended Solids	24	30	38	10	2	2
TDS	940	1250-1500	127	1250	500	650
Sodium	--	--	--	--	88	50
Chloride	--	315	70	345	82	20
pH (units)	7.7	7.5	7.2	7.3	7.0-7.2	6.9
Coliforms (MPN/100 ml)	10	--	16,000	--	2 to 62	225
Total Hardness	--	--	--	250	160	240
Phosphate	19	--	--	21	20	1
Organic Nitrogen	1.0	--	--	--	39	1-5
Heavy Metals	--	--	trace	trace	trace	trace
Color (units)	--	--	--	--	1	5
MBAS	--	--	--	--	0.5	0.15
Ammonia	--	--	--	--	6	27
Nitrate	1.0-3.4	--	--	--	8	0.5
Total Average Reuse[b] (mgd)	5.2	4.3	varies	20	6	2.4

[a]Concentrations in mg/l except as noted.
[b]Not all use is for cooling; some is reused for irrigation and other purposes.

Table XIII. Comparison of Phoenix Secondary Effluent Characteristics and Desired Quality for Cooling Water at Palo Verde Nuclear Power Plant [55]

	Effluent	Desired Quality
Ca	82	28
Mg	36	b
Na	200	b
K	80	b
SO_4	107	200
Cl	213	b
NO_3	0.7	b
pH (units)	7.9	7.5-8.0
Electrical Conductivity (μmhos)	1320	b
Total Alkalinity (as $CaCO_3$)	290	100
Total Hardness (as $CaCO_3$)	220	b
Silica (Total as SiO_2)	30	10
Fe (Total)	0.15	b
Suspended Solids	20-100	10
TDS	1000	b
NH_3-N	25	5
NO_3-N	0.5	b
Total Phosphorus (P)	9.0	0.5
TOC	25	b
COD	50	b
BOD	15	10
Fecal Coliform (MPN/100 ml)	10^5	b
Virus (PFU/liter)	21	b

[a]Concentrations in mg/l except as noted.

[b]Parameters not specified are either interrelated, not considered critical, or will be controlled with on-site treatment.

DIRECT POTABLE REUSE OF SEWAGE EFFLUENT

Indirect use of sewage effluent for drinking is a common occurrence where municipalities derive water from the same rivers or lakes that receive sewage effluent from other municipalities. Many of these surface waters contain several percent wastewater, with one river in South Carolina reaching 16% [57]. In suburban or rural areas, proximity between septic tanks and shallow domestic wells also leads to indirect (and often unknown to the consumer) reuse of sewage effluent. Direct, planned reuse of sewage effluent for municipal purposes is being considered more and more as an increasing number of cities face water shortages with no possibilities of additional water supply development. Where groundwater recharge is not possible, such cities will then have to treat their sewage effluent so that it can be returned into the water supply system. This type of reuse could also be called recycling, since wastewater basically is directly used again by the entity producing it in a closed system.

The quality standards for reclaimed wastewater to be used for potable purposes are more severe than the normal drinking water quality standards which were developed for water supplies derived from relatively unpolluted surface or ground waters (see, for example, Bouwer [32] and references therein). Trace organics must be completely removed or proved harmless to the public health. The final product water must also be routinely tested for viruses (see also Dean and Convery) [58].

An example of a city planning direct reuse or recycling of its sewage effluent is Denver. Additional water supplies for this city must be transported over great distances from the west slope of the Rocky Mountains at costs that are estimated to reach $4-$5/m^3 ($5000-$6000/ac-ft) by the year 2000 [59]. Since Denver neither has the type nor volume of industry required to make industrial reuse of water feasible from either an economic or a net yield standpoint, and since groundwater recharge is not feasible because aquifers are deep and marginal in permeability, potable reuse is the only form of successive use that would substantially augment present water supplies. The objective then is to treat the wastewater to a potable quality which could without reservation or restriction be supplied through the potable system to the residents of the Denver area. Based on pilot and full-scale studies and on concerns regarding harmful products in the water, the following treatment train will be used in a demonstration plant [59]:

- Phosphorus will be removed by high lime precipitation utilizing single-stage recarbonation. The side benefits of this process are removal of turbidity, suspended solids, and some heavy metals; disinfection; viral inactivation; and some softening.
- Primary suspended solids will be removed by chemically aided dual media pressure filtration which can provide additional benefits of bacterial and virus removal, and additional reductions in phosphorus.
- Nitrogen will be removed by a clinoptilolite filter for the selective ion exchange of ammonia. The process will provide additional filtration.
- Soluble organics will be removed by two-stage granular activated carbon filtration with ozonation between the two stages. The process will provide additional removal of viruses, bacteria, and heavy metals, and oxidation of residual ammonia nitrogen to nitrate.
- Dissolved salts will be removed by reverse osmosis. This process will provide additional removal of bacteria and viruses, trace metals, soluble organics, and ammonia and nitrate nitrogen.
- The disinfection process will involve the onsite generation and addition of chlorine dioxide and should provide final and residual disinfection of the product water.

The demonstration plant will have a capacity of 3780 m^3/day (1 mgd), except for the second-stage activated carbon filter and the reverse-osmosis unit, which will have a capacity of 378 m^3/day (0.1 mgd). The plant is scheduled to be completed in about 1982 and will be operated for 4.5 years of normal use after a shakedown period of about one-half year. The product water will undergo intensive analytical testing. Health aspects will be studied by evaluating acute and genetic toxicity and chronic and subchronic effects on warm blooded animals. Eventually, Denver hopes to have a 378,000 m^3/day (100 mgd) reuse program to provide 15% of its municipal water needs by the year 2000 [59].

The city of Windhoek in South West Africa, plagued for many years by chronic water shortages, has reclaimed 4536 m^3/day (1.2 mgd) of sewage effluent for direct potable use since 1969 [60]. The domestic effluent in that city is kept separate from the industrial wastewater and first receives conventional primary and secondary treatment (trickling filter). It is then passed through three lagoons where algae fix nitrogen and phosphorus in their biomass. The algal activity increases the pH to about 9, which is reduced to 7 by recarbonating the water with CO_2 produced by a submerged propane burner in another tank. The water then goes through an algae flotation tank where a flocculant like alum is added and algae particles float due to entrapped oxygen. The next step is foam fractionation to remove detergents, followed by breakpoint chlorination to remove most of the remaining nitrogen and other impurities. Rapid gravity sand filters are used to remove most of the remaining suspended solids. The final step is activated-carbon filtration to remove residual dissolved organics, including pesticides. Subsequent testing of the reclaimed wastewater from Windhoek and water from other places in South Africa showed that "All the drinking water supplies tested conformed to international quality criteria. Organic material extracted from drinking waters had no adverse effects on the health of laboratory animals. The quality of reclaimed drinking water was better than that of many supplies derived from surface sources. Consumption of reclaimed water had no detectable effect on the incidence or epidemiology of diseases in the community" [61].

Almost all uses of water increase its salt content. Crop irrigation, which normally is the biggest consumer of water in arid areas with an active agriculture, increases the salt concentration of the water by a factor of 2-10 as some of the irrigation water seeps through the root zone to leach out the salts and becomes deep percolation water that eventually recharges underlying aquifers. Municipal use of water typically adds 200-400 mg/l to the water. Cooling water in steam-electric generating plants that goes through cooling towers for recycling becomes more and more concentrated, until it becomes a brine. A portion of the water must then be continuously blown down and replaced by make-up water. Repeated reuse or recycling of water thus will lead to a salt buildup in the water that eventually requires demineralization by reverse osmosis or electrodialysis if reuse is to be continued. Other than that, water indeed is indestructible.

SOCIAL AND INSTITUTIONAL CONSTRAINTS

Public attitudes and possible legal, environmental, and other barriers can profoundly affect the adoption of wastewater reuse projects. Where water supplies become scarce, decreased use and more industrial recycling are the first obvious adjustment to make. Domestic use is quite elastic; increases in the price of water immediately produce significant reductions in water use [62].

Public education, citizen participation and demonstration projects are important factors in overcoming social constraints in wastewater reuse. A recent California study [63], however, indicated a remarkable stability in public attitudes toward wastewater reuse for the 10-year period 1970-1980. Of 972 respondents from ten cities, 56.4% opposed use of reclaimed water for drinking, 56% were against use for food preparation in restaurants, 54.5% against use for cooking in the home, 54% against use for canned vegetables, and 38.7% against use for bathing in the home. The farther away the reuse was from body or home, the lower the percentages opposed became. The percentages against were still 23.7 for swimming and 23.2 for well injection, but decreased to 14 for irrigation of vegetables, 13.3 for groundwater recharge with rapid-infiltration systems, 7.3 for pleasure boating, 3.8 for home toilet flushing, 2.7 for residential lawn irrigation, 1.6 for golf course irrigation, and 1.2 for freeway landscaping irrigation.

Institutional constraints can be quite severe. A new water reuse project can involve numerous agencies, which do not always give uniform recommendations or judgements. In California, for example, 22 different agencies had to be dealt with for one water reuse project [64]. Under those circumstances, it is often difficult to make progress toward the goal of beneficial reuse of wastewater.

REFERENCES

1. Culp-Wesner-Culp and M. V. Hughes, Jr. "Water Reuse and Recycling," Office of Water Research and Technology, U.S. Department of the Interior (1979).
2. Wesner, G. M., and M. V. Hughes, Jr. "The Potential for Wastewater Reuse in the United States," in *Proceedings of the 1979 AWWA Water Reuse Symposium, Vol. I* (Denver, CO: American Water Works Association Research Foundation, 1979), pp. 85-102.
3. Municipal Wastewater Reuse News, "Water Reuse Plans and Demonstrations" (Denver, CO: American Water Works Association Research Foundation, 1980), pp. 4-21.
4. Bouwer, H. "Deep Percolation and Groundwater Management," in *Proceedings of the Deep Percolation Symposium*, Department of Water

Resources Report No. 1 (Phoenix, AZ: Arizona Water Commission, 1980), pp. 13-19.
5. Foster, D. H., and R. S. Engelbrecht. "Microbial Hazards of Disposing of Wastewater on Soil," in *Recycling Treated Municipal Wastewater and Sludge through Forest and Cropland,* W. E. Sopper and L. T. Kardos, eds. (University Park and London: The Pennsylvania State University Press, 1973), pp. 247-270.
6. "Water Quality Criteria," National Academy of Sciences-National Academy of Engineering, No. 5501-00520 (Washington, DC: Superintendent of Documents, 1972).
7. California Administrative Code, Chapter 4, Title 22, Division 4, Environmental Health.
8. Arizona Department of Health Services, Rules and Regulations, Chapter 20, Title 9, Article 4.
9. Biological Standards for Effluent Reuse in Arizona. Technical Report by the Bureau of Water Quality Control, Arizona Department of Health Services, Phoenix, AZ, 21 March 1980.
10. "Process Design Manual for Land Treatment of Wastewater," U.S. EPA Report 625/1-77-0087 (1977).
11. Weinberger, L. W., D. G. Stephan and F. M. Middleton. "Solving our Water Problems—Water Renovation and Reuse." *Ann. N.Y. Acad. Sci.* 136:131-154 (1966).
12. Mytelka, A. I., J. S. Czachor, W. R. Guggiono and H. Golub. "Heavy Metals in Wastewater and Treatment Plant Effluents," *J. Water Poll. Control Fed.* 45:1859-1864 (1973).
13. Blakeslee, P. A. "Monitoring Considerations for Municipal Wastewater Effluent and Sludge Application to the Land," *Proceedings of the Conference on Recycling Municipal Sludge and Effluent on Land,* U.S. EPA, USDA and National Association of State Universities and Land-Grant Colleges, Washington, DC (1973), pp. 193-198.
14. Bouwer, H., J. C. Lance and M. S. Riggs. "High-Rate Land Treatment II. Water Quality and Economic Aspects of the Flushing Meadows Project," *J. Water Poll. Control Fed.* 46(5):844-859 (1974).
15. Ayers, R. S. "Quality of Water for Irrigation," *Proceedings of the Irrigation and Drainage Division, Specialty Conference,* Am. Soc. Civil Eng. Logan, UT, August 13-15, 1975, pp. 24-56.
16. Bernstein, L. "Salt Tolerance of Plants," USDA Information Bulletin No. 283, 24 pp. (1964).
17. "Diagnosis and Improvement of Saline and Alkaline Soil," U.S. Salinity Laboratory, USDA Handbook No. 60, 160 pp.
18. Miley, W. N., and R. Maples. "Why Nitrate Testing Pays Off for Cotton Consultants," *Agri-Fieldman and Consultant J.* 35(5):46-47 (1979).
19. Baier, D. C., and Fryer, W. B. "Undesirable Plant Responses with Sewage Irrigation," *J. Irrig. and Drain. Div., Am. Soc. Civ. Eng.* 99(IR2):133-141 (1973).
20. Lau, L. S., R. M. Young, P. C. Loh, V. F. Bralts and E. K. F. Lu. "Recycling of Sewage Effluent by Sugarcane Irrigation," Techn. Report No. 121, Water Resources Research Center, University of Hawaii, Honolulu (1978).
21. Dallaire, G. "First U.S. Carrousel 'Racetrack' Sewage Plant: Simple, Economical, Excellent Removals," *Civil Eng.* 49(5):65-67 (1979).
22. Iskander, I. K. "Overview of Existing Land Treatment Systems," in *Proceedings of the International Symposium on the State of Knowledge in*

Land Treatment of Wastewater, Vol. 1, H. L. McKim, Ed. (USAEC, 1978), pp. 193-200.

23. Jewell, W., and B. Seabrook. "Historical Review of Land Application as an Alternative Treatment Process for Wastewater," U.S. EPA Report No. 43019-77-008 (1978).

24. Rafter, G. W. "Sewage Irrigation, Part II," U.S. Geol. Survey Water-Supply and Irrrigation Paper 22, 100 pp. (1899).

25. Hutchins, W. A. "Sewage Irrigation as Practiced in the Western States," USDA Techn. Bull. No. 675, 59 pp. (1939).

26. Wasserman, K., and J. Radimsky. "Water Reclamation Efforts in California," *Proceedings of the 1979 AWWA Water Reuse Symposium, Vol. I* (Denver, CO: American Water Works Association Research Foundation, 1979), pp. 69-84.

27. "Irvine Ranch Water District—Total Water Management System," *Municipal Wastewater Reuse News* (Denver, CO: American Water Works Association Research Foundation, 1978), pp. 13-23.

28. Hurst, W. F., G. E. Lee and J. Scherfig. "Irvine Ranch Water District Turns Research to Application," *Proceedings of the 1979 AWWA Water Reuse Symposium, Vol. 3* (Denver, CO: American Water Works Association Research Foundation, 1979), pp. 1657-1662.

29. Harvey, C., and R. Cantrell. "Use of Sewage Effluent for Production of Agricultural Crops," Texas Water Development Board, Report No. 9 (1965).

30. Gray, J. F. "Irrigation Process Using Reclaimed Water Described," *West Texas Today,* January 1965, pp. 18-23 (1965).

31. Law, J. P., Jr. "Agricultural Utilization of Sewage Effluent and Sludge—an Annotated Bibliography," U.S. EPA, Robert S. Kerr Water Research Center, Ada, OK, 89 pp. (1968).

32. Bouwer, H. *Groundwater Hydrology* (New York: McGraw-Hill Book Company, 1978).

33. Bouwer, H., R. C. Rice, J. C. Lance and R. G. Gilbert. "Rapid Infiltration Research at Flushing Meadows Project, Arizona," *J. Water Poll. Control Fed.* 52:2457-2470 (1980).

34. Bouwer, H., R. C. Rice, J. C. Lance and R. G. Gilbert. "Renovation of Sewage Effluent with Rapid-Infiltration Land-Treatment Systems," *Proceedings of the Symposium on Wastewater Reuse for Groundwater Recharge,* Pomona, CA, September 1979. Office of Water Recycling, California State Water Resources Control Board, Sacramento (May 1980), pp. 265-282.

35. Lance, J. C., F. D. Whisler and R. C. Rice. "Maximizing Denitrification During Soil Filtration of Sewage Water," *J. Environ. Qual.* 5(1):102-107 (1976).

36. Gilbert, R. G., J. C. Lance and J. B. Miller. "Denitrifying Bacteria Populations and Nitrogen Removal in Soil Columns Intermittently Flooded with Secondary Sewage Effluent," *J. Environ. Qual.* 8(1):101-104 (1979).

37. Bouwer, H., R. C. Rice and E. D. Escarcega. "Infiltration and Hydraulic Aspects of the Flushing Meadows Project," *J. Water Poll. Control Fed.* 46(5):835-843 (1974).

38. Gilbert, R. G., C. P. Gerba, R. C. Rice and H. Bouwer. "Virus and Bacteria Removal from Wastewater by Land Treatment," *Appl. Environ. Microbiol.* 32(3):333-338 (1976).

39. Lance, J. C., C. P. Gerba and J. L. Melnick. "Virus Movement in Soil Columns Flooded with Secondary Sewage Effluent," *Appl. Environ. Microbiol.* 32(4):520-526 (1976).
40. Bouwer, H. "Ground Water Recharge Design for Renovating Wastewater," *J. San. Eng. Div., Am. Soc. Civil Eng. Proc.* 96(SA1):59-74 (1970).
41. Bouwer, H. "Design and Operation of Land Treatment Systems for Minimum Contamination of Ground Water," *Ground Water* 12(3):140-147 (1974).
42. Bouwer, H. "Urbanizing Irrigated Valleys for Optimum Water Use," *J. Urban Planning Devel. Div. Proc.*, Am. Soc. Civil Eng. 105(UP1):41-50 (1979).
43. Bouwer, E. J., P. L. McCarty and J. C. Lance. "Trace Organic Behavior in Soil Columns Inundated with Secondary Sewage," *Water Res.* 15:151-159 (1981).
44. Horvath, R. S. "Microbial Cometabolism and the Degradation of Organic Compounds in Nature," *Bacteriol. Rev.* 36:2, 146 (1972).
45. McCarty, P. L., B. E. Rittmann and M. Reinhard. "Processes Affecting the Movement and Fate of Trace Organics in the Subsurface Environment," in *Wastewater Reuse for Groundwater Recharge,* T. Asano and P. V. Roberts, Eds. (California State Water Resources Control Board, May 1980), pp. 93-227.
46. Rittmann, B. E. "The Kinetics of Trace-Organic Utilization by Bacterial Films," Ph.D. Thesis, Department of Civil Engineering, Stanford University (May 1979).
47. Cline, N. M. "Groundwater Recharge at Water Factory 21," *Proceedings of the 1979 AWWA Water Reuse Symposium, Vol. I* (Denver, CO: American Water Works Association Research Foundation, 1979), pp. 139-166.
48. Knorr, D. B. "Direct Recharge for El Paso, Texas," *Proceedings of the 1979 AWWA Water Reuse Symposium, Vol. I* (Denver CO: American Water Works Association Research Foundation, 1979), pp. 212-223.
49. Romero, J. C. "The Movement of Bacteria and Viruses Through Porous Media," *Ground Water* 8:37-39 (1970).
50. Anderson, M. P. "Using Models to Simulate the Movement of Contaminants Through Groundwater Flow Systems," *CRC Critical Reviews in Environmental Control* 9(2):97-156 (1979).
51. Roberts, P. V., P. L. McCarty, M. Reinhard and J. Schreiner. "Organic Contaminant Behavior During Groundwater Recharge," *J. Water Poll. Control Fed.* 52:161-172 (1980).
52. Roberts, P. V. "Removal of Trace Contaminants from Reclaimed Water During Aquifer Passage," paper presented at NATO-CCMS Conference on Oxidation Techniques in Drinking Water Treatment, Karlsruhe, Fed. Republic of Germany, 11-13 Sept. 1978.
53. Idelovitch, E., R. Terkeltoub, M. Butbul, R. Friedman and M. Michail. "Dan Region Sewage Reclamation Project—Groundwater Recharge with Municipal Effluent," Annual Report 1977, Tahal-Water Planning for Israel Ltd., Tel Aviv, Israel (1977).
54. Burdick, C. R., and G. Dallaire. "Florida Sewage Plant First to Remove Nutrients with Bacteria Alone—No Need for Costly Chemicals," *Civil Eng.* 48(10):51-56 (1978).
55. "Wastewater Reuse in the Salt River Valley, Arizona," *Municipal Wastewater Reuse News* (Denver, CO: American Water Works Association Research Foundation, 1978), pp. 13-22.

56. Bouwer, H. "Geothermal Power Production with Irrigation Wastewater," *Ground Water* 17:375-384 (1979).
57. Swayne, M. D., G. H. Boone, D. H. Bauer, J. S. Lee, J. T. Morgan and J. English. "Wastewater in Drinking Water Supplies," *Proceedings of the 1979 AWWA Water Reuse Symposium, Vol. I* (Denver, CO: American Water Works Association Research Foundation, 1979), pp. 167-168.
58. Dean, R. B., and J. J. Convery. "Wastewater Treatment Technology for Potable Reuse," in *Research Needs for the Potable Reuse of Municipal Wastewater*, K. D. Lindstedt and E. R. Bennett, Eds., U.S. EPA Report 600/9-75-007, Cincinnati, OH (1975), pp. 67-77.
59. Rothberg, M. R., S. W. Work, K. D. Lindstedt and E. R. Bennett. "Demonstration of Potable Water Reuse Technology—the Denver Project," *Proceedings of the 1979 AWWA Water Reuse Symposium* (Denver, CO: American Water Works Association Research Foundation, 1979), pp. 105-138.
60. Clayton, A. J., and P. J. Pybus. "Windhoek Reclaiming Sewage for Drinking Water," *Civil Eng.* (1972), pp. 103-106.
61. Nupen, E. M., and W. J. J. Hattingh. "Health Aspects of Reusing Wastewater for Potable Purposes—South African Experiences," in *Research Needs for the Potable Reuse of Municipal Wastewater*, K. D. Lindstedt and E. R. Bennett, Eds., U.S. EPA Report No. 600/9-75-007, Cincinnati, Ohio (1975), pp. 109-119.
62. Haan, C. T., D. I. Carey and O. C. Grunewald. "Water Conservation Via a Variable Pricing Mechanism," *Proceedings of the Symposium on Effects of Urbanization and Industrialization on the Hydrological Regime and on Water Quality*, Amsterdam, The Netherlands, IAHS-AISH Publ. No. 123 (October 1977), pp. 213-225.
63. Bruvold, W. H. "Public Attitudes Toward Community Wastewater Reclamation and Reuse Options," Water Resources Center, University of California at Davis, Contribution No. 179 (1979).
64. Stokes, H. W., and J. Colbaugh. "Who's Calling the Signals?" paper presented at the 52nd Annual Conference of California Water Pollution Control Association, Monterey, May 1980, as summarized in *Wastewater Reuse News* (August 1980).

CHAPTER 7

HUMAN FACTORS ENGINEERING
AND WASTEWATER REUSE

Walter P. Lambert
Roy F. Weston, Inc.
West Chester, Pennsylvania

Patrick Y. Yang
General Electric
Cleveland, Ohio

INTRODUCTION

This chapter provides a simplified introduction to the discipline of human factors engineering (HFE). The presentation is made from a process engineering point of view and represents a number of lessons learned while wrestling with human-oriented problems during design of a potable reuse pilot facility for the U.S. Army. Relevant aspects of that project are presented in the latter portion of the chapter. The major lesson learned from that project was that end-item design and performance would suffer if explicit hardware and software accommodations were not made for actual human beings expected to operate the system. Engineers confronted with the problem of designing complex reuse systems to be operated by minimally trained people can benefit from learning about HFE.

HUMAN FACTORS ENGINEERING

HFE has nothing to do with genetic manipulation or socioeconomic impacts of technology on society. It is a pragmatic discipline, a kind of synthesis of certain aspects of industrial engineering, human physiology and applied experimental psychology that was spawned by man/machine interface requirements of World War II weapon systems. Called ergonomics in England, HFE focuses on the interface between man and the machines men operate. Men plus machines and their environment comprise the man/machine system (MMS). The goal of HFE is minimization of system failures caused by inappropriate man/machine interaction. For a wastewater reuse system, failure means producing water outside specified quality constraints. The approach is to design out of a system predispositions to human error. Hopefully, the result is safe, reliable and efficient system performance.

The presence of a human operator as an active subsystem of the MMS attributes the MMS with biological rather than strictly mechanistic characteristics. Since the human operator is usually imbedded as an integral part of the MMS control system, operational characteristics of the MMS are linked to the behavioral characteristics of the operator. Meister ([1], p. 9) defined the human factors discipline as "... the application of behavioral principles and data (i.e., those describing how people behave) to engineering design to do two things: to maximize the human's contribution to the effectiveness of the system of which he is a part and to reduce the impact of that over-all system on him."

Human factors engineering is divided into two broad fields: one concerned with mating human bodies and machines, and the other with mating minds and machines. The supporting science base of the first field is drawn from anthropometry, biomechanics, work physiology (e.g., biological effects of workload), environmental physiology (e.g., heat and cold stress) and sensory physiology (e.g., visual, auditory and tactile acuity). A portion of this base has been translated into sets of human factor standards and criteria such as MIL-STD-1472B from the Department of Defense [2]. Other portions have been used for development of models of human performance in a wide variety of complex systems and under a number of physiological, psychological and physical stresses [3]. Although applications in this field include design of work place environments, vehicle seats, control handles, protective clothing, work/rest procedures and control panel display devices, a cursory scan of *Consumer Reports* illustrates that applications of HFE to the design of some consumer products are not as widespread as one would hope. Cases of HFE deficiencies in defense systems and industrial facilities have included switches which could not be operated by gloved hands, protective masks which could not be used by persons wearing glasses, test circuits located between floors in elevator shafts and alarm displays located near ceilings and floors rather than at eye level. Results of such deficiencies are increased risk of preventable injury, increased risk of preventable control errors and system operations at levels below design capabilities.

The second broad field of HFE deals with mating minds and machines. This activity requires skills, background and attitudes not commonly found within the design engineering community. Skill psychology (dealing with information processing and decision-making characteristics of operators) and occupational psychology (dealing with training efforts and individual operator differences) are represented in the science base for this field ([3], p. 5), as are such other disciplines as industrial and organizational psychology, clinical and social psychology, and economic and management science ([3], p. 18).

The main concern of this second field of HFE is how operators acquire, process and act upon system-generated information. Operator characteristics important to this field include learning abilities, awareness, motivation, comprehension, memory and psychomotor coordination. Example applications include specification of process control panel display information, distribution of process control decisions between operators and machines, design of operator tasks, and selection and training of personnel. The first broad field of HFE can be said to encompass the biology of work, and this second field to encompass the psychology of work. As such, it differs from the first in that only guiding principles rather than criteria and standardized procedures are available ([3], p. 15). One result is that subjectivity, style and opinion play stronger roles in influencing system design. HFE analyses and decisions are more difficult and effective communication between the HFE specialist and the design engineer may not be as fluent because of greater disciplinary differences between them.

USE OF HUMAN FACTORS ENGINEERING

In a general sense, "technology is effective to the extent that men can operate and maintain the machines they design" [4]. More specifically, reuse system effectiveness depends in large part on system reliability measured as the proportion of total production falling within specified quality limits. System reliability depends on the mechanical performance of the plant equipment and on the way the plant operators manipulate and use the equipment [5]. Machine design, wear and fatigue play roles in the former. Human error plays a role in the latter. An important operating concept in HFE is "that error occurs only when the conditions which predispose the human to make the error exist" [1]. In fact, there is an increasingly detailed and sophisticated literature which is disproving concepts that human error is related to inherent flaws in the human material. Designing to minimize predispositions to human error has been demonstrated possible. There is also evidence that users and operators are no longer willing to accept liability for poor performance of poorly human factored systems [5].

That places a demand on the system designer for more than adequate system performance based on engineering design expectations. It also leads to two corollary incentives for investing in HFE. The first is the selfish concern about being associated with a reuse system which does not perform to specification and which is perceived to be a hazard to the user. System designers can use HFE to avoid such a situation. The second incentive stems from uncertainties in the quality and size of the manpower pool for reuse facility operators and maintenance persons. Scoggins [6] of the Georgia Institute of Government has described trends in user-oriented computer systems as a field in which widely expanding computer applications have not been accompanied by an equally expanding operator pool. The result has been increased design sophistication of software packages so systems can be used effectively by people without traditional computer science training. HFE concepts have played strong roles in that environment. A similar situation can be expected in the case of proliferation of wastewater reuse facilities where, as the number of facilities increases, the quality of available operators can be expected to decrease. Facilities will have to be designed for operation by the people actually available.

It would be reassuring to learn that there exist a substantial number of human factors specialists actively participating in the wastewater treatment industry, but such is not the case. Of the approximately 3000 human factors specialists in the United States, most are located in industry and are primarily involved with defense and aerospace systems development. However, a few other industries have recognized a need and a shortage. In 1976, the Division of Operational Safety, Energy Research and Development Administration, commissioned specially designed introductory courses in HFE for associated design engineers [5].

HFE specialists represent a broad range of talents and draw from a number of associated disciplines such as operations research, experimental psychology, anthropology and physiology. An experimental psychologist on a reuse system design team may be a disturbing concept to some engineers, but other industries, e.g., chemical process and nuclear power, are accepting the idea.

HFE in the wastewater reuse industry may of necessity be left in the hands of cross-trained design engineers and systems analysts. Emphasis is placed on the concept of explicit cross training because review of HFE applications literature (and behavioral studies of design engineers [1, Ch. VII]) reveals that common sense and engineering training are necessary but not sufficient for competence.

Applications of HFE fall into two time categories, the theoretical and the practical. Theoretical applications should be made throughout the entire system design period from concept to startup. If we consider the evolvement of a reuse system as an iterative, time-phased process, it should be expected that HFE would play a role in each iteration. The purpose for early and continuous involvement is to effectively influence system design. This implies high-intensity applications in the concept and predesign phases with continuing lower intensity involvement during detailed design, construction and startup.

Application targets will change with time. For example, early involvement may be directed toward influencing design tradeoff decisions. Midstream involvement may be directed toward influencing detailed design of control panels and behaviorally-based formulations of routine and contingency operation and maintenance procedures. Construction phase involvement may be directed toward operator training and selection. Important justification for continuous HFE participation is the concept of the human subsystem or the Human Unit Process [7]. The operator, including maintenance and on-site supervisory staff, is an integral component of the reuse system. The same care and attention needs to be given to design of the Human Unit Process and its respective interfaces as is given to all other system components.

If design engineers executed design activities according to textbook prescriptions, theoretical time applications of HFE might be possible. Behavioral studies of design engineers, however, imply that if the human factors specialist does not make inputs to the design process within the first few days (or hours) of a new project, few opportunities for influencing design will remain [1, Ch. VII].

HFE APPLIED TO WASTEWATER REUSE SYSTEMS

Standard references on HFE supply little specific guidance on where HFE fits in the design and operation of wastewater reuse plants because most HFE literature deals with weapon systems. Large differences are obvious between weapon system characteristics and those of reuse systems. Response time is an example. Many weapon systems require constant human monitoring and manipulation compared to only periodic human interaction in reuse systems. Response times of weapon systems are often measured in fractions of a second, and interactions may border on human physiological and psychological limits. A reuse system, in contrast, may have a response time measured from hours to months depending on the presence of intermediate storage reservoirs. It can be assumed that psychological and physiological stresses on reuse plant operators do not approach human tolerance limits although no data exist for verification. Thus, not all HFE applications in weapon system design are appropriate for reuse system design, and the extensive data base developed for weapon systems may have limited applicability to reuse systems. There are three areas of application, however, which appear to provide beneficial transfer of concepts and data: anthropometric machine design, outside-in system design, and operator job design.

Anthropometric Machine Design

Anthropometry is the study of human body measurements. Anthropometric machine design is the technology of designing mechanical interfaces between machines and human bodies. This is the "knobs and dials" side of HFE. Extensions of the field include the design of working environments, of information display devices and of certain aspects of test instrumentation and repair tools. A large anthropometric literature exists, much of it formalized into tables, drawings, specifications, standards and routine measurement procedures for use by design engineers and human factors specialists [2,8-10]. Knowledge of this field is of value to reuse system designers as an aid to selection of subsystem components, provision for sampling facilities, selection and placement of emergency equipment, design of maintenance access ways and design of operator environments to include control rooms and service areas. The referenced literature provides specific details about human factoring detail and layout designs. A few examples are presented here to illustrate some anthropometric concepts. System designers cannot control the makeup of the operator population beyond certain bounds. Some percentage of that population has color vision impairment. It would be counterproductive to specify color-coded status and warning displays which confuse recognition by persons with such impairment. It is simply counterproductive to design maintenance access ways which meet local codes but which discriminate against veteran beer drinkers. It can be expected that reuse facilities will require several sample points throughout the plant. These points should be usable by short people as well as tall, and no one should be required to have an aptitude for gymnastics to be able to collect samples. Women can be expected to be part of the operations staff. Manual controls including handles, levers, switches and valve wheels should be operable by one woman rather than three gorillas. Twenty-four and 48-channel strip chart recorders may save a few dollars of initial investment, but certain risks are assumed when the symbols and numbers overlay each other in an undecipherable mass of colored ink. And the list goes on.

Outside-in System Design

This is a jargon phrase which expresses an important HFE attitude. Man comes first as the priority subsystem. A closed-loop relationship exists in which the operating system generates information which is input to the human operator. The operator analyzes this information as a basis for making operational decisions. Process control actions resulting from such decisions provide information flow from man to machine. Machine behavior is modified. New operating system information is generated, and the cycle continues. Presence of the operator within the information loop introduces a behavioral dimension to system performance. Humans have capacity limitations on the acquisition, synthesis and evaluation of information [10,11] and capability limitations on decision making. Capacities and capabilities vary widely [12], but they must be considered as constraints on design if the system is to satisfy performance requirements. This closed-loop interrelationship also precludes effective study of either the operator or the operating system in isolation, an important concept to use when, for example, trying to determine why approximately half the secondary treatment plants in the country do not perform to expectations [13].

Execution of outside-in design follows a more or less linear path from the operator back into the operating system, through unit processes to individual components. The starting point is an understanding of the operator. This understanding is reflected in a detailed decision matrix which relates routine, contingency and emergency operating system status and status changes to a set of specific process control decisions allowed to the operator. Analysis of each element of the decision set results in derived sets of specific information requirements for input to the operator and for output from the operator. This permits specification of content, format and frequency of information to be transmitted. Input requirements to the operator can be translated into sensors, transmission methods and displays which include on-line devices, manual sampling protocols and laboratory procedures. Output requirements from the operator are translated into controls and actuators by type, location and number. Process control decisions not allowed to the operator are assigned to either the operating system or to technical elements of the management organization. The result is allocation of decisions between those subsystems best equipped to make them. What constitutes "best", however, is determined by a set of criteria which is situation dependent and may range in content from economics to local politics.

An interesting implication is that the Process and Instrumentation (P&I) drawings for a reuse facility designed outside-in may be different from the P&I for a traditionally designed facility. Specifications for sensors, displays, controls and actuators to support sets of predefined decisions sound like functional requirements for an information system, and that is exactly what is intended. Strong parallels exist between design concepts for information systems and for outside-in design of reuse facilities. Successful information systems are user oriented [14]. They provide information of proper content and format to support predefined decision making tasks. Certain flexibilities are also incorporated to allow accommodation to unusual or unique situations. The same attitudes apply to the information system supporting reuse facility process control. The user is the operator. The control information system is interactive in the sense that the operator maintains a dialogue with the operating system. This situation exists independent of the presence or absence of computers. It can be expected, therefore, that a P&I designed outside-in as an interactive process control information system would be different from one based on traditional concepts.

A more radical implication is that operator capacities and capabilities could (maybe should) influence unit process selection. Traditionally, wastewater treatment unit processes are selected on the basis of past success, technical performance, cost, availability, compatibility with preceding and succeeding unit processes, and regulatory agency requirements. Human operability and controllability enter the selection process as one component of an experience factor with facility size a major determinant. Large communities with large budgets are encouraged to buy one set of unit processes, and smaller communities with lower budgets are offered a different set. Rules of thumb appear to be common decision criterion. One sometimes suspects that personal preferences of design engineers and technological fads also play significant roles. Reports that more than one-third of the secondary treatment plants built with U.S. EPA construction grant funds do not operate to design and that more than half the factors contributing to poor performance are human related [15,16] should aggravate a gnawing suspicion that traditional concepts of inside-out design may have limited applicability for reuse facilities.

Operator Job Design

The operator job design function is directly related to outside-in design concepts and is the link between man/machine interfacing with the operating system and man/organization interfacing with the management structure. It is also that system design function which has greatest relevance to personnel selection and training. Job design is an explicit function which begins with the concept of task as the design variable. Meister identifies the element of a task as "the stimulus to the operator, which triggers performance of the task, the required response to that stimulus (i.e., the performance criterion), a procedure for performing this response (which includes the equipment to be utilized for performing the task), and a goal or purpose that organizes the whole"

[4]. Tasks are identified by formal analysis of specific operating system configurations.

Task descriptions are usually written in a standardized format which contains a behaviorally relevant verb such as "observes" coupled with a noun which can be either an information display device such as "chlorine pressure meter" or a datum such as "chlorine pressure value". A proper task description incorporates a behavioral action for acquisition of information, an evaluative process explaining what the operator is to do with the information and a motor action which is the result of the previous two processes, e.g., "reads chlorine dosage, compares actual with required, and adjusts chlorine control valve". A process control task is a specific case of the closed loop relationship between operator and operating system.

The result of job design is specification of job content. This is the set of tasks expected of an operator by number, frequency and variety. Job content determines operator capacity and capability requirements. It is, therefore, a performance specification for the human subsystem. As a set of explicitly identified and quantified operator activities, it is the actual starting point for execution of outside-in system design. What is interesting is that job content also contributes to design of the interface between the operator and the supervisory management structure. Job design, job content and task analysis are components of organization design [17]. Interrelationships exist between the operator and the management structure just as they exist between the operator and the operating system. Information flows, decisions are made, responses are solicited, actions are taken and certain competencies are required for operator participation in the management structure.

Application of human factors concepts and information used for job design for the operating system are also applicable to job design for the management organization. Job content for the two are not necessarily the same, however. Job content for operating system design is based on system performance specifications, but job content for organization design has trended toward satisfaction of higher-order human needs such as self esteem and self identity. A conflict arises when job design for an operating system tends toward greater division of labor and decreased skill requirements in the names of efficiency, productivity and system reliability, but organization design for the same operating system calls for expanded job content requiring greater skills and less division of labor [10, pp. 464-470]. The degree of potential conflict for reuse systems is unknown but it is anticipated that economic considerations limiting the size of the work force at any given facility will force a reconciliation. It is important to understand, however, that the same concept applied to the same operator for satisfaction of conflicting goals has the potential of adversely affecting reuse system performance.

Job content has obvious implications for personnel affairs such as selection and training. Personnel selection presumes the existence of a manpower pool large enough to support system demands, and training presumes selected individuals can be brought to acceptable levels of competency within reasonable time and cost. Usually, the manpower pool available to a customer is predetermined by status, pay structure, local competitors, security and career incentives. The pool is outside the influence of either the design engineer or the human factors specialist. Reconciliation is appropriate between task specifications and personnel realities. The human factors specialist should conduct job designs with both system performance and manpower constraints in mind so that acceptably large proportion of the available manpower pool is available to the customer as potential reuse system operators. Job content should be structured so that members from less qualified segments of the available pool can be trained at reasonable cost to the customer. Techniques for making these types of analyses are not intuitive and are not usually available to the design engineer. It is a serious fallacy, for example, for a design engineer to place himself mentally in the role of an operator since the engineer is not an operator, and does not have the same incentive structure, personal values, cognitive styles, reasoning capabilities and information processing capacities. Meister claims that when "we come to the matter of extracting . . . training implications from the task, the [human factors] specialist's methods become important because the engineer almost never performs a comparable analysis" ([1], p. 79). It may be that he cannot. It is a disservice to the customer to design and build a state-of-the-art reuse facility which cannot be operated by local people. One suspects that this has been a common occurrence for secondary treatment plants, and it may be time to add HFE specialists to reuse facility design teams.

A CASE STUDY

In the late 1960s, the U.S. Army published specifications and requirements for a potable reuse system called the Water Processing Element (WPE). The intent was to equip certain mobile field hospitals with a potable reuse capability for operations in water deficient areas. Engineering performance specifications were published. A Basis of Issue Plan with operator staffing levels by number and rank was formulated, but nothing was done about job design, training requirements and capability constraints on the operators. Process development work proceeded into the mid-1970s at a bench and breadboard level until the best candidate process train was identified. In 1976, when it was time to go to full-scale prepilot design [18,19], the question of the operator became important. Success of the system depended on its use by field hospital commanders. Building the system did not automatically mean it would be used, for commanders would use the system only if it worked and if it were operable by the people available. The system was inherently complex and although highly skilled operators could be specified and obtained during peacetime, this system would be needed during wartime when the availability of highly skilled operators would be in doubt. The system had to be operable by people actually on-site during wartime. In lieu of formally specified operator characteristics, it was decided that a crude operator model would be used as a prepilot design target: minimally trained draftee reluctantly doing twice the work originally assigned and under the psychological stress of being in a combat area. It was necessary, therefore, that human factors considerations be incorporated at the prepilot level. This represented a significant conceptual departure from the strictly mechanical performance focus up to that time.

Emphasis on Mind/Machine Interface

For prepilot design, greater needs existed for analyzing and accommodating interfaces between the operator's mind and the WPE than between his body and the operating system. Decision analysis for allocation of tasks between man and machine played an important role. Sensors and unit processes were carefully screened and selected for their operability and maintainability [20]. Lacking an adequate operator profile, a minimum skill level was assumed, and more machine intelligence was specified to simplify operation, reduce operator errors, and minimize the effects of such errors. Final allocation was for the machine to provide fully automated control, monitor and maintenance functions. These included procedural control (e.g., sequentially switching process hardware configurations from one operating mode to another), parameter control (e.g., maintaining a process parameter at its setpoint for the current operating mode), fault detection, process performance trend analysis, fault prediction, fault isolation and the logic to provide maintenance aid instructions. The operator was allocated setup, takedown and maintenance responsibilities. He was expected to set up influent and effluent plumbing, activate the automatic control system, perform scheduled maintenance, and respond to unscheduled tasks posted as machine-generated messages.

1. Operator/System Interface: The WPE operator/system interface panel was probably the most tangible result of the HFE effort. The panel layout reflected an effort to simplify a complex operation by grouping command pushbuttons, switches and displays according to function. Figure 1 shows the WPE automatic control/monitor instrumentation [21]. Grouped at the right side on the operator panel are the control buttons and indicators for entering system commands, product and source water selections and auxiliary functions.

Versatility and flexibility were among the requirements for the WPE prepilot. Various influent sources had to be treated such as natural fresh and

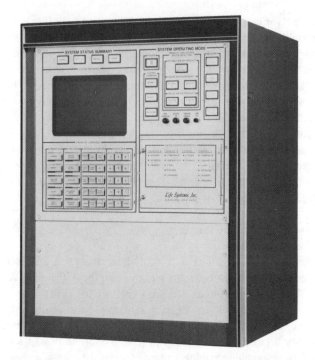

Figure 1. WPE automatic control/monitor instrumentation.

brackish waters or shower, operating room, laundry, medical complex, laboratory, X-ray and kitchen wastewaters. Different water products were required
for different missions. System effluent could be potable water for human consumption derived from natural water sources, wastewater for discharge to
the environment, or potable water renovated from hospital wastewaters. The
combination of several sources and several products posed certain risks dependent on operator competence. Panel design was simplified by aggregating the
three product waters and several source waters into five command buttons
(Figure 1, upper right) and four lists of wastewater sources (Figure 1, right
center). Note that some risk was reduced by including 'unknown' as one
wastewater source in case the operator had doubts concerning the origin of
wastewaters entering his system for potable reuse. Operator selection of one
of the five command buttons activated a different subset of unit processes
within the WPE for best treatment optimized on energy utilization.

2. Operator Command Validation: The WPE was designed to allow certain dual functions concurrently. For example, treating Wastewater Sources
A for surface discharge and processing natural brackish water for potable use
was allowed. However, not all combinations were possible. For example,
treating Sources B and D for two different purposes could not be done con-

currently because of conflicting demands for the same internal unit processes. This type of intelligence was designed into the machine to check operator command validity and to identify and reject erroneous requests or maintenance operations.

3. Color Coded Indicators: Pushbuttons on the control panel both accepted commands and displayed the resulting system action. The buttons were color coded. Product and Source selectors, when activated, were illuminated in green; auxiliary modes were in amber. System mode displays were coded in green and amber, with green indicating a steady-state mode, and amber a transition mode (system in process of startup, shutdown, or changing from one source and product mix to another). This was done so that the operator could readily tell the system's status. The operator's panel also had an additional mode transition indicator which was incorporated to accommodate color blind operators.

4. Monitor Functions: The monitor panel, grouped at the left side of the panel, consisted of a System Status Summary, a video message display and an operator command keyboard.

The System Status Summary displayed the current system condition: Normal, Caution, Warning or Alarm coded in green, amber, flashing red and red, respectively, as separate displays. The video display was partitioned for five types of messages (see Figures 2 and 3): fault diagnostic, on-line display, command echo, interactive data input/output, and communication messages. This type of status summary and partitioned display provided rapid, understandable, interactive communication between the operator and the operating system. The operator command keyboard was designed to require of the operator only one simple syntax rule: operation-function-sensor/actuator-code/data. If an error were made, the machine rejected the input command and output an appropriate message on the video display.

5. Protection Against Unauthorized Use: The instrumentation system was designed with three levels of protection against unauthorized commands from unauthorized personnel. The first level was a hidden "Command Disable" switch. When activated this switch disabled the front panel keyboard and all front panel pushbuttons. Second level protection was a password requirement for anyone changing the operating mode or modifying the system fault-detection and protection setpoints. Only with a correct password entered through the keyboard prior to operating the system could the operator change the system operating mode or modify the setpoints. The third level protection was a higher level password with which a qualified person could change the sensor scale factors, allowable ranges and control constants.

6. Manual Overrides: A manual override panel was provided for bypassing the automatic instrumentation. This function was needed for testing and troubleshooting. The override switches and dials were located on a recessed panel behind a closed door which, when necessary, could be modified to in-

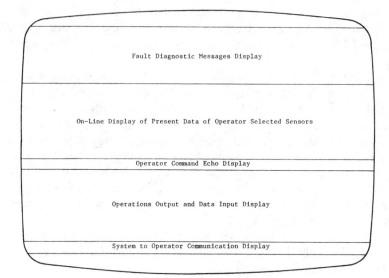

Figure 2. Video message display format.

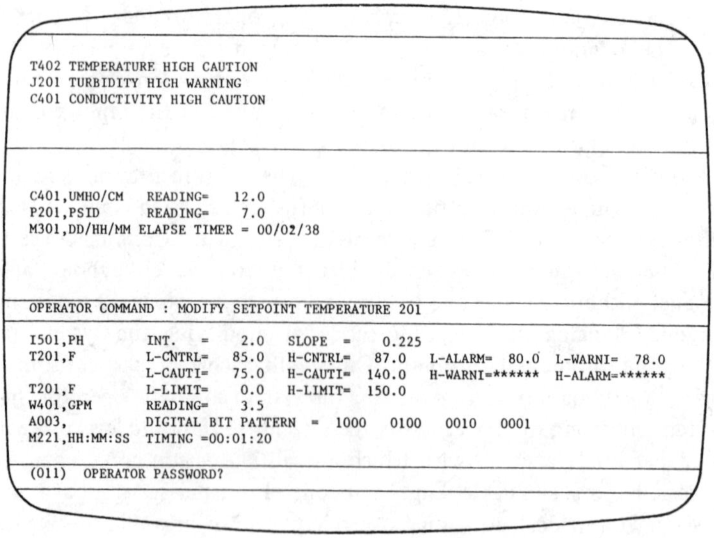

Figure 3. Video message display example.

corporate a lock for more security. Figure 4 shows the manual override panel. Note that the layout of the override switches and dials was made with human factors in mind. Switches and dials were divided horizontally into different unit processes and vertically into types of actuators.

7. Equipment and Sensor Selection: In addition to applications for the control and monitor instrumentation, HFE concepts were employed in the selection of equipment and sensors. Maintainability, operability and factors relevant to operator cognition were the three major considerations. Selections were made after extensive comparative analysis of available components in which engineering performance and human factors criteria carried equal weight.

8. Component Accessibility: Minimizing prepilot downtime was an important design goal. Layout was specifically made for maximum accessibility, with instrumentation components accessible from the top, both sides and the back. Mechanical components such as valves, regulators and pumps were arranged on or close to a common place for easy service. Results can be seen in the valve planes shown in Figures 5 and 6. The operator could easily reach all the valves and inspect, adjust or repair them as required.

Figure 4. Operator/system interface panel.

Figure 5. Valves are arranged on a valve plane for accessibility.

Figure 6. Another example of valve plane arrangement.

Case Study Comments

Prepilot design (which was certainly a lead in to end-item design) of the Army's Water Processing Element potable reuse system required an extensive effort in mind/machine interface analysis and specification. Task allocation between the human operator and the operating system resulted in a heavy burden on the operating system for intelligent decisions. This was primarily caused by the absence of a well-considered, quantified specification for either the prepilot or the end-item operator. Results of HFE concepts applications were simplified operation of an otherwise complex, special-purpose machine and a safer, more reliable wastewater reuse system. Application of HFE concepts was made counter to the best advice learned after the fact. Engineers of several backgrounds, working under considerable time/performance stress, applied their best estimates of human-oriented design in the absence of an HFE specialist with a wastewater treatment orientation. Concepts drawn from information system design and military recruit selection processes helped. The next time around, if things are done consistent with best HFE practice, human factors engineering will be a standard component of reuse system design.

REFERENCES

1. Meister, D. *Human Factors: Theory and Practice* (New York: John Wiley & Sons, Inc., 1971).
2. "Human Engineering Design Criteria for Military Systems, Equipment and Facilities," Department of Defense, MIL-STD-1472B (Washington, DC: Superintendent of Documents, 31 December 1974).
3. Roe, W. T., and D. L. Finley. "Ergonomic Models of Human Performance: Source Materials for the Analyst" (Springfield, VA: Defense Technical Information Center, ADA020-086, August 1975).
4. Meister, D. *Behavioral Foundations of System Development* (New York: John Wiley & Sons, Inc., 1976), p. vii.
5. Nertney, R. J., and M. G. Bullock. "Human Factors in Design," report ERDA-76-45-2 (Springfield, VA: National Technical Information Service, February 1976).
6. Scoggins, J. Paper presented at the Mid-Atlantic Technology Exchange Conference and Exposition, 19-20 March 1980, Baltimore, MD.
7. Miller, R. D. Personal communication (January 1979).
8. "Military Specification, Human Engineering for Military Systems, Equipment, and Facilities," Department of Defense, MIL-H-46855 (Washington, DC: U.S. Government Printing Office, 1968).
9. Roebuck, J. A., K. H. E. Droemer and W. G. Thomson. *Engineering Anthropometry Methods* (New York: John Wiley & Sons, Inc., 1975).
10. McCormick, E. J. *Human Factors in Engineering and Design*, 4th ed. (New York: McGraw-Hill Book Company, 1976).

11. Martin, J. *Design of Man-Computer Dialogues* (Englewood Cliffs, NJ: Prentice-Hall, Inc., 1973), Ch. 19.
12. West, B., and J. A. Clark. "Operator Interaction with a Computer-Controlled Distillation Column," in *The Human Operator in Process Control*, E. Edwards and F. P. Lees, Eds. (New York: Halsted Press, 1974), pp. 206-221.
13. Bender, J. H. "EPA-ORD Researchers Plant Operation and Design," U.S. EPA information paper (Cincinnati, OH: U.S. EPA Municipal Environmental Research Laboratory, 1980).
14. King, W. R., and D. I. Cleland. "Manager-Analyst Teamwork in MIS," in *Interactive Decision Oriented Data Base Systems*, W. C. House, Ed. (New York: Petrocelli/Charter, Inc., 1977), pp. 33-43.
15. Hegg, B. A., K. L. Rakness and J. R. Schultz. "Evaluation of Operation and Maintenance Factors Limiting Municipal Wastewater Treatment Plant Performance," EPA-600/2-79-034 (Cincinnati, OH: Municipal Environmental Research Laboratory, June 1979).
16. Gray, A. C., Jr., P. E. Paul and H. D. Roberts. "Evaluation of Operation and Maintenance Factors Limiting Biological Wastewater Treatment Plant Performance," EPA-600/2-79-078 (Cincinnati, OH: Municipal Environmental Research Laboratory, July 1979).
17. Galbraith, J. R. *Organization Design* (Reading, MA: Addison-Wesley Publishing Company, 1977), p. 24, pp. 28-32, pp. 344-355.
18. Lee, M. K., P. Y. Yang, R. A. Wynveen and G. G. See. *Pilot Plant Development of an Automated, Transportable Water Processing System for Field Army Medical Facilities*, Report LSI ER-314-7-1 under contract DAMD17-76-C-6063 to U.S. Army Medical Research and Development Command (Cleveland, OH: Life Systems, Inc., June 1979).
19. Lee, M. K., and L. H. Reuter. "Multipurpose Water Treatment System Development," *Proceedings of the 1979 AWWA Water Reuse Symposium* (Denver, CO: American Water Works Association Research Foundation, March 1979), pp. 1276-1310.
20. Yang, P. Y., W. P. Lambert, B. W. Peterman and J. D. Powell, Jr. "Instrumentation for Automating Water Processing Systems," *Proceedings of the 1979 AWWA WAter Reuse Symposium* (Denver, CO: American Water Works Association Research Foundation, 1979), pp. 1843-1864.
21. Yang, P. Y., J. Y. Yeh, J. D. Powell and R. A. Wynveen. *Advanced Instrumentation for a Water Processing Pilot Plant for Field Army Medical Facilities*, Report LSI ER-314-7-4 under contract DAMD17-76-C-6063 to U.S. Army Medical Research and Development Command (Cleveland, OH: Life Systems, Inc., May 1978).

CHAPTER 8

POTABLE WATER REUSE

K. Daniel Linstedt
Black & Veatch Consulting Engineers
Denver, Colorado

Michael R. Rothberg*
Denver Water Department
Denver, Colorado

INTRODUCTION

When one considers the complete hydrologic cycle, it is readily apparent that potable water reuse has been practiced for a long time. However, potable water reuse has also been practiced within the cycle between the point at which precipitation reaches the earth and the point where it is discharged into the oceans. This practice has arisen because natural surface water systems have been used historically for both wastewater disposal and water supply purposes. This practice has been called indirect or inadvertent reuse, and should be contrasted to planned or direct reuse practices. In order to minimize confusion about these terms, the definitions developed at a World Health Organization meeting on potable reuse [1] are used.

Wastewater: the spent water of a human settlement, consisting of the water carried wastes from residences, commercial buildings and industrial plants, and surface drainage and groundwater from urban areas and industrial premises.

Renovated
Wastewater: wastewater that has been treated by one or more advanced wastewater treatment processes to render it fit for direct reuse.

*Present address: Culp/Wesner/Culp, Denver, Colorado.

Direct
Reuse: the planned and deliberate use of treated wastewater for some bene-
                ficial purpose, such as irrigation, recreation, industry, the recharging of
                underground aquifers, and human consumption
Indirect
Reuse: reuse when water already used one or more times for domestic or indus-
                trial purposes is discharged into fresh surface or underground waters
                and used again.

Unplanned indirect reuse has been practiced in varying degrees in cities
throughout the world. An indication of the extent of this reuse in the United
States was provided in 1966 by the Federal Water Pollution Control Administration [2]. In 155 cities studied, the percentage of municipal wastewater in
the surface supplies at the low flow condition ranged between 0 and 18%.
The median reuse level was determined to be about 3.5%. These percentages were generally quite low, but they were significant and will undoubtedly
increase as population growth increases the demand on available surface water
sources. Although unplanned indirect reuse has been widely practiced, this
type of reuse practice is not the focus of this chapter. Rather, the emphasis
is directed at reviewing past practice and future plans for implementing potable reuse of a planned nature by both indirect and direct means.

Motivation for Potable Water Reuse

In reviewing the earliest potable reuse practices, it is not surprising to observe that such reuse was generally fostered by an acute water resource problem related to severe drought conditions which limited the available options
for supply water sources [3,4]. However, more recent potable reuse projects
have included rather thorough and extensive planning [5-9]. In these cases,
a course involving potable reuse has been pursued only after considering many
alternatives for dealing with a long-term water supply situation.

The motivations for pursuing this potable reuse option are varied. Perhaps
the most often mentioned consideration relates to the economics of developing new sources of supply for major metropolitan areas. In many cases, the
readily available sources of supply have been fully utilized and only distant
sources of supply offer potential for satisfying future water needs. The development and transmission costs associated with these distant sources of supply
make water reuse economically attractive because of the proximity of the
source to the point of use [9]. Related to these economic considerations is
the recognition by some that as wastewater discharge requirements become
more stringent, the quality requirements and the associated costs of pollution
control make the water too good to discharge [10]. It is certainly true that
the incremental costs of renovating water are significantly reduced as the extent of treatment for pollution control purposes is increased.

Another factor that encourages consideration of water reuse is that the sources and quantities of pristine water are already rare, and diminishing [11]. Some existing water supplies already have substantial components of waste-water that have been introduced into the supply in an indirect way, but di-rectly controlled, planned reuse could provide a more reliable water supply option. In line with this, if potable reuse were to become a widely accepted practice, it would allow us to optimize more closely the management of our total water resource [12].

In many areas, where the major source of supply is ground water, the rate of withdrawal has increased with the population to the point that withdrawals exceed the rate of natural recharge. In these groundwater mining situations, soil subsidence problems may occur, and saltwater intrusion problems may develop where the aquifer is located near the ocean. Recharging with reno-vated water has been proposed to minimize these adverse impacts [6,13]. In the process, the potable supply aquifer serves as a potable recycling system.

Finally, there is motivation for considering planned potable reuse when one recognizes the uncertainties associated with conventional supply sources. The drought conditions which have led to some historic reuse practices can occur in other locations in the future. Reuse that is planned prior to that eventuality would ensure a more consistent and reliable water quality in the event of such an emergency in the future.

Factors Limiting Implementation of Potable Water Reuse

While there are significant reasons for considering implementation of plan-ned potable water reuse, there are also many unknowns in the practice which suggest that a cautious approach is necessary. Among the most significant concerns are those related to the public health of the consumers. It is general-ly accepted that some constituents are refractory to the most complete reno-vation treatment systems which have been developed to date. As such, trace concentrations of these compounds will exist in any renovated water. Further-more, the concentration of these compounds will increase with successive recycles [14,15]. The materials in this refractory fraction are of health con-cern because they have not been adequately characterized. Most of the organic compounds which are resistant to water renovation treatment have not been identified except in rather gross ways as total organic carbon or chemical oxy-gen demand. In the same way, the health effects of these compounds and the synergistic effects of combinations of compounds have not been established [12]. As a result, there is a lack of standards for assessing the acceptability of supply waters derived from a wastewater source, and a difficult task ahead in developing such standards. Until such standards can be established, it is not likely that public health officials will sanction widespread potable reuse.

In addition to the fact that some compounds are refractory to water renovation treatment, concern has also been expressed about the reliability of these treatment operations [16]. If direct recycle is to be practiced, it is necessary that the product quality be maintained at a consistently high level. Such fail-safe operation will have to be demonstrated before concerns about bypassing the perceived natural purification processes will be alleviated. A related concern in the development of such fail-safe systems is the high economic and energy cost associated with the required level of treatment. If such treatment is to be implemented, it must be competitive with that of other supply sources from both the economic and energy standpoints.

The treatment operation must be supported by high-quality monitoring capabilities. If a water renovation plant is to be put on-line, with the product going into a water supply system, there must be capability for monitoring the suitability of this product for potable purposes. This requires a more rapid and complete analytical capability than is currently available, particularly in the areas of organic analyses and virus testing [14].

The concept of potable water reuse also must be acceptable to the public before the practice will be widely implemented. Surveys of public opinion indicate support for water reuse that is not of a potable nature, but there are significant reservations to potable reuse [17]. It will have to be demonstrated to the public that such practice is acceptable. This will likely require an extensive educational effort.

HISTORICAL REUSE PRACTICE

With the unknown areas previously discussed, the experience with intentional potable reuse has been quite limited. However, there have been a few examples, and the experiences at these sites are reviewed to lend historical perspective to the topic of potable water reuse.

Chanute, Kansas

Perhaps the earliest example of direct reuse reported in the literature is that which occurred in Chanute, Kansas. Emergency reuse at Chanute was necessitated by water shortages associated with a severe drought in Kansas over the period of 1952-1957 [3]. During a part of this period, the source of supply for Chanute, the Neosho River, actually ceased to flow. When this occurred, a decision was made to reuse water through the recirculation pattern shown in Figure 1. The treatment applied to the wastewater discharged to the supply reservoir consisted of primary and secondary treatment incorporating a high-rate trickling filter system. The water withdrawn from the intake system was subjected to chemical treatment in conjunction with solids-contact softening, recarbonation, sedimentation, filtration and chlorination.

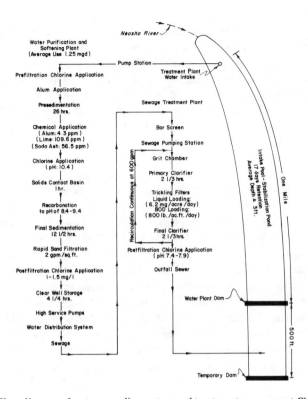

Figure 1. Flow diagram of water recycling system and treatment processes at Chanute [3].

Operation in this recirculation mode was continued for five months, amounting to approximately seven cycles through the community. During this period, the treated water contained taste, odor and color associated with undesirable levels of dissolved inorganic and organic constituents. With these adverse aesthetic qualities and the publicity surrounding the recirculation practice, public acceptance of the reuse program was reported to be poor. Nevertheless, there were no known cases of waterborne disease or other adverse health effects resulting from the use of the recirculated water.

Windhoek, South Africa

The city of Windhoek is situated in a very arid area of southwest South Africa where the rate of population growth had made it impossible to meet future water demands from the available conventional supplies. As a result, water reclamation studies were performed, and a 1.2 mgd plant was constructed to augment the existing conventional supply sources with reclaimed wastewater. This plant was originally put into operation in 1968 with the treatment components shown in the schematic diagram of Figure 2 [4,18,19, 20]. As shown, the wastewater was subjected to a series of conventional and advanced wastewater treatment processes. The advanced wastewater treatment included maturation ponds, recarbonation, flotation for removal of algae, foam fractionation, chemical coagulation, breakpoint chlorination, rapid sand filters and activated carbon filters. The product water from this treatment sequence was commingled with water from an adjacent surface source that had been subjected to conventional water treatment comprised of flocculation, chlorination, sedimentation, and sand and carbon filtration. During the initial two-year phase of reclamation plant operation, the plant contributed an average of 13.5% of the total water consumption in Windhoek [21]. The quality of the product consistently complied with standards recommended by the World Health Organization [4,21]. Review of hospital and laboratory records for Windhoek did not reveal any significant change in the disease pattern for this area after two years of reuse. Further, the public acceptance of the reclaimed water was generally favorable.

As a result of some easing in the drought conditions, and the existence of several inherent inefficiencies in the original Windhoek design, the operation of the Windhoek facility had decreased to some 70 days per year by 1973, compared to more than 8 months per year immediately following commissioning of the facility in 1968. In order to facilitate better utilization of the

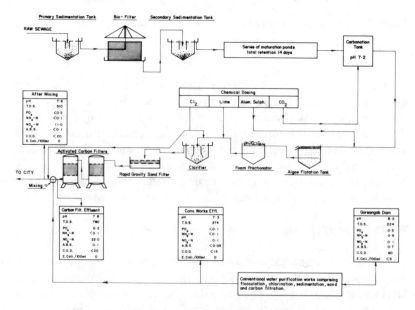

Figure 2. Schematic of Windhoek Water Reclamation [4].

plant, facility modifications were designed and constructed into the treatment system. These modifications included the following components:

1. addition of high lime treatment utilizing solids contact type clarifiers,
2. installation of an ammonia stripping tower,
3. modification of the algae flotation and foam fractionation tanks to provide recarbonation facilities, and
4. modification of the granular activated carbon system and installation of a regeneration furnace.

Following implementation of these process changes the facility was operated about 200 days in 1977 and 1978. During the time of this operation the water reclamation facility provided reclaimed water which constituted as much as 50% of the total water in the potable supply system [22].

Occoquan River, Virginia

The Occoquan River project presents a somewhat different approach to potable reuse than that practiced in either Chanute or Windhoek. In fact, it would most appropriately be characterized as indirect reuse. However, it is appropriate for discussion in the context of this chapter because of the planning and treatment associated with the project.

The Occoquan River watershed is situated in Fairfax and Prince William Counties of northern Virginia. These counties comprise a portion of the Washington, DC metropolitan area, and have been heavily impacted by the population growth associated with that city. This location is shown in the map of Figure 3 [7]. The Occoquan watershed covers a 600 mile^2 area which drains into the Occoquan Reservoir, a water supply reservoir serving approximately 600,000 people. By 1969, the discharge of wastewaters of secondary effluent quality had contributed to the development of significant algal blooms in the reservoir and accompanying taste and odor problems in the water supply. In addition, the water in the reservoir was devoid of oxygen at depths below 10 feet. It was estimated that a future wastewater flow of 40 mgd into the Occoquan watershed would represent 12% of the total annual inflow to the reservoir, and 46% of the July low flow. With recognition of the potential increased adverse impact on the Occoquan supply reservoir, the Virginia State Water Control Board adopted the Occoquan policy in 1971 to correct this situation [8]. This policy mandated that a watershed monitoring program be instituted, and that a regional water reclamation plant be established to replace the existing wastewater treatment plants. The reclamation plant was to include the best available treatment technology, and include extensive fail-safe features so that stringent effluent requirements, as shown in Table I, could be met on a consistent basis.

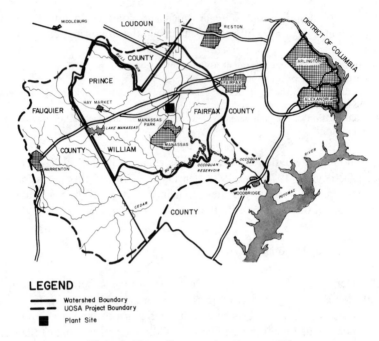

Figure 3. Upper Occoquan location map [8].

The treatment system chosen for meeting the indicated effluent quality requirements contained the components shown in the schematic diagram of Figure 4. The plant was designed to treat an average flow of 15 mgd. Unique features of the plant design for providing fail-safe operation included redundant units for all mainstream processes. For example, three primary clarifiers were provided in the design, each with a capacity of 7.5 mgd. Another fail-safe design feature in the Occoquan plant was the capability to divert a portion of the plant influent or primary effluent to an emergency retention pond

Table I. Upper Occoquan Sewage Authority Effluent Requirements [8]

Constituent	Required Effluent Quality
BOD_5 (mg/l)	1.0
COD (mg/l)	10.0
Suspended Solids (mg/l)	<1.0
Phosphorus (Total) mg/l	0.1
Nitrogen (Total Unoxidized) (mg/l)	1.0
Methylene Blue Active Substances (mg/l)	0.1
Coliform Bacteria MPN[a]	<2/100 ml

[a]MPN = most probable number.

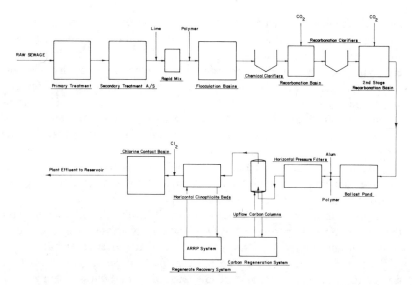

Figure 4. Upper Occoquan Sewage Authority Reclamation Plant schematic.

having a capacity of 45 million gallons. This water would be directed back into the plant headworks when the emergency situation was corrected. Another feature of the plant design which lent reliability to the operation was a ballast pond ahead of the column treatment operations of filtration, carbon adsorption and selective ion exchange. This pond provided for relatively constant loading and quality onto these units. A final fail-safe provision involved continuous effluent monitoring of the permit discharge parameters. If any parameter exceeded the permit limitation, the unacceptable water would be automatically recycled for additional treatment within the plant.

Operation of the reclamation plant began in 1978 and the quality of water discharged since then to the Occoquan Reservoir is reported to have shown substantial improvement over that discharged during any of the six years of the monitoring program [8]. This is particularly important in view of the fact that as much as 80% of the flow to the reservoir during the 1977 drought was from wastewater discharges.

PLANNING FOR FUTURE POTABLE REUSE

While no direct potable reuse operations now exist in the United States, the Denver Water Department has been actively evaluating the potential for such reuse in order to augment the conventional water supply for Denver, Colorado [9]. The Denver program has been an ongoing effort since the late 1960s when it was initiated with an evaluation of the various forms of water reuse which might be practical in the context of the Denver situation. The data developed in this early evaluation indicated that total reuse, including potable water reuse, would provide the only type of reuse capable of making a significant impact on the future water resource requirements of that city. The conclusion that direct potable reuse should be given the highest consideration was based on several factors: the location of Denver and the methods by which water is obtained for that city; the legal factors associated with the ability to reuse water in the context of local water law; the quantity of water available for various forms of reuse, and the associated demand; the likelihood of potable water reuse acceptance by the Denver population; and the comparative economics associated with development of a potable reuse system and more conventional forms of supply [23].

In planning a water supply program which involves potable water reuse, the Denver Water Department was faced with many areas of technical and institutional uncertainty. To address these uncertainties, Denver embarked on a demonstration project to evaluate the capability of current treatment technology for producing a safe and reliable reuse product. Following ten years of pilot plant research and careful review of advanced wastewater treatment literature, processes were selected for inclusion in the demonstration plant facility at a 1 mgd capacity. The flow schematic of this plant is shown in Figure 4 [24]. Considerations involved in the selection of these treatment processes included: An assessment of the ability of the unit processes to meet projected removal efficiency requirements; the operating history and reliability of the unit processes; potential operational difficulties or other drawbacks associated with the unit processes; and the potential of the unit processes for providing constituent removals in addition to those of the major contaminants for which the processes were included.

Analysis of each of the treatment processes incorporated into the demonstration plant facility would reveal that provision was made for redundant removal processes to eliminate each of the major categories of water and wastewater contaminants. The first process, that of high lime precipitation, was included to provide one means of removing phosphorus and most heavy metals. The process has also been observed to be highly effective in reducing the concentrations of virus and bacteria because of the adverse environmental conditions which develop at pH levels of 11.2 or greater. The filtration process provided polishing of the lime-treated effluent for removal of suspended solids, metal precipitates and microbial organisms. Selective ion exchange was intended primarily for ammonium ion removal. However, it has also been shown to provide some removal of other cationic species and an additional filtration step. The carbon and ozone portion of the treatment sequence was intended to remove the bulk of the organic material from the plant influent. Additionally, inactivation of virus and other pathogenic organisms was expected, together with some heavy metal removal. The use of reverse osmosis

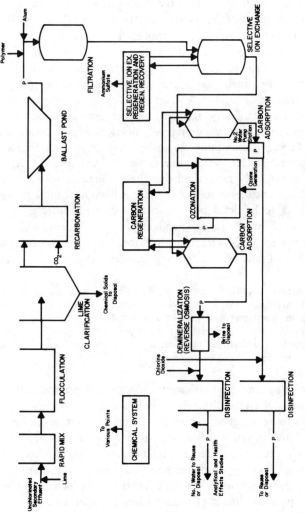

Figure 5. Denver Water Reuse Demonstration Plant process flow diagram.

as the demineralization process provided a physical barrier limiting the passage of most wastewater contaminants. As such, it has been shown to provide a high degree of removal for virus and bacteria, phosphorus, nitrogen compounds, heavy metals and other dissolved salts, and most organic materials [24]. With these capabilities, the reverse-osmosis unit provided redundancy for most of the other processes in the treatment sequence. The decarbonator, which was included with the reverse-osmosis unit for pH and stability control, has been shown to be effective in stripping low molecular weight volatile organic compounds from solution [25]. The final process, chlorine dioxide disinfection, provided a final means of dealing with residual pathogenic organisms.

Related to the capabilities of renovation treatment, the questions associated with quality of reclaimed water have been paramount in the minds of health officials, regulatory agencies and the general public. Conventional drinking water standards were originally developed from the assumption that water for human consumption would be drawn from groundwater sources or from protected surface water sources which have received little contamination. It is generally recognized that these standards were not intended to apply in direct reuse situations. As such, it is likely that more stringent standards will have to be developed to define acceptable potable reuse practice. With the lack of water quality standards for potable reuse water, two alternative paths were considered to be available for addressing the questions of the suitability of renovated water for potable reuse:

1. to conduct in-depth toxocological and epidemiological studies over a period of years to identify any health problems and to provide background criteria to assist in the development of meaningful standards, and
2. to provide renovation treatment capabilities which would remove all the contaminants added between the water supply and the wastewater treatment facility, or the use-increment of contaminants.

The Denver program contains elements of both of these strategies.

The analytical and health effects program supporting the demonstration project was to be designed to provide a complete characterization of the product water, and to verify that all constituent concentrations were no greater than those in the existing Denver supply water. A health effects program involving animal toxicity testing has been provided in addition to the extensive analytical program to ensure that those compounds which cannot be measured would not pose the threat of negative health impacts to the Denver water consumer [26].

If the Denver Water Department reuse demonstration program is successful, the Department intends to construct and operate a 100-mgd reuse plant for treatment of water to potable quality. This water would be commingled with existing water supplies to increase the total available potable water resource. The program involves a demonstration of at least eight years, followed by design and construction of the full-scale facility. With these program elements, Denver would not begin to rely on water reclamation for potable use as an increment to the existing water supply system until about the turn of the century.

REFERENCES

1. World Health Organization, "Report of the International Working Meeting on Health Effects Relating to Direct and Indirect Reuse of Wastewater for Human Consumption," Technical Paper No. 7, International Reference Center for Community Water Supply, P.O. Box 140; Leidschendam, The Netherlands (1975).
2. Koenig, L. "Studies Relating to Market Projections for Advanced Waste Treatment," Federal Water Pollution Control Administration Publication WP-20-AWTR-17, Washington, DC (1966).
3. Metzler, D. F., et al. "Emergency Use of Reclaimed Water for Potable Supply at Chanute, Kansas," *J. Am. Water Works Assoc.* 50:1021-1060 (1958).
4. Stander, G. J., and L. R. J. van Vuuren. "Municipal Reuse of Water," *Water Quality Improvements by Physical and Chemical Processes*, E. F. Gloyna and W. W. Eckenfelder, Jr., Eds. (Austin: University of Texas Press, 1970), pp. 31-48.
5. Parkhurst, J. D., and W. E. Garrison. "Water Reclamation at Whittier Narrows," *J. Water Poll. Control Fed.* 35(9):1094-1104 (1963).
6. Cline, N. M., and D. G. Argo. "Wastewater Reuse in Orange County, California," in *Research Needs for the Potable Reuse of Municipal Wastewater*, K. Daniel Linstedt and E. R. Bennett, Eds., U.S. Environmental Protection Agency EPA-600/9-75-007 (Cincinnati, OH, 1975).
7. Culp, G. L., R. L. Culp, and C. L. Hamann. "Water Resource Preservation by Planned Recycling of Treated Wastewater," *J. Am. Water Works Assoc.* 65:641-647 (1973).
8. Robbins, M. H., and G. A. Gunn. "Water Reclamation for Reuse in Northern Virginia," *Proceedings of the 1979 AWWA Water Reuse Symposium* (Denver, CO: American Water Works Association), pp. 1311-1327.
9. Linstedt, K. D., K. J. Miller and E. R. Bennett. "Development of Successive Water Use in a Metropolitan Area," *J. Am. Water Works Assoc.* 63:610-615 (1971).
10. Stephan, D. G., and L. W. Weinberger. "Wastewater Reuse—Has it 'Arrived'?" *J. Water Poll. Control Fed.* 40:529-539 (1968).
11. Graeser, J. H. "Water Reuse: Resource of the Future," *J. Am. Water Works Assoc.* 66:575-578 (1974).
12. AWWA-WPCF. "Statement by the Joint Committee on Wastewater Reclamation Policy," *J. Am. Water Works Assoc.* 65:700 (1973).
13. Oliva, J. A., F. J. Flood and J. S. Gruen. "Observations on the Start-up of the Water Reclamation-Recharge Project in Nassau County, N.Y.," *Proceedings of the 1979 AWWA Water Reuse Symposium* (Denver, CO: American Water Works Association, 1979), pp. 199-211.
14. Suhr, L. G. "Some Notes on Reuse," *J. Am. Water Works Assoc.* 63:630-633 (1971).

15. Dick, R. I., and V. L. Snoeyink. "Equilibrium Concentrations in Reuse Systems," *J. Am. Water Works Assoc.* 65:504-505 (1973).
16. Linstedt, K. D., and E. R. Bennett. "Research Needs for the Potable Reuse of Municipal Wastewater," U.S. Environmental Protection Agency EPA-600/9-75-007 (Cincinnati, OH, 1975).
17. Bruvold, W. H., and P. C. Ward. "Using Reclaimed Wastewater—Public Opinion," *J. Water Poll. Control Fed.* 44:1690-1696 (1972).
18. van Vuuren, L. R. J., M. R. Henzen, G. J. Stander and A. J. Clayton. "The Full-Scale Reclamation of Purified Sewage Effluent for the Augmentation of the Domestic Supplies of the City of Winhoek," *Proceedings of the 5th International Conference on Water Pollution Research* I-31:1-6 (1970).
19. Clayton, A. J., and P. J. Pybus. "Windhoek Reclaiming Sewage for Drinking Water," *Civil Eng.* 42:103-106 (1972).
20. van Vuuren, L. R. J., and M. P. Taljano. "The Reclamation of Industrial/Domestic Wastewaters," *Proceedings of the 1979 AWWA Water Reuse Symposium* (Denver, CO: American Water Works Association, 1979), pp. 906-924.
21. Hart, O. O., and L. R. J. van Vuuren. "Water Reuse in South Africa," in *Water Renovation and Reuse,* H. I. Shuval, Ed. (New York: Academic Press, Inc., 1977), pp. 355-395.
22. van Vuuren, L. R. J., A. J. Clayton and D. C. van der Post. "Current Status of Water Reclamation at Windhoek," presented at the Annual Conference of the Water Pollution Control Federation, Anaheim, California (1978).
23. Work, S. W., and N. Hobbs. "Management Goals and Successive Water Use," *J. Am. Water Works Assoc.* 68:86-92 (1976).
24. Rothberg, M. R., S. W. Work, K. D. Linstedt and E. R. Bennett. "Demonstration of Potable Water Reuse Technology: The Denver Project," *Proceedings of the 1979 AWWA Water Reuse Symposium* (Denver, CO: American Water Works Association, 1979), pp. 105-138.
25. McCarty, P. L., and M. Reinhard. "Statistical Evaluation of Trace Organics Removal by Advanced Wastewater Treatment," presented at the Annual Conference of the Water Pollution Control Federation, Anaheim, California (1978).
26. Work, S. W, M. R. Rothberg and K. J. Miller. "Water Reuse—Safety and Quality," *Proceedings of the AWWA 1979 Annual Conference* (Denver, CO: American Water Works Association, 1979), pp. 1035-1050.

CHAPTER 9

DUAL WATER SYSTEMS IN WATER REUSE

Arun K. Deb
Environmental Systems
Roy F. Weston, Inc.
Designers-Consultants
West Chester, Pennsylvania

INTRODUCTION

A fundamental need of any community is an adequate supply of biologically and chemically safe, palatable water of good mineral quality. While demand for this good quality water is high, its availability is limited.

Technological advances, changes in lifestyle, and increases in population during the past decades have caused both the demand for fresh water and the discharge of wastewater to increase. If the present rate of growth of population and industry continues, the quality of natural water will deteriorate, and it will be difficult to obtain a high quality bulk water supply for domestic, industrial and commercial use.

Every year our industries produce over 1000 new chemicals. The development of these new chemical compounds for the increasing demands of the consumer market and the growth of chemical use in agriculture and industry have allowed new micropollutants to enter the natural water courses. In addition to the chemicals discharged into streams, many other chemicals are formed, primarily through chemical reactions with chlorine. The carcinogenic or otherwise toxic behavior of these chemicals is not clearly known, but Harris has stated "there is a relationship between drinking water quality and cancer" [1]. DDT and other substances, now proved dangerous to man, were of little concern two decades ago. The latency period of these substances is on the order of 20 years. If, after 20 years or so it is concluded that these sub-

213

stances in drinking water do cause cancer, adverse effects on society cannot be overcome by short-term measures.

Good quality water for potable uses to meet primary and secondary drinking water standards (if available) should be obtained from protected natural sources. However, it has become increasingly difficult and expensive to bring up to potable quality the water found in natural lakes, streams, ponds and subsurface locations. This consideration and the fact that only a small fraction of a community's water needs require potable quality water have made questionable the supplying of only a high-quality water.

The requirements for water pollution abatement as mandated by the Water Pollution Control Act Amendment (PL 92-500) and the Clean Water Act have resulted in the production of wastewater effluents of such quality that, in many instances, these effluents are too valuable to be discarded but can be useful as resources for nonpotable purposes. The National Interim Primary Drinking Water Regulations, under the Safe Drinking Water Act (PL 93-523), established a maximum contamination level (MCL) of trace metals and organics in drinking water. The cost of production of water of single potable quality from natural sources to meet primary drinking water regulations will be very high in many communities.

Water quality management in this country is now taking a new direction from technological and management points of view. Adequate pollution control measures, along with conservation and reclamation of water resources, are necessary in long-term urban water planning and management. In some areas of the country, there is a growing imbalance between demand and available water resources. If wastewater effluent can be reused for purposes which do not warrant high-quality water (such as urban irrigation, industrial cooling and process water, agriculture, recreation and groundwater recharge), the demand on good quality natural water sources can be reduced.

The 1976 drought in California focused the attention of the whole country on the need for water conservation and reuse. The future costs of producing potable water, particularly from polluted sources, will be very high; with the availability of good quality effluent, in many cases, there will be an economic incentive to reuse wastewater for nonpotable purposes. Therefore, water reuse should be an essential consideration in developing a long-term water management plan for a city or a region studying conservation and optimum uses of water.

The average total urban water usage for Americans today is approximately 160 gallons per capita per day (gpcd). Nationally, the approximate categorization for the 160 gallons of per capita average daily use is as follows:

Use	Percentage
Residential	40
Commercial	15

Industrial   25
Unaccounted for   20

Vast amounts of potable quality water for industrial, commercial and public sectors are not required. Of the 60-gpcd average interior residential usage, 40% is required for toilet flushing, 30% for bathing, 15% for laundering, 6% for dishwashing, 5% for drinking and cooking, and 4% for other miscellaneous uses. In fact, of the average amount of water consumed per person, only about one-half gallon per day is required to be of high potable quality. The U.S. EPA and the National Academy of Science, in their studies to determine the MCL for primary drinking water regulations [2], assumed water consumption of 2 liters/capita/day. From a health standpoint, therefore, only 5% of water for interior residential use (drinking and cooking) actually needs to be of the highest quality. Okun [3] suggests that a hierarchy of water supply should be established, with the quality of water being adapted for the use to which it is put.

CONCEPT OF DUAL SUPPLY SYSTEMS

If it is assumed that recycled water will not be employed for potable uses, a separate distribution system will be necessary for supplying water for non-potable uses. In fact, dual supply systems would be a necessary feature of water management in areas where wastewater reuse will be considered.

A dual water supply can also be adapted in cases where raw water source quality is poor and has high organic, heavy metal, and total dissolved solids (TDS) content. The cost of removing these materials is high, and dual supply systems may be economically viable over the conventional systems. Again, this depends on the savings from eliminating additional treatment versus the additional distribution cost.

In dual water supply systems, two qualities of water (one potable, the other nonpotable) would be supplied to consumers through separate distribution systems. Since only a small fraction of municipal water must be of primary drinking water quality, the volume of water to be treated by an expensive and sophisticated process also would be small, and the bulk portion of the nonpotable water supply could be obtained from high-quality treated effluent. Thus, two qualities of water, potable and nonpotable, could be supplied economically through separate distribution systems.

The quality of the potable water would have to conform to the primary and secondary drinking water regulations of the Safe Drinking Water Act. The nonpotable water would be disinfected and would be bacteriologically safe, but might not meet primary drinking water regulations for trace heavy metals and organic materials. The effluent water would be filtered and chlorinated with large dosages, and a long contact period would be provided to meet the bacteriological standard of the Safe Drinking Water Act. The quality of non-potable water should be such that it is free from harmful bacteria and viruses below detection level. It would be relatively clear and might contain nutrients, but is not intended for and may not be safe for human ingestion. It can be used for irrigation of public parks and golf courses, air conditioning, industrial cooling, toilet flushing, exterior household uses and other low-grade water uses.

The following sources of water can be considered for the various dual supply options:

- Potable Supply
    a. good quality protected groundwater,
    b. protected upland reservoir,
    c. unprotected groundwater after extensive treatment, and
    d. unprotected surface sources (polluted streams, rivers, or lakes) after extensive treatment.
- Nonpotable Supply
    a. secondary effluent for reuse, and
    b. advanced wastewater effluent for reuse.

The benefits of a dual-water supply are twofold: by providing small quantity of high quality potable water, the risk of health hazard from continuous ingestion of low levels of toxic contaminants over a period of years would be eliminated; and the good quality water, which otherwise would have been used for nonpotable purposes not warranting high-quality water, would be conserved for potable use only.

Water Quality Criteria and Treatment

The National Interim Primary Drinking Water Regulations prescribed maximum contamination levels for ten inorganic contaminants. Removal of inorganic ions from drinking water is done in most cases by conventional coagulation and lime softening. Recent EPA research for removal of inorganic contaminants is based on coagulation and lime treatment. These EPA studies show that no one treatment is effective for all contaminants. Most of the methods listed in EPA studies [4] are conventional coagulation and lime softening. Other treatment techniques such as ion exchange and reverse osmosis may be equally effective.

In developing treatment systems for water containing trace metals, either chemical coagulation or lime treatment may be adopted. In the removal of high total dissolved solids, reverse osmosis may be adopted. If it is necessary to remove all the inorganic pollutants described in the Interim Primary Drinking Water Regulations, more than one unit process of treatment will be required.

The granular activated carbon adsorption method is most effective for trace organics removal.

Analysis of Dual Water Systems

Increased demands for good quality water increase the necessity for better management and planning of water systems. Water pollution abatement requirements have resulted in the production of wastewater effluents of such quality that they may be considered sources of water for nonpotable uses. Where good quality water is not available, potable water would need to be produced from unprotected sources after extensive treatment to meet primary and secondary drinking water regulations.

Therefore, in areas where good quality water is not available, or is available in limited quantity, or where there is an over-all shortage of water, the development of a long-term economic water management plan will be complex. It might be necessary to consider more than one source of water (including reuse), and more than one distribution system to supply water for potable and other uses. A comprehensive and easy-to-use systems model (WATMAN) for analysis of long-term urban water supply planning using single- or dual-supply concepts has been developed by Deb [5].

A conceptual dual supply urban water system model is shown in Figure 1. The model outline may have numerous options, depending on the quality and quantity of sources, potable and nonpotable water demand, and institutional and political considerations. This model provided a methodology of physical and economical analyses of urban water systems. It also provided the user with a tool to analyze various water supply system alternatives, including dual distribution, depending on the composition of sources and demands. This model can only be used as a planning tool rather than design tool and can be used to analyze up to:

- 3 qualities of water and supply systems,
- 10 communities/cities in a region, or
- 50 unit processes or elements.

In analyzing a dual supply system, based on the sources and treatment requirements, the potable and nonpotable system may be completely or partially extended to all water users in an urban community. Total present worth of collection, transmission, treatment, and distribution of a water system of any configuration for dual supply can be calculated using this model.

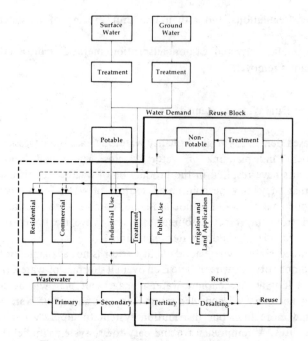

Figure 1. Urban water reuse management concept.

In formulating the mathematical model, parameters such as potable-to-total flow ratio, interest rate, annual capital and operation and maintenance (O&M) cost increase rates, and cost functions are considered as variables. In treatment system cost formulation, the quality of water from a single supply is assumed to be the same as that of the potable supply from a dual system.

The WATMAN model is a powerful tool for analyzing the various alternatives of long-term urban water management. The model is also very flexible and can accommodate any configuration of water supply system. It can handle a conventional system, and a dual system with more than one source of water (including reuse), and it can supply more than one grade of water to various demand centers.

The model has also been designed to accommodate regional management of water, with up to ten cities in the region. In a regional water supply management analysis, one limited good quality source may be considered to supply all potable demands of the region, and nonpotable demands can be drawn from local unprotected sources or from effluents of sewage treatment plants.

Distribution System Analysis

A distribution main system consists of all pipes and appurtenances in the water distribution system from service reservoirs to consumers. If the lengths and diameters of all pipes in a system are known, capital cost of the total distribution system can be calculated. If the lengths and diameters of pipes are not known, as in the case of a new city, a simplified procedure has been developed to estimate the total lengths of pipes and average cost diameter for the city. This average cost diameter can be defined as the diameter of a pipe the length of which is equal to the total length of distribution main, and the cost is the same as for a total distribution system.

The average cost diameter can be expressed as follows:

$$\text{Capital Cost} = 1.01 \, d_{av}^{1.29} L_m \, 5280$$

where d_{av} = average cost diameter, and
L_m = total length of distribution main in miles.

A simplified method of estimating capital and O&M costs of a distribution system has been developed [5].

Relationship of Population Density With Water Main Length

The length of water main required in a community water supply is an important parameter in the dual water supply cost analysis. It is expected that the length of water main in a community will be dependent on population density of the community. By analysis of the data of population density per square mile and lengths of water mains in mile per thousand of population of various water utilities, it was found that the larger the population density, the smaller is the main length per thousand population [5]. By analysis of existing data, the following relationship was given by Deb [5].

$$L_m = K_{Lm} P_d^{-0.458} \text{POP} \tag{1}$$

where L_m = main length in miles,
K_{Lm} = coefficient for length (by regression analysis K_{Lm} = 125.39),
P_d = population density (people/square mile), and
POP = population in base year in thousands.

If the population density and population of a town are known, the total length of distribution main can be estimated.

Average cost diameter of the distribution system of a city is found to increase with the population of the city. By regression analysis of average cost diameter of distribution systems with size of cities, the following relationship has been established [5] :

$$d_{av} = K_d \, POP^{\,0.065} \tag{2}$$

where d_{av} = average cost diameter in inches,
  K_d = coefficient for average diameter (by regression analysis K_d = 6.2), and
  POP = population in thousands.

Capital Cost of Distribution Mains

From known values of population and population density of a city and using Equations 1 and 2, the total capital cost of a distribution system can be calculated using the relationship:

$$\text{Capital Cost} = 1.01 \times 5280 \, (K_d \, POP^{\,0.065})^{1.29} K_{Lm} P_d^{\,-0.458} POP$$

$$= 5332.8 \, K_d^{\,1.29} \, K_{Lm} P_d^{\,-0.458} \, POP^{\,1.084} \tag{3}$$

By regression analysis of existing data, the values of K_{Lm} and K_d were found to be 125.39 and 6.2, respectively.

Water mains from service reservoirs to consumers are assumed to be subject to the same hydraulic gradients in potable and nonpotable water supply systems. Using the Hazen-Williams equation of pipe flow, it can be shown that for constant hydraulic gradient the diameter of the pipe is proportional to $Q^{0.38}$.

For conventional system, once the average cost diameter is calculated using Equation 2, the average cost diameter for potable and nonpotable systems can be calculated as proportional to respective flows. It is also assumed that, for a complete dual supply system, the length of potable mains is equal to the length of nonpotable mains. Once the lengths and average cost diameters are calculated, the total cost of the distribution system can be calculated using the pipe cost function.

However, if it is intended to have a partial dual distribution system in a city, as in the case of supplying nonpotable water for selected public, commercial and industrial uses, the lengths and sizes of various pipes required should be used to calculate the capital cost of the distribution system.

Retrofitting Cost of Distribution

The retrofitting cost of laying distribution mains for a dual system in an existing city will be higher than the cost of a new system. To calculate the capital cost of retrofitting distribution mains, the pipe cost function is multiplied by a retrofitting factor. The value of this factor depends on the complexity of development of the area and should be given as input to the model. In England, it has been found that the cost of laying a pipe in developed areas is equal to twice the cost of laying the same pipe in open areas [6]. The model (WATMAN) incorporates a retrofitting factor and can be used to analyze a complete, partial or dual distribution system in a new or retrofitting condition.

Operation and Maintenance Cost of Distribution Mains

Operation and maintenance cost per mile of a distribution system is found to vary with the average cost diameter of the system. The following relationship was obtained using regression analysis of data from water utility companies [5] :

$$\text{O\&M Cost (\$) per year} = 33.63 \, K_d^{1.29} \, K_{Lm} \, P_d^{-0.458} \, POP^{1.084} \tag{4}$$

where P_d = population density (people/square mile), and
         POP = population in thousands.

Once the total length and average cost diameter of any supply system are known, the capital and O&M cost of distribution mains can be calculated. The O&M cost equation (Equation 4) is essentially of the same form as the capital cost equation (Equation 3). In fact, O&M cost is 0.0063 times the capital cost.

In a dual water supply analysis, capital and O&M costs of distribution systems of potable and nonpotable systems can be converted to present worth of the base year of the planning period, considering proper salvage value and inflation rates.

Cost Functions

To analyze cost effectiveness of water reuse and dual water systems over the conventional system, capital and O&M cost functions of various treatment processes are required. The capital and O&M cost functions developed by the Environmental Protection Agency can be used for cost comparison of dual supply over the conventional system [7-9].

Present Worth Method of Cost Calculation

In calculating the costs of a single supply and a corresponding dual supply system, all costs incurred during the planning period should be converted to the present worth cost. The total cost includes capital costs, O&M costs, replacement costs, and salvage value. This procedure converts these figures over the project life into an equivalent cost representing the current investment required to satisfy all of the identified project costs for the planning period.

The present worth of a system unit cost, Y_t, that would be installed t years after the base year can be given as:

$$Y_{pw} = \frac{Y_t}{(1 + i)^t} \tag{5}$$

where Y_{pw} = present worth (base year) of the cost Y_t that would be incurred t years after the base year, and

        i   = the annual rate of interest.

To include the effects of inflation, assuming a capital cost inflation rate of C_c per year, the cost of the unit after t years would be:

$$Y_t = Y_o (1 + C_c)^t \tag{6}$$

where Y_o = present cost of this unit. Replacing the express for Y_t of Equation 6 in Equation 5, the present worth of a future unit cost can be expressed as:

$$Y_{pw} = Y_o \left(\frac{1 + C_c}{1 + i}\right)^t \tag{7}$$

$$= Y_o (IF_c)^t$$

where IF_c = the inflation factor for capital cost.

If the O&M annual cost increase rate is C_o, a similar expression for one future year's O&M cost can be converted to present worth as:

$$Y_{pw} = Y_o \left(\frac{1 + C_o}{1 + i}\right)^t \tag{8}$$

$$= Y_o \, IF_o^{\,t}$$

where IF_o = inflation factor for O&M cost. The present worth of the total O&M cost during the planning period can be given as:

$$Y_{pw} = \sum_{t=0}^{n} Y_o \left(\frac{1 + C_o}{1 + i} \right)^t \qquad (9)$$

where n = planning period.

The salvage value represents the value remaining for all capital at the end of the planning period. Considering inflation, the salvage value of a unit cost converted to present worth can be given as:

$$SV_{pw} = (1 - \frac{UL}{DL}) Y_o (IF_c)^n \qquad (10)$$

where SV_{pw} = present worth of salvage value of a unit of which present day (base year) cost is Y_o,

n   = planning period in years,
DL = design life period of the unit in years, and
UL = used life period of the unit in years.

All equipment found within a water supply system has a finite service life. This service life represents a period of time when a particular equipment item must be replaced. Mechanical equipment such as pumps, chlorinators, chemical feeders, etc., tend to have a low service life, whereas structural equipment such as sedimentation tanks, filtration units and buildings tend to have a long service life. Appropriate service life should be established for all unit processes. The present worth of replacement cost of a unit considering inflation can be calculated using Equation 7. This applies only for equipment with a service life shorter than the planning period.

Applications

This systems methodology can be used for any urban water supply management and planning effort where a large number of alternatives are to be analyzed. Specifically, this systems model can be applied to evaluate various alternative plans using conventional and dual supply concepts in areas where:

* the quality of the raw water source is poor,
* the availability of good quality water is limited, or
* there is an over-all shortage of water.

In the design of new water supply systems, it is advisable to consider the water reuse and dual supply concept as a viable alternative for a long-term safe water supply plan. From social, engineering, economic and institutional points of view, it would be easier to establish a dual supply in a new city. Depending on the source and quality of raw water, there may be many cases where a dual supply can be economically adopted.

With the introduction of Primary Drinking Water Regulations, many existing water systems will have to provide further treatment in order to meet requirements. In cases where the cost of additional treatment for the bulk water is high, a dual supply system, using either the same source or an alternate source for potable supply, might prove to be more desirable. Figures 2 and 3 describe the typical existing systems and outline conventional or dual supply alternatives. In cases where existing source quality is poor, it might be less expensive to treat a small fraction of the water to meet drinking water regulations and distribute it through a separate distribution system.

Common examples of reuse of wastewater for conservation of good quality sources are shown in Figures 4 and 5. In an industrial city, where a large portion of the municipal water demand is for industrial needs, a dual supply can

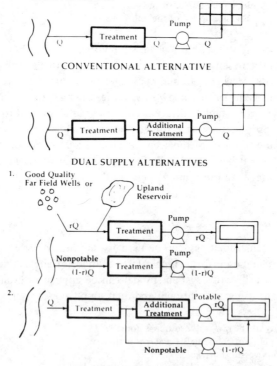

Figure 2. Existing system 1.

be effectively used with one nonpotable supply for industries and a potable supply for other uses. The industrial nonpotable supply can be derived from the same source with only existing treatment, or from reclaimed municipal wastewater with further treatment.

Figure 6 shows a schematic of alternatives for an existing system on a polluted river which has been analyzed with this model.

In regional management of water, good-quality-limited sources can be conserved and used for potable supply only through a regional water treatment and distribution system. The water for nonpotable uses can be obtained from reclaimed water (Figure 7). Where a limited quantity of protected water is available, existing communities obtaining their water supply from nearby polluted sources will have the option of having additional treatment for removing trace chemicals, or using a limited protected source for potable supply to all communities in the region using a dual distribution system. The nonpotable supply can be obtained from polluted river source or from reclaimed wastewater.

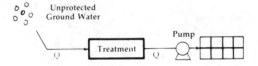

CONVENTIONAL ALTERNATIVE

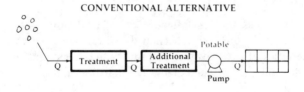

DUAL SUPPLY ALTERNATIVE

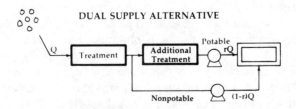

Figure 3. Existing system 2.

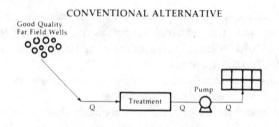

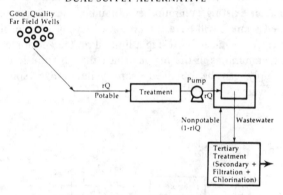

Figure 4. Systems 1-2; city: 100,000 population.

Results

In order to check the cost effectiveness of dual systems, the two water systems shown in Figure 2 have been analyzed for various potable/total flow ratios. The ratios of total costs of dual supply to conventional supply for the two systems and for various potable to total flow ratios are presented in Figures 8 and 9.

In System 1, the source of water for the conventional system is an unprotected river. In order to produce potable water from this source, granular activated carbon (GAC) has been added with the standard treatment of coagulation and filtration. In the corresponding dual system, the potable source is good quality groundwater, and the unprotected river source furnishes nonpotable water.

It is apparent from Figure 8 that the dual water supply cost is less than 85% of that of a conventional supply in a new system. However, if it is assumed that the potable well source for dual supply is 30 miles away, dual supply becomes more expensive then conventional supply when the flow ratio exceeds 0.25.

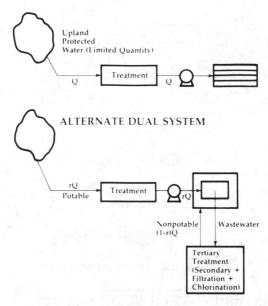

Figure 5. Systems 1-2; city: 10,000 population.

In System 2, unprotected river water is the only source for either a conventional or a dual system. Activated carbon treatment has been considered along with the standard treatment to produce potable water in the dual and conventional systems, whereas only standard treatment is given to produce nonpotable water in the dual system. The result, presented in Figure 9, shows that conventional system will be more economical when the flow ratio exceeds 0.2.

REGULATIONS AND GUIDELINES

The use of reclaimed water for various nonpotable uses in the community is gaining importance. Several states are already using reclaimed water for various beneficial uses and have developed some form of regulations and guidelines.

To compile nationwide regulations and guidelines regarding dual water supply and water reuse, a survey was conducted of all the State Health Departments [5]. Of 37 states replying, about 10 have some form of guidelines or regulations regarding reuse of wastewater effluent, mainly for irrigation. California, Colorado, Kansas, Maryland and Utah have detailed regulations regarding reuse of wastewater effluent for various irrigation uses.

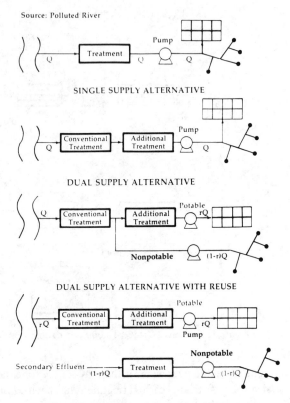

Figure 6. Existing system for industrial uses.

Four states (California, Colorado, Kansas, and Utah) have regulations regarding treatment of wastewater and quality of effluent to be used for urban landscape irrigation. All the states as a minimum require secondary treatment and disinfection of effluent for landscape irrigation. However, Utah requires AWT effluent for irrigation of public parks and lawns with BOD and suspended solids less than 5 mg/l and total coliform of 1/100 ml. Kansas does not have any requirement for quality of effluent to be used for urban landscape irrigation, but does require secondary treatment of wastewater followed by filtration and disinfection.

Virus standards have not been adopted. Where the reclaimed waters are accessible to the public, there is a need to establish virus limits and virus monitoring. However, technology is not now adequate, and standards have not yet been developed.

The coliform limit, similar to those for drinking water (1/100 ml), is probably adequate.

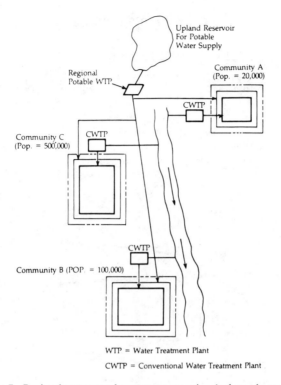

WTP = Water Treatment Plant

CWTP = Conventional Water Treatment Plant

Figure 7. Regional water supply management using dual supply concept.

Cross-Connection Control

The use of reclaimed water for various nonpotable water uses within a community requires the installation of a dual distribution system. In a dual distribution system, there is always concern about the possibility of cross connection of the potable system with the nonpotable system. All states have cross connection prevention regulations and programs with the conventional single pipe distribution system. Because the possibility of cross connection would be increased with the dual system, installation of such a dual system would require a rigid cross connection prevention program and regulations.

A detailed regulation for the construction and operation of a distribution system conveying reclaimed water has been developed by the Irvine Ranch Water District (IRWD) of California [10,11]. These regulations can be considered by any water authority as a good starting point for a new dual distribution system. The specific IRWD regulations regarding the reclaimed water systems are as follows:

CITY: 100,000 POPULATION

Conventional System
_____

Source: Unprotected River
Treatment: Coag. + Sed. + Filt. + Gac + Disin.

Dual System
_____

Potable Source: Protected Ground Water
Treatment: Disinfection

Nonpotable Source: Unprotected River
Treatment: Coag. + Sed. + Filt. + Disin.

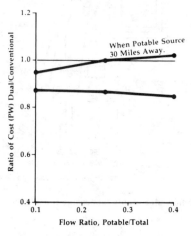

Figure 8. Dual water system model no. 1.

1. Reclaimed Water Pipeline Identification: All reclaimed water pipes shall be identified and marked in the following manner:
   • All reclaimed water mains shall have the words "Reclaimed Water" stenciled in two-inch letters on both sides of the pipe in at least three places in a 13-foot section of pipe. The other acceptable alternative is placing a warning tape directly on the top of the pipe identifying it as Reclaimed Water.
   • The stenciling on irrigation pipes shall appear on both sides of the pipe with the marking "Reclaimed Water" in 5/8-inch letters repeated every 12 inches.
2. Control and Regulating Valves: All gate valves, manual control valves, electrical control valves, and pressure relief valves shall be housed in a valve box with a locking green cover.
3. Quick-Coupling Valves: Quick-coupling valves shall be installed with green cover and shall be operated only with a special coupler key for opening and closing the valves.
4. Identification of Potable Pipelines: All potable water piping installed in the same premises as the on-site irrigation piping shall be installed in accordance with uniform plumbing code and shall be blue in color.

CITY: 100,000 POPULATION

Conventional System
_____

Source: Unprotected River
Treatment: Coag. - Sed. - Filt. + Gac + Disinf.

Dual System
_____

Potable: Unprotected River
Treatment: Coag. - Sed. - Filt. - Gac - Disint.

Nonpotable: Same Source
Treatment: Coag. - Sed. - Filt. - Chlorination

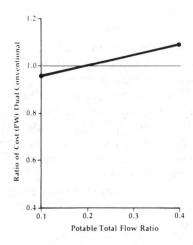

Figure 9. Dual water system model no. 2.

5. Intersection of Potable and Reclaimed Water Lines: Intersection of the reclaimed water and potable water systems is treated the same as water-sewer system crossings.
6. Identification of Valve Covers for Reclaimed Water System: Valve box covers shall be specially marked to identify the system as tertiary reclaimed water.
7. Back Flow Prevention: The on-site reclaimed water system is a separate and controlled nonpotable system. Under normal conditions, protective devices will not be required on the potable water services. However, the authority may require back flow prevention devices on the potable water supply in certain cases and shall review each service on a case-by-case basis. The installation of back flow prevention devices may be required in the potable line: where reclaimed water is used on individually- and privately-owned premises and is under individual and private control; where reclaimed water is used in schools and public parks with extensive systems; and where there is high public exposure and use. One additional measure of back flow prevention would be to keep the pressure in nonpotable pipeline less than the pressure in potable supply in the same vicinity.
8. Inspection of Protective Devices: It shall be the duty of the water user on any premises on which back flow prevention devices are installed to have competent inspections made at least once a year.

9. Marking Potable and Nonpotable Water Lines: Where the premises contain more than one water supply system, the exposed portions of the pipelines shall be painted, banded, or marked at sufficient intervals to distinguish clearly which water is potable and which is nonpotable. All outlets from reclaimed water systems shall be posted as being unsafe for drinking purposes.

10. Water Supervisor: For control of cross connections, a permanent education program is sponsored. At each premise (school, community, association, etc.) supplied with reclaimed water, a water supervisor shall be designated and trained. This supervisor shall be responsible for the installation and use of pipelines and equipment, and for the prevention of cross connections.

Monitoring and Reliability

Monitoring of reclaimed water in the treatment plant and in the distribution system is very important to make the dual water supply system reliable. The performance of the tertiary treatment plant should be monitored regularly to check the performance of each unit process and to check the quality of reclaimed water in conformance with the requirements. With the distribution of reclaimed wastewater throughout a city, and into and in the vicinity of homes, commercial establishments, parks, etc., monitoring is advisable, particularly in the early years, even if somewhat costly. With experience, the monitoring requirement might be relaxed.

The following quality and monitoring measures are suggested for the distribution of reclaimed wastewater for nonpotable uses:

- Secondary maximum contaminant levels should be met, with some relaxation of limits for dissolved solids as appropriate.
- Routine viral and bacterial monitoring should be done. Coliform should not be greater than 1/100 ml (primary drinking water standard) and virus standards will need to be established.
- Turbidity before post chlorination should not be more than 1 unit (primary drinking water standard).
- Plant effluent free chlorine should be more than 3.0 mg/l, and system residual should be more than 0.5 mg/l.

It is expected that reliably operated secondary treatment followed by polishing and chlorination would be adequate to meet these criteria.

To increase the reliability of the performance of the system, backup units for each unit process are desirable. Some of the critical unit operations (power supply, pumps, disinfection, storage for diverted reclaimed water, etc.) must have backup units in order to supply reclaimed water of reliable quality.

SUMMARY AND CONCLUSIONS

Due to Federal regulations, a considerable change of degree of treatment of potable water and wastewater has taken place. Treatment-related costs of water and wastewater utilities will constitute a larger share of total costs. The cost of producing high quality water from a poor quality source and cost of producing a high quality effluent are high. All these factors necessitate reevaluation of economics of conventional water and wastewater management systems. Dual water systems and water reuse would be cost-effective alternatives under many circumstances.

Good quality water sources are limited and should be protected and conserved for uses requiring high quality water. Water for other uses can be from lower quality sources, including reused water. Where recycled water will not be employed for potable uses, a separate distribution system will be necessary for supplying recycled water for nonpotable uses. In fact, dual supply systems would be a necessary feature of water management in areas where wastewater reuse will be considered.

A dual water supply can also be adopted in cases where raw water source quality is poor and has high organic, heavy metal and total dissolved solids (TDS) content. The cost of removal of these materials is high and dual supply systems may be economically viable over the conventional system. Again, this depends on the savings of additional treatment costs over the additional distribution costs.

A dual water supply should be an essential consideration in a long-term water management plan for a city or region considering conservation and optimum uses of water.

These results undoubtedly reflect the assumptions made in the analysis regarding the sources of water for the potable and nonpotable supplies. Although certain general conclusions are possible, it would be unwise to accept these conclusions for all circumstances. It must also be stressed that the results that have been presented assume the development of new supplies.

A general methodology has been developed by Deb [5] for comparing the costs of single and dual supply. This method can be used in any specific case by using the proper values of the various costs and other economic parameters.

REFERENCES

1. Dellaire, G., Ed. "Are Cities Doing Enough to Remove Cancer-Causing Chemicals from Drinking Water?," *Civil Eng.* (September 1977).
2. "Drinking Water and Health," Recommendations of the National Academy of Science, *Federal Register* (July 11, 1977).

3. Okun, D. A. "New Directions of Wastewater Collection and Disposal," *J. Water Poll. Control Fed.* 43:1582 (1968).
4. "Manual of Treatment Techniques for Meeting the Interim Primary Drinking Water Regulations," U.S. EPA Report 600/8-77-005 (May 1977).
5. Deb, A. K. "Multiple Water Supply Approach for Urban Water Management," Final Report on research supported by the National Science Foundation Grant ENV 76-18499 (November 1978).
6. Jackson, J. K. "Dual Water Supplies in the Trent River Basin," presented at the Symposium on Advanced Techniques in River Basin Management, Institution of Water Engineering, London, England, 1972.
7. Culp Wesner Culp. "Estimating Costs of Water Treatment as a Function of Size and Treatment Efficiency," U.S. EPA Contract No. CI-76-0288 (July 1979).
8. Dames & Moore. "Construction Costs For Municipal Wastewater Treatment Plants," U.S. EPA Report 430/9-77-013 (January 1978).
9. Dames & Moore. "Analysis of Operations and Maintenance Costs for Municipal Wastewater Treatment Systems," U.S. EPA Report 430/9-77-015 (May 1978).
10. AWWA Research Foundation. "Irvine Ranch Water District—Total Water Management System," *Munic. Wastewater Reuse News* (April 1978).
11. Irvine Ranch Water District. "Rules and Regulations for Water, Sewer, and Reclaimed Water Services" (June 1977).

CHAPTER 10

REVERSE OSMOSIS IN WATER REUSE SYSTEMS

Dennis B. George
Division of Environmental Engineering
Utah State University
Logan, Utah 84322

Membrane processes have been employed successfully by industry and municipalities to remove dissolved inorganic and organic compounds. Membrane processes used in water and wastewater systems are reverse osmosis, ultrafiltration and electrodialysis. Each process uses a semipermeable membrane to separate impurities from the liquid stream. Historically, reverse osmosis, electrodialysis and ultrafiltration have been used in the fields of food processing, industrial wastewater treatment, clinical medicine and saline water desalting. Membrane technology also has been used to remove toxins from metal streams and dye wastes. Radioactive substances contained in wastewater derived from the washing of contaminated garments worn by nuclear power plant personnel have also been effectively treated by membrane processes.

In addition to dissolved inorganic salt removal, reverse osmosis effectively removes dissolved organics with molecular weight ≥ 150 [1]. Reverse osmosis processes, however, have been shown to be ineffective in removing trihalomethanes and other halogenated aliphatics [2,3]. The ultrafiltration process separates suspended solids and dissolved solutes with a molecular weight greater then approximately 15,000 from the liquid stream. Figure 1 indicates the useful range of each process.

Currently Director, Lubbock Christian College Institute of Water Research, Lubbock, Texas 79407.

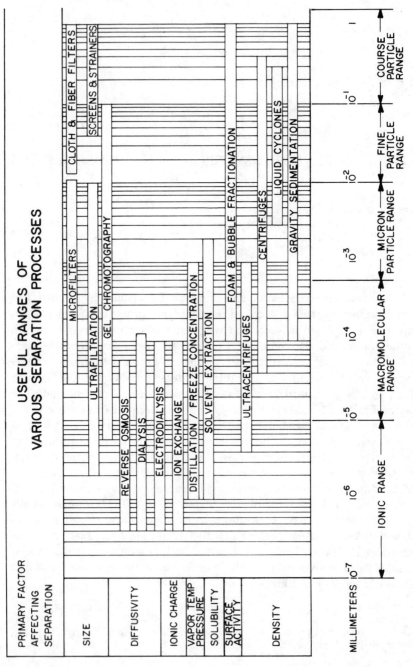

Figure 1. Useful ranges of separation processes [4].

Each use of water by municipalities and industry adds $\sim$ 200-500 ppm of salinity to the wastewater stream [5]. Conventional treatment processes are unable to effectively reduce these dissolved inorganic salts. Consequently, in water reuse cases, membrane processes may be required following advanced wastewater treatment. Reverse osmosis is the major membrane process employed in the reuse of municipal and industrial wastewaters. The focus of this chapter is the description and use of reverse osmosis in water reuse systems. Various references are available which provide a more detailed discussion of membrane process [4,6-10].

REVERSE OSMOSIS

When a solution of a solute is separated from pure solvent by a semipermeable membrane which allows the flow of solvent but not solute, osmosis occurs. The concentration of solvent molecules in the solute solution is less than in the pure solvent. The system will compensate for this difference by a net transport of solvent across the membrane into the solute solution until the concentration of the solvent is equal on both sides of the membrane. The flow of solvent into the solute solution causes a rise in the hydrostatic head of the solute solution (Figure 2). At equilibrium the hydrostatic head counterbalances the osmotic pressure. Osmotic pressure is the force that must be applied to exactly oppose the flow of solvent across the membrane. Osmotic pressure is a property of the solution and not of the membrane. An increase in the pressure above the osmotic pressure on the concentrated solute side of the membrane will reverse the flow of the solvent through the membrane from the concentrated solute to the dilute solution. This is the basic theoretical foundation for reverse osmosis (RO). A basic RO system is depicted in Figure 3.

Reverse osmosis systems normally employ some form of pretreatment or posttreatment depending on the application. Pretreatment can be as simple as a cartridge filter which removes particulate matter which can damage the high pressure pump. However, when employed on highly turbid industrial waste water, pretreatment may consist of coagulation, sedimentation and filtration followed by pH adjustment [11]. Posttreatment can consist of aeration, degasification, chlorination and pH adjustment [11]. When complete demineralization of the waste water is required, ion exchange can be used as a posttreatment process.

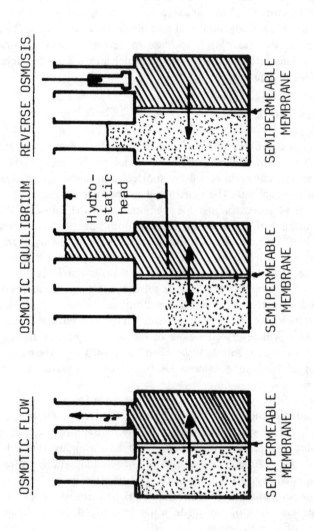

Figure 2. Principle of reverse osmosis.

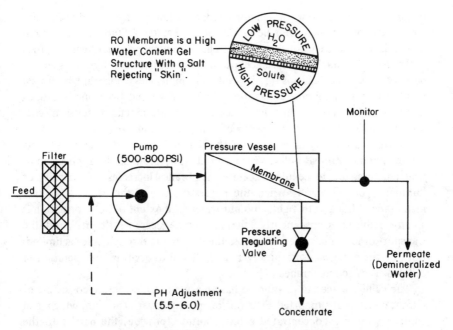

Figure 3. Basic reverse osmosis system [11].

Membranes

Membranes used in RO systems must meet such stringent requirements as: (1) hydrophilic membrane which selects water molecules over ions (governs the number of pressurized stages needed to produce potable water), (2) permeation rate of water per unit pressure gradient which determines the size of equipment per unit production rate of potable water, and (3) membrane durability which determines the replacement rate of the membrane [12]. Some factors which affect membrane durability are chemical agents, heat and bacteria.

Reverse osmosis systems generally employ semipermeable membranes made of cellulose acetate or polyamide. Membranes made of sulfate polysulfones, polyphenylene-oxides and polyureas are being developed [5]. The most widely used membrane is the cellulose acetate membrane.

Cellulose-Acetate Membrane

Cellulose-type membranes are composed of polymer structural units $\sim$ 15,000 Å in length and 10-20 Å in width [13]. These polymers are randomly oriented. When polymer chains are parallel to each other, crystalline regions are created. However, an enlargement of holes or spaces between polymers occurs when the chains lie in a disordered state. Cellulose acetate maintains a large percentage of the polymer chains in highly ordered regions [14].

Transport through cellulose acetate membranes has been hypothesized to occur by two mechanisms [15]. Ions and molecules which are hydrogen bonded to the membrane are assumed to be transferred from one hydrogen bonding site to another. In "tight" cellulose acetate membranes, this mechanism is considered the major means of water diffusion [16]. Ions and molecules which do not hydrogen bond with the membrane are transported across the membrane through holes. Vincent et al. [16] indicate the application of heat and pressure to cellulose acetate membranes increases the quantity of "primary" bound water. When the pore structure of the cellulose acetate membrane is filled with tightly bound water, the extent of hole formation is greatly reduced. Consequently, the membrane becomes "tight," and the transport of ions and molecules across the membrane through holes is limited. Both of these mechanisms, however, are believed to occur in the operation of cellulose acetate membranes [7].

The cellulose acetate membrane has an asymmetric structure consisting of a thin, dense skin supported by a thicker porous layer. The skin, which is in contact with the concentrated solute solution, provides the barrier to the transport of salts and impurities across the membrane. The porous support layer permits diffusion of the product water into a collection system. A typical membrane is approximately 100-μ thick and contains a surface skin which is 0.2-μ thick [4]. The porous layer contains about two-thirds water by weight and normally must be kept wet at all times.

Cellulose acetate membranes are susceptible to biological degradation and have a limited operating pH range. Salt rejection decreases with time, which is attributed to hydrolysis of the membrane. The hydrolysis rate is minimized between pH 4 and 5 [4].

The process should be operated at a pH range of 3-7 to prevent rapid deterioration of the membrane. Temperature increase will accentuate the membrane hydrolysis rate; however, the water flux across cellulose acetate membranes increases about 3.5%/°C in the temperature range of 15-30°C.

Polyamide Membrane

Polyamide membranes have high chemical and physical stability, therefore, longer membrane life. Certain aromatic polyamide membranes (Dupont's "Permasep") are stable during continuous operation over a pH range of 4-11 [17]. For limited periods of time, these aromatic polyamide membranes can maintain stability at pH levels between 2 and 12. Furthermore, polyamide membranes are more resistant to higher temperatures and essentially immune to biological degradation.

A polyamide membrane which has attained commercial acceptance is PA-300. The PA-300 membrane barrier is formed by the following chemical polymerization reaction of epiamine and isophthaloyl chloride (IPC):

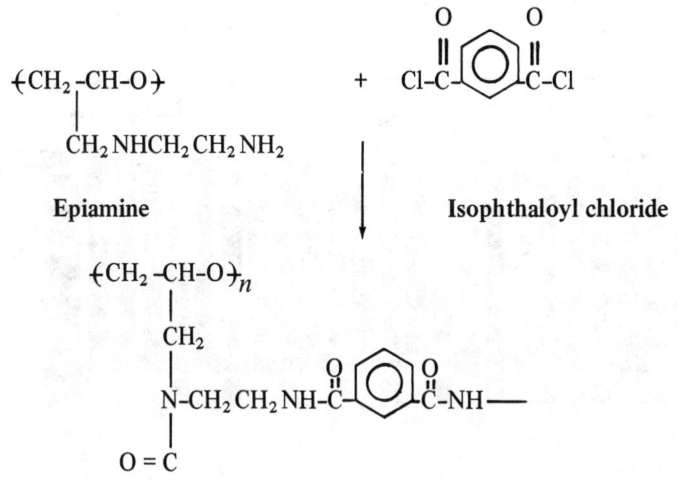

PA-300 Barrier

The interfacial condensation of epiamine and IPC is formed on a polysulfone porous support [18]. PA-300 and other polyamide membranes are susceptible to degradation by chlorine and/or other oxidizing agents in the feedwater. It has been hypothesized that the low tolerance of the membrane to chlorinated feedstreams is associated with the presence of secondary amine in the backbone of the barrier polymer [18].

Membrane Modules

Reverse osmosis membranes are prepared to achieve a thin, dense, surface layer (skin) supported by a thicker, porous layer. The skin is the barrier to the passage of the solute. The product water is diffused through the porous layer into a collecting system. There are four common configurations in which the membrane can be contained: plate-and-frame (flat membrane), spiral-wound, tubular and hollow-fiber.

Plate-and-Frame

Plate-and-frame design consists of membranes attached to both sides of a rigid plate (Figure 4). The membranes are sealed around the edge to prevent leaks. The most common plates are constructed of thin plastic; however,

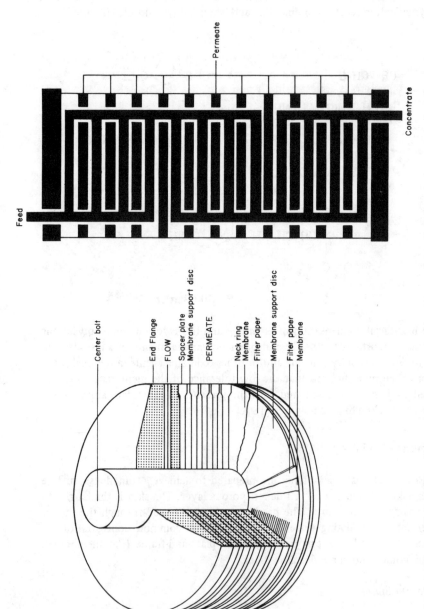

Figure 4. Plates and frames in section with membranes mounted in sandwich fashion in the module and held together by a center bolt [11].

plates may be constructed of porous fiberglass materials or reinforced porous paper. The flat surfaces of the plates contain grooved channels through which the permeate flows after passing through the membrane. The plate/membrane modules are contained in pressurized vessels which are designed to allow the flow of the feed stream on both sides of the plates. The feed water is introduced under pressure into the vessel and concentrated brine is removed from the opposite end. Each plate is at low pressure; therefore, water passes through the membrane and is collected in the porous media. Product water is collected from all the plates. Due to the complexity of plate-and-frame RO systems, they are very expensive to operate for large-scale operations [11]. Consequently, plate-and-frame systems are not being applied to water and/or wastewater treatment.

Spiral Wound

The spiral-wound configuration consists of a large flat sheet of membrane which covers each side of a flat sheet of porous, woven fabric (Figure 5). The two long edges and one end of the membrane are sealed to form a flexible envelope. The open end of the membrane envelope is sealed to a perforated tube. The perforated tube collects the product water. A sheet of plastic netting is placed adjacent to one side of the membrane envelope. The plastic netting separates the membrane layers during assembly and promotes turbulence in the feedwater during process operation. The membrane envelope and netting are wrapped around the central tube and form a spiral configuration. The spiral-wound unit is, subsequently, inserted into a pressure vessel. Pressurized feedwater is introduced through the open mesh netting and flows through the membrane into the porous, woven fabric. The permeate proceeds to the porous central tube and is removed from the RO system.

Advantages of spiral-wound configurations include its high packing density, low manufacturing cost and relative ease of cleaning both chemically and hydraulically. The process cannot be used, however, on highly turbid feedwaters without extensive pretreatment because the passages of the feedflow are very small and very susceptible to clogging. Spiral-wound configurations, tubular and hollow-fiber designs are currently being employed in water and/or wastewater treatment.

Tubular

The tubular design is the simplest of all the configurations. Tubular configurations consist of parallel bundles or a series of small diameter porous or perforated, high-strength, pressure-resisting tubes (Figure 6). The tubes are 1.3-2.5 cm in diameter. The membrane is either cast on the inside of, or placed within, a porous tube and sealed into place. Perforated, copper tubing

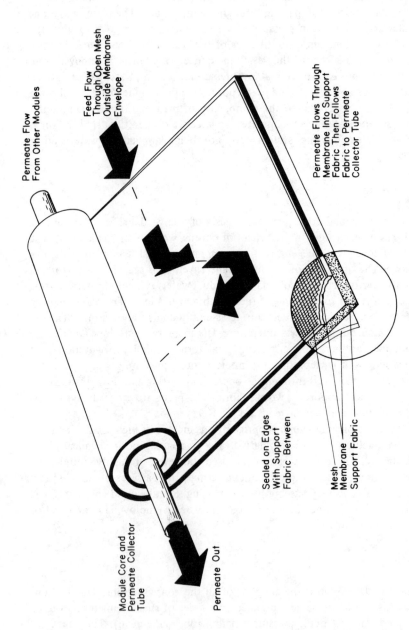

Figure 5. Spiral wound membrane element [11].

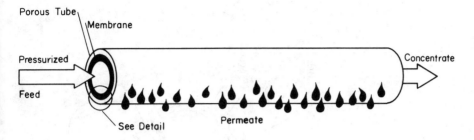

Multiple tubes normally connected in series
or parallel to form complete module

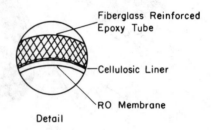

Figure 6. Tubular membrane element [11].

was used to make early porous tubes. Filter paper or woven fabric was used
as a backing material between the membrane and tube wall so that permeate
could be transported to the holes in the copper tube. Recent porous tubing is
fabricated from glass fiber-plastic combinations. Woven fabric backing is used
(although not required) to help prevent damage of the membrane. Pressurized
feedwater flows inside the tube. High quality water passes through the membrane, through the porous tube and drips off the outside surfaces. The permeate is collected in troughs or vessels.

   Tubular systems can be used on extremely turbid feedstreams. Furthermore, the system can be easily cleaned mechanically or hydraulically. The
high capital costs associated with tubular RO systems, however, limit their
ability to compete with spiral-wound or hollow-fiber RO systems for water
treatment application.

Hollow-Fiber

   The hollow-fiber membrane unit (Figure 7) consists of large numbers of
microfibers with an outside diameter of 25-250 microns and a wall thickness
of 5-50 microns. The ends of the fibers are potted into an epoxy tube sheet
so that the bores are exposed. The membrane fibers are packaged in a cylindrical bundle configuration, evenly spaced around a central feed distribution

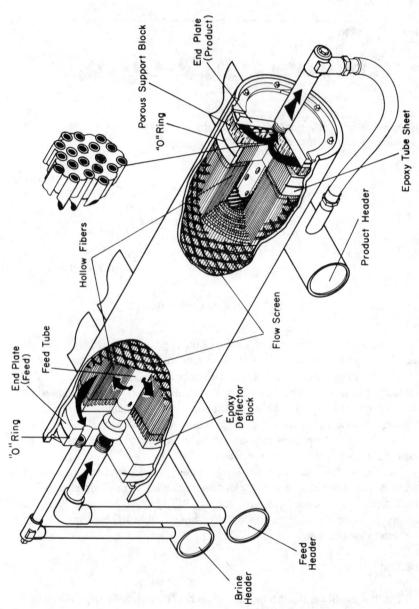

Figure 7. Hollow fine fiber membrane element (Dupont's "Permasep").

tube. The bundle of microfibers is inserted into a pressure vessel. During operation, pressurized feedwater is introduced through the distribution tube and flows over the outside membrane-surface of the hollow-fibers. Purified water passes through the outside membrane into the bore of each of the hollow-fibers and out through the epoxy tube sheet.

Hollow-fiber configurations have extremely high packing density. Reverse osmosis systems utilizing hollow-fiber design are compact and have very little space requirements, and manufacturing costs are quite low compared to other RO configurations. However, the very small spacing between fibers in the bundle makes the hollow-fiber systems extremely susceptible to fouling and difficult to clean effectively. Therefore, the system is limited to treating relatively clear waters which are extensively pretreated.

Water and Solute Diffusion

The reduced forms of the equations which describe the diffusion of water and solute through membranes are:

$$M_w = K_w(\Delta P - \Delta \pi) \qquad (1)$$

$$M_s = K_s(\Delta C) \qquad (2)$$

where
M_w = water mass flux through the membrane,
M_s = salt mass through the membrane,
K_w = coefficient of permeability for the water,
K_s = coefficient of permeability for the solute,
ΔP = applied pressure differential across the membrane,
$\Delta \pi$ = osmotic pressure differential across the membrane, and
ΔC = solute concentration differential across the membrane.

Equations 1 and 2 indicate that water flux across the membrane is a function of the net pressure difference and solute flux is only a function of the concentration differential across the membrane. Therefore, an increase in feedwater pressure will cause an increase in water flux across the membrane; whereas, solute flux will remain relatively constant.

The mass flux of water, M_w, decreases with time. The reduction of water flux is due to the degradation of the permeability of water across the membrane (K_w). Reduction of the membrane water permeability is due to [19]:

1. membrane fouling, which is caused by deposition of soluble material (i.e., calcium carbonates, calcium sulfate and hydrous metals), biological growth, deposition of organics, suspended solids, colloidal particles and other contaminants;

2. membrane compaction and compression, which results from the sustained high pressure fluid crushing the porous support structure of the membrane; and
3. hydrolytic deterioration of the membrane, which is related to hydrolysis of the membrane skin polymer (i.e., cellulose acetate, polyamide, etc.).

Membrane fouling can be controlled by proper pretreatment of the feedstream. Pretreatment may consist of filtration, pH adjustment, carbon adsorption, chlorination or a combination of the preceding processes, and others. Methods for controlling membrane compaction have not been very successful. As previously mentioned, the rate of hydrolytic deterioration of the membrane can be controlled by adjusting the pH of the feedwater.

The rejection of solute by a membrane also deteriorates with time. Factors contributing to a decrease in salt rejection are:

1. membrane compaction, which is caused by high feedwater pressure altering the thin membrane (skin) pore size;
2. coupling, which is a small coupling effect between the flow of pure water and solute;
3. hydrolytic deterioration of the membrane; and
4. concentration polarization, which results from an accumulation of solute on the membrane surface which is greater than the solute concentration in the bulk solution.

Concentration polarization increases the amount of solute present in the product water as predicted by Equation 2. Furthermore, the increase of solute at the membrane surface will increase the local osmotic pressure, which reduces the net pressure difference across the membrane. The reduction of net pressure difference will decrease the water recovery flux (Equation 1). Concentration polarization can also cause a deterioration of the membrane. In addition, solutes with low solubility may precipitate on the membrane surface and create fouling problems.

The decline in solute rejection and water flux determines the useful lifetime of the membrane. Proper pretreatment of feed water and cleaning of membranes are essential to the prolonged lifetime of the membrane.

Water Reuse Applications

Direct reuse of wastewater has been practiced primarily by industry and agriculture. Limited implementation of direct wastewater reuse systems for domestic drinking water has occurred in the United States. A severe problem associated with the direct reuse of wastewater is the buildup of dissolved solids in the recycled water. Reverse osmosis has been used in full-scale and pilot water reuse systems to demineralize recycled water.

In addition, water produced from wastewater reclamation facilities utilizing the RO process has been injected into groundwater aquifers to augment groundwater supplies and control seawater intrusion. Groundwater injection of reclaimed wastewater constitutes an indirect water reuse. Several case studies are provided to illustrate the use of the reverse osmosis process in the direct and indirect reuse of industrial and municipal wastewater.

Water Factory 21 [20]

In southern California, groundwater resources are the major component of water supply for municipalities and industry. Increasing population and industrial growth in the area have caused the demand for water to exceed the safe yield of the local watersheds. Furthermore, severe overdrafting of the ground water has promoted the intrusion of sea water into the freshwater zones of the aquifer. The Orange County Water District developed a wastewater reclamation system to produce a source of water to hydraulically prevent seawater intrusion into groundwater aquifers. Reclaimed water will eventually co-mingle with native ground water and subsequently be used for domestic, agricultural and industrial purposes. Therefore, very stringent water quality standards were imposed on the reclaimed water by the California Regional Water Quality Control Board and the California State Department of Health. The coastal barrier system consists of Water Factory 21 (advanced wastewater treatment facility), 23 injection wells, 31 monitoring wells, five supplementary deep wells and seven extraction wells.

Water Factory 21 is designed to treat 39 m³/min (15 mgd) of secondary effluent produced by activated sludge treatment systems owned by the Orange County Sanitation District. The advanced wastewater treatment system includes lime clarification, ammonia stripping, recarbonation, mixed media filtration, activated carbon adsorption, RO and chlorination (Figure 8).

Wastewaters produced in Orange County typically contain approximately 1000 mg/ℓ total dissolved solids (TDS). A 13 m³/min (5 mgd) RO demineralization plant has been operating at Water Factory 21 since July 1977 to achieve a reduction in TDS levels in the product water to satisfy injection quality standards. In addition, water from deep wells can be blended with the effluent from Water Factory 21 to meet mineral requirements. The RO plant is designed to provide 90% rejection of all salts and achieve an overall product water recovery of 85%. The flow schematic for the RO plant is presented in Figure 9. The activated carbon effluent is treated prior to entering the RO process by chlorination and filtration.

Table I presents typical performance data for the RO plant. The electrical conductivity (EC) of the influent stream to the RO plant is reduced by > 90%. The blended injection water does not violate the California State

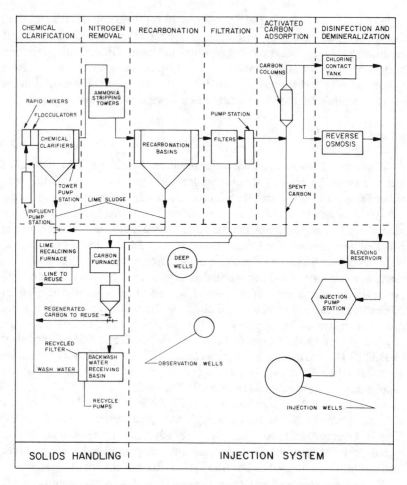

Figure 8. Flow schematic of 15-mgd wastewater reclamation plant [20].

maximum contamination level for EC (900 μmhos, cm). McCarty et al. [21] determined the percent probability of water meeting various chemical oxygen demand (COD) concentrations after different stages of treatment at Water Factory 21 (Table II). As COD requirements become more stringent, a high probability exists that reverse osmosis will be able to achieve the necessary degree of treatment.

Since October 1976, Water Factory 21 has produced over 40 million m³ (33,000 acre-ft) of high quality water which have been injected into the groundwater aquifer. The reliability of the advanced wastewater treatment plant to produce high quality effluent is enhanced by the physical chemical system's ability to adapt to varying operating conditions.

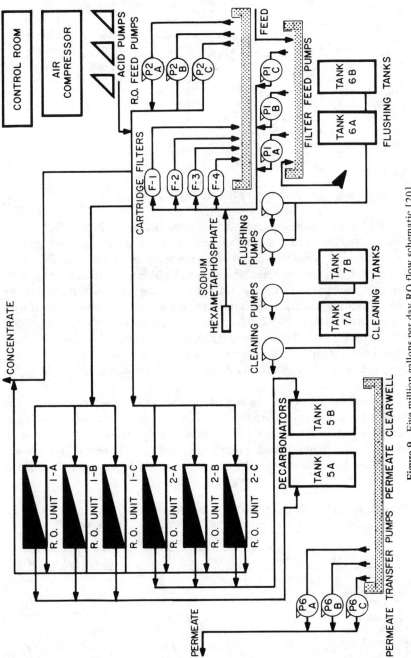

Figure 9.   Five million gallons per day RO flow schematic [20].

Table I. Typical Reverse Osmosis Plant Performance

Constituent	Inflow Quality (Activated Carbon Effluent)	Product Quality	Blended Injection Water Quality
Sodium (mg/ℓ)	210	11.0	108.0
Total hardness (mg/ℓ)	300	Trace	–
Sulfate (mg/ℓ)	280	0.8	121.0
Chloride (mg/ℓ)	240	16.0	103.0
Ammonium-nitrogen (mg/ℓ)	45	Trace	0.86
Chemical oxygen demand (mg/ℓ)	15	1.5	10.0
Electrical conductivity (μmhos/cm)	1460	70	784.0

Oxnard Wastewater Treatment Plant [22]

The Ventura Regional County Sanitation District (VRCSD) in Southern California was confronted with the problem of seawater intrusion into the Oxnard aquifer. The Oxnard aquifer is a major source of water for agricultural activities in the Oxnard Plain. The most viable alternative to control seawater intrusion and provide irrigation water for farming was reclamation of waste water from the Oxnard Treatment Plant. The advanced wastewater treatment system that will treat the secondary effluent from the Oxnard Treatment Plant will include coagulation, multimedia filtration, RO and chlorination (Figure 10). A pilot plant system is being developed to determine the ability of the proposed system to meet quality requirements for injection and public health protection.

Approximately 46% of the reclaimed water will be transported to the RO facility for desalination. The RO process will employ spiral wound cellulose acetate membrane modules. The filtered secondary effluent will be pressurized to about 500 psi and passed through the membranes. Chemical pretreatment of influent stream will be used for pH control and to prevent scaling on the

Table II. Probability (in %) of Meeting Various Hypothetical COD Criteria at Different Sampling Points at Water Factory 21 During the Thrid Period [21]

Sampling Point	Hypothetical COD Criteria				
	2 mg/ℓ	5 mg/ℓ	10 mg/ℓ	20 mg/ℓ	30 mg/ℓ
Plant Influent	0.0	0.0	0.0	0.0	1.0
Lime Effluent	0.0	0.0	0.0	0.5	82
Filter Effluent	0.0	0.0	0.0	9.2	96
Activated Carbon Effluent	0.0	0.2	25	94	99.8
RO Effluent	69	95	99.8	99.97	>99.99
Injection Water	1.2	35	85	99.3	99.9

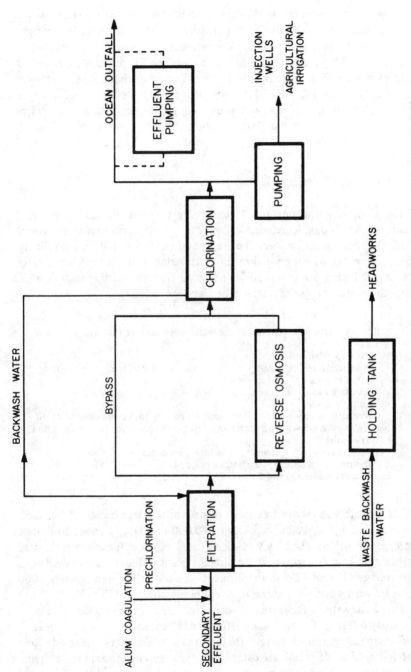

Figure 10.   Process schematic of injection/extraction well system seawater intrusion control barrier [22].

membrane. Approximately 85% of the influent water will be recovered with a residual pressure of 30-50 psi; the remaining 15% brine solution will be rejected at a pressure ranging up to 450 psi. A turbine generator or pump may be used to possibly recover some of the energy. The RO product water will be blended with the filtered secondary effluent. The resultant reclaimed water will have dissolved solids < 1200 mg/ℓ. The quality of the product water from the advanced treatment processes will be low enough to permit use for crop irrigation or injection into the Oxnard aquifer for seawater intrusion control. The waste brine from the RO process will be discharged into the ocean.

Petromin of Saudi Arabia [23]

The General Petroleum and Minerals Organization (Petromin) of Saudi Arabia was expanding its refinery capacity in Riyadh to process an additional 100,000 barrels of oil per day. The increase in water required for the refinery expansion was to be supplied through reclamation of the Riyadh wastewater effluent. Drinking water was to be obtained from an existing shallow well. The refinery water requirements can be classified as follows:

1. utility water for hose station and fire water which has the following characteristics:
   a. low suspended solids;
   b. biologically and chemically stable for storage in steel tanks to prevent corrosion and odor problems;
   c. positive Langlier Index; and
   d. nonpathogenic.
2. intermediate quality water which is used for cooling tower makeup and desalting (refinery process). This water has a total dissolved solids concentration (TDS) less than 500 mg/ℓ.
3. boil feed water which is subject to the following conditions:
   a. electrical conductivity < 20 μmhos/cm; and
   b. silica concentration < 0.5 mg/ℓ.

The Riyadh wastewater treatment plant is a two-stage trickling filter facility with a design hydraulic capacity of 40,000 m³/day. Plans are to double the capacity of the plant by converting the two-stage filtration process to single-stage. Aerated lagoons will polish the plant effluent prior to discharge. The average value of TDS in the influent wastewater is approximately 3000 mg/ℓ. Influent wastewater characteristics are presented in Table III.

The Wastewater Reclamation Facility will treat up to 20,000 m³/day of secondary effluent. The secondary effluent will be initially chemically treated. The chemical treatment system (Figure 11) consists of two aerated, completely-mixed surge ponds, first-stage lime treatment and second-stage chemical treatment. The first-stage lime treatment and the second-stage chemical

Table III. Influent Wastewater Characteristics [23]

Parameter	Value
Flow Rate (m^3/d)	20,000
pH	7.0-8.0
Temperature (°C)	43 maximum
Dissolved Oxygen (mg/ℓ)	0
Ammonia (mg/ℓ as Nitrogen)	55 maximum
SiO_2 (mg/ℓ)	35 maximum
Suspended Solids (mg/ℓ)	194 maximum
Biochemical Oxygen Demand (mg/ℓ)	225 maximum
Chemical Oxygen Demand (mg/ℓ)	275 maximum
Total Dissolved Solids (mg/ℓ)	3300 maximum
Total Hardness (mg/ℓ as $CaCO_3$)	1060 maximum
Total Alkalinity (mg/ℓ as $CaCO_3$)	230 minimum
Calcium (mg/ℓ)	268 maximum
Magnesium (mg/ℓ)	103 maximum
Sodium (mg/ℓ)	995 maximum
Iron (mg/ℓ)	2 maximum
Sulfate (mg/ℓ)	1000 maximum
Chloride (mg/ℓ)	650
Phosphate (mg/ℓ)	15

treatment are designed to soften the water and reduce the particulate and colloidal solids. The pH in the two second-stage recarbonation basins will be adjusted from ∿ 9.5 to 8.5. Sulfuric acid is added to reduce the alkalinity to stabilize the water and prevent precipitation of calcium carbonate in water storage tanks and subsequent treatment processes. Following chemical treatment, the water is stored in utility water storage tanks.

The chemically treated and chlorinated waste water is pumped from the storage tanks to a cooling tower which reduces the water temperature from ∿ 45°C to 25°C (Figure 12). The temperature reduction is required to minimize degradation and compaction of the membrane. The pH of the water into the tower can be adjusted from 8.5 to 6.0 with sulfuric acid to limit calcium carbonate scaling and prevent delignification and weakening of weed packing. Polymers will be added to the discharge water from the cooling towers. Polymers will aid in removal of colloidal or particulate solids. The water will be periodically chlorinated to prevent algal growth on the gravity filters.

The wastewater will be transported through activated carbon columns after gravity filtration. The carbon adsorption system will reduce the organic loading to the RO system, thereby reducing downtime and prolonging the life of the membranes. After filtration and carbon adsorption, the discharge water will have a COD of 10-20 mg/ℓ, and a turbidity of 0.5 turbidity units.

The effluent from the physical treatment system will be dechlorinated (Figure 13). Secondary RO system brine reject water will be mixed with

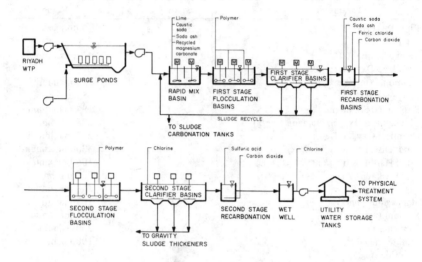

Figure 11. Chemical treatment plant [23].

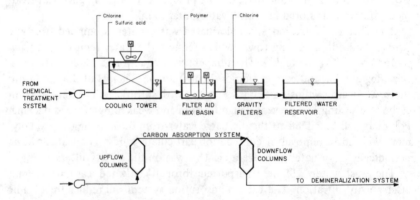

Figure 12. Physical treatment system [23].

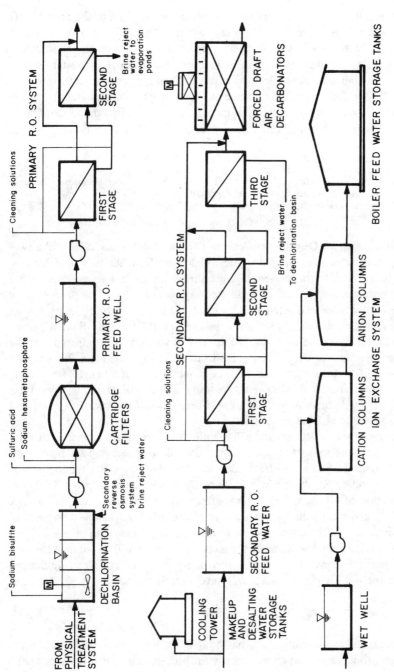

Figure 13. Demineralization system [23].

dechlorinated water and pumped through cartridge filters to the primary RO feed wet well. Sulfuric acid will be used to depress the pH to approx. 5.7 which will reduce the potential for calcium carbonate sealing on the membranes.

Spiral wound membranes will be incorporated into the RO system. Spiral wound membranes were selected because of their greater resistance to fouling and the extensive testing of these modules in pilot wastewater treatment plants. Polyamides were selected as membrane material. Polyamide membranes were selected rather than cellulose acetate for the following reasons:

1. increased water permeability rate;
2. increased chemical stability;
3. increased operating pH range (3-12 compared to 4-7 for cellulose acetate);
4. increased resistance to biological degradation of membrane;
5. increased temperature range; and
6. increased resistance to compaction.

The primary RO system consists of five treatment trains, each with a two-stage arrangement of pressure tubes. The reject from the first-stage tubes is the feed stream to the second-stage pressure tubes. Each is designed to achieve 50% water recovery with an overall recovery of 75%.

The hydraulic feed rate to each of the RO trains will be 173.3 m^3/hr. At 75% recovery, the water production rate will be 129.9 m^3/hr. Four RO treatment trains will produce $\sim$ 12,470 m^3/day of water. It is anticipated that the system will reduce the influent TDS from 2600 to $<$ 200 mg/ℓ. The product water TDS concentration is below the required TDS of 500 mg/ℓ for cooling tower and refinery desalter makeup water. A total of 4160 m^3/day with a TDS of approximately 9800 mg/ℓ will be rejected from four RO trains and transported to evaporation ponds.

The secondary RO system will include three treatment trains. Each train will have three pressure vessels in series. Overall water recovery will be $\sim$ 90%. Each train will be able to treat 2380 m^3/day of primary RO product water. The secondary RO system will produce water with an anticipated TDS concentration of 40 mg/ℓ. Reject brines from each train will contain 1600 mg/ℓ of TDS and will be discharged at a rate of 240 m^3/day. The reject stream will be blended with the influent stream to the primary RO system.

The RO product water from the secondary system will be pumped through two parallel trains of two bed ion exchange columns. The water discharged from the ion exchange process will have a specific conductivity $<$ 20 μmhos.

Beckman Instruments Plating Plant [24]

Beckman Instruments, Inc., has built a facility at Porterville, California, which produces printed-circuit boards for process and scientific instruments.

The facility processes $\sim$ 1300 m^2 of material per month. Water is obtained from Porterville city water and from recirculated, contaminated rinse water. The water requirement of the plant was $\sim$ 5700 m^3/month. However, water requirements were increasing and the availability of additional water was extremely limited. In order to reduce supplemental water requirements, the firm decided to: (1) reduce overall water usage, (2) improve the quality of the recirculated rinse water, and (3) minimize or eliminate waste discharge.

Wasted rinse water from the plating shop is divided into two streams. A flow of $\sim$ 9.5-11.4 m^3/day is transported directly to solar ponds. This waste stream consists mainly of rinse water containing tin, lead and palladium salts, pumice, silicates or methylene-chloride. The other stream, which has a flow of $\sim$ 45-57 m^3/day, is pumped to a RO system (Figure 14). Contaminants in the waste stream transported to the solar ponds would either degrade the RO membranes or pass through the membranes unaffected. Therefore, no attempt is made to reclaim this wastewater.

The plating-wastewater being reclaimed proceeds to a pretreatment operation prior to the RO unit. The average pH of the wasted rinse water is $\sim$ 3.2; therefore, caustic soda is added to adjust the pH of the waste stream to between 4.5 and 5.1. Following pH adjustment, the waste stream passed through diatomaceous-earth filters and subsequently 5-μm and 0.5-μm cotton cartridge filters.

The RO unit consists of two 20-cm polyamide membranes in series, followed by two 10-cm cellulose triacetate membranes in parallel. Hydraulic handling capacity of the RO unit is 136 m^3/day at a design efficiency of 90%. The influent stream is supplied to the membranes at 2758 kN/m^2 (400 psi). Wastewater temperature must be $<$ 32°C.

The membranes have an estimated life span of three years. One membrane is removed from service each week for maintenance. The membrane is flushed with a circulating rinse of a solution of 2% ethylenediaminetetraacetic acid and 0.1% cleansing agent. The membrane is then rinsed with demineralized water. If bacterial growth is observed on the membranes, it is submerged in a 1% formaldehyde solution prior to flushing.

The conductivity of the influent stream to the RO unit is $\sim$ 800-1000 μmhos. The concentrate from the RO unit is recycled to the pretreatment cycle. The RO permeate conductivity is $\sim$ 35 μmhos. City water which has been treated by anion, cation and mixed-bed columns is added to the RO product water. The reclaimed plating-wastewater and makeup water are polished by carbon adsorption and mixed-bed exchange columns. The waste concentration from the RO unit is disposed of in a solar pond.

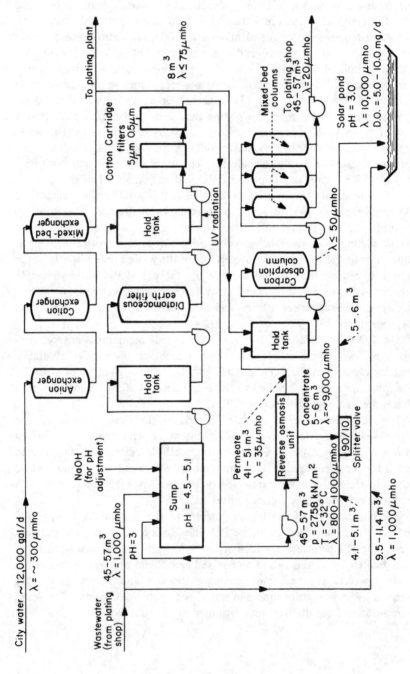

Figure 14. Reclamation system extracts full benefit of its sunny locale [24].

Denver Plant [25]

Natural water supplies serving the Denver, Colorado, metropolitan area are insufficient to meet the current and future water demands of the area. Therefore, the city of Denver has initiated a program to evaluate the reuse of secondary effluent from the Metropolitan Denver Sewage Disposal District Number 1 facility for potable water demands. A potable reuse demonstration facility is currently under construction. The treatment train is illustrated in Figure 15. Unchlorinated secondary effluent will be treated with a high lime dose to promote flocculation of colloidal and particulate solids, heavy metals, pathogenic organisms and phosphorus. Solids-liquids separation of the chemical floc particles will then occur followed by recarbonation with carbon dioxide. The discharge from the lime treatment will flow to an equalization basin to equalize pump withdrawals.

From the basins the waste stream will be pumped to chemically aided dual media pressure filters for polishing. Ammonia in the waste stream will then be removed by passing the flow through selective ion exchange beds containing the natural zeolite, clinoptilolite.

Following ion exchange, the waste stream will pass through two carbon adsorption columns in series. Ozone will be introduced between the adsorption columns to enhance the second adsorption step.

The water will be split after the second carbon adsorption column and 3400 m^3/day will be made available for industrial use. The remaining portion of flow (378 m^3/day) will be further treated by reverse osmosis for removal of dissolved solids. The permeate from the RO unit will then be disinfected with chlorine dioxide.

The design of the facility began in July 1979 and is scheduled to be ready for operation in the fall of 1982. The plant will be operated and evaluated continuously for four and a half years.

REFERENCES

1. Lee, M. K., and L. H. Reuter. "Multipurpose Water Treatment System Development," in *Water Reuse Symposium* (Denver, CO: AWWA Research Foundation, 1979), pp. 1276-1311.
2. Wojcik, C. K., J. G. Lopez and J. W. McCutchan. "Renovation of Municipal Wastewater by Reverse Osmosis: A Case Study," in *Water Reuse Symposium* (Denver, CO: AWWA Research Foundation, 1979).
3. Hrubec, J., J. C. Schippers and B. C. J. Zoeteman. "Studies on Water Reuse in the Netherlands," in *Water Reuse Symposium* (Denver, CO: AWWA Research Foundation, 1979), pp. 785-807.
4. Weber, W. J., Jr. *Physicochemical Processes for Water Quality Control* (New York, NY: Wiley-Interscience, 1972).

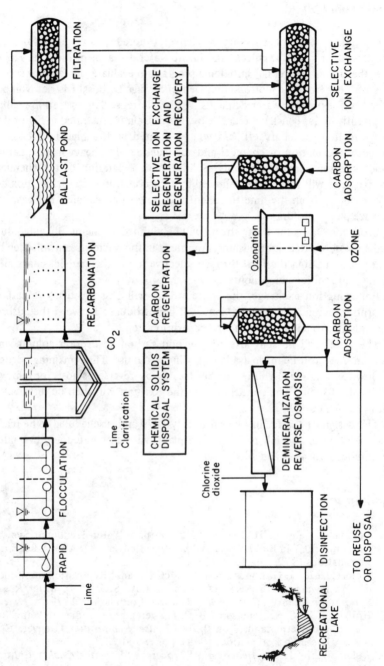

Figure 15. Process flow diagram of the proposed Denver plant [25].

5. "Desalting Handbook for Planners," 2nd ed. U.S. Department of the Interior, Office of Water Research and Technology, OWRT TT/80.3 (1979).
6. Friedlander, H. F., and R. N. Rickles. "Membrane Separation Processes, Parts I-VI," *Chem. Eng.* 73(5):111-116, 73(7):121-124, 73(9):163-168, 73(11):153-156, 73(12):145-148, 73(13):217-224 (1966).
7. Merten, U., Ed. *Desalination by Reverse Osmosis* (Cambridge, MA: The MIT Press, 1966).
8. Spiegler, K. S., Ed. *Principles of Desalination* (New York, NY: Academic Press, 1966).
9. Tuwiner, S. B., L. P. Miller and W. E. Brown. *Diffusion and Membrane Technology* (New York, NY: Reinhold, 1962).
10. Rickles, R. N. *Membranes–Technology and Economics* (Park Ridge, NJ: Noyes Development, 1967).
11. "Reverse Osmosis," Office of Water Research and Technology. U.S. Department of the Interior (1977).
12. Perry, R. H., and C. H. Chilton. *Chemical Engineer's Handbook,* 5th ed. (New York, NY: McGraw-Hill Book Co., 1973).
13. Ott, E., H. M. Spurlin and M. W. Grafflin, Ed. *High Polymers, Vol. V, Cellulose and Cellulose Derivatives* (New York, NY: Interscience Publishers, Inc., 1954).
14. Baker, W. O., C. S. Fuller and N. R. Pape. "Effects of Heat, Solvents, and Hydrogen-bonding Agents on the Crystallinity of Cellulose Esters," *J. Am. Chem. Soc.* 64:776-782 (1942).
15. Reid, C. E., and E. J. Breton. "Water and Iron Flow Across Cellulose Membranes," *J. Appl. Polymer Sci.* 1(2):133-143 (1959).
16. Vincent, A. L., M. K. Barsh and R. E. Kesting. "Semipermeable Membranes of Cellulose Acetate for Desalination in the Process of Reverse Osmosis. III. Bound Water Relationships," *J. Appl. Polymer Sci.* 7:2363-2378 (1965).
17. Shields, C. P. "Five Years Experience with Reverse Osmosis Systems using Dupont "Permasep" Permeators," in *Desalination* (Amsterdam, Netherlands: Elsevier Scientific Publishing Co., 1979), pp. 157-1979.
18. Sudak, R. G., M. E. McKee, R. L. Fox and J. B. Bott. "Development of a Chlorine Resistant Reverse Osmosis Membrane and Porous Membrane Supports," Office of Water Research and Technology, U.S. Department of the Interior, Contract No. 14-34-0001-8510 (1979).
19. Curran, H. M., D. Dykstra and D. Kenkeremath. "State-of-the-Art of Membrane and Ion Exchange Desalting Processes," Office of Water Research and Technology, U.S. Department of the Interior, Contract No. 14-34-0001-6705 (1976).
20. Cline, N. M. "Groundwater Recharge at Water Factory 21," in *Water Reuse Symposium* (Denver, CO: AWWA Research Foundation, 1979), pp. 139-166.
21. McCarty, P. L., D. G. Argo and M. Reinhard. "Reliability of Advanced Wastewater Treatment," in *Water Reuse Symposium* (Denver, CO: AWWA Research Foundation, 1979), pp. 1249-1269.
22. Wedding, J. J., M. A. Fong and C. M. Crafts. "Use of Reclaimed Wastewater to Operate a Seawater Intrusion Control Barrier," in *Water Reuse Symposium* (Denver, CO: AWWA Research Foundation, 1979), pp. 639-661.

23. Kalinske, A. A. "Reclamation of Wastewater Treatment Plant Effluent for High Quality Industrial Reuse in Saudi Arabia," in *Water Reuse Symposium* (Denver, CO: AWWA Research Foundation, 1979), pp. 958-992.
24. Warnke, J. E., K. G. Thomas and S. C. Creason. "Wastewater Reclamation System Ups Productivity, Cuts Water Use," *Chem. Eng.* 84(7):75-77 (1977).
25. Work, S. W., M. R. Rothberg and K. J. Miller. "Denver's Potable Reuse Project: Pathway to Public Acceptance," *J. Am. Water Works Assoc.* 72(8):435-440 (1980).

CHAPTER 11

IMPLEMENTING DIRECT WATER REUSE

Ju-Chang Huang
Department of Civil Engineering
Environmental Research Center
University of Missouri-Rolla
Rolla, Missouri 65401

It has been more than a decade since the need for direct water reuse was recognized in the United States. However, the matter has not been pursued seriously and vigorously until the recent severe drought in California. The Federal Clean Water Act of 1977 (PL 95-217) also gave impetus to wastewater reclamation and reuse by offering incentives of federal sharings of the waste treatment construction grants.

Actually, water reuse is not new in this nation although most reuse has been limited to industrial and agricultural applications. The largest industrial reuse is for cooling, but reuse for process water is somewhat limited. Agricultural reuse for supplemental irrigation has been increasing rapidly in recent years and reuse of treated municipal waste water for groundwater recharge is also being practiced today in some parts of the nation. However, direct reuse of treated water for potable purposes has not been attempted so far, although its need has been recognized for many years, particularly in the semi-arid southwest United States.

Perhaps many people even doubted in the past whether we really had a need for practicing direct water reuse. The recent episode of severe water shortage in California, which finally led to strict water rationing, has hopefully convinced these people that it is time now for us to seriously consider direct water reuse. It must be pointed out, however, that direct water reuse is not needed in all parts of the nation. As long as there is enough natural water of suitable quality, there is no need to practice direct reuse. But there are regions where the amount of naturally available waters is not adequate to meet the local demand. Water shortages have been experienced in the southwest and other places either perennially or occasionally. Inadequate water supplies reduce the quality of life, and often are major factors in limiting municipal, industrial and agricultural growth. Under such situations, direct water reuse should be implemented to augment the inadequate natural water resources. This chapter discusses various steps and considerations which must be taken in order to speed up the progress toward implementation of direct water reuse.

TECHNICAL FEASIBILITY

Is direct water reuse technically feasible today? The answer is generally "yes". In the last two decades, significant advances have been made in the technology of water and wastewater treatment. The conventional types of pollutants such as BOD, SS, coliforms, pathogens, nitrogen and phosphorus, etc., can all be removed effectively from waste water. Many treated effluents even have lower concentrations of these pollutants than are commonly found in natural surface waters. Specific examples cited in the following paragraphs illustrate the high quality of effluents that have been produced by several full-scale or pilot-scale advanced waste treatment plants.

Perhaps the first and most publicized high-quality effluent is one produced by the South Tahoe Advanced Waste Treatment Plant in California. Through the use of chemical clarification, ammonia stripping, activated sludge biological process, sand filtration, carbon adsorption and chlorine disinfection, the plant reduced its effluent BOD_5 to 1 mg/l, SS to 0 mg/l, total coliforms to less than 2.0 MPN/100 ml and virus to 0 PFU, as shown in Table I [1].

Table I. Treatment Efficiencies Achievable by Two Early Advanced Plants

Effluent Parameters	South Tahoe Plant (7.5 mgd)		Ahuimanu Plant (1.4 mgd)	
	Effluent Concentrations	Efficiency (%)	Effluent Concentrations	Efficiency (%)
pH	6.8-8.7	–	7.0-7.8	–
BOD_5 (mg/l)	1	99.6	3	98.7
SS (mg/l)	0	100	1	99.6
Total Kjeldahl Nitrogen (mg/l N)	–	–	0.09	97.0
Total Nitrates (mg/l N)	–	–	3.3	–
Total Phosphorus (mg/l P)	0.6	99.6	1.5	90.3
Total Coliform (MPN/100 ml)	2.0	99.9+	43	99.99+
Virus Recovered (PFU)	0	100	–	–
Dissolved Oxygen	–	–	6	–

Similar high-quality effluent had also been produced by another early full-scale advanced waste treatment plant, Ahuimanu Plant in Hawaii. This writer was personally involved in its start-up operation for a period of six months in 1972 and 1973 [2]. The plant employed a treatment train similar to that of the South Tahoe Plant, but had no sand filtration or carbon adsorption. During the six-month start-up period, the treated effluent was also of high quality, as shown in Table I. That is, the effluent BOD$_5$ was reduced to 3 mg/l and SS to 1 mg/l. Although neither of these plants had any intention of reusing their treated effluents for domestic purposes, the produced effluents were qualified for serving as raw water sources according to the standards promulgated by the U.S. Public Health Service in 1962 [3]. Of course, the National Interim Primary Drinking Water (NIPD) Regulations, issued by the U.S. Environmental Protection Agency (EPA) in 1976, would require expansion of the effluent quality monitorings if these two plants would ever intend to practice direct reuse for drinking purpose.

A more recent example of high-quality effluent is the one produced by Water Factory 21 in Orange County, California. The plant has a capacity of 15 mgd, and employs a treatment train of lime precipitation/clarification, air stripping, recarbonation, filtration, activated carbon adsorption, disinfection and reverse osmosis (for one-third of the flow) to improve the quality of secondary biological effluent so that it can be used to provide the injection water for a hydraulic seawater barrier system. In 1976-1978, without the use of reverse osmosis, the geometric mean concentrations of all contaminants in the treated effluent met the 1976 EPA National Interim Primary Drinking Water Regulations, as shown in Table II [4]. Only chromium and mercury did not meet the regulations more than 98% of the time. If reverse osmosis were used, there is little doubt that these two metals would be further reduced to more acceptable levels.

Current drinking water regulations cover few organic materials, such as the pesticides listed in Table II and a current proposed maximum level of 100 μg/l for trihalomethane. However, there is always a legitimate concern over other trace organic substances that may be present in the reclaimed wastewater. McCarty [4] carried out extensive monitoring of organic distributions at various sampling locations at Water Factory 21. The probabilities of different water samples meeting various COD criteria are shown in Table III. It is evident that lime treatment, activated carbon adsorption and reverse osmosis (RO) are most effective in removing COD. Although the RO process is expensive, it produces water with a geometric mean COD of about 2 mg/l (equivalent to about 0.8 mg/l organic carbon), which is as low as a value found in many water supplies in the United States [5]. In addition to the COD concentrations, McCarty et al. [4] also analyzed the treated effluent for many trace organics which are contained in the current EPA list of priority pollutants. Since the maximum contaminant levels (MCL) for these trace organics

Table II. Comparison of National Interim Primary Drinking Water (NIPD) Regulations
with Water Factory 21 Effluent Quality

Contaminant	NIPD Limits	Water Factory 21 Effluent	
		Geometric Mean	98% of Time Less Than
Arsenic (mg/l)	0.05	0.005	
Barium (mg/l)	1.0	0.03	0.08
Cadmium (mg/l)	0.01	0.001	0.006
Chromium (mg/l)	0.05	0.02	0.09
Lead (mg/l)	0.05	0.002	0.036
Mercury (mg/l)	0.002	0.0017	0.012
Nitrate (mg/l)	10	10	10
Selenium (mg/l)	0.01	0.004	?
Silver (mg/l)	0.05	0.003	0.007
Fluoride (mg/l)	1.4	0.6	1.2
Coliforms (MPN/100 ml)	1	0.01	14
Endrin ($\mu g/l$)	0.2	0.01	0.01
Lindane ($\mu g/l$)	4	0.05	0.05
Toxaphene ($\mu g/l$)	5	0.01	0.01
2,4,5-TP ($\mu g/l$)	10	0.01	0.01
Methoxyclor ($\mu g/l$)	100	0.1	0.1
Turbidity (TU)	1	0.4	1.1

in drinking water have not been established, they used a strict, hypothetical MCL of 1 $\mu g/l$ for nonchlorinated trace organics and 0.5 $\mu g/l$ for the chlorinated compounds as a basis for judging the fitness of the reclaimed water for potable purpose. Based on such a criterion, McCarty et al. found that after treatment through activated carbon adsorption and disinfection, only some of the chlorinated methanes and ethanes and some of the phthalates exceeded the hypothetical MCL values more than 10% of the time. The chlorobenzenes, aromatic hydrocarbons and pesticides were all efficiently and reliably removed by treatment. From the operational experience obtained at Water Factory 21, it appears that current waste treatment technology has advanced to the point where it is capable of producing an effluent, on a full-scale plant, of sufficiently high quality for potable reuse.

In addition to the full-scale establishment, numerous pilot-scale and smaller laboratory-scale studies have attempted to establish a suitable treatment scheme to produce an effluent of potable quality [6-8]. The Denver Water Department initiated a demonstration project in 1975 to evaluate the effectiveness and reliability of a certain treatment train to produce reusable water for drinking purposes [6]. After the demonstration study, a full-scale water reclamation plant using a process train as shown in Figure 1 was proposed, and its operation is targeted in 1983. By the year 2000, it is expected that the reclaimed water will provide up to 15% of Denver's municipal water supply.

Table III. Percent Probability of Different Water Samples Meeting Various COD Criteria at Water Factory 21

Sampling Point	Hypothetical COD Criteria				
	2 mg/l	5 mg/l	10 mg/l	20 mg/l	30 mg/l
Plant Influent	0.0	0.0	0.0	0.0	1.0
Lime Effluent	0.0	0.0	0.0	0.5	82
Filter Effluent	0.0	0.0	0.0	9.2	96
Activated Carbon Effluent	0.0	0.2	25	94	99.8
Reverse Osmosis Effluent	69	95	99.8	99.97	99.99

A smaller-scale wastewater recycling system was developed in Boulder, Colorado, by the Purecycle Corporation [7]. The treatment system employs anaerobic/aerobic biological processes followed by membrane ultrafiltration (using a pore size of 50 Å), activated carbon adsorption and UV disinfection with an irradiation energy of over 300,000 Ultrads (microwatt seconds/cm^2). The treated water is reported to have a quality surpassing the EPA NIPD Regulations, as shown in Table IV [7].

In a laboratory-scale exploratory study using alum coagulation/clarification, rapid sand filtration, reverse osmosis and ozonation to treat a municipal secondary effluent, a final effluent with TDS of 17 mg/l, TOC of 2 mg/l was produced [8]. The product water was claimed to have a quality comparable to or better than that of distilled water in terms of TDS, TOC, turbidity and color.

These examples indicate that, based on the 1976 EPA NIPD regulations, many of the full-scale and pilot-scale advanced waste-treatment facilities have been able to produce effluents either nearly or already meeting the NIPD drinking water requirement. However, it must be realized that the NIPD regulations have not been developed to address the special problems associated with the direct reuse of treated municipal effluents. The main concern here is the potential presence of trace organic materials and virus particles which

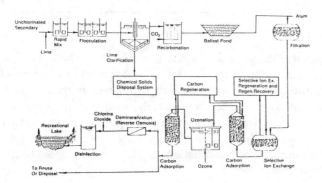

Figure 1. Proposed treatment scheme for the new Denver Water Reclamation Plant.

Table IV. Reclaimed Water Qualities by the Purecycle System

Parameter	Purecycle Operating Levels
Nitrate (as mg/l N)	0.05
Turbidity (NTU)	0.15
Chloride (mg/l)	1
Fluoride (mg/l)	0.01
Iron (mg/l)	0.01
Magnesium (mg/l)	0.01
Sulphate (mg/l)	0.01
Total Dissolved Solids (mg/l)	2
Zinc (mg/l)	0.01
Total Organic Carbon (mg/l)	1
Calcium (mg/l)	0.1
Sodium (mg/l)	1
Ammonia (mg/l)	0.01
pH	6
Phosphorus (mg/l)	0.01
Conductivity (micromhos/cm)	3
Fecal Coliforms (MPN/100 ml)	0
Silica (mg/l)	0.01
Arsenic (mg/l)	0.01
Barium (mg/l)	0.01
Cadmium (mg/l)	0.01
Chromium (mg/l)	0.01
Lead (mg/l)	0.01
Mercury (mg/l)	0.002
Selenium (mg/l)	0.01
Silver (mg/l)	0.01
Copper (mg/l)	0.01
Manganese (mg/l)	0.01
Endrin (mg/l)	0.0002
Lindane (mg/l)	0.004
Methoxychlor (mg/l)	0.1
Toxaphene (mg/l)	0.005
Chlorophenoxys (2,4-D) (mg/l)	0.1
2,4,5-TP Silvex (mg/l)	0.01

these treated effluents may contain beyond safety limits. In light of this, determinations of the "appropriateness" of direct water reuse must be based on comparisons of the treated effluent with a new set of criteria developed for such a special purpose. Unfortunately, such criteria are not available at this time. But if the waste water is treated by a scheme similar to that used by the Water Factory 21 and polished by reverse osmosis with a fine pore size plus further disinfection by ozonation, the final effluent should be safe enough for direct potable reuse.

PSYCHOLOGICAL PROBLEMS

Slow advancement toward direct water reuse is not the result of delayed progress in treatment technologies, but rather of psychological unreadiness among the general public.

Waste water has been invariably linked with a psychological implication of "filth" and "dirtiness," no matter how thoroughly it has been treated. Although there is concrete evidence that harmful impurities in waste water can be reduced to an acceptable level by today's treatment technologies, the public has not been convinced (and the sanitary engineering profession has not actively been trying to convince them either) that the properly treated effluent can be reused for domestic purposes. In fact, many people have been deeply imbedded with such an attitude that "if no other people drink the treated sewage effluent first, I am not going to be the first one to try!" But if some daredevil is willing to try the reclaimed effluent, others will follow him. Three specific examples of this attitude follow:

Example 1: On June 29, 1973, when Hawaii's first advanced waste treatment plant, the Ahuimanu Plant, was dedicated, about 100 individuals representing different sectors of the public were invited to participate in the event. Before the dedication ceremony, all participants were overwhelmingly impressed with the absolute clarity of the treated effluent—just like tap water —but no one was willing to drink it when asked to. At the end of the ceremonial address, the design engineer of the plant took a bold step by making a toast to the participants with the plant effluent. Because of his voluntary move, more than 20 people, including TV reporters and cameramen, were either curious or brave enough to try the same effluent that day. Most important of all, none of these individuals showed any subsequent illness.

Example 2: In April 1975, this writer was invited to give a talk to the Science Club of West Plains High School in Missouri. The topic was on the "Feasibility and Risks of Drinking Sewage Effluent!" The writer brought with him two large bottles of tap water, but one was intentionally mislabeled as "sewage effluent" while the other was correctly identified as "tap water". At the end of his talk, the author emphatically pointed out to the 40 high school students that the "sewage effluent" had been thoroughly treated in the Sanitary Engineering Research Laboratories at the University of Missouri-Rolla, and that there was absolutely no risk or danger involved in drinking such a sewage effluent sample. He then asked for any volunteer from his audience who was willing to try such an effluent first. For 5 minutes the students looked at each other and jokingly tried to volunteer other fellow classmates to make such a first move, but no individual was willing to step out and give it a try. Finally, the writer had to "volunteer" himself to take the first drink. Within 15 minutes, all of the 40 students had tried the "sewage effluent," at least half of a glassful each. This episode clearly reflects that the psychological barrier, even among our "liberal" youngsters, is the major obstacle delaying the progress of direct water reuse.

Example 3: The writer participated in an international conference on advanced treatment and reuse of wastewater, held in Johannesburg, South Africa, in June 1977. The conference participants had an opportunity to visit the Stander Water Reclamation Plant on the outskirts of Johannesburg. The treatment plant had 4,500 m^3/day full-scale facilities consisting of conventional secondary treatment plus ammonia stripping, recarbonation, chemical clarification, chlorination/ozonation and carbon adsorption. The treated effluent was claimed to be of potable quality. It was a "traditional custom" for visitors to try the plant effluent and log their names down on a registration book. Since the book had already carried thousands of names in it, none of the 300 conference participants seemed to show any hesitation in trying the plant effluent. It is hoped that the willingness of this international professional group to drink the treated effluent is a reflection of their confidence in today's technical readiness for direct potable reuse.

These examples suggest that a considerable psychological barrier exists in today's society toward direct water reuse. A recent survey study conducted in Irvine and Anaheim, California, to evaluate the educational and social factors affecting the public acceptance of reclaimed water [9] found formal education to be the most important factor in public awareness of water resources issues and in acceptance of direct water reuse. Individuals formally educated in the sciences (biology, health/medical sciences, physical sciences and social sciences) had a more favorable attitude toward reclaimed water than those educated in the humanities or business. Common psychological attitudes which affect an individual's acceptance of reclaimed water include aversion to the unclean, warning of health risks in childhood and faith in technology. To overcome these psychological difficulties, a proper health education program should be developed to reduce the negative impacts of childhood health risks as well as to increase awareness of today's advanced technology which is able to remove all harmful pollutants from waste water. Perhaps the most effective way of convincing the general public that we are now ready to practice direct water reuse would be to have some persons volunteer to practice direct reuse under broad coverage of TV and newspaper media. Logically, the volunteer should be the professional group actively involved in the treatment of water and wastewater.

ACTIONS TO BE TAKEN

The public is not aware that almost all of the impurities in waste water can be removed by today's advanced waste treatment technologies. Intensive public education plus an enthusiastic professional campaign are necessary so that the public can be fully convinced that waste water, like a piece of equipment or other merchandise, can be "repaired" or "renovated" to make it clean and wholesome again.

In carrying out the public education and professional campaign, the following actions warrant serious consideration:

1. Consideration should be given to change some of the common terminology used today which either explicitly or implicitly links waste water with a sense of filth and dirtiness, such as:

Existing Terms	Suggested New Terms
wastewater or sewage	used water (making it synonymous to used car or other merchandise that can be fixed and reused)
waste treatment plant	water renovation plant or water reclamation center
treated sewage effluent	renovated water, or reclaimed water
pollutant or contaminant in water	impurities in water
sewage sludge	municipal humus
sewage treatment plant operator	water renovator

The main objective of the new terms is to impress on the public that "waste water" is just like a "used car" or a "used typewriter" that can be fixed by removing its impurities and reused. It is hoped that with these new terminologies, the public will gradually be convinced that waste water is really not that awful, that it has only been "used" and can be cleaned up for subsequent reuse.

2. The public must be informed through education and public news media that "water reuse" is not new at all. We have been reusing our wastewaters for many decades. For an American city experiencing only average reuse of its surface supply, half of its water has passed through a municipal or industrial outfall. For other cities in the upper 10% reuse bracket, supplies have been

used at least three times previously. Of course, such reuse is neither deliberate nor direct; it merely occurs from one community to another as water flows downstream through surface or underground drainage systems. However, in a dry season, most of the water in a natural stream may be from municipal sewage effluents. Under such a situation, direct reuse is already in existence.

3. The environmental and sanitary engineering profession should actively publicize the fact that today's advanced water reclamation technologies are adequate to renovate wastewater to potable qualities. More importantly, they should take a lead in demonstrating their willingness to drink the reclaimed water.

4. More extensive research must be undertaken immediately to assess the real toxicological implications of the various impurities present in wastewater effluent. At present our knowledge in this field is still in its infancy; therefore, many new standards established today may be unduly conservative. This definitely makes direct water reuse more difficult. It must be pointed out here that in many underdeveloped countries, people are drinking water from fairly polluted sources without adequate treatment. Yet there is still no concrete evidence that these people have serious chronic toxicity problems (except bacterial and viral infections) that can be linked directly with the chemical impurities present in their polluted drinking waters. Since the toxicological implications of various impurities may not be able to be defined precisely in a foreseeable future, a thorough scientific investigation should be initiated by the government to make a statistical correlation between the chemical qualities of drinking water and the resultant chronic toxicity, if any and if identifiable, in some underdeveloped countries. This will help us to determine the advisability of establishing more stringent chemical qualities of drinking water in the future.

5. Since direct reuse of reclaimed water for drinking purpose is somewhat personally sensitive for either health or aesthetic reason, a complete closed system for such a reuse may not be easily accepted by the water user in the immediate future. It is recommended that the reclaimed water be first used to feed swimming pools or man-made ponds and reservoirs. In the latter case, sensitive game fish may be maintained to provide a reassuring indicator that the pond water is indeed safe for human consumption. This would actually introduce a "natural barrier" in the reuse approach. Such a natural barrier, if its detention time is two weeks or more, can serve at least three major functions in safeguarding the public health:

   a. Exposure of the reclaimed water to an open environment allows the natural purification process to proceed. This includes a combination of sunlight disinfection of harmful organisms and viruses and adsorption of trace substances and virus particles by soil materials. A recent laboratory study found that of 40 selected representative soils and minerals, most of them were good adsorbents for Poliovirus type 2. Only a few have shown an adsorption efficiency of less than 95%. Therefore, placing reclaimed water in an earth basin for a certain period of time before its reuse will likely further reduce the health risk associated with viral infection.

b. The storage basin minimizes the health risk associated with fluctuations of the reclaimed water qualities. This is important because water reclamation facilities always show certain variations in their performance. With an equalization basin, the extent of water quality variations will be greatly dampened.

c. If any pollutant exists at a harmful level, a first warning sign would be indicated by the fish kills in the storage basin. Under such a situation, an emergency measure can be taken immediately to protect the public health.

Of course, after the water is withdrawn from the holding pond, it has to be treated through a conventional water treatment plant before it is sent to a distribution system.

SUMMARY

The subject of direct water reuse has been traditionally more sensitive and touchy than many other scientific topics. As long as there is enough natural water in a local area to meet its demand, there is no need to practice such a reuse. However, in some arid or semiarid areas, or in some island countries in the world, direct water reuse is often more economical than desalination of ocean water. Under such circumstances, serious consideration must be given to direct water reuse.

In the United States, the public is not psychologically ready today to accept such a reuse. But the currently available treatment technology appears to produce effluents of sufficiently high quality that they are more or less harmless to drink despite some "uncertainties" that no one really can know for sure. At present, the public psychological barrier outweighs the technical credibility toward the reuse application; thus intensive public health education and a professional campaign with broad coverage by news media are definitely needed. Safeguarding of public health can be ensured by use of a receiving stream to serve as a natural barrier and give a first warning sign of any potential toxicity in the public water supply.

REFERENCES

1. Culp, R. L., and G. L. Culp. *Advanced Wastewater Treatment* (New York: Van Nostrand Reinhold Company, 1971).
2. Huang, J. C., R. L. Smith and T. S. Otaguro. "Hawaii Gives Leis to New Tertiary Waste Plant," *Water Wastes Eng.* 10(5):24 (1973).
3. McKee, J. E., and H. W. Wolf. *Water Quality Criteria,* California State Water Quality Control Board (1963).

4. McCarty, P. L., D. G. Argo and M. Reinhard. "Reliability of Advanced Wastewater Treatment," *Proceedings of the Water Reuse Symposium, Vol. 2,* sponsored by AWWA Research Foundation, Washington, DC (March 1979).
5. Symons, J. M., et al. "National Organics Reconnaissance Survey for Halogenated Organics," *J. Am. Water Works Assoc.* 67:634 (1975).
6. Rothberg, M. R., et al. "Demonstration of Potable Water Reuse Technology—The Denver Project," *Proceedings of Water Reuse Symposium, Vol. 1,* sponsored by AWWA Research Foundation, Washington, DC (March 1979).
7. Mankes, R. O. "Purecycle Corporation—A Domestic Wastewater Recycling System," *Proceedings of the Water Reuse Symposium, Vol. 1,* sponsored by AWWA Research Foundation, Washington, DC (March 1979).
8. Roy, D., and E. S. K. Chian. "Treatment of Municipal Wastewater Effluent for Potable Water Reuse," *Proceedings of the Water Reuse Symposium, Vol. 1,* sponsored by AWWA Research Foundation, Washington, DC (March 1979).
9. Olson, B. H., et al. "Educational and Social Factors Affecting Public Acceptance of Reclaimed Water," *Proceedings of the Water Reuse Symposium, Vol. 2,* sponsored by AWWA Research Foundation, Washington, DC (March 1979).
10. Fuhs, G. W., et al. "A Laboratory Study of Virus Uptake By Minerals and Soils," *Proceedings of the Water Reuse Symposium, Vol. 3,* sponsored by AWWA Research Foundation, Washington, DC (March 1979).

CHAPTER 12

WATER RECLAMATION EFFORTS
IN THE UNITED STATES

Takashi Asano
   Office of Water Recycling
   California State Water Resources Control Board
   Sacramento, California

Robert S. Madancy
   Office of Water Research and Technology
   U.S. Department of the Interior
   Washington, DC

INTRODUCTION

Renovation of wastewater for beneficial use has been practiced in many parts of the world for many years. In highly industrialized nations, there are growing problems of providing adequate water supply, and municipal and industrial wastewater disposal. In developing countries, particularly those in arid parts of the world, there is a need to develop low cost, low technology methods of developing new water supplies and protecting existing water sources from pollution. In the United States, the 1976-1977 drought of the western states and later droughts in midwestern and southwestern states focused special attention on several water resources management options to meet the water needs of agriculture, industry and urban areas. These options include measures to reduce water consumption, water exchanges and transfers, conjunctive use of surface and ground waters, crop selection and wastewater reclamation and reuse.

Future freshwater supplies are expected to be higher in cost than existing supplies because of the remoteness of new water sources, escalating energy and delivery costs, environmental considerations and increasing competition for available supplies.

As the demand for water increases, wastewater reclamation and reuse have become increasingly important sources for meeting some of this demand. Because wastewater reclamation and the planned use of reclaimed water are so closely linked to the freshwater supply of a region, significant wastewater reuse projects are typically implemented in water-short areas of the country. However, because of the very nature of the waste water used for augmentation of freshwater supplies, significant public health, social, legal, economic and institutional issues must be addressed and carefully evaluated.

Although water supplies generally are sufficient to meet the requirements for all beneficial purposes, certain areas have major water supply problems as shown in Figures 1 and 2. These problems include shortages resulting from inadequate distribution systems, groundwater overdrafts and quality degradation of both surface and underground supplies [1]. In addition, there is a rather large area with future water demands that exceed surface water flows even in an average year (Figure 1). The area includes: southern California, Arizona, southeastern Nevada, New Mexico, Colorado, southeastern Wyoming

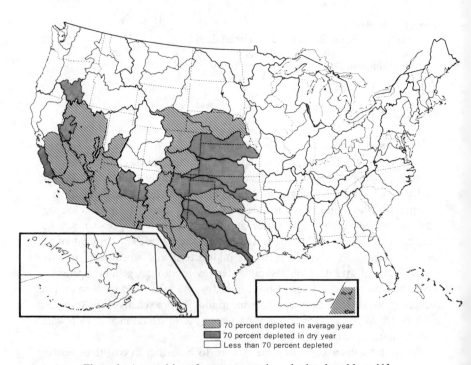

70 percent depleted in average year
70 percent depleted in dry year
Less than 70 percent depleted

Figure 1. Areas with surface-water supply and related problems [1].

and northern Nebraska. With respect to groundwater mining (Figure 2), areas with more than 1.9×10^6 m³/day (500 mgd) of groundwater overdraft include: central California, Arizona, extreme eastern and western New Mexico, southern Nebraska, western Kansas, western Oklahoma and central Texas.

Federal water pollution control regulations [2] now generally require a minimum of secondary treatment and, in some cases, advanced treatment to meet municipal discharge standards, thus creating the unique opportunity for reuse with higher quality treated water. The degree of wastewater treatment and requirements for process performance reliability will depend on the categories of planned reuse encompassing both present and future demand. The categories of planned reuse of municipal wastewater are listed in Table I in descending order of anticipated future volume of reuse.

In 1975 total municipal wastewater discharges in the United States amounted to about 90.8×10^6 m³/day (24 billion gallons per day (bgd)). Table II presents a summary of current and projected wastewater reuse (municipal) and recycle (industrial) in the United States [3]. The data in this summary indicate that the total reuse/recycle quantity may be expected to increase from 139.8 bgd in 1975 to 870.3 bgd in the year 2000. These data also help somewhat in placing municipal wastewater reuse in perspective on a

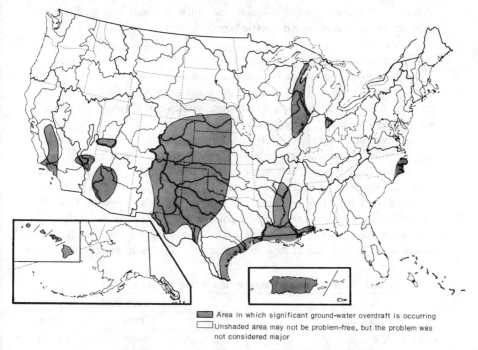

Area in which significant ground-water overdraft is occurring
Unshaded area may not be problem-free, but the problem was not considered major

Figure 2. Areas with groundwater overdraft and related problems [1].

Table I. Categories of Municipal Wastewater Reuse
(in descending order of volume of reuse)

1. Agricultural and Landscape Irrigation
2. Industrial Process and Cooling Water
3. Impoundment for Recreational Facilities
4. Stream Flow Augmentation
5. Groundwater Recharge
6. Direct Consumptive Use

national scale. The relatively small municipal reuse fraction in 1975 is expected to remain about the same in the future even though the actual quantity could increase sevenfold.

Table III presents a more detailed breakdown of the present and potential markets for industrial recycling and municipal wastewater reuse. The data presented in Tables II and III are based on realistic evaluations of available wastewaters, future water needs, availability of other water sources, geographic and climatic factors and conformance to national discharge standards [4]. In many water-short areas industrial process and cooling water recycling may provide an alternative to potable and nonpotable municipal reuse, that could have a considerably greater impact on supplementing existing water supplies. Process and/or cooling water recycling may allow available high-quality water supplies to be reserved for domestic purposes, thereby reducing the need for municipal reuse.

The dominant functional uses by geographic areas are shown in Figure 3. These functional uses give some idea as to how reclaimed water may be used throughout the country. In comparison, approximately 2.6×10^6 m^3/day (0.7 bgd), almost all of which is municipal wastewater effluent, was directly reused in 536 separate projects [3]. Most of the water volume (62%) and the largest number of projects (470) are for irrigation; only 29 large projects supply industrial cooling and process waters. The majority of reuse projects, accounting for the largest volume of reclaimed water, are located in the southwest and southcentral regions of the country, primarily in Arizona, California and Texas. A summary of municipal reuse projects is shown in Table IV.

Among water-short regions of the country, California has the most active and ambitious wastewater reclamation and reuse program. As a result of the severe drought, the use of reclaimed water has steadily increased and today

Table II. Summary of Wastewater Ruse and Recycling in the United States

Category	Quantity (bgd)		
	1975	1985	2000
Freshwater Withdrawals	362.7	356.3	330.9
Wastewater Recycle (industrial)	139.1	386.7	865.5
Wastewater Reuse (municipal)	0.7	2.1	4.8

Table III. Breakdown of Industrial Recycling and Municipal Wastewater Reuse

Category	Quantity (bgd)	
	1975	2000
Steam Electric	57	517.3
Manufacturing Industries	61	316.2
Mineral Industries	21	32.0
Municipal	0.7	4.8
Total	139.7	870.3

some 266×10^6 m^3/yr (215,200 ac-ft/yr) is used in California for various purposes as shown in Table V.

The Federal Clean Water Construction Grant Program has provided most construction costs of wastewater reclamation projects consisting of 75% by the U.S. Environmental Protection Agency (EPA), and in the case of California, 12.5% each by the State of California and by the municipality. Table VI shows current status of wastewater reclamation projects in California. However, due to the limited financial resources available in the Construction Grants Program, only wastewater reclamation projects which also proved to be cost-effective water pollution control options are being funded. Because the need to expand the use of reclaimed water is clearly evident, several financial assistance programs are being planned or implemented including: the California Clean Water and Water Conservation Bond Law of 1978, a State Assistance Program (16 million dollars financing five wastewater reclamation projects), the Renewable Resources Bill, the State Water Project, and local and regional financing of wastewater reclamation projects.

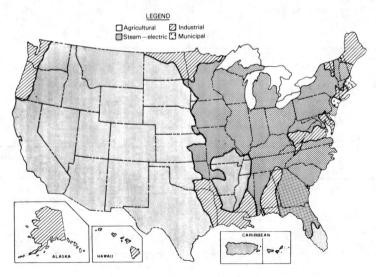

Figure 3. Dominant functional water uses [3].

Table IV. Existing Municipal Wastewater Reuse Projects in the United States

Category	Number of Projects	Reclaimed Water (mgd)
Irrigation	470	420
Agriculture	(150)	(199)
Landscape	(60)	(33)
Not Defined	(260)	(188)
Industrial	29	215
Process		(66)
Cooling		(142)
Boiler Feed		(7)
Groundwater Recharge	11	34
Other (recreation, etc.)	26	10
Totals	536	679

Table V. Present Municipal Wastewater Reuse in California

Type of Use	Number of Operations	Reclaimed Water (ac-ft/yr)	Percent of Volume
Agricultural Irrigation	232	147,920	69
Landscape Irrigation	104	28,700	13
Groundwater Recharge	5	26,069	12
Industry	22	10,480	5
Recreational Impoundments	13	880	<1
Other	3	1,180	<1
Totals	379	215,220	100

Table VI. Current Status of Wastewater Reclamation Projects in California

Status	Number of Projects	Reclaimed Water (ac-ft/yr)
Planning	72	292,160
Design	14	36,920
Construction	30	54,510
Construction Completed	21	130,130
Totals	137	513,720

PLANNING METHODOLOGY FOR WASTEWATER RECLAMATION AND REUSE

Although a number of factors affect wastewater reclamation and reuse decisions, historically, the impetus for wastewater reuse has resulted from three prime motivating factors:

1. availability of high quality effluent,
2. increasing cost of freshwater development and
3. desirability of establishing comprehensive water resources planning including wastewater reuse and water conservation.

As mentioned previously, federal water pollution control regulations generally require secondary treatment, and in some cases advanced treatment to meet effluent discharge standards; thus creating the unique opportunity for reuse with higher-quality reclaimed water. Consequently, discharging high-quality effluents in water-short areas may be considered to be a waste of limited water resources.

While the once-in-five year or once-in-twenty year drought highlights the need for additional water resources development, the next large increment of freshwater supplies will cost much more than existing supplies, mainly because of the remoteness of new water sources, higher energy and delivery costs, and environmental considerations.

The availability of reclaimed water for reuse at relatively low incremental cost, the increasing cost of fresh water, and the demonstrated need for additional water, form a basis for incorporating wastewater reclamation and reuse as an integral and complementary source of water. Reclaimed water is a reliable water supply source even in the drought year.

The general factors affecting water reuse decisions include: (1) local and regional water supply conditions, (2) water quality requirements for various water reuse applications, (3) degree of existing or proposed wastewater treatment facilities and process reliability, (4) potential health risks mitigation and (5) environmental issues and public acceptance.

Facilities Planning

Facilities planning for wastewater reclamation and reuse projects that involve primary benefits in the area of water supply require additional steps necessary for water pollution control planning. Although wastewater reclamation and reuse may be justified on the basis of the least-cost alternative to water pollution control projects, much of the effort in wastewater reclamation projects focuses on the market assessment or actual marketing of reclaimed water as the basis for facilities development. Therefore, the facilities planning should consist of three primary steps:

1. preliminary market assessment,
2. engineering and economic analyses and
3. detailed market analysis.

These steps should result in development of a recommended facilities plan for wastewater reclamation and reuse including reclaimed water pricing policy, and financial plan for the facilities. Table VII summarizes a checklist for the key elements in wastewater reclamation and reuse planning.

Table VII. Checklist for Wastewater Reclamation and Reuse Planning

I. Market Assessment/Feasibility Report
  1. Survey of potential users
  2. Water quality requirements and reuse requirements
  3. Estimate of future freshwater supply costs:
    (Data base: specific potential uses, quantity, time of needs, water quality needs, reliability needs, capital investment need to accept reclaimed water, desired pay back period and rate of return on investment, desired savings for using reclaimed water and plans for developing land or abandoning farm operation.)
  4. Feasibility report
    (Data base: study area boundaries and topographic maps, water and wastewater agency boundaries, wholesale and retail water supply entity boundaries, groundwater basins boundaries, major streams, receiving waters and summary tables of potential reclaimed water users and data.)

II. Engineering and Economic Analyses
  1. study area characteristics
  2. water supply characteristics and facilities
  3. wastewater characteristics and facilities
  4. treatment requirements for discharge and reuse
  5. water reclamation potential
  6. wastewater reclamation feasibility
  7. viability of wastewater reclamation and reuse
    (Data base: Source and availability of fresh water, price structure and comparison to reclaimed water, regulatory requirements, future changes in market and institutional and water rights assessment.)

III. Detailed Market Analysis/Project Report
  1. description of market assessment procedures
  2. descriptions of all users or categories of potential users
  3. define logical service area boundaries
  4. planning and design assumptions
  5. wastewater treatment and alternatives
  6. required water qualities for potential uses
  7. pipeline route and alternatives
  8. alternative reuse markets
  9. storage locations and capacity
  10. recommended plan
  11. financial plan and revenue program

CATEGORIES OF WASTEWATER RECLAMATION
AND REUSE

The categories of direct reuse of municipal wastewater effluents include agricultural and landscape irrigation, industrial processes or cooling, recreation and groundwater recharge. Direct reuse implies the existence of a pipe or other conveyance facilities for delivering the first user's effluent to the second users or uses.

Indirect reuse, through discharge of an effluent to a receiving water and withdrawal downstream, is recognized to be important but is not a primary focus of this discussion. In addition, direct, potable reuse is not considered here because further research and demonstration are required to provide additional assurances of safety, and these programs would take at least several years before implementation can be seriously contemplated. (The reader is referred to the Working Meeting of Experts for this issue, "EPA Protocol Development: Criteria and Standards for Potable Reuse and Feasible Alternatives," July 29-31, 1980, Airlie House, Warrenton, Virginia, and to the Proceedings to be published by the U.S. Environmental Protection Agency.)

In contrast to direct reuse, recycling or recirculation involves only one user or use and the effluent from the user is captured and redirected back into that use scheme. Water recycling is practiced predominantly in steam electric, manufacturing and mineral industries. In this context, reclaimed municipal wastewater is considered to be only reusable and not recyclable.

The municipal wastewater reuse categories in Table I may be expanded further to indicate additional reuse applications and their relative sensitivity in terms of quality and treatment requirements and public health considerations, as shown in Table VIII. Local circumstances could result in some reordering of the sensitivity ranking and possibly some degree of expansion or consolidation of the reuse applications.

More often than not nonpotable alternatives should prove to be more feasible than potable reuse. In many instances availability of low-quality waters will permit the use of municipal wastewater for low sensitivity reuse applications. The lower sensitivity applications shown in Table VIII will normally require a lesser quality water than potable reuse. This lower quality requirement will generally result in a simpler treatment system along with its associated lower capital and operation and maintenance costs and lower energy requirements. A potable reuse treatment system might typically consist of ten or more unit processes or unit operations such as primary sedimentation, biological treatment, secondary sedimentation, chemical clarification, recarbonation, multimedia filtration, selective ion-exchange, breakpoint chlorination, carbon adsorption, reverse osmosis and disinfection along with ancillary operations such as recalcining and carbon regeneration. In contrast a nonpotable irrigation alternative might only require two-stage aerated ponds, disinfection and the necessary distribution system.

Table VIII. Ranking of Reuse Applications in Order of Increasing Sensitivity

Rank	Reuse Application
1	Agricultural Irrigation, Forage Crops
2	Agricultural Irrigation, Truck Crops
3	Urban Irrigation, Landscape
4	Livestock and Wildlife Watering
5	Cooling, Once Through
6	Cooling, Recirculation
7	Industrial Process and Boiler
8	Fisheries
9	Recreation
10	Public Water Supply, Ground water
11	Public Water Supply, Surface water
12	Potable

Potential Substitution for Freshwater Withdrawals

Reclaimed water available for reuse can be divided into treated and untreated categories which is capable of substituting for freshwater withdrawals. Treated effluent includes: (1) primary, secondary and advanced treatment effluents from municipal wastewater treatment facilities; (2) industrial discharges meeting regulatory requirements and (3) discharges from steam electric plants. Untreated wastewaters are primarily agricultural irrigation return flows that could be collected and reused. Table IX summarizes reclaimed water use which is capable of substituting for freshwater withdrawals.

Several factors determine the quantity of reclaimed water to be actually used: (1) the geographic location of discharges and point of potential users, (2) the timing of reclaimed water discharges and the requirements of potential users (necessity of holding reservoirs, e.g., irrigation requirements vary with the season and climate but municipal discharges are relatively constant throughout the year) and (3) the availability and cost of alternative supplies. In addition, there are large regional differences in wastewater availability and reuse potential.

Table IX. Potential Use Categories and Reuse/Recycle Water Quality

Use Category	Required Water Quality
Agricultural Irrigation (return flows)	Untreated
Landscape Irrigation	Treated
Steam Electric	Treated or untreated
Industrial Cooling	Treated or untreated
Other Industrial	Treated

CONSTRAINTS TO WASTEWATER RECLAMATION AND REUSE

While the need to expand the use of reclaimed water is clearly evident, major constraints must be resolved before large-scale wastewater reclamation and reuse can be implemented. The complex constraints to be overcome involve economic and environmental costs, institutional and legal constraints and social impact. Independently they pose significant questions which must be answered. Collectively their interrelationship further complicate attempts to find solutions to the questions posed.

Economics

Major economic constraints to increased water reuse are:

1. The cost of most present freshwater supplies is relatively low due mainly to the supplies coming from the most easily developed sources.
2. The price the consumer pays for fresh water may be highly subsidized (a number of agencies which operate development facilities charge customers only a fraction of the cost of water; indeed, some purveyors use property taxes to reduce the price charged customers).
3. The cost of using reclaimed water in many cases is substantial.

While treatment of wastewater to levels required by discharge permits can be written off as a pollution control expenditure and subsidized by the federal and state construction grant programs, any additional treatment and the delivery system required for reclaimed water of a lower quality than fresh water may be incurred in terms of yield decrements of lower production efficiency. Since the great majority of the reuse potential lies in replacing fresh water, reclaimed water usually must be price competitive as well as cost competitive with fresh water. When all three constraints are present in a reuse situation, a formidable barrier often presents itself.

However, many areas of the nation are water short and need to expand their water supply. From federal and state viewpoints, the total cost of reclaiming waste water in most cases is less than the total cost of developing new freshwater supplies. For example, in the southern California service area of the California State Water Project, the price of water will increase because of the expiration of long-term power contracts. The new contracts will reflect actual power costs, which are estimated to be more than ten times higher than present rates. The water supply cost analysis for the Orange County Water District [4], as shown in Table X, illustrates how reclamation energy needs and actual over-all costs compare favorably with energy needs and costs of imported water in this area. Table X indicates that the present cost of reclaimed water is high in comparison with imported water. However, the actual cost, reflecting the present power costs and without the ad valorem

Table X. Cost of Future Water Supply Alternatives for Orange County Water District [5]

Alternative Water Supply	Present Cost ($/mil gal)	Actual Cost[a] ($/mil gal)	Energy (kwh/1000 gal)	TDS (mg/l)
Reclaimed Water				
1. Lime/Reverse osmosis treated	780	780	8.5	50
2. Present Blend of Activated Carbon/ Reverse osmosis plus well water	836	836	9.2	680
Colorado River Water		832	6.4	750
State Water Project		933	10.2	300

[a]Cost reflecting actual power cost and without ad valorem tax subsidy.

tax subsidy, is lower than the cost of imported water supplies. The Colorado River supply to this area is the least energy consuming alternative; however, its quality is marginal and delivery of this water to California will be drastically reduced upon completion in Arizona of the Central Arizona Project.

Public Health and Health Risks

Technology is available today to treat waste water so that it meets traditional drinking water standards with respect to microbiological, chemical and physical quality. Not enough is yet known about virus removal, the long-term carcinogenic and mutagenic effects of ingesting trace organics chemicals, and the cumulative effects of heavy metals to determine if these standards are in fact appropriate for reclaimed water supplies. An overriding consideration is the reliability of present technology and processes to provide reclaimed water that continuously meets accepted standards. The reliability question has, to a large extent, been answered at a cost by requiring redundancies, back-ups and fail-safe systems in treatment plant operations. Further research is needed to evaluate and improve the capability and reliability of treatment processes and equipment to consistently produce a uniform quality of reclaimed water.

Because of the major potential for using reclaimed water for groundwater recharge, health criteria must be developed where reuse involves recharge of groundwater basins serving domestic water supplies. In California, an expert consulting panel on health aspects of wastewater reclamation was established to guide state agencies in developing a research program to establish criteria and to plan and implement programs for the use of reclaimed water for groundwater recharge [5]. A Groundwater Recharge Symposium [6] was held in September 1979 to discuss technical and policy issues related to water reclamation for the purpose of groundwater recharge. The health related research is underway and hopefully should provide, over the next few years, a foundation of information and safeguards to allow increased water reuse without endangering public health or degrading water quality.

The farming community has been reluctant to use reclaimed water because of concern that the marketability of crops irrigated with reclaimed water may be reduced. Health regulations allow the use of highly treated wastewater for irrigation of food crops but in many cases the cost of such highly treated water is not competitive with other irrigation water. The bulk of reclaimed water used in agriculture is for irrigation of pasture, fodder, fiber and seed crops where health concerns are minimum and where lower levels of water quality can be provided by less costly treatment processes.

Institutional/Legal

In most instances, the agencies that develop, deliver and regulate water supply are separate from those that collect, treat and discharge wastewater.

This is true for agencies at the local level as well as at the state and federal levels. At the local level, this presents a constraint because it is often difficult to bring together the wastewater treatment agency that produces the reclaimed water, and the water purveyor who is usually the logical seller of reclaimed water. The constraint presented by such separation at the state and federal levels is more subtle, but just as significant. For example, in the arid western states where the quantity and quality of the water are strongly related, such separation presents a definite constraint.

There are laws in many states that were originally instituted to protect water purveyors, but now present a constraint to reuse of waste water, e.g., one law prohibits the installation of a duplicate distribution system, required for delivering reclaimed water, without reimbursing the original water purveyor. Another institutional constraint to reuse is the adverse financial impact on water suppliers which may be caused by the use of reclaimed water in the supplier's service area.

There are three basic types of water rights laws:

1. Riparian right: The right of a property owner to take water from a natural stream or lake bordering that property for use on the riparian right (a right determined solely by location of the land with respect to water supply).
2. Groundwater rights: The rights of owners of land overlying a groundwater basin to withdraw water for reasonable beneficial use on their overlying lands (another right determined by location of land with respect to water supply).
3. Appropriation of surface water: The right to take surface water and apply it to a beneficial use (a right granted according to the "first in time, first in right" doctrine by which priority of water use, in time of shortage, is given to the user who has had the right the longest).

Under various laws, it has become apparent that the sale and distribution of reclaimed water may raise water rights questions regarding the ownership of this water resource. These problems will arise both prior to the treatment of the water and subsequent to its discharge. Prior to treatment, a wastewater treatment facility may receive waste water from local sanitation districts. These districts normally convey the water through a sewage collection system after it has been discharged by local municipal and industrial users. These local users receive their water from a municipal water supply system, a private water company or through their own diversions. The water may be used on the basis of groundwater rights, surface water rights or contract rights with the Federal Water and Power Resources Service (formerly the U.S. Bureau of Reclamation) or a state water agency. It may not be clear under existing law who—the owner of the wastewater treatment facility or the water supplier— may rightfully claim ownership of the reclaimed water.

Parties have commonly settled such questions through private agreements. In order to encourage the sale and distribution of reclaimed water, it would be desirable to concentrate the ownership of the resource in one entity rather than in multiple entities. The subsequent reuse of reclaimed water raises a different set of ownership issues. Commonly, downstream users will have obtained rights to the return flow that upstream users have discharged into the stream. Generally, upstream discharges must respect the rights of downstream users to the return flow.

Two exceptions to this rule occur: (1) where the owner of a wastewater treatment plant initially discharges treated effluent with the prior intent of recapturing the water (i.e., a reclamation program is planned for implementation at some future date), (2) or where the source of the water is imported water and the water is recaptured within the plant boundaries or the boundaries of the district, then the treatment plant owner may be able to market that water to the detriment of downstream users. Existing law now provides substantial judicial consideration of downstream rights to return flow, thus creating a potential problem where the treatment plant owner proposes to produce and market reclaimed water.

All of these constraints to wastewater reclamation and reuse exist because up to now there has been no accommodation for reclaimed water in the water supply industry. As institutional and legal arrangements are changed to bring reclaimed water into the total water resources picture, these constraints should disappear.

Public Acceptance

Any program that considers the use of reclaimed water must take into account public attitudes toward such reuse and questions concerning protection of public health. A recent study [7] measured public attitudes toward

25 general uses of reclaimed water in California. As expected, the use of reclaimed water for drinking and food preparation received the strongest opposition. The lowest level of opposition was directed to such applications as irrigation of golf courses and highway greenbelts, and road construction. The study concluded that the extent of opposition is correlated with the likelihood or extent of close personal contact. Of the respondents in the study, 50% opposed the use of reclaimed water because the water was considered psychologically repugnant or lacking in purity. The public perceives that foul smelling, bad-tasting, dirty water is likely to contain harmful organisms or substances, and that clear water, having no odor and a good taste, is likely not to be harmful [7]. The goal of scientists, engineers and health professionals in controlling the use of reclaimed water is to provide a reclaimed water of suitable water quality for the intended use that is free from unacceptable risks posed by disease-causing organisms or substances.

SUMMARY

This chapter summarizes the current status of municipal wastewater reclamation and reuse as practiced in the United States. Those categories of wastewater reuse comprise present and future large volumes of municipal wastewater reuse. Special attention is directed to the planning methodology, categories of wastewater reclamation and reuse and constraints to wastewater reclamation and reuse.

REFERENCES

1. "The Nation's Water Resources, 1975-2000, Vol. 1: Summary," U.S. Water Resources Council, U.S. Government Printing Office (December 1978).
2. "EPA Secondary Treatment Information," Federal Register, Vol. 38, No. 159 (August 17, 1973).
3. Culp/Wesner/Culp, "Water Reuse and Recycling," Vol. 1, Office of Water Research and Technology, U.S. Department of the Interior (July 1979).
4. Argo, D. G. "The Cost of Water Reclamation by Advanced Waste-Water Treatment," WPCF 51st Annual Conference, Anaheim, California (October 1978).
5. "Report of the Consulting Panel on Health Aspects of Wastewater Reclamation for Groundwater Recharge," State of California, State Water Resources Control Board (June 1976).
6. Asano, T., and P. V. Roberts, Eds. "Wastewater Reuse for Groundwater Recharge," State of California Office of Water Recycling, State Water Resources Control Board (May 1980).
7. Bruvold, W. H. "Public Attitudes Toward Community Wastewater Reclamation and Reuse Options," California Water Resources Center, University of California, Contribution No. 179, Davis, CA (August 1979).

CHAPTER 13

RISK ASSESSMENT IN WATER REUSE

Neil J. Hutzler
Department of Civil Engineering
Michigan Technological University
Houghton, Michigan

William C. Boyle
Department of Civil and Environmental Engineering
University of Wisconsin
Madison, Wisconsin

INTRODUCTION

The United States enjoys one of the safest water supplies in the world. Any suggestion to compromise this safety by increased recycle of wastewater must be viewed with suspicion. However, as unpolluted water resources become scarcer, the risks of disease transmission must be weighed against the risk of having no water.

Waste recycle is not new. The Chinese directly recycle human excrement (night soil) back onto land used for food production with no apparent health problems. Sewage farming has been practiced in Europe for centuries, yet there is no evidence to suggest that the Europeans are less healthy than we (they appear to be more protective and selective of their water supplies than we are). Only recently, however, has it been suggested that wastewater can be directly recycled back to a community's water supply.

This chapter discusses the general procedure of risk assessment with special emphasis on biological diseases commonly associated with waterborne transmission. Although waste water can be reused for many purposes such as irrigation, recreation and industrial processing, this report focuses on direct recycle—water consumption—because it is felt that this represents the greatest risk. The process of risk assessment is demonstrated by estimating the risk of contracting infectious hepatitis (hepatitis A) by consuming recycled waste water that has had a minimum amount of treatment. This chapter presents methods which can be used to deduce the risks to water consumers, to explain risks to consumers and to place risks in perspective.

GENERAL PROCEDURE FOR RISK ASSESSMENT

Risk assessment comprises three broad and overlapping phases: hazard identification, risk estimation and social evaluation [1]. Hazard identification defines the potential for harm or injury, risk estimation attempts to quantify the potential for harm or injury and social evaluation judges whether or not the risk is acceptable.

The general procedure for assessing the risks resulting from pollutant discharges can be outlined as follows:

A. Hazard identification
1. describe properties of hazardous substance,
2. identify harmful effects, both acute and chronic, and
3. identify potential benefits.
B. Risk estimation
1. base estimates on disease statistics,
2. base estimates on epidemiological studies, or
3. base estimates on disease transmission models:
   a. define dose-response relationships,
   b. estimate or predict production, use, and disposal of hazardous substances,
   c. predict or monitor fate of hazardous substances in the environment,
   d. predict or measure exposure to transport vehicle and estimate actual dosage, and
   e. calculate overall risk =
      (experimental response)(actual dose/experimental dose).
C. Social evaluation
1. adhere to guidelines for effective risk assessments,
2. describe consequences of risk, and
3. outline costs of disease.

Hazard Identification

Hazards can be identified in a variety of ways including studies of disease statistics, epidemiological studies, animal studies, short-term screening with nonmammalian systems and analogy with known hazards [2,3]. When the effect of a suspected disease agent is acute and seen immediately, it is fairly easy to show the cause and effect between the agent and susceptible populations. Chronic or delayed effects are increasingly more difficult to study [2]. The problem of identifying hazards is compounded by the fact that many diseases have multiple causes [4].

During this phase of risk assessment, several decisions might be made. A substance may obviously be nonhazardous and therefore require no further analysis or a substance may be well-characterized as being hazardous so one could proceed to the next step of the assessment. A third decision might be to look for further evidence, in which case the type of required evidence should be specified. Two primary types of errors could be made during this phase of assessment: first, to decide that a substance is safe when in fact it is not; and second, to declare a substance hazardous when it is not. Errors of the first type are likely to be detected by epidemiological studies of public health, but errors of the second type are less likely to be known since this decision would restrict or prohibit use of the substance. This decision should also be viewed as a risk since it also deprives society of the substance's potential benefits. Therefore, decisions to ban the use of substances should be considered as thoroughly as those to accept risks.

Waste water undoubtedly contains many hazardous substances, but as unpolluted water resources become more scarce the pressure to reuse waste water will increase. The reluctance of the public to accept recycled water must be balanced against the dangers of having insufficient water supplies and of retarding industrial and agricultural growth.

Wastewater Hazards

Waste water can be hazardous to health for several reasons with the possible disease agents being classified as biological or chemical. In the past, the most important disease agents identified in waste water were pathogens—the cause of infectious disease—but recently the focus has shifted to the presence of toxic chemicals.

Infectious diseases can be classified according to the type of pathogen (bacterial, viral, parasitical), the site of infection (enteric, respiratory, dermal) or the primary mode of transmission (foodborne, airborne, anthropodborne). While many infectious diseases are potentially waterborne, enteric (intestinal) diseases are most often assumed to be transmitted by waste water since the pathogens primarily exit the body with feces, sometimes in large numbers. Other groups of disease that may have a relationship to sewage but have not been adequately evaluated are those affecting the eyes, ears, nose, throat and skin. Considering the greater number of microbial species, surprisingly few are typically pathogenic to man [5].

Craun et al. [6] have reviewed the causes of disease outbreaks involving water supplies. Illnesses associated with these outbreaks included: amebiasis, gastroenteritis, giardiasis, hepatitis A, salmonellosis, shigellosis and typhoid fever—all enteric diseases. Other diseases which can be identified as potentially waterborne by virtue of being enteric are: cholera, *Escherichia coli* enteropathy, aseptic meningitis, leptospirosis and yersiniosis [7]. Of the known viral diseases, only hepatitis A has been demonstrated to be clearly waterborne [8]; however, occasional cases of viral gastroenteritis can be attributed to contaminated drinking water [6]. Because enterovirus can be transported by water, there is concern about their presence in water supplies [9]. Feachem [10] has classified most water-related infections as either being waterborne (contracted by direct ingestion of water), water-washed (contracted by direct immersion in water), water-based (pathogen spends part of life cycle in water) or insect-related. Table I summarizes his classification scheme.

There is also increasing concern about the presence of chemical agents in wastewater, probably because less is known about their disease potential. Both inorganic substances as arsenic, heavy metals and nitrate and organic materials such as chlorinated hydrocarbons are implicated [11]. Organics found in U.S. water supplies based on a U.S. Environmental Protection Agency (EPA) study in 1975 [12] appear in Table II. This table presents the concentration ranges of those compounds and the relative frequency in which they were found. In a recent survey [13] of potential toxic compounds from U.S. households, the EPA listed the most frequently occurring compounds used and wasted into domestic wastewater (see Table III).

The National Academy of Sciences (NAS) recently completed an extensive review of the health effects of more than 80 inorganic and organic chemicals that could potentially be found in drinking water [14]. The national interim primary drinking water standards summarized in Table IV are, in part, based on this review. These standards were established primarily by identifying concentrations that have had no observed effect on exposed populations and by applying orders-of-magnitude safety factors that reflect the reliability of the data—a procedure that has long been used to set standards of this type [14]. Considering the great number of chemicals, these standards constitute a very short list. As more becomes known about the degree of hazard posed by the many chemicals known to be in water, this list will undoubtedly expand. Very few disease outbreaks are reported as being due to chemical poisoning [6,15]. Chemicals are more often associated with chronic or long-term effects rather than with acute symptoms, which makes finding a link between their presence in drinking water and disease much more difficult. It is not possible to make detailed risk assessments at this time but the prospect is improving.

Risk Estimation

The fact that a substance is potentially harmful is not a sufficient cause to declare that its use is unsafe. Safety relates to the probability of adverse effects and risk estimation attempts to evaluate this probability. Defining the

Table I. Water-Related Infections [10]

Disease	Pathogen	Water Association
Bacterial		
Shigellosis	*Shigella* spp.	Waterborne, water-washed
Salmonellosis	*Salmonella* spp.	Waterborne, water-washed
Cholera	*Vibrio cholerae*	Waterborne, water-washed
Leprosy	*Mycobacterium leprae*	Water-washed
Tularaemia	*Brucella tularensis*	Waterborne, insect-related
Typhoid, Paratyphoid	*Salmonella* spp.	Waterborne, water-washed
Viral		
Denque	Denque virus	Insect-related
Yellow fever	Yellow fever virus	Insect-related
Poliomyelitis	Poliovirus	Waterborne, water-washed
Arboviral diseases	Arboviruses	Insect-related
Hepatitis A	Hepatitis A virus	Waterborne, water-washed
Protozoa		
Amebiasis	*Entamebia histolytica*	Water-washed, waterborne
Balantidiasis	*Balantidium coli*	Water-washed, waterborne
Giardiasis	*Giardia lamblia*	Water-washed, waterborne
Malaria	*Plasmodium* spp.	Insect-related
Trypanosomiasis	*Trypanosomium* spp.	Insect-related
Helminths		
Ascariasis	*Ascari lumbricoides*	Water-washed, waterborne
Schistosomiasis	*Schistosoma* spp.	Water-washed
Spirochaetes		
Leptospirosis	*Leptospira* spp.	Water-washed, waterborne
Yaws	*Treponema pertenue*	Water-washed
Fungal		
Ringworm	*Trichophyton concentricum*	Water-washed
Miscellaneous		
Conjunctivitis	Infection of the eye caused by several organisms	Water-washed
Gastroentenitis	Enteric infection by several organisms	Water-washed, waterborne
Skin Sepsis	Infection of the skin by bacteria	Water-washed

most likely disease risk under various conditions of exposure requires evaluation of several factors: (1) dosage of substance required to produce an adverse response (the higher the dosage, the greater the risk); (2) the concentration of hazard in suspending media (the closer to the hazard source, the higher the concentration and the greater the risk); (3) the amount of medium

Table II. The Amounts and Expected Percent Distribution of Selected Constituents in the U.S. Drinking Water Supplies [12]

Constituents	Amounts ($\mu g/l$)	Percent distribution[a]
Carbon tetrachloride	$<$2-3	10
Chloroform	$<$0.3-311	100
Other halogenated C_1 and C_2[b]	$<$0.3-229	100
Bis (2-chloroethyl) ether	0.02-0.12	Low
β-chloroethylmethylether	Unknown	Low
Acetylenedichloride	$<$1	Low
Hexachlorobutadiene	$\sim$0.2	Low
Benzene[c]	10	High
Octadecane	$\sim$0.1	High
C_8-C_{30} hydrocarbons	$<$1	High
Phthalate esters	$\sim$1	50
Phthalic anhydride	$<$0.1	Low
Polynuclear aromatics[d]	0.001-1	High

[a]100% distribution means that tests of all drinking waters (24) showed the presence of the listed constituents. Percent values are rounded to the nearest 10%. Where insufficient sites have been sampled, low or high estimates have been made.

[b]Includes summation of all C_1 and C_2 halogenated hydrocarbons except carbon tetrachloride and chloroform.

[c]Whereas benzene has not been frequently reported, its distribution is probably widespread. The amount and distribution columns here refer to benzene and the alkylated benzenes up to C_6 which have been reported in many drinking waters.

[d]The listed amounts are a summation of the concentrations of individual compounds.

Table III. Predicted Priority Pollutants in Household Wastewater [13]

Organics	Inorganics
Benzene	Arsenic
Phenol	Cadmium
2,4,6-trichlorophenol	Chromium
2-chlorophenol	Copper
1,2-dichlorobenzene	Lead
1,4-dichlorobenzene	Mercury
1,1,1-trichloroethane	Zinc
Naphthalene	Antimony
Toluene	Silver
Diethylphthalate	
Dimethylphthalate	
Trichloroethylene	
Aldrin	
Dieldrin	

Table IV. National Interim Primary Drinking Water Standards (40CFR141)

Contaminant	Level (mg/l)
Inorganic	
Arsenic	0.05
Barium	1.
Cadmium	0.01
Chromium	0.05
Lead	0.05
Mercury	0.002
Nitrate (as N)	10.
Selenium	0.01
Silver	0.05
Fluoride	1.4-2.4
Organic	
Endrin	0.0002
Lindane	0.004
Methoxychlor	0.1
Toxaphene	0.005
2,4-D	0.1
2,4,5-TP	0.01
Trihalomethane (proposed)	0.01
Coliform bacteria	$< 1/100$ ml

taken in (the greater the amount, the greater the risk); (4) the duration of exposure (the longer the exposure, the greater the amount taken in) and (5) the characteristics of the exposed population (the greater the number of susceptible persons exposed, the greater the risk).

The risks associated with recycling of waste water could be deduced several ways including analysis of general disease statistics, epidemiological studies of exposed populations and mathematical modeling based on dose/response information. Risk should be expressed as the probability of experiencing an adverse effect (e.g., death, cases of disease, etc.) per unit time of exposure or, as in the case of drinking water which is more or less a continuous exposure, as probability per unit time. Table V, for example, presents some risks commonly faced by the U.S. population.

From a practical viewpoint, the effective risk assessment must be able to catalog and analyze large amounts of diverse information, make appropriate assumptions and reduce data to uniform units of expression, and yet be clear enough to undergo public scrutiny. Table VI, for example, lists the types of information necessary to predict the mobility of chemicals. The analyst should not be overwhelmed by the apparent magnitude of necessary information because risk assessment represents at least a procedure by which the data can be organized. One of the major benefits of a systematic risk estimate is that it points to needs for additional research.

Table V. Risks Faced by U.S. Population [16]

Event/Activity	Risk of Dying (per person per year)
All causes of death	9.15×10^{-3}
Smoking (males, ages 30-75)	2.32×10^{-2}
Motor vehicle accidents	2.2×10^{-4}
Homocide	1.02×10^{-4}
Falls	7.6×10^{-5}
Poisoning	1.1×10^{-5}
Fires	3.0×10^{-5}
Drowning	3.3×10^{-5}
Firearms	1.2×10^{-5}
General aviation	6.8×10^{-6}
Water transport	8.0×10^{-6}
Floods	2.0×10^{-6}
Boating	6.0×10^{-6}
Lighting	5.9×10^{-7}
Bites and stings	2.2×10^{-7}
Enteritis and diarrheal diseases	1.7×10^{-5}

Risk Estimates Based on Disease Statistics

Disease statistics can be used to reveal general trends of risk, but they usually underestimate the actual risk for several reasons: not all types of disease are reported to health agencies; not all cases of a specific disease are reported although the percentage of cases reported tends to increase as the severity of the disease increases; and often the mode of disease transmission is unreported. Nonetheless, crude estimates of risk can be made from existing data.

Table VII summarizes the incidence of notifiable enteric diseases in the United States for the period from 1970 through 1979. Presently, hepatitis A, salmonellosis and shigellosis represent the greatest health risks as far as enteric diseases are concerned, although, diseases such as giardiasis are receiving increasing attention. The reporting efficiencies for these diseases are typically low. For example, several studies have indicated that from 60 to 90% of the clinical cases of hepatitis are not reported to health authorities [19-23]. During nonepidemic periods, as few as 10% of the actual cases may be reported [24]. It also should be recognized that the incidence of enteric disease varies greatly between states. This variability may be attributed to the aggressiveness of the particular state agency as much as to any difference in factors such as the level of sanitation, population age, population density or the level of population immunity. If the diseases listed in Table IV represent the major enteric diseases in the United States and the reporting efficiencies seen for hepatitis A apply to all diseases, then there are approximately 100-420 annual cases of enteric disease per 100,000 persons. This compares to the

Table VI. Processes that Affect the Mobility of Chemicals[a]

Introductory Processes	Water Transport
Production	Bulk Flow Transport
Use	Mixing
Disposal	Diffusion
	Sedimentation
Physical Transformations	Scouring
Physical Adsorption/Desorption	
Vaporization/Condensation	Air Transport
Solution/Precipitation	Particle and Vapor
	Transport by Wind
Chemical Transformations	Mixing and Diffusion
Ion Exchange	Sedimentation
Dissociation	Impaction
Oxidation	Wind Erosion
	Precipitation
Microbial Transformations	
Oxidation	Water/Soil Transport
Reduction	Leaching
Metabolism	Erosion
Plant Transformations	Plant Transport
Metabolism	Adsorption on Leaves
Bioaccumulation	Root Uptake
	Plant Decay
Animal Transformations	Excretion
Metabolism	
Bioaccumulation	Food and Web Transport
	Aquatic Organism Uptake
Human Intervention/Contact	Terrestrial Organism Uptake
Water Supply/Water Consumption	Organism Decay
Irrigation/Agriculture/Food Consumption	Excretion
Inhalation	
Harvest Aquatic Organisms	
Occupational Exposure	
Consumer Goods	
Recreation	

[a]Adapted from Stolzenberg [17].

Table VII. Summary of Notifiable Enteric Diseases, 1970-1979 [18]

Disease	Incidence (cases/100,000 pop./yr)	Deaths (number/yr)
Amebiasis	1.4	40
Cholera	0	0
Hepatitis A	20	759
Poliomyelitis	0.01	11
Salmonellosis	12	73
Shigellosis	8.6	28
Typhoid Fever	0.21	4

rates observed in eastern Europe, where reporting is much more complete [25]. In 1974, for example, Czechoslovakia reported a rate of about 290 cases per 100,000 population; Poland, 245; and Yugoslavia, 360 [26]. Translated into risk, the probability of contracting a case of enteric disease in a given year may be 0.1-0.4%.

Not all enteric diseases are transmitted through water (Figure 1). Other possible routes of transmission include: personal contact, food, air, inanimate objects, transfusions and animal vectors. The relative importance of each route varies with the disease. Salmonellosis, for example, is considered to be primarily foodborne, while some parasitic diseases such as giardiasis are more likely to be waterborne. Water is known to play a role in the transmission of enteric diseases as evidenced by the occurrence of occasional waterborne outbreaks [6,27,28]. Some estimates of the proportion of waterborne cases of hepatitis A range as high as 33% [29], but it is likely to be much less than this, at least in developed countries [8,26]. Part of the confusion lies with the definition of "waterborne". In most of the documented waterborne outbreaks, the contamination of a water supply had to be demonstrated [28]. Table VIII summarizes the cases of waterborne disease in the United States from 1961 to 1976. About 60% of the documented cases were due to gastroenteritis of unknown etiology. If the previous reporting efficiencies were applied to this data then the estimated risk of contracting waterborne disease is approximately 0.006-0.025%/yr.

Risk Estimates Based on Epidemiological Studies

Some humans are inadvertently or intentionally exposed to infectious agents. They form the basis for epidemiological studies. Epidemiological studies have been invaluable in deducing many of the causes of disease and will continue to provide perspective as to the relationship between man and his pathogens. Epidemiology is a broad field of study with many branches

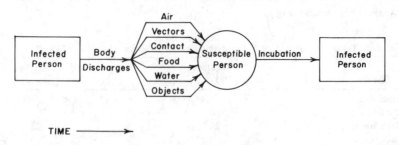

Figure 1. Infectious disease transmission.

including: (1) descriptive epidemiology, which is a general collection of disease data (e.g., [18]); (2) surveillance activities, which are more detailed observations of important diseases (e.g., [28]); (3) field epidemiology, which includes the specific studies of outbreaks (e.g., [31,32]); and (4) analytic epidemiology, which involves controlled studies of specific populations (e.g., [33-35]). Epidemiology has revealed many of the hazards involved with the discharge of wastewater effluents but it is very difficult to obtain specific measures of risk since the information needed most—the magnitude of the exposure—is usually not available. Some of the other weaknesses of epidemiologic study are the difficulty of separating the "causes" of disease, problems in obtaining representative samples and problems in data collection and analysis.

A possible situation within which to study the relationship between wastewater, drinking water and health is to observe a community that recycles part or all of its waste water back into its drinking water system. It is extremely difficult to implicate the discharge of treated waste water with disease contracted as a result of consuming contaminated drinking water because communities that knowingly take their drinking water supply from contaminated (or potentially contaminated) sources normally expend greater efforts to remove presumed microbial pathogens. Although potential study areas have been identified (Windhoek and Pretoria, South Africa), no reports on the health effects have been noted.

Metzler et al. [36] reported an unusual instance where a community was forced to recycle its own waste water back to its water supply during a pro-

Table VIII. Summary of Cases from Reported Waterborne Disease Outbreaks, 1961-1975 [6,15,30]

Disease	Number of cases	Incidence[a] (cases/100,000/yr)
Amebiasis	39	0.0001
Cholera	0	0
Hepatitis A	1,310	0.04
Poliomyelitis	0	0
Salmonellosis	17,493	0.54
Shigellosis	6,644	0.21
Typhoid Fever	326	0.01
Gastroenteritis	46,955	1.46
Giardiasis	5,951	0.19
Chemical Poisoning	592	0.02
Totals	79,310	2.47

[a] Based on an average population of 201 million.

longed drought. For 2 months, waste water was discharged into a stabilization pond with a 17-day detention time and within which the water treatment plant located its water intake. During this time the most notable change in treatment methods was an increase in chlorine dosage from 1 to 7 mg/l at the water treatment plant. Initially, public acceptance of the recycled water was good because the city was accustomed to receiving treated sewage from upstream communities; however, as the practice gained more publicity, the public greatly reduced their consumption of the water as evidenced by a sudden increase in the purchase of bottled water. During the entire episode, no unusual disease patterns were noted by local physicians.

Risk Estimates Based on Modeling

Disease statistics and epidemiology are useful for defining historical risk but tell us nothing about new hazards or the risk of known hazards under new circumstances. When the direct monitoring of risk is not possible, it is necessary to model risk. Information on the production, use, disposal and transport of suspected hazards can be coupled with information on exposure and dose-response to make estimates of risk for various disease agents and various situations [37,38]. The over-all risk resulting from the production, use and disposal of a hazardous substance is the product of the probabilities that: (1) it is being discharged into transport media such as air, water or food (determined from data on production, use and disposal); (2) it is transported intact to zones of human activity such as the home or public places (predicted by modeling or monitoring); (3) it is taken in by susceptible people through inhalation, ingestion or immersion (determined from population data and physiology); and (4) the amount taken in is sufficient to elicit an undesirable response (extrapolated from dose-response data). This concept as applied to infectious disease is depicted in Figure 1. For there to be disease transmission by direct water recycle the following events must occur in order: (1) the disease agent must be discharged into waste water, (2) it must not be removed by wastewater and water treatment processes (it is not clear where wastewater treatment ends and water treatment begins), (3) it must be taken in by susceptible persons by ingestion or through contact, and (4) the amount of disease agent taken in must cause disease or produce an infection which results in the subsequent production of disease agent.

Dose-Response. The basis for a quantitative risk estimate is a well-defined dose-response relationship between the hazard and the populations exposed to it. Unfortunately, the necessary information to develop dose-response relations is often either nonexistent or available only in a crude form. Not only is there a wide variety of disease agents and undesirable responses but there is a wide range of tolerance between individuals within a population.

Even if sound experiments can be designed there is still the problem of extrapolating from experimental dosages, which must be high to ensure response, to environmental dosages, which are typically low. If risk must be deduced from animal experiments, then there is the additional problem of extrapolation from the test species to man.

The wide range in tolerance to a given disease agent leads to a characteristic sigmoid shape for the dose-response curve when dosage is plotted against the percentage of subjects exhibiting adverse symptoms at that dosage (Figure 2). In principle, there should be a low dosage to which no one will respond and a high dosage to which all will respond. Since individuals are normally exposed to low concentrations of hazardous substances, low dosages and the associated risk are of greatest interest. It is tempting to try to define a threshold or "no-adverse-effect" dosages and there are some biological reasons to expect that true threshold dosages exist for some disease agents; however, a true no-effect level cannot be observed experimentally because of statistical interpretation of zero responses for small samples [2]. At best, a "no-observed-effect" level can be determined and it should be qualified by a statement of the size of the group in which no effect was observed [2].

Uncertainty about the shape of the dose-response curve at low dosages requires conservative extrapolations from existing data. Several methods have been suggested but the easiest to apply and understand is a linear extrapolation from a known point on the curve to zero [39]. An added degree of conservatism is provided by starting from the upper confidence limit, UCL, of the response at an experimental dosage, d_e (Figure 2). The maximum risk at a dosage, d, can be estimated by the equation:

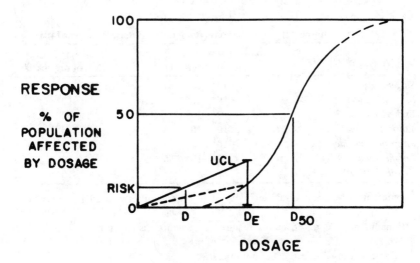

Figure 2. Estimating risks from dose-response data.

$$\text{maximal risk} = \text{UCL} \times \frac{d}{d_e} \qquad (1)$$

This analysis ignores the possibility of a threshold or minimum infectious dose in that a zero response occurs only in a zero dose. This also provides a conservative risk estimate.

Bryan [40] has summarized the results of the limited number of studies where enteric pathogens were fed to human volunteers. These data are further summarized in Table IX. While the data are not complete enough to develop curves like that shown in Figure 2, they do indicate which microorganisms are the most pathogenic. *Shigella dysenteriae, Giardia lamblia* and *Entamoebia coli* may cause infection when fewer than 10 organisms are ingested. The data on virus are incomplete. It is suspected that very low numbers are required to produce infection [41], but it is also known that large numbers are sometimes required to ensure infection [42].

Most feeding studies have been conducted on healthy, adult males so there is some question as to the applicability of this work to the general population, the feeling being that the very young, the aged and the infirm are at greater risks. Therefore, for the time being, conservative safety factors are indicated.

Production and Disposal of Hazards. Risk begins with the production and improper disposal of wastewater hazards. Chemicals can originate in residential [13], institutional, commercial and industrial [43] wastes. In addition, some chemicals including chlorinated hydrocarbons and the trihalomethanes (THM), can be generated during the processing of wastewaters [12,44,45].

Table IX. Approximate Dose-Response for Selected Enteric Pathogens [40]

Organism	Challenge Dosage[a] (No. of organisms)	Minimum Dosage[b] (No. of organisms)
Shigella spp.	10^2 -10^5	10^1
Salmonella spp.	10^5 -10^9	10^4
Escherichia coli	10^6 -10^{10}	10^6
Vibrio cholerae	10^3 -10^{11}	10^3
Streptococcus faecalis	$> 10^{10}$	10^{10}
Entamoeba coli[c]	10^1 -10^3	10^1
Giardia lamblia[c]	—	10^1

[a]Approximate range of organisms required to produce disease in 25-75% of subjects tested.

[b]Minimum number of organisms required to produce disease in any of the subjects tested.

[c]The dosages caused infection but not disease.

It is difficult to assess the quantities of toxic and mutagenic chemical present in raw municipal wastewaters. Raw wastewater levels of some of these compounds found in San Francisco municipal wastewaters are listed in Table X. Production of chlorinated hydrocarbon residues is dependent on specific wastewater treatment plant unit operations, presence of precursors and chlorine dosage.

Pathogens are shed primarily from infected persons and enteric pathogens are shed in the feces. During peak shedding, a gram of feces may contain 10^8-10^{10} pathogenic bacteria or virus and, since the average fecal production is 100-200 g/day, an infected individual may shed as many as 10^{12} pathogens in a single day [47]. Pathogen shedding may last from a few weeks (e.g., hepatitis A, salmonellosis) to several months (e.g., amebiasis). For some diseases such as typhoid fever, a permanent carrier state may be established in a small portion of the population. In an urban setting, most fecal matter is flushed to public sewage systems for disposal. A small percentage may enter solid wastes on disposable diapers [48].

Table X. Raw Wastewater Characterization for Selected Components[a] [46]

Constituent	Domestic + Commercial and Industrial	Domestic	Domestic + Industrial
pH (pH units)	7.0	7.2	–
Arsenic	0.003	0.004	0.008
Barium	0.136	0.041	0.141
Cadmium	0.004	0.006	0.002
Chrome	0.560	0.213	0.029
Copper	0.156	0.217	0.067
Fluoride	1.36	0.90	0.60
Iron	3.54	2.78	1.24
Lead	0.206	0.203	0.100
Manganese	0.111	0.195	0.044
Mercury	0.004	0.003	0.003
Selenium	0.005	0.001	0.014
Silver	0.044	0.034	0.013
Zinc	0.641	1.195	0.193
Endrin (μg/l)	–	0.007	0.003
Lindane (μg/l)	0.021	0.70	0.06
Methoxychlor (μg/l)	0.20	0.08	0.027
Toxaphene (μg/l)	0.001	0.002	0.001
2,4 D (μg/l)	0.017	0.4	0.04
2,4,5-TP Silvex (μg/l)	0.120	0.14	0.07
TOC	100	171	73
TDS	1918	5259	554

[a] mg/l except as noted.

Fate of Hazard in the Environment. The next step in risk estimation by modeling is to approximate or measure the level of the hazardous material under a particular set of conditions. The fate of a substance, once discharged to the environment, is dependent upon its chemical or biological characteristics, and on the environment in which it is placed. Chemicals may undergo a variety of separations, including volatilization, precipitation, exchange or adsorption, or may be chemically or biochemically transformed to a separable or innocuous material (Table VI). As mentioned previously, certain hazardous chemicals may actually be produced through these transformations.

In analyzing risks associated with wastewater reuse, the major line of defense would be the treatment facility although all steps along the transport vector from source to host must be considered in estimating the actual dose. Detailed transport models can be formulated by treating transport and transformation processes as discrete steps [17,49]. The most likely transport processes (e.g., flow through sewers) and transformation processes (e.g., substance decay or precipitation) affecting a given substance can be selected and placed in order. Probability expressions (e.g., the fraction of a substance destroyed) based on process mechanism (e.g., decay kinetics) and environmental conditions (e.g., temperature, pH) are developed for each process at each step. The over-all simulation of substance movement is then based on a materials balance for the specific environmental system. At the very least, a modeling effort like this forces the analyst to collect and catalog the most important information and allows one to identify the most critical pathways of substance movement.

The fate of chemical hazards in wastewater treatment facilities has been reported by several investigators. McCarty et al. [50] presented data on the fate of both heavy metals and trace organics at an advanced wastewater treatment (AWT) facility in Orange County, CA (Water Factor 21) (Tables XI and XII). Pressley et al. [51] found that biological/physical/chemical treatment of a municipal raw wastewater produced final effluent concentrations of halogenated methanes and ethanes, other volatile organics and extractable materials to be similar to those found in finished drinking water as reported by the EPA National Reconnaissance Survey of 1975 [12] (Table II). Secondary and AWT effluent showed significant mutagenesis in base pair substitution mutants of *S. typhimurium* based on studies at the Piscataway high-lime AWT plant [52]. Fewer mutagenic samples were reported for AWT plants at Dallas, Lake Tahoe and Pomona, however, based on *Salmonella* microsome assays [53]. Englande and Reimers [54] indicated that, in general, AWT and physical-chemical facilities produced effluents of excellent heavy metal quality as compared with potable water criteria. Metals that did persist through the treatment schemes studied included boron, mercury, selenium and zinc.

Table XI. Comparison Between National Interim Primary Drinking Water (NIPD) Regulations and Effluent Water Quality at Water Factory 21 [50]

Contaminant	NIPDWS	Effluent Water Geometric Mean	98% of Time Less Than
Arsenic (mg/l)	0.05	< 0.005	
Barium (mg/l)	1.0	0.03	0.08
Cadmium (mg/l)	0.01	0.001	0.006
Chromium (mg/l)	0.05	0.02	0.09
Lead (mg/l)	0.05	0.002	0.036
Mercury (mg/l)	0.002	0.0017	0.012
Nitrate (mg/l)	10	< 10	< 10
Selenium (mg/l)	0.01	< 0.004	?
Silver (mg/l)	0.05	0.003	0.007
Fluoride (mg/l)	1.4	0.6	1.2
Coliforms (MPN/100 ml)	1	0.01	14
Endrin (μg/l)	0.2	< 0.01	< 0.01
Lindane (μg/l)	4	< 0.05	< 0.05
Toxaphene (μg/l)	5	< 0.01	< 0.01
2,4,5-TP (μg/l)	10	< 0.01	< 0.01
Methoxyclor (μg/l)	100	< 0.1	< 0.1
Turbidity (μg/l)	1	0.4	1.1

There are two basic approaches to reducing the chloroform or trihalomethanes resulting from wastewater chlorination: one is to employ alternative disinfectants, such as ozone or ultraviolet, the other is to remove either the organic precursor or the chlorinated by-product. Currently, the most complete elimination of total trihalomethanes appears to occur through removal of the precursor [55].

The fate of selected biological pathogens and indicators is summarized in Table XIII. A review of this table suggests that organism disappearance or destruction is highly variable. Processes that are effective for one organism may be completely ineffective for another.

Disinfection processes have been considered the most effective control measures for biological pathogens. Chlorine applied to secondary effluent at a dose of 5 mg/l can effect bacterial kills of four logs in 30 minutes contact. Yet, this dosage may be ineffective against a number of virus and protozoan cysts. Table XIV provides comparative information on the effectiveness of several disinfectants on three biological pathogen groups. Effectiveness of the disinfectant is dependent on organism type, disinfectant dose and contact time, presence of reactive substances in the waste (disinfectant demand), pH, temperature and shielding of the organisms within solids or fat globules. Current wastewater disinfection practice, employing a minimum chlorine residual (e.g., 0.5-2.0 mg/l) or residual coliform count (e.g., 200/100 ml) will not necessarily guarantee an effluent free of virus or infectious protozoan cysts.

Table XII. Percentage of Time Hypothetical Maximum Contaminant Levels (MCL) for Various Trace Organics Were Exceeded at Water Factory 21 [50]

Contaminant	Hypothetical MCL (μg/l)	Geometric Mean Influent Concentration (μg/l)	Percent of Time Hypothetical MCL Exceeded		
			Influent	AWT Effluent	RO Effluent
Carbon Tetrachloride	0.5	0.03	0.1	6	5
1,1,1-Trichloroethane	0.5	3.2	92	33	28
Trichloroethylene	0.5	1.0	71	38	40
Tetrachloroethylene	0.5	1.7	92	27	17
Chlorobenzene	0.5	0.14	8	2	0.8
1,2-Dichlorobenzene	0.5	0.64	60	0.03	0.8
1,3-Dichlorobenzene	0.5	0.16	8	<3	0.1
1,4-Dichlorobenzene	0.5	1.8	99.5	1	0.03
1,2,4-Trichlorobenzene	0.5	0.11	12	<3	<6
Heptaldehyde	1.0	0.1	1	2	–
Heptylcyanide	1.0	0.002	0.2	0.1	–
Ethylbenzene	1.0	0.04	0.4	10^{-5}	10^{-5}
m-Xylene	1.0	0.04	0.003	0.003	0.002
p-Xylene	1.0	0.02	10^{-8}	10^{-10}	10^{-9}
Naphthalene	1.0	0.03	0.2	0.001	0.001
1-Methylnaphthalene	1.0	0.01	10^{-5}	<3	0.8
2-Methylnaphthalene	1.0	0.01	0.1	<3	10^{-5}
Styrene	1.0	0.05	0.1	0.5	–
Dimethylphthalate	1.0	4.8	99	26	50
Diethylphthalate	1.0	0.10	15	<4	<8
Di-n-burylphthalate	1.0	0.79	39	14	56
Diisobutylphthalate	1.0	4.7	99.7	21	25
Bis-(2-ethylhexyl) phthalate	1.0	11	99.9	>99.9	85
PCB as Aroclor 1242	0.5	0.47	45	<5	<7
Lindane	0.5	0.14	10^{-7}	<5	<7

Exposure. People are exposed to wastewater hazards by a variety of mechanisms including ingestion of contaminated water or food, inhalation of wastewater aerosols and immersion in polluted waters. An important part of defining exposure to wastewater hazards is the determination of the concentration of hazard. Of several possible approaches, the most direct is to monitor areas of human activity for the disease agents of interest [47]. This is not feasible in many cases because of the prohibitive cost of measurement. The next approach might be to relate disease risk to the presence of pollution indicators such as fecal coliforms [60]. However, acceptable indicator systems are not presently available for many disease agents such as viruses [61]. If the suspected hazard cannot be monitored or is not currently being produced, then the only way to deduce environmental concentrations is to couple estimates or measurements of total production to projected movements through the environment as previously discussed.

Table XIII. Removal of Selected Organisms by Various Treatment Processes [56-58]

Organism	Primary Treatment	Activated Sludge	Trickling Filter	Lime Coagulation	Alum Coagulation	Rapid Sand Filter	Chemical Sand Filter	Adsorption
				Unit Process				
Protozoan	nil	var.	var.	–	–	–	–	–
E. histolytica	nil	0-99	10-99.9	–	–	Sig	Sig	–
Eggs	~50	0-99	~30	–	–	Sig	Sig	–
Bacteria	20-50	>90	var.	–	–	–	–	–
Coliform	30-95	90-99	60-95	99	90	–	–	–
Salmonella	>50	85-99	70-99	99	90	–	–	–
Strep. faecalis	<50	85-95	–	99	90	–	–	–
Virus	0-85	60-99	0-85	98-99.9	95-99.9	0-90	90-99.99	35-99
Poliovirus	0-10	75-99	~85	70-99	–	–	–	–
Coxsackie	<50	0-50	~95	–	–	–	–	–

Determination of dosages to hazardous substances also requires knowledge of the duration of exposure and typical values for vehicle intake. Waterborne risks, for example, are related to average water consumption, airborne to average respiration rates and foodborne to average caloric intake. Although there should be some concern for those persons working around wastewater processes, the direct ingestion of recycled water and, to a lesser extent, contaminated food represents the highest probability for exposure because everyone has a minimum daily requirement for water and food.

Table XIV. Performance of Halogens and Ozone at 25°C [59]

Disinfectant	Necessary Residual After 10 min. to Achieve 99.999% Destruction (mg/l)		
	Amoebic Cysts	Enteric Bacteria	Enteric Virus
HOCl (predominates @ pH <7.5)	3.5	0.02	0.4
OCl- (predominates @ pH >7.5)	40	1.5	100
NH_2Cl[a]	20	4	20
I_2 (predominates @ pH <7.0)	3.5	0.2	15
HIO/IO- (predominates @ 8.0 > pH > 7.0)	7	0.05	0.5
O_3	0.3->1.8	0.2-0.3	0.2-0.3

[a]$NHCl_2$:NH_2Cl Efficiency = 3.5:1.

The National Academy of Sciences in its recent study [14] summarized estimates of average water consumption and found that the average American ingests about 1.6 liters of water per day. Although daily water consumption by an individual is a function of temperature, humidity and physical activity as well as other factors and can vary widely, the Academy adopted a value of 2 liters/day as being a representative intake of the majority of consumers.

Overall Risk. The last step in modeling risks is to calculate risks under a set of critical conditions. To demonstrate the procedure of risk estimation, the maximal probability of contracting hepatitis A as the result of drinking recycled wastewater will be calculated. Hepatitis A is used as an example because of its relatively high incidence and its large economic impact [24]. The focus is on direct exposure by the intake of recycled waste water because it represents the greatest risk.

Social Evaluation

The last phase of risk assessment is social evaluation—judging the acceptability of risk. Many methods could be used but the most common ones are: the traditional cost/benefit or risk/benefit analysis, comparison to accepted risks as revealed by a study of the marketplace, comparison to accepted risks as expressed by public opinion and comparison to historical background levels of pollution [62].

No single method is acceptable for all hazards or to all interest groups, so a combination of methods is necessary. It is clear, however, that any effective risk assessment should adhere to a few basic guidelines [2,62-64]:

1. The uncertainty of risks and the risk assessment process should be clearly acknowledged.
2. The steps taken and the assumptions made should be clearly presented.
3. The analysis should be flexible enough to incorporate new information.
4. Only evidence or facts that are easily referenced should be used.
5. Data should be condensed to common units of expression that are easily understood (e.g., risk expressed as cases of disease per unit of time of exposure).
6. When numbers are meaningless or untrustworthy, a digitless structuring of the problem is required (e.g., relative risks versus absolute risks).
7. Unclear or complex concepts should be presented a variety of ways.

The challenge of an effective risk assessment is to focus the complex issues of safety onto the major issues. At the very least, the analyst should be able to determine what questions should be answered first. It is not likely that a risk analysis can be complete with a single iteration. While a good analysis should be insightful, it does not need to be conclusive. It should increase understanding, clarify issues and, perhaps, even create debate.

WATERBORNE RISK OF CONTRACTING
INFECTIOUS HEPATITIS

Infectious hepatitis or hepatitis A is a common viral disease that is transmitted by the fecal-oral route. It is important economically because, though seldom fatal, it is quite debilitating in its acute stages [24]. Most infections, however, do not result in clinical symptoms making it more difficult to trace transmission since asymptomatic cases may also shed the hepatitis A virus (HAV) [65]. The prevalence of hepatitis A infections has been shown to be inversely related to social and economic conditions with the incidence being the highest where sanitation and efforts toward personal hygiene are the lowest; a correlation that is common for many intestinal infections. It is not surprising that improved sanitation is listed as the primary weapon in the battle against infectious hepatitis [66].

Dose-Response

The site of HAV invasion is most likely the intestine since the fecal-oral route is the generally accepted mechanism of transmission. This concept is reinforced by the facts that high numbers of virus have been found in the feces of infected persons [67,68] and that infection by the oral route has been demonstrated by many studies [69,70]. During incubation the virus multiplies in the gastrointestinal tract and the liver. Lesions may form in the liver, intestine and, occasionally, the kidneys. Viral shedding begins and is almost completed before the onset of symptoms [68]. The duration of HAV shedding is about 4 weeks but shedding never occurs after convalescence; i.e., no carrier state for hepatitis A has been identified [65].

Data from several experiments where infective fecal material was intentionally fed to humans are summarized in Figure 3. HAV could not be isolated at the time of the experiments, so the dosages are related to an equivalent amount of infective feces. The high variability of responses can be attributed to several factors: (1) although volunteer subjects were selected on the basis of having no previous exposure to hepatitis, it is likely that some had had subclinical infections and, thus, were immune; (2) the fecal samples were usually pooled specimens often collected after the onset of clinical symptoms and, therefore, were likely to have had a wide range of virus titer; (3) the various researchers used different methods to prepare the fecal extract which was fed to the subjects—thus, there is some uncertainty about comparing dosages between experiments; and (4) a natural variation in responses most likely exists between various groups of people. Even though only 166 subjects are represented by these studies (Krugman et al. [69] administered 0.001 g and 0.00001 g of feces to additional groups of eight subjects each, and observed

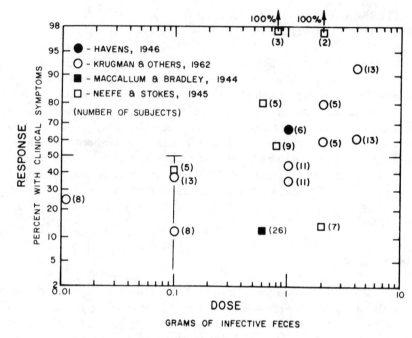

Figure 3. Dose-response of hepatitis A virus in humans.
● — Havens [71]
○ — Krugman et al. [69]
■ — MacCallum and Bradley [72]
□ — Neefe and Stokes [70]

no response) and the results are highly variable, this dose-response relationship can be used in a rough order-of-magnitude estimate of risk.

Because the result of a single dose-response trial can be expressed as a positive or negative result (i.e., either a person becomes diseased or he does not), the results of several trials at a given dosage can be approximated by a binomial distribution and appropriate confidence limits can be computed [73]. The width of the confidence limits is a function of the number of trials and the accepted level of confidence: the greater the number of trials or the lower the accepted level of confidence, the narrower the confidence limits. From the data presented in Figure 3, the 0.1 gram dosage (d_e) was selected for analysis because it represents a low dosage that produced positive responses (21% of subjects became diseased) and because about 15% of the subjects received this dosage. The 95% UCL of the response at this dosage was calculated by the procedure outlined by Davies [73] and was found to be 50%. This response and dosage can be used along with Equation 1.

The great uncertainty in interpreting these data points to the need for better dose-response information. The ethics of human experimentation, however, are currently under question so it is not likely that research of this type will be conducted in the near future. The alternative is to study the dose-response using chimpanzees or marmosets, the only animals known to be susceptible to HAV infection [74].

Production and Disposal

The first step toward estimating an environmental dosage of HAV is estimating the production of virus and, since they are only produced by infected humans, this means estimating the incidence of hepatitis A infections. Because hepatitis is a reportable disease, statistics on its incidence can be obtained from the Center for Disease Control as well as from state and local health departments. However, the incidence of hepatitis cannot be used to estimate directly the number of virus shedders for several reasons: (1) there are at least three types of hepatitis (A, B, and non-A/non-B), (2) not all clinical cases are reported, and (3) not all shedders exhibit clinical symptoms. The reported rate of hepatitis A for the United States from 1970-1979 ranged from 14 to 29 cases/100,000 persons/yr with an average incidence of 20 [18].

Several studies [19-23] have indicated that from 60 to 90% of hepatitis A cases are not reported to health authorities, and if these percentages are applied to the national average, the actual annual incidence of hepatitis A is probably 50-200 cases/100,000. However, only about 5-20% of hepatitis A infections result in disease and if it is assumed that all infections result in pathogen shedding then there may be 250-4000 HAV shedders/100,000 persons in a given year. Over the course of one generation in a stable population, the average rate of HAV infection cannot exceed the average birth rate since infection confers life-long immunity. Therefore, the infection rate in the United States is probably less the 1400/100,000/yr, the current birth rate. A tentative estimate is that there might be about 1000 infections and 100 cases of infectious hepatitis in a population of 100,000 per year or, in other words, the risk of being infected by HAV in a given year is 1% while the risk of becoming diseased is 0.1%.

The average duration of HAV shedding is about 1 month and if all infections were to result in shedding, then, based on the aforementioned tentative estimate, there would be about 80 HAV shedders in a population of 100,000 people at any given time or there would be about one HAV shedder per 1250 people. This number may be fairly constant throughout the year because the seasonal fluctuation for the reported incidence of hepatitis A has disappeared in recent years [75]. The average fecal output from an adult ranges from 100 to 200 g (wet weight)/day [76] and average wastewater production is about 450 1 (120 gal)/person/day [77]. In a large population (greater than 1250 people) the concentration of HAV-infected feces may be approximately 0.3 mg/l of sewage. This estimate can be refined by better information on a particular community's hepatitis A incidence and wastewater production.

Most HAV exit an infected person in feces, although urine and blood can also contain viruses. In an urban setting, most feces and urine will be flushed into sewers and ultimately will travel to a sewage treatment plant where the majority of virus will either be inactivated or removed.

Transport and Inactivation

Little is known about the fate of HAV once it is discharged into the environment because there are no methods for culturing it in the laboratory, but recent studies have shown that HAV is similar in many respects to enteroviruses [26] for which there is much information. Because of the epidemiologic similarities between hepatitis A and poliomyelitis, one could assume that HAV and poliovirus share similar fates after being discharged into the environment. The assumption that HAV behaves as other enteroviruses is questionable, however, and should be applied with caution until more information is available.

Although a certain amount of fecal material enters solid waste streams (primarily via disposable diapers), most of it will be flushed to public sewage systems. Sewage treatment can be accomplished by many methods but the most common system includes an initial settling period, followed by biological treatment and sludge digestion [78]. Primary sedimentation may remove more than 50% of the influent virus [79] and since the detention time through sedimentation basins is usually short, inactivation is not very important. Viruses are more likely to be inactivated by activated-sludge treatment where more than 90% of the virus may be removed by sorption as well as by inactivation [80,81]. Wastewater chlorination often follows secondary treatment, but chlorination as currently practiced is not very effective in reducing the level of viruses in effluents [82]. In general, higher dosages and longer contact times are required. Of the enteroviruses produced by a given community, fewer than 4.8% are likely to be discharged with the effluent of a conventional secondary sewage treatment plant (see the Appendix). Conceptually, this reduces the concentration of HAV-infected feces from 0.3 mg/l to less than 0.014 mg/l.

Water treatment processes reduce the concentration of HAV even further. Conventional water treatment with coagulation, sedimentation, filtration and chlorination should be capable of inactivating well over 99.95% of the virus entering the water treatment plant (see the Appendix). This reduces the concentration of HAV-infected feces to much less than 7 ng/L.

Overall Risk

The estimation of risk is completed by coupling the estimate of HAV concentration with data on average drinking water consumption and the dose-response relation. If the majority of people consume less than 2 liters/day for a dosage less than 14 ng and fewer than 50% of the persons that ingest 0.1 g of infective feces contract disease as suggested by the existing dose-response data, then the daily risk of contracting hepatitis A by drinking recycled waste water with a minimum of treatment should be less (and probably much less) than 7×10^{-8} $(0.5 \times 1.4 \times 10^{-8}/0.1)$. This risk is equivalent to 2.6 cases/100,000 persons/yr or 2.6% of the total incidence estimated previously. This can be compared to the incidence of reported waterborne incidence in Table V.

This level of risk would probably be unacceptable to most consumers but this estimate represents an upper bound to waterborne risk of hepatitis A—a risk that could be directly reduced by several methods such as improved disinfection, dilution with water from other sources and reduction of water consumption (e.g., substitution with bottled water). This estimate also represents an upper bound when considering other types of exposure. For example, swimming risks must be smaller because: (1) a small percentage of the population swims every day, (2) effluents are diluted by receiving streams, and (3) the ingestion of water is small (these counterbalance the probability that the HAV concentrations are higher).

CONCLUSIONS

- Even though the risk estimates presented here are only crude order-of-magnitude estimates, they do serve to put the role of water resources into perspective with other possible routes of disease transmission.
- Several events are required to transmit diseases. Although the probabilities of some of these events are more uncertain than others, a general statement about disease risk can be made: as the number of events increase or as the probability of any event decreases the over-all risk decreases. Thus any effort to control infectious disease will probably reduce the risk, but not all control measures have the same effect—those occurring nearest the source will have the greatest effect. Eliminating the source will eliminate risk completely. In addition, some control measures pose risks in themselves (e.g., chlorination can produce chlorinated hydrocarbons which are known carcinogens).
- While it may not be possible to reduce risks to zero, the engineer should calculate risks as objectively and completely as possible with the goal of minimizing the over-all risks as much as possible.
- A good design should include safeguards, especially disinfection capable of inactivating viruses and protozoan cysts.

APPENDIX—RISK ASSESSMENT EXAMPLE

Suspected hazard: hepatitis A (infectious hepatitis) virus.
Safety question: risk of contracting hepatitis A as the result of drinking
            recycled wastewater.
A. Hazard Identification
  1. Description of hazard
     a. an enteric virus similar to poliovirus [26]
     b. virus replicates in the intestine of infected human and is shed in
        feces and, occasionally, urine [83]
  2. Identification of harmful effects
     a. acute stage of disease results in inflammation of the liver and causes
        fever, fatigue, headache, nausea and abdominal pain
     b. disease normally lasts a few weeks
     c. disease is seldom fatal [84]
  3. Identification of benefits
     a. infection appears to produce life-long immunity [85]
B. Risk Estimation
  1. Dose-response
     a. data from limited human feeding studies indicate that fewer than
        50% of subjects fed about 100 mg of infective fecal material con-
        tracted hepatitis A [86]
        Assumption: data from a limited number of studies (4) and a limited
                    number of subjects (166) can be extrapolated to the
                    general population
     b. to be conservative assume a linear relationship between dose and
        response [39]
        Maximal risk = (experimental response) (actual dose)/(exp. dose)
  2. Production of hazard
     a. average reported incidence of hepatitis A = 20 cases/100,000 persons/
        year [75]
     b. estimated reporting efficiency = 20% of total cases [21]
     c. estimated incidence of hepatitis A = 20/0.2 = 100 cases/100,000
        persons/year
        or the risk of contracting a case of hepatitis in a given year is 0.001
     d. ratio of cases to infections = 0.1 [34]
     e. estimated hepatitis A infections = 100/0.1 = 1000/100,000 persons/
        yr or the risk of becoming infected by hepatitis A virus in a given
        year is 0.01
     f. infection and replication occurs in intestinal tissue.
        Most, if not all, viruses are shed in feces.
        Viral shedding occurs for an average of 1 month [68]

g. estimated number of shedders at any given time =
(1000 infections/100,000 persons/yr) $\times$ (1/12 yr) =
80 shedders/100,000 persons
Assumptions: all persons infected with virus shed viruses
h. probability that a person is shedding virus = 0.0008
3. Use of hazard—not applicable
4. Disposal of hazard
a. mass of infective feces per shedder = 150 g/person/d [76]
b. average wastewater production = 450 L/person/d [77]
c. average concentration of infective feces in municipal wastewater =
(80 shedders) (150 g feces/shedder/d)/(100,000 persons)
(450 L/person/d) =

$$C_0 = 0.3 \text{ mg infective feces/L}$$

Assumption: all feces are disposed of in wastewater. A small
amount (about 2%) is probably disposed of on
diapers in solid waste [48]
5. Transport and Transformation of Hazard
a. inactivation rate, $\log (C_t/C_0) = -kt$
where t = time in days
C_t = concentration at t
C_0 = initial concentration
k = first-order rate constant = 1.5×10^8 exp (−6200/
(T+273))
T = temperature in °C [49]
Assumption: hepatitis A virus are inactivated at a first-order rate
lower than most other enteroviruses
b. inactivation during flow in sewers, $C_1/C_0 = 0.95$
Assumption: t = 0.25d, T = 20°C
c. removal by sedimentation, $C_2/C_1 = 0.5$ [79]
Assumption: viruses are removed at same rate as suspended solids
d. removal and inactivation by activated sludge, $C_3/C_2 = 0.1$ [81]
e. removal by coagulation/flocculation/sedimentation, $C_4/C_3 = 0.1$
[79]
f. removal by filtration, $C_5/C_4 = 0.5$ [79]
g. inactivation by chlorination, $C_6/C_5 = 0.01$ [11]
6. Environmental exposure to hazard
a. estimated concentration of infective feces in recycled wastewater =
$C_6 = C_0(C_1/C_0)(C_2/C_1)(C_3/C_2)(C_4/C_3)(C_5/C_4)(C_6/C_5)$
= (0.3 mg/l)(0.95)(0.5)(0.1)(0.1)(0.5)(0.01) = 7 ng/l
b. estimated water consumption = 2 l/person/day [14]
c. estimated dosage = (7 ng/l)(2 l/person/day) = 14 ng/person/day

7. Maximal risk $= (0.5 \text{ cases})(0.000014 \text{ mg/person/day})/(100 \text{ mg})$
$= 7 \times 10^{-8} \text{ cases/person/day}$
$= 0.007 \text{ cases}/100{,}000 \text{ persons/day}$
$= 2.6 \text{ cases}/100{,}000 \text{ persons/yr}$

C. Social Evaluation
1. Relative risk: assuming the current incidence of hepatitis A = 100 cases/100,000 persons/yr, recycling wastewater under the conditions outlined above would increase the incidence less than 2.6%.
2. Consequence of risk
   a. as many as 1 case out of 100 may result in death
   b. acute disease lasts about 1 month
   c. estimated cost of disease = $1500 [24].

REFERENCES

1. Kates, R. W. "Assessing the Assessors: The Art and Ideology of Risk Management," *Ambio* 6(5):247 (1977).
2. "Principles for Evaluating Chemicals in the Environment," National Academy of Sciences, Washington, DC (1975).
3. "Environmental Assessment: Short-Term Tests," U.S. EPA Report No. 625/9-79-003, Research Triangle Park, NC (1979).
4. Jacobs, K. H., A. H. Goodman and G. J. Armelagos. "Disease and the Ecological Perspective," *The Ecologist* 6(2):40 (1976).
5. Hoeprich, P. D. *Infectious Diseases* (Hagerstown, MD: Harper and Row, 1977).
6. Craun, G. F. and L. J. McCabe. "Review of the Causes of Waterborne Disease Outbreaks," *J. Am. Water Works Assoc.* 65(1):74 (1973).
7. Benenson, A. S., Ed. *Control of Communicable Disease in Man* (New York: American Public Health Association, 1975).
8. Mosley, J. W. "Transmission of Viral Diseases by Drinking Water," in *Transmission of Viruses by the Water Route,* G. Berg, Ed. (New York: John Wiley and Sons, 1967).
9. Berg, G., et al., Eds. *Viruses in Water* (Washington, DC: American Public Health Association, 1976).
10. Feachem, R. G. "Water Supplies for Low-Income Communities in Developing Countries," *J. Environ. Eng. Div. ASCE* 101(EE5):687 (1975).
11. Sproul, O. J. "The Efficiency of Wastewater Unit Processes in Risk Reduction," in *Risk Assessment and Health Effects of Land Application of Municipal Wastewater and Sludges,* University of Texas, San Antonio, TX (1978).
12. "Assessment of Health Risk from Organics in Drinking Water," A Report to the Hazardous Materials Advisory Committee, Science Advisory Board, U.S. EPA, Washington, DC (April 30, 1975).
13. Hathaway, S. W. "Sources of Toxic Compounds in Household Wastewater," U.S. EPA Report No. EPA-600/2-80-128, Cincinnati, OH (1980).

14. "Drinking Water and Health," National Academy of Sciences, Washington, DC (1977).
15. Craun, G. F., L. J. McCabe and J. M. Hughes. "Waterborne Disease Outbreaks in the U.S.—1971-1974," *J. Am. Water Works Assoc.* 68(8): 420 (1976).
16. Pipes, W. O. "Water Quality and Health: Significance of Bacterial Indicators of Pollution," National Science Foundation Workshop Proceedings, Drexel University, Philadelphia, PA, April 17-18, 1978.
17. Stolzenberg, J. E. "A Stochastic Model for the Transport of a Trace Chemical in a Regional Environment," PhD Thesis, University of Wisconsin, Madison, WI (1975).
18. Center for Disease Control. "Reported Morbidity & Mortality in the United States: Annual Summary—1979," *Morbidity Mortality Weekly Rep.* 28(54):11-13 (1980).
19. Liao, S. J., F. P. Berg and R. J. Bonchard. "Epidemiology of Infectious Hepatitis in an Urban Population Group," *Yale J. Biol. Medicine* 26:512 (1954).
20. Koff, R. S., T. C. Chambers, P. O. Culhane and F. L. Iber. "Underreporting of Viral Hepatitis," *Gastroenterology* 64(6):1194 (1973).
21. Bernier, R., et al. "Viral Hepatitis Reporting," *Morbidity Mortality Weekly Rep.* 24(19):165 (1975).
22. Levy, B. S., J. Mature and J. W. Washburn. "Intensive Hepatitis Surveillance in Minnesota: Methods and Results," *Am. J. Epidemiol.* 105(2): 127 (1977).
23. Marier, R. "The Reporting of Communicable Diseases," *Am. J. Epidemiol.* 105(6):587 (1977).
24. Tolsma, D. D. and J. A. Bryan. "The Economic Impact of Viral Hepatitis in the United States," *Public Health Rep.* 91(4):349 (1976).
25. Szmuness, W. and A. M. Prince. "Epidemiologic Patterns of Viral Hepatitis in Eastern Europe in Light of Recent Findings Concerning the Serum Hepatitis Antigen," *J. Infect. Dis.* 123(2):200 (1971).
26. "Advance in Viral Hepatitis," World Health Organization, Technical Report Series No. 602, Geneva, Switzerland (1977).
27. Mosley, J. W. "Water-Borne Infectious Hepatitis," *New England J. Med.* 261(14):703, 261(15):748 (1959).
28. "Foodborne and Waterborne Disease Outbreaks," Center for Disease Control (CDC) 76-8185, U.S. Department of Health, Education and Welfare, Atlanta, GA (1976).
29. Singley, J. E., A. W. Hoadley and H. E. Hudson. "A Benefit Cost Evaluation of Drinking Water Hygiene Programs," U.S. EPA, NTIS Report No. PB 249-891 (1975).
30. Craun, G. F. and R. A. Gunn. "Outbreaks of Waterborne Disease in the United States: 1975-1976," *J. Am. Water Works Assoc.* 71(8):422 (1979).
31. Aach, R. D., J. Evans and J. Losee. "An Epidemic of Infectious Hepatitis Possibly Due to Airborne Transmission," *Am. J. Epidemiol.* 87(1): 99 (1968).
32. Rosenberg, M. L., et al. "Shigellosis From Swimming," *J. Am. Med. Assoc.* 236(16):1849. (1976).
33. Cabelli, V. J., et al. "Relationship of Microbial Indicators to Health Effects of Marine Bathing Beaches," *Am. J. Public Health* 69(7):690 (1979).

34. Szmuness, W., et al. "Distribution of Antibody to Hepatitis A Antigen in Urban Adult Populations," *New England J. Med.* 295(14):755 (1976).
35. Mosley, J. W. "Epidemiology," in *Infectious Diseases,* P. D. Hoeprich, Ed. (Hagerstown, MD: Harper and Row, 1977).
36. Metzler, P. F., et al. "Emergency Use of Reclaimed Water for Potable Supply in Chanute, Kansas," *J. Am. Water Works Assoc.* 50(8):1021 (1958).
37. Crites, R. W. and A. Uiga. "An Approach for Comparing Health Risks of Wastewater Treatment Alternatives," U.S. EPA Report No. 430/9-79-009, Washington, DC (1979).
38. Dudley, R. H., K. K. Hekiman and B. J. Mechalas. "A Scientific Basis for Determining Recreational Water Quality Criteria," *J. Water Poll. Control Fed.* 48(12):2761 (1976).
39. Interagency Regulatory Liaison Group. "Scientific Basis for Identification of Potential Carcinogens and Estimation of Risks," *Federal Register* 44(131):39858 (1979).
40. Bryan, F. L. "Diseases Transmitted by Foods Contaminated by Wastewater," *J. Food Protection* 40(1):45 (1977).
41. Plotkin, S. A. and M. Katz. "Minimal Infective Dose of Viruses for Man by the Oral Route," in *Transmission of Viruses by the Water Route,* G. Berg, Ed. (New York: John Wiley & Sons, 1967), p. 151.
42. Sabin, A. B. "Behavior of Chimpanzee—Avirulent Poliomyelitis Viruses in Experimentally Infected Human Volunteers," *Am. J. Med. Sci.* 230:1 (1955).
43. Dyer, J. C. "Impact of Pretreatment Regulations on Reuse Potential of Municipal Wastewaters," Water Reuse Symposium, American Water Works Assn. Research Foundation, Vol. 3 (1979), p. 1648.
44. Arguello, M. D., et al. "Trihalomethanes in Water: A Report on the Occurrence, Seasonal Variation in Concentration, and Precursors of Trihalomethane," *J. Am. Water Works Assoc.* 71(9):504 (1979).
45. Kavanaugh, M. C., et al. "An Empirical Kinetic Model of Trihalomethane Formation: Applications to Meet Proposed THM Standard," *J. Am. Water Works Assoc.* 72(10):578 (1980).
46. CH$_2$M Hill, Inc. "San Francisco Wastewater Treatment Pilot Plant Study," San Francisco, CA (1974).
47. Geldreich, E. E. "Bacterial Populations and Indicator Concepts in Feces, Sewage, Stormwater and Solid Wastes," in *Indicators of Viruses in Water and Food,* G. Berg, Ed. (Ann Arbor, MI: Ann Arbor Science Publishers, Inc., 1978).
48. Klein, S. A., C. G. Golucke, P. H. McGauhey and W. J. Kaufman. "Environmental Evaluation of Disposable Diapers," SERL Report No. 72-1, Sanitary Engineering Research Lab, University of California, Berkeley, CA (1972).
49. Hutzler, N. J. "Risk Assessment of Wastewater Discharges and the Effect of Wastewater Disinfection," Ph.D. Thesis, University of Wisconsin, Madison, WI (1978).
50. McCarty, P. L., et al. "Reliability of Advanced Wastewater Treatment," Water Reuse Symposium, American Water Works Association Research Foundation, Vol. 2 (1979), p. 1249.

51. Pressley, T. A. "Behavior of Volatile and Extractable Organics in Combined Biological-Physical/Chemical Treatment of Municipal Wastewater," Water Reuse Symposium, American Water Works Association Research Foundation, Vol. 3 (1979), p. 2298.

52. Saxena, J., et al. "Occurrence of Mutagens/Carcinogens in Municipal Wastewaters and Their Removal During Advanced Wastewater Treatment," Water Reuse Symposium, American Water Works Association Research Foundation, Vol. 2 (1979), p. 2209.

53. Pahren, H. R. and R. G. Melton. "Mutagenic Activity and Trace Organics in Concentrates from Advanced Wastewater Treatment Plant Effluents," Water Reuse Symposium, American Water Works Association Research Foundation, Vol. 2 (1979), p. 2170.

54. Englande, A. J., Jr. and R. S. Reimers III. "Wastewater Reuse—Persistence of Chemical Pollutants," Water Reuse Symposium, American Water Works Association Research Foundation, Vol. 2 (1979), p. 1368.

55. Gummerman, R. C., et al. "Estimating Costs for Water Treatment as a Function of Size and Treatment Efficiency," U.S. EPA Contract No. C1-76-0288 (1976).

56. SCS Engineers. "Contaminants Associated with Direct and Indirect Reuse of Municipal Wastewater," U.S. EPA Report No. EPA 600/1-78-019, Cincinnati, OH (1978).

57. Culp/Wesner/Culp. "Water Reuse and Recycling—Evaluation of Treatment Technology," Report No. OWRT/RU-79/2, Dept. of Interior, Office of Water Research Technology (1979).

58. Bitton, G. Introduction to Environmental Virology (New York: John Wiley & Sons, Inc., 1980).

59. Chang, S. L. "Modern Concepts of Disinfection," J. San. Eng. Div., ASCE, 97(SA5):689 (1971).

60. Cabelli, V. J. "New Standards for Enteric Bacteria," in Water Pollution Microbiology, R. Mitchell, Ed. (New York: John Wiley & Sons, 1978), pp. 233-271.

61. Berg, G., Ed. Indicators of Viruses in Water and Food (Ann Arbor, MI: Ann Arbor Science Publishers, Inc., 1978).

62. Fischoff, B., P. Slovic and S. Lichtenstein. "Weighing the Risks," Environment 21 (4): 17 (1979).

63. Lowrence, W. W. Of Acceptable Risk (Los Angeles, CA: William Kaufman, Inc., 1976).

64. Rowe, W. D. An Anatomy of Risk (New York: John Wiley and Sons, 1977).

65. Dienstag, J. L., W. Szmuness, C. E. Stevens and R. H. Purcell. "Hepatitis A Virus Infection: New Insights from Seroepidemiologic Studies," J. Infec. Dis. 137(3):328 (1978).

66. "Principles and Methods for Evaluating the Toxicity of Chemicals. Part I," Environmental Health Criteria 6, World Health Organization, Geneva, Switzerland (1978), p. 27.

67. Frosner, G. G., et al. "Seroepidemiological Investigation of Patients and Family Contacts in an Epidemic of Hepatitis A," J. Medical Virol. 1(3):163 (1977).

68. Rakela, J. and J. W. Mosley. "Fecal Excretion of Hepatitis—A Virus in Humans," J. Infect. Dis. 135(6):933 (1977).

69. Krugman, S., R. Ward and J. P. Giles. "The Natural History of Infectious Hepatitis," *Am. J. Medicine* 32(5):717 (1962).
70. Neefe, J. R. and J. Stokes. "An Epidemic of Infectious Hepatitis Apparently Due to a Waterborne Agent," *J. Am. Medic. Assoc.* 128(15): 1063.
71. Havens, W. P. "Period of Infectivity of Patients With Experimentally Induced Infectious Hepatitis," *J. Experim. Medicine* 83(3):251 (1946).
72. MacCallum, F. O. and W. H. Bradley. "Transmission of Infective Hepatitis to Human Volunteers," *Lancet* 2:228 (1944).
73. Davies, O. L., Ed. *Statistical Methods in Research and Production* (New York: Hafner Publishing Co., 1961).
74. Dienstag, J. L., et al. "Experimental Infection of Chimpanzees With Hepatitis A Virus," *J. Infect. Dis.* 132(5):532 (1975).
75. "Hepatitis Surveillance," Center for Disease Control Report No. 41, Phoenix, AZ (1977).
76. Ligman, K., N. Hutzler and W. C. Boyle. "Household Wastewater Characterization," *J. Environ. Eng. Div., ASCE* 100(EE1):201 (1974).
77. Metcalf and Eddy, Inc. *Wastewater Engineering: Treatment, Disposal, Reuse* (New York: McGraw-Hill Book Company, 1979), p. 19.
78. "Cost Estimates for Construction of Publicly Owned Wastewater Treatment Facilities—Summaries of Technical Data," U.S. EPA Report No. 430/9-76-011, Washington, DC (1977).
79. Berg, G. "Removal of Viruses from Sewage Effluents and Waters," Bulletin of the World Health Organization 49(5):451 (1973).
80. Balluz, S. A., H. H. Jones and M. Butler. "The Persistence of Poliovirus in Activated Sludge Treatment," *J. Hyg.* 78:165 (1977).
81. Malina, J. F., K. R. Ranganathan, B. P. Sagik and B. E. Moore. "Poliovirus Inactivation by Activated Sludge," *J. Water Poll. Control Fed.* 47(8):2178 (1975).
82. Water Pollution Control Federation. "Chlorination of Wastewater, Manual of Practice No. 4," Washington, DC (1976).
83. Zuckerman, A. J. *Human Viral Hepatitis* (New York: American Elsevier Publishing Co., 1975).
84. McCollum, R. W. "Viral Hepatitis," in *Virus Infections in Humans,* A. S. Evans, Ed. (New York: Plenum Medical Book Co., 1976).
85. Szmuness, W., et al. "Hepatitis A and Hemodialysis," *Ann. Internal Med.* 87(1):8 (1977).
86. Hutzler, N. J. and W. C. Boyle. "Wastewater Risk Assessment," *J. Environ. Eng. Div. ASCE* 106(EE5):919 (1980).

CHAPTER 14

DISINFECTION FOR REUSE

V. Dean Adams
  Division of Environmental Engineering
  College of Engineering
  Utah State University
  Logan, Utah

As discharge regulations become more stringent and our water resources become more limited, water reuse using high level sophisticated wastewater treatment technology to augment our domestic water supplies is becoming increasingly important. These systems will require relatively high capital expenditure and sophisticated operation, control and process procedures beyond the conventional treatment systems. The most important aspect of the treatment process to control will definitely be the disinfection process. Potential health hazards and social, political and economic implications also will be very important in the development of a viable working process for domestic water reuse.

Disinfection can be achieved by a variety of methods including chlorine, ozone, bromine, ultraviolet light radiation and chlorine dioxide. Other means which have been utilized to a lesser degree are metal ions (silver), iodine, bromine chloride, potassium permanganate, hydrogen peroxide, heat and sulfurous acid. This chapter focuses primarily on chlorination, ozonation, ultraviolet light radiation and chlorine dioxide disinfection techniques.

The rate of kill in disinfection processes was first formalized in the literature by Chick [1]. Chick's law simplistically stated is that for a given concentration of disinfectant, the longer the contact time, the greater the kill. This is shown in differential form by Equation 1:

$$\frac{dN}{dt} = -kN_t \tag{1}$$

where dN/dt =   rate of kill of organisms
$\quad\quad\quad k$ =   rate constant, time^{-1}, and
$\quad\quad\quad N_t$ =   number of living organisms at time t.

Integrating Equation 1 from $t = 0$ to $t = t$ and N_0 to N_t gives:

$$\frac{N_t}{N_0} = e^{-kt} \quad \text{or} \quad \ell n \frac{N_t}{N_0} = -kt \tag{2}$$

where N_0 =   number of organisms living initially and
$\quad\quad\quad N_t$ =   number of organisms living at time t.

It is often more convenient to state Equation 2 in terms of base 10 instead of base e with inversion of N_0 and N_t to eliminate the negative sign.

$$t = \frac{2.3}{k} \log \frac{N_0}{N_t} \tag{3}$$

The rate constant k can be determined (slope = $2.3/k$) by plotting t versus $\log(N_0/N_t)$ which yields a straight line on semilog paper. The rate constant k is dependent on concentration and type of chemical disinfectant, temperature, pH and the number and types of organisms.

Equation 4 empirically relates the effects of the concentration of a particular disinfectant and time [2,3]:

$$tC_{dis}^{n} = k_1 \tag{4}$$

where t =   time required to achieve a constant percentage kill,
$\quad\quad\quad C_{dis}$ =   disinfectant concentration,
$\quad\quad\quad n$ =   coefficient of dilution, and
$\quad\quad\quad k_1$ =   constant.

The constants (n and k_1) can be determined by plotting concentration versus time required to achieve a constant percentage kill on log-log paper (slope corresponds to $-1/n$). If n exceeds unity, concentration strongly influences effectiveness, but if n is less than one, contact time is more important than disinfectant concentration. When n is near one, contact time and concentration have about the same influence on disinfection effectiveness [3]. The relationship of the temperature to the rate of kill can be developed in a form of the van Hoff-Arrhenius relationship [4]:

$$\ell n \, \frac{t_1}{t_2} = \frac{E(T_2 - T_1)}{RT_1 T_2} \tag{5}$$

where t_1, t_2 =   time required for given percentage kill at temperatures T_1 and T_2, °K, respectively,

   E =   energy of activation, and

   R =   universal gas constant, 1.99 cal/°C·mol.

As can be seen from an examination of Equation 5, increasing temperature results in a shorter period of time being required for an equivalent kill.

Fair et al. [3] have proposed a relationship described in Equation 6 which indicates the more microorganisms present, the longer the time required for a given kill:

$$C^x N_p = k_2 \tag{6}$$

where C =   concentration of killing agent,

   N_p =   concentration of microorganisms reduced to a given percentage in a given time,

   x =   constant related to strength of killing agent, and

   k_2 =   constant.

The effects of pH, types of organisms and other environmental factors are not as well defined and need to be developed or established for specific sets of conditions.

CHLORINATION

Chlorine in a variety of forms is the most commonly used chemical disinfectant throughout the world. Chlorine gas (Cl_2) and calcium or sodium hypochlorite (OCl^-) are the most common chlorine compounds used in wastewater treatment plants. Chlorine gas is very soluble in water and its solubility can be calculated using Henry's law constants [5] at given temperatures and partial pressures. Chlorine hydrolyzes rapidly in water forming hypochlorous acid and hydrochloric acid as follows:

$$Cl_2 + H_2O \rightleftharpoons HOCl + H^+ + Cl^- \tag{7}$$

$$k_H = 3.94 \times 10^{-4} @ 25°C \, [6]$$

At low Cl_2 concentration (< 1000 mg/ℓ) and a pH value > 3, virtually no Cl_2 is present in solution as hydrolysis is essentially complete [2].

Hypochlorous acid partially ionizes (dissociates)

$$HOCl \rightleftharpoons OCl^- + H^+ \tag{8}$$

$$k_i = 2.9 \times 10^{-8} \text{ @ } 25°C$$

to produce a hydrogen ion and a hypochlorite ion [7]. The relative amounts of OCl^- and $HOCl$ (free available chlorine) in solution are pH dependent and are shown in Figure 1. The relative distribution of these species is very important due to the higher killing efficiency of $HOCl$.

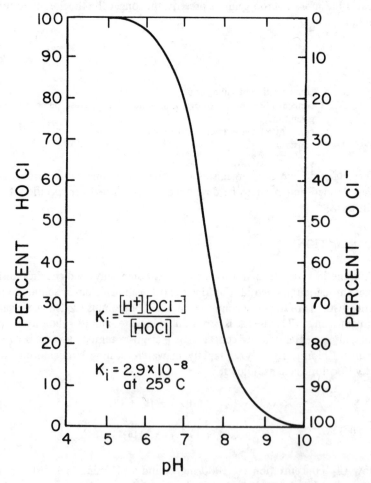

Figure 1. The effect of pH (at $25°$ C) on the relative concentrations of $HOCl$ and OCl^- in aqueous solution.

Hypochlorites (calcium and sodium), when dissolved in water, ionize to yield hypochlorite ion:

$$Ca(OCl)_2 \rightarrow Ca^{++} + 2OCl^- \tag{9}$$

$$NaOCl \rightarrow Na^+ + OCl^- \tag{10}$$

The hypochlorite ions establish an equilibrium with hypochlorous acid as illustrated in Equation 8.

Chlorine, HOCl and OCl⁻ are very active and strong oxidizing agents and react with many chemicals found in wastewater. Ammonia reacts with hypochlorous acid and forms monochloramines, dichloramines, and trichloramines as shown in the following equations:

$$NH_3 + HOCl \rightarrow NH_2Cl\,(\text{monochloramine}) + H_2O \tag{11}$$

$$NH_2Cl + HOCl \rightarrow NHCl_2\,(\text{dichloramine}) + H_2O \tag{12}$$

$$NHCl_2 + HOCl \rightarrow NCl_3\,(\text{trichloramine}) + H_2O \tag{13}$$

The chloramines which result and the relative amounts of each depend on pH, temperature and the concentrations of ammonia and hypochlorous acid. The most predominant species are the mono- and dichloramines. HOCl and OCl⁻ are referred to as free available chlorine, and chlorine in the form of chloramines is referred to as combined available chlorine.

The establishment of a residual (combined or free) for the purpose of wastewater disinfection is complicated by the reaction of many of the materials in the waste stream with the free chlorine. These side reactions complicate the use of chlorine as a disinfection agent. Many substances such as Fe^{++}, Mn^{++}, H_2S and NO_2^- (reduced forms) are oxidized very rapidly and convert the chlorine to chloride, which is ineffective as a disinfection agent. It is thus important to add sufficient amounts of chlorine to a waste water to produce a chlorine residual.

The resulting process—in which the chlorine demand is satisfied and ammonia is removed—often is referred to as breakpoint chlorination (Figure 2). After meeting the initial demand of readily reactive species (the area between 1 and 2 in Figure 2), chlorine reacts with ammonia to produce chloramines (the area between 2 and 3). As point 3 is reached, the ammonia and organic nitrogen compounds have reacted with the chlorine and free available chlorine reaches a critical stage and is sufficient to begin oxidation of the chloramines formed.

$$NH_2Cl + NHCl_2 + HOCl \rightarrow N_2O + 4HCl \tag{14}$$

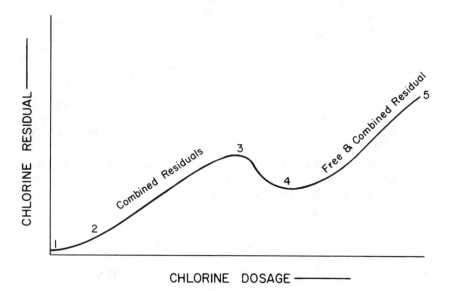

Figure 2. Generalized breakpoint chlorination curve.

$$HNH_2Cl + 3Cl_2 + H_2O \rightarrow N_2 + N_2O + 10HCl \qquad (15)$$

$$2NH_2Cl + HOCl \rightarrow N_2 + H_2O + 3HCl \qquad (16)$$

$$NH_2Cl + NHCl_2 \rightarrow N_2 + 3HCl \qquad (17)$$

With sufficient chlorine, the chloramine oxidation is essentially completed between points 3 and 4 in Figure 2. Point 4 is referred to as breakpoint. Past point 4, continued addition of chlorine results in greater amounts of free available chlorine in addition to the somewhat resistant or nonoxidizable chlorination products. The amount of chlorine required to meet a disinfection efficiency with a desired residual level is called chlorine demand.

The use of chlorine as a disinfectant for controlling microorganisms in the treatment of waste water for reuse has some potentially serious problems. When chlorine is used as the disinfectant, chloramines, simple and complex chlorinated organics and the free chlorine residual need to be removed prior to reuse as many of the products of chlorination are toxic, mutagenic or carcinogenic [8-10]. Other potentially harmful long-term adverse effects may still be unknown. Thus the removal of "chlorination by-products of waste treatment" needs to be accomplished by further treatment (i.e., physical or chemical techniques). Chlorine and chloramine residuals have been removed by the use of sulfur compounds, primarily Na_2SO_3 and SO_2 [4,11-13].

Activated carbon also has been used for dechlorination [4,13]. Granular activated carbon has the potential for adsorption of the chlorine or chlorinated products or for serving as a catalyst for chemical reaction of the chlorine (free and combined) [14]. There are also some additional benefits of granular activated carbon with the removal of many soluble organics from the chlorine treated waste water. Activated carbon is generally quite costly and is used only when dechlorination and soluble organic removal is required.

Although chlorination is a very effective and acceptable means of disinfection, beneficial uses of the waters for reuse may be limited without some highly sophisticated and costly techniques for further purification. Even then wastewater chlorination followed by reuse should be approached with caution with respect to unknown harmful constituents which remain in the water.

OZONATION

Ozone is a triatomic allotropic form of oxygen with a pale blue color and a characteristic pungent odor. Ozone is a very powerful oxidizing agent and reacts with many oxidizable substances that O_2 will not oxidize under similar conditions. Ozone is a much better oxidant than chlorine and appears to be the most powerful oxidizing agent that can be used on a practical basis for wastewater disinfection and treatment.

Ozone was first discovered in 1785 but was not formulated until 1867 [11,13]. After some experimentation in France and Holland in the 1880s [15], the first full-scale municipal water supply system using ozone as the disinfectant was placed in operation in 1906 in Nice, France. The ozone treatment was designed to disinfect and to remove the earthy odors associated with high runoff periods [16]. With the successful operation of the Nice installation, rapid development of ozonation for municipal water treatment occurred in Europe. There are now more than a thousand water treatment plants using ozone as part of the chemical treatment process [11,15]. The ozone process in these treatment plants is primarily used for taste and odor control and color removal.

Use of ozone in North America has not been widespread in the past due to the very good quality of the water to be treated and the much lower cost of the use of chlorine. Also in the United States there is the requirement of a disinfectant residual at the consumer tap which cannot be met with ozone. Even so, within the past 10 to 15 years there has been increasing interest in ozonation spawned by more stringent pollution control requirements, trends in wastewater reuse requiring additional treatment, toxicity and hazardous materials resulting from chlorine, the additional benefits derived from ozone (high DO, reduction in TOC, COD, etc.) and advances in ozone technology.

Ozone cannot be stored in bulk containers like chlorine, and must be produced on site in low concentrations. It is most practically produced by passing air or oxygen through a high voltage discharge gap. Ozone is toxic and corrosive but with the use of proper construction materials and a properly designed system, relatively few safety or handling problems occur.

Ozone is approximately 10 times more soluble in water than oxygen [11] and it follows Henry's law, but because it is produced in very low concentrations in the gas phase, its partial pressure never gets very high and so it is not found in high concentration in aqueous solution. This is one of the factors that limits the practical use of ozone for wastewater treatment.

When ozone is placed in an aqueous solution, it quickly decomposes to oxygen as follows:

$$2O_3 \rightarrow 3O_2 \tag{18}$$

In the decomposition process, intermediate species are formed which are very reactive. Ozone oxidizes ferrous and manganous ions, sulfides and sulfites and nitrites, but it does not oxidize ammonium ion under normal wastewater conditions. Unsaturated organic compounds react readily with ozone-producing aldehydes, ketones and organic acids [17]. The pH of the wastewater has very little influence on the ozonation process. Because of the highly reactive nature of ozone and its rapid decomposition, no chemical residual persists in the treated effluent that would require removal and it does not add dissolved solids to the treated waste water.

A number of commercial ozone generators are available in many designs. In general, however, they all follow the same basic principles. Gas, usually dry oxygen in air, or dry pure oxygen, is passed between two parallel surfaces separated by a space (discharge gap) and subjected to a high-voltage alternating current. A variety of dielectric materials may be used to obtain a uniform current over the surface areas. The oxygen molecules are excited to the point of splitting some of the O_2 in two; these then react with other O_2 in the system to form O_3. Ozone production is a function of many variables. The surface separation or the air gap must be sufficient to allow relatively free gas flow and a uniform current flow without requiring excessive voltage. The absolute gas pressure within the discharge gap must be sufficient to allow gas movement through the gap at the desired flow to overcome the pressure losses within the generator. This is necessary since as the gas pressure increases, the electrical resistance of the gas changes, which in turn affects the air gap and optimum voltage. The dielectric material used between the electrode surfaces should have high electrical resistance and high thermal conductivity which is difficult to obtain and is usually optimized by varying dielectric thickness, voltage and high frequency. Ozone decomposition rate back to O_2 is temperature dependent and is greatly accelerated with increasing temperature. Since much of the energy supplied to the generator is lost as heat, a significant factor is the efficient removal of heat from the generator. Commercial generators produce ozone varying in concentration from 0.5 to 3.0% by weight of the gas flow using air as the supply gas. Using oxygen, the concentration is about twice as much.

The ozone wastewater treatment system requires the establishment of intimate contact between the waste water and the ozone gas. The mixing of the ozone gas and the waste water is very important for successful use and has received considerable attention. A maximum amount of ozone must be transferred from the gas phase into the liquid phase to meet the ozone demand and to provide contact with the microorganisms in order to disinfect effectively. Packed towers, diffuser columns, pressure injectors and atomizers —all designed to optimize gas transfer—have been used as ozone contactors. Selection of the appropriate contacting system depends on the application and design required (flow rate, contact, gas-liquid exchange surface and contact time to achieve disinfection).

The effectiveness of ozone is also a function of contact time and dosage. Ozone reacts very rapidly as compared to chlorine and is classified as an oxidant and as a germicidal agent. The germicidal properties of ozone seem to be associated with its powerful oxidizing properties, and it appears to disintegrate the bacterial cell wall during disinfection [18]. This phenomenon is slightly different than the more generally accepted mode of disinfection for chlorine which seems to be cell wall penetration and destruction of an enzyme which results in the death of the organism [19].

Ozone is recognized as a very strong oxidant and an effective disinfectant, but quantitative measures of its actual disinfection potential are limited due to its highly reactive nature and the difficulty in obtaining dose-time dependent relationships. There are also interferences from the oxidizable compounds in the waste streams (which create an ozone demand) that complicate the characteristics of ozone with respect to its germicidal efficiency in waste water. For wastewater treatment, doses of 6-50 ppm have been used for disinfection and oxidation with contact times up to 10 minutes [20].

Use of ozone has been established in Europe as an effective water treatment process, and it is gaining momentum in the treatment of municipal and industrial effluents. Ozone has been shown to be an excellent disinfectant and provides other benefits such as color and odor removal; BOD, COD and TOC degradation; oxidation of many inorganics (Fe^{++}, Mn^{++}, $SO_3^=$, NO_2^-, etc.); and no additional TDS is added to the system. The effluents are oxygen saturated and could be beneficial to aquatic life in receiving streams. In terms of reuse its most effective use may be in combination with chlorination as a polishing agent for tertiary effluents or possible pretreatment.

The ozonation of rather simple compounds such as 2-propanol, acetic acid and oxalic acid yielded formaldehyde, formic acid and glyoxylic acid, most of which was oxidized to carbon dioxide [21]. The formation of epoxides and organic peroxides which may be hazardous is suggested by the reaction of ozone with many organic compounds [22-26]. At present, little has been reported as to organic end products of ozonation of waste water for reuse or what their potential health effects may be. Extensive research in the areas of reaction products and health effects from ozonated sewage is certainly needed if ozone is to be used as a disinfectant for potable water reuse.

ULTRAVIOLET RADIATION

Ultraviolet (UV) light is one of the physical methods by which water can be disinfected. In 1878 it was discovered that sunlight could destroy some bacteria and in 1893 it was shown that the UV radiation of the sunlight spectrum was responsible for the germicidal bacterial action [27,28]. To kill microorganisms, the UV electromagnetic radiation energy of the appropriate wavelength must actually strike the microorganisms [13]. The microbiocidal action seems to be most effective at a wavelength of 254 nm [29]. The actual germicidal mechanism of UV light is not known but is speculated to be associated with the UV radiation absorbed by the nucleic acids. The photochemical changes in these compounds which are essential to cell metabolism result in biological changes that are lethal to the microorganism. The effective UV radiation is that energy which reaches the organism. Water is relatively transparent to UV radiation at 250-260 nm, but penetration into waste water is affected by turbidity, color and soluble organic matter which absorbs UV radiation at this particular wavelength. Thus, the water needs to be free of absorbing materials to allow UV transmissibility for disinfection. Ultraviolet treatment does not add chemicals which is an advantage to the water, but it also does not provide a disinfecting residual.

Ultraviolet demand may be described as those components which affect the rate of absorption of the UV energy in treatment waste water in a disinfection chamber. It is desirable to have no more than 90% of the UV light expended prior to reaching the bottom of the chamber [13]. Transmission of UV light through water has been studied with respect to impurities which reduce absorptivity of the UV energy. A coefficient of absorption can be calculated from the following equation [13,30-32]:

$$T = e^{-kd} \qquad (19)$$

where T = transmission at a given wavelength (254 nm),
       e = base of natural logarithm,
       k = coefficient of absorption, and
      d = depth of water (cm).

Because each water source is different, the coefficient of absorption must be determined experimentally. There does not seem to be a good correlation between UV transmissibility and parameters such as suspended solids, turbidity, color, dissolved organic carbon, etc.

Two basic types of units are used for UV disinfection, a suspended system and a submerged system. The suspended system consists of UV lights suspended (perpendicular to flow) over a long trough through which water flows. The lamps are generally placed 10-20 cm above the water surface. The water depth is 5-8 cm and an ultraviolet monitor is placed in the bottom of the trough to measure the UV dosage. Baffles are placed in the trough perpendicular to flow to provide agitation. The major design parameters to be considered are water depth, water quality, lamp-to-water distance, lamp spacing, lamp-reflector combination and water flow rate. After evaluation of the water to be treated with respect to the above parameters and after the required dosage is determined, the physical size of the disinfecting unit can be selected.

The submerged or flow-through UV treatment system utilizes a UV lamp or lamps submerged or surrounded by water. It consists of a cylindrical quartz housing in which a lamp is placed in the center parallel to flow. The quartz chamber and tube keep the water from contacting the lamp but allow UV radiation to be absorbed by the water and the microorganisms in the water. As the water passes through the cylindrical quartz housing, particles, space and debris collect on the glass surface blocking the UV radiation. Either manual or automatic cleaning is required and is an indispensable part of the system. Cleaning frequencies depend on the quality of the water, with more frequent cleaning required of poorer quality water.

The most significant problems associated with UV systems are the cleaning frequency and the UV intensity monitoring. As UV transmission is reduced, the disinfection efficiency declines. The submerged system has the advantage of using all the lamp output other than the slight loss to the quartz glass. Submerged systems also can be installed directly in line provided there is not excessive pressure in the line.

Although UV radiation has excellent disinfection qualities on rather good quality water, the reliability of ultraviolet treatment as a wastewater disinfection process does not look promising for treatment of waste water for reuse.

CHLORINE DIOXIDE

Because the harmful side effects of chlorination are of considerable concern, other disinfectants have stimulated interest. Chlorine dioxide has been considered a rather expensive treatment alternative and has been used primarily for removal of tastes, odors, color, iron and manganese in potable water treatment. It was first used in the United States in 1944 for taste and odor control at a water treatment plant in Niagara Falls, New York [23,33]. As the benefits of chlorine dioxide have been recognized, usage has increased. In 1977, a survey [23,33] indicated that 84 water treatment plants were using chlorine dioxide in treatment processes (taste and odor control, disinfection, oxidation of organics and removal of iron, manganese and color). It also indicated that at least 495 water treatment plants in Europe were using chlorine dioxide for a variety of treatment processes.

Research has shown that treatment of water with chlorine dioxide does not produce chloroform under similar conditions to chlorination [34]. Chlorine dioxide has another advantage over chlorine in that it does not react with ammonia nitrogen in waste water [13]. Thus, chlorine dioxide has become an interesting alternative to chlorine with respect to disinfection of waste water for reuse.

In 1811, Sir Humphrey Davy first discovered chlorine dioxide by reacting potassium chlorate with hydrochloric acid [23]. Chlorine dioxide is a yellow-green, unpleasant-smelling gas which is very unstable and explosive [35]. Concentrations > 10% at atmospheric pressure are easily detonated by sunlight, heat or contact with organic materials [35]. Therefore, it must be generated at the point of use and then handled only in aqueous solution. Although it is a gas at room temperature, it can be compressed to a liquid with a density of 2.4, a boiling point of 11°C and a melting point of -59°C [35-37]. Chlorine dioxide is soluble in water to 3 g/ℓ at 25°C and 34.5 mm Hg [35]. It is normally used in an aqueous solution but does not react with the water as does chlorine. Thus it is quite volatile and is stripped rather easily from an aqueous solution by mild aeration. Because of its instability and volatility, the handling of chlorine dioxide solutions requires extreme care in design so that there is no possibility of chlorine dioxide gas coming out of solution. Because of its disagreeable odor (similar to that of chlorine gas), it is detectable at 14-17 ppm in air by the human nose [13,23]. It is distinctly irritating to the respiratory tract at concentrations of 45 ppm in air [13,23].

Chlorine Dioxide Preparation

Like ozone, the instability of the chlorine dioxide gas requires on-site generation of the chlorine dioxide. There are two principal methods for the preparation of chlorine dioxide in aqueous solution for water or wastewater treatment. One method uses sodium chlorite ($NaClO_2$) to generate chlorine dioxide.

Acid and sodium chlorite

$$5NaClO_2 + 4HCl \rightarrow 4ClO_2 + 5NaCl + 2H_2O \qquad (20)$$

Gaseous chlorine and sodium chlorite (Olin chlorine-chlorite system)

$$2NaClO_2 + Cl_2 \rightarrow 2ClO_2 + 2NaCl \qquad (21)$$

The other method uses sodium chlorate ($NaClO_3$) [13,23,37,38].

Sulfur dioxide process (Mathieson process)

$$2NaClO_3 + H_2SO_4 + SO_2 \rightarrow 2ClO_2 + H_2SO_4 + Na_2SO_4 \qquad (22)$$

Methanol process (Solvay)

$$2NaClO_3 + CH_3OH + H_2SO_4 \rightarrow 2ClO_2 + HCHO + Na_2SO_4 + 2H_2O \quad (23)$$

Chloride reduction

$$2NaClO_3 + 2NaCl + 2H_2SO_4 \rightarrow 2ClO_2 + Cl_2 + 2Na_2SO_4 + 2H_2O \quad (24)$$

For small operations such as potable water, wastewater or industrial processes, chlorine dioxide is generated from $NaClO_2$. For large production such as bleaching of paper pulp and textiles, chlorine dioxide is generated from $NaClO_3$ due to better economics for large-scale operations.

Modifications and continued refinement in the above-described processes for chlorine dioxide production will continue as increased interest is generated in the use of chlorine dioxide for water and wastewater treatment. If chlorine dioxide onset production can proceed without contamination of free chlorine in the process, a very bright outlook is forecast.

Chlorine Dioxide Reactivity

Although the chemistry of chlorine dioxide in water and wastewater treatment is not well understood, it is considered to have a greater oxidative capacity than chlorine, and thus is a more effective oxidant at lower concentrations than chlorine. At low pH, depending on the conditions in the system and on the reducing agent, the chlorine atom in the chlorine dioxide can undergo a five-valence change to yield the chloride ion [37].

$$ClO_2 + 4H^+ + 5e^- \rightleftharpoons Cl^- + 2H_2O \qquad (25)$$

In the pH range associated with water and wastewater treatment, only a one electron change generally occurs, forming chlorite ion [37].

$$ClO_2 + e^- \rightleftharpoons ClO_2^- \qquad (26)$$

Oxidation of organic materials results in the formation of aldehydes, ketones, carboxylic acids and quinones, but is not known to produce trihalomethanes [13,38]. Chlorine dioxide seems to be very effective in the destruction of phenols and can be used for removal of iron and manganese in potable water supplies [13,23,37].

Although discussion in the literature of its disinfection efficiency is rather sketchy, it appears that the unaltered chlorine dioxide molecule is responsible for bactericidal action [39]. The mechanism of kill by chlorine dioxide also was investigated by Bernarde et al. [40] who concluded that the chlorine dioxide inhibited protein synthesis which resulted in destruction of the micro-organisms. Hettche and Ehlbeck [41] reported that chlorine dioxide is more effective than either ozone or chlorine in the deactivation of poliomyelitis virus. This was shown also by Aieta et al. [42].

In summary, the most important aspects associated with the use of chlorine dioxide as a wastewater disinfectant and in water reuse are:

1. it does not ionize in aqueous solution as does chlorine, thus lower concentrations and shorter contact times are more effective in disinfection when compared to chlorine;
2. it is effective in oxidizing organics (iron and manganese) but does not react with ammonia; and
3. it can provide a very stable and long-lasting residual and is germicidally effective in the pH range of 6-10.

At present there are no known studies which indicate the production of trihalomethanes from the interaction of chlorine-free chlorine dioxide with either natural or manmade organic precursors.

The main disadvantages of chlorine dioxide are its higher cost when compared to chlorine, the unknown potential health hazard associated with residual oxidants and the formation of unknown products that potentially may be harmful. Although there may be some disadvantages and much research to be done with the use of chlorine dioxide, it may be the disinfectant of choice for water reuse systems.

DISINFECTION–REUSE OR POTENTIAL
POTABLE REUSE STUDIES

1. *Denver's Potable Reuse Project* [43,44]. In an area such as Denver, Colorado, where current water supplies are insufficient, a potable reuse supply may be an attractive alternative to meet increasing water needs. A potable reuse demonstration plant (a $22 million program over an 8-year period beginning in 1979) is in progress. The proposed treatment scheme is shown in Figure 3. A combination of ozone and chlorine dioxide will be used in the process. Ozone will be used as an oxidant to enhance biological adsorption and chlorine dioxide will be used as the final disinfectant prior to discharge. Analytical quality and health effects will be thoroughly tested throughout the project. The investigation should provide some in-depth answers to many important technical questions regarding potable reuse technology. This project with respect to potable reuse is on the forefront and will be watched by regulatory agencies, health authorities and the general public with much interest and enthusiasm.

2. Dallas Water Reclamation [31]. The project was to evaluate UV light irradiation as an alternative disinfection technique for secondary effluent.

Using a Kelly-Purdy system (shallow tray exposure system), solids depositions, large surface area requirements and inefficient uses of light emissions caused major problems, and it was suggested by this research that the system should not be used for secondary effluents. A submerged lamp system was the UV system preferred. The UV light treatment was effective in killing a type I poliovirus and an F2 coliphage and, due to better utilization of radiation emitted by the lamps, this system was more energy efficient than the shallow-tray design. It was also shown that 200 fecal coliforms/100 mℓ could be easily achieved when disinfecting a secondary effluent with UV irradiation.

It was also demonstrated that UV light is more effective on a nitrified activated sludge effluent than on a nonnitrified effluent. Typical effluent dosages for secondary effluent were in the range of 30,000-35,000 μW·sec/cm^2. It was noted that it was absolutely necessary to monitor the UV intensity for process control. From this study, UV light appears to be a potentially effective disinfectant method for reuse systems but may need to be used in combination with a chemical addition technique to maintain an active residual.

3. *Ultraviolet Teflon*®* *Tube Flow System* [45]. Another system using Teflon tubes to transport the fluid to be sterilized (Teflon transmits UV

---

*Registered trademark of E. I. du Pont de Nemours & Company, Inc., Wilmington, Delaware.

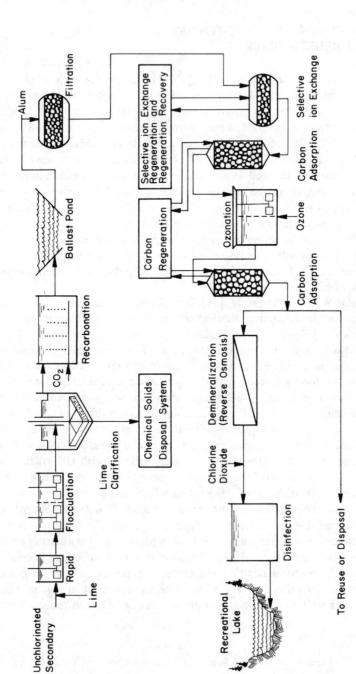

Figure 3. Process flow diagram of the proposed Denver plant [44].

light very well) has been introduced to the field of UV systems and may have some advantages over the submerged-tube systems [45]. The Teflon is chemically inert, nonwetting and virtually unaffected by UV rays. Germicidal lamps are placed between the Teflon tubes so that each tube is exposed to UV light from all sides. A reflecting system has been designed to efficiently use the UV energy emitted by the lamps. The water to be treated flows inside the Teflon tubes and varied configurations of the tubes allows a variety of flow capacities and contact times. Although data are somewhat limited, the system seems to achieve very good coliform kills (i.e., 99.86-99.998%). Problems associated with quartz-tube UV sterilizers such as fouling, leaks, frequent cleaning and breakage have been eliminated. More full-scale data are needed to further evaluate this system.

4. *Northwest Bergen County Water Pollution Control Plant Using UV Disinfection* [12]. The wastewater treatment plant is a conventional air-activated sludge plant with a design capacity of 30,000 m^3/day (8 mgd) and an average yearly flow (as of 1978) of $\sim$18,900 m^3/day (5 mgd). It utilizes four hundred 6-ft germicidal lamps each jacketed in quartz. The total power consumption by the unit is 45 KVA at an operating voltage of 480 V. The quartz tubes can be cleaned with a mechanical wiper mechanism. The influent waste water was relatively stable with mean densities of total and fecal coliforms of 3.6×10^5 and 9.5×10^4, respectively, the COD averaged 26 mg/ℓ, and the turbidity and suspended solids were at 4 FTU and 6 mg/ℓ, respectively. The water temperature averaged 22°C. During treatment the UV transmittance averaged 67%. To obtain a 99.9% removal, a dosage of 35 kW·sec/m^2 under average flow conditions and an exposure time of $\sim$2-2.5 sec would be required.

Based on preliminary costs at that time (1978), the cost associated with obtaining an effluent fecal coliform of < 200 organisms/100 mℓ in a 5 mgd plant was estimated to be 1.2¢/1000 gal and for a 10 mgd plant, 0.9¢/1000 gal. It should be stressed that these were only preliminary cost estimates.

A chlorine diffuser was located downstream of the UV unit to be in compliance with New Jersey law but was not a part of the UV disinfection process.

5. *Ozonation—Estes Park, Colorado* [12]. The plant has a design flow of 5680 m^3/day (1.5 mgd) and a flow range of from 1140 m^3/day (0.3 mgd) to 3790 m^3/day (1.0 mgd) based on tourist influx and outflux. Several unit processes were incorporated in the plant to handle the flow variations and discharge a high-quality plant effluent. Among these unit processes were flow equalization, activated sludge, attached growth nitrification, tri-media filtration and ozone disinfection. The ozone system was one of the first full-scale ozone wastewater disinfection processes in the United States.

At design capacity the ozone contact basin would give an ozone contact time of 14 min. Ozone is injected into the waste water in the contact basin through porous stone diffusers. The ozone generators are air-fed units which can provide a maximum ozone/liquid dosage of 6 mg/ℓ/unit at the plant design flow of 1.5 mgd. After many problems associated with plant startup, the effluent fecal coliform concentrations were reduced to < 200 organisms/ 100 mℓ and COD was reduced from 7.5 to 12%. Although performance data are limited, the data do indicate that good disinfection can be achieved with an ozone system.

6. *Water Factory 21* [33]. McCarty et al. [33] evaluated advanced waste-water treatment plant Water Factory 21 for a variety of parameters. They were looking at plant performance with respect to reliability. In anticipation of the time when direct potable reuse of reclaimed waste waters becomes acceptable, they wanted to determine if Water Factory 21 could produce water that meets water quality requirements. The study was primarily concerned with a variety of organics and heavy metals. The treatment processes involved were lime clarification, ammonia stripping, breakpoint chlorination, filtration, activated carbon adsorption, reverse osmosis demineralization and final chlorination. The capability of the system for reliable removal of many of the contaminants from biologically treated municipal waste water was noted. When compared to the National Interim Primary Drinking Water (NIPD) regulations of 1 coliform, MPN/100 mℓ, the coliform geometric mean was 0.01 and 9.8% of the time was < 14. It should be noted that many chlorinated organic compounds were quantified but many remain uncharacterized and the health significance of these materials still needs to be evaluated prior to direct reuse of municipal wastes for potable purposes.

Disinfection Cost

In evaluating disinfection treatment alternatives, cost is an important factor. Vogt and Regli [46] have reported costs for water treatment using EPA's computer model of unit treatment costs, capital expenditures, annual operation and maintenance expenses and annual per capita costs (Table I). From Table I, chlorination/ammoniation is less costly, followed by chlorine dioxide which is three to four times more costly and then ozone which is seven to ten times more costly than chlorination/ammoniation. All costs are in 1980 U.S. dollars and are for water treatment. Cost comparisons reported by Clark [47] are similar.

DISINFECTION—WATER REUSE

In the past few years many texts have been written, symposia proceedings published, scientific articles and reviews published, reports and documents distributed, and theses and dissertations written concerning many significant and important aspects of water and wastewater disinfection. Recent literature

Table I. Costs for a Typical Water System for Selected Treatments[a] [46]

Cost Item	Population Range Served by System					
	10,000-25,000	25,000-50,000	50,000-75,000	75,000-100,000	100,000-1 million	More Than 1 million
Ozone						
Capital expenditures	$341,000	$ 625,000	$1,080,000	$1,471,000	$2,729,000	$ 7,161,000
Anual O & M expense	17,000	27,000	44,000	60,000	148,000	943,000
Annual per capita cost	3.00	2.40	2.40	2.30	1.60	1.40
Chlorine dioxide						
Capital expenditures	$ 30,000	$ 31,000	$ 34,000	$ 38,000	$ 76,000	$ 362,000
Annual O & M expense	17,000	26,000	41,000	56,000	148,000	680,000
Annual per capita cost	1.20	0.80	0.70	0.70	0.60	0.60
Chlorination/ammoniation						
Capital expenditures	$ 12,000	$ 15,000	$ 19,000	$ 22,000	$ 31,000	$ 61,000
Annual O & M expense	4,000	8,000	13,000	18,000	51,000	253,000
Annual per capita cost	0.30	0.20	0.20	0.20	0.20	0.20
GAC						
Capital expenditures	$435,000	$ 760,000	$1,264,000	$1,733,000	$5,240,000	$21,063,000
Annual O & M expense	20,000	36,000	63,000	90,000	269,000	756,000
Annual per capita cost	3.70	3.10	3.00	2.90	3.00	2.40
GAC and ozone (BAC)						
Capital expenditures	$760,000	$1,357,000	$2,299,000	$3,141,000	$7,753,000	$27,259,000
Annual O & M expense	29,000	48,000	79,000	104,000	307,000	1,107,000
Annual per capita cost	6.20	5.00	5.00	4.50	4.20	3.20

[a]Costs are in 1980 U.S. dollars.

reviews give a good indication of ongoing current activities with respect to research and practical applications of modern techniques as applied to water and wastewater treatment.

A workshop on "Protocol Development: Criteria and Standards for Potable Reuse and Feasible Alternatives" sponsored by the U.S. Environmental Protection Agency in 1980 brought together many technical experts to consider the subject. Some of the issues discussed were: need, quality, public health effects and risks, site-specific water limited areas, sub-potable options, sources, economics, risk assessment, criteria, regulations and standards of potable reuse, consumer acceptance, adequate chemical, toxicological and microbiological tests and indicators, monitoring, disinfection, etc. There are many different views regarding the use of treated waste water for potable reuse. One of the key (and a most important) issues is the presence, generation, elimination and reduction of pathogens during treatment of waste water. Appropriate research in the disinfection of waste water with respect to new and better disinfectants and engineering technology should continue to provide information and achievable solutions to problems associated with potable reuse.

REFERENCES

1. Chick, H. "Investigation of the Laws of Disinfection," *J. Hyg.* 8:92 (1908).
2. Fair, G. M., J. C. Morris, S. L. Chong, I. Weil and R. P. Burden. "The Behavior of Chlorine as Water Disinfectant," *J. Am. Water Works Assoc.* 40:1051 (1948).
3. Fair, G. M., J. C. Geyer and D. A. Okun. *Water and Wastewater Engineering, Vol. 2* (New York, NY: John Wiley & Sons, Inc., 1968).
4. Metcalf & Eddy, Inc. *Wastewater Engineering: Treatment, Disposal, Reuse,* 2nd ed. (New York, NY: McGraw-Hill Book Co., 1979).
5. Loomis, A. G. "Solubilities of Gases in Water," in *International Critical Tables,* E. W. Washburn, Ed. (New York, NY: McGraw-Hill Book Co., 1928), pp. 255-261.
6. Jolley, R. L. "Chlorination Effects on Organic Constituents in Effluents from Domestic Sanitary Sewage Treatment Plants," ORNL-TM-4290 (1973).
7. Morris, J. C. "The Acid Ionization Constant of HOCl from 5 to 35°C," *J. Phys. Chem.* 70:3798 (1966).
8. Jolley, R. L. *Water Chlorination—Environmental Impact and Health Effects, Vol. 1* (Ann Arbor, MI: Ann Arbor Science Publishers, Inc., 1975).
9. Jolley, R. L. *Water Chlorination—Environmental Impact and Health Effects, Vol. 2* (Ann Arbor, MI: Ann Arbor Science Publishers, Inc., 1977).
10. Jolley, R. L. *Water Chlorination—Environmental Impact and Health Effects, Vol. 3* (Ann Arbor, MI: Ann Arbor Science Publishers, Inc., 1979).

11. Johnson, D. J., Ed. *Disinfection Water and Wastewater* (Ann Arbor, MI: Ann Arbor Science Publishers, Inc., 1975).
12. Venosa, A. D., Ed. "Progress in Wastewater Technology," in *Proceedings of the National Symposium*, U.S. EPA Report-600/9-79-018 (1978).
13. White, G. C. *Disinfection of Wastewater and Water for Reuse* (New York, NY: Van Nostrand Reinhold Environmental Engineering Series, 1978).
14. Stasiuk, W. N., L. J. Hetling and W. W. Shuster. "Removal of Nitrogen by Breakpoint Chlorination Using Activated Carbon Catalyst," New York State Dept. of Environmental Conservation, Tech. Paper #26 (1973).
15. McCarthy, J. J., and C. M. Smith. "A Review of Ozone and its Application to Domestic Wastewater Treatment," *J. Am. Water Works Assoc.* 66:718 (1974).
16. Richards, W. N., and B. Shaw. "Developments in the Microbiology of Water Supplies," *J. Inst. Water Eng. Sci.* 30:191 (1976).
17. Morrison, R. T., and R. N. Boyd. *Organic Chemistry*, 3rd ed. (Boston, MA: Allyn and Bacon, Inc., 1973).
18. Rosen, H. M. "Wastewater Ozonation. A Process Whose Time Has Come," *Civil Eng.*, ASCE 46(3):65 (1976).
19. White, G. C. *Handbook of Chlorination* (New York, NY: Van Nostrand Reinhold Co., 1972).
20. Evans, F. L., Ed. *Ozone in Water and Wastewater Treatment* (Ann Arbor, MI: Ann Arbor Science Publishers, Inc., 1972).
21. Kuo, P. P. K., E. S. T. Chian and B. J. Chang. "Identification of End Products Resulting from Ozonation and Chlorination of Organic Compounds Commonly Found in Water," *Environ. Sci. Technol.* 11(13): 1177 (1977).
22. Carlson, R. M., and R. Caple. "Chemical/Biological Implications of Using Chlorine and Ozone for Disinfection," U.S. EPA Report-600/3-77-066 (1977), NTIS NO. PB270:694.
23. Miller, G. W. et al. "An Assessment of Ozone and Chlorine Dioxide Technologies for Treatment of Municipal Water Supplies," U.S. EPA Report-600/8-78-018 (1978).
24. Rice, R. G., and G. W. Miller. "Reaction Products of Organic Materials with Ozone and with Chlorine Dioxide in Water," paper presented at IOI Symposium on Advanced Ozone Technology, Toronto, Ontario (1977).
25. Richard, Y., and L. Brener. "Organic Materials Produced upon Ozonation of Water Ozone–Chlorine Dioxide Oxidation of Organic Materials," International Ozone Association, Vienna, VA (1978).
26. Simmon, V. F., and R. J. Spanggord. *The Effects of Ozonation Reactions in Water, Vol. 1* (Menlo Park, CA: SRI International, 1979). NTIS No. PB 294,795.
27. Downs, A., and T. P. Blunt. "Research on the Effects of Light upon Bacteria and Other Organisms," in *Proceedings of the Royal Society of London* (London, 1878), p. 488.
28. Ward, H. M. "The Action of Light on Bacteria," in *Proceedings of the Royal Society of London* (London, 1893), p. 472.
29. Venosa, A. D. "State of the Art–Alternatives to Chlorination," paper presented at the Disinfection Seminar, Seattle, WA, 1976.

30. Koller, L. R. *Ultraviolet Radiation,* 2nd ed. (New York, NY: John Wiley & Sons, Inc., 1965), p. 312.
31. Petrasek, A. C., H. W. Wolf, S. E. Esmond and D. C. Andrews. "Ultraviolet Disinfection of Municipal Wastewater Effluents," U.S. EPA Report-600/2-80-102 (1980), p. 262.
32. Roeber, J. A., and F. M. Hoot. "Ultraviolet Disinfection of Activated Sludge Effluent Discharging to Shellfish Waters," Municipal Environmental Research Laboratory, U.S. EPA Report-600/2-75-060 (1975).
33. McCarty, *Proceedings of the Water Reuse Symposium* (Washington, DC: AWWA Research Foundation, 1979).
34. Symons, J. M. "Interim Treatment Guide for the Control of Chloroform and other Trihalomethanes," U.S. EPA Municipal Environmental Research Laboratory (1976).
35. Windhole, M., Ed. *The Merck Index, An Encyclopedia of Chemicals and Drugs,* 9th Ed. (Rathway, NJ: Merck and Co., Inc., 1976).
36. Gordon, G., R. G. Kieffer and D. H. Rosenblatt. "The Chemistry of Chlorine Dioxide," in *Progress in Inorganic Chemistry, Vol. 15,* S. J. Kippard, Ed. (New York, NY: John Wiley and Sons, Inc., 1972), pp. 201-286.
37. Weber, W. J., Jr. *Physicochemical Processes for Water Quality Control* (New York, NY: John Wiley & Sons, Inc., 1972).
38. *The Chemistry of Disinfectants in Water Reactions and Products* (Washington, DC: National Academy of Sciences, 1979). Subcommittee on Disinfectants and Products of the Safe Drinking Water Committee, NTIS No. PB 292776.
39. Bernarde, M. A., B. M. Israel, V. P. Oliveri and M. L. Granstrom. "Efficiency of Chlorine Dioxide as a Bactericide," *Appl. Microbiol.* 13:776 (1965).
40. Bernarde, M. A., W. B. Snow, V. P. Olivier and B. Davidson. "Kinetics and Mechanism of Bacterial Disinfection by Chlorine Dioxide," *Appl. Microbiol.* 15(2):257 (1967).
41. Hettche, O., and H. W. S. Ehlbeck. "Epidemiology and Prophylaxes of Poliomyelitis in Respect of the Role of Water in Transfer," *Arch Hyg. Berlin* 137:440 (1953).
42. Aieta, E. M., J. D. Berg, P. V. Roberts and R. C. Cooper. "Comparison of Chlorine Dioxide and Chlorine in Wastewater Disinfection," *J. Water Poll. Control Fed.* 52:810 (1980).
43. Rothberg, M. R., S. W. Work, K. D. Linstedt and E. R. Bennett. "Demonstration of Potable Water Reuse Technology: The Denver Project," paper presented at the Water Reuse Symposium, Washington, DC, March 25-30, 1979.
44. Work, S. W., M. R. Rothberg and K. J. Miller. "Denver's Potable Reuse Project: Pathway to Public Acceptance," *J. Am. Water Works Assoc.* 72(8):435.
45. Cruver, J. E. "Ultraviolet Disinfection of Wastewater," paper presented at the 1981 Annual Meeting of the Utah Water Pollution Control Association, Park City, UT, May 1, 1981.
46. Vogt, G., and S. Regli. "Controlling Trihalomethanes While Attaining Disinfection," *J. Am. Water Works Assoc.* 83:33 (1981).
47. Clark, R. M. "Evaluating Costs and Benefits of Alternative Disinfectants," *J. Am. Water Works Assoc.* 73:89 (1981).

SECTION 2

AQUACULTURE AND WETLANDS

CHAPTER 15

ASSESSMENT OF AQUACULTURE FOR RECLAMATION OF WASTEWATER

William R. Duffer
U.S. Environmental Protection Agency
Robert S. Kerr Environmental Research Laboratory
Ada, Oklahoma

INTRODUCTION

Wastewater aquaculture is a relatively new field in the United States, but it has a large potential for application to environmental problems. Wastewater aquaculture is broad in scope for it involves a variety of organisms, both freshwater and marine environments, organized production and treatment processes, and wastewater recycling by inputs into natural aquatic habitats. In order to develop processes for wastewater treatment or utilization by the aquaculture alternative, it will be necessary to evaluate criteria such as nature of wastewater source, system loading, land area availability and suitability, pretreatment requirements, stocking and harvesting rates, product processing and utilization, acceptability of species, climatic factors, energy requirements, discharge requirements, and economy of operation.

The major areas of wastewater aquaculture to be considered, include aquatic plants, natural and artificial wetlands, aquatic invertebrates, finfish, and highly structured or integrated systems. The types of wastewaters included are municipal and industrial effluents.

The developmental status of the major areas of treatment and reuse of municipal wastewater was reviewed by Duffer and Moyer in 1978 [1]. This review placed major emphasis on the reduction or fate of pollutants such as organics, solids, nutrients, heavy metal, residual hydrocarbons and potentially pathogenic organisms. The overall conclusion was that while exploratory studies demonstrate a definite potential for development of aquaculture processes, information was not adequate at that time for design of operational aquaculture systems to treat or utilize inputs of municipal wastewater.

The potential for development of aquaculture was discussed by Duffer and Harlin [2]. This study identified those aquaculture areas where technological progress is sufficient to proceed with developmental or relatively low risk activities through performance testing to establish process design criteria. Research needs, current status of development, and research resource requirements were considered in determining the potential of major areas of aquaculture for reclamation of municipal wastewater. Based on available technology and potential for early pay off, it was concluded that the current research and development effort should be concentrated on two major aquaculture areas: aquatic macrophytes and wetlands.

New national goals in municipal wastewater treatment place strong emphasis on the use of natural systems and improved biological processes [3-4]. The Federal Water Pollution Control Act Amendments of 1972 [PL 92-500 § 201 (d) (1)] encourage the recycling of potential sewage pollutants through the production of agriculture, silviculture or aquaculture products. More recently, the 1977 Amendments to the Act provide special incentive for the use of alternative and innovative technology for wastewater treatment management, processes and techniques which will reduce total energy requirements. To further ensure the development and utilization of these alternative and innovative technologies, the amendments authorize special research fundings for the evaluation of these systems. These amendments clearly give impetus to the development and use of aquaculture and other natural systems for treatment and management of municipal wastewater.

Aquacultural wastewater treatment processes could provide a simple and effective alternative to conventional municipal systems for treatment and management of wastewaters. Two factors must be considered in assessing the value of these processes: the role of aquaculture in the overall management of municipal wastewater and progress toward developing design criteria for operational systems. Smith et al. [4] stress that design criteria for aquacultural systems are very site specific. Areas cited as having the greatest technology base are culture of water hyacinth, natural and artificial wetlands and finfish. Municipal treatment technology fact sheets, containing two-page capsule summaries of processes and techniques commonly used for treatment and disposal of wastewater, are available for wetlands and water hyacinth culture [5]. It is expected that developed aquacultural processes will be preceded by conventional secondary treatment with current emphasis being nutrient removal and polishing treated effluents. Consideration is being given, however, to treatment of conventional primary effluents in some instances. As new processes are developed, pretreatment requirements will need to be established in order to effectively position the aquacultural component within the overall treatment of management system.

A recent engineering assessment of aquacultural wastewater treatment systems concludes that aquatic plant systems using water hyacinth and wetlands systems have a potential for removal of pollutants that can equal or may exceed that of mechanical treatment systems [6]. Water hyacinth systems were identified as having reliable design criteria for treating a wide range of conventional wastewater effluents within the geographical range where these plants grow naturally. It was concluded that although optimum, cost effective criteria were not yet available for routine design of wetland-type systems, the present technology base was sufficient to qualify these systems as innovative technology under current EPA definitions. Additionally, artificial wetlands were specified as having more promises for process development than natural wetlands.

While most of the research efforts relative to aquaculture techniques for the treatment of industrial wastewater have involved exploratory or proof-of-concept studies, aquacultural processes may offer the best opportunity for reclamation of industrial wastewater. These exploratory efforts are sufficient to indicate that use of several types of aquatic organisms is feasible for economically and effectively removing low levels of hazardous chemicals from industrial wastewater.

It is the purpose of this assessment to review recent developmental progress, identify specific areas of research needed, and project benefits expected from cost and energy savings associated with aquacultural wastewater treatment processes. In addition, aquacultural processes being utilized as well as those being considered for pilot-scale demonstration or full-scale application are addressed.

AQUATIC MACROPHYTE SYSTEMS

The municipal wastewater treatment process utilizing water hyacinth is the most advanced. The largest technology base for water hyacinth systems exists for low levels of organic loading. Several studies at diverse locations in the southern latitudes of the United States have contributed toward the development of the water hyacinth process [7-10]. Middlebrooks [11], in assessing the developmental status of aquatic plant processes, concluded that available data were adequate for design of water hyacinth treatment systems in warm temperate or tropical climates capable of producing advanced secondary effluent. The recommended design criteria for treating secondary effluent are summarized in Table I. These design criteria emphasize nutrient removal and control of algae.

Table I. Design Criteria for Water Hyacinth Wastewater Treatment Systems Receiving Conventional Secondary Effluent[a]

Parameter	Design Value		Expected Effluent Quality
	Metric	English	
Hydraulic Residence Time	>6 days	>6 days	$BOD_5 \leq 10$ mg/l
			$SS \leq 10$ mg/l
Hydraulic Loading Rate	800 m³/ha day	0.0855 mgd	$TP \leq 5$ mg/l
Depth, Maximum	0.91 meter	3 ft	$TN \leq 5$ mg/l
Area of Individual Basins	0.4 ha	1 acre	
Organic Loading	≤ 50 kg BOD₅/ ha day	≤ 44.5 lb BOD₅/ac day	
Length to Width Ratio of Hyacinth Basin	>3:1	>3:1	
Water Temperature	>20°C	>68°F	
Mosquito Control	Essential	Essential	
Diffuser at Inlet	Essential	Essential	
Dual Systems, Each Designed to Treat Total Flow	Essential	Essential	
Nitrogen Loading Rate	≤ 15 kg TKN/ ha day	≤ 13.4 lb TKN/ac day	

[a]Adapted from Ref. 11, p. 57.

In an engineering assessment by O'Brien [12] the economic incentive for including water hyacinth ponds in a wastewater treatment facility was emphasized. Comparative cost estimates for a 1 mgd municipal facility indicate that water hyacinth ponds are very attractive for achieving both advanced secondary and advanced waste treatment when compared to alternative treatment methods (Tables II and III). Based on less extensive economic analysis, the author indicates that the cost advantage of water hyacinth treatment systems continues to be favorable for municipal facilities with hydraulic capacities up to at least 10 mgd. Additional process development was believed necessary to

Table II. Comparison of Total Costs, Alternative Methods for Achieving Advanced Secondary Treatment [12] (Plant capacity: 3785 m³/day)

Treatment System	Total Cost ¢/3.785 m³ [a]	
	Favorable Conditions	Less Favorable Conditions
Oxidation pond plus hyacinths	45	74
Overland flow land treatment	96	115
Conventional advanced secondary treatment	130	130

[a]Cost includes amortized capital, operation maintenance and land.

Table III. Comparison of Total Costs Alternative Methods for Achieving Advanced Wastewater Treatment [12] (Plant capacity: 3785 m^3/day)

Treatment System	Total Cost ¢/3.785 m$^{3\,a}$
Overland flow plus hyacinths	79
Slow rate land treatment	110
Conventional advanced waste treatment	240

^{a}Cost includes amortized capital, operation, maintenance, and land.

further define conditions under which water hyacinth can be effective for advanced secondary and advanced waste treatment.

A recently prepared manual for water hyacinth wastewater treatment [13] provides system design criteria for upgrading secondary sewage lagoons and for secondary and tertiary municipal treatment systems. Example design problems and description of several case study systems are also included.

Water hyacinth wastewater treatment systems are being considered by at least 15 municipalities in the United States at this time for innovative applications in connection with the design of new or modified facilities to treat sewage. The system planned for the city of San Diego is one of these innovative facilities. A technical advisory committee, consisting of scientists and engineers having specialties in wastewater aquaculture, has provided recommendations and guidelines for design of this 1-mgd pilot-scale facility. The objective of the facility is to achieve secondary wastewater treatment utilizing water hyacinth culture units in combination with other treatment components. Cost, energy consumption, reliability and by-product utilization will be evaluated and compared with conventional wastewater treatment techniques. The San Diego aquaculture project is scheduled for EPA Construction Grants funding under the 201 Program.

Additional developmental activities which are needed for municipal water hyacinth systems include evaluation of environmental control systems for use in northern climates, testing of high levels of organic loading and processing and utilization of harvested products. Research studies currently being conducted are oriented toward addressing these developmental needs.

A study is being conducted by the Reedy Creek Improvement District at Walt Disney World, Florida, to determine the treatment effectiveness of a water hyacinth system receiving relatively high levels of organic loading. The experimental facility consists of three channels, each of which is 29 feet wide and 360 feet long. In addition to level of pretreatment studies, tests are being performed relative to biomass production, harvesting methods and frequency and composting of water hyacinth. Preliminary results indicate that the water hyacinth system can treat conventional municipal primary effluent to secondary treatment standards [14]. During the first few months of operation with each channel receiving 17,000 gpd of primary effluent, 80-90% removals were achieved for BOD_5 and total suspended solids. The maximum biomass production rate observed was 72 dry ton/acre/year. Experiments with primary effluent are continuing in order to establish critical design criteria.

Future operational phases for the water hyacinth system at Walt Disney World include tests for further refinement of water hyacinth culture as a tertiary treatment process and evaluation of the use of a protective cover during the winter months. In addition, relationships for production and utilization of energy will be examined. Energy consumed by the water hyacinth system will be compared to conventional systems, and anaerobic digesters will be operated for the conversion of water hyacinth to methane.

Experiments being conducted by NASA at the National Space Technology Laboratories (NSTL), Mississippi, are designed to evaluate methods for production of methane from water hyacinth juices using anaerobic digestion filters. This research involves laboratory experiments to determine optimum size, configuration, microbial surface area and detention time for anaerobic digestion filters to provide biomass containing a high percentage of methane.

Very little developmental effort has been directed toward using vascular aquatic plants other than water hyacinths for municipal wastewater treatment processes. Exploratory or proof-of-concept experiments indicate that a process utilizing duckweed has good potential for development [15-17]. In an engineering assessment of aquatic plant processes, O'Brien [12] recommended that additional developmental research effort be directed toward use of duckweed in wastewater treatment systems. Wolverton and McDonald [18] studied a system receiving wastewater from a small residential development. The system was composed of an aerated algal pond followed by a duckweed pond. Good removals for total suspended solids and BOD_5 were achieved with the discharge attaining mean annual levels of 18 and 15 mg/l, respectively.

A large variety of aquatic vascular plants should be tested in order to assess their potential for use in developing wastewater treatment processes. An annotated bibliography on the environmental requirements of aquatic plants [19] provides extensive background information on emergent and floating species likely to be considered for testing. Emergent aquatic plants discussed include species of catttails, reeds, rushes, sedges and grasses. Water fern, duckweed and water hyacinth are among the species of floating plants discussed.

A study was recently initiated by Roseville, California, to determine the treatment effectiveness of aquatic plant culture systems. The bacterial populations associated with the root systems of species cultured also will be quantified. Water primrose, water hyacinth, duckweed, cattails and bullrush are among the aquatic species that will be cultured utilizing municipal sewage.

The treatment effectiveness of several species of aquatic and nonaquatic vascular plants is being determined by Cornell University researchers using a specialized hydroponic culture system. The objective of the study is to determine the feasibility of using plants grown in a nutrient film technique (NFT) system for treatment of municipal wastewater. The approach involves testing several plant species to establish the effectiveness of removal of gross and trace inorganic and organic chemicals relative to pretreatment of municipal sewage. Biomass production rates using the NFT system will also be determined. Water cress, cattails, reed canarygrass, Bermuda grass and napier grass are among the species being evaluated.

Although aquatic macrophytes systems for treatment and use of industrial wastewaters are in the very early stages of development, this approach may offer one of the best opportunities for beneficial use of this type of wastewater. O'Brien [12] suggests that developmental progress in this area has good application potential due to the low costs associated with aquatic plant system.

Research studies are needed to determine the feasibility of using aquatic plants to economically and effectively remove low levels of hazardous chemicals from industrial wastewater, and to develop and demonstrate practical systems and techniques for industrial wastewater treatment. Many industrial wastewaters contain toxic organic and inorganic chemicals. Often these contaminated industrial waters are released in small quantities into the environment, or stored in earthen lagoons. Neither disposal method is entirely satisfactory since aquatic organisms may biomagnify some chemicals even though they are released in very dilute concentrations, and wastewater stored in lagoons may seep into and contaminate ground waters. Ion exchange resins are expensive and sometimes not effective. An inexpensive and effective alternative has been demonstrated. Some vascular plants have been shown to remove low concentrations of some organics such as phenol, mirex, toxaphene, kepone, PCBs and others, and heavy metals such as cadmium, lead, chromium, copper, zinc, mercury and cobalt [20-27]. A small water hyacinth chemical wastewater treatment system has been successfully operating at NASA's NSTL in Mississippi for several years.

One of the most serious water pollution problems facing the United States today is groundwater contamination with hazardous chemicals. The prevention of these problems is an area where aquaculture techniques using vascular aquatic plants may have very promising potential. The use of aquacultural methods to treat wastewaters from industry is much more complicated than treating domestic wastewater. Industrial wastewaters may contain hundreds of different types of chemicals at varying concentration levels. Aquacultural methods using vascular aquatic plants have the greatest potential as final filtration systems for treating industrial wastewaters before discharge into rivers and streams. This method shows promise in treating leachates from hazardous chemical dump areas. Research efforts should be concentrated on chemical disposal areas and on industrial effluents containing low levels of chemical pollutants.

WETLAND SYSTEMS

Research studies in several regions of the United States indicate that wetlands can be an effective wastewater treatment mechanism. Nearly all of the technology developed for wetland systems to treat municipal wastewater has been oriented toward low levels of organic loading. In a recent engineering assessment of the status of these processes Tchobanoglous and Culp [28] point out that both natural and artificial wetland treatment systems represent an extremely attractive alternative to conventional secondary treatment and advanced treatment of municipal wastewater. The authors feel that while these systems should be considered within the scope of the innovative and alternative technology provisions of Public Law 95-217, additional technology is needed before their use can be considered routine. Artificial wetland systems were identified as having the greatest potential for development of usable design criteria for general process application.

The University of Michigan conducted research studies over a period of five years (1972-1977) involving discharge of secondary wastewater into peat wetlands [29]. Wastewater was stored for eight months and discharged for four months during the warm season. Research results indicated that nitrogen and phosphorus were removed from 100,000 gal/day within a five-acre area. Suspended solids were deposited close to the point of discharge and odor problems were slight. Based on these research results, wetlands treatment came under consideration for the required 1978 plant expansion at Houghton Lake, Michigan. Williams and Works, Inc., designed a full-scale system for wetlands treatment as a plant expansion alternative [30]. The concept and design were ultimately approved by regulatory agencies, and the construction phase of the 201 Facilities Planning process was completed during 1978. The Houghton Lake Wetlands Treatment System operated successfully in the treatment of 65 million gallons during the summer of 1979. Approximately $700,000 in capital costs savings relative to upland irrigation was realized for the Houghton Lake system.

The University of Florida's Center for Wetlands conducted studies on cypress domes near Gainesville, Florida, which received secondary effluent from a small residential area [31]. Results of this work indicate that cypress dome treatment was feasible for treatment of secondary municipal effluents. From monitoring of ground water it was determined that 98% of the total nitrogen and 97% of the total phosphorus was removed. Cypress wetlands were shown to be cost effective compared to spray irrigation and physical/chemical treatment facilities for the combination of variables utilized in the analysis.

The treatment effectiveness of a wetland system composed of shallow shrub swamp and deep marsh was determined for inputs of secondary wastewater from the town of Concord, Massachusetts [32]. The shrub swamp and deep marsh had surface areas of 6 and 48 acres, respectively. The mean discharge to the wetland system during the period of study was 0.61 mgd and detention time was calculated to be about 5 days. Rates of removal for nutrients and BOD$_5$ displayed significant variations which were attributed to seasonal changes and difference in wetland subtypes. Mean annual removal of BOD$_5$ from the shrub swamp and deep marsh sections of the wetlands were 20.8 and 2.6 lb/acre/day, respectively. Overall removal rates for the wetland for nitrogen and phosphorus were in the range of 0.05 lb/acre/day for each of these nutrients.

During the fall of 1974, Mt. View Sanitary District at Martinez, California, established eight acres of artificial wetlands [33]. In 1978, the wetlands system was expanded to cover a total of 21 acres. The primary purpose of the system was to demonstrate substantial environmental benefits derived from the use of secondary municipal effluents (1.6 mgd) for creating wildlife habitat. The system has been very successful, with 86 species of birds, 63 plant species, 34 species of aquatic invertebrates, and 22 species of other animals having been identified. Improved water quality has been an additional benefit. Capital costs for the 21-acre system were $94,000 and the District funded the entire project. Operation and maintenance costs are expected to average about $1200 per year.

An artificial wetland system established at Vermontville, Michigan (population 975), treats municipal wastewater effluent from two facultative sewage lagoons with a surface area of 10.9 acres. Williams and Sutherland [30] describe the system as a seepage wetland which was designed for phosphorus removal. Four diked wetland plots having a total surface area of 11.5 acres were constructed in an upland area and allowed to colonize naturally. Several species of emergent aquatic plants became established, the predominant type being cattail. The observed seepage rate of wastewater averages 4 in./week with some surface water being discharged from the system. Water quality determinations were made for the influent, lagoon effluent, wetland water, groundwater and final surface overflow. Results of this study indicate that nutrient removals for the system were very effective (Table IV). Although final surface overflow from the system slightly exceeded the strict limit (0.5 mg/l) for phosphorus, groundwater levels are well below discharge requirements. The slight increase in total phosphorus levels from the lagoon discharge to the wetland water is attributed to decomposing detrital organic matter. It was concluded that 95% of the phosphorus removal occurred in the upper three feet of soil.

In comparing capital costs for seepage wetlands and spray irrigation, cost differences noted between the two methods were in requirements for chlorination, isolation land, site grading, electrical power and irrigation structures (Table V). Capital cost comparisons for system components having cost differences favor the seepage wetland process by nearly constant amounts within

Table IV. Mean Changes in Water Quality for the Municipal Wastewater Treatment Facility at Vermontville, Michigan[a]

		Concentration (mg/l)			
Parameter	Influent	Lagoon Effluent	Wetland Water	Groundwater	Surface Overflow
Chloride	280	207	157	124	123
NH_3N	37	2.5	2.0	0.7	
NO_3N	1.3	1.0	1.2	1.4	
TKN	81	6.5	5.0	3.7	
P	5.3	1.8	2.1	0.04	0.64

[a]Adapted from Ref. 30.

the range of flows examined. An assumed land acquisition cost of $1154/acre was considered to be the same for both treatment methods at application rates of 4 in./wk. Costs for transmission pipelines as well as pumping stations were also considered to be the same for both treatment methods. Taking into account system component areas having the same costs and the nearly constant amount of cost differences for increased daily flow, it becomes apparent that the greatest percentage of overall cost advantage exists for the smaller capacity seepage wetland systems.

Wetland wastewater treatment systems are being considered at this time by 12 municipalities in the United States for innovative applications in connection with the design of new or modified facilities to treat sewage. The system pro-

Table V. Unique Capital Costs Comparison for Seepage Wetlands and Spray Irrigation[a]

Treatment Condition and System Components	Total Daily Flows (mgd)[b]			
	0.031	0.124	0.28	0.50
Spray Irrigation				
Treatment Acreage	4	16	36	64
Isolation Land Costs	$ 90,012	$116,554	$143,096	$170,792
Spray Equipment Costs	13,840	55,360	124,560	221,440
Chlorination Costs	28,838	28,838	28,838	28,838
Power Facility Costs	11,535	11,535	11,535	11,535
Subtotal Spray Cost (A)	144,225	175,373	308,029	432,605
Seepage Wetland				
Treatment Acreage	4	16	36	64
Gated Pipe and Fixtures	10,100	20,200	30,300	40,400
Site Preparation Costs	16,000	64,000	144,000	256,000
Subtotal Wetland Cost (B)	26,100	84,200	174,300	296,400
Cost Differences				
A-B	118,125	128,087	133,729	136,205

[a]Adapted from Ref. 30.
[b]Application rates are 4 in./wk for 6 months each year.

posed for Riverside, Iowa, provides an example of one of these innovative facilities. Design plans and specifications for this facility are in the final stages. A riverine wetland area of 25 acres will be used to treat effluent from conventional sewage lagoons serving a population of about 800 people. Most of the innovative applications are being considered for relatively small-sized systems, but one large-scale application has been suggested. Ward [34], in a discussion of wastewater treatment for Las Vegas, Nevada, indicates that a 90 mgd marsh treatment system has been proposed for the Las Vegas Wash as an alternative to advanced wastewater treatment.

Additional research and development relative to natural wetland wastewater treatment processes is needed to determine pretreatment requirements, hydraulic loadings, fate of nutrients and suspended solids, mechanism of microbial mediated processes and effects of changes in wetland hydrology. The effects of inputs of wastewater on natural populations of vegetation and wildlife should be assessed. In addition to establishing enhancement and degradation criteria for wastewater inputs into natural wetlands, the fate of materials believed to be toxic or of concern to human health should be determined. Heavy metals, pathogenic organisms, chlorinated hydrocarbons and pesticides are among the important public health considerations.

Regarding artificial wetland systems, it will be necessary to compare and evaluate various wetland plants and system types in terms of climatic variations for treatment effectiveness and system stability. Pretreatment requirements and hydraulic loadings will be very important in establishing design criteria for artificial wetland wastewater treatment processes. The feasibility of commercial utilization of natural and artificial wetland products should be determined. Critical factors in this regard are harvesting techniques, product processing and product evaluation.

Current efforts by researchers at The University of Michigan are aimed at establishing design criteria for wetland wastewater treatment. The objective of this study is to collect, organize, and interpret the existing data on the design and performance of wetland wastewater treatment systems. Available information on aesthetics, economics, products and values will be included. Outputs are to include an annotated bibliography of wetland wastewater treatment information sources, a technical summary and analysis of quantitative performance predictions, and a preliminary design manual that presents an interpretation and summary of design and operating conditions, economics, products, values, constraints and impacts.

In experiments being conducted in the artificial wetland area, emphasis is placed on cultural practices necessary for establishment of systems and treatment effectiveness for application of municipal secondary wastewater effluent. The San Diego Regional Water Reclamation Agency has established several artificial wetlands plots at the Padre Dam Municipal Water District in Santee, California. The site consists of 14 experimental plots covering an area of about one acre. A variety of emergent aquatic plants as well as wastewater application rates are being tested during the first phase of the project. After the most suitable and best application rates are determined, the second phase will consist of demonstrating the treatment effectiveness of the most promising combination.

The University of Wisconsin, Stevens Point, is investigating the effectiveness of a natural peat wetlands in the treatment of wastewater from primary and secondary sewage lagoons. A full-scale peat bog system which receives sewage lagoon effluent during the summer and fall months from the town of Drummond, Wisconsin, is being evaluated in terms of water quality changes through the system. Smaller-scale experimental plots are being utilized to test the treatment effectiveness of peat wetlands for inputs of primary sewage lagoon effluent.

OTHER AQUATIC SYSTEMS

Experiments involving exotic filter feeding finfish have been conducted by the Arkansas Game and Fish Commission at Benton, Arkansas [35]. The system consists of six ponds, each having a surface area of about 4 acres. The ponds are operated in series with the first two ponds serving as stabilization and plankton culture units. The remaining four ponds of the series are stocked with silver and bighead carp. The influent flow rate is 0.45 mgd and the residence time for the entire pond series is about 72 days. The average influent concentration for BOD_5 is 260 mg/l and that for total suspended solids is 140 mg/l. Results indicate that this process has good potential for municipal application achieving a BOD_5 reduction of 96%, and a total suspended solids reduction of 86%. In addition, an annual fish production rate of 6000 lb/acre was achieved. A health effects study is being conducted at this time in order to determine the feasibility of a variety of uses for the large amounts of fish produced. This study is concentrating on levels of enteroviruses and pathogenic bacteria. Due to the successful operation of the system at Benton, at least three municipalities in Arkansas are considering this process in connection with the design of new or modified facilities to treat sewage.

Annual fish production rates in Arkansas are comparable to those achieved in China at the Pearl River Fisheries Research Institute. Culliton [36] reports that the annual average fish production in still water ponds was 2900 lb/acre with maximum fish production reaching 14,500 lb/acre. The four species grown in the ponds at Pearl River were grass carp, silver carp, bighead carp and mud carp. Ponds were fertilized with animal and human excrement following a fermentation process for removal of pathogens.

Additional species of finfish should be tested to determine their potential for use in municipal wastewater treatment processes. An annotated bibliography on the environmental requirements of fish provides extensive background information on five species: channel catfish, rainbow trout, common carp, tilapia and Sacramento blackfish [37]. Effects of specific environmental parameters on mortality, physiology and growth of each species are emphasized.

A wastewater treatment process which utilizes aquatic organisms along with other components was developed by Solar Aquasystems, Inc., Encinitas, California [38]. The process consists of an anaerobic cell followed by aerobic lagoons. A greenhouse pond cover, a diffused aeration system and high surface area fixed-film substrates are components utilized in the process. Floating aquatic plants such as water hyacinth and duckweed are used for polishing effluent and nutrient removal. Based on impressive reductions for BOD_5 and suspended solids in small scale experimental units, a 0.35 mgd municipal wastewater treatment system utilizing this process was constructed at Hercules, California.

An aquatic process to treat industrial wastewater was developed by researchers at the Texas Division of the Dow Chemical Company [39]. The process was designed for removing organic contamination from a hypersaline (55 to 70 g/l NaCl) wastestream. Two separate but sequential steps are employed in the process: culturing of planktonic algae and culturing of brine shrimp and rotifers. The pilot plant system tested had a surface area of 0.3 ha and received 129 m^3/day of effluent from a 6-ha biological waste treatment aeration lagoon. Operational results indicate that removals from the aerated lagoon effluent are up to 89% for total suspended solids, 89% for BOD_5 and 88% for ammonia. Analysis of brine shrimp cultured in the industrial wastewater treatment system indicated that there were no significant accumulations of industrial waste chemicals in these herbivores. Researchers estimated that a full scale operation utilizing this type of wastewater treatment facility could produce over 1430 metric tons of brine shrimp and 3100 metric tons of rotifers per year.

ENERGY AND COST REQUIREMENTS

Aquatic treatment systems utilizing natural processes conserve a significant amount of energy when compared to conventional wastewater treatment systems. Energy production utilizing biomass from aquatic plants cultured in municipal and industrial wastewater treatment systems should be considered for contribution to energy reserves in the United States. Benemann [40,41] estimated that long-term aquatic biomass production from wastewater aquaculture systems would have an annual fuel value of about 0.1 quads (1 quad = 1 $\times$ 10^{15} Btu). Current annual energy use in the United States is about 80 quads. It was concluded that while the total energy contribution potential of these systems was limited, wastewater aquaculture processes offered the most immediate application of aquatic plants in energy production and conservation. In discussing the potential of wastewater aquaculture processes, Benemann states that "Although still not perfected, these technologies have considerable potential in the treatment of municipal and industrial wastes, particularly where nutrient removal may be required for installations of small to medium scale (up to 10 mgd). If developed, these technologies may be capable of net energy production in a cost-effective manner" [40].

Selected land and aquatic treatment systems were compared to commonly used conventional treatment facilities by Tchobanoglous et al. [42] to assess the consumption of energy and resources. Initial construction costs for facilities (Table VI) as well as annual energy and labor requirements and annual costs for parts and supplies (Table VII) were compared for systems that can

Table VI. Land Requirements and Construction Costs for Alternative and Conventional Treatment Systems with a Plant Capacity of 1 mgd [a]

System Description	Designation	Land Required (acre)	Total Construction Cost ($ \times 10^{-6}$)[b]
Conventional Treatment			
1. Activated sludge + chlorination	As+c	4.0	1.600
2. High rate trickling filter + chlorination	TF+c	5.0	1.700
Land Treatment			
3. Primary + overland flow + chlorination	P+OF+c	61.5	1.050
4. Primary + rapid infiltration	P+RI	21.5	0.790
5. Primary + slow rate (solid set sprinklers)	P+SR (sss)	161.5	1.500
6. Primary + slow rate (ridge and furrow)	P+SR (r+f)	161.5	1.130
7. Facultative pond + overland flow	FP+OF	90.0	1.950
8. Facultative pond + rapid infiltration	FR+RI	50.0	1.690
Aquatic Treatment			
9. Primary + artificial wetland + chlorination	P+AW+c	41.5	0.900
10. Primary + water hyacinths + chlorination	P+WH+c	21.5	0.830
11. Facultative pond + artificial wetland	FP+AW	70.0	1.800
12. Facultative pond + water hyacinths	FP+WH	50.0	1.730

[a] Adapted from Ref. 42.
[b] The cost of land is not included.

be used to achieve secondary treatment of municipal wastewater (BOD_5 = 10 mg/l and TSS = 30 mg/l). The activated sludge and trickling filter processes were used as a basis of comparison for conventional systems. Land treatment processes evaluated were overland flow, rapid infiltration and slow rate application. Aquatic treatment processes evaluated were artificial wetlands and water hyacinth covered basins. All of the land and aquatic treatment processes were combined with pretreatment components to achieve the desired level of treatment. From examination of values presented in Tables VI and VII, it is clear that several land and aquatic treatment alternatives achieve very significant savings in costs and energy consumption for the facility capacity evaluated.

CONCLUSIONS

Technology development in connection with natural processes to treat municipal wastewater has been substantial during recent years in the United States. This development is consistent with new national attitudes which are oriented beyond resource utilization and environmental protection to the broader concepts of resource recycling and environmental enhancement. Wastewater treatment processes utilizing water hyacinth, natural and artificial wetlands, and finfish are in varying stages of development. Due to treatment effectiveness and the potential for cost and energy savings, aquacultural pro-

Table VII. Annual Energy Requirements, Labor Requirements, and Parts and Supplies Costs for Conventional and Alternative Treatment Systems with a Plant Capacity of 1.0 mgd[a]

System	Labor Required (person hr/yr)	Parts & Supplies ($/yr)	Total Energy[b] (Btu/y × 10⁻⁶)
AS+c	5500	16,000	7480
TF+c	4200	14,000	6084
P+OF+c	4200	13,000	3795
P+RI	4000	12,000	3124
P+SR (sss)	4200	13,000	5405
P+SR (r+f)	4200	13,000	3485
FP+OF	5700	18,000	3765
FR+RI	5500	17,000	3470
P+AW+c	3000	6,000	3479
P+WH+c	4000	8,000	3950
FP+AW	4500	11,000	3450
FP+WH	5500	13,000	3921

[a]Adapted from Ref. 42.
[b]Includes primary energy as electricity and fuel and secondary energy relative to plant construction, parts and supplies and chemicals, etc.

cesses are being considered by at least 30 municipalities in the United States as favored alternatives to conventional wastewater treatment facilities. Additional research is needed for development of new processes to treat municipal wastewater as well as for refinement of existing processes. In view of the encouraging results from exploratory or proof-of-concept studies involving aquacultural techniques for treatment of industrial wastewater, the research effort in this area should be expanded. Energy from plant biomass produced in municipal and industrial aquacultural wastewater treatment systems may constitute an additional benefit of these processes.

REFERENCES

1. Duffer, W. R., and J. E. Moyer. "Municipal Wastewater Aquaculture," U.S. EPA Report 600/2-78-110 (1978).
2. Duffer, W. R., and C. C. Harlin, Jr. "Potential of Aquaculture for Reclamation of Municipal Wastewater," in *Proceedings of the 1979 AWWA Water Reuse Symposium* (Denver, CO: American Water Works Association, 1979), pp. 740-746.
3. Hais, A. B., J. M. Smith and L. K. Lim. "An Overview of EPA's Innovative Alternative Technology Program," paper presented at the 52nd Annual Conference of the Water Pollution Control Federation, Houston, TX, October 8-12, 1979.
4. Smith, J. M., J. J. McCarthy and H. L. Longest II. "Impact of Innovative and Alternative Technology in the United States in the 1980's," paper presented at the 7th United States/Japan Conference on Sewage Treatment Technology, Tokyo, Japan, May 1980.
5. "Innovative and Alternative Technology Assessment Manual," MCD-53, U.S. EPA Report 430/9-78-009 (1978).
6. Reed, S. C., R. K. Bastian and W. J. Jewell. "Engineering Assessment of Aquaculture Systems for Wastewater Treatment: An Overview," in *Aquaculture Systems for Wastewater Treatment; Seminar Proceedings and Engineering Assessment,* MCD-67, U.S. EPA Report 430/9-80-006, University of California, Davis, September 11-12, 1979.
7. Wolverton, B. C. "Engineering Design Data for Small Vascular Aquatic Plant Wastewater Treatment Systems," in *Aquaculture Systems for Wastewater Treatment; Seminar Proceedings and Engineering Assessment,* MCD-67, U.S. EPA Report 430/9-80-006, University of California, Davis, September 11-12, 1979.
8. Dinges, R. "Development of Hyacinth Wastewater Treatment Systems in Texas," in *Aquaculture Systems for Wastewater Treatment; Seminar Proceedings and Engineering Assessment,* MCD-67, U.S. EPA Report 430/9-80-006, University of California, Davis, September 11-12, 1979.
9. Swett, D. "A Water Hyacinth Advanced Wastewater Treatment System," in *Aquaculture Systems for Wastewater Treatment; Seminar Proceedings and Engineering Assessment,* MCD-67, U.S. EPA Report 430/9-80-006, University of California, Davis, September 11-12, 1979.

10. Stewart, III, E. A. "Utilization of Water Hyacinth for Control of Nutrients in Domestic Wastewater—Lakeland, Florida," in *Aquaculture Systems for Wastewater Treatment; Seminar Proceedings and Engineering Assessment,* MCD-67, U.S. EPA Report 430/9-80-006, University of California, Davis, September 11-12, 1979.

11. Middlebrooks, E. J. "Aquatic Plant Processes Assessment," in *Aquaculture Systems for Wastewater Treatment; Seminar Proceedings and Engineering Assessment,* MCD-67, U.S.EPA Report 430/9-80-006, University of California, Davis, September 11-12, 1979.

12. O'Brien, W. J. "Engineering Assessment: Use of Aquatic Plant Systems for Wastewater Treatment," in *Aquaculture Systems for Wastewater Treatment; Seminar Proceedings and Engineering Assessment,* MCD-67, U.S. EPA Report 430/9-80-006, University of California, Davis, September 11-12, 1979.

13. Gee and Jenson, Engineers-Architects-Planners, Inc. "Water Hyacinth Wastewater Treatment Design Manual," Project Report for NASA/ National Space Technology Laboratories, NSTL Station, MS (1980).

14. Lee, C. "Water Hyacinth Wastewater Treatment System at Walt Disney World, Florida," Project report for NASA Contract No. NAS 13-76 (1980).

15. Sutton, D. L., and W. H. Ornes. "Phosphorus Removal from Static Sewage Effluents Using Duckweed," *J. Environ. Qual.* 4:367-370 (1975).

16. Harvey, R. M., and J. L. Fox. "Nutrient Removal Using Lemna Minor," *J. Water Poll. Control Fed.* 45:1928-1938 (1973).

17. Culley, Jr., D. D., and E. A. Epps. "Use of Duckweed for Waste Treatment and Animal Feed," *J. Water Poll. Control Fed.* 45:337-347 (1973).

18. Wolverton, B. C., and R. C. McDonald. "Secondary Domestic Wastewater Treatment Using a Combination of Duckweed and Natural Processes," (submitted to: *J. Water Poll. Control Fed.*) (1980).

19. Stephenson, M. et al. "The Environmental Requirements of Aquatic Plants," Publication No. 65, University of California, Davis (1980).

20. Rhynes, S., E. Garrity and F. Shore. "Water Hyacinths for Removal of DDT from Water," NASA Progress Report, National Space Technology Laboratory, NSTL Station, MS (1978).

21. Smith, R. W., and F. Shore. "Water Hyacinths for Removal of Mirex from Water," NASA Progress Report, National Space Technology Laboratory, NSTL Station, MS (1977).

22. Smith, R. W., and F. Shore. "Water Hyacinths for Removal of Kepone from Water," NASA Progress Report, National Space Technology Laboratory, NSTL Station, MS (1978).

23. Shore, F., W. VanZandt and W. Smith. "Water Hyacinths for Removal of Toxaphene from Water," *J. Missi. Acad. Sci.* 23:17-23 (1978).

24. Wolverton, B. C., and R. C. McDonald. "Wastewater Treatment Utilizing Water Hyacinths (*Eichnornia crassipes*) (Mort Solms)," Proceedings of the National Conference on Treatment and Disposal of Industrial Wastewaters and Residues, April 26-28, 1977, Houston, TX, pp. 205-208.

25. Wolverton, B. C., and M. M. McKown. "Water Hyacinths for Removal of Phenols from Polluted Waters," *Aquatic Bot.* 2:191-201 (1976).

26. Wolverton, B. C., and D. D. Harrison. "Aquatic Plants for Removal of Mevinophos from the Aquatic Environment," *J. Miss. Acad. Sci.* 19:84-88 (1973).

27. Wolverton, B. C., R. M. Barlow and R. C. McDonald. "Application of Vascular Aquatic Plants for Pollution Removal, Energy, and Food Production in a Biological System," in *Biological Control of Water Pollution,* J. Toubier and R. W. Pierson, Eds. (Philadelphia, PA: University of Pennsylvania, 1976, pp. 141-149.
28. Tchobanoglous, G., and G. L. Culp. "Wetland Systems for Wastewater Treatment; an Engineering Assessment," in *Aquaculture Systems for Wastewater Treatment; Seminar Proceedings and Engineering Assessment,* MCD-67, U.S. EPA Report 430/9-80-006, University of California, Davis, September 11-12, 1979.
29. Kadlec, R. H. "Wetland Tertiary Treatment at Houghton Lake, Michigan," in *Aquaculture Systems for Wastewater Treatment; Seminar Proceedings and Engineering Assessment,* MCD-67, U.S. EPA Report 430/9-80-006, University of California, Davis, September 11-12, 1979.
30. Williams, T. C., and J. C. Sutherland. "Engineering, Energy and Effectiveness Features of Michigan Wetland Tertiary Wastewater Treatment Systems," in *Aquaculture Systems for Wastewater Treatment; Seminar Proceedings and Engineering Assessment,* MCD-67, U.S. EPA Report 430/9-80-006, University of California, Davis, September 11-12, 1979.
31. Fritz, W. R., and S. C. Helle. "Cypress Wetlands as a Natural Tertiary Treatment Method for Secondary Effluents," in *Aquaculture Systems for Wastewater Treatment; Seminar Proceedings and Engineering Assessment,* MCD-67, U.S. EPA Report 430/9-80-006, University of California, Davis, September 11-12, 1979.
32. Yonika, D. A. "Effectiveness of a Wetland in Eastern Massachusetts in Improvement of Municipal Wastewater," in *Aquaculture Systems for Wastewater Treatment; Seminar Proceedings and Engineering Assessment,* MCD-67, U.S. EPA 430/9-80-006, University of California, Davis, September 11-12, 1979.
33. Demgen, F. C. "Wetlands Creation for Habitat and Treatment At Mt. View Sanitary District, California," in *Aquaculture Systems for Wastewater Treatment; Seminar Proceedings and Engineering Assessment,* MCD-67, U.S. EPA Report 430/9-80-006, University of California, Davis, September 11-12, 1979.
34. Ward, P. S. "Las Vegas Goes to Battle over AWT," *J. Water Poll. Control Fed.* 50(11):2432-2435 (1978).
35. Henderson, S. "Utilization of Silver and Bighead Carp for Water Quality Improvement," in *Aquaculture Systems for Wastewater Treatment; Seminar Proceedings and Engineering Assessment,* MCD-67, U.S. EPA Report 430/9-80-006, University of California, Davis, September 11-12, 1979.
36. Culliton, B. "Aquaculture: Appropriate Technology in China," *Science* 206:539 (1979).
37. Colt, J. et al. "The Environmental Requirements of Fish," Department of Civil Engineering Report. University of California, Davis (1979).
38. Stewart, W. C., and S. A. Serfling. "The Solar Aquacell System for Primary, Secondary or Advanced Treatment of Wastewaters," in *Aquaculture Systems for Wastewater Treatment; Seminar Proceedings and Engineering Assessment,* MCD-67, U.S. EPA Report 430/9-80-006, University of California, Davis, September 11-12, 1979.
39. Milligan, D. J. et al. "Sequential Use of Bacteria, Algae and Brine Shrimp to Treat Industrial Wastewater at Pilot Plant Scale," presented at the

International Symposium on Brine Shrimp, *Artemia salina*, Corpus Christi, TX, August 20-23, 1979.

40. Benemann, J. R. "Energy from Aquaculture," Office of Technology Assessment. U.S. Congress (in press).
41. Benemann, J. R. "Energy from Wastewater Aquaculture Systems," in *Aquaculture Systems for Wastewater Treatment; Seminar Proceedings and Engineering Assessment*, MCD-67, U.S. EPA Report 430/9-80-006, University of California, Davis, September 11-12, 1979.
42. Tchobanoglous, G., J. E. Colt and R. W. Crites. "Energy and Resource Consumption in Land and Aquatic Treatment Systems," presented at the Energy and Optimization of Water and Wastewater Management for Municipal and Industrial Applications Conference, New Orleans, LA, December 10-13, 1979.

CHAPTER 16

MICHIGAN WETLAND WASTEWATER
TERTIARY TREATMENT SYSTEMS

Jeffrey C. Sutherland
Williams & Works
Grand Rapids, Michigan

Between Michigan's larger cities and her uninhabited forest and waterlands lie several hundred well-established communities with populations ranging from fewer than 100 to more than 1000. Typical of the older small communities is a minimal commercial row including a grocery, hardware, coffee shop and service station, and up to a score of bordering residential blocks which give immediate way to woods, farms and recreational waters. Within such communities, potable water is provided either by household wells or community wells through a public distribution system. Wastewater is typically treated and disposed through the use of passive individual systems, or stabilization ponds followed by discharge to a watercourse.

Methods of managing sanitary wastewater from small Michigan communities have been evolving since the early 1960s under the impetus of concern for health and water quality. Conventional mechanical-biological-chemical treatment plants have not proved economically affordable and for that reason play no significant role. Instead, land application approaches have developed through deliberate accommodation to the rural surroundings, varying geological settings and treatment needs. Over the past ten years, Michigan's small communities have been increasingly served by stabilization ponds and upland irrigation systems in response to phosphorus and nitrogen removal requirements for Lake Michigan Basin point discharges.

Agricultural vegetation is a "living" filter [1] which removes a percentage of applied wastewater and nutrients through plant growth. Surprisingly though, the agricultural productivity of the land is seldom an essential factor either in locating or designing upland wastewater irrigation systems for very small communities. Concern for health has eliminated most row crops other than field corn from consideration and has favored forage grasses and legumes. Further, minimal land is usually consistent with over-all economy, which for very small communities may mean little or no profit potential in agricultural operations. Also, removal of phosphorus is usually accomplished by reaction of wastewater with soil particles within the upper several feet of soils. Reactive soils have proved to be much more efficient than plants in removing phosphorus [1,2].

The location of upland irrigation sites has instead been determined by the availability of affordable land having acceptably permeable agricultural soils. The design of upland irrigation systems has depended on topography, agricultural soil permeability, permeability of underlying glacial soils, depth to ground water and proximity to discharging streams. The idea of living systems in association with (now) conventional land disposal has been downplayed. Yet, covering vegetation of some kind is desirable for controlling erosion and for keeping the soil surface open to infiltration. A well-run irrigation operation will be favorably reflected in the pleasing appearance of a vigorous field cover. Deliberate effort is exerted to maintain soil openness and field appearance. Effort is applied to maintain soil openness and field appearance.

Evolution in wastewater irrigation field cover is occurring fortuitously by design of nature rather than by engineering intent, in a small number of Michigan communities. Highly productive wetlands are establishing themselves where irrigation field drainage is favorably slow. In other Michigan communities, natural wetlands are being used for wastewater renovation. Wetlands provide wastewater treatment advantages and benefits for wildlife that are not available with conventional land treatment. This chapter is concerned with highlighting some of these beneficial factors, through the examples of systems serving the village of Vermontville and the Houghton Lake area.

THE VERMONTVILLE POND AND IRRIGATION SYSTEM

The village of Vertmontville is a community served by one of the first pond and irrigation systems designed and constructed in the state (Figure 1). Vermontville's service area population of 975 generates average wastewater flows of 70 gallons per capita per day (gpcpd), which is typical for small communities. The treatment system consists of two facultative ponds and four flood irrigation fields located on a hill bordering a minor stream tributary of the Thornapple River, a first-quality warm water stream located approximately one-half mile to the south [3].

Figure 1. Vermontville pond and irrigation system.

In most respects, the system behaves like a conventional flood irrigation operation. Slowly-permeable soils favor irrigation by flooding methods. Pond effluent is released into the fields by gravity flow, and wastewater is applied through a header with multiple stand-pipe outlets in each field. Runoff control is provided by berms at the perimeter of each of the four fields. The hill site slopes directly to a discharging stream, providing a depth of unsaturated soils above the water table adequate to assimilate the applied wastewater. The stream discharges ground water from the site, preventing it from reaching household wells. There is a constant surface overflow of water from the final field into the nearby stream. This water is comprised mainly of irrigated water which, after seeping through subsoils beneath the upper fields, emerges into the final field. Wastewater from field 3 will occasionally overflow into the final field during prolonged or intense rainfall.

While the conventional aspects of a flood irrigation system are operative mainly at and below the land surface, the surface features of the irrigation system are highly unconventional. The fields are heavily overgrown with aquatic vegetation, mainly two different cattail species, with subordinate willows and duckweed (Figure 2). This wetland vegetation is entirely "voluntary," having first come in during the rainy 1972 summer, one full year ahead of the initial irrigation season [4]. The fields are not entirely taken over by marsh vegetation. Interspersed with cattail stands and open water areas in fields 1 and 3 are poorly drained to moderately well-drained "uplands" dominantly covered with terrestrial grasses and goldenrod. The contrast in vegetation is owed mainly to contrasting depths of water. The nonuniform water depths result from rough grading, differential compaction of extensively filled

Figure 2. Wetland vegetation (photo courtesy of Professor F. B. Bevis, Grand Valley State Colleges, Allendale, MI).

surfaces situated side by side with down-cut areas, and erosion at some of the irrigation inlet standpipes.

What features and values do the Vermontville wetlands possess for wildlife and wastewater management? These wetlands came into existence to take advantage of flood conditions and nutrients provided through irrigation of municipal wastewater. Approximately 4 in. of pond-treated wastewater containing around 10 mg/l total nitrogen and 1.8 mg/l total phosphorus are applied weekly to the 11-acre marshlands during five or six months of the year (June forward). The vegetation takes up approximately 20% of the applied phosphorus in spring and early summer. The vegetation takes up approximately 20% of the applied phosphorus in spring and early summer. Between early summer and early September, much of this liberated phosphorus is returned to the wetland surface waters to await seepage and interaction with soils before it is "permanently" removed [3]. The wetland vegetation, therefore, provides only temporary storage space for phosphorus, and this behavior must be respected in situations where 90% and greater removal of phosphorus is a goal of wastewater treatment.

Thousands of acres of inadvertent cattail wetlands have become established in floodings resulting from highway and railroad construction in the United States, providing habitat for one of the most easily recognized marsh denizens, the redwing blackbird. Similarly, hundreds of redwing blackbirds rest, feed, nest, and raise their young within the wastewater-grown cattail stands at Vermontville. The Vermontville wetlands provide better habitat for wildlife than is offered by other, natural wetlands marginal to the nearby Thornapple River. Whitetail deer, including does with their young, have used field 3 for rearing habitat, shelter, and resting area. The small mammals which inhabit or feed around the wetland edges include muskrats, voles, shrews, opposums and raccoons.

The variety of plants and water habitats available to plant species, together with the fact of public ownership and the practice of discouraging trespass on lands set aside for wastewater treatment, prompted the idea that such wetlands might be suitable as sanctuary and conservancy for certain sensitive and threatened plant species. A preliminary test of this idea has been highly encouraging. Several blue flag iris plants, a sensitive species, were transplanted from natural wetlands near Lake Michigan to the Vermontville field 3 in the spring of 1979. The transplants were set in nearly saturated but unflooded ground, and by late summer they had spread and given rise to new green shoots. The transplants survived the winter and are showing spring shoot growth comparable to that of their wild counterparts [5].

An intriguing feature of the volunteer wetlands is that, unlike exposed upland mineral soil surfaces which become plugged through resuspension of silt and clay during times of irrigation, the wetland soil surface openness is apparently preserved, allowing seepage to continue year after year, without the need for soil surface maintenance. Lack of mineral fines in the irrigated wastewater, the several inches of accumulated cattail straw covering the mineral soil substrate (impure sand, sandy-gravelly clay), and the decomposition-recycling of fine, dark organic matter, may all have a role in preserving the openness of the wetlands soils [4]. The natural counterparts of these wetlands include isolated marshes situated over drainage-restrictive glacial soils, from which seepage has been continuous for many centuries.

Operation and maintenance (O&M) costs of the Vermontville system are low.   The annual labor costs with wetland irrigation are $1500-1600. Of these costs, 30% is incurred in mowing the long, grassed slopes facing the wetland fields. Most of the labor involves manual operation of irrigation valves on a 5-day/week schedule. Use of a pickup truck, a tractor with sickle-bar cutter and fuel totals around $900/yr. There is virtually no repair or replacement. Concerning electrical energy, there is a 22-ft elevation difference (average) between the ponds and the fields, which allows irrigation by gravity flow. The 22-ft pumping lift consumes around $340/yr. Of the total wetland O&M cost of $4800/yr, 43% ($2100) is spent on state-required environmental monitoring [6].

Volunteer seepage wetlands can be purposefully designed for irrigation treatment of municipal wastewater. Seepage rapid enough to avoid uncontrolled or undesirable runoff, but slow enough to maintain flooded or saturated soils for wetland establishment, must be provided by the natural setting and any necessary engineering improvements. Silty soils and very fine sand have the slow drainage characteristics desired (10^{-3} to 10^{-5} cm/sec hydraulic conductivity). Also, the mineral soils with permeabilities in the given range tend to be readily compactible. Mineral soils with permeability in the higher range, approaching 10^{-3} cm/sec, could be ideal: the area intended for wetland establishment could be graded for any desired water depth variations using equipment heavy enough to compact the graded soil surface. Add to this an uncompacted, berm-enclosed perimeter area wide enough to absorb any over-flow from the irrigated interior and the needs for slow seepage and prevention of runoff will be met. Coarser sands with greater permeability could be used, but in lieu of compaction, a soil sealant could be injected, spread or disked into the area intended for marsh habitat [3].

The costs associated with wetland construction are lower than for upland spray irrigation systems, and are comparable to those for flood irrigation. As indicated in Figure 3, the estimated costs for both seepage wetlands and upland spray irrigation systems tend to increase in parallel manner as system flows increase from below 0.1 mgd to 0.5 mgd. At and below 0.1 mgd flows, that is for very small communities, savings of greater than 25% in capital costs may be anticipated [3]. Dollar figures should be increased by 20% over those given in Figure 3 to adjust to current estimated construction costs.

Natural wetlands in Michigan are being irrigated with municipal waste-water as a deliberate final step in achieving removal of nutrients. The most outstanding of these natural wetland projects serves the Houghton Lake resort/recreation community of 5300 permanent and 13,500 seasonal residents, under management by the Houghton Lake Sewer Authority (HLSA).

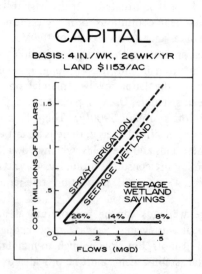

Figure 3. Capital cost comparison for spray irrigation and seepage wetland wastewater treatment systems.

In 1971, Williams & Works (W&W), as engineering consultant for the three rural townships bordering the lake, developed collection and treatment concepts including stabilization ponds and upland irrigation on community-owned land to serve through 1978. But additional uplands for expansion to serve the 1990 and 2000 design year populations were not readily affordable by this typically rural, low-tax based community.

Late in 1971, W&W met with Michigan Department of Natural Resources (MDNR) personnel to discuss using a state-owned peat wetland in the nearby headwaters of the Muskegon River in place of a more expensive irrigation system to accomplish tertiary treatment. The peatland offered poor to marginal habitat for waterfowl and game animals, owing partly to its headwaters location and consequent absence of nutrient-bearing runoff. A comprehensive research program was begun in 1972 by The University of Michigan (U of M), with support from the National Science Foundation (NSF) and the Rockefeller Foundation. Under continuing NSF support, the researchers demonstrated that phosphorus and nitrogen species are removed from applied wastewater after several hundred feet of surface flow across the vegetated peat substrate [7].

An irrigation system for the peatland was designed and constructed between 1975 and 1978 by W&W through the U.S. EPA facility planning program [8]. The design includes an elevated wooden walkway suspended on two-inch pilings, extending one half mile into the wetlands (Figures 4 and 5). The walkway supports aluminum transmission and gated irrigation pipelines up to 12 in. in diameter, and allows operation and maintenance access with virtually no physical disturbance of the fragile peatland surface [9].

Nearly 170 million gallons of wastewater, containing 4 mg/l phosphorus, 0.6 mg/l ammonia-N and 5.4 mg/l nitrate-N, were applied to the peatland during the summers of 1978 and 1979 [10,11]. Concurrently, proof of concept studies to document and interpret everything from environmental water quality to small mammal populations were made under direction of the U of M with grants from the NSF and the HLSA. Similar studies will continue through 1981 and perhaps beyond.

The 500-acre peatland is extensive compared to the amount of wastewater being applied—approximately 6 in. over a four-month growing season. Environmental responses to the applied wastewater are modest, to date, including prominent increases in the growth of plants close to the irrigation pipeline [11]. But the economic value of the wetland system is being realized by the Houghton Lake community now, both in capital savings of $1 million compared to the upland irrigation alternative and in operation and maintenance costs. Costs borne by HLSA related to wetland O&M in 1979 were approximately $111 per million gallons irrigated, including pumping energy, repairs, new laboratory equipment and a $5800 grant for ecological monitoring [6,9].

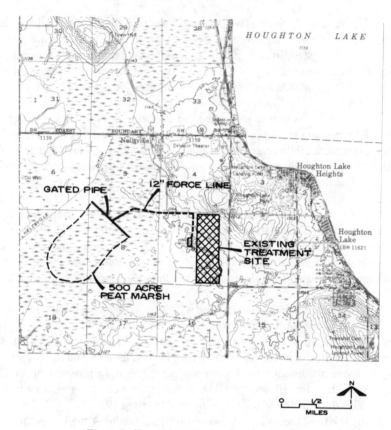

Figure 4. Houghton Lake area treatment facility.

Natural wetlands, unaffected or only lightly affected by human culture, tend to be fragile and irreversibly alterable when their water and nutrient budgets are changed through human activity. Adventitious or capricious changes may be to the disadvantage of wildlife. Therefore, a conservative approach to wetland-wastewater management is needed, beginning with thorough background studies to identify wetland hydrologic characteristics, and wetland ecological attributes and vulnerabilities. Joint efforts by qualified engineers and ecologists can help assure objective and sensitive design to achieve favorable tradeoffs between economical wastewater treatment and wetland modification. The establishment of new, volunteer wetlands on wastewater-saturated surface soils can be done purposefully to the economic advantage of small communities, as we have seen at Vermontville. In strip-mined or other barren lands on fine-textured soils, small amounts of wastewater could transform large unproductive tracts into productive marshes. In

Figure 5. Elevated walkway and pipelines on fragile peatland surface.

regions that are lacking viable marsh habitat, new wastewater wetlands could significantly increase the recreational game base, and would provide new educational opportunities for potential young naturalists and environmental engineers.

REFERENCES

1. Parizek, R. R. et al. "Waste Water Renovation and Conservation," Pennsylvania State University Studies No. 23, 71 pp. (1967).
2. Myers, E. A., and J. C. Sutherland. "Reuse and Renovation of Sewage Stabilization Pond Effluent through Irrigation," U.S. EPA Project S-803807-01 (1978).
3. Sutherland, J. C., and F. B. Bevis. "Reuse of Wastewater by Volunteer Fresh-Water Wetlands," *Proceedings of the 1979 AWWA Water Reuse Symposium* (Denver, CO: American Water Works Association, 1979).
4. Williams & Works. "Reuse of Municipal Wastewater by Volunteer Fresh-Water Wetlands" (ENV-20273), Interim Report to the National Science Foundation (April 1979).
5. Bevis, F. B., Chairman, Department of Natural Resources Management, Grand Valley State Colleges, Allendale, MI, Personal communication.

6. Williams, T. C., and J. C. Sutherland. "Engineering, Energy, and Effectiveness Features of Michigan Wetland Tertiary Wastewater Treatment Systems," *Aquaculture Systems for Wastewater Treatment: Seminar Proceedings and Engineering Assessment,* (Davis, CA: University of California, 1979).

7. Kadlec, R. H., D. L. Tilton and J. A. Kadlec. "Feasibility of Utilization of Wetland Ecosystems for Nutrient Removal from Secondary Municipal Wastewater Treatment Plant Effluent," Semiannual Report No. 5 to the National Science Foundation (1977).

8. Williams & Works. "Houghton Lake Area Facilities Plan, Step I," Federal Project C-262768 (April 1976).

9. Williams, T. C., and J. C. Sutherland. "Technical Aspects of the Tri-Township (Houghton Lake, Michigan) Peat Wetland Tertiary Treatment System," Water Pollution Control Federal Conference, Houston, TX (October 1979).

10. Kadlec, R. H. "1978 Operations Summary, Houghton Lake Wetlands Treatment Project, Wetland Utilization for Management of Community Wastewater," Report to the National Science Foundation (March 1979).

11. Kadlec, R. H., and D. E. Hammer. "1979 Operations Summary, Houghton Lake Wetlands Treatment Project, Wetland Utilization for Management of Community Wastewater," Report to the National Science Foundation (February 1980).

CHAPTER 17

FRESHWATER FISH CULTURE
WATER REUSE SYSTEMS

Dennis B. George
Institute of Water Research
Lubbock Christian College
Lubbock, Texas

The optimization of fish production has been the goal of fish culturists since the inception of the science of aquaculture. The requirements to achieve such a goal are optimum water temperature, water chemistry and an adequate supply of food resources. Supplying the food resource is easily accomplished and in fact is a well-refined science. The management of water temperature and water chemistry has long been left to natural environmental factors. Management and control of these parameters is achieved in a large-scale rearing facility where up to 5662 liters (200 ft^3)/sec of water may be recirculating. The most successful and productive rearing facilities for trout have been those using high-volume isothermal spring sources. Trout have stringent water quality requirements. One facility in Idaho has produced up to 60,000 kg of trout per acre [1]. Such success is due not only to the quality of the water but also to the quality of management of the facility.

Since suitable natural springs are rare, most fish facilities must operate under less than optimal conditions or install expensive heating and chilling units and various other devices to treat the large amounts of raw water coming into a rearing facility. Water pretreatment capital and operating costs can be reduced substantially by reducing the volume of water to be treated. This reduction can be accomplished through water reuse systems in which water requirements can be reduced over 95% [2]. Reduction in water consumption significantly reduces the costs of controlling water quality to attain maximum production.

Production rates in controlled environments can be very impressive. In 1951, a Japanese recirculating system produced 400 kg of carp/m^2, the highest level ever achieved in Japan to that time. Growth rates of fish in German controlled-reuse systems were 500 to 600 times those for fish maintained in uncontrolled pond culture [1].

Reuse systems also allow development of hatcheries and rearing facilities in areas where water is in short supply or existing water quality is substandard for fish production. The reuse of a substantial percentage of the hatchery water also assists the facility in complying with water pollution regulations by reducing the effluent output.

In principle, the recirculating system seems to be the ultimate answer to providing the controlled environment sought after for efficient fish production. Reuse systems at the present, however, are expensive from the standpoints of initial investment and operation.

Many existing reuse systems depend on biological fixed-film filters (biofilters) for water treatment. The biofilter is that portion of the system which, through bacterial action, removes metabolic wastes produced by the fish and decomposes uneaten food. The biofilter is a highly complex and sensitive combination of mechanical filtration and bacterial decomposition.

The waste product of most concern to the fish culturist is ammonia because of its toxic effect at very low concentrations [3,4]. Biofilters handle ammonia through the nitrification process in which toxic ammonia is oxidized through bacterial action to the relatively harmless compound, nitrate (NO_3). For reuse facilities biological nitrification is considered to be the most practical and economical method of ammonia removal [2].

THE NITRIFICATION PROCESS

The nitrification process is carried out primarily by the chemoautotrophic bacteria, *Nitrosomonas* (Ns) and *Nitrobacter* (Nb). These two genera are the most common and important representatives of the nitrifying bacteria (Family Nitrobacteraceae). Ns and Nb are considered "primary autotrophic nitrifiers" [5]. Five other genera of nitrifying bacteria are in the same family but are considered "secondary autotrophic nitrifiers". Many "primary" nitrifiers are able to grow heterotrophically or at least to be stimulated by the presence of organic matter, but "secondary" nitrifiers show strictly autotrophic tendencies. The result is that the primary nitrifiers tend to dominate most systems because rarely is organic matter absent.

The growth rate of nitrifiers is very slow and inefficient by comparison to that of heterotrophic bacteria. Generation times for Ns and Nb have been reported to be in the range of 20-40 hr [6,7]. The yield of cells per unit of substrate oxidized is also low: 1×10^4 to 4×10^4 cells Nb/μg nitrogen [8-10] and 3×10^4 to 12×10^4 cells Ns/μg nitrogen [11].

Heterotrophic nitrification has been demonstrated to occur in a number of bacteria and fungi species [5]. Painter [12] reported that 104 species can produce nitrite from ammonia, the first stage of the two-step nitrification process. Rates of nitrification by heterotrophs, however, are much lower than for autotrophs. Rates measured have been from 10^3 to 10^4 lower even under ideal pure culture conditions [5]. As a result, heterotrophic nitrification has been considered to play a very limited role in nitrogen oxidation [13]. There is, however, speculation that heterotrophic nitrification may be of importance in acidic soils [14,15] and in highly alkaline nitrogen rich aquatic environments where autotrophic nitrification does not occur [5,16].

The stoichiometric reactions for the nitrification process can be summarized in two equations:

$$NH_4^+ + 1.5\ O_2 \rightarrow 2H^+ + H_2O + NO_2^- + 58 \text{ to } 84 \text{ kcal/mole} \qquad (1)$$

$$NO_2^- + 0.5\ O_2 \rightarrow NO_3^- + 15.4 \text{ to } 20.9 \text{ kcal/mole} \qquad (2)$$

The first reaction is carried out by Ns and results in the production of nitrite which serves as a substrate for oxidation by Nb. Nb then converts nitrite to nitrate. It is common practice to refer to each reaction singly and collectively by the term "nitrification".

Examination of these equations yields valuable information about the nitrification process. First, it is aerobic, requiring two molecules of oxygen for each molecule of ammonia oxidized to nitrate. Theoretically the process should require 4.57 mg O_2 for each mg of ammonia oxidized. The first stage conversion of ammonia to nitrite utilizes 3.43 mg O_2 and the balance is used in the final oxidation of nitrite to nitrate [17]. Experimentally measured values frequently differ from this stoichiometric prediction [18]. Several hypotheses exist to explain the differences but none have been proven.

The reaction equations also indicate the production of energy from each reaction. This energy is used by the bacteria for growth and metabolic activity.

The production of Ns and Nb biomass can be incorporated into equations of the form [19]:

$$15\ CO_2 + 12\ NH_4^+ \rightarrow 10\ NO_2^- + 3\ C_5H_7NO_2 + 23\ H^+ + 4\ H_2O \qquad (3)$$
$$\text{Ns Biomass}$$

$$5\ CO_2 + NH_4^+ + 10\ NO_2^- + 2\ H_2O \rightarrow C_5H_7NO_2 + H^+ \qquad (4)$$
$$\text{Nb Biomass}$$

The CO_2 used and the H^+ produced react within the constraints of the carbonate buffer system if the reaction is occurring in water. At pH 8.3 and below where nitrification generally occurs, the hydrogen ion reacts immediately with bicarbonate (HCO_2^-) to produce carbonic acid (H_2CO_3) which will lower the pH of a poorly buffered system. This reaction in conjunction with the oxidation of ammonia and nitrate consumes a total of 7.14 mg of alkalinity as $CaCO_3$ for each mg of ammonia oxidized [19]. It is, therefore, important to maintain an adequate buffer system to sustain nitrification. If the pH is to remain above 6.0, the alkalinity should be ten times greater than the amount of ammonia oxidized [19].

Under most conditions, the growth limiting substrate (S) for Ns will be ammonia and for Nb it will be nitrite. The concentrations at which these substrates become limiting is a function of temperature and values have been reported as 0.06-5.6 mg $NH_3 \cdot N/\ell$ for Ns and 0.06-8.4 mg $NO_2 \cdot N/\ell$ for Nb [18]. Maximum oxidation rates measured for Ns and Nb vary widely with temperature following Arrhenius' law [13]. Values measured for Ns and Nb range from 0.25 day^{-1} at $8°C$ to 1.08 day^{-1} at $23°C$ and 0.28 day^{-1} at $15°C$ to 1.44 day^{-1} at $23°C$, respectively. The optimum temperature range for nitrification is 30-36°C [13].

Dissolved oxygen is also an important parameter in nitrification kinetics. Several investigators have determined concentrations at which oxygen becomes a limiting factor. Saturation constants for nitrification are generally about 1.0 mg/ℓ for both stages of the process and it is observed that in nitrifying systems, oxygen levels above 3 to 4 mg O_2/ℓ are required to ensure complete nitrification [17]. At $30°C$ oxygen concentrations below 2 mg/ℓ reduce growth by 60.6% of maximum [19] and at 0.2 mg O_2/ℓ no oxidation occurs [12]. Under differing situations, inhibitory levels may be different; there are situations which indicate complete nitrification may occur at 0.5 mg/ℓ dissolved oxygen [19].

The effects of pH on nitrification are also well documented. As pH decreases into the acid range, the rate of ammonia oxidation declines. This decrease in pH can occur naturally from the nitrification of ammonia on poorly buffered systems and eventually inhibit the bacterial oxidation of ammonia. Data indicate that through acclimation, nitrification can continue at pH levels as low as 5.8, although values between 7.2 and 8.0 are considered optimal [20]. Christensen and Harremoës [17] suggest that pH 7.8-9.2 is optimal for Ns and pH 8.5-9.2 for Nb.

Nitrifying bacteria need certain micronutrients for growth and metabolism Table I). Most notable are the effects of molybdenum and biotin on *Nitrobacter*. Molybdenum supplied at concentrations of only 10^{-9} M resulted in an 11-fold increase in growth and activity of *Nitrobacter*, while small quantities of biotin produced two- to four-fold increases in activity and an impressive

Table I. Substances Required or Stimulatory for Nitrification [18]

Substance	Concentration[a,b]	Effect[c]	Reference[d]
Phosphate		Required for Ns G and Nb G	[21]
	310 as P	Required for Ns G and Nb G	[22]
	5 as P	Required for Nb G	[23]
Magnesium		Required for Ns G and Nb G	[21]
	5	Required for Nb G	[23]
	10.5-50.5	No effect on Ns A	[24]
	(as $MgSO_4 \cdot 7H_2O$)	Ns A +	[25]
	12.5-50	Slight Ns A −	
	50-100		
Molybdenum	−	Nb A +	[23]
	$10^{-9} M$ (0.0001)	11-fold increase in Nb A and G	[26]
	$10^{-2} M$ (1000)	Slight Nb A, G −	
Iron	−	Required for Ns G and Nb G	[21]
	0.5-0.6	Ns G +	[24]
	7	Required for Nb G	[23]
Calcium	−	Required for Nb G	[21]
	0.5-10	No effect by itself on Ns A;	
	10.5-50.5	+ in presence of 5 mg/ℓ EDTA	
	(as $CaCl_2 \cdot 2H_2O$)	No effect on Ns A	[24]
Copper	−	Required for Nb G	[21]
	0-0.06	Ns A +; Added	[25]
		Ns A + along with 5 mg/ℓ	
		EDTA	
	0.1	Slight Ns A +; with higher	[27]
		concentrations Ns A −	
	0.1-0.5	Increasing Ns A	[24]
Sodium	0.6-1.5	Ns A +; Ns G −	[25]
	1.5-7.0	Ns A −; Ns G +	
Marine Salts	−	Required by some estuarine	[28]
		or littoral cultures of ammonia	
		oxidizers	
Vitamins			
A-Palmitate	50,000 USP/mℓ	Ns G +; Nb G+	[29]
Pantothenic Acid	0.05 mg/mℓ	Nb G+	
	0.0025 μg/mℓ	Ns A +	[30]
Nicotinic Acid	0.05 mg/mℓ	Nb G +	[29]
Ascorbic Acid	0.05 mg/mℓ	Nb G +	
Biotin	0-150 mμg	2- to 4-fold Nb A +;	[31]
		100- to 1000-fold Nb G +	
	2	Slight Ns A, G +	[32]
Adenine Sulfate	0.05 mg/mℓ	Nb G +	[29]
Sodium Glutamate	1720 mg/mℓ	Ns G +; Nb G +	
Yeast Extract	2 mg/mℓ	Nb G +	
1-Serine	4 μg/mℓ	Ns A +; Ns G +	[32,33]
	1050 mg/mℓ	Nb G +	[29]
1-Glutamate	4 μg/mℓ	Ns G +; Ns A +	[32]
	1450 mg/mℓ	Nb G +	[29]
1-Glutamic Acid	4 μg/mℓ	Nb G +; Ns A +	[32]

Table I, continued

Substance	Concentration[a,b]	Effect[c]	Reference[d]
1-Aspartic Acid	4 μg/mℓ	Ns G +; Ns A +	
Ash of corn steep liquor		Ns G +	[34]
Glucose, p-amino-benzoic acid	2-5	Ns A +; Nb A +; impure, mixed culture	Cooper & Catchpole (cited in [35])

[a] All results are for pure cultures unless indicated otherwise.
[b] In mg/ℓ unless specified otherwise.
[c] NS = *Nitrosomonas*; Nb = Nitrobacter; G = Growth; A = Activity; + = Stimulation; − = Inhibition, e.g., Ns A + = stimulation of *Nitrosomonas* activity = stimulation of nitrification.
[d] All references are included in Sharma and Ahlert [18].

100- to 1000-fold increase in growth. Bacteria supplied with an adequate source of micronutrients may function more efficiently over a wider range of environmental conditions than those with deficiencies.

Inhibitory compounds also must be given consideration in development of a nitrifying system. High levels of organic matter and ammonia may result in an accumulation of nitrite through inhibition of Nb by free ammonia (FA) and free nitrous acid (FNA). Ionized nitrite does not act to inhibit Nb until very high levels are reached. Meyerhof [36] found 1400 mg of nitrite were required to reduce respiration in Nb by 26% and 4700 mg to achieve 100% inhibition.

The concentrations of FNA and FA are linked to the pH of the solution with higher levels of FA at higher pH. Since FA begins to inhibit Nb at concentrations of 0.1-1.0 mg/ℓ [37], a build-up of nitrite can easily result under higher pH. Ns can continue to oxidize ammonia and produce nitrite until FA levels reach 10-150 mg/ℓ [37].

FNA concentrations increase as pH decreases. As mentioned earlier, pH will decrease when nitrification occurs in poorly buffered environments. Again Nb is the more sensitive species and as a result its activity is inhibited before that of Ns so an accumulation of nitrite again results. FNA begins to exhibit inhibitory effects between 0.22 and 2.8 mg/ℓ [37].

Another possible source of inhibitory nitrite accumulation is denitrification. Belser [38] has proposed that an additional source of nitrite in soils might come from the reduction of nitrate in anaerobic environments. Under anaerobic conditions, a variety of microorganisms including Nb are able to reverse this nitrification sequence in a process known as denitrification. Anaerobic conditions may occur at microsites in a filter which might otherwise show high oxygen levels [13,17]. These microsites could easily provide the anaerobic conditions conducive to denitrification and nitrite formation.

Consideration has been given to the possibility of different tolerances and environmental optima of different bacterial strains, which may explain differences in measurements of pH optima for Nb from those expected [39]. There is also some evidence to suggest different temperature responses by bacteria from warm and cold soils [5].

AMMONIA AND NITRITE TOXICITY TO FISH

Ammonia and nitrite may cause death at relatively low concentrations and a reduction in growth at sublethal levels. A review of the literature indicates that salmonids and channel catfish have been the most studied fishes in the U.S. with regard to ammonia and nitrite toxicity because of their great popularity and high economic value.

The toxicity of ammonia in fishes is determined by the amount of un-ionized or free ammonia (FA) in solution [40]. The ratio of FA to total ammonia is a function of pH with higher FA values corresponding to increasing pH. Values of FA as a percentage of total ammonia are presented in Table II.

The median lethal threshold concentration (LC_{50}) of FA has been reported for rainbow trout (*Salmo gairdneri*) to be 0.45 mg $NH_3 \cdot N/\ell$ [41]. Smart found that this concentration was reached at a total ammonia concentration of 205 mg/ℓ at pH 6.9 and at only 11 mg/ℓ at pH 8.2. Other LC_{50} for rainbow trout are 0.39 mg/ℓ FA [42], 0.41 mg/ℓ FA [43], and 0.07 mg/ℓ FA [44]. Free ammonia concentration of 90 μg/ℓ, however, may cause reduction of growth in rainbow trout if the condition persists for a period of six months or longer [45]. Reduction of total ammonia levels below 1 mg/ℓ should reduce free ammonia levels below stress conditions.

Juvenile coho salmon (*Oncorhynchus kisutch*) have a LC_{50} of 0.45 mg/ℓ FA, similar to that of rainbow trout [3]. Buckley observed acute toxic effects only after FA concentrations surpassed 0.178 mg/ℓ. An FA concentration of 0.023 mg/ℓ was a safe level for coho salmon.

Channel catfish (*Ictalurus punctatus*) are apparently more tolerant of FA and have a reported 24-hr LC_{50} of 2.36 mg/ℓ [4]. Robinette was careful to point out that toxicity values determined in conditions where free carbon dioxide is lacking may be much greater than under high carbon dioxide, low pH conditions. He observed catfish growth to be completely inhibited at FA levels of 0.12 and 0.13 mg/ℓ. Colt [46] found complete inhibition of growth in channel catfish at 0.97 mg/ℓ FA and a 50% reduction in growth at 0.52 mg/ℓ. He recorded a 96-hr LC_{50} of 1.61 mg/ℓ FA.

The accumulation of nitrite resulting from incomplete nitrification can be toxic to fish. A build-up of nitrite results in methemoglobinemia, a condition in which the blood pigment hemoglobin is oxidized to methemoglobin and is

Table II. Percentage of Free Ammonia (as NH_3) in Fresh Water (FW) and Sea Water (SW) at Varying pH and Temperature [a]

pH	10°C		15°C		20°C		25°C	
	FW	SW	FW	SW	FW	SW	FW	SW
7.0	0.19		0.27		0.40		0.55	
7.1	0.23		0.34		0.50		0.70	
7.2	0.29		0.43		0.63		0.88	
7.3	0.37		0.54		0.79		1.10	
7.4	0.47		0.68		0.99		1.38	
7.5	0.59	0.459	0.85	0.665	1.24	0.963	1.73	1.39
7.6	0.74	0.577	1.07	0.836	1.56	1.21	2.17	1.75
7.7	0.92	0.726	1.35	1.05	1.05	1.52	2.72	2.19
7.8	1.16	0.912	1.69	1.32	2.45	1.90	3.39	2.74
7.9	1.46	1.15	2.12	1.66	3.06	2.39	4.24	3.43
8.0	1.83	1.44	2.65	2.07	3.83	2.98	5.28	4.28
8.1	2.29	1.80	3.32	2.60	4.77	3.73	6.55	5.32
8.2	2.86	2.26	4.14	3.25	5.94	4.65	8.11	6.61
8.3	3.58	2.83	5.16	4.06	7.36	5.78	10.00	8.18
8.4	4.46	3.54	6.41	5.05	9.09	7.17	12.27	10.10
8.5	5.55	4.41	7.98	6.28	11.18	8.87	14.97	12.40

[a] Salinity = 34% at an ionic strength of 0.701 m for the seawater values [40].

prevented from functioning in its capacity as an oxygen transport mechanism. The result is that the fish experience hypoxia, cyanosis and eventual death. Smith and Russo [47] reported that methemoglobin was produced in rainbow trout exposed to as little as 0.096 mg $NO_2 \cdot N/\ell$. Brown and McLeay [48] reported similar results at levels of 0.100 mg $NO_2 \cdot N/\ell$. They estimated 96-hr LC_{50} of 0.230 mg $NO_2 \cdot N/\ell$ and recommended levels below 0.200 mg/ℓ be maintained for rainbow trout to survive. Russo et al. [49] found that younger fish seem to have a greater tolerance for nitrite than older, larger fish. A 96-hr LC_{50} for 2.0-gm fish was calculated to be 0.39 mg $NO_2 \cdot N/\ell$ while 235-gm fish had a 96-hr LC_{50} of 0.20 mg $NO_2 \cdot N/\ell$. The nitrite level for zero mortality also varied from 0.14 mg/ℓ for the smaller fish to 0.06 mg/ℓ for the larger fish.

Channel catfish appear to be more tolerant of nitrite than trout although the data are difficult to interpret. Data presented by Konikoff [50] as total nitrite rather than nitrite-nitrogen indicate 96-hr median tolerance limits of 24.8 mg/ℓ. This value would be roughly equivalent to 7.5 mg $NO_2 \cdot N/\ell$ [40].

There is evidence that nitrite tolerance can be increased in fish by alteration of the ionic composition of the water. Crawford and Allen [51] showed increased tolerance to nitrite of Chinook salmon (*Oncorhynchus tschawytscha*) fingerlings in sea water and concluded that calcium ion was somehow involved. Meade [52] demonstrated that nitrite toxicity is related to temperature, oxygen concentration and chloride ion concentration.

Toxicity of nitrite may also be linked to fish activity. Fish in environments where they must swim constantly to maintain position seem to have a lower threshold tolerance to nitrite [51,53].

The resultant product of the nitrification process is nitrate. As far as has been determined, nitrate does not affect fish until concentrations become very high [40]. A nitrate concentration of 1000 mg $NO_3 \cdot N/\ell$ produced trauma in Chinook salmon only after five to eight days of exposure [54]. It is, therefore, unlikely that nitrate would ever build up to a concentration that would create problems in a reuse system since periodic inputs of raw water would dilute and export nitrate [2].

The goal of nitrification in a rearing facility then becomes one of maintaining low ammonia and nitrite levels. Based upon available data, these levels must be below 0.03 mg/ℓ FA and 0.100 mg $NO_2 \cdot N/\ell$ for trout and salmon. At these levels growth should not be limited and mortality due to these factors should be negligible. Channel catfish seem more tolerant of FA and nitrite but the available data are insufficient to set reliable limits for exposure in a reuse systems.

DESIGN

Biological filters are fixed film reactors which contain a porous solid phase support media. Nutrients, oxygen, carbon and other required trace minerals are extracted from the wastewater stream by a heterogeneous culture of bacteria grown on the solid phase. The liquid phase passes through the pores and is either recycled to the fish rearing ponds or wasted.

Two types of filters are currently used in fish hatchery reuse systems: submerged filters, trickling filters. Submerged filters may be classified as upflow or downflow depending on the manner in which wastewater enters and exits the filter. The upflow submerged filter is the most commonly used process in reuse systems. The upflow filter does not clog as rapidly as downflow submerged filters [55]. Trickling filters are operated intermittently or continuously. A liquid film flows along the surface of the solid phase support media.

Biofilters have been designed to reduce ammonia concentrations below fish toxicity levels. No consideration has been given to lethal or sublethal effects of nitrite concentrations in the process water.

Aeration of the water can occur naturally at the air-liquid interface within the pores of a trickling filter. Dissolved oxygen concentrations in submerged filters must be maintained primarily by mechanical aeration.

The design procedures available were developed from data obtained from research on trout and/or salmon. Burrows and Combs [2] pioneered the initial work concerning controlled environments for fish propagation. The biofilter consisted of a 1.2-m (4-ft) layer of crushed oyster shell. The size of the oyster shells varied from 0.6 to 1.9 cm (0.25 to 0.75 in.). The size of the filter was based on 2.4 m³/hr/m² of surface area (1 gal/min/ft²). The filter system was designed as a downflow submerged filter.

Calcium carbonate present in the oyster shells is solubilized and reacts with nitrate to form calcium nitrate. Failure to maintain an adequate base concentration will result in the accumulation of nitrous and nitric acid [2].

An advantage of the water reuse system used in fish culture is the tremendous reduction in new water required to operate the system. Supplemental make-up water is added in amounts ranging from 2 to 10% of the total water used in the recirculating system. Without temperature control equipment 5 to 10% make-up water is required.

Burrows and Combs [2] recommend that in large hatcheries a series of 10 rearing ponds be used (Figure 1). This configuration will allow isolation of different size fish groups and species and permit more precise temperature control. The size of the biofilter for 10 or more rearing ponds should be 6.1 × 23 m (20 × 75 ft) as shown in Figure 2. Fish hatcheries containing fewer than 10 rearing ponds should have filters 4.6 × 15.2 m (15 × 50 ft).

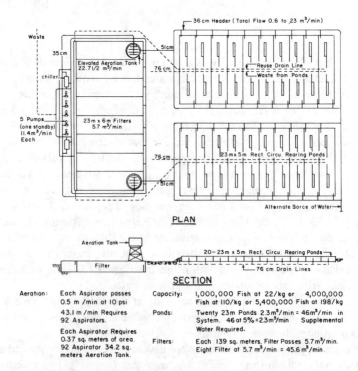

Figure 1. Schematic drawing of typical controlled environment system for rearing salmon (after Burrows and Combs [21]).

Filters are cleaned by air scour followed by backwashing with raw water [2]. The raw water inlet valve is first closed, the filtered water outlet valve is closed, and the water level is lowered to 30 cm (1 ft) above the oyster shell media. An air blower is actuated and produces an air upflow rate through the filter of 24.3 m³/hr/m² of filter area (1.33 ft³/min/ft²). The air turbulence agitates the oyster shell and scrubs the biomass and other debris from the media. The blower is operated for one hour. At the end of 40 min the raw water valve is opened which flushes the loose debris out of the filter. Backwashing the filter with raw water requires 20 min. The cleaning of the filters is a function of the fish loading. Fish culturists at hatcheries operating at full capacity may be required to clean the biofilters every other day.

Based on data obtained primarily from trout hatchery reuse systems, Liao and Mayo [56] and Speece [57] developed design procedures for hatchery reuse systems. The design procedures are reported to be applicable for designing nitrification filters for salmonids.

A basic parameter which governs the production of metabolites (i.e., ammonia, organic matter, phosphorus, suspended solids, etc.) is the feeding level used by the hatchery. Liao and Mayo [56] and Speece [57] use the feeding rate (mass of food/mass of fish per day) as the foundation to their respective design techniques. Speece [57] utilized the work of Haskell [58,59],

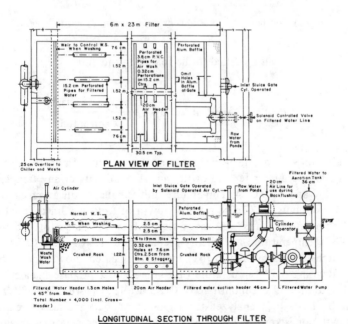

Figure 2. Design drawing of a filter for water reclamation (after Burrows and Combs [2]).

Piper [60], Bowen [61] and Buterbaugh and Willoughby [62] to develop graphical procedures to determine the feeding rate as a function of water temperature and fish length.

Haskell [59] stated that the growth rate of immature trout is the same regardless of size if the fish are contained under comparable environmental conditions (e.g., temperature, density, food, etc.). Feeding levels above a certain minimum have little effect on fish growth [63]. The greatest influence on the growth of fish is water temperature.

Haskell [59] presented the following mathematical definition of feeding levels for trout:

Percent of body weight fed/day =

$$\frac{3 \times \text{feed conversion} \times \text{daily length increase} \times 100}{\text{length of fish}} \tag{5}$$

The constant factors in Equation 5 were consolidated by Buterbaugh and Willoughby [62] in a single "Hatchery Constant". The Hatchery Constant was defined as:

$$\text{Hatchery Constant} = 10 \times \text{feed conversion} \times \text{monthly length increase} \tag{6}$$

Haskell's equation was modified to the following equation:

$$\text{Percent of body weight fed/day} = \frac{\text{Hatchery Constant}}{\text{Length of fish}} \tag{7}$$

In addition, Haskell [59] proposed a "Temperature Unit" theory which states that a definite rate of growth can be predicted for any temperature between 2.7 and 15.6°C. A temperature unit is equal to the average monthly temperature minus 3.7°C. The rate of growth of trout is a linear function of the "Temperature Units" in the water within the range of 3.6-15.6°C. Haskell's "Temperature Unit" theory predicts zero growth at 3.6°C. An analysis of hatchery data by Bowen [61] indicated that zero growth rate occurred at 0°C. Revising the "Temperature Unit" theory to a 0°C base and assuming a growth rate of 1.52 cm/month at 10°C, Speece [57] plotted Haskell's results (Figure 3). Piper [60] analyzed trout growth rate data and discovered that if the growth rate is known at 10°C, the growth rate at temperatures above 10°C will increase 7.2%/°C. Below 10°C the growth rate will decrease 9%/°C. Piper's relationship is shown in Figure 3.

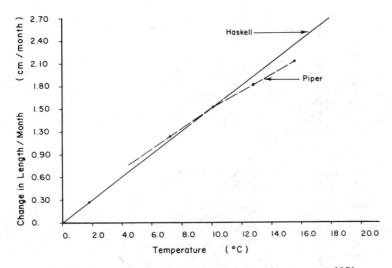

Figure 3. Growth response of trout as a function of temperature [57].

Speece [57] assumed the growth rate followed Piper's prediction model. Furthermore, Speece assumed the growth rate of fish at 10°C was 1.52 cm/month and the feed conversion was 1.5. Speece calculated Hatchery Constants at various temperatures using the growth rates obtained from Figure 3 and Equation 6. The Hatchery Constants at various degrees Celsius are presented in Table III.

Substituting the Hatchery Constant at a particular temperature into Equation 7, Speece constructed a feeding rate chart of percent of body weight fed daily as a function of fish length and water temperature (Figure 4). Therefore, the feeding rate can be predicted for a specified length fish and water temperature.

Knowledge of the feeding rate will enable prediction of metabolite production. Design techniques proposed by Liao and Mayo [56] and Speece [57] are based on the fish feeding rate.

Table III. Predicted Hatchery Constants as a Function of Temperature [57]

	Temperature (°C)				
	5.0	7.0	10	12.5	15.5
Hatchery Constant	12.6	16.7	22.0	26.9	31.8

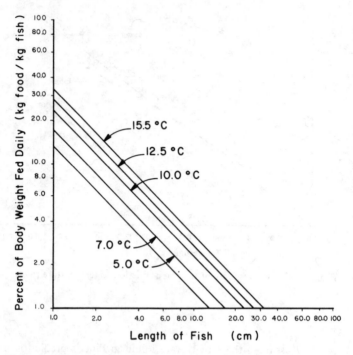

Figure 4. Feeding rate as a function of temperature and length of fish [57].

Design Procedure I: Liao and Mayo [55]

Liao and Mayo utilized data obtained from rearing of trout to develop the following equations which define metabolite production:

$$N_A = 0.0289F \tag{8}$$

$$N_n = 0.024F \tag{9}$$

$$P = 0.0162F \tag{10}$$

$$SS = 0.52F \tag{11}$$

$$BOD = 0.60F \tag{12}$$

$$COD = 1.89F \tag{13}$$

where N_A = ammonia production rate (kg $NH_4 \cdot N/100$ kg fish/day),
$\qquad\quad N_n$ = nitrate production rate (kg $NO_3 \cdot N/100$ kg fish/day),
$\qquad\quad P$ = phosphate production rate (kg $PO_4 \cdot P/100$ kg fish/day),
$\qquad\quad SS$ = suspended solids production rate (kg SS/100 kg fish/day),
$\qquad\quad BOD$ = BOD production rate (kg BOD/100 kg fish/day),
$\qquad\quad COD$ = COD production rate (kg COD/100 kg fish/day), and
$\qquad\quad F$ = feed rate (kg food/100 kg fish/day).

Equations 8-13 are applicable at water temperatures of 10-15°C. Up to 90% recycling of water was used in the trout culture system. Trout densities were maintained up to 28.4 kg/m^3 (1.77 lb/ft^3) of water in the tanks.

Oxygen consumption by the fish is important in the design of reuse systems. The mathematical expression used to determine oxygen consumption rates is [64] :

$$O_c = K_2 T^a W^b \qquad (14)$$

where O_c = oxygen consumption rate (lb $O_2/100$ lb fish/day),
$\qquad\quad K_2$ = rate constant,
$\qquad\quad T$ = temperature (°F),
$\qquad\quad a,b$ = coefficient constants, and
$\qquad\quad W$ = fish size (lb/fish)

Wheaton [55] presents the values of constants, K_2, a, and b as obtained from Liao [64] (Table IV). Liao et al. [65] proposed the fish carrying capacity of the reuse system can be predicted by the following equation:

$$L_c = \frac{0.14 (DO_i - DO_m)}{O_c} \qquad (15)$$

where L_c = carrying capacity (kg fish/ℓ min),
$\qquad\quad DO_i$ = dissolved oxygen concentration at a specific temperature and altitude (mg/ℓ),
$\qquad\quad DO_m$ = minimum dissolved oxygen concentration allowable in the rearing ponds (mg/ℓ), and
$\qquad\quad O_c$ = oxygen uptake rate (kg $O_2/100$ kg fish).

Burrows and Comb [2] stated that the minimum dissolved oxygen concentration (DO_m) of 6 mg/ℓ should be maintained in the ponds. Equation 15 fails to account for a potential accumulation of metabolic produces in the system which will affect the carrying capacity [55].

Table IV. Oxygen Consumption Contants[a] [55]

Species	Temperature	K_2	a	b
Salmon	$\leqslant 50°F$	7.2×10^{-7}	3.200	-0.194
	$> 50°F$	4.9×10^{-5}	2.120	-0.194
Trout	$\leqslant 50°F$	1.90×10^{-6}	3.130	-0.138
	$> 50°F$	3.05×10^{-4}	1.855	-0.138

[a]Insufficient data available to convert the constants to SI units.

The concentration of metabolite (e.g., ammonia) contained in the discharge from a fish culture water reuse system can be predicted by the following expression [66];

$$C = \frac{C_s}{1 - R + RE} \tag{16}$$

where
C = concentration of metabolite in water discharged from a fish rearing unit,
C_s = concentration of metabolite in water discharged from a fish rearing unit in a single pass system,
R = percentage (as a decimal) of water recycled, and
E = metabolite removal efficiency (as a decimal fraction) of a single pass through the biological filter.

Equation 16 can be used to determine the desired removal efficiency and recycle rate necessary to maintain a specified level of metabolite, C.

Ammonia removal in biofilters was primarily dependent on organic, nutrient and hydraulic loading, temperature, pH, dissolved oxygen concentration and retention time [65]. Liao et al. [65] developed empirical relationships between ammonia removal and ammonia loading rate for a trickling filter and two upflow biofilters (Table V). The media retention time is defined by the following expression:

$$\text{media retention time} = V_m \, \epsilon / Q \tag{17}$$

where
V_m = volume of media (m³),
ϵ = void fraction (volume of voids in filter media/total volume of filter media), and
Q = flow rate (m³/hr).

The mathematical expressions presented in Table V are restricted by the following limitations:

Table V. Relationships Describing Ammonia Removal Rates
for Various Biological Filters (after Wheaton [55])

Filter Type	Media Retention Time (hr)	Ammonia Removal Rate $(kg\ NH_4\text{-}N/m^2 \cdot day) \times 10^{-5}$
Trickling filter	0.46	$0.489A_L$ [a]
	0.294	$0.258A_L$
Upflow biological filter number 1	0.33	$0.2533A_L$
	0.294	$0.2227A_L$
Upflow biological filter number 2	0.206	$0.1811A_L$

[a] A_L = ammonia loading rate $[(kg\ NH_4\text{-}N/m^2 \cdot day) \times 10^{-5}]$.

1. temperature range of 10-15°C;
2. hydraulic loading rates of 1.0-1.7 $\ell/s \cdot m^2$ of horizontal filter surface;
3. pH from 7.5 to 8.0;
4. filter media 9-cm (3.5-in.) Koch rings;
5. maximum ammonia loading rate not to exceed 97.6×10^{-5} kg NH_4-N/ $m^2 \cdot day$.

Utilizing the data of Liao et al. [65], Wheaton [55] presented an expression relating ammonia removal as a function of media retention time (t_m) and ammonia loading rates.

$$N_{AR} = 0.96A_L t_m \qquad (18)$$

where N_{AR} = ammonia removal rate of rate of filter (kg NH_4-N/m$^2 \cdot$day),
m^2 = square meters of specific medium surface area,
A_L = ammonia loading rate (kg NH_4-N/m$^2 \cdot$day), and
t_m = media retention time (hr).

Equation 18 is constrained by the following conditions:

1. media retention time between 0.206 and 0.46 hr;
2. hydraulic loading < 101 $\ell/min \cdot m^2$;
3. water temperature range of 10 to 15°C;
4. ammonia-nitrogen concentration approximately ≤ 1 mg NH_4-N/ℓ; and
5. ammonia loading < 97.6 kg NH_4-N/m$^2 \cdot$day.

Haug and McCarty [67] observed that the nitrification rate constant (K) varied linearly with temperature. Based on the linear relationship discovered by Haug and McCarty, Liao et al. [65] developed the following equation to mathematically define the nitrification rate constant (K) as a function of temperature:

$$K = 0.097T - 0.215 \tag{19}$$

where T = water temperature (°C). Equation 19 was developed from two data points; one obtained at 1.67°C and one at 12°C [55]. Substitution of Equation 19 into Equation 18 yields the following expression:

$$N_{AR} = (0.098T - 0.217)A_L t_m \tag{20}$$

The coefficients have changed due to rounding differences or addition of data in the derivations of the equations. The ammonia removal efficiency of the biofilter is the ratio of the ammonia removal rate (N_{AR}) to the ammonia loading rate (A_L). Therefore, the ammonia removal efficiency (E) of the filter may be mathematically defined as:

$$E = \frac{N_{AR}}{A_L} = (0.098\,T - 0.217)t_m \tag{21}$$

The data used in the development of the design equation were obtained primarily from nitrification filters used in trout hatchery reuse systems. Similarly, Speece [57] used data obtained from trout hatcheries to develop a graphical technique for the design of nitrification biofilters.

Design Procedure II: Speece [57]

Speece compiled available information concerning trout metabolism and nitrification to develop a rational graphical approach to the design of nitrification processes for use in trout hatchery water reuse systems. The amount of ammonia production is a function of the fish feeding rate. The graphical approach for determining the feeding rate for a specific size fish was previously discussed. An "ammonia factor" was developed from data obtained at the Bozeman Fish Cultural Development Center. The ammonia factor is mathematically defined as:

$$\text{Ammonia factor} = \frac{8.34CQ}{F} \tag{22}$$

where C =   ammonia-nitrogen concentration (ppm),
       Q =   volumetric water flow rate (ℓ/min), and
       F =   feeding rate (kg).

The studies at Bozeman showed that at normal feeding rates the average ammonia-nitrogen concentration in water at 16.7°C was 0.6 ppm. When the water temperature was 9.4°C, the average ammonia-nitrogen concentration was 0.3 ppm. The feeding rates were based on the water temperature. The ammonia factor was computed to be 22.5 for the 16.7°C water and 18.3 for the 9.4°C water. Equation 22 can be solved to determine the ratio of the amount of ammonia-nitrogen produced per kilogram of feed,

$$\text{Ammonia factor } (1.44 \times 10^{-3}) = \text{kg NH}_3\text{-N/kg feed} \qquad (23)$$

Therefore, the ammonia production per kilogram of feed is 0.032 kg NH_3-N/kg feed at 16.7°C and 0.026 kg NH_3-N/kg feed at 9.4°C. Ammonia production is a function of the amount of food fed. Speece assumed the temperature effect on the ratio of ammonia production per kilogram of feed was linear as shown in Figure 5. Utilizing Figure 5 Speece determined rates of ammonia production at various temperatures (Table VI).

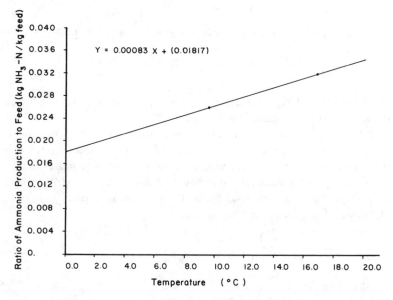

Figure 5. Ratio of ammonia production to feed as a function of temperature [57].

Table VI. The Effect of Temperature of Ammonia-Nitrogen Production[a] [57]

	Temperature (°C)				
	5	7	10	12.5	15.5
Ammonia production, kg NH$_3$-N/kg food	0.022	0.024	0.026	0.029	0.031

Utilizing the information obtained in Table VI and presented in Figure 4, the rate of ammonia production per day for a given fish length and temperature was calculated by Speece by the following formula:

$$N_A = F \times \text{(ammonia conversion)} \tag{24}$$

For example, the ammonia production for 100 kg of 15.2 cm trout at 10°C would be determined as follows:

1. From Figure 4 the feed rate (F) for 15.2 cm trout at 10°C is 1.5 kg feed/100 kg fish.
2. The ammonia conversion at 10°C is 0.026 kg NH$_3$-N/kg feed (Table VI).
3. Therefore, the ammonia production rate (N_A) is

$$N_A = \text{(1.5 kg feed/100 kg fish)} \times \text{(0.026 kg NH}_3\text{-N/kg feed)}$$

$$= 0.039 \text{ kg NH}_3\text{-N/100 kg of 15.2 cm trout per day.}$$

Speece constructed Figure 6 showing the ammonia-nitrogen production as a function of water temperature and fish length.

Data from Bozeman Fish Cultural Development Center indicated that the growth of fish was reduced when the dissolved oxygen concentration was less than 5 mg/ℓ and the ammonia-nitrogen concentration was greater than 0.5 mg/ℓ [57]. Speece assumed the ammonia-nitrogen level in the effluent would be 0.5 mg/ℓ. Based on this assumption, the required flow per 100 kg of fish was calculated from the following equation:

$$Q/100 \text{ kg of fish} = N_A/C'_e\, \rho_w \text{ (1440 min/day)} \tag{25}$$

where Q = required flow (ℓ/min),

N_A = kg NH$_3$-N produced per 100 kg of fish per day,

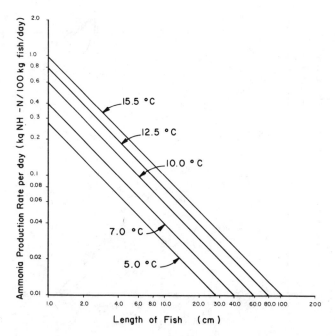

Figure 6. Ammonia production rate as a function of length of fish and temperature [57].

C'_e = fraction of ammonia-nitrogen in effluent (0.5 ppm x 10^{-6}), and
ρ_w = density of water (1 kg/ℓ).

The use of Equation 25 is demonstrated in the following example. The required flow rate for 100 kg of 15.2 cm trout at 10°C can be determined by:

1. The ammonia production rate for a given fish length and water temperature is obtained from Figure 6 (0.039 kg NH_3-N/100 kg fish-day).
2. The required flow rate is calculated using Equation 25,

$$Q/100 \text{ kg of fish} = \frac{0.039 \text{ kg } NH_3\text{-N}/100 \text{ kg fish-day}}{(0.5 \text{ ppm})(10^{-6})(1440 \text{ min/day})(1 \text{ kg/ℓ})}$$

$$= 54.2 \text{ ℓ/min}/100 \text{ kg of } 15.2 \text{ cm trout}$$

Using this procedure, Speece developed the graph presented in Figure 7. The ammonia-nitrogen limit was 0.5 mg/ℓ. Having obtained the required flow of water, Speece constructed Figure 8 which enables the determination of the

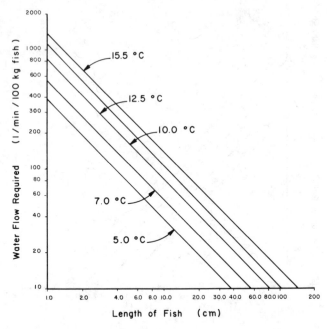

Figure 7. Water flow requirement per 100 kg of fish to limit ammonia-nitrogen to 0.5 ppm as a function of length of fish and temperature [57].

allowable mass loading of fish which would produce 0.5 mg NH_3-N/ℓ in the effluent.

In a fish hatchery water reuse system there are primarily two processes which exert a significant oxygen demand: fish metabolism of food, and nitrification of ammonia within the biofilters. Willoughby [68] determined that 100 g of oxygen were used to metabolize 454 g of trout pellets. This ratio of oxygen to trout feed is relatively constant over the temperature range 4.4-15.6°C. Therefore, Speece assumed this ratio of 0.22 kg oxygen/kg feed was valid and determined the oxygen required for fish metabolism for a given length and a specified temperature as follows:

$$\text{Oxygen requirement} = 0.22 \; \frac{\text{kg oxygen}}{\text{kg feed}} \times F \qquad (26)$$

Thus, the oxygen requirement for 100 kg of 15.2 cm trout at 10°C can be determined as follows:

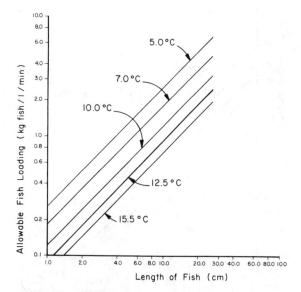

Figure 8. Allowable fish loading to limit ammonia-nitrogen to 0.5 ppm as a function of length and temperature [57].

1. The feed rate (F) for 15.2 cm trout at $10°C$ is 1.5 kg feed/100 kg fish (Figure 4).
2. The oxygen requirement for fish metabolism is:

$$\text{Oxygen requirement} = 0.22 \; \frac{\text{kg oxygen}}{\text{kg feed}} \; (1.5 \text{ kg feed}/100 \text{ kg fish per day})$$

$$= 0.33 \text{ kg oxygen}/100 \text{ kg fish per day}$$

Figure 9 presents the oxygen requirement for fish metabolism for various fish lengths at several temperatures. The change in dissolved oxygen (DO) through the rearing ponds resulting from fish metabolism can be determined as follows:

$$\Delta DO_M = \frac{(\text{kg oxygen}/100 \text{ kg fish-day})}{Q/100 \text{ kg fish } (1440 \text{ min/day})(10^{-6})(\rho_w)} \tag{27}$$

where ΔDO_M = reduction in dissolved oxygen by fish metabolism (ppm).

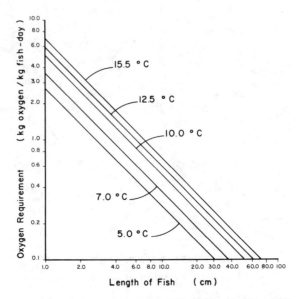

Figure 9. Oxygen requirements for fish metabolism as a function of length of fish and temperature [57].

Therefore, a 15.2 cm trout at 10°C would reduce the DO in the water by the following amount:

1. From Figure 9 the oxygen requirement for 15.2 cm trout at 10°C is 0.33 kg oxygen/100 kg fish-day.
2. The water flow required is 54.3 ℓ/min/100 kg fish (Figure 7).
3. Therefore the DO reduction would be:

$$\Delta DO = \frac{(0.33 \text{ kg oxygen}/100 \text{ kg fish-day})}{54.2 \text{ ℓ/min}/100 \text{ kg fish } (1400 \text{ min/day})(10^{-6})(1 \text{ kg/ℓ})}$$

$$= 4.2 \text{ ppm}$$

Speece noted that the reduction in DO resulting from fish metabolism is constant at any temperature and not a function of fish length. The same quantity of food is added to the rearing pond for a specified water flow and temperature, and the fish weight is varied. The dissolved oxygen reduction is higher at lower temperature due to less ammonia produced per kg of food fed (Figure 10). Therefore, less water is required to transport the ammonia away at 0.5 ppm while the oxygen requirement is assumed constant with temperature [57].

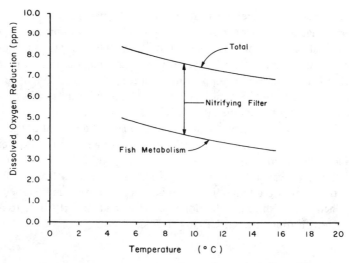

Figure 10. DO reduction from water for fish metabolism and by nitrifying filter as a function of temperature [57].

The nitrification process in the water reuse system also exerts an oxygen demand. Gigger and Speece [69] discovered the total oxygen requirement of the nitrifying filter to be 150% of the stoichiometric oxygen demand (4.57 ppm dissolved oxygen to oxidize 1 ppm NH_3-N). Based on a limited 0.5 ppm NH_3-N in rearing pond effluent, the additional oxygen demand of the nitrifying filter (ΔDO_N) would be

$$\Delta DO_N = 150\% \, (4.57 \text{ ppm } O_2/\text{ppm } NH_3\text{-N})(0.5 \text{ ppm } NH_3\text{-N})$$

$$= 3.4 \text{ ppm DO}$$

Consequently, if a 5-ppm minimum dissolved oxygen is required in the rearing ponds to prevent reduction in fish growth, the total dissolved oxygen entering the rearing ponds must be 5 ppm (minimum level) plus 4.2 ppm (O_2 demand by fish metabolism) or 9.2 ppm.

The nitrifying process will reduce the DO in the final recycled flow to 1.6 ppm (5 ppm O_2 - 3.4 ppm DO). Therefore, since the recycled water is the bulk of the water passing through the over-all system, the influent water to the rearing ponds must be reaerated to meet the oxygen demands of the system.

Organic matter present in the hatchery effluent also exerts an oxygen demand. The biochemical oxygen demand (BOD) of fish hatchery effluents is associated primarily with suspended solids. Speece found that 1 kg of dry fecal matter suspended solids (SS) exerts approximately 1 kg of ultimate

BOD. In a study conducted on a catfish culture in Speece's laboratory, it was discovered that approximately 0.4 kg of dry fecal matter was produced per kg of trout feed fed to the fish. Speece also noted that this ratio fit salmonid BOD production data obtained by Liao [70]. Assuming this ratio is independent of fish size and water temperature, Speece constructed graphs relating the kilograms BOD and SS production per 100 kg of fish per day as a function of fish length and water temperature (Figure 11).

Nitrification in a submerged filter is independent of dissolved oxygen concentration as long as the stoichiometric oxygen requirement for nitrification of ammonia is satisfied [67]. Haug and McCarty [67] discovered that the nitrification rate was independent of pH in the range from 6.5 to 8.0. The rate of nitrification is a function of ammonia concentration and temperature. The mathematical expression developed by Haug and McCarty [67] to determine the nitrification rate is a 2.5 cm gravel biofilter is:

$$\frac{dC}{dt} = (0.11T - 0.2)(C/10)^{1.2} \tag{28}$$

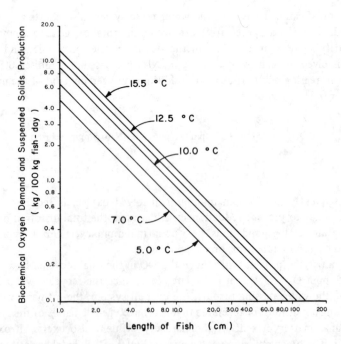

Figure 11. Biochemical oxygen demand and suspended solids production from fish metabolism as a function of length of fish and temperature [57].

where $\dfrac{dC}{dt}$ = rate of nitrification (ppm NH_3-N oxidized/min),

T = temperature ($^\circ$C), and

C = ammonia-nitrogen concentration (ppm).

The relationship between the nitrification rate and temperature was evaluated from 1 to 25°C. Gigger and Speece [69] determined that at 20°C the nitrification rate in a submerged filter containing 1.9 cm gravel was approximately 1 g NH_3-N/m^2 day (or 5.3 g/m^3-hour). The ammonia-nitrogen concentration into the biofilter was 0.5 ppm. Figure 12 shows the effect of temperature on the nitrification capacity of submerged filters. The specific nitrification surface area can be determined from the ammonia production rate (Figure 6) and the nitrification rate (Figure 12). The specific nitrification surface area is the total exposed surface area of the aggregate within the filter upon which the nitrifying bacterial population grows. Speece computed the specific nitrification area required for a given weight and size of fish at a specified temperature from the following relationship:

$$a = N_A / r_N \qquad\qquad (29)$$

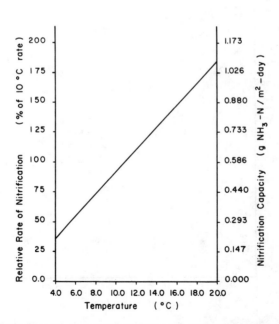

Figure 12. Effect of temperature on nitrification capacity of submerged biofilter [57].

where a =   specific nitrification surface area (m²/100 kg fish), and
r_n =   nitrification rate (kg NH_3-N/m²-day).

Therefore, the specific nitrification surface area required to oxidize ammonia produced from 100 kg of 15.2 cm trout at 10°C would be:

1. From Figure 6 the ammonia production rate of 15.2 cm trout at 10°C is 0.039 kg NH_3-N/100 kg fish-day.
2. The nitrification capacity of a submerged filter at 10°C is 0.527 g NH_3-N/m²-day or 5.27 x 10⁻⁴ kg NH_3-N/m²-day (Figure 12).
3. The specific nitrification surface area is:

$$a = \frac{(0.039 \text{ kg } NH_3\text{-N}/100 \text{ kg fish-day})}{(5.27 \times 10^{-4} \text{ kg } NH_3\text{-N}/m^2\text{-day})}$$

$$= 74 \text{ m}^2/100 \text{ kg of 15.2 cm trout.}$$

The specific nitrification surface area for various lengths of fish (Figure 13) is independent of temperature from 4.4 to 15.6°C.

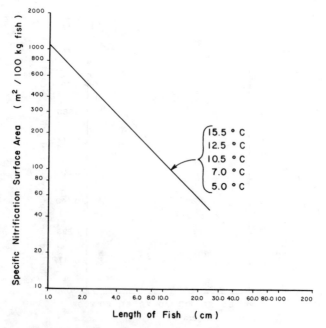

Figure 13. Specific surface area of nitrifying filter required per 100 kg of fish versus length of fish and temperature [57].

Speece developed a graph showing the specific surface area of crushed rock per unit volume as a function of the size of rock (Figure 14). Figure 14 only holds for the filter medium used by Speece and probably would not be valid for other filter media. Consequently, the specific surface area of the medium per unit volume as a function of the size of the medium must be determined for each medium under consideration.

The volume of the nitrifying filter for a given length of fish is calculated by dividing the specific nitrification surface area (a) (Figure 13) by the specific surface area per unit volume of filter medium as follows:

$$V = a/S_s \qquad\qquad (30)$$

where V = required filter volume (m^3), and
S_s = specific surface area per unit volume of medium (m^2/m^3).

The volume required in a nitrifying filter of 5-cm crushed rock for 100 kg of 15.2 cm trout would be:

1. The specific nitrification surface area (a) is 74 m^2/100 kg of 15.2 cm trout (Figure 13).

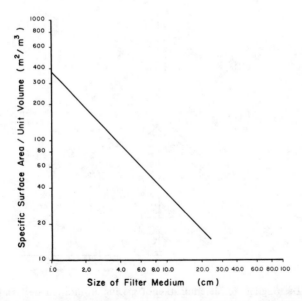

Figure 14. Specific surface area per unit volume of medium versus size of filter rock [57].

2. The specific surface area per unit volume (S_s) of medium is 73.8 m^2/m^3 of 5 cm crushed rock (Figure 14).
3. The required volume (V) for the nitrifying filter is:

$$V = \frac{(74\ m^2/100\ kg\ fish)}{(73.8\ m^2/m^3\ of\ 5\ cm\ crushed\ rock)}$$

$$= 1\ m^3\ of\ 5\ cm\ crushed\ rock/100\ kg\ of\ 15.2\ cm\ fish$$

Speece did not include additional volume to provide the filter with a margin of safety. Furthermore, the entrapment of suspended solids in the filter media is not accounted for in this design procedure.

Speece provided a composite of the preceding graphs from which the volume of a submerged nitrifying filter for a particular trout hatchery can be determined (Figure 15).

The Hatchery Constant relationships are as proposed by Piper [60] (Figure 3). Having selected a size of fish to be reared in the hatchery water system, a vertical line is extended until it intersects the Hatchery Constant line (A). The feeding rate (B) is determined from the particular Hatchery Constant. The feed rate line can be extended until it intersects with a specified water temperature line (C). From the intersection at point C a vertical line is

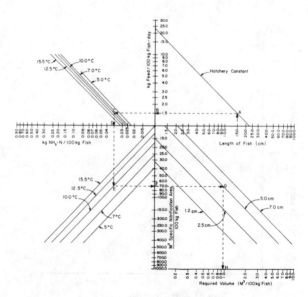

Figure 15. Design graph for determination of specific nitrification area as a function of length and temperature [57].

drawn downward to the corresponding ammonia production rate (D). The line is extended downward until it intersects the given water temperature line (E) which will also delineate a corresponding specific nitrification surface area (F). By extending the line from point F to the line indicating the size of filter medium to be used (G) and dropping the line to the horizontal axis, the required filter volume can be determined (H).

In addition, Speece developed a composite graph which permits the determination of the waste load produced by trout and the required water flow rate (Figure 16). The Hatchery Constant must be determined for the particular hatchery in question and located on the vertical and horizontal axes. The length of fish is selected. A line is projected vertically from the selected fish size to the Hatchery Constant line (A). Extending the line from the point of intersection horizontally to the right indicates the associated oxygen requirement (B) and BOD and SS production rates (C). Extending the line from point A horizontally to the left delineates the feeding rate (D). The feeding rate line is extended to the hatchery water temperature line (E) and dropped downward to intersect with the ammonia production rate value (F). The vertical line is projected downward from point F until it intersects the slanted line (G). A horizontal line is extended from point G to the vertical axis which indicates the water flow required (H).

Speece's design approach for hatchery water reuse system is applicable to the culturing of trout and possibly salmonids in general. The design equations

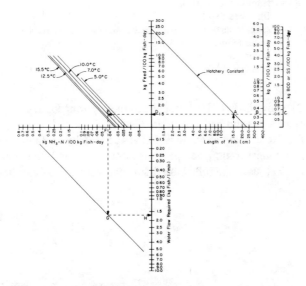

Figure 16. Trout metabolism characteristics as a function of length and temperature [57].

and graphs are subject to the limitations of the assumptions upon which they are based (i.e., growth rate of fish is 1.52 cm/month at 10°C; amount of oxygen required by fish to metabolize 1 kg of feed is 0.22 kg; maximum concentration of ammonia-nitrogen in system effluent is 0.5 ppm; etc.).

Biofilter Predictive Models

As previously discussed, Haug and McCarty [67] mathematically described nitrification rates in fixed-film biofilters, utilizing 2.5-3.8 cm diameter smooth quartzite filter medium (Equation 28). Tests conducted by Haug and McCarty were at ammonia-nitrogen concentrations of 2-14 mg/ℓ. Ammonia-nitrogen levels existing in trout rearing pond effluents are normally less than 2 mg/ℓ. In addition, the data presented by Haug and McCarty failed to show the importance of media characteristics on the reaction kinetics. Hess [71] developed the following equation to delineate the importance of media characteristics on biofilter performance:

$$E = \frac{4.8 A_B (D_B)^{0.8} (S_s)^{0.8} (0.604 T - 2.15)}{1000 \, (Q) \epsilon} \tag{31}$$

where E = percent removal of ammonia,
A_B = filter bed area (ft^2),
D_B = filter bed depth (ft),
S_s = specific surface area per unit volume of medium (ft^2/ft^3),
T = water temperature (°F),
Q = water flow (gpm), and
ϵ = void fraction.

Cooley [72] showed that Hess' model produces erroneous results when high specific surface areas and bed depths are tested.

In an effort to develop a relationship between nitrifying filter performance and ammonia concentration, temperature, flow rates, and media characteristics, Cooley [72] used the following first-order reaction to describe ammonia oxidation:

$$\frac{dc}{dt} = KC' \tag{32}$$

where C = ammonia-nitrogen concentration (mg/ℓ), and
K = reaction rate constant (time^{-1}).

Integrating Equation 32 and solving for percent removal (E) yields:

$$E = 1 - e^{-kt} \tag{33}$$

The actual bed retention time (t) was defined by Cooley as:

$$t = A_B D_B \epsilon / Q \tag{34}$$

where all symbols are as previously defined.

Cooley indicated that the reaction rate coefficient, K, is dependent on:

1. biological equilibrium,
2. pH and alkalinity,
3. dissolved oxygen,
4. temperature, and
5. media characteristics.

Cooley assumed that the bacterial population is in equilibrium and the pH, alkalinity, and dissolved oxygen are in the acceptable ranges defined by Haug and McCarty. Therefore, the reaction rate coefficient is affected only by temperature and the medium characteristics.

Hess [71] developed a temperature correction factor (T_c) from work published by Haug and McCarty. The correction factor corrects reaction rates to 11.2°C and is mathematically defined as:

$$T_c = 0.1087T - 0.2172 \tag{35}$$

where T = water temperature (°C). Equation 35 can be applied directly to the reaction rate coefficient.

Cooley noted that an increase in the specific surface area and decrease in void fraction caused an increase in the reactor rate coefficient, K. This is obvious since an increase in the specific surface area and a decrease in void fraction results in an actual decrease in the distance between medium surfaces, thereby enhancing the availability of ammonia to the attached bacterial film. The reaction rate coefficient corrected for temperature, K_{T_c}, was linearly related to the ratio of the specific surface area to the void fraction as follows:

$$K_{T_c} = 0.00328 S_s / \epsilon \tag{36}$$

where S_s = specific surface area per unit volume of medium (m^2/m^3), and
        ϵ = void fraction.

Combining Equations 35 and 36, the corrected reaction rate coefficient is defined as:

$$K_c = 0.003285_s/\epsilon \, (0.1087T - 0.2172) \tag{37}$$

Substituting Equations 34 and 37 into Equation 33 yields the following expression:

$$E = 1 - e^{-(0.003285_s/\epsilon)(0.1098T - 0.2172)\dfrac{A_B D_B \epsilon}{Q}} \tag{38}$$

where S_s = specific surface area per unit volume of medium (m^2/m^3),
        ϵ = void fraction,
        T = temperature ($^\circ$C),
        A_B = area of filter bed (m^2),
        D_B = depth of filter bed (m), and
        Q = water flow rate (m^3/min).

Cooley compared predicted ammonia removals against actual removals reported by Haug and McCarty. Close agreement was observed between actual and predicted removals (Figure 17). Equation 38 is based on the following assumptions:

1. stable pH within acceptable range (6.5-8.0).
2. dissolved oxygen and alkalinity equal to or above the stoichiometric requirement.
3. plug flow reaction with no short circuiting.
4. no secondary ammonia production within the filter.

The ammonia oxidation model developed by Cooley can be employed to predict the performance of a nitrifying filter and in designing a filter to meet a specified ammonia removal efficiency.

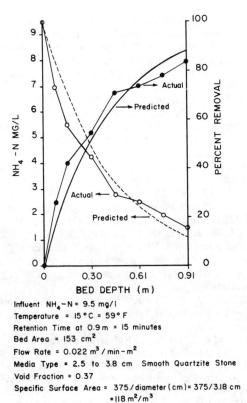

Influent NH$_4$-N = 9.5 mg/l
Temperature = 15 °C = 59° F
Retention Time at 0.9 m = 15 minutes
Bed Area = 153 cm^2
Flow Rate = 0.022 m^3 / min - m^2
Media Type = 2.5 to 3.8 cm Smooth Quartzite Stone
Void Fraction = 0.37
Specific Surface Area = 375 / diameter (cm) = 375/3.18 cm
= 118 m^2/m^3

Figure 17. Comparison of predicted and actual ammonia-nitrogen removal as a function of biofilter depth (after Cooley [72]).

REFERENCES

1. Baradach, J. E., J. H. Ryther and W. O. McLarney. *Aquaculture: The Farming and Husbandry of Freshwater and Marine Organisms* (New York: Wiley-Interscience, 1972), p. 868.
2. Burrows, R. E., and B. D. Combs. "Controlled Environments for Salmon Propagation," *Prog. Fish-Cult.* 30:123-136 (1968).
3. Buckley, J. A. "Acute Toxicity of Un-ionized Ammonia to Fingerling Coho Salmon," *Prog. Fish-Cult.* 40:30-32 (1978).
4. Robinette, H. R. "Effect of Selected Sublethal Levels of Ammonia on the Growth of Channel Catfish (*Ictalurus punetatus*)," *Prog. Fish-Cult.* 38-26-29 (1976).

5. Focht, D. D., and W. Verstraete. "Biochemical Ecology of Nitrification and Denitrification," *Adv. Microbial Ecol.* 1:135-214 (1977).
6. Buswell, A. M., T. Shiota, N. Lawrence and I. V. Meter. "Laboratory Studies on the Kinetics of the Growth of *Nitrosomonas* with Relation to the Nitrification Phase of the BOD Test," *Appl. Microbiol.* 2:21-25 (1954).
7. Knowles, G., A. L. Downing and M. J. Barrett. "Determination of Kinetic Constants for Nitrifying Bacteria in Mixed Culture, with the Aid of an Electronic Computer," *J. Gen. Microbiol.* 38:263-278 (1965).
8. Belser, L. W. "The Ecology of Nitrifying Bacteria," PhD Thesis, University of California, Berkeley, CA (1974).
9. Ardakani, M. S., R. K. Schulz and A. D. McLaren. "A Kinetic Study of Ammonium and Nitrite Oxidation in a Soil Field Plot," *Soil Sci Soc. Amer. Proc.* 38:273 (1974).
10. Schmidt, E. L. "Quantitative Autecological Study of Microorganisms in Soil by Immunofluorescence," *Soil Sci.* 118:141 (1974).
11. Volz, M. G., L. W. Belser, M. S. Ardakani and A. D. McLaren. "Nitrate Reduction and Associated Microbial Populations in a Ponded Hanford Sandy Loam," *J. Environ. Qual.* 4:99 (1975).
12. Painter, H. A. "A Review of Literature of Inorganic Nitrogen Metabolism in Microorganisms," *Water Res.* 4:393 (1970).
13. Focht, D. D., and A. C. Chang. "Nitrification and Denitrification Processes Related to Wastewater Treatment," *Adv. Appl. Microbiol.* 19:153-186 (1976).
14. Weber, D. F., and P. L. Gainey. "Relative Sensitivity of Nitrifying Organisms to Hydrogen Ions in Soils and in Solutions," *Soil Sci.* 94:138 (1962).
15. Becker, G. E., and E. L. Schmidt. "B-Nitropropionic Acid and Nitrite in Relation to Nitrate Formation by *Aspergillus flavus*," *Arch. Microbiol.* 49:167-176 (1964).
16. Verstraete, W., and M. Alexander. "Heterotrophic Nitrification in Samples of Natural Ecosystems," *Environ. Sci. Technol.* 7:39 (1973).
17. Christensen, M. H., and P. Harremoës. "Nitrification and Denitrification in Wastewater Treatment," in *Water Pollution Microbiology. Vol. 2*, R. Mitchell, Ed. (New York: Wiley-Interscience, 1978), pp. 391-414.
18. Sharma, B., and R. C. Ahlert. "Nitrification and Nitrogen Removal," *Water Res.* 11:897-925 (1977).
19. "Process Design Manual for Nitrogen Control," U.S. EPA Office of Technology Transfer, Cincinnati, OH (1975).
20. Poduska, R. A., and J. F. Andrews. "Dynamics of Nitrification in the Activated Sludge Process," in *Proceedings of the 29th Ind. Waste Conference* (Lafayette, IN: Purdue University, 1974), pp. 1005-1025.
21. Lees, H. *Biochemistry of the Autotrophic Bacteria* (London, England: Butterworths Scientific Publications, 1955).
22. Van Droogenbroek, R., and H. Laudelout. "Phosphate Requirements of the Nitrifying Bacteria," *Antonie Van Leeuwenhock* 33:287 (1967).
23. Aleem, M. I. H. "The Physiology and Chemoautotrophic Metabolism of *Nitrobacter agilis*," PhD Thesis, Cornell University, Ithaca, NY (1959).
24. Skinner, F. A., and N. Walker. "Growth of *Nitrosomonas europea* in Batch and Continuous Culture," *Arch. Mikrobiol.* 38:339 (1961).
25. Loveless, J. E., and H. A. Painter. "The Influence of Metal Ion Concentration and pH Value on the Growth of a Nitrosomonas Strain Isolated from Activated Sludge," *J. Gen. Microbiol.* 52:1 (1968).

26. Finstein, M. S., and C. C. Delwiche. "Molybdenum as a Micronutrient for *Nitrobacter*," *J. Bacteriol.* 89:123 (1965).
27. Tomlinson, T. G., A. G. Boon and C. N. A. Trotman. "Inhibition of Nitrification in the Activated Sludge Process of Sewage Disposal," *J. Appl. Bacteriol.* 29:266 (1966).
28. Finstein, M. S., and M. R. Butzky. "Relationships of Autotrophic Ammonium-Oxidizing Bacteria to Marine Salts," *Water Res.* 6:31-40 (1972).
29. Pan, P. C. "Basis of Obligate Autotrophy," PhD Thesis, Rutgers University, New Brunswick, NJ (1971).
30. Gunderson, K. "Effects of B Vitamins and Amino Acids on Nitrification," *Physiol. Plant.* 8:136-141 (1955).
31. Krulwich, T. A., and H. B. Funk. "Simulation of *Nitrobacter agilis* by Biotin," *J. Bacteriol.* 90:729-733 (1965).
32. Clark, C., and E. L. Schmidt. "Growth Response of *Nitrosomonas europaea* to Amino Acids," *J. Bacteriol.* 93:1302-1308 (1967).
33. Clark, C., and E. L. Schmidt. "Uptake and Utilization of Amino Acids by Resting Cells of *Nitrosomonas europaea*," *J. Bacteriol.* 93:1309-1315 (1967).
34. Gundersen, K. "Influence of Corn-Steep Liquor on the Oxidation of Ammonia to Nitrite by *Nitrosomonas europaea*," *J. Gen. Microbiol.* 19: 190-197 (1958).
35. Painter, H. A. "Microbial Transformations of Inorganic Nitrogen," in *Progressive Water Technology,* S. H. Jenkins, Ed. (New York: Pergamon Press, 1977).
36. Meyerhof, O. "Untersuchungen uber den Atmungsvorgang nitrifizieren der Bakterien," *Arch. Ges. Physiol.* 164:229 (1916)(cited in Focht and Verstraete, 1977).
37. Anthonisen, A. C., R. C. Loehr, T. B. S. Prakasam and E. G. Srinath. "Inhibition of Nitrification by Ammonia and Nitrous Acid," *J. Water Poll. Control Fed.* 48:835-852 (1976).
38. Belser, L. W. "Nitrate Reduction to Nitrite: A Possible Source of Nitrite-Oxidizing Bacteria," *Appl. Environ. Microbiol.* 34:403-410 (1977).
39. Kholdebarin, B., and J. J. Oertili. "Effect of pH and Ammonia on the Rate of Nitrification in Surface Water," *J. Water Poll. Control Fed.* 49: 1688-1692 (1977).
40. Spotte, S. H. *Fish and Invertebrate Culture.* 2nd ed. (New York: John Wiley and Sons, Inc., 1979), p. 145.
41. Smart, G. R. "Investigations of the Toxic Mechanisms of Ammonia to Fish-Gas Exchange in Rainbow Trout (*Salmo gairdneri*) Exposed to Acutely Lethal Concentrations," *J. Fish Biol.* 12:93-104 (1978).
42. Lloyd, R., and L. Orr. "The Diuretic Response by Rainbow Trout to Sub-Lethal Concentrations of Ammonia," *Water Res.* 3:335-344 (1969).
43. Ball, I. R. "The Relative Susceptibilities of Some Species of Freshwater Fish to Poisons. I. Ammonia," *Water Res.* 1:767-775 (1967).
44. Rice, S. D., and R. M. Stokes. "Acute Toxicity of Ammonia to Several Developmental Stages of Rainbow Trout, *Salmo gairdneri*," *U.S. Natl. Mar. Fish. Serv. Fish. Bull.* 73:207-211 (1975).
45. Smith, C. E. "Effects of Metabolic Products on the Quality of Rainbow Trout," *Am. Fish. U.S. Trout News* 17:1-3 (1972).
46. Colt, J. E. "The Effects of Ammonia on the Growth of Channel Catfish, *Ictalurus punctatus*," PhD Thesis, Auburn University, Auburn, AL (1978).

47. Smith, C. E., and R. C. Russo. "Nitrite-Induced Methemoglobinemia in Rainbow Trout," *Prog. Fish-Cult.* 37:150-152 (1975).
48. Brown, D. A., and D. J. McLeay. "Effect of Nitrite on Methemoglobin and Total Hemoglobin of Juvenile Rainbow Trout," *Prog. Fish-Cult.* 37: 36-38 (1975).
49. Russo, R. C., C. E. Smith and R. V. Thurston. "Acute Toxicity of Nitrite to Rainbow Trout (*Salmo gairdneri*)," *J. Fish. Res. Board Can.* 31:1653-1655 (1974).
50. Konikoff, M. "Toxicity of Nitrite to Channel Catfish," *Prog. Fish-Cult.* 37:96-98 (1975).
51. Crawford, R. E., and G. E. Allen. "Seawater Inhibition of Nitrite Toxicity to Chinook Salmon," *Trans. Amer. Fish. Soc.* 106:105-109 (1977).
52. Meade, T. L. "Environmental Parameters Affecting Fish Physiology in Water Reuse Systems," *Marine Fish Rev.* Paper No. 1349. 40:46-67 (1978).
53. Smith, C. E., and W. G. Williams. "Experimental Nitrite Toxicity in Rainbow Trout and Chinook Salmon," *Trans. Amer. Fish. Soc.* 103:389-390 (1974).
54. Westin, D. T. "Nitrate and Nitrite Toxicity to Salmonid Fishes," *Prog. Fish-Cult.* 36:86-89 (1974).
55. Wheaton, F. W. *Aquacultural Engineering* (New York: Wiley-Interscience, 1977), p. 708.
56. Liao, P. B., and R. O. Mayo. "Intensified Fish Culture Combining Water Reconditioning with Pollution Abatement," *Aquaculture* 3:61-85 (1974).
57. Speece, R. E. "Trout Metabolism Characteristics and the Rational Design of Nitrification Facilities for Water Reuse in Hatcheries," *Trans. Amer. Fish. Soc.* 102:323-334 (1973).
58. Haskell, D. C. "Weight of Fish per Cubic Foot of Water in Hatchery Troughs and Ponds," *Prog. Fish-Cult.* 17:117-118 (1955).
59. Haskell, D. C. "Trout Growth in Hatcheries," *N.Y. Fish Game J.* 6: 204-237 (1959).
60. Piper, R. G. "Know the Proper Carrying Capacities of your Farm," *Am. Fishes U.S. Trout News* 4-6 (1970).
61. Bowen, J. T. Personal communication with Speece. U. S. Bureau of Sport Fisheries, Washington, DC (1971).
62. Buterbaugh, G. L., and H. Willoughby. "A Feeding Guide for Brook, Brown, and Rainbow Trout," *Prog. Fish-Cult.* 29:210-215.
63. Freeman, R. I., D. C. Haskell, D. L. Longacre and E. W. Stiles. "Calculations of Amounts to Feed in Trout Hatcheries," *Prog. Fish-Cult.* 29:194-209 (1967).
64. Liao, P. B. "Water Requirements of Salmonids," *Prog. Fish-Cult.* 34(4): 210-224 (1971).
65. Liao, P. B., R. D. Mayo and S. W. Williams. "A Study for Development of Fish Hatchery Water Treatment Systems," Report prepared for U.S. Department of Army Corps of Engineers, Walla Walla District, Walla Walla, WA (1972).
66. Liao, P. B., and R. D. Mayo. "Salmonid Hatchery Water Reuse Systems," *Aquaculture* 1(1):317-335 (1972).
67. Haug, R. T., and P. C. McCarty. "Nitrification with the Submerged Filter," Technical Report, Stanford University, Stanford, CA (1971).

68. Willoughby, H. "A Method for Calculating Carrying Capacities of Hatchery Troughs and Ponds," *Prog. Fish-Cult.* 30:173-174 (1968).
69. Gigger, R. P., and R. E. Speece. "Treatment of Fish Hatchery Effluent for Recycle," Technical Report No. 67, New Mexico State University, Las Cruces, NM (1970).
70. Liao, P. B. "Pollution Potential of Salmonid Fish Hatcheries," *Water Sewage Works* 291-297 (1970).
71. Hess, J. W. "Aids to Evaluation and Design Sizing of Submerged Nitrification Biofilters in Fish Rearing Water," MS Thesis, University of Idaho, Moscow, ID (1976).
72. Cooley, P. E. "Summary of Biofilter Testing," Prepared for Corps of Engineers, Walla Walla District, Walla Walla, WA (1979).

CHAPTER 18

PISCICULTURE IN WASTEWATER

Scott Henderson
Assistant Chief, Fisheries Division
Arkansas Game and Fish Commission
Lonoke, Arkansas

Pisciculture, the controlled production of finfish, in waste water has been a subject of interest in the United States for a relatively brief period of time and remains largely conceptual from a practical standpoint. The broader field of wastewater aquaculture has been around longer but efforts have been concentrated mainly toward utilizing plant species. It is ironic that fish culture has only recently evolved as a serious consideration in the field of wastewater aquaculture since it has been practiced in some parts of the world for hundreds or even thousands of years. This is one wheel, however, that is being reinvented with an almost totally new set of design specifications.

Because of the many variables involved with water quality, culture methods and species used, it is beyond the scope of a single chapter to consider all aspects of wastewater utilization in fish culture. Most of the possibilities must be considered on an individual-site basis, but certain constraints pervade the whole issue.

FISH CULTURE METHODS

Fish culture as practiced today involves many variations of two basic methods:

1. Raceways. Raceways are relatively small rearing units that are characterized by very high stocking densities and rapid flow-through of large volumes of water. These factors prevent the buildup of toxic metabolic wastes and

maintain sufficient oxygen levels. A prodigious, continual supply of high-quality water is usually the limiting factor in raceway production. Raceways are typically thought of as synonymous with the production of trout species even though the method is applicable to any species. Densities of fish in raceways vary greatly in practice but, as a point of reference, a continuous flow of 450 gpm of good quality water is necessary for the production of 10,000 lb of trout annually with the raceway designed to allow a complete water exchange every hour [1].

Wastewater use in the raceway culture of salmonids as presently practiced is limited in scope. Raceways could utilize large volumes of wastewater but due to water quality requirements this water would have to be already reclaimed waste water. Also, since large amounts of metabolic wastes will result from the high-density production of a raceway system, raceway effluent will require further treatment before final discharge. Under proper circumstances, this system would provide the benefit of reuse of the water and the production of a valuable product. However, it will also result in the need for further water treatment, instead of solving the original treatment problem. The economics might be improved as a result of fish production but until further advances are made or a new wrinkle in recycling water from raceways is devised, wastewater utilization in this method of fish culture will remain limited.

2. Pond Culture. Pond culture is characterized by a more or less static water supply throughout the production period and is generally more consistent in meeting the needs of warm water species. Land area and type are usually the most limiting factors for pond culture, with water supply following closely. As a point of reference here, one surface acre of water with an average depth of three feet (about 1,000,000 gal) would allow for the production of approximately 3500-4000 lb of channel catfish annually. Pond production levels vary greatly depending on the species being raised.

Pond fish culture offers many more alternatives for wastewater use than does the raceway type. In fact, the culture of certain selected species or combinations of species can in itself provide water quality improvement. Fertilization of warm water fish ponds has long been recognized by the fish culturist as a method of increasing production by stimulating the lower end of the food chain to provide increased natural food supplies. Most species presently cultured commercially in the United States benefit from additional nutrients but the preferred species typically require artificial diets for maximum, economical production rates.

OPPORTUNITIES

It can be reasoned intuitively that if nutrients (fertilizers) are added to fish ponds and fish production increases, that these nutrients travel through the food chain and are finally exhibited in the stable and useful form of fish flesh. It is also known that these nutrients must be photosynthetically transformed into plant material to be usable by fish. It follows that those fish feeding at the lowest trophic level would be the most efficient at converting these nutrients. Therefore, any species of fish used in pond fish culture must meet certain biological as well as economic criteria. It is paradoxical that those fish most preferred as food in the United States do not lend themselves to traditional culture methods [2]. Due to this lack of economic incentive, those lower trophic level feeders have received little attention and have often been considered a problem because of their productive capacity. This is not the case in other parts of the world where these species are prized food items. Only with the advent of higher production costs (mainly for pelleted food and energy) and the emphasis on better water quality discharges have these previously considered "trash fish" been looked upon with any promise.

There are a variety of ways waste water could be applied to pond culture. It could be metered into ponds previously filled with fresh water to provide the optimum nutrient level for the species cultured. In many cases it could also be used as "makeup" water throughout the year to prevent water level fluctuations due to evaporation of seepage. The "strength" of the waste water would dictate many such uses. Water suitable for growing small fish to a larger size might not be suitable for spawning or fry rearing where water quality is much more critical. The more the water quality of a pond varies from the optimum, the greater the stress placed on the fish within that pond. This stress may manifest itself in a number of ways ranging from slowed growth to a greater susceptibility to disease, to death. There are so many variables involved and so many interactions possible that no one parameter can be measured to provide the basis for definitive management decisions. This problem is not unique to the application of waste water in fish culture. Inability to control variables has kept pond fish culture as much in the realm of art as in science.

Pond fish culture has been practiced for centuries in China and parts of Southeast Asia using the same principles as those involved with nutrient-laden waste waters. In these cases, protein production was the goal. Human, livestock and agricultural wastes were added directly to the fish pond to provide the increased nutrient levels [3]. The fish cultured were primarily species belonging to the carp family. This long history has provided a wealth of information concerning how much waste of what type is necessary for suitable production. These systems also combined several noncompeting species to utilize all natural food sources in the pond, thereby increasing overall production. These combinations generally included a filter feeding planktivore, a bottom feeding omnivore and a herbivore feeding on aquatic macrophytes. Species ratios, stocking densities and application rates of various types of wastes necessary for a given production level are well documented and reported throughout the literature.

This ancient method of fish culture has been rediscovered in recent years for a totally different reason. Although there is a wealth of information concerning production enhancement by adding nutrients, little or nothing is known about the amount of nutrient removal that can be expected as a result of the growth of the fish, even though the two processes are inseparable. As United States water quality standards have become more stringent and conventional treatment methods have become cost prohibitive, alternative treatment methods became more attractive. Water quality improvement through fish culture is only one method presently receiving attention.

Many species of fish ranging from the lowly esteemed common carp, *Cyprinus carpio,* to such highly prized sport fish as the muskellunge, *Esox masquinongy,* have been produced in wastewater-fed ponds. Most instances reported are isolated, one-time trials that were aimed primarily at the production of the fish and little other information is available. While aquaculturists have used waste nutrients for production purposes in other parts of the world, the first attempts in the United States stemmed from the need to remove the nutrients from waste water.

In attempts to design totally self-sufficient ecosystems for extended space travel, B. C. Wolverton of NASA was among the first to discover the constructive utilization of water hyacinths for improving the quality of wastewater [4]. Since this somewhat unlikely beginning, researchers have tried a variety of species for use in upgrading water quality. Unicellular algae are highly efficient at nutrient removal but they themselves become a problem since they cannot be economically removed from the water. Attention, therefore, has been turned to organisms that can do a similar job and be more easily harvested. Both shellfish and finfish have been tested in these studies.

Shellfish

Though technically not included in the term pisciculture, shellfish were among the first organisms evaluated for nutrient removal from a closed system. Dr. John Ryther and associates at the Woods Hole Oceanographic Institute began work in the early 1970s utilizing oysters, clams, mussels and scallops [5]. Their system mixed nutrient-laden waste water with sea water to stimulate plankton growth, which in turn was used as a food supply for the cultured species. In later years, lobsters and other marine organisms were added to the system. This pilot-scale project has shown promise. However, the length of time necessary to culture these species to a usable size, the rather advanced technology necessary and the specialized problems involved with dilution rates and plankton culturing, have cast some doubts on the cost effectiveness of a full-scale plant of this type.

Finfish

Waste treatment lagoons have been utilized widely as a simple, inexpensive method of treating a wide variety of wastes, particularly domestic waste from small communities. These ponds attracted attention from the fish culturist because of the possibility of managing them as highly fertilized fish ponds.

Salmon

Dr. George Allen and his students at Humboldt State University in Arcata, California, established a pilot project to test the capacity of sea water fertilized with treated domestic sewage to produce food for juvenile salmon [6]. Both earthen ponds and cages suspended in the ponds were used to rear small Chinook salmon, *Oncorhynchus tschawytscha*, and Coho salmon, *Oncorhynchus kisutch*. Because high-quality water is needed for salmonid culture, their early attempts were promising but not highly successful. Water quality fluctuations resulted in low survival and low productivity levels. However, those fish that did survive exhibited suitable growth. Cost effectiveness and production levels were not thought to be competitive with conventional culture methods without further refinements to the system.

Channel Catfish and Baitfish

In 1970, Mark Coleman and Dr. LeRoy Carpenter of the Oklahoma Department of Health began what was probably the first controlled attempt at fish culture in full-scale sewage lagoons at the one million gallon per day Quail Creek Treatment Plant at Oklahoma City [7]. The plant consisted of six lagoons arranged in a serial flow-through pattern with a forced air aeration system in the first two ponds. Channel catfish, *Ictalurus punctatus*; golden shiners, *Notemigonus crysoleucas*; fathead minnows, *Pimephales promelas*; and *Tilapia nilotica* were introduced into the ponds.

The food habitats of these species are particularly noteworthy. The tilapia, golden shiners and fathead minnows are primarily plankton feeders and the channel catfish is an omnivore which, like all fish, begins life as a plankton feeder but becomes more dependent upon higher organisms in the food chain as it increases in size. *Tilapia nilotica* is notorious for its prolific reproductive habits but since it evolved in the tropics, it cannot survive low water temperatures. Golden shiners and fathead minnows do not usually produce a high biomass even under ideal hatchery conditions.

The Quail Creek experiment saw a tremendous increase in the tilapia population before all fish were lost due to low water temperatures in November. This loss prevented an accurate biomass measurement. The total production of golden shiners and fathead minnows was disappointing even though an increase six times the original stocking weight of 85 lb occurred. The biomass of channel catfish increased from the initial 600 lb to an estimated 4400 lb. This increase, however, took place in a short six-week period after stocking, since as the catfish increased in size, their food habits changed and growth stopped. Growth cessation resulted in stressed catfish, disease problems and the failure of this species to further remove significant amounts of nutrients from the water.

The importance of the Quail Creek Project is that it was the forerunner of this simplistic type of culture and wastewater treatment. Water quality at the discharge was improved, probably due in part to the ability of the new ponds to accept the initial impact of the biochemical oxygen demand and nutrient loads and undoubtedly in part to the presence of the fish. Although the species used at Quail Creek were probably not the ideal ones for the situation, this project set the stage for further work in this area.

Chinese Carps

In 1974, the Arkansas Game and Fish Commission, with funding support from the EPA, began a development and demonstration project based on the results of Quail Creek and a conceptual model proposed by Dr. S. Y. Lin [8], a long time fisheries expert from Taiwan. The Arkansas Game and Fish Commission's interest in this project evolved from the importation to Arkansas of two species of Chinese carps by a private fish farmer. These species, the silver carp, *Hypopthalmichthyes molitrix,* and the bighead carp, *Aristichthyes nobilis,* were two of the species used in the classical Chinese polyculture scheme of fish culture previously mentioned.

The feeding mechanisms of these fishes seemed ideally suited for a waste treatment project. Both are filter feeders, with the silver carp being the more efficient of the two since it possesses a gill raker apparatus capable of filtering particles as small as four microns [9]. The silver carp ingests predominately phytoplankton, and the bighead carp ingests primarily zooplankton and the larger blue-green algae species. Also aiding their waste removal capabilities is the fact that the feeding habits of these fish do not change throughout their life span nor is their filtering capability reduced. These fish are fast growing, reach a maximum size of 40-60 lb, are tolerant of low oxygen levels (typical of sewage lagoons), withstand low water temperatures, are resistant to most diseases and do not spawn in static water which allows total control of stocking densities.

The fish were stocked in the last four ponds of a serially arranged six-cell lagoon treatment plant at the Benton Services Center at Benton, Arkansas. The 500,000 gal/day plant passes waste water through a grinder and clarifier which removes the larger solids before the water is introduced into the ponds. All water introduced into the system is waste water and no fresh water is provided for dilution. The six interconnecting lagoons, which total 24 acres in area, have a daily loading rate of 976 lb of BOD and 460 lb of suspended solids with a residence time of approximately 70 days for all six ponds. Ponds 1 and 2 serve as stabilization and algae culture ponds; the remaining four ponds are utilized for fish culture.

The plant has been in operation in this manner for almost two years and the experiment is continuing. Results of operation during the first year show a reduction of BOD by 96.4% and of suspended solids by 86%. The presence of the fish continually grazing the plankton bloom has caused increased reduction in total inorganic nitrogen and total phosphorus. With these species and the typical 235-day growing season in central Arkansas, this system is capable of producing approximately 4500 lb of fish per acre, per year. The plant has consistently met the discharge requirements for a secondary treatment permit since the introduction of the fish. However, the application of these fish in advanced wastewater treatment has not yet been fully explored.

Chinese carp have been distributed to a variety of research institutions by the Arkansas Game and Fish Commission for work in this area, although most have been for small bench scale projects [10]. Probably the most notable of these smaller projects has been research by Dr. Homer Buck at the Illinois Natural History Survey who has grown a mixture of the Chinese carps and native species in ponds receiving swine wastes [11].

COST EFFECTIVENESS

According to U.S. EPA report 600/2-76-293, only when finfish aquaculture was not capable of meeting water quality objectives was it deemed not to be cost effective when compared to conventional systems [12]. The report went further to state that aquaculture wastewater alternatives appear to be economically attractive regardless of the market for products if water quality goals are met.

Although there are several possibilities and likely many useful fishery products yet to be developed, it appears that the long and the short of the present market for wastewater-reared fishes lie with the use of the product as a processed food item for direct consumption or with rendering them into meal and oil for use as livestock feed supplements. With a dwindling supply of ocean catches and a growing market for fisheries products, the economics of the industry will likely change drastically in the future. It should be understood that in present-day freshwater pond aquaculture, the greatest overhead costs are land, feed, fertilizer and water. By utilizing a system of wastewater aquaculture, these costs would be borne by the primary function of water treatment. By accepting this and some other rather basic assumptions within the framework of present markets, some cursory economic projections have been made.

The Quail Creek study assumed the sale of catfish and tilapia as a food item and of shiners and minnows as baitfish and calculated a net profit of 2.3 cents/1000 gal of wastewater passing through the plant [7]. Based on the production potential of the Chinese carp, the Arkansas study projected a gross return of $180-240/ac/yr if the fish were used in the meal industry and from $1500-1800/ac/yr should they be sold directly as a food item [13].

SOCIAL AND LEGAL CONSTRAINTS

Those species of fish amenable to wastewater aquaculture are not popular food items in most of the western world [14] and the sale of a fish grown in "sewage" has its obvious problems. However, the ability to sell these products at low cost, coupled with proper processing and marketing, could likely overcome that stigma, especially in light of present economic developments. In fact, the channel catfish is one of the few successful examples of commercial freshwater aquaculture even though it is limited to the southern United States [2]. Marketing tests by Auburn University have shown the Chinese carps to be successful sellers when attractively presented and sold at reasonable prices [15].

The major constraints on the sale of wastewater-reared fish stem from public health laws [16,17]. These laws and their intent of protecting the consumer's health are certainly of major importance. Shellfish are not considered marketable as a direct food item because of the custom of eating them raw and their known affinity for concentrating chemical and biological contaminants. There are no apparent problems with the use of finfish rendered to a meal product primarily because of the heat used in processing and the existence of other suitable quality control measures. It also appears that finfish grown in wastewater could be used for direct consumption, if acceptable guidelines could be established. Since contaminants in the fish would result from contaminants in the water, the decision for safe use would have to be made on a site-by-site basis and acceptable levels would have to be determined as in other fisheries products. The Arkansas study has monitored viruses, pathogenic bacteria, heavy metals and pesticides in both the water and the fish flesh. To date, no contaminant has been found in the fish flesh that would prevent its being used under present FDA guidelines [18]. Work being done in Israel [19] is aimed at this same objective. Preliminary findings indicate that they have identified a threshold level for bacteria in water that is a reasonable indicator that no bacteria will be found in the fish flesh. While quality control guidelines are far from being established for wastewater-reared fish, there is no reason to believe that acceptable levels cannot be determined as they have been for wild caught fish.

OTHER POSSIBILITIES

This chapter has dealt primarily with domestic wastes due to the emphasis placed on utilizing this most abundant wastewater "resource". There are many types of waste water that could be more easily adapted for use in fish culture. Cannery wastes, for example, would offer the abundance of nutrients without the associated problems of viral and bacterial contaminants. Many industrial wastewaters would require removal of substances toxic to the cultured organism. Heated effluents from power generation have received considerable attention [20] because the ability to use the water directly or mix the warm water with other sources offers great potential in maintaining optimum growth temperature and extending the natural growing season for any culturable fish species.

Because of the necessity of a plentiful water supply, the varied applications for the use of waste water in aquaculture, and the growing demand for fisheries products, fish culture will remain high on the list of opportunities for recycling waste water into a useful product.

REFERENCES

1. Scheffer, P., and D. Marriage. "Trout Farming," USDA Soil Conservation Service, Leaflet 552, U.S. Government Printing Office (1969).
2. "Aquaculture in the United States: Constraints and Opportunities," National Academy of Sciences (1978).
3. Ryther, J. H. "Aquaculture in China," *Oceanus* 22 (1979).
4. Wolverton, B. C., and R. C. McDonald. "Water Hyacinths For Upgrading Sewage Lagoons to Meet Advanced Wastewater Standards," NASA Technical Memorandum TM-X-72729 (1976).
5. Huguenin, J. E., and J. H. Ryther. "Experiences With a Marine Aquaculture-Tertiary Sewage Treatment Complex," in *Wastewater Use In The Production of Food and Fiber—Proceedings*, U.S. EPA Report-660/2-74-041 (1974), pp. 377-386.

6. Allen, G. H., and L. Dennis. "Report On Pilot Aquaculture System Using Domestic Wastewaters For Rearing Pacific Salmon Smolts," in *Wastewater Use In The Production of Food and Fiber—Proceedings*, U.S. EPA Report-660/2-74-041 (1974), pp. 162-198.
7. Coleman, M. S., J. P. Henderson, H. G. Chichester and R. L. Carpenter. "Aquaculture As A Means To Achieve Effluent Standards," in *Wastewater Use In The Production of Food and Fiber—Proceedings*, U.S. EPA Report-660/2-74-041 (1974), pp. 199-212.
8. Lin, S. Y. Personal communication, Lonoke, Arkansas (1974).
9. Henderson, S. "An Evaluation of The Filter Feeding Fishes, Silver and Bighead Carp, For Water Quality Improvement," in *Culture of Exotic Fishes Symposium Proceedings*, R. O. Smitherman, L. Shelton and John Grover, Eds. (Fish Culture Section, American Fisheries Society, 1978), pp. 121-136.
10. Arkansas Game and Fish Commission, Stocking and Fish Distribution Records, Lonoke, Arkansas.
11. Buck, D. H., R. J. Baur and C. R. Rose. "Experiments in the Recycling of Swine Manure Using A Polyculture of Asian and North American Fishes," in *Agriculture and Energy* (New York: Academic Press, Inc., 1977).
12. Henderson, U. B., and F. S. Wert. "Economic Assessment of Wastewater Aquaculture Treatment Systems," U.S. EPA Report-600/2-76-293 (1976).
13. Henderson, S. "Utilization of Silver and Bighead Carp For Water Quality Improvement," *Proceedings of Symposium on Aquaculture Systems for Wastewater Treatment, September 1979* (in press).
14. Ryther, J. H. "Waste Recycling in Warm Water Aquaculture in the United States," *Proceedings of Symposium on Aquaculture in Wastewater, November 1980*. Pretoria, S. Africa, in press.
15. Dunseth, D. R., and R. O Smitherman. "Polyculture of Catfish, Tilapia, and Silver Carp," excerpt from PhD Dissertation, Auburn University (1976), and in *"Proceedings of Sixth Annual Commercial Fisheries Workshop"* (1977), pp. 25-26.
16. Walker, W. R., and I. E. Cox. "Legal Constraints On The Use Of Wastewater For Food and Fiber," in *Wastewater Use In The Production of Food and Fiber—Proceedings*, U.S. EPA Report-660/2-74-041 (1974), pp. 330-343.
17. Environmental Protection Agency. "Quality Criteria for Water," U.S. Government Printing Office (1978).
18. Henderson, S. "An Evaluation of Filter Feeding Fishes for Removing Excessive Nutrients and Algae From Wastewater," U.S. EPA Project # R80545301, Interim report, unpublished, 1979.
19. Buras, N., B. Hepher and E. Sandbank. "Public Health Aspects of Fish Culture In Wastewater, First Progress Report, January 1-June 30, 1980," IDRC Sponsored Project, Israel (1980), In progress.
20. TVA and EPRI, Eds. "State-of-the-Art: Waste Heat Utilization for Agriculture and Aquaculture," Technical Report B-12, Tennessee Valley Authority, Norris, TN (1978).

SECTION 3

MUNICIPAL AND INDUSTRIAL REUSE

CHAPTER 19

MUNICIPAL WASTEWATER REUSE IN POWER PLANT COOLING SYSTEMS

D. J. Goldstein and John G. Casana
  Water Purification Associates
  Cambridge, Massachusetts

INTRODUCTION

Maintaining an adequate water supply is becoming an increasingly difficult problem in many areas of the United States. This fact, coupled with the high cooling water requirement for electric power generation (0.45 gal/kWh) and other industrial processes, suggests the use of alternative water sources for cooling purposes. Biologically treated municipal sewage is a reliable source of cooling water. Several electric power plants primarily in the southwest are currently using treated municipal sewage for cooling purposes. This chapter discusses the potential pitfalls associated with this application and evaluates design procedures, using the electric power industry for illustrative purposes. Sewage chemical characteristics required for recycle are determined. Specific cases where sewage is being used as cooling water in the electric power and other industries are reviewed. Finally, methods of conditioning sewage to render it more amenable for cooling purposes are presented.

If between one half and three fourths of the wastewater fed to the cooling tower is evaporated and the rest is blown down (the blowdown is concentrated two to four times over the makeup), wastewater can be used for cooling with no problems. Foaming is no longer a problem because biodegradable detergents are used, scale from calcium phosphate can be prevented by lowering the pH, and organic carbon is usually reduced by biological oxidation. Ammonia is either stripped or nitrified, the latter being preferable because nitrification yields nitric acid, which prevents calcium carbonate and calcium phosphate scale. Biological slime, which can interfere with heat transfer, can be controlled by shock chlorination without killing biological oxidation, and by mechanical cleaning. Chlorine consumption varies from a high level of ten times the consumption with fresh water, if biological oxidation is prevented, to one or four times the consumption with fresh water when biological oxidation is encouraged, to zero if mechanical cleaning is used.

In parts of the western United States where water is scarce and the use of wastewater is particularly attractive, discharge is forbidden and blowdown is impounded in solar evaporation ponds. It is desirable to reduce the blowdown rate by concentrating the circulating cooling water as much as possible. To prevent scale formation at high concentrations, phosphate must always be removed, calcium must be removed from many wastewaters, and magnesium and silica from some wastewaters. Currently cold lime treatment is used to remove phosphate with high lime doses to a pH > 11 if magnesium or silica must be precipitated. This treatment increases calcium, and to remove calcium the water must be recarbonated to pH 10 and treated with soda ash. There is evidence that cold lime treatment is more difficult with sewage effluent than with fresh water and an alternative procedure is suggested. Phosphate can be removed biologically or by cold lime. Calcium, magnesium and silica can be removed by lime-soda treatment on a sidestream taken from the circulating cooling water. Administrative means may also be employed to reduce phosphorus concentrations in sewage.

THE NEED FOR COOLING IN POWER PLANTS

In a fuel-powered (not hydroelectric) electric generating cycle the fuel is burned to convert water to steam in a boiler. Steam is produced at temperatures as high as 1000°F and pressures as high as 3000 psig. The steam is let down in pressure and temperature through a turbine which drives the generator. Steam leaving the turbine is saturated at a low pressure and no further work can be extracted from it. Furthermore, it is not practical to compress the steam to boiler pressure so that it can be reheated and reused. Instead, the steam is condensed to water and the water is pumped into the boiler for reconversion to high-pressure, high-temperature steam.

About one-third of the heat of combustion of the fuel is recovered as electrical energy, about one-half is dissipated in the condenser and the balance is lost in various ways. A typical disposition of energy for a coal-fired, steam-cycle generating plant is shown in Table I.

Water is used to cool the steam condenser, which is more than 90% of the entire cooling load, and for most other cooling requirements in the plant. It

Table I. Typical Disposition of Energy in a Coal-Fired Generating Plant

	10^9 Btu/hr for 1000 MWe Generated	Percent
Coal	10.33	100
Electricity	3.41	33
Stack losses	1.03	10
Flue Gas Reheating and Other Plant Losses	0.72	7
Condenser Cooling Load	5.17	50

is easily possible to cool the condenser with atmospheric air, but air is not used because it is expensive. For maximum power plant efficiency the condenser should be as cold as possible. The usual design temperature is in the range of 100-140°F (which corresponds to a pressure of 0.95-2.9 psi absolute, or 50-150 cm of mercury absolute). Air is usually warmer than water and so a smaller temperature difference must be used across the condenser, necessitating more heat transfer surface. Also, the heat transfer rate to air is very much less than to water, necessitating still more heat transfer surface. In short, air cooling results in larger condensers and less efficient generating cycles than water cooling.

Of course water costs money. It has been estimated that if water costs in the range of $3.50-4.50/1000 gallons treated and ready for cooling, it would pay to transfer about 85-90% of the plant cooling load to air [1]. In practice, air cooling is not used in generating plants today, although plans exist.

Water cooling means that there must be a supply of cold water. On large rivers, lakes, estuaries and oceans many plants take in water, pass it once through the condenser and return the water 15-40°F hotter. The heat is then dissipated as the returned water is cooled by conduction, mixing and evaporation. Evaporation, which is a consumption of water, is low in once-through cooling. However, once-through cooling requires a larger source of water than is usually available, and in many existing plants and most new plants, recycled cooling water will be used. When cooling water is recycled, it is cooled by evaporation, and water consumption is high.

In a cooling tower, cycled water is cooled by letting it splash down against a rising current of air which is heated and humidified. About 1200-1400 Btu are dissipated to the atmosphere for each pound of water evaporated. Water consumption is about 0.45 gal/kWh generated. The average per capita consumption of electricity is about 30 kWh/day and is increasing. If half of this electricity were generated in fuel-powered plants using cooling towers, the per capita consumption of cooling water would be 6.8 gal. Per capita consumption of domestic water is about 80-140 gal/day. Consumption of water for power plant cooling is not trivial.

Cooling towers, although usually used in recycled cooling systems, are not the only possibility. Large ponds or reservoirs cool water by evaporation to the wind with or without water sprays to enhance evaporation. Where solar evaporation is larger than precipitation, ponds consume more water than cooling towers.

SEWAGE AS A SOURCE OF COOLING WATER

Municipal wastewater can be used in circulating cooling systems using towers. The wastewater is sewage which has received biological treatment (called secondary treatment), but no other treatment. In arid but industrialized parts of the world, this water has been used for cooling for at least twenty years. In the Vaal River Triangle in South Africa (the area includes Johannesburg) 50% of all sewage effluent is reused; 14.4% of it for industrial cooling [2].

In the United States, municipal wastewater has been used for cooling by a few plants for many years. It is used in the power plant of Burbank, California, and is thought to be a valuable future source of cooling water in California [3]. McKee [4] recommends cooling as the best way to reuse municipal wastewater in North Central Texas. Wastewater is already used in Texas by Southwestern Public Service at Amarillo and Lubbock [5], and by El Paso Products at Odessa [6]. Hanssen [7] finds that municipal wastewater will be increasingly used for cooling in many parts of the country. Other current users include Nevada Power Company [4,7] and Los Alamos National Laboratory [4].

At least three power stations in England use municipal wastewater: Oldham, Stoke-on-Trent [8] and Croydon [9].

To find how best to treat and use municipal wastewater in circulating cooling systems, we first compare the quality of wastewater and fresh water to try to determine what problems may arise. Then published experience is examined, and current practice determined. Finally, treatments are considered with emphasis on obtaining high concentrations in the circulating water.

COMPARISON OF WASTEWATER AND
FRESH WATER FOR COOLING

The problems of using fresh water for cooling can be grouped into four areas:

1. Biological Growth. Circulating cooling water is warm and oxygenated. Bacterial growth is rampant and where the water is sunlit, as on the sides of the cooling tower, algal growth occurs as well.

2. Fouling. Suspended solids in the intake water, plus rust and other corrosion products, plus dust scrubbed from the atmosphere, plus sloughed off growth, all concentrate as water is evaporated and tend to settle out at stagnant points in the piping and generally to foul the system.
3. Corrosion. Water, particularly oxygenated water, is corrosive.
4. Scale Formation. As water is evaporated, dissolved salts concentrate and tend to precipitate. Commonly found precipitates are calcium carbonate, calcium sulfate (gypsum), silica and magnesium silicate. The calcium and magnesium salts decrease in solubility as the temperature is raised and tend to precipitate on the condenser surface where they settle to form a hard scale which interferes with heat transfer. Scale is often a mixture of salts and bacterial slime.

Most circulating cooling systems are treated with a little biocide and all are blown down. Blowing down prevents the accumulation of nonevaporated matter, both dissolved and in suspension. In cooling tower language, the ratio of makeup rate to blowdown rate is called the cycles of concentration. It is often measured by determining the concentration of a nonprecipitating ion such as chloride. If a tower operates at 2 cycles of concentration, 50% of the makeup is evaporated and 50% is blown down. If a tower operates at 15 cycles, 93% of the makeup is evaporated and 7% is blown down.

High cycles of concentration and low blowdown rates lead to scale formation. Scale cannot be allowed and condensers are regularly chemically cleaned to dissolve accumulated scale. Chemical cleaning is expensive and much effort is made to prevent scale formation. Fresh water and wastewater must be compared at the same cycles of concentration.

Sewage differs from the municipal drinking water supply (the carriage water) in having an important organic content, an increased inorganic content and the specific additions of ammonia and phosphate. Problems that can arise when treated sewage is substituted for fresh water are:

• Foaming. Foaming has been a problem in the past, but as biodegradable detergents are now required, the problem is much reduced. Linear alkyl sulfonates should be below 0.5 mg/l in the treated waste. Periodic use of antifoaming agents may be required during periods of upset, but no special precautions are needed today.

• Ammonia. Ammonia is corrosive to copper alloys and reacts with chlorine to lessen its effect as a biocide. However, repeated experience is that ammonia does not accumulate in circulating cooling water and problems do not arise. Grutsch and Griffin [10] have reported on a tower using treated refinery sour water as makeup and operating at 6-14 cycles of concentration. While ammonia in the makeup varied from 2-9 mg/l, ammonia in the circulating water varied from 0.4-9 mg/l with a 50% probability that ammonia would be reduced by stripping to half of the makeup concentration. When municipal wastewater is used, both stripping and nitrification have been reported [9,11]. The disappearance of ammonia is further described in the discussion of published experience.

   · Bacterial Slime. Wastewater contains carbon, nitrogen and phosphorus, so slime and algal growths are regularly found to be problems. Biocide requirements have been found to vary from zero to ten times the level required for fresh water makeup. As long as scale does not form, soft slime can be easily cleared from condenser tubes by mechanical means and, as described later, mechanical cleaning is recommended.

   · Calcium Phosphate Scale. As less phosphate is used in detergents, this problem may disappear. Today, however, the possibility of calcium phosphate scale formation is high and must be controlled by acidification or removal of phosphorus. Control methods are given in detail below.

   · Other Scale Formation. If the community is softening its drinking water, then the sewage may be less scale-forming than alternative water sources available for cooling. In the usual situation, however, sewage contains more alkalinity (bicarbonate and carbonate) and may contain more calcium, magnesium and sulfate than alternative water sources and will be more prone to cause scale. The national average mineral pickup is shown in Table II. Treatment is the same as for a freshwater source. In spite of the higher dissolved salt content, particularly including chloride, sewage is not found to be more corrosive than other waters.

   · Health Hazard. Drift from a cooling tower circulating sewage or any contaminated water might cause spread of infection [13]. Although pathogenic organisms are partly destroyed by warm temperature, by chlorination and by sunlight, there is no guarantee that all infectious organisms will be destroyed. On the other hand, there is no evidence that an epidemic has been started. A small increase in the occurrence of diseases which are spread through water will be very hard to prove or disprove. Adams et al. [14,15] found no added danger to health. Indeed in one case studied, the average number of bacteria-bearing particles emitted by a tower using secondary sewage which had been chlorinated was less than the number emitted by a tower using polluted river water.

Table II. National Average Mineral Pickup from Domestic Water Usage [12]

Constituent	Incremental Concentration Increase (mg/l)
Chloride	20-50
Sulfate	15-30
Nitrate	20-40
Phosphate	20-40
Calcium	6-16
Magnesium	7-19

PUBLISHED EXPERIENCE

1. South Africa. Flook [11] has described how purified sewage has been used in Johannesburg for cooling in three power plants commissioned in 1942, 1957 and 1963. For many years nitrification occurred in the cooling tower. Ammonia entered at 8.9 mg/l and was found to be circulating at 0.4-0.6 mg/l in spite of concentrations of 3.6 cycles at one plant and 2.6 cycles at another plant. Nitrification turned ammonia into nitric acid, so the pH of the circulating water was 5.9-6.8 and there was no scale formation at the modest cycles of concentration used. Neither was corrosion noticed even when the Langelier Index was negative.

At a more recent date the alkalinity of the sewage increased and the pH rose from a previous 7.0 to 7.2. At the same time the ammonia nitrogen dropped to 5.5 mg/l, nitrification was lost. Ammonia was still reduced to 0.3 mg/l in the circulating water, but this was by stripping because the nitrate content was not increased and pH rose to 7.8. Serious phosphate and carbonate scale occurred and acid had to be added to lower the pH by 0.2 units.

At one plant a spray pond was used to store the water. Growth of algae in the pond raised the pH (presumably by adsorption of carbon dioxide) and caused the precipitation of calcium phosphate. The water from the pond was acidified in the circulating loop and no further precipitation occurred.

2. England. Humphris [9] has described the use of sewage effluent since 1950 at the Croydon Power Station in England. From the start this plant was plagued by calcium phosphate scale. Neither filtration nor chlorrination helped. The scale was not due to biofouling or suspended solids; it was calcium phosphate. Since 1957 acidification has been used for control. Chlorine was used in large amounts, being considered preferable to sulfuric acid which might cause calcium sulfate scale. Chlorine, although successful, was too expensive and was changed from continuous to intermittent application in 1960. With no chlorination, nitrification was fully established in seven weeks and when Humphris reported in 1975, ammonium sulfate was being added to control scale. The ammonium was nitrified to nitric acid which lowered the pH enough to prevent precipitation. Oxidation of BOD gave a BOD in the blowdown which was 40 to 20% of that of the makeup. Chlorination has not been practiced at all since 1974. Even with no chlorination, acid washing of the condensers is not required and only occasional mechanical cleaning by bulleting (firing nylon and wire brushes with compressed air and water) is used at shutdowns. What slight condenser fouling is experienced appears to be due to an organic film which can be removed by drying out the condenser. It is thought that the organic film forms only under stagnant conditions; under normal operating conditions the flow rates are too high.

3. Southwestern Public Service, Texas. Experience at the Nichols Station in Amarillo and at the Jones station in Lubbock has been described by Ladd [5] and by Ladd and Terry [16] (the two descriptions are the same). The Nichols Station has been operating since 1961 and the Jones stations since 1972. Effluent from sewage treatment is given a cold lime treatment to pH 10-10.5, reacidified to pH 9.2 for storage and circulated at pH 7.0. Some analyses from Amarillo are given in Table III, where it can be seen that fresh water and sewage effluent both have high dissolved solids (the fresh water is over 1000 ppm TDS). The cold lime treatment reduces the Mg, alkalinity, SiO_2, PO_4 and BOD, but does not reduce the Ca. The cooling tower is operating at just over 5 cycles of concentration and we estimate (in a later section) that the circulating water is just saturated in calcium phosphate. In the tower, nitrification is occurring and there is some oxidation of BOD.

The cold lime treatment is necessitated by the poor quality of the local water supply and the medium high cycles of concentration wanted. The equipment is designed for a rise rate of 0.7 to 1 gpm/ft^2 and works well provided the influent has a BOD and suspended solids both below 25 mg/l. Chemical use rates per 1000 gallons are 2.5-3.0 lb lime and less than 0.25 lb alum.

Foam has not been a problem. Chlorination has been required at a rate of 135 lb/10^6 gallons makeup compared to 10-15 lb/10^6 gallons if well water is used. Chlorination can be much reduced or stopped if mechanical cleaning using the Amertap system is used, and this is preferred. In the Amertap system small, compressible balls are passed through the condenser tubes to clean the surface.

Table III. Typical Water Analyses[a] at Amarillo, Texas

As $CaCO_3$	Amarillo Fresh Water	Sewage Effluent	C.T. Make-up	Cooling Water
Ca	170	185	180	940
Mg	120	150	42	210
Na	484	584	584	2994
K	7	20	20	100
NH_3	0	73	73	8
HCO_3	170	220	40	32
CO_3	0	0	60	0
SO_4	270	293	350	1800
Cl	341	443	443	2215
NO_3	0	5	4	145
PO_4	0	51	2	10
SiO_2	9	28	10	50
pH	8.1	7.3	9.2	7.0
B.O.D.	0	15	2	6

[a]Analysis results corrected for calculated cation and anion balance.

On one occasion 300-400 mg/l nitrate was found with a pH of 6.4-6.0. This was corrosive. Apparently atmospheric nitrogen was being fixed and chlorination mended the situation.

4. El Paso Products, Odessa, Texas. This petrochemical complex has used treated sewage from the city of Odessa for all purposes in the plant since about 1957. Reports have been published by Smythe [6] and Cummings [17]. As in the preceding example from Texas, the sewage effluent has more than 1000 ppm TDS. Analyses from Smythe are presented in Table IV with the units changed. The analyses are not in electrobalance and we suspect that the sodium analysis is too low.

Received sewage is given a very complicated treatment. First a cold lime treatment is used which reduces phosphate and suspended solids, the water is recarbonated to prevent plugging of the filters and then it is filtered. The stream is then split in two. Half is sodium ion exchange softened and the rest is treated with hydrogen ion exchange. The streams are then blended and the carbon dioxide released by the hydrogen ion exchange is removed in a degasifier.

The treated water is used in cooling towers and concentrated seven times. A side stream filter is necessary to remove sand scrubbed from the atmosphere. There is no deliberate blowdown; drift losses, side stream filter backwash, etc., keep the concentration at 7 cycles.

Cummings, in 1964, reported serious foaming trouble, but Smythe, in 1971, did not mention foam as a problem. A complex biocide program is used with intermittent and not simultaneous use of quaternary ammonium compounds, nitrogen-based compounds, pentachlorophenate, trichlorophenate, peracetic acid and chlorine. Heat exchangers have been kept free of slime. It must be remembered that this is a chemical plant with a multitude of heat exchangers, not a power plant with one large heat exchanger; mechanical cleaning in a chemical plant is very difficult.

Table IV. Analyses of Sewage and Treated Effluent from Odessa, Texas

Hardness (mg/l as $CaCO_3$)	Sewage Effluent	Cold Lime Effluent	Recarbonator Effluent	Split Stream
Ca	128	118	128	0
Mg	42	42	42	0
Na	200	170	200	254
HCO_3	137	159	159	64
CO_3	0	85	0	0
SO_4	105	101	105	101
Cl	206	206	213	220
SiO_2	19	19	19	19
PO_4	40	4	–	–
pH	7.2	10.2	7.9	7.1

5. Burbank, California. Treated municipal wastewater is used for cooling in the municipal power plant at Burbank, California [18,19]. (An analysis of the circulating water is included in [19].)

Sewage plant effluent is used without additional treatment. The cycles of concentration were 4 and are now 2.5. Blowdown is blended with unused sewage plant effluent and discharged. The cycles of concentration are set by the legal maximum of total dissolved solids in the blended discharge. There have been no corrosion problems. When the sewage plant has been upset and linear alkyl sulfonates have exceeded 0.5 ppm, there have been foam problems, but not otherwise. Chlorine is used in slugs to keep 1 ppm residual after the condenser. The amount of chlorine used is two to four times as much as with surface water.

6. Other Experience. A summary of the reported experience described above is given in Table V. Additional information on this table is taken from data given by Schmidt et al. [20], Hoppe [21] and Goldman and Kelleher [22]. (Hoppe gives a list of plants operating without phosphate removal.) The Comanche station of the Public Service Company of Oklahoma is a 230 MW combined cycle station with 110 MW generated by the steam cycle [23]. Cooling is from a 250-acre lake. Makeup to the lake is 3.5 mgd of treated sewage from the city of Lawton. About 2.4 mgd are discharged from the lake, with the rest of the losses being from evaporation and seepage. High concentrations are not reached in the water. The lake is used to cool the condenser and an auxiliary closed cooling system is used for the rest of the plant. Both heat exchangers use Amertap mechanical cleaning, and chlorination has been found to be unnecessary.

There are many examples of industries using their wastewater as part or all of the makeup to their own cooling towers. Of particular pertinence to our discussion is the use of organically contaminated wastewater in cooling towers by refineries and petrochemical plants. Grutsch and Griffin [10] found ammonia to be stripped and not to accumulate in the circulating water, and they found the same with organic carbon. When the makeup varied from 60 to 90 mg/l total organic carbon and the system operated between 5.5 and 13.9 cycles of concentration, half of the carbon fed was removed. Phenol in the makeup varied from 0.2 to 0.5 mg/l and more than 80% was removed. When a slug of phenol was deliberately added to raise the concentration to 5000 mg/l, 80% of the phenol was removed in three hours. Removal occurred whether or not the system was sterilized, proving that bio-oxidation was not the means of removal.

However, in other cases bio-oxidation is the cause of removal of organics, particularly phenol. Hart [24] reported on tests in Mobil Oil Corporation's East Chicago refinery where, in a cooling tower receiving phenol in its makeup at rates of 85-140 lb/day, between 70 and 97% was removed by bio-oxidation.

Hart states: "Phenol removal was adversely affected by overchlorination. Careful control of chlorine-injection rates is needed to efficiently oxidize phenolics and yet prevent slime growth."

The best documented use of wastewater for cooling is at the Toledo Refinery of Sun Oil. This system has been running since 1954 and has been documented by Mohler and Clere [25,26]. The makeup water is stripped of hydrogen sulfide to less than 2 mg/l, which is essential because sulfide is toxic. Phenol in the makeup varies from 10 to 70 mg/l (equivalent to at least 20-140 mg BOD/l) and over 99.4% is biologically oxidized. Sixty to seventy percent of the total organic carbon is oxidized. The tower operates at 2 cycles of concentration. Chlorination is not used and suspended solids (which are mostly biological) build up to 110 mg/l. The flow through each heat exchanger is reversed daily (they are back flushed) and this is enough to keep them clean. Heat exchange is not impaired. Blowdown is clarified and used in other cooling towers.

CONCLUSIONS FROM CURRENT PRACTICE

The information reported leads to the following conclusions: (1) sewage effluent can be used as makeup to cooling towers in power plants and concentrated up to about four-fold with little trouble; (2) foaming, which has occurred in the past, is no longer occurring; (3) calcium phosphate scale can be controlled by acid addition which may be needed anyway to control calcium carbonate scale; (4) ammonia is not a problem; if not nitrified, it will be stripped, although nitrification is preferable because the nitric acid produced can control phosphate scale; and (5) slime can be controlled by chlorination or the use of other biocides while maintaining the biological activity which nitrifies ammonia and reduces BOD in the makeup. Chlorination should be in slugs and is most often reported to be daily, though the dose seems to be somewhat arbitrary. Petrey has suggested controlling chlorination to maintain a total bacterial count [27] and, indeed, this may result in a saving in the cost of chlorine. Mechanical cleaning of the condenser tubes to remove slime will reduce the need for chlorine and no chlorination may be needed. Mechanical cleaning is recommended; it will probably only be effective if hard salt scale is not formed. Drying out the condenser is also a successful means of cleaning it.

At four cycles of concentration, 25% of the makeup will be blown down. The blowdown will have a higher salt concentration, a higher suspended solids and a lower biological oxygen demand than the makeup. There are circumstances where blowdown will be no more environmentally harmful than sewage effluent and the use of sewage effluent to save other waters will be environmentally and economically desirable.

Table V. Sewage Effluent Quality and Treatment

Company/Plant	Effluent Sewage Water Quality					Treatment	No. of Cycles of Concentration
	BOD (mg/l)	O-PO$_4$ (mg/l)	TDS (mg/l)	NH$_3$ as N (mg/l)	Ca as CaCO$_3$ (mg/l)		
Nichols P.S. Amarillo, TX [5,16,22]	10-15	50	1400	73	185	Cold lime treatment, pH adjustment, shock chlorination, mechanical cleaning of condenser	3-5
Jones P.S. Lubbock, TX [5,16,22]	25	29					
Clark P.S. and Sunrise P.S. Las Vegas, NV [4,20,22]	20	35	1000-1500			Shock chlorination, lime clarification, pH adjustment, corrosion inhibitor	
Los Alamos Scientific Los Alamos, NM [4,21,22]		35				Shock chlorination, pH adjustment, corrosion inhibitors	4-5
El Paso Products Odessa, TX [6,17]	2-10	40	1300	3	130	Lime clarification, recarbonation, pH adjustment, filtration, ion exchange softening	7-7.4
Burbank, CA [18-20,22]	2	36	555		118	pH control, shock chlorination	2.5-4
Denton P.S. Denton, TX [20]	30		130			Shock chlorination, pH adjustment, corrosion inhibitor (treatment inadequate)	
Croydon, P.S. England [7]	21	15		29	350	Filtration, biological oxidation of NH$_3$ in tower for pH control (NH$_4$-salt sometimes added)	

Kelvin P.S. Johannesburg, S.A. [11]	25	8.9-5.5	Shock chlorination, pH adjustment nitrification preferred	2-4
Orlando P.S. Johannesburg, S.A. [11]			Shock chlorination (phosphate precipitation occurs in a storage pond)	2-4

In much of the western United States, where water is scarce and municipal wastewater is the first choice as a reliable supply, discharge to a surface stream is forbidden and all blowdown is impounded and evaporated in solar ponds. Here it is most desirable to circulate water at more than 4 cycles of concentration. At high cycles of concentration, scale will form and will mix with bacterial slime to decrease heat transfer intolerably. Treatment to prevent salt scale is necessary.

The first problem is phosphate scale. In the following section we describe how the onset of phosphate scale can be predicted.

Phosphate can be removed by treatment with lime, alum or ferric chloride. Lime has the potential advantage of removing other scale-forming chemicals such as magnesium and silica, and currently is the only treatment used. Phosphate can also be removed by modification of the sewage treatment without the use of chemicals. Biological removal seems to be desirable provided removal of other scale-forming chemicals is not required or can be handled in some other way. To determine the best treatment, we must be able to predict the onset of all forms of salt scale and these predictions are given later.

PREDICTION OF PHOSPHATE SCALE FORMATION

Phosphate in solution is mostly present in four forms [28]: H_3PO_4, $H_2PO_4^-$, $HPO_4^=$ and $PO_4^\equiv$. An analysis for phosphate gives the sum of the four forms, that is:

$$[PO_4]_{Total} = [H_3PO_4] + [H_2PO_4^-] + [HPO_4^=] + [PO_4^\equiv] \qquad (1)$$

The scale-forming salt is dependent on pH and may be $CaHPO_4$, $Ca_3(PO_4)_2$ or hydroxyapatite, $Ca_5(PO_4)_3OH$. The distribution of total phosphate between the four forms depends on pH.

$$K_1 = [H^+] \times [H_2PO_4^-]/[H_3PO_4] \qquad (2)$$

$$K_2 = [H^+] \times [HPO_4^=]/[H_2PO_4^-] \qquad (3)$$

$$K_3 = [H^+] \times [PO_4^\equiv]/[HPO_4^=] \qquad (4)$$

So:

$$[HPO_4^=] = [H^+] \times [PO_4^\equiv]/K_3 \qquad (5)$$

$$[H_2PO_4^-] = [H^+]^2 \times [PO_4^\equiv]/K_2K_3 \qquad (6)$$

$$[H_3PO_4] = [H^+]^3 \times [PO_4^{\equiv}]/K_1K_2K_3 \qquad (7)$$

and:

$$[PO_4^{\equiv}] = [PO_4]_T \times K_1K_2K_3/F \qquad (8)$$

$$[HPO_4^{=}] = [PO_4]_T \times K_1K_2 \times [H^+]/F \qquad (9)$$

where:

$$F = [H^+]^3 + K_1[H^+]^2 + K_1K_2[H^+] + K_1K_2K_3$$

$$= 10^{-3pH} + K_1 \times 10^{-2pH} + K_1K_2 \times 10^{-pH} + K_1K_2K_3 \qquad (10)$$

If the solubility product of calcium monohydrogen phosphate, $CaHPO_4$, is written as:

$$[Ca^{++}] \times [HPO_4^{=}] = K_{S2} \qquad (11)$$

then:

$$Log[Ca^{++}] + Log[PO_4]_T - LogF - pH = LogK_{S2} - LogK_1K_2 \qquad (12)$$

If the solubility product of calcium orthophosphate, $Ca_3(PO_4)_2$, is written as:

$$[Ca^{++}]^3 \times [PO_4^{\equiv}]^2 = K_{S3} \qquad (13)$$

then:

$$1.5 \, Log[Ca^{++}] + Log[PO_4]_T - LogF = 0.5 \, LogK_{S3} - Log \, K_1K_2K_3 \qquad (14)$$

If we consider hydroxyapatite we have, from Truesdell and Jones [29]:

$$Ca_5[PO_4]_3OH + 3H_2O \rightleftharpoons 5Ca^{++} + 3HPO_4^{=} + 4OH^-, \qquad (15)$$

so:

$$5 \, Log[Ca^{++}] + 3 \, Log[HPO_4^{=}] + 4 \, Log \, K_w + 4 \, pH = LogK_{S.Ap} \qquad (16)$$

and, by manipulation similar to that used above:

$$1.667 \, \text{Log}[Ca^{++}] + \text{Log}[PO_4]_T - \text{Log} \, F + pH/3 =$$

$$0.33 \, \text{Log} \, K_{S.Ap} - \text{Log} \, K_1 K_2 - 1.33 \, \text{Log} \, K_w \tag{17}$$

Various values of the equilibrium constants have been given as shown on Table VI. The variations in the solubility product of calcium orthophosphate are large. McCoy's value is a laboratory measurement from pure solutions. McCoy points out that the rate of crystal growth is slow and that this constant may not measure the concentrations when a hard scale is formed. The value of DeBoice and Thomas [31] is from lime treatment of sewage and seems particularly applicable to the present purpose. The value given by Chen et al. [32] is their suggestion for a practical solubility product which they use regardless of pH. The value of Humphris [33] was measured on a simulated cooling system and is the most directly applicable.

In the preceding equations the concentrations, written in brackets, are in moles per liter. If the concentrations are expressed as: calcium in mg/l as $CaCO_3$, written CaH, and phosphate in mg/l as PO_4, written PO_4, then:

$$10^5 [Ca^{++}] = CaH \text{ and } 0.95 \times 10^5 [PO_4]_T = PO_4 \tag{18}$$

Equation 12 using Humphris' first set of constants [9] becomes:

$$\text{Log CaH} + \text{Log PO}_4 - \text{Log} \, F - pH = 13.45 \tag{19}$$

Equation 12 using Humphris' second set of constants [33] at 35°C and TDS = 1000 becomes:

$$\text{Log CaH} + \text{Log PO}_4 - \text{Log} \, F - pH = 12.47 \tag{20}$$

Equation 14, using McCoy's constants, becomes:

$$1.5 \, \text{Log CaH} + \text{Log PO}_4 - \text{Log} \, F = 19.48 \tag{21}$$

Equation 14 using McCoy's values for K_1, K_2 and K_3 and Chen's value for K_S becomes:

$$1.5 \, \text{Log CaH} + \text{Log PO}_4 - \text{Log} \, F = 21.63 \tag{22}$$

Equation 17 using McCoy's values for K_1 and K_2 and Truesdall and Jones' value for $K_{S.Ap}$, becomes:

$$1.667 \, \text{Log CaH} + \text{Log PO}_4 - \text{Log} \, F + pH/3 = 21.82 \tag{23}$$

Table VI. Equilibrium Constants for the Prediction of Calcium Phosphate Scale[a]

	(1) McCoy [30]	(2) DeBoice and Thomas [31]	(3) Chen et al. [32]	(4) Humphris [9]	Humphris [28,33][b] (5) 25°C	Humphris [28,33][b] (6) 35°C	(7) Truesdell and Jones [29] (at 30°C)
Log K_1	-2.125	-2.143			-2.123	-2.171	
Log K_2	-7.208	-7.201		-7.7	-7.201	-7.185	
Log K_3	-12.319	-12.347			-12.335	-12.195	
Log K_{S2}				-6.36	-7.1		
Log K_{S3}	-29.301	-25.46	-24.98		-24.635	-25.365	
Log $K_{S.Ap.}$							-58.92

[a]The tabulated values are for zero ionic strength. At finite TDS, Humphris [33] uses:

$$\text{Log } K_1' = \text{Log } K_1 + (TDS)^{1/2}/200$$
$$\text{Log } K_2' = \text{Log } K_2 + (TDS)^{1/2}/100$$
$$\text{Log } K_3' = \text{Log } K_3 + (TDS)^{1/2}/66.6$$
$$\text{Log } K_{S3}' = \text{Log } K_{S3} + (TDS)^{1/2}/13.33$$
$$\text{Log } K_{S2}' = \text{Log } K_{S2} + (TDS)^{1/2}/50.$$

[b]Humphris also uses $-\text{Log } K_{S3} = 22.81 + 0.073 \text{ t°C}$.

Equation 14, using DeBoice and Thomas' constants, becomes:

$$1.5 \, \text{Log CaH} + \text{Log PO}_4 - \text{Log F} = 21.43 \qquad (24)$$

where

$$F = 10^{-3\,pH} + 7.2 \times 10^{-3-2\,pH} + 4.7 \times 10^{-10-pH} + 2.1 \times 10^{-22} \qquad (25)$$

Equation 14 using Humphris' constants [33] at 35°C and TDS = 1000 becomes:

$$1.5 \, \text{Log CaH} + \text{Log PO}_4 - \text{Log F} = 21.58 \qquad (26)$$

where:

$$F = 10^{-3\,pH} + 9.7 \times 10^{-3-2\,pH} + 1.3 \times 10^{-9-pH} + 2.5 \times 10^{-21} \qquad (27)$$

In the range of pH 6 to 10 only the middle two terms in F are necessary, i.e.:

$$F = (approx.) \, K_1 \times 10^{-2\,pH} + K_1 K_2 \times 10^{-pH} \qquad (28)$$

So Equation 19 can be written:

$$\text{CaH} \times \text{PO}_4 = 10^{3.62} + 10^{11.32 - pH} \qquad (29)$$

which is Humphris' form of the equation with the units changed [9]. Similarly, Equation 20 can be written:

$$\text{CaH} \times \text{PO}_4 = 10^{3.59} + 10^{10.46-pH} \qquad (30)$$

Equation 24 can be written:

$$\text{CaH}^{1.5} \times \text{PO}_4 = 10^{19.29-2\,pH} + 10^{12.09-pH} \qquad (31)$$

and Equation 26 can be written:

$$\text{CaH}^{1.5} \times \text{PO}_4 = 10^{19.57-2\,pH} + 10^{12.70-pH} \qquad (32)$$

When Equations 19-27 are used to compute the permissible concentration of phosphate for a given concentration of calcium and pH, it is found that

Equations 21 and 23 give about one tenth the concentration of the rest of the equations. Equations 21 and 23 give very similar concentrations, all of which are much lower than is found to be nonscaling in practice.

Some calculations are presented on Table VII. The equations vary, but there is an indication that $CaHPO_4$ precipitates at pH below 7 to 7.5 and $Ca_3(PO_4)_2$ precipitates at pH above 7 to 7.5. Equation 29 was used to control the Croydon Power Station for many years, but it was found to be biased toward the safe side at medium pH and to be in error at high and low pH. Humphris has supplied Table VIII which shows conditions which were scale forming, marginally scale-forming and nonscale-forming at the Croydon Power Station. The equilibrium pH, shown in the right-hand column, is below the measured pH when scale is forming and above the measured pH when scale is not forming. The calculations were made using Equation 14 for the precipitation of $Ca_3(PO_4)_2$, and the constants in the fifth column of Table VI with corrections for TDS as shown on Table VI. Note that a change in TDS from 200 to 2000 ppm causes the permissible phosphate concentration at constant calcium to be multiplied by 3.5 to 4. The TDS correction cannot be ignored.

An attempt to correlate the data on Table VIII using Equation 12 for the precipitation of $CaHPO_4$ failed. The calculation shows that $CaHPO_4$ should always be scaling even when no scale is formed. Apparently, the solubility products listed for $CaHPO_4$ on Table VI are measurements of precipitation rather than scale formation, but the solubility products given for $Ca_3(PO_4)_2$ in all the columns of Table VI except Column 1 are for scaling rather than

Table VII. Permitted Concentrations of Phosphate (mg/l as PO_4) for Various Concentrations of Calcium and Various pH

Precipitant	pH						
	6	6.5	7	7.5	8	8.5	9
Calcium 300 (mg/l as $CaCO_3$)							
$CaHPO_4$ (Eq. 29)	710	234	84	36	21	16	15
$CaHPO_4$ (Eq. 30)	109	43	23	16	14	13	13
$Ca_3(PO_4)_2$ (Eq. 31)	4000	450	61	11	3	0.8	
$Ca(PO_4)_2$ (Eq. 32)	8100	1020	168	38	10	3	1
Calcium 750 (mg/l as $CaCO_3$)							
$CaHPO_4$ (Eq. 29)	284	94	33	14	8	6	6
$CaHPO_4$ (Eq. 30)	44	17	9	6	6	5	5
$Ca_3(PO_4)_2$ (Eq. 31)	1000	114	15	3	0.7		
$Ca_3(PO_4)_2$ (Eq. 32)	2050	258	42	10	3	0.8	

precipitation. The solubility product in Column 1 is for precipitation. We recommended calculating the pH of scaling by assuming that only $Ca_3(PO_4)_2$ scales.

Table VIII. Calcium Phosphate Scale Formation at the Croydon Power Station

	Alkalinity (mg/l CaCO$_3$)	pH	Phosphate (mg/l CaCO$_3$)	Calcium (mg/l CaCO$_3$)	Total Dissolved Solids (mg/l)	Calculated Equilibrium pH (at 25°C)
	92	7.9	40	688	1380	7.55
	100	7.9	37	693	1400	7.60
	104	7.9	36	665	1370	7.60
	130	8.0	38	512	1130	7.65
	132	8.0	32	516	1110	7.75
Scaling	140	8.0	33	512	1130	7.70
Conditions	148	8.1	32	488	1130	7.75
	144	8.1	41	504	1120	7.61
	164	8.1	44	552	1200	7.59
	128	8.0	30	428	990	7.83
	134	8.0	36	456	1060	7.75
	140	8.1	42	468	1050	7.65
	136	8.0	45	436	1080	7.70
	120	8.0	39	442	930	7.59
	184	8.2	42	444	1310	7.75
	164	8.1	23	504	1060	7.87
	43	7.3	72	584	631	7.24
	48	7.4	72	588	642	7.25
	45	7.4	72	600	642	7.24
	43	7.3	76	600	653	7.17
Marginal	51	7.4	76	588	612	7.16
Scaling	38	7.4	76	616	612	7.13
	42	7.3	72	576	642	7.20
	35	7.3	70	596	653	7.20
	45	7.5	67	580	631	7.21
	44	7.4	67	572	620	7.21
	47	7.5	66	568	612	7.28
	47	7.4	66	568	620	7.28
	36	7.2	66	556	612	7.28
	32	7.3	64	536	590	7.30
	29	7.1	67	528	600	7.30
	18	6.7	73	616	714	7.20
Nonscaling	22	7.0	61	492	540	7.35
	28	7.2	63	516	600	7.32
	26	7.0	61	516	570	7.32
	24	7.0	61	516	581	7.33
	31	7.1	61	500	559	7.34
	34	7.1	60	488	550	7.35
	26	6.9	60	504	570	7.34
	22	6.8	60	484	550	7.35
	33	7.1	58	472	550	7.37

TREATMENTS FOR HIGH CYCLES
OF CONCENTRATION

Specification of Makeup Quality for Fifteen Cycles
of Concentration

Studies have been published [34,35] on the use of sewage effluent from Phoenix, Arizona, in a proposed nuclear generating station at Palo Alto. High cycles of concentration are required because the blowdown must be impounded. Based on the experience of Bechtel, Inc. with fresh water sources, Weddle and Rogers [34] state the quality of makeup required for 15 cycles of concentration. Their recommendations and ours are shown in Table IX.

Apart from ammonia and BOD, these are widely used specifications with an experimental and calculational basis. The sources of these concentration limits are explained briefly below.

Calcium concentration is limited by the formation of calcium sulfate scale. Using data given by Radian Corporation [36], the solubility product of calcium sulfate was calculated at 40°C and zero ionic concentration:

$$K_s = [Ca^{++}] \times [SO_4^{=}] = 1.62 \times 10^{-5}, (moles/l)^2$$

From the same data the equilibrium constant for unionized $CaSO_4^0$ in solution was calculated to be:

$$K = \frac{[Ca^{++}] \times [SO_4^{=}]}{[CaSO_4^0]} = 4.19 \times 10^{-3}, (moles/l)$$

Although there are many other complexes of calcium existing in water, our calculations show them to be of secondary importance and it is sufficiently accurate to write:

Table IX. Recommended Makeup Water Concentration
Limits for 15 Cycles of Concentration

mg/l	Weddle & Rogers [34]	Authors
Ca	28	25
Mg	2.4 [a]	15
SiO$_2$	10	10
PO$_4$	1.5	1.5
Ammonia	5	[b]
BOD	10	[b]
Suspended solids	10	[b]

[a] Concentration achieved from two-stage lime softening higher concentration acceptable in makeup.

[b] See text.

$$[Ca]_{Total} = [Ca^{++}] + [CaSO_4^0]$$

from which one may derive:

$$[Ca]_{Total} \times [SO_4^=] = K_s \left\{ 1 + \frac{[SO_4^=]}{K} \right\}$$

To complete the calculation it is necessary to compute the activity coefficients which, in the concentrated waters of concern here, cannot be taken to be one. We assumed the solution to be sodium sulfate (which is the most important salt in circulating cooling water in most of the western United States) and computed the activity coefficients at 40°C by the method given by Radian [36] and by Truesdell and Jones [29]. The calculation is shown on Figure 1 which is a graph of the concentration of total calcium against the concentration of sulfate ion at which precipitation just begins. As long as calcium is below 250 mg/l, calcium sulfate will not precipitate at any concentration of sulfate.

In practice, the onset of precipitation does not mean scale formation, and higher concentrations of calcium are permissible. Klen and Johnson [37] have found in laboratory experiments that scale will not form at any concentration of sulfate as long as calcium is below 380 mg/l, and Stearns-Rogers Inc. use a similar limit in their designs. If water is circulating at 15 cycles of concentration with 380 mg/l calcium, then the makeup must be limited to 25.3 mg/l calcium, a little less than recommended by Weddle and Rogers [34].

From Equation 32 the permissible concentration of phosphate in the circulating water can be calculated when the calcium concentration is 380 mg/l (950 mg/l as $CaCO_3$).

pH in Circulating Water	Phosphate in Circulating Water	Phosphate in Makeup
6.8	59	3.9
7	30	2.0
7.2	16	1.0
7.5	6.7	0.4

Weddle and Rogers recommend 1.5 mg PO_4/l.

The limit on silica of 150 mg/l in circulating water is empirical, but widely demonstrated. Norman et al. [38], at Public Service of New Mexico, managed to circulate 240 mg/l SiO_2 by using quite large doses of antiscalant chemicals and by rigorously avoiding the formation of calcium carbonate solids. No higher concentration has been reported.

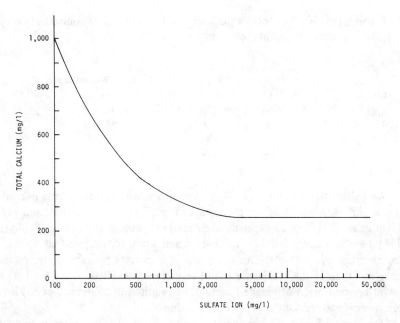

Figure 1. Concentration of total calcium at which calcium sulfate just precipitates.

The magnesium silicate which forms scale is the mineral sepiolite. The solubility of sepiolite has been measured by Christ et al. [39] by dissolving. Correcting for temperature as described by Truesdell and Jones [29] we calculate their results at 40°C to be:

$$Log(Mg) + 1.5 \, Log(SiO_2) + 2 \, pH = 20.62$$

and at 20°C to be

$$Log(Mg) + 1.5 \, Log(SiO_2) + 2 \, pH = 19.57$$

where (Mg) and (SiO_2) are in mg/l.

Wollast et al. [40] have measured the solubility of sepiolite by precipitation. Their results at 40°C may be written

$$Log(Mg) + 1.5 \, Log(SiO_2) + 2 \, pH = 21.67$$

and at 20°C

$$Log(Mg) + 1.5 \, Log(SiO_2) + 2 \, pH = 20.15$$

If $(SiO_2) = 150$, then these equations give the permissible concentration of magnesium in the circulating water, in mg/l:

pH	Christ et al. [39]		Wollast et al. [40]	
	40°C	20°C	40°C	20°C
6.8	5,700	500	64,000	1,900
7.0	2,300	200	25,500	770
7.2	900	80	10,000	300
7.5	230	20	2,500	77

We believe that the results of Wollast et al. are the more applicable to cooling water. However, there is clearly a strong effect of temperature and pH. Chen et al. [32] have recommended 57 mg/l magnesium and Crits and Glover [41] have recommended 233-400 mg/l magnesium. We suggest that 15 mg/l in the makeup is suitable. This difference is important because many waters have less than 15 mg/l magnesium, but very few have as little as 2.4 mg/l. The lower concentration of magnesium may be required in cold weather but it is also possible to reduce the pH in cold weather.

The limit of 150 mg/l suspended solids in circulating water is widely recommended to avoid fouling. The ammonia and BOD limit suggested by Weddle and Rogers is in part because biological oxidation of these contaminants will increase the suspended solids. But it will be difficult to reduce suspended solids in the makeup to 10 mg/l, and a clarifier or filter on a side stream probably will be necessary. For this reason we do not specify suspended solids, BOD or ammonia.

Kleusener et al. [35] found that a trickling filter which reduced BOD and ammonia in normal sewage plant effluent much improved the cold lime treatment which followed. This is discussed under Cold Lime Treatment.

It is clear that to reach 15 cycles of concentration, phosphate must be removed from all wastewaters, calcium from many wastewaters and magnesium and silica from some wastewaters.

Actual choice of treatment depends on the salt content of the sewage and therefore of the municipal drinking water. One treatment is modification of the sewage plant to remove phosphate—a treatment available to the municipality, but not usually to the power company. The usual treatment is cold lime. Data on these treatments are summarized in the following subsections, as is the method of treating a side stream from the circulating cooling water with lime and soda ash. Finally, administrative means of reducing phosphate discharge to treatment plants are mentioned.

Biological Removal of Phosphate

It has been found that phosphate can be removed to the sludge in biological treatment. Bernard [42] and Davellaar et al. [43] have summarized past findings and their own laboratory experiments. The exact mechanism of removal is not fully understood, but the most probable explanation is that bacteria can be induced to adsorb more phosphorus than they need for metabolism (called a "luxury uptake"). To cause the adsorption, the bacteria must first be subject to anaerobic conditions, where they release phosphorus, and then to aerobic conditions, where they adsorb phosphorus. The anerobic conditions must be truly anaerobic. If nitrate is present, called "anoxic condition," the adsorption of phosphorus will not be induced. The aerobic condition requires 3-4 mg/l dissolved oxygen.

Several designs are available for plants to remove phosphorus. Campbell et al. [44] have reported on Union Carbide's "Phostrip" process pictured in Figure 2. Recycled sludge is held in an anaerobic tank where phosphorus is released. The high phosphorus supernatant (16-25 mg/l P or 48-75 mg/l PO_4) is treated with low doses of lime to reduce the concentration to 2-5 mg/l P (6-15 mg/l PO_4) and then recycled. This is a very efficient use of lime. This system can conveniently be added to an existing sewage plant.

Goldieri [45] has described a system for phosphorus removal which also removes ammonia by nitrification and denitrification. This is the Air Products and Chemicals, Inc. A/O System pictured in Figure 3. Influent wastewater and recycled sludge are mixed in a plug flow anaerobic section where phosphate is released by the sludge as required for later uptake of phosphorus. In the anoxic section, recycled mixed liquor containing nitrate is added and the nitrate is denitrified using sewage as the carbon source. Finally in the aerobic plug flow section the remaining BOD is oxidized, ammonia is nitrified and phosphate is adsorbed. Phosphorus is removed from the system in the waste

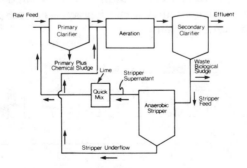

Figure 2. "Phostrip" process.

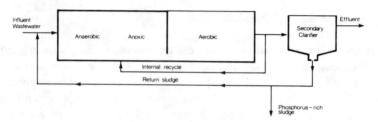

Figure 3. "A/O" process.

sludge which, in the plant in Largo, Florida, is dried and sold as a fertilizer. Goldieri suggests that the A/O System can be put into existing plug flow basins because of the short times required. Barnard [42] proposes a very similar process which he calls "Phoredox."

Barnard [46] has devised a most interesting process for high removal of ammonia in a single sludge system, which also results in phosphorus removal. This is the Bardenpho process pictured in Figure 4 (U.S. patent rights held by Environtech Corp., Salt Lake City, Utah). Recycled mixed liquor at four to five times the flow rate of the influent sewage is mixed with sludge and raw sewage in an anoxic basin where nitrate is denitrified using sewage BOD as the carbon source. In the following aerated basin the balance of the BOD is oxidized and ammonia is nitrified. The flow then goes to a secondary anoxic basin where denitrification occurs and a true anerobic state can be reached, thus conditioning the sludge for later phosphate adsorption. In the final aerobic basin the BOD and ammonia are oxidized to low concentrations and quite a high dissolved oxygen can be maintained, which helps to achieve a high adsorption of phosphorus. Provided the sludge in the secondary clarifier is not allowed to go anaerobic, phosphate will be wasted from the system in the waste sludge. This system is probably cheaper than a two-sludge or three-sludge process for nitrification and denitrification [47], but is more expensive than a standard one-sludge-activated sludge process. For our purpose nitrification and denitrification are not required.

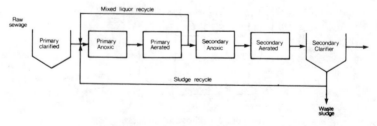

Figure 4. "Bardenpho" process.

All of the processes described appear to reduce phosphorus to < 1 mg/l (< 3 mg/l PO_4). This is not low enough to be able to concentrate fifteen times in a cooling tower, but as described below it is quite low enough to make a side-stream treatment feasible on the cooling tower.

Cold Lime Treatment of Sewage

Cummings [17], Kluesener et al. [35], O'Farrell and Menke [48], Ladd and Terry [16] and Dvorn and Wilcox [49] have studied the addition of lime to secondary treated sewage. Merrill and Jordan [50] have studied the addition of lime to degritted raw sewage. Some of these results have been replotted in matching units on Figures 5 and 6. The authors found, and the figures show, that:

1. Adequate phosphorus removal occurs at a pH greater than 10.5. Kluesener found that treated water was low in eutrofication and gave low algae growth in storage or in the tower.

2. To reach a low level of silica Kluesener et al. had to raise the pH to 11 or higher and even then effluent of the specified quality was uncertain. If, however, the wastewater was passed through a trickle filter to nitrify it and remove some BOD, an effluent of 10 mg/l Mg (as $CaCO_3$) was reliably obtained at pH 11.2. The single point from Ladd and Terry of 42 mg/l Mg (as $CaCO_3$) at pH 10 to 10.5 is close to the curve from Kluesener.

Ladd and Terry were satisfied with this reduction in magnesium. They obtained reliable removal provided both BOD and suspended solids in the influent were both less than 25 mg/l. Cummings, with an influent of 42 mg/l Mg (as $CaCO_3$) got no reduction, agreeing with the other results. Note that our recommendation of 15 mg/l Mg in the cooling tower makeup is 62 mg/l Mg (as $CaCO_3$).

3. Calcium usually increases when the pH is raised enough to remove phosphate and increases a great deal if a pH above 10.5 is required. Kluesener et al. and O'Farrell and Menke got satisfactorily low calcium by recarbonating the lime treater effluent to pH 10.0 and adding soda ash. O'Farrell and Menke used ferric chloride as a coagulant in the second-stage reactor-clarifier.

4. Merrill and Jordan [50] reported dissolved carbonate in the effluent as 140 (mg/l as $CaCO_3$) at pH 11 and as 200 at pH 10.5. At pH 10, Cummings [17] reported the carbonate as 110 and the total alkalinity as 216 (mg/l as $CaCO_3$).

5. Kluesener [35] reported unexpectedly unsatisfactory silica removal. For best removal, the reaction zone residence time should be at least 20 minutes and the reaction zone suspended solids at least 10,000 mg/l. Then

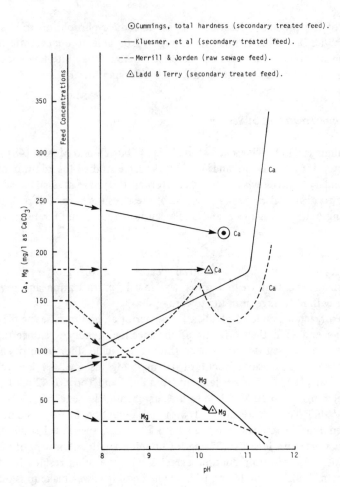

Figure 5. Concentrations of calcium and magnesium in effluent from cold lime treatment of sewage.

silica was removed at the rate of 1 mg SiO_2/7 mg Mg hardness. Ladd and Terry got a reduction of SiO_2 from 28 to 10 mg/l; i.e., silica was reduced at the rate of 1 mg SiO_2/6 mg Mg hardness.

It seems that cold lime treatment is certainly satisfactory for the removal of phosphate and is probably satisfactory for the removal of magnesium. Whether or not removal of silica is adequate will depend on the magnesium to silica ratio in the treated sewage and on the silica concentration. Removal of silica may or may not be adequate. Calcium is added rather than removed, and a second stage is needed to remove it. Cold lime treatment is not the ideal

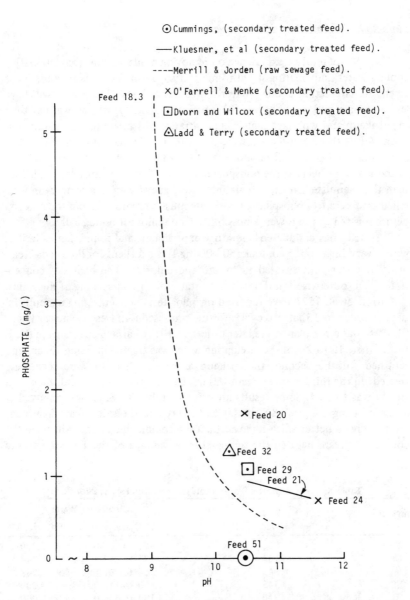

Figure 6. Concentrations of phosphate in effluent from cold lime treatment of sewage.

treatment, but it can usually be expected to work and it is the only treatment backed by experience. We suggest that a side stream treatment with lime and soda ash should be tried, particularly when silica removal is required, and its cost compared to cold lime treatment. The side-stream treatment will usually require that phosphate be separately removed from the makeup.

Lime-Soda Treatment of a Side Stream

There is not a lot of experience on removing a side stream from circulating cooling water and treating it with lime and soda ash. There is no experience when the circulating cooling water is wastewater. The side stream will have less ammonia and possibly less BOD than the makeup, will be warmer than the makeup and will have higher concentrations of calcium, magnesium and silica. All of these considerations lead one to believe that removal of calcium, magnesium and silica will be more reliable at lower pH in a side stream than in the makeup. Because the phosphate content of many wastewaters is higher than that permitted in the circulating water, side stream treatment cannot be relied on to control phosphate. Phosphate must be removed from the makeup before it is fed to the tower; a possibility is to remove it biologically.

Darji [51] has conducted laboratory experiments and found that a clarifier with a very high solids content (50,000 mg/l) is excellent for lime-soda treatment of water concentrated in sulfate and chloride. The high salt concentration can otherwise interfere with crystal growth and settling of the solids.

Carroll et al. [52] have reported on pilot tests concentrating agricultural run-off water to 15 and 30 cycles for the proposed Sun Desert Nuclear Plant. (Plans to build the plant were later cancelled.) Three arrangements were tried. In the first, 12 to 2% of the circulating water was treated in a side stream and returned. In the second, the makeup and a side stream were separately treated. In the third, a side stream was mixed with the makeup and the mixed stream was treated. Some results are given in Table X. (Silica was removed at a rate of 1 mg SiO_2/5.7-8.8 mg Mg as $CaCO_3$ and the side stream treatment did not give a better silica removal than the combined stream.) Although the combined stream negated the temperature advantage of the side stream, the

Table X. Representative Water Analyses from Sun Desert Pilot Plant

mg/l as Ion	Raw Water	Circulating Water	
		15 Cycles[a]	30 Cycles[b]
Na	360	6,900 to 9,100	12,300
Ca	134	173 to 363	344
Mg	43	64 to 263	248
SO_4	545	9,400 to 11,500	20,000
Cl	340	5,100 to 5,800	10,000
HCO_3	315	90 to 252	112
SiO_2	21	28 to 47	39
TDS	1,780	23,200 to 27,600	45,400
pH	8.0	8.2 to 8.3	8.2

[a]Range of results from three arrangements: side stream only, separate side stream and makeup, combined sidestream and makeup.
[b]Combined side stream and makeup.

lower concentration apparently led to a better performance with less chemical cost. The combined stream arrangement was also operated successfully at 30 cycles. A combined stream arrangement, if operable, seems particularly suited for sewage makeup where the phosphate concentration makes side-stream-only useless.

Administrative Means of Phosphorus Control

Phosphorus enters domestic wastewater through human wastes, primarily urine, and through phosphate compounds used as builders in detergent formulations. Each of these sources contributes significantly to the total amount of phosphorus, which is typically 20-30 mg/l in sewage.

During the late 1960s phosphorus was recognized as contributing to accelerated eutrophication of lakes. Numerous states and municipalities responded with partial or complete bans on the sale of phosphate detergents. Among these were Connecticut, Florida, Indiana, New York, Detroit and Chicago [53]. These bans often resulted in litigation claiming that the reduction in phosphorus from the area affected by the ban would not significantly affect receiving water quality.

If such bans were employed in a municipality primarily for the purpose of facilitating the use of sewage for cooling purposes, the above argument would become invalid since the desired result of effectively increasing the available water supply would be achieved. Where sewage is being considered for use as cooling water, administrative means of reducing phosphorus could be investigated.

ACKNOWLEDGMENT

We would like to express our gratitude to T. H. Humphris, Croyden Power Station, Great Britain, for his discussion with us of the section on phosphate scale formation.

REFERENCES

1. Gold, H., and D. J. Goldstein. "Wet/Dry Cooling and Cooling Tower Blowdown Disposal in Synthetic Fuel and Steam-Electric Power Plants," U.S. EPA Report-600/7-79-085 (March 1979).
2. Hart, O. O., and L. R. J. vanVuuren. "Water Reuse in South Africa," in *Water Renovation and Reuse*, H. I. Shuval, Ed. (New York: Academic Press, 1977).

3. "Water for Power Plant Cooling, California State Department of Water Resources, Sacramento, Bulletin No. 204 (July 1977). NTIS Catalog PB 285-126.
4. McKee, J. E. "Potentials for Reuse of Wastewater in North Central Texas," *Water Resources Bulletin* 7(4):740-749 (1971).
5. Ladd, K. "City Wastewater Re-Used for Power Plant Cooling and Boiler Make-up," in *Water Management by the Electric Power Industry*, Gloyna, Woodson and Drew, Eds. (Austin, TX: University of Texas, 1975), pp. 165-171.
6. Smythe, F. "Multiple Water Reuse," *J. Am. Water Works Assoc.* 63(10): 623-625 (1971).
7. Hanssen, N. S. "The Capacity of Thermal Electric Power Plants to Accept Municipal Wastewater," *Complete WateReuse*, AIChE Publication (1973), pp. 220-225.
8. Eden, G. E., D. A. Bailey and K. J. Jones. "Water Reuse in the United Kingdom," in *Water Renovation and Reuse*, H. I. Shuval, Ed. (New York: Academic Press, 1977).
9. Humphris, T. H. "The Use of Sewage Effluent as Power Station Cooling Water," *Water Res.* 11:217-223 (1977).
10. Grutsch, J. F., and R. J. Griffin. "Water Reuse Studies Sponsored by the Petroleum Industry," 85th National Meeting, AIChE, Philadelphia, Pa., Paper 5e, June 1978.
11. Flook, R. A. "Problems Associated with Re-Use of Purified Sewage Effluents for Power Station Cooling Purposes," presented at the International Conference on Advanced Treatment and Reclamation of Wastewater, NTIS Catalog N78-26027, 1978.
12. "Wastewater Reclamation," Bureau of Sanitary Engineering, California Department of Public Health, Berkeley, CA (November 1967).
13. Lewis, B. G. "On the Question of Airborne Transmission of Pathogenic Organisms in Cooling Tower Drift," presented at the Cooling Tower Institute, Annual Meeting, New Orleans, LA, January 1974.
14. Adams, A. P., M. Garbett, H. B. Rees and B. G. Lewis. "Bacterial Aerosols from Cooling Towers," *J. Water Poll. Control Fed.* (October 1978), pp. 2362-2369.
15. Adams, A. P., M. Garbett, H. B. Rees and B. G. Lewis. "Bacterial Aerosols Produced from a Cooling Tower Using Wastewater Effluent as Makeup Water," *J. Water Poll. Control Fed.* (March 1980), pp. 498-501.
16. Ladd, K., and S. L. Terry. "City Wastewater Reused for Power Plant Cooling and Boiler Makeup," *Complete WateReuse*, AIChE Publication (1973), pp. 226-231.
17. Cummings, R. O. "The Use of Municipal Sewage Effluent in Cooling Towers," presented at the Cooling Tower Institute Annual Meeting No. 41, June 1964.
18. Gray, H. J., C. V. McGuigan and H. W. Rowland. "Treated Sewage Serves as Tower Makeup," *Power* (May 1973), pp. 75-77.
19. Gray, H. J., C. V. McGuigan and H. W. Rowland. "Sewage Plant Effluent as Cooling Tower Makeup—A Continuing Case History," presented at the International Water Conference, 34th Annual Meeting, October 1973.
20. Schmidt, C. J., E. V. Clements and I. Kugelman. "Current Municipal Wastewater Reuse Practices," 2nd National Conference on *Complete WateReuse*, Chicago, IL, AIChE (May 1975), pp. 12-21.

21. Hoppe, T. C. "Secondary Effluent Without Phosphate Removal Used for Cooling Water Makeup," *Water Sew. Works* (February 1976), pp. 62-65.
22. Goldman, E., and P. J. Kelleher. "Water Reuse in Fossil-Fueled Power Stations," in *Complete WateReuse*, AIChE Publication (1973), pp. 240-249.
23. McGilbra, A. F. Personal communication (1980).
24. Hart, J. A. "Wastewater Recycle for Use in Refinery Cooling Towers," *Oil Gas J.* (June 11, 1973), pp. 82-86.
25. Mohler, E. F., and L. T. Clere. "Bio-Oxidation Process Saves H2O," *Hydrocarb. Process.* (October 1973), pp. 84-88.
26. Mohler, E. F., and L. T. Clere. "Development of Extensive WateReuse and Bio-Oxidation in a Large Oil Refinery," in *Proceedings of the Conference on Complete WateReuse*, AIChE (1973), pp. 425-447.
27. Petrey, E. Q., Jr. "The Role of Cooling Water Systems and Water Treatment in Achieving Zero Discharge," paper no. 118a, Cooling Tower Institute, Houston, TX (1973).
28. Humphris, T. H. Personal communication (1980).
29. Truesdell, A. H., and B. F. Jones. "WATEQ, A Computer Program for Calculating Chemical Equilibria of Natural Waters," *J. Research U.S. Geological Survey* 2(2):223-248 (1974).
30. McCoy, J. W. *The Chemical Treatment of Cooling Water* (New York: Chemical Publishing Company, 1974).
31. DeBoice, J. N., and J. F. Thomas. "Chemical Treatment for Phosphate Control," *J. Water Poll. Control Fed.* 47(9):2246-2255 (1975).
32. Chen, Y. S., J. L. Petrillo and F. B. Kaylor. "Optimal Water Reuse in Recirculating Cooling Water Systems for Steam Electric-Generating Stations," *Proceedings of the 2nd National Conference on Complete Water Reuse*, AIChE (1975), pp. 528-541.
33. Humphris, T. H. "The Control of Phosphate Scaling in Power Station Condensers," Central Electric Research Laboratories, England, Note RD/L/N 168/79, September 1979; presented at the International Conference on the Fouling of Heat Transfer Equipment, Rensselaer Polytechnic Institute, Troy, NY, August 1979.
34. Weddle, C. L., and A. C. Rogers. "Water Reclamation Process Evaluation for the Arizona Nuclear Power Project," presented at the AIChE Water Reuse Conference, Chicago, IL (May 1975).
35. Kluesener, J., J. Heist and R. H. Van Note. "A Demonstration of Wastewater Treatment for Reuse in Cooling Towers at 15 Cycles of Concentration," *Proceedings of the Second National Conference on Complete Water Reuse*, AIChE (1975), pp. 554-563.
36. Radian Corporation, "A Theoretical Description of the Limestone Injection—Wet Scrubbing Process. Vol. 1," NTIS Catalog PB-193 029 (June 1970).
37. Klen, E. F., and D. A. Johnson. "Calcium Sulfate Solubility in Dynamic Cooling Tower Systems: Zero Blowdown," presented at the W.W.E.M.A. Industrial Pollution Conference, Houston, TX, 1976.
38. Norman, L. R., W. S. Midkiff and J. A. Baumbach. "Team Approach Saves Water for PNM," presented at the International Water Conference, 40th Annual Meeting, Pittsburgh, PA, October 1979.
39. Christ, C. L., P. B. Hostetler and R. M. Siebert. "Studies in the System MgO-SiO2-CO2-H2O) (III): The Activity-Product Constant of Sepiolite," *Am. J. Sci.* 273:65-83 (1973).

40. Wollast, R., F. T. Mackenzie and O. P. Bricker. "Experimental Precipitation and Genesis of Sepiolite at Earth-Surface Conditions," *Am. Min.* 53:1645-1662 (1968).
41. Crits, G. J., and G. Glover. "Zero Blowdown from Cooling Tower—the Problems and Some Answers," *Proceedings of the 34th International Water Conference*, Pittsburgh, PA (1973), pp. 15-22.
42. Barnard, J. L. "A Review of Biological Phosphorus Removal in the Activated Sludge Process," *Water SA* 2(3):136-144 (1976).
43. Davelaar, D., T. R. Davies and S. G. Wiechers. "The Significance of an Anaerobic Zone for the Biological Removal of Phosphate from Wastewaters," *Water SA* 4(2):54-60 (1978).
44. Campbell, T. L., et al. "Biological Phosphorus Removal at Brocton, Massachusetts," *J. New England Water Poll. Control Fed.* 12(1) (1978).
45. Goldieri, J. V. "Biological Phosphorus Removal," *Chem. Eng.* (December 31, 1979), pp. 34-35.
46. Barnard, J. L. "Cut P and N Without Chemicals," *Water Wastes Eng.* Part 1 (July 1974), pp. 33-36; Part 2 (August 1974), pp. 41-44.
47. Goldstein, D. J., I. W. Wei and R. E. Hicks. "Reuse of Municipal Wastewater as Makeup to Circulating Cooling Systems," *Ind. Water Eng.* 16(4):. 20-29 (1979).
48. O'Farrell, T. P., and R. A. Menke. "Operational Results for the Piscataway Model 5 MGD AWT Plant," U.S. EPA Report-600/2-78-172 (September 1978).
49. Dvorn, R., and R. Wilcox. "Treated Sewage for Power Plant Makeup Water," *Power Eng.* (November 1972), pp. 40-41.
50. Merrill, D. T., and R. M. Jordan. "Lime-Induced Reactions in Municipal Wastewaters," *J. Water Poll. Control Fed.* 47(12):2783-2808 (1975).
51. Darji, J. "Reducing Blowdown from Cooling Tower by Sidestream Treatment," *Proceedings of the 1977 Waste Water Equipment Manufacturing Association Conference*, Atlanta, GA, April 1977.
52. Carroll, J. W., L. R. Lepage and B. A. Milnes. "Use of Agricultural Runoff Water for Power Plant Cooling," Paper No. IWC-78-30, 39th Annual Meeting of the International Water Conference, Pittsburgh, PA, October 1978.
53. Porcella, D. B., and A. B. Bishop. *Comprehensive Management of Phosphorus Water Pollution* (Ann Arbor, MI: Ann Arbor Science Publishers, Inc., 1975).

CHAPTER 20

WASTEWATER RECLAMATION AND REUSE
AT MILITARY INSTALLATIONS

Curtis J. Schmidt, Marsha Gilbert, and Ernest V. Clements III
SCS Engineers, Inc.
Long Beach, California

Stephen P. Shelton
University of South Carolina
Columbia, South Carolina

INTRODUCTION

Military bases are often excellent candidates for on-site wastewater reclamation and reuse. Reasons for this include:

- A large percentage of the fresh water supply is used for nonpotable purposes (i.e., cooling, air scrubbing, vehicle washing, irrigation, etc.).
- There is often relatively close proximity of wastewater sources to potential water uses.
- There are usually well-controlled, discrete areas where public access can be limited and potential risks to health minimized.
- Military bases often have the capability to implement reuse systems relatively quickly compared to the lenghty procedures required for private industry and communities to go through the many levels of regulatory agencies, public hearings for approvals and financing.

This is not to say that all, or even most, military bases can reuse wastewater in a cost-effective manner. Recent experience does indicate, however, that wastewater reclamation and reuse should be evaluated as part of the total analysis whenever future improvements to water or wastewater facilities are being contemplated.

Although the primary mission of military facilities is to secure our national defense, they are also charged with maintaining the quality of the national environment through compliance with federal and sometimes state environmental regulations. The military encourages water conservation and reuse as a part of their environmental programs. One means of achieving this is through wastewater reclamation and reuse.

Wastewater reuse can have several benefits: freshwater supplies are conserved by substituting the reclaimed water for subpotable uses; some pollution control problems can be alleviated by internal recycling and reuse at specific activities; treatment plant performance can be enhanced by reusing water, thereby reducing the wastewater flow to the treatment plant; nutrients in the wastewater can be utilized as fertilizer in irrigation waters; zero discharge reuse schemes can eliminate problems with meeting stringent effluent or pretreatment criteria; and by reducing wastewater volume through reuse, bases can reduce their expenditures on sewer discharge fees.

Both manual and computerized models have been developed to assess the technical and economic feasibility of wastewater reuse at fixed military installations and to conduct field investigations of those showing good potential [1-3]. Several U.S. military bases now have reclamation/reuse systems in operation. The procedures developed for: performing these preliminary evaluations, conducting the field investigations, developing conceptual reuse systems, and performing economic analysis are described in this chapter. The types of military installations and their similarities to civilian counterparts are included, and a concise summary of pertinent water and wastewater characteristics is provided for military activities with good reuse/recycle potential.

WATER-RELATED ACTIVITIES AT MILITARY INSTALLATIONS

The approximately 400 major fixed military installations in the United States can be divided into two broad categories: training-oriented and maintenance/storage-oriented facilities.

Training-oriented facilities have water- and wastewater-related activities that are similar to small civilian communities, with a few scattered industrial, business, recreational and commercial establishments. Typically, they provide several types of post housing, a hospital, post office, industrial laundry, laundromats, cafeterias, restaurants, mess halls, various community facilities (e.g., school, church, gymnasium, swimming pool, auditorium, hobby shops, golf course), protective facilities (police, fire), commercial facilities (commissary, post exchange, gas station), office buildings, a photographic lab, and wastewater treatment plant. They commonly have their own heating and cooling and sometimes electricity-generating plants. Industrial-type activity is usually limited to minor or intermediate level vehicle, ship, and aircraft maintenance and cleaning. Thus, wash racks, vapor degreasers, steam cleaners and paint booths are common.

Maintenance-oriented installations (e.g., Army depots, Air Force logistics centers and Navy shipyards) perform major repair and maintenance of tanks, weapons, planes, ships, heavy equipment, electronics, and engines. A few select bases work only with munitions, others are specialty bases and storage depots. Most maintenance-oriented installations are predominantly industrial in character, and metal cleaning, electroplating and metal finishing, wash racks, steam cleaning, and paint booths are common. For this reason, they often operate independent industrial waste treatment facilities, as well as sewage treatment facilities.

Table I summarizes the prevalence of various water-related activities at Navy bases [1]. The Army [2] and Air Force [3] have similar patterns of activities. As shown, most posts have many activities that use and generate water and wastewater, including domestic, industrial, recreational, and pollution control. Activities which have excellent potential for reuse are:

- sewage and industrial wastewater treatment plants,
- irrigated golf courses and landscape,
- large cooling towers (including those associated with engine and transmission test cells),
- plating shops,
- large wash racks and steam cleaning facilities, and
- other large industrial water users.

These activities are prevalent at most installations.

Table II summarizes the estimated effluent quality from major sources of wastewater, and Table III shows the tolerable water supply quality required by the potential users of reclaimed water. Data for metal electroplating and finishing are not included in Table II, although this activity may be an important source of reclaimed water for a particular installation. The quality of metal plating and finishing wastewater varies so widely, depending on the plating process and rinses used, that no typical values for these wastewaters can be assigned.

Many activities at military installations are similar to civilian operations (e.g., laundromats, car washes, recreational activities, office buildings, maintenance shops, sewage treatment plants, etc.). However, populations at military installations can fluctuate greatly between day and night, since a large number of civilians may come onto an installation to work during the day. Also, the populations often decline significantly on weekends or vary depending on

Table I. Prevalence of Water-Related Activities at Navy Bases

Activities	Norfolk NAS	Norfolk NSY	Norfolk NARF	Corpus Christie NAS	Alameda NAS	Alameda NARF	Concord NWS	Mare Island NSY	Moffett NAS
1. Aircraft Washrack	X	-	X	X	X	X	-	-	X
2. Base Housing	X	-	-	X	X	-	-	X	X
3. Boiler	X	X	X	X	X	X	X	X	X
4. Cafeteria, Mess Halls, Restaurants	X	X	X	X	X	X	X	X	X
5. Cooling Tower	-	X	X	X	-	X	-	X	-
6. Engine Test Cell	X	-	X	-	X	X	-	-	-
7. Fire Protection/Spill Washdown Reservoir	-	X	-	-	-	-	-	-	-
8. Golf Course	-	-	-	X	-	X	-	-	X
9. Heat Treatment	-	X	X	-	-	-	X	X	-
10. Hospital/Clinic	X	X	-	X	X	X	-	X	X
11. Industrial Laundry	X	X	X	X	X	-	-	-	-
12. Industrial Waste Treatment Plant	X	-	X	X	X	X	-	X	X
13. Irrigation	X	-	-	X	X	-	X	X	X
14. Laundromat	-	-	-	-	X	-	-	-	X
15. Metal Finishing/Plating	-	X	X	-	X	X	-	X	-
16. Metal Cleaning	-	X	X	-	X	X	-	X	-
17. Motor Pool	X	X	X	X	X	X	X	X	X
18. Paint Booth	-	X	X	-	X	X	-	X	X
19. Photography Lab	X	X	-	X	-	-	-	X	X
20. Sewage Treatment Plant	X	X	-	X	-	-	-	X	X
21. Steam Cleaning	-	X	X	-	X	X	-	X	-
22. Swimming Pool	-	-	-	-	-	-	X	-	X
23. Vehicle/Washrack	X	X	-	X	X	X	-	X	-
24. Water Treatment Plant	X	X	-	-	X	-	X	X	X

Table II. Sources of Reclaimed Water—Typical Effluent Quality

Concentration (mg/l)

Constituent	Base Housing[a]	Hospitals/ Clinics	Industrial Laundries	Laundro-mats	Boilers	Cooling Waters[b]	Air Pollution Wet Scrubbers	Vehicle Wash Racks	Aircraft Wash Racks	Steam Cleaning	Metal Cleaning[c]	Paint Booths	Photographic Laboratories
BOD_5	200	250	450	200	5.0	7.0	10	60	5700	1300	3000	8100	300
COD	300	850	2000	400	15	35	720	900	8400	2800		13600	500
pH (pH units)		7.6	11.2	8.2	10.0	7.4	4.0		8.5	9.7	9.0		7.8
TDS	300+SWC[d]	1400	2000	360	3500	5xSWC	5000						2900
O&G	50-100	45	300	750	0.5	30	0.3	60	280	245	350	280	4.0
SS	300	200	1000	130	50	0-0.1	3270	2000	470	1000	300	2800	225
PHNL	0.15				0.5		0.001	0.01	8.5	8.0	70	1.2	0.001
Ag		0.3											0.5
B	1.0+SWC				10	0.1	0.1	0.01	0.1			0.1	18
Ca		15	740										
$CaCO_3$	80+SWC			250	50		200	31					
Cd		0.02	0.04		0.005	0.05	0.01			0.5	0.5		
Cr		1.1	0.06			0-2	0.005			0.3	25	13	
Cu			0.3		3.0					0.2		0.005	
CN	0.01				2.5			0.005	0.005	<0.01	0.6		4.8
Cl^-	100+SWC				1000	5xSWC	400						
Fe	1.0+SWC	0.3	1.0		2.5	0.6	5.3	4.7	1.1			3.2	2.0
K		34											
MN					2.5								
Mg		16	6.4					15					
Na	50+SWC	360			1000		72						
Ni			2.1		0-0.1					<0.05			
NH_4	30				2.0		0.1	0.01	0.1			0.1	16

Table II, continued

Constituent	Base Housing[a]	Hospitals/ Clinics	Industrial Laundries	Laundro- mats	Boilers	Cooling Waters[b]	Air Pollution Wet Scrubbers	Vehicle Wash Racks	Aircraft Wash Racks	Steam Cleaning	Metal Cleaning[c]	Paint Booths	Photographic Laboratories
Concentration (mg/l)													
NO$_3$				1	150		28	3.3	0.8			28	8.8
P (Total)						2.1		33					
Pb	10	0.3	0.7					2.5		0.6	0.4		
PO$_4$		170	130	220	60		5.4	12	80	65	40	3.0	9.3
Si					2.5								
SO$_4$		35			0-1	5xSWC							
Zn			0.5		50	5wSWC		2.9		2.0	6.0		
Hardness (as CaCO$_3$)													
Detergents (as MBAS or ABS)		75		60							3.0		
Hexane Sol.				32								4900	
ALK (as CaCO3)	50-150	125	500	182	500	5xSWC		115			400		
Level of Confidence[e]	1	1	1	1	1	1	3	1	2	1	1	2	1

[a] Applies to officers' quarters, BOQ, VOQ, barracks, and unclassified office space, as well as family housing.
[b] Recirculating type.
[c] Rinse waters.
[d] SWC = Source water concentration.
[e] Level of Confidence denoted confidence in data shown 1 = High level of confidence. Well documented. 2 = Moderate level of confidence. Some documentation, some values may be engineering estimates. 3 = Low level of confidence. Limited documentation, most values are engineering estimates.

NOTE: Metal electroplating and finishing is a potential source of reclaimed water. The effluent contains high concentrations of various metal ions and cyanides, depending on the processes used, and small amounts of BOD, P, O&G, and SS.

Table III. Users of Reclaimed Water–Tolerable Water Supply Quality

Constituent	Laundries[a]	Recreational Lakes[b]	Fire Protection/Spill Washdown Reservoirs	Irrigation[c]	Boilers	Cooling Waters[d]	Wet Scrubbers	Wash Racks and Steam Cleaning[e]	Metal Electroplating and Finishing[f]	Metal Cleaning	Paint Booths	Photographic Laboratories
						Concentration (mg/l)						
BOD_5	45	10	10	30	1.0		100	10	1	1	30	0.1
COD	500	60	22	60	3.0	75	200	25	3	3	60	1.0
pH (pH units)	6.0-6.8	5.0-9.0	5.0-9.0	4.5-9.0	>9.0							
TDS	3300	2000	2000	2000	2000	500-1500	2000	2000	500	500		700
O&G	10	5	1.0	30	0.0		50	5	1	1	30	0.2
SS	30	10	10	50	10		100	10			60	1.0
PHNL	0.05	0.01	0.01	0.5	0.1				0.001	0.001		0.001
As	0.5								0.05	0.05		
B	0.1	0.1	0.1	3	2.0		2.0	2.0	1.0	1.0		0.1
$CaCO_3$					20		300	500				400
Cd									0.01	0.01		
Cr	0.5								0.05	0.05		
Cu	1.0								1.00	1.00		
CN	0.2	0.1	0.1	0.01	0.5	0.5	0.5	0.5	0.2	0.2	0.5	0.5
Cl^-		300		350	200	500	600	600				0.01
Fe	1.0	5.0	5.0	10	0.5	0.5	20	40	0.3	0.3		0.3
Mn	1.0				0.5	0.5			0.05	0.05		0.5
Mg							200					
Na	250	250	10	350	200		600	600	0.5	0.5	15	100
NH_4	0.1	0.1		20	2.5		20	5				0.1
NH_3-N	1.5		5.0									
NO_3		2.5					50		10	10		20
Pb	0.5					0.3		1	0.05	0.05		

Table III, continued

Constituent	Concentration (mg/l)											
	Laundries[a]	Recreational Lakes[b]	Fire Protection/Spill Reservoirs	Washdown Irrigation[c]	Boilers	Cooling Waters[d]	Wet Scrubbers	Wash Racks and Steam Cleaning[e]	Metal Electroplating and Finishing[f]	Metal Cleaning	Paint Booths	Photographic Laboratories
PO_4	0.3	0.3	1		0.3								3.0
Si					50	50						
SO_4						200						
Zn	0.5									5.0	5.0		
Hardness (as $CaCO_3$)	50				10	50				10	10		
ALK (as $CaCO_3$)	60				100	350						
DO	>0	5	3	>0	>0	>0	>0	>0	>0			
Median coliform (No./100 ml)	≤2.2	≤2.2		g				≤2.2				
Level of Confidence	1	2	3	1	1	1	3	3	1	1	3	3

a Applies to both industrial laundries and coin-operated laundromats.
b Limited body contact.
c Note that some plants are much less tolerant than others to various constituent concentrations and may require water of higher quality than that shown in this table; BOD, TDS, chlorides, and Boron are particularly important in this regard.
d Applies to cooling waters for boilers, dynamometers, air compressors, vapor degreasers, and any operation that has a cooling tower.
e Applies to both vehicle and aircraft wash racks.
f Rinse waters.
g Agricultural irrigation – 2.2/100 ml; landscape irrigation – 23/100 ml.

staffing levels, training programs, maneuvers, etc. An important consideration is that military personnel are a captive audience in that they must follow all commands and rules, including any that may pertain to water reuse. Thus, water conservation/reuse strategies can be more easily and quickly implemented, and more directly enforced than among civilian populations.

Preliminary Evaluation of Wastewater Reuse

A weighted comparison evaluation model has been developed to provide an inexpensive, rapid method for quantifying the potential for wastewater reuse at fixed military installations [1-3]. By evaluating the results of this analysis, called Tier I, it is possible to eliminate installations with little potential for reuse, and concentrate on those that show good potential. The principal evaluation criteria are summarized below.

Positive factors for reuse:

- water supply shortages, problems or high cost,
- direct discharge of treated or untreated effluent to surface water or land [National Pollution Discharge Elimination System (NPDES) permits required],
- wastewater treatment/discharge problems or high cost,
- presence of high water use activities that could use reclaimed water, such as: golf course irrigation, landscape irrigation, cooling towers, plating shop, industrial laundry, wash and steam cleaning racks, artificially filled recreational lakes, dynamometers, air pollution scrubbers, and engine test cells.
- arid climate,
- positive attitudes toward reuse by key military personnel, and
- absence of legal/instituional constraints to reuse.

Negative factors for reuse include:

- ample, reliable water supplies,
- long-term water purchase or wastewater discharge (to municipal or regional sewer system) agreements that constrain water or wastewater reductions,
- absence of irrigation potential, and
- lack of enthusiasm for reuse among key personnel.

Generally, the types of installations with the greatest potential for reuse are: those with many major industrial activities and training installations in water-short areas with large irrigation demands. The industry-oriented posts use large amounts of water, often have several direct discharges, and may have significant pollution control problems. In contrast, the training installations are troop-oriented facilities that rarely have many large industrial activities. Their wastewater is primarily domestic with few problem contaminants and, therefore, is excellent for reuse for irrigation and cooling after secondary treatment and filtration. Note that these are very broad categories and reuse considerations should not be limited to these types.

Field Investigations

If the preliminary technical and economic analysis (Tier I) shows that wastewater reuse is potentially feasible, a detailed field investigation, called a Tier II evaluation [1-3], can be conducted. Basically, the evaluation model provides an orderly method to obtain and analyze the information necessary to make logical engineering and economic judgments regarding potential for reuse/recycle. The four parts of the model are: (1) activities information, (2) spatial relationship, (3) diagrams for conceptual reuse systems, and (4) economic analysis. Information pertinent to Parts 1 and 2 is collected during the field investigations.

Activities information provides background information on activities and contains standardized formats for collecting data specific to each installation. Significant water use and/or wastewater generating activities are investigated. Typical flow rates and effluent quality data, and the tolerable water supply quality required by various activities, are determined or estimated.

The characteristics of a good potential wastewater source are high volume, reliable flow and low contamination. The wastewater volume should be high enough for reuse to be economical due to freshwater savings and/or wastewater discharge reduction, the flow should be predictable so that storage needs can be determined and a reliable supply provided to the user activities, and the wastewater should not be so highly contaminated that prohibitively expensive treatment would be needed before reuse.

Sanitary and domestic-type wastewaters from housing, community, protective, administrative/institutional, and commercial activities have excellent reuse potential. Some industrial effluents (e.g., some plating shop rinse waters and cooling system blowdowns) also meet the above criteria and are good candidates for reuse. However, others, such as steam cleaning and metal cleaning wastes, have very low reuse potential because extensive treatment is required to remove or neutralize such diverse contaminants as oils, grease, cyanides, phenols, heavy metals, phosphates, acids and caustics.

The economics of reuse can be appreciably improved if domestic-type wastewaters are collected through the installation's sanitary sewer system and are centrally treated prior to reuse. Industrial waste treatment plant effluents can also be reused if the treatment provided is complete and effective. However, additional treatment to remove oils and dissolved metals and salts from industrial wastewaters may be required before reuse is possible.

A good potential user of reclaimed water requires a high volume of low quality water. A good secondary effluent is usually available and not expensive to produce. Activities that can make use of filtered secondary effluent also have good potential, because most secondary plants can be readily upgraded with filtration, although this process is more expensive. Activities that require very high water quality (e.g., boilers) are generally poor users because the processes required to sufficiently purify wastewater are too expensive in most cases.

Although we are considering only nonpotable reuse, bacteria and viruses in the reclaimed water must still be considered hazardous in activities that involve human contact with water, sprays and aerosols (e.g., wash racks, paint water walls and, to a lesser extent, spray irrigation of golf courses and landscape. Reclaimed water for these activities must meet required limits on bacterial and viral levels. Good potential users also are those that use significant amounts of water on a regular basis. Activities such as irrigation can use tremendous volumes of reclaimed water, but frequently require large storage basins to hold the water during high precipitation periods. Activities that use a great deal of water continuously, such as industrial cooling towers, are optimal.

Activities that are best suited for internal recycling usually are those with a fairly clean discharge or a discharge that is simply treated. Activities using large volumes of water offer economies of scale for pretreatment systems. Small water savings gained by internal treatment and recycling are rarely economically feasible.

Table IV lists those activities with the greatest potential as sources and users of reclaimed water, and with potential for internal reuse.

The second element of the Tier II model involves consideration of spatial relationships. The proximity of the source to user activities is important in any reuse network because piping and pumping are usually expensive components of the system. Typically, the evaluator obtains a large map of the installation that shows all buildings, roads, grounds, etc.; he marks all important potential source, user, and internal recycling activities, and locates all treatment facilities, irrigated areas and large industrial activities. He generally eliminates isolated, small flow volume activities in locations remote from the main area of activity, for it may not be economically feasible to build a separate pipeline to carry a small volume of reclaimed water over long distances.

Areas of major potential reuse activity, and areas that will be difficult to reach with reclaimed water lines (e.g., those cut off from the main center of activity by runways or highways), are noted. A determination is made as to whether the facility is essentially troop-oriented, with reuse of secondary effluent for irrigation the main possibility, or if it is a heavily industrial depot with possible reclaimed water cascades from one activity to another.

Conceptual Reuse Schemes

The next step in the Tier II evaluation is to develop diagrams for conceptual reuse systems. This is the most important step in the evaluation process and requires the highest level of engineering judgment to complete.

Table IV.   Activities with Greatest Potential as Sources or Users of Reclaimed Water, or with Potential for Internal Recycle

Sources of Reclaimed Water

Sewage Treatment Plant Effluent	Cooling Towers
Industrial Waste Treatment Plant Effluent	Dynamometers
Vehicle Wash Racks	Industrial Laundries
Metal Plating and Finishing	Boilers

Users of Reclaimed Water

Industrial
    Cooling Towers
    Paint Booth Water Walls
    Air Pollution Wet Scrubbers
    Autoclaves
    Dynamometers
    Vehicle Wash Rack
    Aircraft Wash Rack
    Steam Cleaning
    Ash Handling System Water
    Maintenance Wash Downs

Community
    Golf Course Irrigation
    Landscape Irrigation
    Athletic Field, Playground,
     Park Irrigation
    Recreational Lakes and Ponds

Commercial
    Laundry

Internal Recycling

Metal Plating and Finishing	Large Industrial Autoclaves
Vehicle Wash Racks	Cooling Towers
Aircraft Wash Racks	Paint Booth Water Walls
Dynamometers	Air Pollution Wet Scrubbers

Its purpose is to assist the evaluator in selecting conceptual reuse schemes for the military installation. All the data gathered during the field investigations on activities and spatial relationships are used to develop feasible reuse networks or systems. These networks are essentially schematic diagrams showing the distribution of fresh and reclaimed water throughout the base, as well as the collection, treatment, reuse, and disposal of wastewaters.

Several basic types of water reuse are possible for military bases depending on the presence of treatment facilities, major industrial or irrigation activities, and other factors. Conceptual diagrams are presented in Figure 1 for four basic reuse schemes:

1. Treated effluent reuse: the direct reuse of secondary or tertiary effluent from a wastewater treatment plant by activities, e.g., irrigation and cooling towers (see Figure 2).
2. Direct cascade reuse: the direct reuse of the untreated discharge from one activity as the water supply for another activity. The donor activity usually has a fairly clean discharge, and the user activity can tolerate low quality water (see Figure 3).
3. Cascade reuse with pretreatment: the same as direct cascade reuse with an intervening treatment step to bring the donor's wastewater up to the recipient's quality requirements. These pairings are generally feasible only where a simple treatment can do the job (see Figure 4).

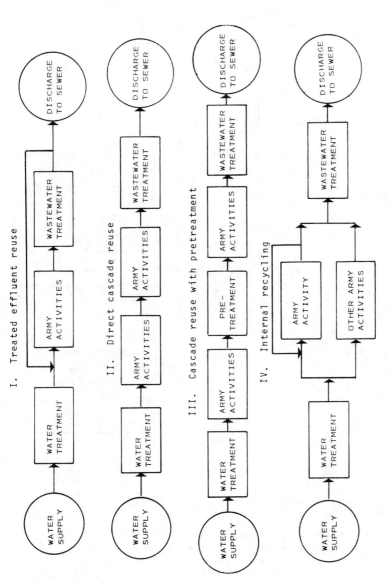

Figure 1. Basic types of reuse schemes [1,2].

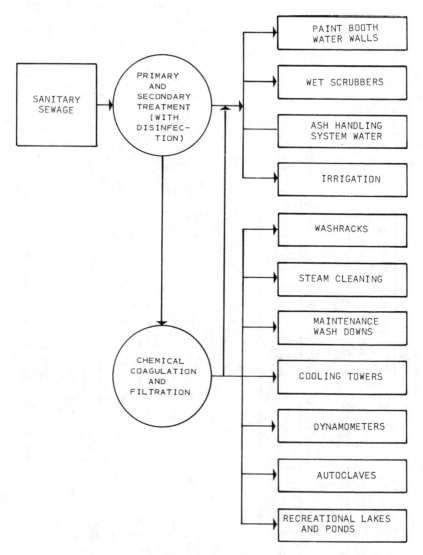

Figure 2. Reuse of reclaimed sewage treatment plant effluent [1,2].

4. Internal recycling: the reuse of wastewater as new source water for the same activity (e.g., a recirculating water system used in spray paint booth water walls, and also in air pollution scrubbers). The recirculated water can be continually bled off and water volume made up with fresh water; or it can be periodically dumped and completely replaced with fresh water. It is possible for some activities to accept internally recycled water by treating their wastewater and then mixing it in with a freshwater supply. Candidate activities are shown in Figure 5.

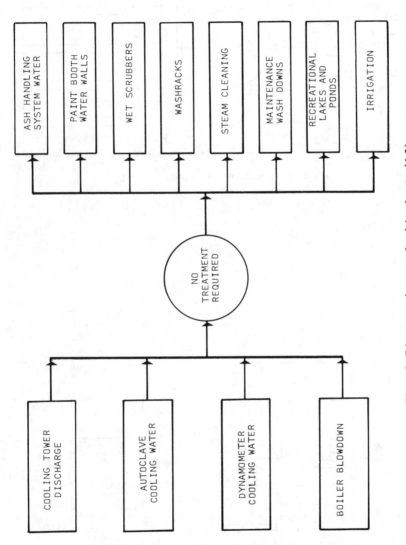

Figure 3. Direct cascade reuse of reclaimed water [1,2].

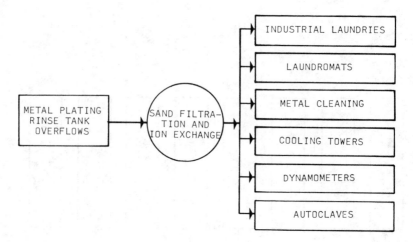

Figure 4. Cascade reuse with pretreatment [1,2].

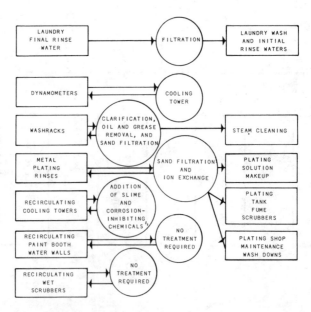

Figure 5. Internal treatment and recycling [1,2].

Some bases may be able to incorporate more than one of the basic conceptual reuse systems in a total reuse scheme.

At this point, the evaluator must compare activities and existing treatment facilities, plan possible reuse modes, and develop feasible reuse schemes. The most efficient method of laying out these networks is to develop a basic system for large sources and users, and then to integrate smaller users where it appears cost-effective.

Required new treatment facilities, modification to existing treatment facilities, storage, pumping, and distribution requirements are all shown on the conceptual networks. A general volume and quality mass balance for each network is developed to ensure that reclaimed water supply requirements can be met. At a large complex installation, it may take many weeks to develop a reuse network(s). At other installations, reuse network(s) may be immediately obvious.

Economic Analysis

At this point anywhere from one to five technically feasible reuse networks have been conceptually developed. The engineer now moves to the final step of the Tier II evaluation and uses life-cycle cost estimating techniques to determine which of the possible reuse networks is most cost-effective, and to compare the cost benefits of that network with the existing system.

The important capital, operation and maintenance costs that will be computed are:

- existing water supply cost,
- existing wastewater treatment and/or disposal cost,
- cost for new reclaimed water pipelines,
- cost for reclaimed water storage,
- cost for reclaimed water pumping, and
- cost for new treatment facilities for reclaimed water.

The model developed for military use provides a step-by-step method of cost analysis which can be done manually, or preferably with an existing computer program.

DESIGN OF THE REUSE SYSTEM

The engineering analyses and field investigations will yield the following basic information:

1. A description of activities which could accept subpotable reclaimed wastewater, including their locations, minimum acceptable supply

water quality, water supply volume required (seasonal, weekly, daily, hourly, and peak flow variations), potential human exposure through direct contact or aerosols, and plans showing the methods needed to ensure positive isolation of reclaimed water systems from potable water systems.

2. A description of potential sources of reclaimed water, including their locations, quality variations, generated volumes, and reliabilities.

3. A complete set of existing, pertinent drawings of the military base, including a base map, topography, inplace utilities, construction drawings for existing wastewater treatment systems (if any), drawings of existing wastewater disposal facilities, soil reports, and other information necessary to begin preliminary design.

4. Information from local state and federal regulatory agencies, and the military agencies which advise what standards and criteria will have to be met by the wastewater reclamation and reuse system designed.

With this information, the engineering staff can begin design. A significant design consideration is that any possibility for cross-connection with the potable water system, either during construction or later, must be positively prevented. All reclaimed water piping and appurtenances must be clearly marked by using different color pipe, labeling the pipe with marked tape throughout its length, labeling valve box covers, attaching permanent signs to exposed piping at frequent intervals, placing signs above underground pippline alignments, etc. Backflow preventers and/or air breaks should be used as needed. As portions of the system are completed, colored dye tests can be made to ensure that no nonpotable water is entering the potable system; thereafter, dye tests should be made annually.

Where public health aspects are a major consideration, some reclaimed water will require disinfection. The designer will consider the necessity for dual disinfection and, in many cases, filtration as well. In these cases, automatic chlorine residual monitoring and alarms will usually be required.

The best applications for reclaimed water are in areas where public contact is easily controlled (e.g., areas associated primarily with industrial/commercial activities). Use of reclaimed water in family housing areas, playground areas, and other difficult to control areas is not recommended without special consideration.

During the life of the system, the reclaimed water produced may not always be of adequate quality. Therefore, provisions will be made to:

1. store and test the reclaimed water before it enters the distribution system;
2. maintain facilities for storing reclaimed water not meeting standards;
3. arrange for alternative disposal or retreatment of substandard reclaimed water (e.g., a "safety valve" plan for dumping the system into a public sewer), divert-

ing to a less restrictive use, recirculating inadequately treated water back through the plant; and

4. furnish an alternate potable water supply to critical activities using reclaimed water.

Education in the use of reclaimed water is very important. At a typical military installation, the department responsible for water supply operates separately from the groups responsible for the various activities using the reclaimed water (e.g., irrigation, cooling tower maintenance, etc.). The user groups must be educated in the proper control and handling of reclaimed water versus the potable water which they have been accustomed to using. Comprehensive operation and maintenance manuals will be needed for the groups responsible for supplying reclaimed water, and for each of the user activity groups.

Basically, the designer should recognize that wastewater reclamation and reuse is a relatively new design area. He must be conservative in his design by providing many safeguards, avoiding reuse where improper public contact cannot be easily controlled, and providing emergency alternative disposal means for periods when the system is not functioning properly. An accident causing adverse health effects at any military base as a result of improper use of reclaimed water would severely set back wastewater reclamation and reuse.

REFERENCES

1. SCS Engineers. "Subpotable Water Reuse at Navy Fixed Installations; A Systems Approach," U.S. Army Medical Research and Development Command, Fort Detrick, Maryland, Volumes I, II, and III, (January 1980).
2. SCS Engineers. "Subpotable Water Reuse at Army Fixed Installations; A Systems Approach," U.S. Army Medical Research and Development Command, Fort Detrick, Maryland, Volumes I, II, and III (August 1979).
3. SCS Engineers. "Wastewater Management Alternatives in Reuse Study," McClellan Air Force Base, Sacramento, California, Volumes I, II, III (August-September 1977).

CHAPTER 21

SEAWATER INTRUSION CONTROL WITH WASTEWATER

Jerome J. Wedding and Harsha V. Kondru
Ventura Regional County Sanitation District
Ventura, California

Seawater intrusion into fresh groundwater aquifers has been a documented problem in various coastal regions of the United States. In hydraulically connected fresh and saltwater zones, as freshwater levels decrease, the salt water intrudes thereby deteriorating the water quality of the aquifer. Such contamination can affect not only a community's water supply but the economic and social structure as well. There are relief measures which may be implemented once this phenomena has occurred. A seawater intrusion control barrier utilizing reclaimed water is one possible solution.

Various factors lead to the seawater intrusion phenomena. General background information, hydrologic conditions, and some specific project area characteristics are described here.

BACKGROUND

Under natural conditions, fresh water is discharged into the ocean where coastal aquifers come in contact with the sea. Demands on the ground water can alter this situation and cause the salt water to penetrate inland in aquifers. This action may directly affect natural underground flows, permeability, and other physical parameters in addition to causing indirect environmental, economic and sociological concerns.

An aquifer is like a large underground storage reservoir. Water enters the aquifer by means of natural or artificial recharge and leaves by natural flows or is extracted by wells. There are two basic classifications for aquifers: unconfined and confined. An unconfined aquifer has a water table as its upper surface; a confined aquifer is under pressure beneath relatively impermeable strata. It is the unconfined aquifer which is discussed in the following sections.

A diagram of a coastal groundwater basin not subject to seawater intrusion is shown in Figure 1. The piezometric surface on the diagram indicates the level at which water stands in wells. Under equilibrium conditions, there is no energy gradient within the saline wedge to provide movement. The seawater front assumes the shape of an inclined wedge which slopes landward, advancing or receding in response to changes in the hydraulic gradient [1].

Salt water and fresh water often share the same formation, yet there are few naturally occurring instances of saltwater intrusion. It is man's activities—primarily pumping more water from an aquifer than can be naturally replenished—that are responsible for destroying the hydraulic continuity between fresh and saline waters [2]. The sea water will continue to move inland as long as the water table remains below sea level and slopes toward land. This situation can be reversed only by returning groundwater levels to near natural conditions. Once saline water has entered a basin, the fresh water in the degraded portion is no longer available as a source of potable water. Restoration of the water to acceptable levels of quality may require many years [3].

HYDROLOGICAL CONDITIONS

Permeable geologic formations which allow water to move under ordinary field conditions are defined as aquifers. Aquitards are confining layers of relatively low permeability such as clays and silts. An aquifer may be bounded by impermeable or permeable formations.

In coastal areas where extraction from the underground aquifers exceed inflow rates, seawater intrusion may occur. To maintain equilibrium between

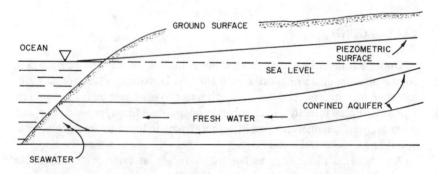

Figure 1. Confined groundwater basin not subject to seawater intrusion.

fresh water and sea water (due to differences in density), 1.025 ft of fresh water is required for every foot of sea water. Variations in the protective elevations are determined by the varying depth of the aquifer bottom. The maintenance of a protective elevation can be computed from an equilibrium equation, assuming that the pressure of the barrier must equal that exerted by sea water to halt its progression inland. The overdraft of an aquifer by extraction wells can affect the equilibrium.

Groundwater supplies in coastal areas are rapidly being intruded by sea water because of the overdraft of existing groundwater supplies; thus sea water is intruding at an increased rate and contaminating the groundwater. A case study of Oxnard, California, illustrates this.

In 1974, a seawater intrusion monitoring program was initiated on the Oxnard Plain. By the June-September 1974 sampling period, the seawater intrusion front had moved inland to a point where the intruded area covered approximately 17.9 mi^2. Based on water resources studies, well logs and other survey data, the onshore portions of the Oxnard aquifer contained 833,900 ac-ft of water in the fall of 1975. During the 1977 seawater intrusion sampling program, 75 well-water samples were collected in June and July. Results of the survey indicated that the total onshore intruded area in the Oxnard aquifer zone covered 20.6 mi^2, an increase of 2.9 mi^2 over the intruded area of the previous year [3]. During the 27-yr period since seawater intrusion in the Oxnard aquifer zone was first detected, the intrusion front has moved inland during every monitoring period except 1976 when a small reversal was noted in the vicinity of Port Hueneme, California. Existing and projected seawater intrusion fronts in the study area are illustrated in Figure 2.

The annual overdraft of an aquifer is determined by adding the long-term decrease in storage to the landward underflow of groundwater at the coastline. The annual overdraft is approximately 8700 ac-ft/yr for the Oxnard aquifer.

The study categorized the basin as an area with unabated seawater intrusion, and concluded that continued controlled pumping could possibly overdraft the basin to an extent that irreparable damage would be done, not only to the Oxnard aquifer zone, but also to underlying water-bearing zones. If this were to occur, a significant water resource would be lost, resulting in extreme economic consequences.

The loss of pumping wells because of seawater intrusion has necessitated the drilling of new wells into deeper aquifer zones, obtaining ground or surface water supplies from inland areas, or using imported water. These options mean that water users must bear one or more of the following costs: (1) construction costs of new wells to deeper aquifer zones; (2) increased power costs for pumping water from greater depths; and (3) costs of acquiring more distant water supplies and the related distribution system.

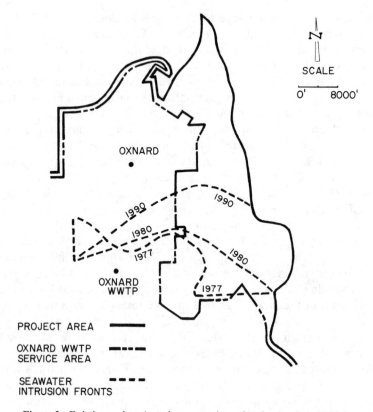

Figure 2. Existing and projected seawater intrusion fronts, Oxnard Plain.

Another water resource problem is that offshore and onshore sea water can intrude portions of the top aquifer zone. This presents a water quality threat to the underlying aquifer because salt water in the intruded aquifer zone can leak downward through improperly sealed wells, areas of aquifer mergence, and through aquitards. Continued seawater intrusion in the upper aquifer system may also adversely affect the lower aquifer system. If this occurs, fresh water stored beneath the plains will not be recoverable. To avoid this situation, seawater intrusion should be controlled.

CONTROL MEASURES

Seawater intrusion can be effectively controlled by groundwater recharge. Since using fresh water as a hydraulic barrier to stabilize and repulse the sea-water intrusion is expensive, use of reclaimed water is a better choice economically, and has the added advantage of conserving water resources.

Historical Precedence

The Los Angeles County Barrier Project, based primarily on an injection system, has proved capable of controlling seawater intrusion by increasing the piezometric surface elevation of the ground water in confined aquifers. The maintenance of a protective elevation can be computed from an equilibrium equation, assuming that the pressure of the barrier must equal that exerted by sea water to halt its progression inland. This equilibrium is satisfied by the expression:

$$H_f D_f = H_s D_s \qquad (1)$$

where
D_f = density of fresh water (62.4 lb/ft^3),
D_s = density of sea water (64.0 lb/ft^3),
H_f = height of fresh water above the bottom of the aquifer, and
H_s = height of saline water above the bottom of the aquifer.

Therefore, 1.025 ft of fresh water is required for every foot of sea water to maintain equilibrium. Variations in the protective elevations are determined by the varying depths of the aquifer bottom. Monitoring of the freshwater elevation is critical and is achieved by the use of observation wells placed midway between wells along the barrier under the assumption that this location corresponds to a piezometric low point.

Theoretically, the processes of injection and extraction can be mathematically defined; whereas practical, efficient operation of the system can be monitored by internodal observation wells. Where pumping troughs are necessary to curtail the inland flow of reclaimed water, a properly designed observation well system can function as a direct measurement of the effectiveness of the method.

The Alamitos Barrier Project, conducted by the Los Angeles County Flood Control District, was initially conceived as a freshwater pressure barrier. The fresh water was to encircle intruded seawater.

Seawater intrusion has been averted by a hydraulic barrier system consisting of an extraction/injection system in the Orange County Coastal Barrier Project. Seven extraction wells, approximately two miles from the coast, intercept the sea water and return it to the ocean through open channels. A freshwater mound is formed by a series of 23 multipoint injection wells into underground aquifers. The injection wells are located on an average of four miles inland. Beginning in October 1976 and continuing through 1977, approximately 13,000 ac-ft of water was injected through the system. The fresh water is a blended combination of reclaimed water, imported water and deep well water. This system has been an effective method of controlling seawater intrusion [4].

An injection/extraction system is currently being studied in a pilot program in Palo Alto, California. This system is effectively controlling seawater intrusion, as well as providing mathematical modeling information and data for injection/extraction well systems.

Artificial Recharging

An obvious method for controlling seawater intrusion is that of artificially recharging an intruded aquifer through the use of spreading areas or recharge wells. Overdraft would be eliminated and water levels and gradients would be properly maintained. The method is technically feasible, spreading areas being best suited for recharging unconfined aquifers and recharge wells being best suited for confined aquifers. However, it may not be economically sound to construct, operate, and maintain a recharge system without reducing the basin pumpage. Importing high-quality supplemental water, recharging it into the ground, and finally pumping it out in nearby areas forms an expensive cycle which, on a continuous basis, would lead to unduly high water costs. In California, consideration has been given to recharging wastewaters (otherwise discharged directly into the ocean) for this purpose.

Pumping Troughs

The theoretical bases for pumping trough operations are presented in *Groundwater Hydrology* [5] and *Geohydrology* [6]. The following sections present assumptions developed in these references.

Drawdown curves for well flows in the practical sense are indicated by a sloping piezometric surface where a pumping well penetrates a confined aquifer with a uniform flow field. This situation is illustrated in Figure 3. The feature of interest in this diagram is the boundary of the region producing inflow to the well. An expression for this boundary has been derived based on the Dupuit assumptions as follows:

$$-\frac{y}{x} = \tan\left(\frac{2\pi Kbi}{Q}y\right) \tag{2}$$

When the limits extend to a stagnation point, the following expression is obtained:

$$X = \frac{Q}{2\pi Kbi} \tag{3}$$

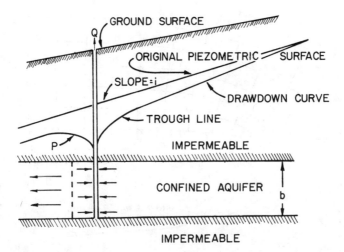

Figure 3. Flow to a well penetrating a confined aquifer having a sloping plane piezometric surface; vertical section.

where   Q = flow rate at a distance r,
            K = coefficient of permeability
            b = aquifer thickness,
             i = slope, and
            x = (refer to Figure 4).

The Dupuit assumptions are: (1) that the velocity of flow is proportional to the tangent of the hydraulic gradient, and (2) the flow is horizontal and uniform everywhere in a vertical section. These assumptions must be made because the shape of the water table determines the flow distribution and vice versa. Although the circular area of influence associated with a radial flow pattern becomes distorted, for most relatively flat natural slopes, the Dupuit radial flow assumptions are applicable.

Although not physically demonstrated, a pumping trough by itself is considered to be effective in preventing seawater and reclaimed water movement past the trough as shown in Figure 5.

This procedure is technically capable of controlling seawater intrusion and is also effective in preventing reclaimed water passage. Extraction troughs are being effectively used in several areas in South Texas and Wyoming for the control of injection fluids in in-situ mining projects.

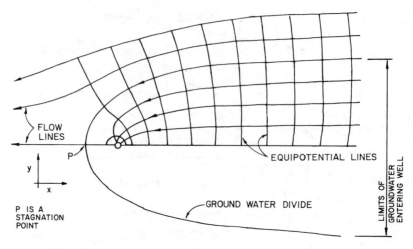

Figure 4. Flow to a well penetrating a confined aquifer having a sloping plane piezometric surface; plan view.

Pressure Ridge

Another important method for control of intrusion is formation and maintenance of a freshwater pressure ridge adjacent to and paralleling the coast. In an unconfined aquifer surface, spreading could create a water table ridge; in a confined aquifer, a line of recharge wells could form a ridge in the piezometric surface. A ridge must be of sufficient height above sea level to repel sea water. A small amount of recharged water would waste to the ocean, the remainder would move landward to supply part of the pumping draft. With

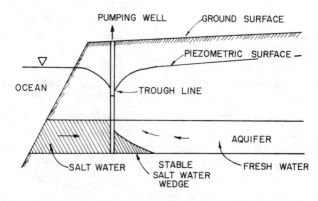

Figure 5. Control of seawater intrusion by a pumping trough paralleling the coast.

recharge wells, the ridge would consist of a series of peaks at each well with saddles between. The necessary elevation of the saddles to displace ocean water would govern the well spacing and the recharge rates required. The ridge should be located inland from the saline front, otherwise sea water inland of the ridge will be driven further inland.

FLOW NETS FOR INJECTION/EXTRACTION SYSTEMS

As illustrated in Figures 6 and 7 where the center of the extraction well is distance X_O and of the recharge well is distance $- X_O$ from the origin on the X axis, the flow rate is assumed to be Q for the extraction well and $-Q$ for the extraction well and $-Q$ for the injection well.

For recharge into a confined aquifer:

$$h = \frac{Q}{4 \pi Kb} \ln \frac{(x-x_O)^2 \times y^2}{(x+x_O)^2 + y^2} + H \tag{4}$$

where
h = piezometric head after Q is pumped,
Q = flow rate,
b = thickness of the aquifer,
K = hydraulic conductivity or coefficient of permeability, and
H = initial piezometric head.

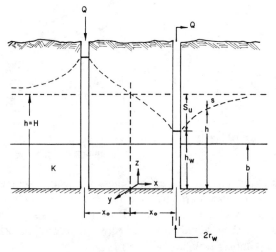

Figure 6. Well and recharge well (confined aquifer).

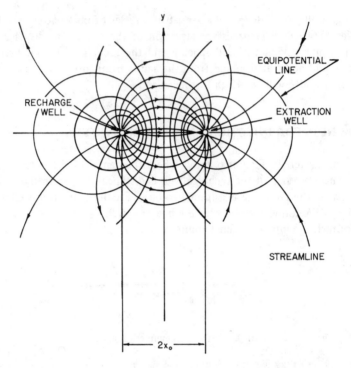

Figure 7. Streamlines and equipotential lines for system of well and recharge well.

The drawdown at the discharging well, S_w, is:

$$S_w = H - h_w \tag{5}$$

where: $H - h_w = \dfrac{Q}{4\pi Kb} \ln \dfrac{(2x_o - rw)^2}{r_w}$ or $\dfrac{Q}{2\pi Kb} \ln \dfrac{2x_o}{r_w}$ (6)

where r_w = actual radius of the pumped well, and
h_w = head of the pumped well corresponding to r_w.

These equations demonstrate the effectiveness of injection wells in preventing inland movement of sea water. Since injection systems are hydraulically equivalent (although flow direction is reversed) to pumping trough systems, it can be concluded that they would also be effective in preventing reclaimed water from passing a pumping trough.

Steady radial flow to a well penetrating a confined aquifer may be described by the Thiem equation as follows:

$$H-h = \frac{Q}{2\pi Kb} \ln \frac{r_e}{r} \tag{7}$$

where r_e = radius of influence, and
        r = radius emanating from the center of the extraction well.

This Thiem equation utilizes the mechanics of steady radial flow to a well and is also known as the equilibrium equation. It is based on the assumption that flow has radial symmetry and that the head must be constant along the perimeter of any circle concentric with the well. Hence, from Figure 8:

$$Q = K \frac{dh}{dr} 2 rb \tag{8}$$

where r = radius of the flow.

A comparison between the Thiem equation and the equation describing drawdown for an extraction well in an injection/extraction system shows that the extraction drawdown is equal to the drawdown at the face of a well following the Thiem assumptions of a radius $2x_o$.

Injection systems have been effective in preventing seawater flow across an injection ridge. Since the hydraulic aspects of injection wells are essentially identical to extraction wells, it can be concluded that extraction troughs are effective in preventing flow from passing the trough.

INJECTION WELLS ARE HYDRAULICALLY EQUIVALENT TO EXTRACTION WELLS

Hydrodynamic concepts have their counterparts in groundwater flow, or in this case, in the flow to and from a well. Assuming a recharge well is the source of a cylindrical flow, the stream function (ψ) is:

$$\psi = -\frac{q}{2\pi} \theta + C \tag{9}$$

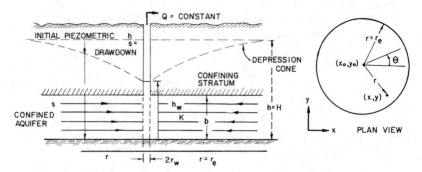

Figure 8. Radial flow to an aquifer completely penetrating a confined aquifer.

and the velocity potential (ϕ) would be:

$$\phi = \frac{q}{2\pi} \ln r + C \tag{10}$$

Assuming that an extraction well is a sink cylindrical flow, the stream function can be written as:

$$\psi = -\frac{q}{2\pi} \theta + C \tag{11}$$

and the velocity potential for the extraction well:

$$\phi = \frac{q}{2\pi} \ln r + C \tag{12}$$

An expression for stream lines may be expressed when $r_1 - r_2$ is constant. The values of θ_1 and θ_2 may be expressed by the geometry of Figure 1. For simplicity, $\tan(\theta_1 - \theta_2)$ equals a constant that is used to derive equations for the stream lines as follows:

$$\tan(\theta_1 - \theta_2) = \frac{\tan\theta_1 - \tan\theta_2}{1 + \tan\theta_1 \tan\theta_2} = \frac{\dfrac{Y}{x - x_o} - \dfrac{Y}{x - x_o}}{1 + \dfrac{y^2}{x^2 - x_o^2}} = \text{constant} = n \tag{13}$$

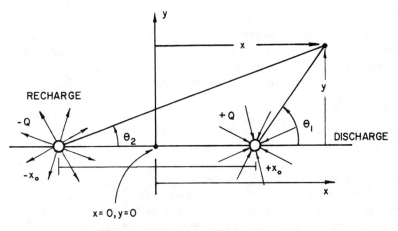

Figure 9. Extraction well and recharge well.

The above equation may be rearranged to represent the circles in Figure 9 with their center at $(0, x_0/n)$ and radius $x_0/1 + (1/n^2)$. The equation of these circles would be:

$$x^2 + y^2 - 2y \frac{x_0}{n} - x_0^2 = 0 \tag{14}$$

These equations are therefore representative of an extraction well and recharge well of equal status.

Review of injection/extraction equations shows that the character of water is not a factor in the hydraulics of the system. Since seawater movement, which has substantial character differences from native ground waters, is controlled, it is concluded that movement of reclaimed water would be effectively controlled also.

CLOGGING CONSIDERATIONS

Discharge equations for pumping and recharge wells are functionally and theoretically similar. Differences in practice are due to clogging of the recharge wells. Periodic surging and pumping of recharge wells was found to be necessary to clear the wells of fine particles which clog casing perforations and gravel packs with resultant high injection heads. Long-term clogging differs from initial high rate clogging, which usually occurs within the first few months of operation due to gravel pack instability.

High injection rates in relation to the transmissivity of an aquifer unit usually result in initial clogging problems. Mechanisms of long-term clogging have been attributed to corrosion products of distribution and supply systems, bacterial growth of sulphate-reducing bacteria, chemical precipitation on sand and gravel particles of calcium carbonate, and physical deposition or rearrangement of particles.

The solution to the problem of clogging has been approached in two ways: by physical redevelopment of the injection well or by chemical conditioning of the injected water. Both alternatives are expensive. Redevelopment becomes necessary when excessive pressure on a well requires substantial reduction of flow rate. The redevelopment process has been undergoing rapid improvements resulting in increased efficiency and ease of redevelopment as well as shorter intervals and improved operating characteristics, but redevelopment is usually expensive, making this alternative undesirable.

Chlorination of injection waters prior to injection in recharge wells has been a common treatment for clogging. Until 1958, chlorination rates of up to 8 mg/l were maintained. However, evidence of the corrosive effects of chlorine caused levels to be changed. Prevention of rapid bacterial growth now is considered possible by operating with a 1.5 mg/l chlorine residual. Another approach, shock chlorination, consists of adding high levels of chlorine during a limited period of time rather than maintaining a constant low level. But shock chlorination is regressive; its effect declines with each application until there is no improvement. Treatment with acid, however, has the opposite impact. Continuous application of acid is not as effective in reducing injection head as shock dosages. Reduction of pH by acid results in the inhibition of chemical precipitation by adjustment of the saturation index.

Redevelopment of wells may not be necessary where water is highly treated, to the extent that the clogging potentials are eliminated. An example of this method is used in Orange County, California, where reclaimed water is injected into wells after treatment and blending. During its entire injection phase of operation, no redevelopment of wells has been required; hence, clogging has not been a problem at this facility [4].

Improvements in redevelopment techniques and design has resulted in a lowering of costs for the process but redevelopment is no longer a significant factor. Information concerning clogging and clogging mechanisms has made a major impact on quality characteristics necessary to control clogging in injection wells.

RECLAIMED WATER CHARACTERISTICS

Wastewater reclamation technology has now advanced to the point where water can be reclaimed to almost any quality desired.

Water characteristics which may affect injection capacities and agricultural crops include Langelier Index (scaling), turbidity (clogging), total bacterial (slime growths), oxygen concentration (corrosion), hydrogen sulfide concentration (corrosion), particle size and distribution (clogging), and total dissolved minerals (potential irrigation use).

Reclaimed water quality requirements for injection in a groundwater basin must meet the Basin Plan groundwater objectives. In order to meet these objectives, the reclaimed water must be adequately disinfected and oxidized with concentration levels not exeeding 1200 mg/l total dissolved solids, 150 mg/l chlorides, 600 mg/l sulfate and 1.5 mg/l boron.

To produce a reclaimed water of suitable quality for injection, various advanced process combinations are being used. These include: (1) multimedia filtration followed by ozonation; (2) multimedia filtration, reverse osmosis, and chlorination; (3) chlorination, diatomaceous earth filtration, and ozonation; and (4) coagulation, multimedia filtration, reverse osmosis, and chlorination.

The process train determined to be most cost effective, and technically feasible includes coagulation, multimedia filtration, reverse osmosis, and chlorination [7,8].

METHODOLOGY OF INJECTION WELL SYSTEM

An injection well system requires the creation and maintenance of a reclaimed water ridge above sea level to act as a barrier to seawater intrusion. Extraction is not part of this system; consequently, frequent well rehabilitation may be necessary to maintain required injection levels. This process provides little control of reclaimed water.

SUMMARY

A primary benefit of using reclaimed wastewater to control seawater intrusion would be increased use of groundwater storage. The beneficial effects of the increased use of groundwater storage include: use and storage of surplus water, possible improvement of groundwater quality by recharging with good quality surface water, and postponement of the need to construct additional surface storage facilities.

Groundwater storage has several advantages when compared with surface storage reservoirs. There is an elimination of the evaporation that is a significant fact in surface storage reservoirs. Biological quality problems are minimized because of the absence of light required for aquatic growth. In addition, groundwater storage is less subject to damage through sabotage and spills than is surface storage.

The use of reclaimed water as a hydraulic barrier to stabilize and repulse the seawater intrusion front is a system for conserving water resources. Disposal of treated effluent to the ocean is a water wasting method in areas where the water supply is limited.

REFERENCES

1. "Santa Ana Gap Salinity Barrier, Orange County." State of California Department of Water Resources, Groundwater Basin Protection Projects, Bulletin No. 147-1 (1966).
2. Newport, B. D., and R. S. Kerr. "Saltwater Intrusion in the United States," U.S. EPA-600 8-77-011. Research Reporting Series (1977).
3. "History of Seawater Intrusion on the Oxnard Plain," Ventura County Public Works Agency, Flood Control District, Task 3, Subtask B (Part 3) of 208 Areawide Waste Treatment Management Planning Study (1977).
4. "The Annual Report," Orange County Water District (1978).
5. Todd, D. K. *Groundwater Hydrology* (New York: John Wiley and Sons, Inc., 1959).
6. DeWiest, R. J. M. *Geohydrology* (New York: John Wiley and Sons, Inc., 1965).
7. PRC Toups Corporation. "Oxnard Wastewater Reclamation Facilities Plan," VRCSD (1979).
8. Wedding, J. J., M. A. Fong and C. M. Crafts. "Use of Reclaimed Wastewater to Operate a Seawater Intrusion Control Barrier," *Proceedings of the 1979 AWWA Water Reuse Symposium* (Denver, CO: American Water Works Association, 1979).

CHAPTER 22

PROBLEMS OF WATER REUSE
IN COKE PRODUCTION

Richard G. Luthy
Department of Civil Engineering
Carnegie-Mellon University
Pittsburgh, Pennsylvania

This chapter discusses water use and wastewater management for coke making and allied operations. Wastewater treatment technologies and alternatives are described. The state-of-the-art for feasibility of water reuse in coal coking and related manufacturing processes is summarized.

COKE MANUFACTURING

Coke manufacturing is an integral segment of the iron making process. Coke is consumed as a basic raw material in the blast furnace wherein the carbon acts as fuel and as a reducing agent for the iron ore. There are two generally accepted processes for manufacturing: the beehive, or nonrecovery process, and the by-product, or chemical recovery process.

In the beehive process air is admitted to the coking chamber in order to achieve controlled burning and distillation of coal-derived volatile components. This process results in loss of all the gaseous and liquid by-products. Consequently beehive coke manufacturing is very dirty; it discharges combustion gases through the charging hole vent, and it releases smoke to the atmosphere when the coking process is completed and the coke oven doors are opened. The beehive coke-making process reached its peak production in 1916 [1]. Since that time the by-product process has largely supplanted the beehive operations. Today the by-product process accounts for more than 99% of all coke production. Even though economic factors governing by-product recovery have changed since the plants were originally installed, coke oven gas remains a valuable by-product for use in coke making and in blast furnace operations. The recovery of light oils, phenolics, tars and ammonia are not usually considered profitable [2]. However, the profitability of recovery of these by-products may change as the result of dramatic increases in the prices of petroleum and natural gas as petrochemical feed stocks.

In the by-product process coke is produced by the destructive distillation of bituminous coal in the absence of air or added water at temperatures as high as 1100°C (2000°F). This process drives off volatile components which are withdrawn from the coke oven under suction and are condensed in the gas main and primary coolers. Tar is decanted from the condensate, and the wastewater which results is called flushing liquor or waste ammonia liquor. This liquor is a major source of wastewater from coke production; other important wastewater streams result from processes to clean the coke oven gas and to recover ammonia, light oil and other products. Specific details on the by-product coke manufacturing process are provided by Denig [1]. Information on processing of coke oven gas and by-product recovery is given in Volume II of *Chemistry of Coal Utilization* [3-5].

Figure 1 shows a schematic of a by-product coke plant process flow diagram [2] including estimates of wastewater discharge flows expressed on a unit mass emission basis, e.g., kg water/kkg coke. Figure 2 shows a schematic

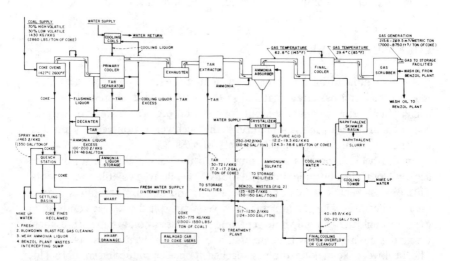

Figure 1. Flow diagram for by-product coke manufacturing process (redrawn from Ref. 2).

of a by-product coke plant light oil recovery and process flow diagram [2].
Major wastewater streams from by-product coking and light oil recovery are
waste ammonia liquor, ammonium sulfate crystallizer effluent, blowdown
from the final cooler and blowdown from the benzol plant [6,7].

By-product coke plant wastewaters vary in composition depending on the
nature of the coking operation and extent of by-product recovery and types
of products recovered. Table I lists general characteristics of coke plant waste
ammonia liquor and compares these values with the characteristics of ammonia still effluent used in various biological oxidation studies. Significant
pollutants in coke plant wastewater include ammonia, BOD, cyanide, phenolics, solvent extractable material (oil), sulfide, thiocyanate, as well as various
acid, base and neutral hydrocarbons including derivatives of benzene, pyridine and polycyclic aromatic hydrocarbons.

The problem of thiocyanate in coke plant wastewater has received insufficient attention in the past [6]. Data in Table I show that thiocyanate is typically in the range of several hundred mg/l or more in coke plant ammonia
still liquor, and that this stream may contain more nitrogen-equivalent in
thiocyanate than in ammonia. Recent investigations [11] have demonstrated
the importance of following the fate of thiocyanate-nitrogen in obtaining a
correct understanding of nitrogen transformations which may occur during
treatment of coke plant effluents.

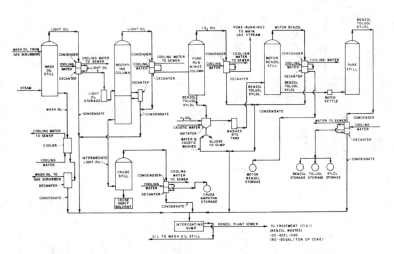

Figure 2. Flow diagram for by-product coke plant light oil recovery and refining (redrawn from Ref. 2).

Table I. Comparison of Coke Plant Effluent Characteristics

Parameter	Waste Ammonia Liquor (Flushing Liquor)[a]	Characteristics of Feeds Used in Biological Oxidation Studies			
		Luthy and Jones [11][b]	Adams et al. [12][c]	Ganczarczyk and Elion [13][d]	Fisher et al. [14][e]
BOD (mg/l)	--	1700-3500	1330	--	4200
COD (mg/l)	2500-10,000	3400-5700	3200	--	6000
Phenolics (mg/l)	400-3000	620-1150	1135	219-288	1150
NH_3-N (mg/l)	1800-6500	22-100	86	97-123	300
NO_3^--N (mg/l)	--	<0.2	--	--	--
Organic-N (mg/l)	--	21-27	260	--	--
P (mg/l)	<1	0.9	0.44	--	--
CN^-, Total (mg/l)	10-100	2-6	4	--	25
SCN^- (mg/l)	100-1500	230-590	--	194-223	550
S^{2-} (mg/l)	200-600	8	--	--	--
SO_4^{2-} (mg/l)	--	325-350	--	--	--
Solvent Extractable Material, Total (mg/l)	100-240	--	25		170
Alkalinity (as $CaCO_3$)	3800-4300	525-920	370	--	--
Conductivity (μmhos/cm)	--	3500-6000	--	--	--
pH (units)	7.5-9.1	9.3-9.8	--	9.0-9.6	10.7

[a] Data Sources: References 2, 6, 8-10.

[b] Ammonia still effluent comprised of flushing and cooling liquors, and tar and benzene plant effluents.

[c] Plant wastewater comprised of stripped flushing liquor, light oil separator underflow, and final cooler and boiler blowdowns. Soluble BOD and COD values reported here.

[d] Stripped wastewater comprised of excess flushing liquor; and collection mains, primary coolers, and light oil plant condensates.

[e] Ammonia still effluent. This waste was diluted to 25% strength prior to biological treatment.

COKE PLANT WATER APPLICATION
AND DISCHARGE RATES

Water use and wastewater loads resulting from coke production vary according to the nature of the process, water use systems, moisture and volatility of the coal, and carbonization temperatures in the coke ovens [2]. These factors result in variance of quality and quantity of coke plant wastewaters.

Water application requirements include water for coke quenching and water for processing coke oven gases. Coke quenching is generally practiced such that quench water is recycled in a closed loop with no discharge. The amount of water applied during coke quenching operations has been estimated to range from 1460 to 6250 l/kkg (350-1500 gal/ton) in which approximately 35% of this water is evaporated by the hot coke and discharged as a steam cloud.

Water application rates for this and other operations in a coke plant are given in Table II. It is estimated that approximately 70-95% of this total is used for indirect cooling and for condensing of steam. Thus the majority of this water receives no contamination from coke production.

Wastewater flow from by-product coal coking operations have been estimated to be in the range of 517-1250 l/kkg (124-300 gal/ton) for plants employing the semidirect ammonia recovery process, indirect (noncontact) cooling, well-designed final cooler recirculation systems and cooling water recirculation in light oil recovery. Effluent limitation guidelines for best practicable control technology currently available (BPCTA) were established on an effluent flow basis of 730 l/kkg (175 gal/ton). Plants operating a gas desulfurization facility, such as potash or soda ash scrubbing, were estimated to generate an additional 104 l/kkg (25 gal/ton) of wastewater. Effluent limitations guidelines for best available technology currently available (BATCA) were based on effluent discharge of 417 l/kkg (100 gal/ton). Three plants achieved this average by employing internal recycle followed by minimal blowdown on such systems as the barometric condenser and final cooler waters.

Table II. Water Application Rates for Coke Plants [7]

Location	Water Application Quantities	
	(l/kkg)	(gal/ton)
Primary coolers	6250-18750	1500-4500
Final coolers	2100-8330	500-2000
Benzol plant	2100-6250	500-1500
Desulfurization plant	2100-8330	500-2000
Total	14600-47900	3500-11500

COKE PLANT WASTEWATER TREATMENT

Effluent limitation guidelines for the by-product coke category of the iron and steel industry [2,15] are summarized in Table III. An assessment of effluent treatment technology required to satisfy these guidelines was made by Wong-Chong et al. [6]. The assessment examined: (1) ammonia removal by steam distillation, (2) physico-chemical dephenolization and (3) biological treatment for control of ammonia, cyanide, phenol and sulfide.

Although solvent extraction and vapor recirculation have been used for physico-chemical dephenolization of coke plant wastewaters, it was concluded [6] that current practice could not reliably and consistently reduce phenol to acceptably low levels for discharge. Furthermore, phenol recovery processes are not presently considered economical for producing a salable by-product from treatment of coke plant effluents. For this reason and because the combination of ammonia recovery by steam distillation in conjunction with physico-chemical dephenolization may still leave residual pollutant concentrations too high for discharge, there has been increased reliance on in-

Table III. Effluent Limitation Guidelines for By-Product Coke Subcategory of the Iron and Steel Industry [2]

Pollutant Parameter	Discharge Load [a] (kg/kkg; lb/1000 lb)	Discharge Conc. [b] (mg/l)
1977 BPCTCA Limits		
BOD$_5$	0.1095	150
Cyanide, Total	0.0219	30
Phenol	0.0015	2
Ammonia (as NH$_3$)	0.0912	125
Oil and Grease	0.0109	15
Suspended Solids	0.0365	50
pH (units)	6.0-9.0	
1983 BATEA Limits		
BOD$_5$	0.0083	20
Cyanide, Amenable	0.00010	0.25
Phenol	0.00021	0.5
Ammonia (as NH$_3$)	0.0042	10
Sulfide	0.0012	0.3
Oil and Grease	0.0042	10
Suspended Solids	0.0042	10
pH (units)	6.0-9.0	

[a]The discharge load limits are 30 consecutive days averages, with a one day maximum permissible discharge of 3 times the average.

[b]The concentrations are based on effluent flows of 730 l/kkg (175 gal/ton) of coke (BPCTCA) and 417 l/kkg (100 gal/ton) of coke BATEA.

plant biological treatment systems to oxidize coke plant wastewater contaminants.

Much of the currently available published research on investigation of treatment of coke plant effluents has addressed issues associated with aerobic biological oxidation of phenolic wastes. A significant amount of earlier work [16,17] evaluated treatment of coke plant and gas works wastes by dilution with municipal sewage. In recent years laboratory and pilot studies have been performed with coke plant wastes in order to determine appropriate criteria for rational, nonempirical design of biological systems for treatment of coke plant effluents without blending with municipal wastewater.

Kostenbader and Flecksteiner [9] presented results of pilot and full-scale treatment studies on a coke plant waste and reported that an ammonia concentration greater than 2000 mg/l will inhibit biological treatment, a claim which has been contradicted by others. In a continuous flow-through system, acceptable influent ammonia concentrations are dependent on the permissible concentration which can be sustained in the mixed liquor. Biological decomposition of thiocyanate and nitrification reactions can cause mixed liquor ammonia concentrations to be significantly different from influent values. Other factors which impact acceptable ammonia levels include mixed liquor pH value and the presence of potentially inhibitory compounds [18].

Barker et al. [19] evaluated biological removal of carbonaceous and nitrogenous compounds in coke plant wastewater. A three-stage biological treatment process was used for the respective removal of carbonaceous compounds, oxidation of ammonia to nitrate (nitrification) and removal of nitrate (denitrification), and it was found that substantial nitrification was difficult to achieve and that insufficient information existed about the control of factors affecting nitrification in coke plant effluents. Biological oxidation studies have been performed [20] on coke plant evaporative condensate containing phenol and ammonia but little other contamination. An important finding from this study was that single-stage nitrification may be feasible as long as the food-to-microorganism ratio, expressed as mg BOD applied per day to mg MLVSS (mixed liquor volatile suspended solids), was on the order of 0.2 or less. However, it must be recognized that this waste was an evaporative condensate and not a typical effluent. Studies on coke plant effluents [12] showed that ammonia increased as a result of biological treatment. It was concluded that nitrogen may be the most difficult parameter to control in the treatment of coke plant effluents.

It has been proposed [19,21,22] that nitrification of coke plant effluents requires multistage treatment. Two-stage activated sludge treatment studies [23] showed 90% ammonia oxidation of second-stage influent. It was believed that two-stage treatment offered greater buffering capacity against fluctuations in effluent characteristics. Luthy and Jones [11] showed that single-stage nitrification of coke plant wastewater was feasible and that it was possible to oxidize ammonia- and cyanogen-nitrogen at COD removal rates less than approximately 0.2 mg COD removed/mg MLVSS-day.

It has been found that thiocyanate removal is erratic and not correlated with any specific treatment parameter [9,19,24], although thiocyanate removals of 90-99% from influent values of 280-550 mg/l SCN⁻ have been achieved in controlled laboratory experiments [11].

Extensive research [25-27] performed over a number of years has revealed the complex nature of biological treatment of phenolic effluents. It has been found that the ease of treatment of these effluents shows little or no relationship to the strength of the waste in terms of COD or phenol content, but the efficiency of compound degradation appears to be influenced by the presence of inhibitory constituents such as calcium, thiocyanate, cyanide, sulfide, chloride and unidentified components such as oxidized or polymerized phenolic constituents which persist after biological treatment. These constituents may be associated with the characteristic residual color resulting from treatment of phenolic wastes. Also, the addition of certain compounds closely related to pyruvic acid may act as accelerating agents for biological oxidation of nitrogenous constituents of coking wastes [26,27].

It is believed by some [14] that different coke plant wastes exhibit different treatment characteristics. Hence, despite various investigations there remain uncertainties regarding biological oxidation of coke plant wastes. Further development work is needed in order to evaluate general process reliability on a large scale, and to determine process stability under fluctuations of coke plant operations. Treatment of coke plant effluents by free- and fixed-leg ammonia distillation followed by biological oxidation are key unit operations in meeting BAT effluents limitation guidelines. However, problems with residual ammonia levels, complexed cyanide species and priority organic pollutants will necessitate improved treatment technology.

FEASIBILITY OF COKE PLANT WATER REUSE

In recent years there has been increased emphasis on evaluation of reuse of wastewaters resulting from iron and steel making. Much of the interest in water reuse has been concerned with wastewaters other than coke plant effluents. The primary reason for this is that coke plant wastewater is the most heavily contaminated of all effluents produced in an integrated steel mill. Hence, water reuse studies have focused on treatment of less contaminated wastewaters where the general concept is that water cascades through various systems with the blowdown from one system supplying the next. The systems are sequenced in order of quality with water of highest quality serving as supply to the system of next highest quality. Intermediate treatment steps between these systems may typically be comprised of physico-chemical processes to neutralize acidic or alkaline streams, to precipitate soluble metals, or to remove suspended or emulsified material. Neither these treatment steps nor treatment required to satisfy coke plant BAT effluent discharge standards would have significant effect on removal of principal inorganic salts in coke plant wastewaters (e.g., Na^+, Ca^{2+}, Cl^-, SO_4^{2-}).

High concentrations of residual salts in coke plant effluents have an especially adverse impact on the acceptability of treated coke plant wastewater for reuse. The problem of high chloride ion concentration is of particular concern. Chloride ion concentration in coke plant ammonia liquors is in the range of 2300-6000 mg/l [19,28]. This high chloride ion concentration can be corrosive to steel. Thus, even in the absence of other contaminants, the high salt concentration restricts the water reuse possibilities of coke plant ammonia liquor in applications such as closed-loop water-to-water heat exchanges, cooling water for rolling mills, and water for open hearth and basic oxygen steel making furnaces [7,21]. Excessive chloride loadings may deteriorate structural components of the blast furnace [7]. Hence, coke plant wastewater may be unsuitable for use as makeup water for tight-water recycle blast furnace scrubbers. This is significant because blast furnace scrubbers handle the lowest quality water next to coke plant effluent.

In view of these issues, it is not surprising that conversations with environmental engineers at various coke and iron manufacturing facilities revealed that reuse of coke plant wastewater is perceived to be a very challenging technological problem. Opinions were expressed that problems related to coke plant wastewater reuse may be surmounted with the application of advanced wastewater technologies such as reverse osmosis or evaporation. However, it was believed that this approach would require much development and that the end result would involve large capital expenditures, significant operating costs and substantial energy consumption. Furthermore, the approach would still leave a problem with residual salt disposal.

Opinions were expressed which questioned the need for reuse of effluents containing high concentrations of dissolved solids, such as coke plant wastewater. Since most steel plants are located in the eastern half of the United States where water is abundant, it was speculated that there may not be sufficient motivation for investing heavily in equipment which would facilitate reuse of these types of effluents. Capital expenditures of this nature may be better spent on facilities which promote reuse of less contaminated effluents or on other types of pollution abatement equipment.

One steel company has been considering tests to evaluate the feasibility of using coke plant wastewater as makeup to the blast furnace scrubbers [29, 30]. This is believed to be feasible if the blast furnaces are operated on a very tight recycle with absolutely no oxygen in the blast furnace scrubber system. Tight water recycle would result in lower scrubber water pH values as the result of carbon dioxide absorption and thereby would help control scale formation. The absence of oxygen would retard corrosion of steel in the presence of chloride. This concept would be difficult to evaluate at the laboratory-scale, and thus will require field evaluation. Nonetheless, the problem of disposal of residual brines from the coke plant would not be solved, for there would still be a blowdown stream to handle from the blast furnace scrubber system.

In summary, there appears to be little opportunity for reuse of coke plant effluents in iron and steel making other than by processing through extensive treatment to reduce dissolved solids concentration. Treatment technology to achieve this objective is discussed later in this chapter. Coke quenching is a major consumptive use of water in the coke making process, and issues and problems associated with use of wastewater for coke quenching are discussed below.

DISPOSAL OF COKE PLANT EFFLUENT
BY QUENCHING HOT COKE

It is possible to dispose completely of coke plant ammonia liquor by using this wastewater as makeup at the coke quench station. The EPA has estimated [2] that approximately 85% of the by-product coke plants use the semi-direct ammonia recovery process and that wastewater discharge at these plants may be estimated at 417 l/kg of coke (100 gal/ton) by employing tight recycle on the barometric condensor and final cooler, and by eliminating noncontact cooling water. This flow could be consumed by quenching, since 417-625 l/kkg (100-150 gal/ton) of water is evaporated during the quenching operation.

Evaporation of coke plant wastewater by quenching hot coke had been widely used in the past as a means of disposing of coke plant wastewater [16, 31]. Concern for protection of the Ohio River from coke plant wastes led to the Ohio River Interstate Stream Conservation Agreement of 1929, which emphasized the need for complete elimination of wastes containing phenolics and tarry substances. The Agreement proposed to eliminate discharge of phenolic wastes by coke quenching. A similar recommendation was adopted by the state of Pennsylvania.

In more recent years, disposal of coke plant effluents by quenching has been judged undesirable because of corrosion problems in the quench tower, quench cars and associated equipment [19], and because of air pollution problems which result from use of contaminated quench water. In 1970, concern for air pollution control led to the ban in Allegheny County, Pennsylvania, on use of any wastewater for coke quenching.

Abnormal rusting was experienced at U.S. Steel's Clariton Works, the world's largest coke plant, when waste ammonia liquor was used for quenching [28]. This was attributed to the action of calcium chloride and/or hydrochloric acid in the plume of steam emitted from coke quenching. Mists containing oxides of sulfur also could contribute to the corrosion problem. It has been suggested [28] that coke quenching with raw liquors produces coke with less desirable properties. However, experiments on the use of diluted ammonia liquors for quench showed no harm to the coke for metallurgical, water gas or domestic purposes [32].

It had been supposed once that quenching would destroy pollutants when contacted with hot, incandescent coke. However, instead of being destroyed, volatile components, such as phenol, cyanide, and ammonia, were simply steam distilled and discharged to the atmosphere [33,34].

Experiences in the Federal Republic of Germany with reuse of coke plant effluent showed that although both the high salt content and the odor of biologically treated effluent restrict its reuse possibilities, treated wastewater has been used successfully at various locations for coke quenching [21]. It was claimed that reduction in coke quality would occur if waste ammonia liquor was not effectively stripped of ammonia; otherwise the quenched coke would be covered with a greyish-white coating from which ammonia vapors would gradually evolve. This problem was overcome by use of a two-stage quenching process, or by use of ammonia-stripped biologically treated wastewater. In the two-stage quenching process, the coke is first partially quenched with contaminated effluent. During a 1-2 minute pause in the quench operation, the adsorbed materials are evaporated, and the quenching process is completed using fresh water. Biologically treated effluent has also been used in conjunction with storm waste runoff to quench coke. The wastewater was processed through multistage biological treatment units and then treated by chemical precipitation. Hence, experience in Germany indicates that suitably treated coke plant effluent can be used as coke quench water, thereby reducing freshwater requirements and the amount of effluent for disposal.

Recently, the EPA sponsored a field study at U.S. Steel's Lorain Works to assess quench system emissions when quenching with fresh water and contaminated wastewater as makeup [35]. Initial conclusions from these tests indicated that quench towers may emit more particulate matter than previously documented, and that quench tower particulate emissions are dependent on quench water dissolved solids content.

It is probable that conclusions drawn from the Lorain sampling effort will not resolve controversy regarding air pollution resulting from coke quenching because of the technical difficulties in accurately assessing quench plume composition, and in interpreting the results of such measurements. The coke quenching process creates a short, violent rush of steam, and plume sampling strategies require verification in order to guarantee high efficiencies for capture of both large and small particles, and to ensure true isokinetic sampling with no bounceoff from cyclones used for particulate and droplet sizing. The Lorain study did not allow for careful discrimination between the nature of the material being evolved in the steam plume, nor for component mass balances around the quenching operation. Additional work needs to be performed in order to determine the size fraction and source of material being evolved in the plume. This is necessary to permit differentiation between particulates resulting from thermal fracturing of incandescent coke, salts evolving from the spray water, and gases being emitted from the coke and water.

CROSS MEDIA ANALYSIS OF COKE PLANT EFFLUENT MANAGEMENT

A systematic analysis of the problem of coke quenching has been made by Dunlap and McMichael [36]. They employed the methodology of cross media analysis which assesses environmental strategies by considering environmental impact to the land, water and air collectively. The analysis also inventories utility requirements and takes into account emissions resulting from use of utilities. This latter issue is important because in general, primary emissions from an industrial process can only be controlled completely at the expense of high secondary emissions from utility production.

The methodology of cross media analysis was developed at Battelle Institute by Reiquam et al. [37]. The procedure does not give absolute answers, but it does provide a logical basis for comparing pollution control strategies [38]. The analysis calculates mass emissions of pollutants to air, land and water. Relative weights are assigned for each receiving media, and hierarchical rankings are given for each pollutant. The environmental impact is given as the sum of damage due to adjusted mass emission to all media. The results of this procedure provide a means of rank ordering various control strategies relative to a base strategy. The base level may be taken as that which results from no environmental controls. The analysis is repeated to evaluate sensitivity of the final ranking to the relative weights to media, the weighting of individual pollutants, and parameters associated with the control strategies.

The environmental degradation index (EDI) is the weighted sum of the relative damage of all pollutants to air, land and water:

$$EDI_s = \Sigma_p d_p W_p \tag{1}$$

where d_p is the relative damage of pollutant p under strategy s. The weighting function W_p distributes relative weighting to each individual pollutant in each media. This weighting function is a relative measure of the impact of a pollutant to all others in a given media. Values of EDI_s may be compared with the case of no environmental control, EDI_o. The parameter:

$$\frac{EDI_o - EDI_s}{EDI_o} \times 100 \tag{2}$$

gives the percentage improvement (positive value) or degradation (negative value) for a given control strategy relative to the base case of no control.

Table III lists the control strategies analyzed by Dunlap and McMichael [36,39] for coke plant effluent treatment strategies. Computation of EDIs for each strategy under varying assumptions is relatively straightforward, but the problem becomes complex through inventory of emissions of pollutants to each media for each strategy. Results from this analysis are displayed as a preference plot for comparison of various strategies. In general, for very low water medium values, no control strategy is preferred; for high water medium values, a very high degree of control is preferred.

A thorough analysis of the coke plant waste disposal problem requires that a number of preference plots be generated in order to assess the sensitivity of change of EDI to variables in the cross media analysis. Some observations from the complete analysis were [39] :

1. Quenching with coke plant effluent is generally the preferred strategy regardless of the level of wastewater treatment. This is preferred over discharge to a watercourse.
2. Tight recycle is preferred over loose recycle.

Table IV. Cross Media Analysis of Coke Plant Effluent Treatment Strategies [36,39]

Strategy	Recycle[a]	Wastewater Treatment[b]	Effluent Discharge
0	Loose	None	To watercourse
1	Tight	None	To watercourse
2	Tight	Level I	To watercourse
3	Tight	Level II	To watercourse
4	Loose	None	To quench
5	Tight	None	To quench
6	Tight	Level I	To quench
7	Tight	Level II	To quench

[a]The volumetric flow of blowdown is reduced by recycle from both the final cooler and the benzol plant.

[b]Level I treatment is wholly physico-chemical with cyanide stripping, ammonia distillation and phenol extraction. It approximates EPA BPCTCA guidelines. Level II treatment consists of cyanide stripping, ammonia distillation and biological oxidation. It approximates EPA BATEA guidelines.

3. Level I (physico-chemical) wastewater treatment is preferred over Level II (stripping, distillation, and biological oxidation) wastewater treatment for media weights that appear most reasonable.
4. A high degree of wastewater treatment appears to be preferred only if water medium is viewed as dominant and out of proportion to air and land media.

This analysis is striking in that it suggests that current directions for coke plant water management, i.e., high degree of coke plant wastewater treatment with discharge to a watercourse with clean water quench, are not only less desirable than other approaches, but may cause additional damage to the environment.

Results of cross media analysis will not settle debate on the impact of coke plant wastewater treatment and disposal. However, this approach clearly illustrates the connectivity of air, land and water media [36]. Hence, selection of appropriate control technologies must not emphasize a single medium regulation and solution. Cross media analysis and experience in Germany [21] suggest that the best environmental management strategy may entail a moderate degree of wastewater treatment with wastewater disposal by quenching.

INTEGRATED STEEL MILL TOTAL WATER REUSE

The EPA has sponsored studies in integrated U.S. steel plants to assess the feasibility of achieving total water recycle. One part of this effort consisted of a survey of minimum water quality requirements for different parts of the iron and steel industry. A second, more extensive project prepared conceptual engineering studies to determine facilities needed to achieve total recycle of water at five integrated steel plants.

The survey of minimum water quality requirements attempted to develop data on minimum quality criteria, information needs and necessary research [40]. A major finding from this study was that there is a general lack of knowledge concerning water flow and composition at consumption points. Further, very little information exists to assess accurately the effect of water quality on product quality. In most cases, distributed water is of a few (two to four) basic qualities and managed between clusters of units in close proximity.

Hofstein and Kohlmann [7] studied integrated steel plants to determine facility requirements to achieve total recycle of water. Although there are 50 or more steel plants in the United States, only 18 are classified as integrated in that the plants possess: blast furnaces, coke and by-product plants, sinter plants, steel making which includes BOF, hot forming (primary and secondary), and cold finishing which includes pickling and cold rolling. This list may be reduced to 14 by eliminating from consideration those plants for which at least three basic process elements are not contiguous. Hofstein and

Kohlmann selected 5 of these 14 for detailed study. Several conclusions from this evaluation are discussed below.

There is a lack of typicality among steel plants, and no simplified solution can be developed that would be applicable throughout the industry. However, there are common problems to be encountered at most facilities. These include [7] :

1. Steel plants in the U.S. are from forty to eighty years old and were constructed in response to changes in demand, war requirements or technological change. Many facilities are built on sites previously occupied by older processes.
2. Many sewers in steel plants do not segregate industrial wastewater from storm water. The mere separation of sanitary wastes proved extremely costly when this was performed in the 1950s. Further segregation of industrial effluent would be even more costly and might force facility shutdowns during the construction period.
3. In many instances, production facilities are separated. Often facilities are constrained against expansion of additional support units by geographical terrain, railroad tracks, and the like.
4. Old sewers and sumps are susceptible to groundwater infiltration.
5. Many facilities were partially engineered in the field. There is a lack of information on actual locations of pipes and sewers, and in some cases, pipe sizes.

Among the integrated steel plants investigated [7], the Kaiser Steel Plant at Fontana, California, is closest to maximizing the reuse of water. This plant is somewhat unusual, and its water reuse has been well-publicized. The plant was constructed during World War II on a green field site in a semiarid region with the express intent of minimizing water consumption. Water of highest quality cascades through four systems with blowdown from one system serving supply needs to the system of next lowest quality. Various wastewaters are disposed of in this play by: (1) evaporation from quenching of coke and slag, (2) discharge of pickle liquor to an onsite ferric chloride manufacturer, (3) discharge as bound water in sludges and (4) discharge to a municipal authority.

Table V summarizes general treatment procedures proposed by Hofstein and Kohlmann that may be employed to facilitate steel plant wastewater reuse. Coke plant wastewaters may be handled by ammonia stripping, biological oxidation and carbon adsorption. It was judged that demineralization of coke plant effluent is required if this wastewater is to be reused in other process facilities.

Hofstein and Kohlmann suggest that blast furnace scrubber water may be used for dilution prior to biological oxidation since it contains many of the same contaminants as coke plant wastewater, but at lower concentrations. This would eliminate the alkaline chlorination process widely employed for treatment of blast furnace wastewater. However, since blast furnace scrubber water contains heavy metals [41] which may be toxic to microorganisms, it would be necessary to conduct bench and pilot tests to evaluate this concept.

Table V. Procedures to Maximize Water Quality for Reuse [7]

Procedure	Facility or Type of Wastewater
Improve water recycle in production facility or reduce water use	Blast furnace gas cleaning Pickling rinse Hot forming
Regeneration	Acid at pickling Chrome Plating
Filtration, SS removal	Virtually all wastes
Ultrafiltration	Preceding all membrane treatment processes
Cooling	All noncontact cooling waters and some contact waters
Biological treatment	Coke plant wastes Blast furnace gas cleaning wastes
Carbon adsorption	Coke plant wastes
Chemical treatment	Oil wastes Between successive membrane processes Ash sluice recycle Blast furnace gas cleaning wastes
Membrane treatment	All wastes with high dissolved solids concentrations

Technologies available for removal of inorganic dissolved solids from coke plant wastewater include: evaporation, electrodialysis, reverse osmosis and ion exchange. Analysis showed [7] that total evaporation was approximately three times as costly as membrane or ion exchange processes with evaporation of residual brines. Although the capital costs of membrane systems are more than ion exchange, these processes had lower annual operating expenses and required less land. Hofstein and Kohlmann decided that reverse osmosis was preferred over electrodialysis for dissolved solids removal because reverse osmosis had a broader technical base for treatment of industrial effluents [42-44]. The use of reverse osmosis assumed that the membranes were not subject to deterioration or disintegration due to contact with low concentrations of organic compounds. The residual waste stream from reverse osmosis was expected to be approximately 25% of the total throughout. The capital costs of a demineralization system including granular media filtration, ultrafiltration, reverse osmosis, evaporators, fuel storage and solids collection was $39,000,000 to treat at a flow rate of 2273 m^3/hr (10,000 gpm). Annual operating costs including amortization and solids disposal was estimated at $44,500,000. This size system would handle all blowdown streams and brines containing high levels of total residual solids including coke plant and blast furnace effluents. The cost data do not include in-plant modifications such as sewer construction or installation of basic biological oxidation systems.

Extensive research is required in order to prove the technical feasibility of total wastewater recycle for the iron and steel industry. Areas requiring research include biological oxidation to meet BAT standards, treatment of blast furnace blowdown and treatment of wastewaters to remove dissolved solids. Certain aspects of multistage biological treatment of coke plant effluents, including combined treatment with blast furnace blowdown, will be evaluated by Hofstein and Kohlmann in a follow-on contract with the EPA in 1980. The problem of removal of dissolved solids from coke plant and other process effluents requires thorough study to understand feasibility, to substantiate costs and to evaluate the need.

SUMMARY

High salt content of coke plant wastewater restricts the reuse of this wastewater in other process units in an integrated steel plant. Some form of demineralization is required prior to reuse in order that process units are not corroded and product quality is not impaired. Technologies to achieve demineralization of coke plant effluents are not proven, although reverse osmosis may offer the best possibility on the basis of experience in related industries. Demineralization of coke plant wastewater will be expensive, and it may be argued that those expenditures are not justified for eastern iron and steel manufacturing facilities where there are no severe water shortages. Those monies may be put to better use to solve other pollution abatement problems.

Coke plant wastewater may be disposed by quenching hot incandescent coke. Presently there is controversy regarding the environmental impact of this strategy. Cross media analysis indicates that coke quenching with wastewater is preferred when effects to air, land and water are considered collectively, including impacts of utility requirements. There is need to conduct additional field studies to evaluate the acceptability of using wastewater to quench coke. These studies should evaluate the effects of quenching with various levels of treated wastewater. Such studies need to be designed carefully in order to ensure reliable sampling methodologies and to permit component mass balances around the quench station and reasonable environmental assessment.

NOTE ADDED IN PROOF

The EPA has recently proposed new effluent limitations, pretreatment rules and new source performance standards for iron and steel manufacturing (46 FR 1858, January 7, 1981). Information regarding the basis for the proposed regulations and the supposed impact on coke and iron making is contained in *Development Document for Effluent Limitations Guidelines and Standards for the Iron and Steel Manufacturing Point Source Category, Vols. I and II,* EPA 440/1-80/024-b (1980).

REFERENCES

1. Denig, F. "Industrial Coal Carbonization," Chapter 21 in *Chemistry of Coal Utilization,* H. H. Lowry, Ed. (New York: John Wiley & Sons, Inc., 1945).
2. "Development Document for Effluent Limitation Guidelines and New Source Performance Standards for the Steel Making Segment of the Iron Steel Manufacturing Point Source Category," U.S. EPA Report-440/1-74-024a (1974).
3. Gollmar, H. A. "Removal of Sulfur Compounds from Coal Gas," Chapter 26 in *Chemistry of Coal Utilization,* H. H. Lowry, Ed. (New York: John Wiley & Sons, Inc., 1945).
4. Hill, W. H. "Recovery of Ammonia, Cyanogen, Pyridine, and Other Nitrogenous Compounds from Industrial Gases," Chapter 27 in *Chemistry of Coal Utilization,* H. H. Lowry, Ed. (New York: John Wiley & Sons, Inc., 1945).
5. Glowacki, W. L. "Light Oil from Coke-Oven Gas," Chapter 28 in *Chemistry of Coal Utilization,* H. H. Lowry, Ed. (New York: John Wiley & Sons, Inc., 1945).
6. Wong-Chong, G. M., S. C. Caruso and T. G. Patanlis. "Treatment and Control Technology for Coke Plant Wastewaters," paper presented at the 84th American Institute of Chemical Engineers National Meeting, Atlanta, GA, 1978.
7. Hofstein, H., and H. J. Kohlmann. "Integrated Steel Plant Pollution Study for Total Recycle of Water," U.S. EPA Report-600/2-79-138 (July 1979).
8. Rubin, E. S., and F. C. McMichael. "Impact of Regulation on Coal Conversion Plants," *Environ. Sci. Technol.* 9(2):112 (1975).
9. Kostenbader, P. E., and J. W. Flecksteiner. "Biological Oxidation of Coke Plant Weak Ammonia Liquor," *J. Water Poll. Control Fed.* 41(2): 199-207 (1969).
10. Jablin, R., and G. P. Chanko. "A New Process for Treatment of Coke Plant Wastewater," paper presented at American Institute of Chemical Engineers Meeting: Advances in Water Pollution Control, New York, NY, 1972.
11. Luthy, R. G., and L. D. Jones. "Biological Oxidation of Coke Plant Effluent," *J. Environ. Eng. Div., Am. Soc. Civil Eng.* 106:847-851 (1980).
12. Adams, C. E., R. M. Stein and W. W. Eckenfelder, Jr. "Treatment of Two Coke Plant Wastewaters to Meet Effluent Guideline Criteria," *Proceedings 29th Annual Purdue Industrial Waste Conference* (Ann Arbor, MI: Ann Arbor Science Publishers, Inc., 1974), pp. 864-880.
13. Ganczarczyk, J. J., and D. Elion. "Extended Aeration of Coke Plant Effluents," in *Proceedings of the 33rd Purdue Industrial Waste Conference* (Ann Arbor, MI: Ann Arbor Science Publishers, Inc., 1978), pp. 895-902.
14. Fisher, C. W., R. D. Hepner and G. R. Tallon. "Coke Plant Effluent Treatment Investigations," *Blast Furnace Steel Plant* (May 1970), pp. 315-320.

15. "Iron and Steel Point Source Category. Proposed Effluent Limitations and Standards," *Federal Register,* February 19, 1974.
16. Hodge, W. W. "Waste Disposal Problems in the Coal Mining Industry," in *Industrial Wastes,* W. Rudolfs, Ed., ACS Monograph Series No. 118 (New York: Reinhold Publishing Corp., 1953).
17. Jenkins, S. H., J. A. Slim, G. W. Cook, A. B. Neale and K. Pickett. "The Biological Filtration of Coal Carbonization Plant Effluents," in *Advances in Water Pollution Research,* Vol. 2 (New York: The MacMillan Co., 1964), pp. 159-188.
18. Luthy, R. G., D. J. Sekel and J. T. Tallon. "Biological Treatment of a Synthetic Fuel Wastewater," *J. Environ. Eng. Div., Am. Soc. Civil Eng.* 106(EE3):609-629 (June 1980).
19. Barker, J. E., R. J. Thompson, W. R. Samples and F. C. McMichael. "Biological Removal of Carbon and Nitrogen Compounds from Coke Plant Wastes," U.S. EPA Report-022-73-167 (April 1973).
20. Adams, C. E. "Treatment of High Strength Phenolic and Ammonia Wastestreams by Single and Multi-stage Activated Sludge Process," in *Proceedings of the 29th Annual Purdue Industrial Waste Conference* (Ann Arbor, MI: Ann Arbor Science Publishers, Inc., 1974), pp. 617-630.
21. Bischofsberger, W. "Biological Treatment of Coke Oven Effluents and Reuse of the Waste in the Coke Oven Plant," *Vom Wasser* 15(1):8-13 (1971).
22. Abson, J. W., and K. W. Todhunter. "The Biological Treatment of Carbonization Effluents," *Ind. Chemist,* pp. 303-308 (1958).
23. Ganczarczyk, J. J. "Second-Stage Activated Sludge Treatment of Coke Plant Effluents," *Water Res.* 13:337-342 (1979).
24. Ludberg, J. E., and G. D. Nicks. "Phenols and Thiocyanate Removed from Coke Plant Effluents," *Water Sew. Works,* Industrial Waste Section, IW10 (1969), p. 116.
25. Ashmore, A. G., J. R. Catchpole and R. L. Cooper. "The Biological Treatment of Coke Oven Effluents by the Packed Tower Process," in *The Coke Oven Manager's Yearbook,* Coke Oven Manager's Association (1970), pp. 103-125.
26. Cooper, R. L., and J. R. Catchpole. "Biological Treatment of Phenolic Wastes," *Iron Steel Inst.* 128(27):27 (1970).
27. Catchpole, J. R., and R. L. Cooper. "The Biological Treatment of Carbonization Effluents," *Water Res.* 6:1459-1474 (1972).
28. Wilson, P. J., and J. H. Wells. "Ammonical Liquor," in *Chemistry of Coal Utilization,* Vol. II, H. H. Lowry, Ed. (New York: John Wiley & Sons, 1945).
29. Peck, D. F. Jones and Laughlin Steel Corp. Personal communication (1979).
30. Samples, W. R. Wheeling-Pittsburgh Steel Corp. Personal communication (1979).
31. ORSANCO. "Reducing Phenol Wastes from Coke Plants," report compiled by the Steel Industry Action Committee of the Ohio River Valley Sanitation Commission, Cincinnati, OH (January 1953).
32. Marson, C. B., and H. V. A. Briscoe. "A Note on the Use of Dilute Ammonical Liquor for Coke Quenching," *Fuel Soc. J.* II:152-153 (1932).
33. Tibbetts, C. *Water Poll. Res.* (Brit.) Summary of Current Literature 9(40), (1936).

34. Samples, W. R. "Fate of Phenolics in Coke Quenching," Mellon Institute Report, Carnegie-Mellon University, Pittsburgh, PA (July 1969).
35. Edlund, C., and A. H. Laube. "Effects of Water Quality on Coke Quench Tower Particulate Emissions," U.S. EPA, Office of Enforcement, Washington, DC (April 1977).
36. Dunlap, R. W., and F. C. McMichael. "Reducing Coke Plant Effluent," *Environ. Sci. Technol.* 10(7):654 (1976).
37. Reiquam, H., N. Dee and P. Choi. "Assessing Cross-Media Impacts," *Environ. Sci. Technol.* 9(2):118 (1975).
38. Dunlap, R. W., and R. C. McMichael. "Cross Media Effects of Environmental Regulation: Technology Selection for Optimal Control," paper presented at the Fifth National Symposium of the Air Pollution Control Division, American Society of Mechanical Engineers, Pittsburgh, PA, May 1977.
39. Dunlap, R. W., and F. C. McMichael. "Comparison of Alternative Strategies for Coke Plant Wastewater Disposal," *Polish/U.S. Symposium on Wastewater Treatment and Disposal,* U.S. EPA Report-600/9-76-021 (1976).
40. Bhattacharyya, S. "Process Water Quality Requirements for Iron and Steel Making," IERL, U.S. EPA Report-600/2-79-003 (January 1979).
41. Wong-Chong, G. M., and S. C. Caruso. "Evaluation of EPA Recommended Control Technology for Blast Furnace Wastewater," *J. Environ. Eng. Div., Am. Soc. Civil Eng.* 104(EE2):305-322 (April 1978).
42. Hauch, A. R., and S. Sourirajan. "Performances of Cellulose Acetate Membranes for the Reverse Osmosis Treatment of Hard and Waste Waters," *Environ. Sci. Technol.* 3(12):1269-1275 (1969).
43. Wiley, A. J., et al. "Concentration of Dilute Pulping Waters by Reverse Osmosis and Ultra Filtration," *J. Water Poll. Control Fed.* 42(8):R279-R289 (1970).
44. Williams, R. H., and J. L. Richardson. "Complete Water Reuse with Membranes—Reverse Osmosis for Dissolved Solids Concentration," Proceedings of the Second National Conference on Complete Water Reuse, American Institute of Chemical Engineers, Chicago, IL, May 1975.

CHAPTER 23

INDUSTRIAL REUSE OF WASTEWATER:
QUANTITY, QUALITY AND COST

Gordon P. Treweek
James M. Montgomery, Consulting Engineers, Inc.
Pasadena, California

INTRODUCTION

Two factors will play major roles in the development of municipal waste-water reuse during the next 20 years: the inadequacy of existing freshwater resources in selected areas of the country, and the upgrading of wastewater treatment plant effluents through at least the secondary level of treatment. Because of the shortfall between available freshwater resources and total needs, both water purveyors and lower-quality users are turning to reclaimed wastewater as a reliable water source, thereby freeing available freshwater resources to higher-quality users. The passage of PL 95-217, the Clean Water Act of 1977, shifted the burden of producing a reusable secondary effluent onto the wastewater discharger, thereby reducing the treatment costs to the potential reuser.

This chapter discusses the various needs for industrial water which can be met by reclaimed wastewater. In general, industrial reuse offers two significant advantages over agricultural reuse: (1) industrial users are located close to the supply of wastewater and generally have continuous needs on a 24-hr/day, 365 day/yr basis, whereas agricultural use is distributed both temporally (season of the year) and spatially (outlying regions); (2) many industrial water quality requirements can be met by secondarily-treated wastewater or by adding physical-chemical pretreatment steps to existing secondary facilities.

The water qualities required by potential industrial users are discussed in general terms because of the range of quality requirements within existing industrial categories, and because of the range of process water qualities even within a particular industry. Water quality variables of primary importance to particular industrial applications are presented with the understanding that the determination of reuse potentials for specific industrial applications will require detailed study on a case-by-case basis.

Finally, this chapter considers the costs associated with treating and delivering reclaimed wastewater to the industrial user. The intent is not to identify specific costs for physical items of equipment, which tend to change rapidly with inflation, but rather to identify those areas of cost which must be considered in evaluating the feasibility of adopting reclaimed wastewater use within industry.

Major potential applications for secondarily-treated municipal effluent are: (1) agricultural irrigation, (2) landscape irrigation, (3) steam electric cooling, (4) industrial cooling, (5) other industrial uses and (6) groundwater recharge. Some studies are currently underway for a seventh possible application: direct potable reuse of reclaimed wastewater; however, this use is not considered further here because of the currently unresolved public health aspects of potable use of reclaimed water and because the needs for nonpotable quality water already greatly exceed the available supply of wastewater ([1], p. 2). Of the potential applications for reclaimed wastewater, this chapter concentrates solely on steam electric cooling, industrial cooling, and miscellaneous other industrial (process) uses.

A clear distinction is drawn between the concepts of industrial recycling, which already occurs on a widespread basis, and industrial reuse, which has only begun to recognize its potential. Recycling is the internal use of water by the original user prior to discharge to a treatment system or other point of disposal. In 1975, approximately 139 billion gallons per day (bgd) of fresh water was recycled in the steam electric, manufacturing and minerals industries [2]. This quantity is projected to increase to 865 bgd by the year 2000. The term reuse is applied to wastewaters that are discharged and then withdrawn by a user other than the discharger. Wastewater reuse in 1975 was estimated at 0.7 bgd (0.4% of available wastewater) and is projected to increase to 4.8 bgd by the year 2000 (4.0% of available wastewater). Approximately one-third of this wastewater reuse can be attributed to industrial reuse, either in cooling, process or boiler feed applications (Table I).

Wastewater discharges available for reuse are divided into treated and untreated categories. Treated wastewater includes: (1) secondary effluent from municipal wastewater treatment plants, (2) industrial discharges treated by best available technology, (3) discharges from steam electric plants, and (4) discharges from fish hatcheries. Untreated wastewaters are agricultural

Table I. Existing (1975) Industrial Wastewater Reuse

		Existing Wastewater Reuse (bgd)
Irrigation		0.42
Agricultural	0.20	
Landscape	0.03	
Not Specified	0.19	
Industrial		0.22
Process	0.07	
Cooling	0.14	
Boiler Feed	0.01	
Ground Water Recharge		0.03
Other (recreation, fish & wildlife, etc.)		0.01
Total Wastewater Reuse (1975)		0.68
Total Potential Industrial Reuse (year 2000)[a]		111.1

[a]The potential industrial reuse is more than 500 times greater than current industrial reuse [2].

irrigation return flows that could be collected and reused. Most irrigation return flows which reach streams through groundwater flow or unconcentrated overland flow are not considered available for direct industrial reuse. Emphasis in this chapter is on treated wastewater flows which are readily available for industrial reuse.

WASTEWATER REUSE POTENTIAL

While a great deal of publicity has already attended those industries which utilize reclaimed wastewater, perspective is needed to evaluate the overall potential for reclaimed wastewater use and the magnitude of this reuse in comparison with overall wastewater discharge and wastewater recycling. Existing wastewater reuse in the United States, including irrigation, industrial, groundwater recharge and other reuse applications, is approximately 0.7 bgd. This quantity pales in comparison with total freshwater withdrawals of 362.7 bdg and wastewater discharges of 244.5 bdg ([1], p. 15). As shown in Table II, total United States freshwater withdrawals and wastewater discharges are expected to decrease over the next 20 years, especially in the areas of steam electric power generation and manufacturing. Thus, the requirements for fresh water are decreasing in two areas which have the greatest potential for reclaimed wastewater usage.

In the same vein, two relatively new federal laws: the Safe Drinking Water Act (PL 93-523) and the Clean Water Act (PL 95-217) will have a significant

Table II. Projected Decrease in Total U.S. Freshwater Withdrawals and Wastewater Discharges by the Year 2000 [2]

Area	Water Withdrawal Quantity (bgd) 1975	Water Withdrawal Quantity (bgd) 2000	Percent Change 1975-2000	Wastewater Discharge Quantity (bgd) 1975	Wastewater Discharge Quantity (bgd) 2000	Percent Change 1975-2000
Agriculture	184.7	180.3	-2	84.7	72.0	-15
Steam Electric	89.1	80.1	-10	87.7	69.5	-21
Manufacturing	51.2	19.7	-62	45.2	5.0	-89
Municipal	28.8	37.1	+29	21.4	27.6	+29
Minerals	7.1	11.3	+59	4.9	7.7	+57
Public Lands	1.2	1.7	+42	0.003	0.006	+100
Fish Hatcheries	0.6	0.7	+17	0.6	0.7	+17
Totals	362.7	330.9	-9	244.5	182.5	-25

impact on the quality of wastewater discharges and the extent of wastewater recycling. The Safe Drinking Water Act will indirectly affect the quality of wastewater because many wastewaters are discharged into streams that are also used for public water supplies. The Safe Drinking Water Act will require upstream industries to pretreat their discharges so as to protect the drinking water quality of downstream users. However, once treated, the former wastewater may have new value for internal recycling. The Clean Water Act will also improve the general quality of wastewater through more stringent control of industrial waste discharge. Clearly, major changes are anticipated in the withdrawal of fresh water by industries, and the extent to which that water is then discharged back to the environment.

On a positive note, of the projected 330.9 bgd of total freshwater withdrawal for the year 2000, approximately 290 bgd could be replaced with reclaimed wastewater, at least on a quality basis [2]. However, as shown in Table III, only 121 bgd of wastewater discharge will be readily available for subsequent treatment and reuse. When one subtracts the agricultural irrigation freshwater withdrawal, and the untreated agricultural discharges, one sees that the projected freshwater withdrawals capable of replacement are approximately equal to the treated wastewater discharges available for reuse. Thus, if wastewater discharges from the steam electric, industrial, and municipal sectors (amounting to approximately 110 bgd) were treated to acceptable quality, they could be paired directly with projected freshwater withdrawals for steam electric and industrial users (amounting to 111 bgd). For a variety of reasons, this potential growth in wastewater reuse will not be achieved. Little more than 4.8 bgd of wastewater reuse is projected for the year 2000, or less than 5% of the treated wastewater discharge available for reuse [2].

Table III. Projected Availability of Wastewater Discharge in the Year 2000 [2]

		Quantity (bgd)
Freshwater Withdrawals Capable of Replacement via Wastewater Reuse		290.3
Agricultural irrigation	177.8	
Landscape irrigation	1.4	
Steam electric	80.1	
Industrial cooling	16.9	
Industrial–other	14.1	
Treated Wastewater Discharges Available for Reuse		109.6
Steam electric	69.5	
Industry	12.7	
Municipal	26.7	
Fish hatcheries	0.7	
Untreated Agricultural Discharges for Reuse		11.5
Projected Wastewater Reuse (year 2000)		4.8
Projected Wastewater Recycle (year 2000)		865.5

Projected wastewater recycling will be approximately 865 bgd, or 180 times greater than wastewater reuse.

Three industrial groupings were defined for estimation of wastewater reuse potential: steam electric power industry, manufacturing industries, and mineral industries.

1. Steam Electric Power Industry. Cooling water for steam electric generating plants represents the second largest requirement for fresh water in the United States (89 bgd in 1975 and an estimated 80 bgd in 2000) [2]. Most of the demand is in the eastern United States with the Great Lakes and Ohio regions accounting for over 50% of the total national requirement ([1], p. 20). Both increased recycling and wastewater reuse are expected to supply a significant part of the makeup water. Since salt water is used for steam electric cooling at coastal plants, obviously low quality is acceptable for once-through cooling purposes. With adequate pretreatment, reclaimed wastewater offers the potential for several cycles of concentration through steam electric cooling condensers.

2. Manufacturing Industry. The manufacturing industries consist of primary metals, chemicals and allied products, paper and allied products, petroleum and coal products, food and kindred products, transportation equipment, textile mill products and other manufacturing. The first three categories accounted for about 77% of the total freshwater requirements for manufacturing industries in 1975. Water use in manufacturing industries is primarily for three purposes: cooling, boiler feed and processing. About 70% of all industrial use is for cooling and this represents the greatest potential for wastewater

reuse and recycling. Boiler feed water and most process waters, in general, must be of high quality, so the potential for reuse of wastewater in these purposes is limited. But water quality requirements for industrial cooling are similar to those for steam electric cooling; and reclaimed wastewater can be readily adapted to these quality needs.

3. Minerals Industry. The mineral industries include metal mining, fuels, and nonmetal minerals. The largest freshwater withdrawal is by the nonmetal minerals industry, with over 90% of the water withdrawn applied for coal processing, with very little cooling. Total water quantities in this field are projected to increase from 3.7 bgd in 1975 to 6.1 bgd in 2000 [2]. Reclaimed wastewater could be readily applied in several nonmetal mineral extractions.

The relationship of the projected freshwater withdrawals for these three major industrial categories to the overall freshwater withdrawal in the United States is shown in Figure 1. Together the industries offer a large potential market for reclaimed wastewater.

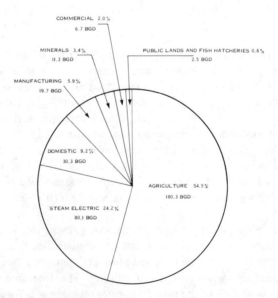

Figure 1. Projected national freshwater withdrawals in the year 2000. The steam electric, manufacturing and minerals industries will account for 111.1 bgd (33.6%) of the total national freshwater withdrawals in the year 2000.

WATER QUANTITY REQUIREMENTS

The basic data on water needs and availability were collected by the U.S. Water Resources Council (WRC) [2]. The WRC data are compiled for 21 major water resource regions and 106 subregions. The data and analyses in the WRC assessment reflect the basic economic, demographic and water and land related conditions in the United States for the years 1975, 1985 and 2000. Information presented in this section summarizes the WRC data, especially as related to industrial reuse.

The total freshwater withdrawal from ground and surface water sources is estimated to decrease from about 363 bgd in 1975 to 331 bgd by the year 2000 (Table II). This reduction will be caused principally by increased recycling of water in the manufacturing and steam electric industries in order to meet water quality requirements prior to discharge. As was shown in Figure 1, by the year 2000, steam electric, manufacturing, and mineral sectors will account for only 34% of the total national freshwater withdrawal (currently 41%). However, because of their proximity to wastewater reclamation plants, and their relatively stable water quality and quantity requirements, these industries remain prime candidates for municipal wastewater reuse.

Requirements by Industry

Steam Electric Power

Steam electric power generation, which is projected to account for 80 bgd of freshwater withdrawals in the year 2000, primarily requires water for the condenser-cooling tower cycle. As less "once-through" cooling is utilized, more cooling towers will be constructed and the actual consumptive use of water by steam electric plants will increase by over seven times between 1975 and 2000 [2]. More than 50% of this water demand for steam electric power generation is in the eastern regions of the United States, principally, the Great Lakes and Ohio regions.

Manufacturing

The Department of Commerce developed water requirements for the manufacturing industries from a comprehensive and confidential set of data on water use in about 10,000 large water using manufacturing plants [3]. That information along with census reports, special reports of the U.S. Environmental Protection Agency (EPA) and consultations with industry specialists provided an accounting of about 95% of the water used by the entire manufacturing sector.

In 1975, the total water use in the United States manufacturing industries was about 136 bgd. However, more than half of this demand was met by recycling. Consequently, as shown in Table IV, only 51.2 bgd was actually freshwater withdrawn to meet manufacturing needs. Because of further recycling within the plant, freshwater withdrawals are projected to decrease to approximately 20 bgd by the year 2000. Of the water used in manufacturing industries, approximately 60% was devoted to cooling tower applications [3]. These cooling uses offer the potential for further reclaimed wastewater usage. As shown in Figure 2, three manufacturing areas: paper, chemical and primary metal, will account for approximately 70% of the freshwater withdrawals in the year 2000. Since these industries use a large percentage of their water for cooling (Table IV)—or in the case of the paper and allied products, for process water which could be replaced with reclaimed wastewater—reuse potential is large. The other industries in the manufacturing group have much smaller freshwater withdrawals, or have requirements for high-quality water, such as the food and kindred products industry. These have limited potential for reclaimed wastewater. Although water quality requirements have been established for process uses, and for boiler makeup waters, in general the costs of reclaiming wastewater to meet these quality constraints are too high to justify planning reclaimed water usage. Most of the water used for manufacturing is in the Great Lakes, Ohio, mid-Atlantic, south Atlantic-Gulf, lower Mississippi and Texas-Gulf regions, with the Great Lakes region alone accounting for over 20% of all industrial use.

Table IV. Projected Decrease in Freshwater Withdrawals in the Manufacturing Sector by the Year 2000 [3]

Manufacturing Industry	Freshwater Withdrawals (bgd)		Percent Usage at Point of Application		
	1975	2000	Cooling	Process	Other
Primary Metals	17.6	3.5	70	25	5
Chemicals and Allied Products	13.5	5.2	80	15	5
Paper and Allied Products	8.4	5.4	35	60	5
Food and Kindred Products	2.6	1.3	50	40	10
Petroleum and Coal	2.5	1.2	65	5	30
Transportation Equipment	1.3	0.5	35	50	15
Textile Mill Products	0.6	0.3	–	–	–
All Other	4.7	2.3	–	–	–
Totals	51.2	19.7	60	25	15

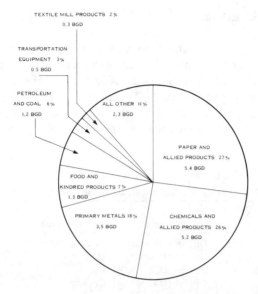

Figure 2. Projected industrial freshwater withdrawals in the year 2000.

Minerals

The mineral industries include metal mining, fuels and nonmetal minerals. As shown in Table V, growth of freshwater withdrawals by the mineral industries through the year 2000 will bring it to within 60% of the water withdrawals by the manufacturing industries ([1], pp. 31-32). Within the mineral industries, most of the water is utilized for processes such as coal washing, sand and gravel washing, subsurface injection and iron and copper ore extractions. Fortunately, the water quality requirements for these process applications are minimal and thus the potential for reclaimed wastewater use in these areas is large.

Table V. Projected Increase in Water Withdrawals by the Minerals Industries over the Next 20 Years

Mineral Industry	Freshwater Withdrawals (bgd)		Percent of Use at Point of Application (%)		
	1975	2000	Cooling	Process	Other
Nonmetals	3.7	6.1	5	90	5
Fuels	2.6	3.8	45	45	10
Metals	0.8	1.4	–	95	5
Totals	7.1	11.3	20	75	5

Requirements by Region

The total estimated wastewater reuse in the United States is approximately 0.7 bgd. Of this amount, industrial wastewater reuse amounts to 0.22 bgd, as shown in Table I. The total wastewater reuse is found at some 500 locations, with the largest amount of use in Arizona and California, and the greatest number of users in Texas and California. Of the approximately 500 reuse locations, only about 30 are industrial reuse sites: the major ones are summarized in Table VI [2].

Table VI. Inventory of Industrial Reuse in the United States ([1], p. 51)

User	Producer	Purpose	Wastewater Reuse[a] (mgd)
Bethlehem Steel Corp.	City of Baltimore	Cooling, Processing	106
Dow Chemical Co.	City of Midland	Cooling	6
Mo. State Park Board	Jefferson City, MO	NA	0.04
City Electric Cio. Martin Drake Plant	City of Colorado Springs	Cooling	21
Champlin Refinery	City of Enid	Cooling	2.0
Southwestern Public Service Company	City of Amarillo	Cooling	10.0
Southwestern Public Service Company	City of Lubbock	Boiler Feed, Cooling	6.5
Municipal Steam Electric Plant	City of Denton	Cooling	3.0
Cosden Oil & Chemical Co.	City of Big Spring	Boiler Feed	0.5
El Paso Products Co.	City of Odessa	Boiler Feed, Cooling	4.8
Bagdad Copper Corp.	City of Odessa	Process	0.2
Kaiser Steel Corp.	Uvalde, Texas	NA	0.5
NA	NA	NA	1.5
Nevada Power Co.	City of Las Vegas	Cooling	27
Nevada Power Co.	Clark County Sanitation District	Cooling	12.5
Phelps Dodge Corp.	Morenci, Az	Process	0.6
NA	Hayden, Az	Process	0.13
NA	NA	NA	0.08
NA	NA	Process	8.23
NA	NA	Cooling	0.96
NA	NA	Cooling	0.26
City Power Generating Station	City of Burbank	Cooling	2.0
Kaiser Steel Corp.	Kaiser Steel	NA	1.45
Total			213.55

[a] Does not include recycling.

The WRC has concluded that United States water supplies are sufficient to meet the requirements for all beneficial uses, especially in light of the anticipated reduction in water withdrawals over the next 20 years. In spite of this overall optimistic assessment, major water supply problems exist in most of the 21 WRC water use regions, and severe local problems exist in 14 of the 106 subregions. These 14 subregions are located mainly in the southwest and midwest. For example, in the lower Colorado region only 0.9 gallons of freshwater are available for each gallon required, in the Rio Grande region only 1.1 gallons are available and in California only 2.5 gallons are available [4]. In each of these cases, the situation is expected to deteriorate even further during the remainder of this century.

Unfortunately, the water shortages cannot be offset directly by wastewater available for reuse. The extent of shortages can only provide an indication of the potential for reuse. Many factors will affect the amount of the potential reuse which can be achieved: (1) the specific geographic relationship between dischargers and users will influence whether economic transfer of water can be made between them, (2) the timing of water needs and discharges will determine storage requirements and costs and (3) the cost of treating and delivering available wastewater will be compared to alternative supplies. In spite of these obstacles, the use of reclaimed wastewater by industry can help to alleviate water imbalances which exist in several parts of the nation. The use of the reclaimed water will enable freshwater resources to be devoted to higher quality needs.

WATER QUALITY REQUIREMENTS

This section describes water quality requirements for the following potential industrial users of reclaimed wastewater: (1) steam electric cooling, (2) industrial cooling and (3) other industrial users. The quality requirements for manufacturing and mineral industries, clustered under "other industrial users," are totally general in nature since the broad industrial categories cover a wide range of individual industrial plant types and processes. For example, the chemical and allied products category includes manufacturing operations for a wide variety of inorganic and organic chemicals, plastics and fertilizers. Similarly, the paper and allied products category includes a variety of paper, paper products and pulp operations. Both manufacturing and mineral industry operating methods vary from one company to another and often from one location to another in the same company. Definite quality requirements for each process water or the character of each wastewater discharge for the broad industrial categories cannot be specified but must be investigated on a case-by-case basis. However, several water quality parameters will be provided to indicate whether the reclaimed wastewater meets the general requirements of various industrial users.

Generally, industries are willing to accept water that meets drinking water standards. Occasionally industrial water quality requirements are more stringent than drinking water criteria, such as water used in high pressure boilers or in selected industrial processes. In these instances, the industry provides additional treatment as required. From the industrial viewpoint, the main criterion is that the primary water supply be of consistent quality so that pretreatment costs can be minimized. Once a consistent water quality is achieved, then pretreatment steps are maintained routinely. Inadequate water quality can cause three major categories of difficulties [5], virtually all of which can be eliminated via good operational control of a typical municipal wastewater reclamation plant:

* Product Degradation
    contamination via biological activity
    staining
    corrosion
    chemical reaction and contamination
* Equipment Deterioration
    corrosion
    erosion
    scale deposition
* Reduction of Efficiency or Capacity
    tuberculation
    sludge formation
    scale deposition
    foaming
    organic growths

Steam Electric Cooling

The basic quality considerations for cooling water are that: (1) scale is not deposited, (2) the cooling system is not corroded, (3) nutrients are not present for slime-forming organisms and (4) delignification does not occur in the cooling towers. Water quality requirements generally meeting these basic considerations are presented in Table VII [6]. Requirements for a recycling makeup system are more stringent than for once-through cooling. Particular problems are associated with the presence of hardness ions such as calcium and magnesium, which can combine with available phosphate to form a hard heat-transfer-limiting scale. Delignification of cooling towers has been studied in considerable depth, and the primary cause found to be sodium carbonate in the cooling water. By maintaining the pH between 6.5 and 7.5, a favorable carbonate-bicarbonate equilibrium can be achieved, thereby minimizing this problem.

Table VII. Water Quality Requirements (mg/l) [6]

Characteristics	Boiler Feedwater (psig)				Cooling Water				Process Water by Industry								
					Once-Through		Makeup for Recirculation						Petroleum and			Bottled and Canned	
	0-150	150-700	700-1500	1500-1500	Fresh	Brackish	Fresh	Brackish	Textile	Lumber	Pulp and Paper	Chem.	Coal Products	Primary Metal	Food Canning	Soft Drinks	Tanning
Silica (SiO_2)	30	10	1.0	0.01	50	25	50	25			50	50	60		50		
Aluminum (Al)	5	0.1	0.01	0.01			0.1										
Iron (Fe)	1	0.3	0.05	0.01			0.5		0.1		0.3	0.1	1.0		0.2	0.3	50
Manganese (Mn)	0.3	0.1	0.01				0.5		0.01		0.1	0.1			0.2	0.05	0.2
Copper (Cu)	0.5	0.05	0.05	0.01					0.05								
Calcium (Ca)		0	0	a	200	520	50	420			20	70	75		100		
Magnesium (Mg)		0	0	a							12	20	30				
Sodium & Potassium (Natr.)													230				
Ammonia (NH_3)	0.1	0.1	0.1	0.7									40				
Bicarbonate (HCO_3)	170	120	50	a	600		25					130	480				
Sulfate (SO_4)					680	2,700	200	2,700				100	600		250	500	250
Chloride (Cl)					600	19,000	500	19,000			200	500	300	500	250	500	250
Fluoride (F)												5	1.2		1	1.7	
Nitrate (NO_3)															10		
Phosphate (PO_4)													10				

Table VII, continued

Characteristics	Boiler Feedwater (psig)				Cooling Water				Process Water by Industry								
					Once-Through		Makeup for Recirculation										
	0-150	150-700	700-1500	1500-1500	Fresh	Brackish	Fresh	Brackish	Textile	Lumber	Pulp and Paper	Chem. Products	Petroleum and Coal Products	Primary Metal	Food Canning	Bottled and Canned Soft Drinks	Tanning
Dissolved Solids	700	500	200	0.5	1000	35,000	500	35,000	100		100	1000	1000	1500	500		
Suspended Solids	10	5	0	0	5000	2,500	100	100	5	<3 mm in dia.	10	5	10	3000	10	10	
Hardness ($CaCO_3$)	20	1.0	0.1	0.07	850	6,250	130	6,250	25		475	250	350	1000	250		150
Alkalinity ($CaCO_3$)	140	100	40	0	500	115	20	115				125	500	200	250	85	
Acidity ($CaCO_3$)														75			
pH (units)	8.0-10.0	8.0-10.0	8.2-9.2	8.8-9.2	5.0-8.3				6.0-8.0	5.0-9.0	4.6-9.4	5.5-9.0	6.0-9.0	5.0-9.0	6.5-8.5		6.0-8.0
Color (units)									5		10	20	25		5	10	5
Organics MBAS							1										
CCl_4							1										
COD	5	5	0.5	0	75	75	75	75						30			
Dissolved Oxygen	<0.03	<0.03	<0.03	<0.005													
Temperature °F	120	120	120	120	100	120	100	120			100			100			
Turbidity (JTU)	10	5	0.5	0.05	5000	100	100				100			100			0

a Determined by treatment of other constituents.

The use of treated wastewater for cooling purposes represents the single largest reuse application by industry and may require special consideration with regard to effluent quality control and water treatment. Fortunately, most of the reclaimed water is used for once-through cooling, and no additional treatment is necessary before use. Where circulating cooling systems are used, industrial users of reclaimed wastewater must provide additional treatment since a much higher quality is required. In general, these users treat the municipal effluent by lime clarification in order to reduce phosphates, organics, silica and suspended solids. Following lime treatment, the effluent is neutralized by acid addition and chlorinated for biological control.

High pressure boilers (pressure > 1500 psig) have the most stringent water quality requirements of any major industrial use; however, as boiler pressures decrease the quality requirements become less demanding, as indicated in Table VII. Pretreatment of existing potable water supplies is expected for boiler feed water. Silica and aluminum are especially critical materials in that they tend to form a hard scale on the boiler heat-exchange tubes. Excessive sodium or potassium can cause foaming of the boiler water.

Manufacturing Industries

Paper and Allied Products

The manufacturing of pulp and paper depends on a large supply of water for cooking and grinding wood chips to make pulp, repeated washings of the pulp and transportation of paper fibers through the bleaching, refining and sheet forming processes.

Water quality criteria for the paper and allied products industry are quite variable, depending on the production process and the desired quality of the finished product. Generally, suspended solids must be minimized since they adversely affect color and brightness of the product. Similarly, turbidity and color can adversely affect the production of the finer grade products. Materials such as silica, aluminum and hardness ions which can corrode or scale the process equipment are kept in low concentrations, as shown in Table VII. With respect to the use of reclaimed wastewater, microorganisms must be removed from the process water since they can result in slime growth, slick spots on the paper and odor development.

Chemicals and Allied Products

As shown in Figure 2, the chemical and allied products industries will be the second major user of water in industry. Nearly 80% of this use is for cooling, and consequently the water quality criteria for cooling waters will apply.

Process temperatures are generally higher than those found in the steam electric generation and petroleum industries, and consequently this requires better water quality. Also, cooling water is usually applied to the shell side of the heat exchangers, thereby increasing the possibility of fouling as a result of low water velocity. For these reasons many chemical systems will require better quality makeup water than listed in Table VII. The suspended solids in the makeup water should not exceed 5-10 mg/l, and high chloride waters cannot be used because of the widespread use of stainless steel in the heat exchangers.

The requirements for process water in the chemical industry are not well defined because of the great variety of processes and products. Generally, the industry requires a water which is moderately soft, has low color and suspended solids and has relatively low silica concentrations. However, because of the diverse nature of the products produced, water quality requirements vary widely and are not easily generalized into specific categories. Even for one specific product, the criteria can vary depending on the process used to manufacture the project. In general, water quality requirements are so restrictive as to eliminate the possibility of application of reclaimed wastewater.

Primary Metals Industry

Water quality requirements for the primary metals industry are variable, depending on whether the water is used for quenching and hot rolling, cold rolling or rinse water. Irrespective of the point of use, the criteria are not demanding, although some rinse waters must be softened. In general, the water quality requirements are as shown in Table VII.

Food and Kindred Products

The water quality requirements for the food canning industry and the bottled and canned soft drink industries are very high, as required for the protection of the products and to maintain quality control during production. Both industries require water of potable quality which must satisfy the U.S. Public Health Service Drinking Water Standards of 1962 and the recommended standards of 1968. While reclaimed municipal effluent can be treated to this extent, other concerns, especially with regard to the removal of pathogenic bacteria and viruses, prohibit the use of reclaimed wastewater for these industries. Complete absence of microorganisms, organic matter, color, turbidity and taste is required, especially for the carbonated beverage industry.

Petroleum and Coal Products

The petroleum and coal products industries represent a potential application point for reclaimed wastewater because the quantity of water required is large and the water quality requirements are nonrestrictive. About 97% of the water used in petroleum refining is for cooling and condensing. An additional 2% of a refinery's water is involved in boiler feed and the final 1% in process use. Water used in refinery cooling should meet the criteria for cooling water presented in Table VII. For process waters, the pH should range between 6 and 9 and suspended solids below 10 mg/l. Because of the generally easily-met water quality requirements, approximately 80% of the refineries water is reused, with only 20% requiring makeup. For example, cooling tower water is recirculated, cooling tower blowdown is used in pump glands, and sour water stripper bottoms are used for crude desalting and cooling tower makeup water.

Textile Mill Products

The textile industry is confronted with three main problems connected with the water supply: (1) suitable quality for processing textile products, specifically the absence of compounds which cause stains, (2) suitable quality for boiler feed in power plants and (3) suitable quality for prevention of corrosion in metal tanks and pipelines.

Impurities in the water are classified under four categories: turbidity and color, iron and manganese, alkalinity and hardness. Turbidity, color, iron and manganese are primarily responsible for stains and their removal is desirable. Hardness adversely affects soaps used in various cleansing operations and causes curd deposits on the textile. The soaps are retained on the fabric, encouraging soiling and impairing the handling of the cloth. Usually soaps are not deposited evenly and are one cause of irregularity in subsequent dyeing. Fortunately, the substitution of detergents for soaps has alleviated this problem. Hardness ions (calcium and magnesium) will precipitate some dyes and increase the breakage of silk during reeling and throwing operations. Alkalinity is important in that iron and manganese may be removed by aeration of water with sufficient alkalinity. If the water lacks alkalinity, insoluble hydroxides of iron and manganese may be formed by the addition of lime or soda. From this discussion one can see that "water good enough to drink is not always good enough for textiles."

Minerals Industry

As shown in Table II, the quantity of freshwater withdrawal for mineral industries is projected to increase by 60% over the next 25 years. Most of the water usage occurs in mining operations that require washing water, froth flotation, leaching for recovery of ores and injection of water into secondary oil recovery formations. No specific water quality requirements have been developed yet for these requirements, primarily because local freshwater resources are still adequate. However, the potential for use of reclaimed waste-water is large because of potential freshwater shortages in many areas which have natural ore materials, and because the water quality requirements can be met by secondary municipal effluent. Some additional treatment may be required to lower the concentration of suspended solids and bacteria, however these steps should only be undertaken on a case-by-case basis.

COSTS

The cost elements which must be considered in determining the economic feasibility of industrial reuse are:

1. reclaimed water supply (WWTP effluent),
2. distribution system,
3. on-site repiping (for systems being converted from potable to reclaimed supplies),
4. engineering analysis of water quality and pretreatment alternatives (including, if applicable, pilot plant analyses),
5. pretreatment capital and O&M,
6. internal treatment (above that associated with potable water use),
7. sludge and brine disposal (above that associated with potable supplies), and
8. institutional, legal and administrative activities:
   a. identification and coordination with supply, use and disposal agencies and regulations,
   b. supply contract negotiation,
   c. permit adjustment or acquisition, and
   d. if appropriate, coordination with other participants in subregional or regional projects.

These elements are evaluated in selecting a most cost-effective reclaimed water system, in comparing potable to reclaimed water usage for a new system, and in determining whether to convert an existing potable system to a reclaimed water system. The following paragraphs will discuss those aspects which have been major cost elements in prior industrial reuse projects.

Reclaimed Wastewater Supply

The base cost to the industrial user of water supplied from the wastewater reclamation plant is usually intended to cover the suppliers' capital and/or O&M costs for secondary treatment or for treatment provided above that required by law. For example, Bethlehem Steel in Maryland required cooling water quality better than that provided by the local treatment plant. They therefore undertook a program to locate and control the source of high salts in the wastewater (seawater intrusion into the sewers), and also provided funding for upgrading secondary trickling filters to activated sludge aeration. By continuing to participate in plant operations, they are assured of a reliable supply of high quality cooling water. Several facilities in Texas and Oklahoma which require activated sludge effluent quality provide the operating cost differential between trickling filters (which satisfy local discharge requirements) and activated sludge aeration at the local reclamation plant. At Southwestern Public Utilities in Amarillo Station, for example, the purchase price of water also covers the city's O&M cost of activated sludge treatment. In addition, Southwestern has a contractual obligation for a minimum amount of water; the power plant must dispose any unused portion. Thus, the city is able to save on its disposal costs.

Distribution System

At many industrial reuse sites, the pipeline for transporting the reclaimed water is constructed, owned, and maintained by the user. Such is the case for most of the industries in Texas, Oklahoma, Nevada, and at Bethlehem Steel in Maryland (see Table VI). These pipelines represent a significant cost to the industries, even though the distance between the local reclamation plant and the industry is generally less than five miles. The industries bear these costs because the value of their product is much greater than the cost of providing usable water.

In those cases where the majority of industries are not located near reclaimed water supplies, the cost of delivery can be consolidated on a regional basis. Within Los Angeles County, for example, the cost of a delivery system to serve 31,700 ac-ft/yr of reclaimed water to about 45 industries is estimated at $32.5 million [7]. Amortized over 20 years, at interest rates of 7.5-12.5% (representing public and private financing, respectively), reclaimed water delivery costs are $100-130/ac-ft. Additional costs are for treatment, operation and maintenance, etc.

Pretreatment, Internal Treatment, and Disposal

Depending on the quality of the treated effluent, its intended use and the relative costs, one of the following pretreatment steps may be required:

1. lime/lime soda clarification,
2. alum treatment,
3. ferric chloride precipitation,
4. ion exchange:
   - sodium ion exchange (NAX),
   - weak acid ion exchange (WAX),
   - split stream strong acid, sodium ion exchange (SANAX), and
5. reverse osmosis.

The usual pretreatment is lime clarification, although alum treatment is being used in one California power plant, and a sodium ion exchange (NAX) system is constructed and slated for operation in northern California.

Associated with each of these processes is either a sludge (lime or alum), or a brine (ion exchange) which must be disposed. The availability of appropriate waste disposal facilities can be a determining factor in the selection of a cost-effective pretreatment system.

Both reclaimed and potable water users require internal chemical addition to inhibit scaling, corrosion, fouling and biological growth within their industrial systems. Internal treatment costs will be somewhat higher for reclaimed water users than for potable water users, although the degree of this difference will depend upon the type of pretreatment given to the reclaimed water, and the final application point of the reclaimed water.

The costs associated with reclaimed water pretreatment, internal chemical treatment and sludge or brine disposal will typically exceed the costs associated with internal treatment of potable water by about 75% ([8], pp. 1-4).

Example ([8], pp. 5-4 to5-6)

A common misconception is that reclaimed water represents a new low-cost water supply. Although the purchase price for reclaimed water is normally significantly lower (75-80%) than for potable supplies, costs associated with reclaimed water transport and treatment will typically more than offset this initial savings.

In Figure 3, two different potable, and three different reclaimed water cost scenarios are shown for a typical 3 mgd cooling use. Cost elements common to both the potable and reclaimed operations are the purchase price of water, and costs of equipment and chemicals to alleviate corrosion, scaling, etc., within the cooling system (i.e., internal chemical treatment). For reclaimed water systems, additional costs are for a distribution line from the

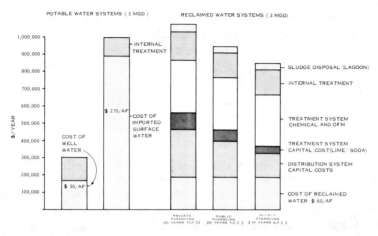

Figure 3. Transport and pretreatment system costs make industrial use with reclaimed water as costly or costlier than with potable supplies.

water reclamation plant to the user's site, pretreatment system capital, chemical and O&M costs, and possibly sludge (or brine) disposal costs.

Price of Water

Within southern California, the retail price of water for large water users (25 AF/yr) ranges between $50/AF and $271/AF. (Fifty dollars reflects the cost of well water, which many industries utilize either alone or mixed with surface supplies. Delivered potable water costs range from a low of $110/AF to $271/AF.) For the typical 3 mgd industry, annual water costs are then between $165,000 and $900,000.

Reclaimed water supplies, on the other hand, show much less variation. Among industries contacted in Texas, Oklahoma, Maryland and southern California, the cost of reclaimed water is around $60/AF. For a 3 mgd industry, total annual water costs would then be about $200,000. This will normally be substantially less than potable water costs (unless an industry has access to 3 mgd of well water).

Internal Treatment Costs

Both reclaimed and potable water users require internal chemical addition to inhibit scaling, corrosion, fouling and biological growth within their indus-·trial systems. These internal chemical addition costs are around $100,000/yr for potable systems and $150,000/yr for reclaimed water systems.

These two costs, water purchase and internal treatment, comprise the total cost for industrial use of potable supplies. Because of the substantial variation in the retail price of water, the total cost for a 3 mgd system ranges from $300,000 to $1,000,000/yr.

Distribution System

The cost of a reclaimed water distribution system represents a major expenditure. Typically, industries locate near their source of natural resources (i.e., water, oil, minerals) or near a shipping port. They do not purposely locate near wastewater treatment plants. A fairly conservative distance of 4 miles between a reclaimed water treatment plant and an industrial user is used to calculate the distribution system capital costs. For a 24 in. pipe, total capital costs would be around $2,000,000.

If the industry finances this cost, as was done in Texas and Oklahoma, the annual costs would be around $270,000. If the industrial reuse system is part of a larger subregional or regional project, then public (municipality) or agency funding may be available. The lower interest rates available would result in lower annual costs of $200,000 and $142,000, respectively. These three costs are reflected in the different reclaimed water system costs shown in Figure 3.

Treatment System Capital

The annual capital costs for lime/soda treatment of an "average" reclaimed water are shown in Figure 3. As for the distribution system, these costs may be publicly financed if the industrial system is part of a larger reuse project. Annual costs for a $600,000 treatment system financed privately, publicly or via agency financing are shown.

Treatment System O&M

Treatment system chemical and O&M costs often represent the largest single cost of reclaimed water operation. Such nonconstruction costs are not usually publicly funded; however, in a regional context, some subsidy of these costs may be possible. The three systems illustrated each include the total annual costs of chemicals and maintenance for lime/soda treatment: $300,000/yr.

Sludge Disposal

Low-cost, on-site, lagoon disposal is depicted. Brine or alum disposal to a sewer would be of comparable cost. Other sludge disposal options would be much more expensive.

Total System Costs

Under the three different reclaimed water system financing alternatives, total system costs range from $850,000 to $1,050,000/yr. The two potable water systems range in annual costs from $300,000 to $1,000,000. Different sludge disposal requirements would increase the reclaimed water costs, whereas better than average reclaimed water quality would lower pretreatment requirements and costs. The length of the distribution system is a significant cost factor. For the general case shown, only the most expensive potable water system would be of comparable cost to a reclaimed water system. More typically, reclaimed water will be a more expensive commodity than potable water.

In spite of the higher costs of operation, three conditions encourage industrial use of reclaimed water for cooling: (1) the uncertainty of future potable supplies, (2) the reliability of locally-developed reclaimed supplies, and (3) a public policy favoring reuse.

Since many benefits from an industry's use of reclaimed water accrue to the public as a whole, the public can reasonably be expected to bear some of the costs. Public sources of funding, perhaps as part of a local or regional project, are thus an appropriate and probably necessary means of financing industrial reclaimed water use.

SUMMARY

The purpose of this chapter has been to provide a realistic appraisal of the reuse potential of reclaimed wastewater in industrial applications. As compared to other reuse applications, such as agricultural irrigation or groundwater injection, industrial reuse offers several distinct advantages: the point of application is often located near the source of supply (industries tend to locate in major urban centers where large secondarily-treated wastewater flows are available); the water quality requirements of major industrial applications (e.g., cooling towers) can often be met by a well-treated secondary effluent with some internal chemical addition to minimize scaling, corrosion, fouling and biological growth; and industrial water quantity needs are uniform temporally, whereas other needs, such as agricultural irrigation, vary widely with season and time of day.

On a quantity basis, the extent of current industrial reuse is very small (0.22 bgd) especially in comparison with total United States freshwater withdrawals (363 bgd) and available wastewater discharges (244 bgd). Over the next 20 years, the extent of freshwater withdrawal is expected to decrease about 10%, as discharge standards become stricter and more industrial recycling is practiced. During this period, the potential for replacement of freshwater withdrawals by reclaimed wastewater will be large (111 bgd), but in fact actual projected wastewater reuse will amount to less than 5% of potential reuse. Actual reuse will occur in selected subregions which are already experiencing deficits in available water supplies, and which have other resources, such as oil or natural gas, whose value as products far exceeds the cost of reclaiming wastewater for reuse.

Industrial categories which probably will increase their extent of reclaimed wastewater reuse are the steam electric power industry (cooling), manufacturing (cooling), and minerals (process). Even with these industries replacing some of their lower quality water needs with reclaimed wastewater, the total projected wastewater reuse (year 2000) of 4.8 bgd is only a small fraction of the total projected wastewater recycle of 866 bgd. Nevertheless, the reuse of reclaimed wastewater will provide a valuable alternative source of water to those subregions experiencing severe shortages, and will free available freshwater resources for higher quality users.

Water quality requirements vary widely between industries, between different plants in the same industry, and between various processes within a single plant. For these reasons, water quality criteria are not generally applicable for a variety of processes or industries. Except for the food processing industry, most industries do not require drinking water quality, although some may require extensive treatment of the reclaimed wastewater before the water is acceptable for reuse. In other situations, the water may be acceptable as received. Table VIII summarizes the major potential industrial applications for reclaimed wastewater, and the general water quality criteria needed to meet reuse applications. In general, water for sanitation, potable uses and processes is a small percentage of the total water used by an industry. Cooling water alone accounts for 50-60% of all water used by industries; this cooling water can often be of relatively poor quality.

Reclaimed wastewater quality differs from that of fresh water in its generally greater potential for corrosion, scaling, biological growth and fouling. The degree of difference between the two supplies depends upon the proportion of industrial to domestic flow into the wastewater reclamation plant and the level of treatment provided at the plant. Raw wastewater is unsuitable for most industrial uses primarily because of organic contamination; however, primary treatment begins the removal of problem constituents and by the end of the secondary activated sludge treatment, the wastewater is relatively low in organic contaminants.

Potential problems from residual contaminants include scaling (the formation of hard deposits, usually on hot surfaces), fouling (the settling out of dirt, debris, organic colloids, scale compounds), corrosion and biological growth. In order to minimize their potential, additional treatment may be provided, either at each industrial site, at one or more sites within or near a cluster of industries, or at the wastewater reclamation plant (external treatment). In addition, chemical additives utilized within each industry help inhibit tendencies of precipitation, corrosion, fouling, etc. (internal treatment).

The reuse of reclaimed wastewater does not represent a new low-cost alternative to freshwater withdrawal or recycling. If additional freshwater resources are available for development or in-house recycling can be practiced, then either of these alternatives will probably be lower in cost than reuse of wastewater. In addition to the basic cost of reclaimed wastewater from the treatment plant, additional costs are new distribution system capital costs, new pretreatment and internal treatment capital and O&M costs, and sludge or brine disposal costs. Furthermore, development of a reclaimed wastewater system will supplant some of the existing demands on the freshwater system, resulting in higher costs to the remaining freshwater users since they will still have to retire existing "sunk" costs in the freshwater system. For these reasons, wastewater reclamation and reuse can usually be justified only under the following circumstances: a scarcity of freshwater supplies exists, reuse is the most cost effective means of effluent disposal, or public policy encourages or mandates reuse as a means of conserving high-quality freshwater supplies.

REFERENCES

1. Office of Water Research and Technology. *Water Reuse and Recycling Volume 1: Evaluation of Needs and Potential,* OWRT/RU-79 1 (April 1979).
2. U.S. Water Resources Council, "The Nation's Water Resources, The Second National Assessment," Draft (January 1978).
3. U.S. Water Resources Council, "The Nation's Water Resources, The Second National Assessment," Appendix B, "Methodologies and Assumptions for Socio-Economic Characteristics and Patterns of Change and Water Use and Water Supply Data," Preliminary Review Draft (April 1978).
4. U.S. Water Resources Council, "Summary Report," The Nation's Water Resources, The Second National Assessment," Draft Review Copy (April 1978), p. 34.
5. McKee, J. E., and H. W. Wolf. *Water Quality Criteria,* Pub. No. 3-A, California State Water Quality Control Board, 1963, p. 92.

Table VIII. Major Potential Industrial Applications for Reclaimed Wastewater and Water Quality Criteria

Industrial Application	Beneficial Use	Water Quality Criteria
Power Plant and Industrial Cooling		
Once-through	Water used by power plants and certain industries for the removal of heat on a once-through basis.	The once-through use of water for the condensation of steam requires large volumes of water and only the removal of contaminants that could clog or settle in the system and requires *no* prior treatment. Other water used at the site will be treated as necessary.
Recirculation	Water used by power plants and certain industries for the removal of heat by recirculating water through cooling facilities.	Water qualities of the raw water supply should be suitably low in contaminants to avoid clogging, scaling, and slime growth accumulation. Contaminant build-up is assumed to be controlled by regular wastage or blowdown. The water should also be noncorrosive. Other water used at the site will be treated as necessary.
Industrial Boiler Make-up Water	Water used for the generation of steam for power or processing needs.	Water quality depend upon the operating pressures of the boiler system; raw water supplies receive extensive onsite treatment. If the steam is used for food processing, water qualities must also meet the requirements for potable use.
Industrial Water Supply Food and Kindred Products	Water used for washing, rinsing, conveying or for the preparation of food products.	Water qualities in general must meet potable use standards. Water used to generate steam must also meet the requirements for boiler make-up water. Water is assumed to be bacteriologically safe. Certain situations which require even more stringent water qualities are not included.

Paper and Allied Products	Water used for cooking and grinding wood chips, washing of pulp and the transport of paper fibers through the various processes.	Contaminants that could clog equipment or cause slime growths, as well as affect the color or texture and uniformity of the pulp must be limited. Hardness which can cause scale on equipment or contaminants causing corrosion or other problems are also assumed to be limited.
Chemicals and Allied Products	Water used for transportation, washing, and mixing the products as well as a chemical reaction media.	Water qualities are assumed to be suitable when the water does not cause unfavorable chemical reactions or when the chemical reactions are not delayed or do not require excessive quantities of chemicals. Water used for cooling or boiler make-up is not included.
Petroleum and Coal Products	Water used in the processes such as refining, desalting and fractionation and also for storage and transportation of the products.	Surface or groundwaters low in dissolved salts and chlorides and iron. Cooling and boiler make-up water are not included.
Primary Metals	Water used for the processing of ferrous and nonferrous metals.	Surface waters that will not corrode or plug equipment or create scale problems. Waters used for cooling or boiler make-up are covered elsewhere.

6. James M. Montgomery, Consulting Engineers, Inc., "Industrial Wastewater Reuse: Cost Analysis and Pricing Strategies" (October 1979), pp. 1-12.
7. James M. Montgomery, Consulting Engineers, Inc., "Engineering Analysis of Reuse Projects Within Areas 4, 5, and 6," Draft Report (January 1980).
8. California Office of Water Recycling, *Evaluation of Industrial Cooling Systems Using Municipal Wastewater Applications for Potential Users* (June 1980).

CHAPTER 24

INTERNATIONAL DEVELOPMENTS AND TRENDS IN WATER REUSE

Jay Messer
Utah Water Research Laboratory and
Division of Environmental Engineering
Utah State University
Logan, Utah

INTRODUCTION

Water must meet a triad of requirements if it is to represent a truly useful resource to a community or society. It must be of an adequate quality and must be available in sufficient quantity and at an appropriate time in order for it to be put to a particular beneficial use. Because of geographical, climatological and socioeconomical factors, many countries suffer a functional deficiency of water resources, because one or more of the above criteria is not met.

Frequently the problem is one of timing. Flood flows associated with spring snowmelt or seasonal weather patterns, such as monsoons and hurricanes, frequently are lost largely to the sea, often causing significant devastation in the process. These flood flows, in turn, often are succeeded by periods of drought or inadequate streamflow, and agricultural and occasionally even municipal and industrial water supplies can be severely stressed. The traditional solution to this problem in timing of resource availability is to construct upstream impoundments to regulate downstream flows and thus to provide a more equitable temporal pattern of water availability. Such projects require large capital expenditures, however, and a long lead time is required before benefits can be realized. Ambroggi [1] recently pointed out that most countries today find it extremely difficult to mobilize the capital required to construct large water development projects, and this in part accounts for his prediction that some 30 countries will face severe water shortages by the end of the century.

In many industrialized countries, sufficient capital for development of such projects has been available in the past, and in some countries the structural regulation of water courses is virtually complete. In these nations, however, many of the same industries that made such projects possible, together with the high density settlements of people they have attracted, have generally reduced the utility of their developed water by reducing its quality to the extent that it cannot be used without extensive treatment. Increasing salinity, corrosive salts and refractory color- and odor-producing organics have left water unfit for many industrial applications, and the specter of carcinogens, heavy metals and pathogenic viruses has engendered grave concern among municipal users. Situations such as these have led many countries to search for effective technologies aimed at the reuse of water that has been degraded by human activities, and for purposes other than merely providing dilution water before it reaches the sea.

The range of technologies employed by different countries to renovate and reuse wastewater is described in the following sections. Because much of the original information is located in in-house reports of government agencies or private consulting firms and is not generally available, this review relies heavily on the reports presented at the Water Reuse Symposium cosponsored by the American Water Works Association [2] and on summaries of earlier experiences collected by Shuval [3]. Additional references can be found in the published proceedings of the United Nations Water Conference at Mar del Plata [4] and in the literature reviews of English [5] and Smith [6-8]. Additional useful information will be found in documents soon to be published by the NATO Committee on the Challenges of Modern Society (CCMS) [9], the Australian Ministry of Water Resources and Water Supply [10], and in the proceedings of the American Water Works Association Second Water Reuse Symposium (1981). These sources can be consulted to obtain specific information on design and operating parameters and detailed experiences with problems and performance.

SOUTH AFRICA

One of the first countries that springs to mind when one thinks of water reuse is South Africa, faced with an annual precipitation excess of only 44 mm (even less if the protectorate of South West Africa is included), and even this scanty runoff is not evenly distributed. Recent estimates place distributable water supplies at 33×10^9 m^3/yr, and consumptive demands are projected to reach 30×10^9 m^3/yr by the end of the century. Present plans are eventually to reduce this demand by 7.2×10^9 m^3/yr (22% of the annual distributable supply) through water renovation and reuse [11].

The rather unusual history of water legislation in South Africa [11,12] has arisen out of the evolution of an agricultural society to an industrial one under conditions of severe water shortage. The parsimony with which water resources are distributed under these legislative mandates has made the development of advanced water treatment virtually requisite. Furthermore, water resources management is kept under tight national control. Research and financial support is provided by the government to put these reclamation schemes on a sound footing.

The cornerstone of water policy in South Africa is the Water Act of 1956. Among other features the Act stipulates that:

1. A permit is required for withdrawal of more than 250 m^3/day, the granting of which may be predicated on reuse, recycling and hierarchical use of water within the operation. Similar provisions apply to the development of municipal works which divert more than 5000 m^3/day.
2. Strict effluent quality standards must be met, thus making reuse schemes economically more attractive.
3. Diversions from a stream are allowed only for beneficial consumptive use, thus discouraging "reuse" schemes based on land disposal.

The result of these regulations is an active, constructive approach to water renovation and reuse in the industrial, municipal and, to a lesser extent, the agricultural sectors.

Agricultural Reuse

The Water Act of 1954 discourages land application of wastewater, and such applications as presently exist are predominantly temporary and provide water to approximately 3.5% of the land area of South Africa. Permanent permits may be approved in situations where high quality waters would otherwise be used, or in lower river reaches where the water, if discharged, would be lost to the ocean and may, indeed, help to prevent saltwater intrusion. Two particular long-term agricultural applications, one of which has apparently grandfathered the Water Act, are of unique interest.

The City of Johannesburg has used its municipal effluent (36% of annual withdrawals) to irrigate pasture and silage crops for its 7500-head cattle herd since 1914. Total crop production averages 9000 metric tons of hay and 7000 of corn silage per year, supporting a stocking density of 4.2 animals/ha, including calves [12]. Although the operation is financially successful, the high stocking density results in some problems with the spread of bacterial infections and parasites (particularly fascioliasis), and the high hydraulic loading (230 cm/yr) results in approximately half of the irrigation water returning to the local watercourse as runoff and seepage. In another application, industrial effluent from the Modderfontein explosives and chemical plant has been applied to pasture at a rate of 3.8 cm/yr, which results in little irrigation but a large application of nitrogen (466 kg/ha·yr). When coupled with fertilization with additional phosphorus and some potassium, an increase in the carrying capacity of the veld to as high as 10 animals/ha has been realized.

Industrial Reuse

In many areas of South Africa, wastewater reuse already exceeds the 22% target discussed above. Table I summarizes effluent reuse in the Pretoria-Witwatersrand-Vereeniging (PWV) complex, where water is in particularly short supply, and in other urban and industrial areas. In the PWV complex alone 23.2% of the effluent is used for power plant cooling and industrial reuse, as opposed to the other urban areas which reuse only 7.8% for these purposes. It is particularly interesting that many industries that were more or less forced into using renovated water have found that such reuse has actually enhanced the quality of their products.

Several pulp and paper plants in the country have enjoyed considerable success with water reuse [11]. The Enstra mill of South African Pulp and Paper Industries Ltd. (SAPPI) supplies 63% of its process water from secondary municipal effluent which receives flocculation with polyelectrolyte, flotation, pH adjustment, chlorination and some filtration. Production of the brightest paper grades has been possible at a 50% savings in cost over full use of culinary water, and the capital costs were saved in the first year of operation. Another paper plant (Mondi at Durban) employing chemical flocculation, foam flotation, sand filtration and activated carbon can produce high-quality paper and newsprint using 100% municipal effluent, although no savings in water costs over culinary water have been realized. Nonetheless, the release of the high-quality water for other uses represents a valuable benefit. Other industries making use of reclaimed municipal wastewater include: slurry processes in Portland Cement, cooling and washing in steel milling, grit washing and slurrying in diamond and gold mining, ash transport in coal conversion and cooling at power stations.

The Croyden Power Station has used sewage effluent as cooling water for more than 20 years and employs a rather unique approach to reducing con-

Table I. Comparison of Daily Sewage Effluent Production and Reuse in the PWV-Complex and Other Urban and Industrial Areas [11]

	Total Effluent Volume (m³/day)	Irrigation (%)	Power Plant Cooling (%)	Industrial Use (%)
PWV-complex [a]	641,200	26.8	14.4	8.8
Other	592,300	4.5	2.6	5.2

[a] Pretoria-Witwatersrand-Vereeniging complex.

denser fouling [13]. Calcium monohydrogen phosphate precipitates were initially found to severely inhibit heat transfer, and means of calcium and phosphate removal from the influent wastewater were found to be too costly. Reduction of pH with HCl to convert $HPO_4^=$ to $H_2PO_4^-$ was prohibitively expensive when the source of the HCl was superchlorination. However, induced biological nitrification (which necessitated some NH_3 feed in the cooling tower) was found to effectively control $CaHPO_4$ scaling at a much lower cost. Nitrification rates remained high in the winter due to icing of the cooling tower resulting in some insulation. Biological slimes represented a problem that has been solved at a reasonably low cost.

Internal water reuse in some other South African industries is shown in Table II. Iscor Steel Mills circulates water at a ratio of 34:1 over its intake rate for a net consumption of only 5.1 m^3 water/ton steel. The South African Electricity Supply Commission (ESCOM) has reduced its specific water consumption by one-third to 2.7 ℓ/kWh, mainly through acid dosing of blowdown water using mine drainage waters and reclamation of ash handling waters. SASOL, the world's only full-scale coal conversion plant, has reduced its potential water needs by 92% through recycling, and the Modderfontein Explosives Plant has similarly reduced its potential needs through recycling.

Municipal Reuse

A South African protectorate, South West Africa, contains the only full-scale plant in the world in its capital, Windhoek, where potable water is reclaimed from municipal wastewater [14]. The Windhoek (Figure 1) consists of a conventional wastewater treatment plant followed by nine maturation ponds with a total hydraulic retention time of 14 days to provide quality equalization, high lime treatment and ammonia stripping, recarbonation and sand filtration followed by breakpoint chlorination, and activated carbon which is regenerated onsite. The effluent stream is then admixed with product water from a conventional potable water treatment plant and post-chlorinated to a free residual of 0.2 mg/ℓ. The proportion of reclaimed water in the municipal system generally ranges from 20 to 50%.

Intensive monitoring of the Windhoek effluent and the community it serves has thus far indicated no significant problems [12,14,15]. Water

Table II. Water Recycling in Four Major South African Industries [11]

	Water Recirculated (m^3/day)	Intake Water (m^3/day)	Intake (%)
Steel (Iscor, Pretoria)	732,000	21,200	2.9
Oil from coal (SASOL)	832,000	63,600	7.8
Chemicals (AECI, Modderfontein)	491,000	17,200	3.5
Thermal Power (Komati)	3,928,000	98,000	2.5

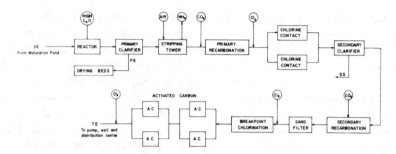

Figure 1. Flow diagram of modified Windhoek water reclamation plant [after 14]. SE = secondary effluent; TE = tertiary effluent; PS = primary sludge; SS = secondary sludge.

quality parameters meet World Health Organization and U.S. Environmental Protection Agency (EPA) standards, neither coliforms nor virus (including reovirus) has been detected in the product water and no epidemiological suggestion of ill effects is evident. Removal of microorganic pollutants is generally greater than 90% and concentrations of polycyclic aromatic hydrocarbons, chlorophenols and trihalomethanes are below EPA limits. Thus far no carcinogenesis has been detected by bioassay techniques. The fact that the nearest perennial river is 720 km from the city has probably contributed significantly to the widespread public acceptance of this project.

Thus far reclamation costs have been approximately $0.32/m³ (1978 dollars) compared to $0.23/m³ for other supply sources in South Africa. The exclusion of industrial wastewater from the input stream, and the proximity of the water and wastewater treatment facilities designed into the system contribute greatly to the cost effectiveness of the system. Biological denitrification and phosphate removal are being added to the system, and further reductions of reclamation costs are anticipated.

The National Institute for Water Research (NIWR) is conducting research at two pilot plants to provide information on the feasibility of reclaiming wastewater with increasing proportions of industrial wastewater input. A plant at Pretoria receives mostly domestic sewage, while another near Cape Town (Athlone) receives a higher proportion of industrial wastewater. These studies have been described by van Vuuren and Talvard [16].

The Pretoria plant (Figure 2) consists of a denitrification reactor where nitrified products of succeeding stages are returned (1:1) to react with the organic carbon in the raw effluent. The product water is drawn to a chemical clarifier, where it is dosed with FeCl₃ and limed to raise the pH. Sludge is continuously drawn from the clarifier and split into three streams. One stream is returned to the denitrification reactor, where it serves as a reserve

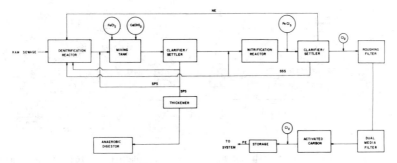

Figure 2. Flow diagram of modified LFB (high lime/biological flotation) pilot plant at Pretoria [16]. RE = raw effluent; SPS = settled primary sludge; SSS = settled secondary sludge; PE = polished effluent; NE = recycled nitrified effluent.

carbon source, especially during low nighttime carbon loading. The second stream is recycled to the chemical clarifier, where it reduces the $FeCl_3$ requirement by more than 50%. The final stream is wasted to the thickener. The effluent from the clarifier is sent to an aeration (nitrification) pond, and a secondary clarifier, where it is again dosed with $FeCl_3$. Further polishing is accomplished by prechlorination and passage through a roughing filter, dual media filter, activated carbon, and by postchlorination, thus minimizing formation of chlorinated hydrocarbons.

The chemical and microbiological quality of the project water was generally good. Removal of COD and nitrogen was approximately 90%, leaving residuals of < 10 mg/ℓ of each. Phosphate-P was reduced to below 0.2 mg/ℓ in the effluent and was not readily released from the sludge, even under anaerobic conditions. Turbidity was less than 0.6 JTU, and neither fecal coliforms nor coliphage was detected in the product water.

The Athlone plant (Figure 3) employs similar processes to treat a poor quality effluent (220 mg/ℓ COD, 36 mg/ℓ NH_3-N) from a trickling filter receiving mixed domestic and industrial wastewater. High lime treatment is being employed rather than $FeCl_3$. Effluent water typically contains 22 mg/ℓ COD, 0.8 mg/ℓ total P, 23 mg/ℓ NO_3^--N, and 1 JTU. Chemical and electrical costs at the Athlone plant ($0.16/m^3) are approximately twice that at the Pretoria plant. This generally points to the much greater economic savings made possible by diverting industrial wastewater prior to municipal wastewater renovation. The higher treatment costs at the full-scale Pretoria operation include labor and capital charges. Chemical and electrical costs at Pretoria total $0.17/m^3 [14], which may reflect the higher quality of the effluent at this plant.

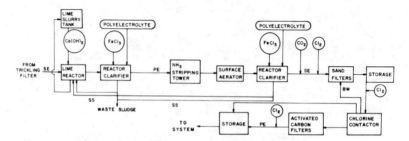

Figure 3. Flow diagram of Athlone water reclamation plant [after 16]. SE = secondary effluent; PE = polished effluent; SS = settled sludge; BW = backwash.

ISRAEL

By the mid-1970s, Israel had reached the point where more than 90% of its "conventional" surface and groundwater resources were being used. To provide for the expected growth in the nation's population and economy, water demand is projected to reach 1980×10^6 m³/yr by 1985, which exceeds the presently available supply (Table III). This demand is expected to be met by increasing wastewater renovation and reuse from 50 to 410×10^6 m³/yr, thus furnishing up to 21% of the 1985 demand. Although desalination may play an increasing role in potable water supply, the technology is presently financially prohibitive in all but a few locations. Direct wastewater reuse in Israel has been practiced in agricultural and industrial settings, but there are presently no plans to depend on renovated wastewater for culinary uses. Israel's water supply problems and the renovation effects being undertaken to confront them have been discussed by Shelef [17], Idelovitch [18], Idelovitch et al. [19] and Farchill [20].

Table III. Water Sources in Israel as of 1974 and Predicted for 1985 [after 17]

Water Source	1974 (10^6 m³)	1985 (est.) (10^6 m³)
Groundwater	910	900
Surface Water	650	670
Renovated Wastewaters		
Domestic	45	260
Industrial	5	50
Desalinated waters[a]	10	100
	1620	1980

[a] Including seawater and brackish water desalination.

Irrigation

Present regulations prohibit the application of renovated watewater to some vegetables, largely because of a cholera outbreak that resulted from the application of untreated sewage to a lettuce crop. However, many crops are allowed to receive treated wastewater effluent, provided certain restrictions are met regarding timing of irrigation with respect to harvest, coliform levels and distance from residential areas. As early as 1971, 20% of the wastewater from sewered communities was being used for irrigated agriculture.

Traditionally, renovation of wastewater for irrigation in Israel has been accomplished using a series of oxidation ponds, which progress from anaerobic to facultative, the final "maturation" pond being essentially fully aerobic. In the rural agricultural settlements (kibbutzim), two parallel anaerobic ponds typically receive BOD_5 loadings of 1000-1500 kg/ha·day and discharge to a single facultative pond with a BOD_5 loading of 150-300 kg/ha·day and finally to a polishing pond and operative irrigation reservoir. Detention times are approximately 1.5-4, 5-15, and 2-4 days, respectively. The final effluent is only of high enough quality for use with wide-nozzle sprinkling irrigation on nonrestricted crops, because pathogens are still present. Although these systems do not produce sufficiently strong odors to offend agricultural communities, the anaerobic ponds are occasionally replaced by facultative ponds in suburban areas.

Shelef [17] has described preliminary plans for wastewater irrigation projects that would collect wastewater from densely populated areas and transport it to water deficient locations. One such project would transport 100-160 km^3 of water 140 km from Netania in the north to Dorot in the Negev desert to provide irrigation for some 33,000 ha of cotton, wheat and sorghum. Aerated lagoons with BOD_5 loadings of 2000-7000 kg/ha·day were recommended in order to produce a high quality effluent for a subsequent polishing pond. Shorter transport distances have been suggested for treated wastewater from Haifa, Jerusalem, Nazareth, Migdal-Ha'emek and Ramat-Ha'sharon. In these systems, impoundments will store wastewater and runoff collected during the rainy season to provide irrigation water during the drier summers.

The Dan River Project

The Dan River project in the highly urbanized Tel Aviv area represents the premier effort to date in Israel for producing a high-quality renovated wastewater for nonpotable use. Following a decade of design and pilot plant experiments, Phase I of the project is now underway [18,19]. Treatment consists of four stages (Figure 4):

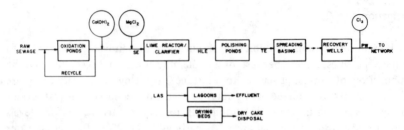

Figure 4. Flow diagram of Phase I of the Dan River Water Recalmation Project [after 18]. SE = secondary effluent; HLE = high lime effluent; TE = tertiary effluent. PW = polished water; LAS = lime-algae sludge.

1. Biological treatment in recirculated oxidation ponds;
2. High-lime-magnesium treatment in a sludge-blanket reactor-clarifier;
3. Detention in polishing ponds for ammonia stripping, natural recarbonation, and additional clarification; and
4. Final polishing by groundwater recharge and infiltration.

In pilot scale studies optimum operation was found to occur with a total dosage of 50 mg Mg/ℓ at a pH of 11.3-11.5. Magnesium was necessary to provide good removal of metals that form insoluble hydroxides (e.g., Fe, Cu, Cd, Pb, Zn) and boron, silica and fluorides, and was readily available from Dead Sea brines. Although natural recarbonation was unable to lower the final pH in the polishing pond to below 9.5-10.5, it was found that mixing with local groundwater pumped from wells near the oxidation ponds, together with percolation through the soil, was sufficient to lower the pH of the reclaimed water to 8.0.

Results of the reclamation process are presented in Table IV for the full-scale plant. Compared to the pilot plant, the full-scale operation (15 × 10⁶ m³/yr) was characterized by slightly lower Mg concentrations (~40 mg/ℓ) with longer detention times in the polishing pond. Nonetheless, the quality is quite good. Infiltration rates in the well field remain high, providing the loading is intermittent (1 day wet to 2-3 days dry), and chemical concentrations appear to have reached equilibrium in the groundwater. Pilot plant studies are presently underway to recycle the Mg in the chemical sludge and to recover algal protein and lime.

At present a large portion of the wastewater from the 1.1 million residents of the Dan Region is being discharged into the sea, where it is causing pollution problems along the beaches. To counteract this loss of two valuable resources, phase II of the Dan Region project is being designed to ultimately treat 160 × 10⁶ m³ of wastewater per year by the turn of the century. A pilot plant which employs a modified, low-rate activated sludge system without primary settling, is presently producing good removal efficiencies (Table V) and the full-scale plant is under the advanced stages of design [20]. Future

Table IV. Performance of Full-Scale Reclamation Plant at the Dan River Project in Israel [19]

Parameter	Raw Sewage		Secondary Effluent		High Lime Effluent		Tertiary Effluent		Reclaimed Water
	Summer	Winter	Summer	Winter	Summer	Winter	Summer	Winter	
pH	7.9	7.9	8.3	8.3	11.7	12.1	10.1	10.2	8.0
Suspended Solids	230	249	212	247	55	154	8	20	0
BOD	230	226	60	53	13	19	7	6	0.2
COD	510	465	355	298	90	125	75	67	10
Ammonia	38	46	21	27	24	30	1	5.5	0.02
Total Nitrogen	58	70	55	55	31	39	6	10.5	5
Phosphorus	11.9	12.0	11.6	11	2.0	1.6	0.5	0.8	0.05
Dissolved Solids	875	778	860	739	830	737	640	588	630

Table V. Performance of Pilot Plant at the Dan River Project in Israel (Phase II) [20]

Parameter (mg/ℓ)	Summer Season Low-Load (15 weeks)			Winter Season High-Load (16 weeks)		
	Infl.	Effl.	% Removal	Infl.	Effl.	% Removal
BOD$_5$	303	N/A	–	357	N/A	–
BOD$_5$–Filt.	133	4.4	96.7	171	4.3	97.5
COD	647	62.1	90.4	716	68.4	90.4
COD–Filt.	205	44.4	78.3	263	49.4	81.2
SS–105°C	322	11.5	96.4	319	9.6	97.0
Total Nitrogen	44.9	3.0	93.3	53.6	3.1	94.2
Ammonia-N	37.0	1.1	97.0	41.8	0.9	97.8
Nitrate-N	Nil	1.0	–	Nil	1.2	–
Phosphorus	14.2	3.2	77.5	12.6	3.7	70.6
Alkalinity (CaCO$_3$)	399	309	22.6	409	311	24.0
pH	7.9	8.2	–	8.0	8.1	–
Liquid Temp. (°C)	–	29.2	–	–	23.3	–

problems now faced by Israel include developing full-cycle reuse in cities (such as Jerusalem) with critical water shortages, overcoming public and institutional obstructions involving dual water systems, and preventing potential contamination of high-quality groundwater aquifers.

THE RHINE BASIN

The Rhine River has been described in Der Spiegel as "the major European sewer". The river annually transports 2.9 Gg (1000 metric tons) of chromium, 1.4 Gg of copper, 11.2 Gg of zinc, 217 metric tons of arsenic, 63 metric tons of mercury, 10,000 Gg of chlorides and over 240 Gg of refractory organic carbon to the sea [21-23]. This same water provides over one-third of the public and municipal water supplies for Germany and The Netherlands, and will be called upon to an even greater extent in the future as groundwater reservoirs are used to their limits [23,24]. Pollution control is difficult along this 1370 km river because of the multinational character, and the several international organizations which police and monitor the river have met with only slight success in reducing pollutant loads [21]. Largely because pollution control has been so difficult in the Rhine Basin, Germany and The Netherlands have become pioneers in the technology of water purification, and various industrial and municipal water reuse schemes have been practiced in the lower Rhine Basin for decades.

Indirect water reuse through natural bank infiltration and artificial groundwater recharge provided over 25% of the public water supply in Germany in 1968, and accounted for over 60% in the heavily industrialized Ruhr basin.

Müller [24] has reviewed the aspects of the individual technologies. In the bank infiltration process, water seeps through the riverbed and subsequently through diluvial or alluvial strata of sand or gravel, whereupon it is pumped from wells or infiltration galleries. However, increasing pollution loads in the lower Rhine have led to decreased infiltration rates caused by increased organic and salt loads, and elevated concentrations of iron, manganese, ammonia and organics have resulted in increasing problems with tastes and odors. In Düsseldorf and Duisburg, combined ozonation-activated carbon treatment has proved necessary to produce a sufficiently high quality, potable water.

Artificial recharge can augment the natural infiltration process and has been practiced using infiltration basins (North Rhine/Westphalia), filter trenches (Hamburg), or infiltration wells (Wiesbaden-Schierstein) [24]. The infiltration basins, which are the most widely used technology, act as slow sand filters, and raw water must be pretreated by settling or biological processes before application if the concentration of suspended solids, bacteria or CO_2/O_2 ratio is too high. Filter trenches require river water to percolate through alluvial strata for 60 days before it is removed via the water works intake system located 85 m away. Infiltration wells essentially act to filter, dechlorinate, deodorize and lower the temperature of pretreated river water from 20°C to 12-13°C as it percolates 180 m through the groundwater alluvium. Pretreatment includes prechlorination, chemical coagulation ($FeCl_3$) and sedimentation, and filtration through rapid gravel filters. Generally, when the reuse ratio (wastewater:natural runoff) can be kept below 0.22 by dilution with reservoir releases, water quality in the Ruhr River is acceptable [9]. During an extreme low flow period in 1959, however, the reuse ratio increased to 0.86, and serious degradation of finished water quality resulted.

Inasmuch as The Netherlands are about to reach their maximum capacity for groundwater withdrawals, Dutch engineers have been evaluating possible direct or indirect reuse of municipal wastewater and river water for groundwater recharge of nonpotable municipal uses, particularly in the coastal Waddeneilanden. Hrubec et al. [23] describe a pilot project at the Dordrecht municipal wastewater facility that receives domestic wastewater. Treatment consists of a conventional activated sludge process with primary and secondary sedimentation, with 10% recycle of product water for scrubbing flue gas resulting from the incineration of household refuse and sewage sludge elsewhere in the plant. The plant effluent consistently exhibits lower levels of metals and organics and no mutagenicity, when compared to untreated Rhine River water.

Two AWT technologies were investigated at the Dordrecht plant, a physical-chemical (PC) and a reverse osmosis (RO) process. The PC process included lime treatment at pH 11.2, recarbonation with CO_2, double-layer filtration using crushed sand and anthracite, ozonation, biological trickling filtration and removal of ammonium ion using clinoptilite. The RO pilot plant consisted of addition of sodium hypochlorite, ferric chloride and polyelectrolytes (Superfloc C_{573}); an upflow contact filter and downflow rapid filter (1.2-2.4 mm sand); HCl addition; an RO unit; activated carbon filter; a dry marble filter and final dosing with sodium hypochlorite. The most effective membranes were found to be cellulose acetate tubular membranes from Patterson Candy International with a salt rejection of 97%. Total membrane area was 2.6 m^2.

The relative effectiveness of the two systems is summarized in Table VI. Removal of phosphate (> 99%), metals (17-92%) and pathogens (> 95%) could be accomplished by PC treatment, although iron, chromium and hardness concentrations tended to increase with treatment because of chemical dosing. Removals of inorganic, organic and pathogenic constituents were likewise high for the RO system, although the trickling filter was necessary for nitrification of ammonium-N, which was not efficiently removed by the RO process. Membrane fouling was controlled by shock applications of 40 mg/ℓ of chlorine, which produced much lower trihalomethane concentrations than did continuous disinfection. Periodic cleaning with enzymes and styrofoam balls was also shown to be effective. Overall, it was concluded that either technique could produce a high quality culinary water.

UNITED KINGDOM

Although the British Isles are usually thought to epitomize dampness, Eden et al. [25] point out that the per capita availability of water (3.9 m^3/day) is among the lowest in Europe. This low availability of high-quality

Table VI. Removal of Wastewater Constituents by Lime Treatment and Reverse Osmosis at the Pilot Plant at Dordrecht, Netherlands [after 23]

Parameter	Lime Treatment		Reverse Osmosis	
	Influent	Effluent	Influent	Effluent
Specific Conductance (μS/cm)	–	–	1220	215
COD (mg/ℓ)	37	29	47	2
Orthophosphate (mg/ℓ)	14	0.1	15	0.04
Calcium (mg/ℓ)	95	79	102	43
Silicate (mg/ℓ)	9.9	7.7	25	5
Color (units)	79	60	–	–
Iron (mg/ℓ)	0.25	0.03	–	–
Cadmium (μg/ℓ)	1.1	0.4	1.2	0.1
Copper (μg/ℓ)	7.5	2.6	3.0	0.5
Lead (μg/ℓ)	7.0	1.8	9.4	0.5
Max. Fecal Streptococci (10^3/mℓ)	5100	160		
Max. Coliphage (10^3/mℓ)	100	1.5	–	–

water has been exacerbated further by urbanization, and the overall demand continues to increase by some 2.5% per year. Present supplies are drawn approximately equally from upland catchments, groundwater and lower river reaches, and the further development of the last source alone appears socially and technically feasible. Increasing pollutant production and decreasing availability of dilution water further exacerbate the need for reuse.

London obviously has the most severe problem, because 70-80% of its water is drawn from the Thames River. Although the quality of water in the Thames reached an abominable state several decades ago, careful management has reduced problems with organics and nitrates to acceptable levels [21,26]. Indirect reuse of Thames River water is reported by Eden et al. [25] to be approximately 13%, based on surface discharges of treated effluents, but may be somewhat higher due to the effluents recharged to groundwater by over 140 wastewater treatment plants.

Direct reuse of wastewater in British industry has generally developed more as a response to individual needs than as a concerted national effort. The Appleby-Frodingham steel works directly integrated sewage effluent from a nearby town into its blast furnace gas cleaning system and (after lime softening) to its coke oven gas cooling and recirculating system [25]. The sewage, though higher in detergents and other pollutants, exhibited a more consistent quality than did the river water supplying the majority of the plant's requirement. Closed cycle recirculation was found to be economically superior to freshwater use if the latter had to be pumped more than 2.5 km.

Other industrial uses of wastewater in England [25] include the use of secondary wastewater from the City of Bristol for gas scrubbing, gypsum slurry and for various other purposes, including quenching the incinerator at the wastewater treatment works. A wool textile mill in Pudsey discovered that secondary effluent from the local treatment plant could be relied on alone for process water, thus saving instream flows in the town's river for other purposes. A limestone slurry pipeline near Dunstable found that secondary effluent from the local treatment plant could save up to 90% of the cost of rail transport without withdrawing valuable, high-quality groundwater. Water from the Mersey River, generally of poorer quality than wastewater effluent, was supplied for years at reduced cost to an industrial complex at Warrington, and millions of cubic meters of used water are used each day for cooling. Research projects have been undertaken on the use of ozone (Redbridge), activated carbon, both with coagulant (Landford) and without (Rye Meads), and reverse osmosis and ultrafiltration (Rye Meads). Water conservation measures, including reservoir control and groundwater recharge and abstraction, have also been under study.

JAPAN

Although Japan receives 650×10^6 m^3 of precipitation annually, much of the usable supply is depleted by the time it reaches the heavily populated urban areas on the coast. Furthermore, the projected 1985 population of some 36 million in the Tokyo area portends severe shortages of industrial water and wastewater disposal problems. The Kinki district (including Osaka, Kobe and Kyoto) will similarly experience growth pains. Kubo and Sugiki [27] and Kashiwaya and Annaka [28] have described research into ameliorative measures.

Between 1963 and 1969, the overall recovery rate of water in Japanese industry rose from 25 to 50% and the water consumption per unit production was reduced by more than two thirds. Reuse tactics ranged from reclaiming wastewater to be used as carriage water in apartment buildings to extensive use of treated sewage in industrial plants in Tokyo, Nagoya and Kawasaki. Treatment processes varied and included both physical-chemical and biological unit operations. The principal problems encountered included corrosion by ammonia and salinity seeping into collection lines and scaling and slime growth.

An unusual technology is being employed to treat night soil in the city of Kiryū. Night soil undergoes aerobic digestion for 10-20 hr to reduce the solids volume (Figure 5). The sludge then is wasted and the supernatant undergoes aerobic decomposition, aided by photosynthetic sulfur bacteria, in three successive digestors which are aerated to provide stirring and suppress the growth of unwanted bacteria. After a total detention time of three days, the solids are harvested and the supernatant is sent to aerobic digestors innoculated with the green alga, *Chlorella,* where it undergoes final polishing for 36 hours. The final effluent exhibits a BOD$_5$ of 15 mg/ℓ and NH$_3$-N of 31 mg/ℓ, representing 99.8% and 99.2% reductions from the night soil, respectively. Some 13 g of cells per liter of effluent are produced, which may have potential for recovery of single cell protein.

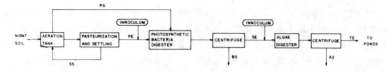

Figure 5. Flow diagram of Kiryū City pilot plant for night-soil treatment [after 27]. SS = settled sludge; PE = primary effluent; SE = secondary effluent; TE = tertiary effluent; PG = product gas; BS = bacterial solids; AS = algal solids.

AUSTRALIA

The Australian continent is the driest on earth and its rivers exhibit severe fluctuations in flow. In order to supply water to its 14 million inhabitants, reservoirs and interbasin transfers are already a necessity. Although reuse of water is not presently widespread, Smith [29] describes some present research efforts.

Wastewater effluent, either treated or untreated, is applied to pasture land by over fifty municipal facilities. The Werribee Farm near Melbourne has been in operation since 1897 and produces over $0.5 million in stock sales annually. Unlike the African experience [12], thus far there has been no evidence of unusual veterinary problems. Effluent has also been applied to silviculture resulting in an enhanced growth rate of native trees, which is contrary to the Rhodesian experience [30] in which native trees withered under irrigation. Trials are also underway with application of chlorinated secondary effluent to vegetable crops and viticulture. Reclaimed water is widely used for municipal landscape watering, and the effluent from the Broken Hills, New South Wales, municipal plants is being used successfully for revegetating mine spoils. Industrial reuse, although not widespread, is slowly increasing.

Much research is presently being undertaken on methods and public health implications of water reuse by the Reclaimed Water Committee of the Ministry of Water Resources and Water Supply. Virological, protozoological and bacterial survival of reuse technologies are being investigated, and thus far lagooning appears to be the safest reclamation technology. Other studies include evaluations of timber and vegetable growth trials, feasibility of groundwater recharge, instream flow augmentation, landscape watering, design and evaluation of facultative ponds, algae harvesting and physical chemical treatment processes [29].

INDIA

The rapidly expanding population in India, coupled with a generally poor groundwater supply, has led to water shortages and wastewater disposal problems for decades. It is thus not surprising that in cities like Calcutta which already supply less than half of their annual water needs, water reuse technologies meet with considerable interest. In addition to water resource considerations, however, the nutrients contained in the wastewater represent an invaluable source of fertilizer in a country struggling to meet its food resource needs. Arceivala [31] has discussed the technological strategies directed toward solving some of these problems.

Bombay has augmented its water supply by treating 5000 m^3/day of municipal wastewater using extended aeration, settling, alum coagulation, sedimentation and rapid filtration. Part of the filtrate is softened using zeolites and blended with the main stream to give approximately 40 mg/ℓ hardness. Approximately 90-93% of the raw sewage inflow can be reclaimed to produce a product water at 15% of the cost of fresh municipal water. In this plant care was taken to obtain raw water from a point in the collection system not heavily contaminated with industrial waste or saline groundwater seepage. Another plant in Bombay has experienced trouble in economically reducing the concentration of yellowish organics resulting from industrial inputs to the raw water.

An innovative method for saving cooling water in tall buildings is the construction of package plants suitable for renovating wastewater directly in their own basements. Extended aeration serves to provide odor free operation, followed by chemical coagulation, sand filtration, partial softening (as necessary), and shock chlorination to prevent growth of slimes and algae in the holding tank. Acid dosing to control scaling of the cooling tower is performed manually, as needed. Excluding the loss of basement rentals, the cost is about 7% of that of fresh municipal water.

In an effort to conserve water, the Bombay Municipal Corporation undertook a survey of industries in the city to determine the potential for water reuse [31]. Considerable savings could be realized in water bottling, automobile manufacture, asbestos cement and clutch lining, food processing, dairy, paper-board, pharmaceutical, chemical, petrochemical and textile industries, and in railway yards, bus depots, docks and harbors. Generally, needs not requiring expensive demineralization (e.g., boiler feed and certain process waters) could be had for a fraction of the cost of municipal water through renovation and reuse. It is particularly significant that one reuse scheme for a textile plant could reduce freshwater demands by more than 50% in an industry that already uses about one third of the design water requirements of those in Western textile mills.

As discussed above, India has been confronted with an insufficient supply of protein, owing to its rapid growth and high population density. To counteract this limited supply of protein, sewage has been used in India as a fertilizer and soil conditioner on sewage farms since the last century. During the mid-1970s, over 1×10^6 m^3 of sewage was applied to over 12,000 ha of farmland each day, and considerable interest has centered on the further development and health aspects of wastewater irrigation. Thus far BOD$_5$ loadings typically range from 25 to 150 kg/ha·day, which represents approximately $1.00/ha·day in fertilizer value at 1975 prices in India. Unfortunately, irrigation of edible crops with raw sewage has led to some contamination of the produce and to a particularly high incidence of disease among sewage farm workers, who wear little protective clothing or footwear [31].

Recent technological efforts have attempted to ameliorate some of these problems. Treatment of sewage prior to irrigation by primary settling, oxidation ponds or trickling filters has significantly reduced contamination. Similarly, digestion of night soil prior to application to the soil results in considerable destruction of microorganisms, and produces biogas as a valuable by-product used in cooking. Application of sewage to nonedible export crops such as citronella and mentha have proven to be economically attractive.

Basu [32] has described an interesting system design proposed to treat sanitary and washing wastewater from a copper ore handling facility at Malanjkhand. Oily wash water is pretreated and combined with sanitary wastes, which are then run through a facultative oxidation pond and a final polishing pond. The effluent from the polishing pond is split between fish polyculture ponds (that are also used for recreational boating) and irrigation water which is applied to an agricultural-horticultural park that surrounds the grounds of the facility. This process provides for final polishing of the runoff together with a park-like setting for local inhabitants, and at least 4000-7000 kg/ha·yr of marketable fish.

CANADA

Although Canada has more than sufficient water to meet its immediate needs, there exist two critical areas for water reuse technologies. The first is in the area of closed-cycle industrial reuse to decrease pollutant discharges. These uses are discussed in many of the papers cited (in English [5] and Smith [6-8]) and will not be dealt with here. The second, and perhaps most unique, area is that of providing water and disposing of wastewater in the virtual deserts of the Arctic.

Bromley and Benedek [33] point out that potable water is relatively unavailable in many Arctic areas due to difficulties in drilling through the permafrost, well freezing and generally low groundwater quality [> 2000 mg/ℓ total dissolved solids (TDS)]. Furthermore, discharge of wastewater into streams and wetlands may overtax ecosystems that already operate under considerable stress. Bromley and Benedek devised a self-contained, heated treatment system for this environment (Figure 6) that includes an innovative provision of chemical coagulant to provide adequate performance of the reactor/clarifier during cold-weather start-up. Pilot tests of a one-man system (45 ℓ/day) showed production of an effluent suitable for toilet flushing over a wide range of loading rates, and that biological processes precluded the need for coagulants after one month. Buildup of COD as a result of continuous recycle was seen as a potential problem.

Canada's Ontario Research Foundation has devised two particularly interesting units for Arctic operations, the Environmental Service Module (ESM)

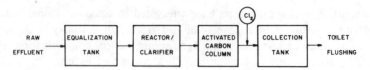

Figure 6. Flow diagram for household wastewater renovator for use in the Arctic [after 33].

and the Canadian Water Energy Loop (CANWEL). The ESM has been described by Smith and Laughton [34]. This module provides extensive recycling for all but potable water, and can be transported in a Hercules C-130 air transport. The heart of the technology (Figure 7) is a wet oxidation process (WETOX), in which macerated garbage and blackwater are oxidized in a violently agitated, pressurized (600 psi), heated (~230°C) chamber. Separate removal of gas and liquid phases provide for a longer liquid retention time (50-60 min), and heat-exchange and the exothermic nature of the oxidation process lead to increased energy efficiency. Removal of 70-75% of the influent COD has been demonstrated in this stage.

Following oxidation, WETOX effluent and graywater are fed to a reactor for lime treatment. At the observed optimum pH of 10, reduction of suspended solids, TOC and COD were 99, 24 and 20%, respectively. Following clarification, the effluent is then filtered and routed to an RO unit employing a tubular membrane, and the permeate is fed to a second hollow fiber membrane RO unit. This permeate is then treated in a UV/ozone contactor which destroys residual organics to yield a potable quality water. The concentrate from the first reactor is evaporated using steam from the WETOX reactor,

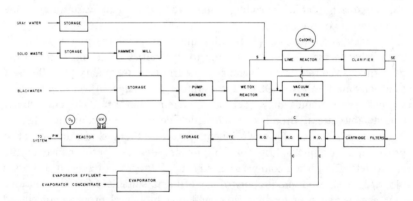

Figure 7. Flow diagram of the ESM total treatment scheme [after 34]. S = stream phase; SE =secondary effluent; TE = tertiary effluent; C = PO concentrate; PW = polished water.

and the concentrate from the second RO unit is recycled to the first RO unit. Overall water requirements are reduced to 47 ℓ/person·day by providing vacuum transport of sewage, a sauna, and low-pressure shower nozzles.

Smith and Laughton also describe the CANWEL (Canadian Water Energy Loop), a system designed to treat all solid and liquid wastes from a community and provide a minimum load of liquid and gaseous discharges. In the CANWEL system (Figure 8) nitrification-denitrification is carried out in a two-stage reactor in which the second stage produces a nitrified effluent which is then recycled to the anaerobic first stage, which supplies organic substrates for denitrification. Recycle is controlled by automated dissolved oxygen monitoring to ensure anoxic conditions in the first stage during low nighttime loading periods. The effluent is passed through a cyclonic sludge settler, made necessary because of the poor settling characteristics of the sludge, then is returned to the first stage. The effluent is treated with alum in a reactor-clarifier, and the final effluent is disinfected with ozone. The effluent is now acceptable for discharge to streams where dilution water is not available. Alternatively, activated carbon and RO treatment can be added to produce a reusable water (Table VII). A demonstration project to serve 500 residents in an apartment building is presently under development.

PROPOSED WATER REUSE INNOVATIONS
IN OTHER COUNTRIES

Many smaller countries, especially in arid and island environments, are finding their freshwater supplies insufficient to meet increasing municipal and industrial demands. Many of these problems are being met with innovative technology imported from countries already experienced in water reuse through past need. A few such projects are described below.

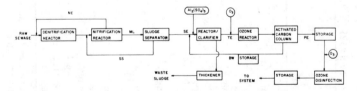

Figure 8. Diagram of CANWEL pilot plant [after 34]. NE = nitrified effluent; ML = mixed liquor; SS = settled sludge; SE = secondary effluent; TE = tertiary effluent; PE = polished effluent; BW = backwash. In the full-scale system, waste sludge and brine from the optional RO unit would be mixed with garbage and burned in a starved air incinerator equipped for heat exchange.

Table VII. Effluent from CANWEL Wastewater Treatment System (reverse osmosis excluded) [34]

Parameter	Acceptable Limits for Canadian Drinking Water Standards	Acceptable Limits for Ontario Public Surface Waters	CANWEL Waste Water Renovation Process	
			Objective	Achieved to Date
Temperature (°C)	15	29.5	30	25
Dissolved oxygen (mg/ℓ)			>5.0	7.0
BOD (mg/ℓ)			3	3
Total solids (mg/ℓ)			500	500
Suspended solids (mg/ℓ)			5	3
Dissolved solids (mg/ℓ)	1000	500		
Total phosphorus (mg/ℓ)	0.07		0.10	0.10
Soluble phosphorus (mg/ℓ)			0.05	0.05
NH_3 as N (mg/ℓ)	0.50	0.50	0.10	0.10
$NO_3 + NO_2$ as N (mg/ℓ)	10.0	10.0	3.0	4.0
Chloride (mg/ℓ)	250	250	150	70
pH	6.5-8.3	6.0-8.5	6.5-8.5	7.0
Turbidity (NTU)	5		.5	<1
Color (TCU)	15	75	15	15
Odor (TON)	4[a]	removable	absent	absent
Total coliform (MF) (org/100 mℓ)		5000	1000	zero[b]
Fecal coliform (MF) (org/100 mℓ)		500	100	zero[b]
Enterococci (MF) (org/100 mℓ)			20	zero[b]
Standard plate (count/ℓ)				<10

[a] 95% of 30-day samples negative; positive counts not to exceed 4/200 mℓ or 10/500 mℓ.
[b] 1 liter samples examined after neutralization of free ozone residual with thiosulphate.

Saudi Arabia

Kalinske et al. [35] describe a wastewater reclamation facility designed to produce additional water for expansion of a petroleum refinery at Riyadh. The source of the 20,000 m^3/day of wastewater is a very low quality effluent from the local wastewater treatment plant. Product water must include utility water for hose station and fire control, cooling tower and desalting use (< 500 mg/ℓ TDS) and boiler feed water (< 20 μmho/cm, < 0.5 mg/ℓ Si). Potable water is drawn from a local well.

The proposed design includes preliminary short detention in surge ponds followed by high-lime, MgOH and $FeCl_3$ coagulation for removal of calcium and silica, flocculation and sludge removal, and recarbonation. The process is repeated in a second set of flocculation basins and subsequently followed by pH adjustment and chlorination. The product water can be used for utility water or as feed for the physical treatment system.

The water is next cooled to 25°C in induced draft cooling towers, flocculated with polymer and passed through dual-media filters. Activated carbon is used to reduce the soluble organic compounds, and following dechlorination, the filtrate is sent to a two-stage RO system equipped with spiral-wound, polyamide filters. Recovery of < 200 mg/ℓ TDS water suitable for process applications should total approximately 75%. The effluent from the second RO unit (also exhibiting 75% recovery) can be treated with two bed ion exchangers for production of high-quality boiler feed water. Reject feeds from the first and second units are wasted to an evaporation pond and recycled to the first unit, respectively. Sludges will be treated to recover Mg, Ca, and CO_2 because of the high cost of chemicals, and activated carbon will be regenerated on site. Costs are conservatively estimated to be \$0.20-0.23/m^3 if the plant were located in the United States.

St. Croix, Virgin Islands

During the 1960s, St. Croix's population roughly doubled as the result of a rapidly expanding tourist industry. This expansion represented a considerable strain on the island's fresh groundwater supply, which is limited by low aquifer capacity. To remedy this situation, the government contracted both desalination and water reuse projects. The water reuse facility consisted of a conventional 2000 m^3/day chemical coagulation, filtration and chlorination facility, followed by recharge to groundwater through spreading basins.

The system initially had problems with insufficient wastewater connections, and subsequently with storm damage to the interceptor system. One of the most significant problems to date, however, has been the use of saline flush- and fire-control water, which provides an unacceptably saline wastewater.

The solution to this problem is seen to be the mixing of groundwater with high-quality desalinated sea water for all municipal use. This would in turn provide a high-quality recharge water, thus helping to close the cycle. Costs of reclamation in 1976 were $0.57/m^3, which compared favorably with costs of desalinated water ranging from $1.30 to $5.00/m^3. Buros [36] provides a detailed discussion.

Indonesia

Snell and Naito [37] discuss a situation in Indonesia in which water reuse was not only necessary, but also profitable. It was discovered that there was insufficient ground water to supply the needs of a proposed 6400-room hotel complex on the Island of Bali. By reducing the size of the complex to 3200 rooms, 55% of the water needs (street washdown, lawn watering, air conditioning make-up water, laundry, etc.) could be met by wastewater renovation and reuse.

In this process raw wastewater will be cycled slowly through four sequential lagoons located on a compacted marl tidal flat. Fresh groundwater seepage will be collected from beneath a ditch separating the last two lagoons, from a porous dike below the fourth lagoon, and in cases of high demand, from the fourth lagoon itself. This ground water then will be treated with alum, mixed, coagulated, settled, filtered through mixed media and chlorinated. Then it will be distributed through a separate water system. Carp and nonbottom feeding fish will be reared in the lagoons as a supplementary protein source for the native staff. As an added bonus, the recycled water will not need to be softened, thus reducing treatment costs below what would be required if the total hotel supply were deep ground water.

Hong Kong

Until twenty years ago, when Hong Kong began to import water from China, this city relied for its water supply on rainfall which was restricted to a few hours daily. Today, with some desalination, the annual consumption is somewhat in excess of 500×10^6 m^3/yr, and is expected to exceed the available supply by 1984. Historically, sea water has been used for toilet flushing, but because it produces a brackish effluent, opportunities for renovation were not good. Lau [38] describes the recent resurgence of interest in reuse brought about by impending demands for additional water.

The two principal sources of reusable water are the lower reaches of streams, which are grossly polluted from agriculture, sewage and light industry effluent (40×10^6 m^3/yr), and the effluent from the municipal wastewater treatment plants (ultimately 140×10^6 m^3/yr) which could be used if reclaimed wastewater replaced sea water in the water carriage system. Pilot flushing trials with chlorinated, clarified and tinted reclaimed water are presently being conducted in an apartment building. Also being investigated is the use of reclaimed wastewater for air conditioning equipment, much of which is presently employing sea water delivered by long lengths of delivery line.

Pilot plants are under investigation for treatment of secondary effluent or river water. Different methods will be evaluated, including coagulation followed by either ammonia stripping or nitrification-denitrification for the secondary effluent, and activated sludge for the surface water. Final polishing could include chlorination, ozonation and activated carbon treatment. Obstacles still to be overcome include establishing reservoir sites to accomplish blending natural water with product water, which is felt to be important in overcoming psychological objections to potable reuse. Also of concern is the present low cost ($0.18/m^3$) of municipal water which tends to discourage industrial reuse.

SUMMARY AND CONCLUSIONS

Given the diverse needs for water reuse technologies among different countries, it is not surprising that the solutions tend to take on characteristic, if not unique, attributes. In water-rich but heavily industrialized areas such as the lower Rhine and Thames River Basins, problems are predominantly associated with heavy reuse cycles. Indirect reuse factors have already reached 22% and 13% in these basins, respectively, and can become much higher during droughts [23]. The heavy industrial base responsible for water degradation is usually capable, however, of assuming the high costs of energy and technology-intensive advanced wastewater treatment techniques. Effluent taxes often serve to internalize the costs of such treatment in European nations.

Another situation often requiring water reuse technologies occurs in extremely arid environments, in which development of freshwater sources is difficult or expensive. Such environments include not only hot deserts, but also oceanic islands and arctic settlements. Frequently reuse in these settings is more attractive economically than desalination of marine waters.

In semiarid, agricultural communities, water represents a sufficiently valuable resource that its application to irrigated agriculture represents a means of simultaneously disposing of a pollutant while increasing productivity. Whether such agriculture is restricted with respect to human consumption depends to a large extent on both the overall availability of both water and protein, and also upon traditional and institutional factors. Energy and protein-rich nations raise low-valued export crops, but less affluent societies raise food, fish and animal crops with lower quality effluents. The relative tradeoffs between increased nutritional status and increased disease transmission appear to be largely unknown.

Perhaps the most salient characteristic shared by most of the reuse technologies described above is their competitiveness with developing additional freshwater supplies. With the cost of a major dam holding approximately 1 km^3 of water at approximately $120 million [1] and the dearth of capital presently available, development of new water sources is difficult even in countries with sufficient undeveloped water. However, reused water is available in many of these countries at a fraction of the cost of fresh water and is adequate for most nonpotable uses. This in turn increases the availability of high-quality water for higher beneficial uses. While water reuse cannot replace conservation and development of new surface and groundwater resources, it can go a long way toward improving the quality of life and the global environment.

REFERENCES

1. Ambroggi, R. "Water," *Scientific Am.* 243:100-116 (1980).
2. *American Water Works Association Water Reuse Symposium* (Denver, CO: AWWA Research Foundation, 1979), p. 2332.
3. Shuval, H. I. *Water Renovation and Reuse* (New York: Academic Press, 1971), p. 463.
4. Biswas, A. K. *Water Development and Management* (New York: Pergamon Press, 1978), p. 2646.
5. English, J. "Water Reclamation and Reuse," *J. Water Poll. Control Fed.* 49:1078-1087 (1977).
6. Smith, D. W. "Water Reclamation and Reuse," *J. Water Poll. Control Fed.* 50:1149-1166 (1978).
7. Smith, D. W. "Water Reclamation and Reuse," *J. Water Poll. Control Fed.* 51:1250-1276 (1979).
8. Smith, D. W. "Water Reclamation and Reuse," *J. Water Poll. Control Fed.* 52:1242-1284 (1980).
9. Richards, R., B. Tunnah, A. Shaikh and K. Stern. "Waste Water Reuse—The CCMS Drinking Water Study," in *Water Reuse Symposium* (Denver, CO: AWWA Research Foundation, 1979), pp. 937-947.
10. Smith, M. A. "Re-use of Water in Australia," in *Water Reuse Symposium* (Denver, CO: AWWA Research Foundation, 1979), pp. 925-936.
11. Odendaal, P., and L. vanVuuren. "Reuse of Wastewater in South Africa—Research and Application," in *Water Reuse Symposium* (Denver, CO: AWWA Research Foundation, 1979), pp. 886-906.
12. Hart, O., and L. vanVuuren. "Water Reuse in South Africa," in *Water Renovation and Reuse,* H. Shuval, Ed. (New York: Academic Press, Inc., 1977), pp. 355-397.
13. Humphries, T. H. "The Use of Sewage Effluent as Power Station Cooling Water," *Water Res.* 11:217-223 (1977).

14. vanVuuren, L., A. Clayton and D. vanderPost. "Current Status of Water Reclamation at Windhoek," *J. Water Poll. Control Fed.* 52:661-671 (1980).
15. Grabow, W., and M. Isaacson. "Microbiological Quality and Epidemiological Aspects of Reclaimed Water," *Prog. Water Technol.* 10:329-335 (1978).
16. vanVuuren, L., and M. Talvard. "The Reclamation of Industrial/Domestic Wastewaters," in *Water Reuse Symposium* (Denver, CO: AWWA Research Foundation, 1979), pp. 907-924.
17. Shelef, G. "Water Reuse in Israel," in *Water Renovation and Reuse,* H. Shuval, Ed. (New York: Academic Press, Inc., 1977), pp. 311-332.
18. Idelovitch, E. "Wastewater Reuse by Biological-Chemical Treatment and Groundwater Recharge," *J. Water Poll. Control Fed.* 50:2723-2740 (1978).
19. Idelovitch, E., T. Roth, M. Michail, A. Cohen and R. Friedman. "Dan Region Project in Israel: from Laboratory Experiments to Full-Scale Wastewater Reuse," in *Water Reuse Symposium* (Denver, CO: AWWA Research Foundation, 1979), pp. 808-833.
20. Farchill, D. "The Dan Region Wastewater Project—Stage II: An Advanced Single Stage Biological Wastewater Treatment and Renovation System," in *Water Reuse Symposium* (Denver, CO: AWWA Research Foundation, 1979), pp. 830-836.
21. Temple, T. "A Tale of Two Rivers," *EPA J.* 6:28-30 (1980).
22. Förstner, U., and G. Wittman. *Metal Pollution in the Aquatic Environment* (New York: Springer-Verlag, 1979), p. 486.
23. Hrubec, J., J. Schippers and B. Zoeteman. "Studies on Water Reuse in the Netherlands," in *Water Reuse Symposium* (Denver, CO: AWWA Research Foundation, 1979), pp. 785-807.
24. Müller, W. J. "Water Reuse in the Federal Republic of Germany," in *Water Renovation and Reuse,* H. Shuval, Ed. (New York: Academic Press, Inc., 1977), pp. 258-276.
25. Eden, G., D. Bailey and K. Jones. "Water Reuse in the United Kingdom," in *Water Renovation and Reuse,* H. Shuval, Ed. (New York: Academic Press, Inc., 1977), pp. 398-428.
26. Onstad, C. A., and J. Blake. "Thames Basin Nitrate and Agricultural Relations," in *Symposium on Watershed Management* (New York: American Society of Civil Engineers, 1980), pp. 961-973.
27. Kubo, T., and A. Sugiki. "Wastewater Reuse in Japan," in *Water Renovation and Reuse,* H. Shuval, Ed. (New York: Academic Press, Inc., 1977), pp. 333-354.
28. Kashiwaya, M., and T. Annaka. "Current Research on Advanced Wastewater Treatment in Japan," *J. Water Poll. Control Fed.* 52:1008-1012 (1980).
29. Smith, M. A. "Re-use of Water in Australia," in *Water Reuse Symposium* (Denver, CO: AWWA Research Foundation, 1979), pp. 925-936.
30. McKendrick, J. "Compulsory Re-use of Water Due to Very Strict Water Pollution Control Regulations in Salisbury, Rhodesia," in *Water Reuse Symposium* (Denver, CO: AWWA Research Foundation, 1979), pp. 1035-1048.
31. Arceivala, S. J. "Water Reuse in India," in *Water Renovation and Reuse,* H. Shuval, Ed. (New York: Academic Press, Inc., 1977), pp. 277-310.

32. Basu, A. "Ecologically Balanced Wastewater Renovation System," in *Water Reuse Symposium* (Denver, CO: AWWA Research Foundation, 1979), pp. 993-1008.
33. Bromley, D., and A. Benedek. "Reduction of Potable Water Demand in the Arctic by Reuse," in *Water Reuse Symposium* (Denver, CO: AWWA Research Foundation, 1979), pp. 1009-1019.
34. Smith, D., and R. Laughton. "Development of Two Sewage and Solid Waste Treatment Systems with Potential for Water Reuse and Energy Recovery," in *Water Reuse Symposium* (Denver, CO: AWWA Research Foundation, 1979), pp. 836-885.
35. Kalinske, A., J. Willis and S. Martin. "Reclamation of Wastewater Treatment Plant Effluent for High Quality Industrial Reuse in Saudi Arabia," in *Water Reuse Symposium* (Denver, CO: AWWA Research Foundation, 1979), pp. 958-992.
36. Buros, O. K. "A Water Reuse Program on St. Croix, Virgin Islands," in *Water Reuse Symposium* (Denver, CO: AWWA Research Foundation, 1979), pp. 1023-1034.
37. Snell, J., and S. Naito. "Water Recycling—Necessity and a Profit," in *Water Reuse Symposium* (Denver, CO: AWWA Research Foundation, 1979), pp. 1020-1022.
38. Lau, T.-H. "The Role of Re-use in Hong Kong as a Potential Future Water Resource," in *Water Reuse Symposium* (Denver, CO: AWWA Research Foundation, 1979), pp. 948-957.

CHAPTER 25

WASTEWATER REUSE IN THE FOOD PROCESSING INDUSTRY

Larry A. Esvelt
Esvelt Environmental Engineering
Spokane, Washington

INTRODUCTION

The food processing industry in the United States is well controlled and regulated by the industry and federal and state agencies, all of whom hold food safety and quality to have paramount importance in processing and production. Federal, state and local agencies are all concerned with public health and safety in the food processing industry and have taken active interest in preservation of food quality and safety through the assurance that food is processed utilizing water of adequate quality.

The food processing industry uses vast quantities of water and this water must be of adequate quality to ensure a safe and wholesome product for human consumption. Within the fruit and vegetable processing segment of the food processing industry alone there are approximately 1600 plants in the United States which process about 30 million tons of raw product per year. The industry uses about 430 billion liter/yr of water, nearly all of which is subsequently discharged as wastewater. About 46% of the wastewater is discharged to publicly owned treatment works (POTW, such as city treatment plants) about 28% is disposed of to surface waters and about 26% is disposed of by land application.

The increasing cost of suitable water supplies, the decreasing availability of such supplies in some regions and the increasing cost of wastewater treatment and disposal will tend to make processing wastewater recycling and processing wastewater effluent treatment, reclamation and reuse increasingly desirable from an economic standpoint. In order for wastewater reuse to become implemented widely in the food processing industry, however, the public must have confidence that the product will be safe and wholesome. It is unlikely that wastewater containing any sanitary sewage will be considered suitable for reclamation for reuse in food processing plants, so the potential source is probably limited to those waste streams emanating from the food processing activity itself.

Many food processing plants are already treating their processing effluents to a quality suitable for discharge to surface waters. Since the requirements for such treatment are relatively high in many places, these plants seem to be the most likely to adapt additional technology for treatment of their wastewater suitable for direct reuse within the processing plants. Dischargers to POTW are being increasingly faced with pretreatment requirements on top of substantial wastewater charges. This may prompt many plants to consider treatment and reuse of their processing wastewaters as an alternative water supply source.

Plants not discharging to surface waters or POTW generally discharge their wastewaters by land disposal. Although this mode of disposal may be considered reuse, the primary objective of the land application is generally wastewater disposal. Some treatment prior to land disposal is often required for odor control, protection of ground waters or compatibility with crops grown on the disposal site. Increases in production, increasing environmental demands on the disposal system and increasing costs of water supplies may lead to an evaluation of processing wastewater reuse by food processors currently using land disposal.

The principal concern among food processors and regulatory agencies is whether and to what degree reuse of wastewaters could introduce contaminants into the processed food. Of primary and most immediate concern are acute toxicants and pathogens which would cause disease or physiological stress to the consumer. Also of great importance are other biological agents which could cause food spoilage, other less toxic chemicals, potential carcinogenic compounds and agents which might alter the food wholesomeness, taste or desirability.

The desirability of reclaiming and reusing processing wastewaters in the food processing industry is thus counterbalanced by the necessity of maintaining product safety and quality. This exhibits itself in a cautiousness and a desire to move in a very deliberate fashion toward implementation of reclamation and reuse while making certain that such practices will not jeopardize the product quality or safety or acceptance by the consumer.

WATER RECYCLING

One method used widely for reducing water consumption, wastewater production and waste discharge is recycling of process waters within the food processing plant. This may be done with or without some form of treatment between the initial and subsequent uses of the wastewater. The treatment, however, does not attempt to reclaim the water quality to its original or water supply quality levels, but merely removes particulate material by such processes as screening on rotary, vibrating or static screens and/or disinfection with chlorine or chlorine dioxide.

Recycling can be made within the particular process module (modular recycle) of a plant such as peeler washwater or product cooling water where the used water is recycled following some appropriate treatment, such as heat removal or disinfection, for use in the same process. Makeup water is added as necessary to replace losses from evaporation and spillage or to maintain the water quality within the process at an acceptable level. The water quality must be maintained at an adequate level to preclude product contamination microbiologically or with "filth".

Another widely used recycling practice within processing plants is countercurrent recycle where used water from one process is recycled counter to the product flow to processes having lower quality requirements. An instance of this type recycle would be utilization of spent cooling waters for raw product washing and conveying. The recycle within a process or modular recycling may be combined with countercurrent recycling to achieve the greatest overall water reduction in a plant.

In-plant water recycling has been practiced by many processors and is well documented [1-4].

WASTEWATER REUSE APPLICATIONS

Direct reuse has been defined as the planned and deliberate use of treated wastewater for some beneficial purpose such as irrigation, recreation, industry or potable reuse. Indirect reuse of wastewater occurs when water used for domestic or industrial purposes is discharged into fresh surface or underground waters and is used again in its diluted form [5].

Indirect reuse is essentially an uncontrollable entity and occurs in many cases without documentation or planning. Direct reuse, however, is usually planned and is indeed deliberate. From the standpoint of the food processing industry recreational reuse is not usually a factor. Direct reuse for irrigation, usually described as land application of wastewater from food processing, is normally justified as a wastewater treatment and disposal means rather than a reuse. The principal benefit from land application of food processing wastewaters is the wastewater disposal. Normally the application site is managed with wastewater disposal as the primary objective, and not optimized for use of the resource as an irrigation supply. Land application of food processing wastewaters has been well documented [1-3,6,7].

Direct reuse of food processing wastewaters as an industrial or potable supply is of particular interest within the food processing industry since it is within these areas that significant reductions in new water supply can be achieved. As a potable supply the interest in reuse is in obtaining sufficient water quality for use in areas of food processing plants where a potable supply is mandated by regulations.

There is currently no interest in the food processing industry for reclaiming and reusing wastewaters containing sanitary sewage. Interest is centered on reclaiming those waters which have been used only for food processing and in the processing plant for nonsanitary purposes.

The reuse of a reclaimed processing effluent in a food processing plant could be directed to specific processing areas believed most suitable for utilizing reclaimed water and thus minimize the risk of product contamination. Industry could alternatively utilize the reclaimed effluent for general purposes interchangeably with and in mixture with new water supply. Using the reclaimed water in selected areas allows for confirmation that the reclaimed effluent is of adequate quality for use in those areas and demonstration that the reclamation process is adequate before its use is expanded to other more critical areas. A dual plumbing requirement and cross-connection control would accompany this reuse philosophy. General use would require that the reclaimed effluent be of a quality unquestioned for the reuse in any area of the plant and would require that it be of a level equal to the plant's freshwater supply. Restricting use of reclaimed water to specific plant areas will probably be most acceptable to regulatory agencies who will have some reservations regarding its general acceptability. It would also allow for a period of demonstration or proving of the effluent quality before it would be introduced into more critical plant areas. This appears to be the most promising method of introducing water reclamation and reuse into the food processing industry.

REGULATION OF WASTEWATER RECLAMATION AND REUSE IN THE FOOD PROCESSING INDUSTRY

Regulation of wastewater reclamation and reuse in food processing is primarily the responsibility of federal agencies. Their regulatory control stems from concern for the health of the consuming public and the wholesomeness of the foods produced by the food processing plants.

Authority

There are several agencies with regulatory control authority over food processing. Each has been directed by Congress to fulfill a particular role in consumer protection and protection of the public health and welfare.

The Food and Drug Administration receives its authority from the Federal Food, Drug, and Cosmetic Act, passed by the 75th Congress in 1938. Section 402 of that Act defined that "A food shall be deemed to be adulterated: (1) if it bears or contains any poisonous or deleterious substance which may render it injurious to health; (2) if it bears or contains any added poisonous or added deleterious substance which is unsafe; (3) if it consists in whole or in part of any filthy, putrid or decomposed substance or if it is otherwise unfit for food; or (4) if it has been prepared, packed or held in unsanitary conditions whereby it may have become contaminated with filth or whereby it may have been rendered injurious to health". The Act provides the Secretary with authority to promulgate regulations to ensure the quality of processed foods entering the interstate commerce.

The Agricultural Marketing Act of 1946 charged the Secretary of Agriculture with duties related to agricultural products. These duties included: (1) determination of methods of processing, packaging, marketing and publication of results; (2) the determination of costs; and (3) the improvement of standards of quality and condition of food products. The Act also included provisions for the inspection and certification of products in interstate commerce. Included within the task of improvement of standards of quality and condition is a direction to develop and improve standards of quality, condition, quantity, grade and packaging and recommendation demonstration of such standards in order to encourage uniformity and consistency in commercial practices.

The Environmental Protection Agency was given authority under Public Law No. 93-523 enacted in 1974, amending the Public Health Service Act, to ensure that the public is provided with safe drinking water. The Safe Drinking Water Act provided for drinking water regulations to apply to each public water system in each state. The term public water system was defined to mean "a system for the provisions to the public of piped water for human consumption if the system has at least 15 service connections or regularly serves at least 25 individuals". It would appear that most, if not all, food processing plants would be covered by the Drinking Water standards under Public Law 93-523 since most use some water in their filling operations which is destined for human consumption by and indeterminate number of people. The Safe Drinking Water Act provides for the establishment of maximum contaminant levels (MCL) meaning the maximum permissible level of a contaminant in water which is delivered to any user of the public water supply system. These MCL would apply to National Primary Drinking Water

Regulations intended to protect the health of the consumer. In addition, the EPA was directed to propose National Secondary Drinking Water Regulations to protect the public welfare.

The EPA was also given authority relating to wastewater reclamation and reuse by Public Law No. 92-500 passed in 1972 which stated an objective to restore and maintain the chemical, physical and biological integrity of the nation's waters. In order to obtain that objective it declared that it is the national goal that the discharge of pollutants into navigable waters be eliminated. The Administrator of EPA was charged to conduct an accelerated effort to develop, refine and achieve practicable application of waste management methods to eliminate the discharge of pollutants and methods for reclaiming and recycling water and confining the pollutants so that they will not migrate to cause water or other environmental pollution.

The EPA then is charged with the dual responsibility of encouraging reclamation and reuse of wastewaters and at the same time protecting the public health of consumers of all public water supplies. The Food and Drug Administration and USDA are charged with protecting the health of consumers of the food products resulting from food processing utilizing waters from any source.

Regulations

The federal agencies with authority to regulate the quality of processed foods or water supplies for food processing accomplish their charges through publication of regulations in the Federal Register and enforcement of the regulations.

Food and Drug Administration

The Food and Drug Administration (FDA) has final authority over the quality of all foods in interstate commerce. In connection with its charge that the food be maintained in a wholesome condition, that it not include any filthy, putrid or decomposed substances and that it has been prepared, packed and held under sanitary conditions and not become contaminated with filth or rendered injurious to health, the FDA has promulgated Good Manufacturing Practices in the manufacturing, processing, packing or holding of human foods. In 1977 these Good Management Practices (GMP) were recodified within Title 21 of the Code of Federal Regulations (CFR) to Part 110. The GMP are to apply in determining whether facilities, methods, practices and controls used in the manufacture, processing, packaging, packing or holding of food are in conformance with procedures to ensure that food for human consumption is safe and has been prepared, packed and held under sanitary conditions. It calls for all reasonable precautions to be taken to assure that production procedures do not contribute to contamination from filth, harmful chemicals, undesirable microorganisms or any other objectionable material.

With regard to water supplies it specifically states in 21 CFR 110.35(a) that "The water supplies shall be sufficient for the operations intended and shall be derived from an adequate source. Any water that contacts foods or food contact surfaces shall be safe and of adequate sanitary quality. Running water at a suitable temperature and under pressure as needed shall be provided in all areas where the processing of food, the cleaning of equipment, utensils, or containers or employee sanitary facilities require."

The FDA considers water to be a food. It regulates the bottling of water for human consumption and has specified the quality of water for sale to the public. The FDA definition of bottled water is "Water which is sealed in bottles or other containers and intended for human consumption." It does not include mineral water or soda water which is defined as a "class of beverages made by adsorbing carbon dioxide in 'potable' water". Bottled water must meet a bacteriological quality according to one of the following: (1) a multiple tube fermentation most probable number (MPN) of not more than 2.2 coliform organisms/100 ml with no analytical unit having an MPN greater than 9.2 coliform organisms/100 ml; or (2) by the membrane filter method, no unit having more than 4.0 coliform organisms/100 ml and the arithmetic mean of the coliform density of the sample not exceeding 1 coliform organism/100 ml. The physical quality requirement for bottled water calls for turbidity not to exceed 5 units, color not to exceed 15 units and odor not to exceed a Threshold Odor Number of 3. The chemical quality requirements for bottled water are that the following constituent concentrations not be exceeded:

Constituent	Concentration (mg/l)
Arsenic	0.05
Barium	1.0
Cadmium	0.01
Chloride	250.0
Chromium	0.05
Copper	1.0
Iron	0.3
Lead	0.05
Manganese	0.05
Mercury	0.002
Nitrate (N)	10.0
Phenols	0.001
Selenium	0.01
Silver	0.05
Sulfate	250.0

Total Dissolved Solids	500.0
Zinc	5.0
Endrin	0.0002
Lindane	0.004
Methoxychlor	0.1
Toxaphene	0.005
2,4-D	0.1
2,4,5-TP Silvex	0.01

Fluoride is also limited according to the average maximum daily air temperature in the region of the United States in which the water is marketed. The limitation is 1.4-2.4 ml/l for water to which no fluoride has been artificially added. A limitation of 0.8-1.7 mg/l applies to water to which fluoride is added. There are limitations for radium, gross alpha activity and beta particle and photon radioactivity also. If a bottled water contains a substance at a level considered injurious to health, it is deemed to be adulterated. These standards for the quality of bottled water are contained in 21 CFR 103.35.

In 21 CFR Part 113 the FDA has issued regulations for thermally processed low-acid foods packaged in hermetically sealed containers. Low-acid foods are defined as "those other than alcoholic beverages with a finished equilibrium pH greater than 4.6 and a water activity greater than 0.85". Tomatoes and tomato products having a finished equilibrium pH less than 4.7 are not classed as low-acid foods. Regulation 21 CFR 113.60(b) specifies that container cooling water shall be chlorinated or otherwise sanitized as necessary for cooling canals and for recirculated water supplies. There should be a measurable residual of the sanitizer employed at the water discharge point of the container cooler. This regulation covers those low-acid foods which were previously retorted or otherwise processed to achieve commercial sterility. Under 21 CFR 113.81 the requirements for blanching by heat and the control of thermophilic growth and contamination of blanchers is covered. It states in that section that if blanched food product is washed before filling, "potable" water should be used.

U.S. Department of Agriculture

The Agricultural Marketing Service branch of the U.S. Department of Agriculture operates a food plant inspection and certification program under which they voluntarily, for a fee, inspect food plants involved in interstate commerce of the processed product to determine and certify they meet good manufacturing practices (GMP). In providing this service an onsite inspector assures the maintenance of a clean and sanitary environment for the processing operation. An ample supply of both hot and cold water of safe and sanitary quality with adequate facilities for its distribution throughout the plant and protection against contamination and pollution is required. They further require that "all ingredients used in the manufacture or processing of any processed product shall be clean and fit for human food." These requirements are included in 7 CFR 52. Inspectors have memoranda of guidance for use during their review of sanitary and sanitation requirements. Memorandum No. 6, "Plant Sanitation Requirements," dated December 1971, provides instructions to help prevent product contamination. Among the instructions are that: "a water potability certificate by state, county or city government is to be obtained at least twice a year".

The Food Safety and Quality Service (FSQS) of the USDA under the Federal Meat Inspection Act and the Poultry Products Inspection Act provides inspection in meat and poultry processing operations to assure the quality of the product. Regulations have been issued under 9 CFR 308 which state that "The water supply shall be ample, clean and potable with adequate facilities for its distribution in the plant and its protection against contamination and pollution." They require that nonpotable water be permitted only in those parts of establishments where no edible product is handled or prepared and then only for limited purposes such as ammonia condensers not connected with the potable water supply, in vapor lines serving inedible product rendering tanks, in connection with equipment used for hashing and washing inedible products preparatory to tanking and in sewer lines for moving heavy solids in the sewage. Nonpotable water is not permitted for washing floors, areas or equipment involved in trucking materials to and from edible product departments nor is it permitted in hog scalding vats, dehairing machines or vapor lines serving edible product rendering equipment or for cleanup of shackling pens, leading areas or runways within the slaughtering department. The regulations state that "reuse of water in vapor lines leading from deodorizers used in the preparation of lard and similar products and in equipment used for the chilling of canned products after retorting may be permitted providing the reuse is for the identical original purpose and that precautions are taken to protect the water that is reused including cleanability of facilities, replacement with fresh potable water at suitable intervals and effective chlorination". In other words recycling within a process as we have defined it in this chapter is permissible under certain circumstances and with approval but reuse as we have defined it is not considered acceptable, unless the reused water is deemed to be "potable".

National Oceanographic and Atmospheric Administration

The National Oceanographic and Atmospheric Administration (NOAA) maintains a seafood processing inspection service for seafood processors similar to the service offered by the Agricultural Marketing Service for fruit and vegetable processors. The service is voluntary and paid for by the processor and the inspector certifies that the processor meets the standards established by NOAA which are basically in compliance with FDA regulations for low acid processed foods and the GMP.

FDA Regulations for Oyster and Shrimp Processors

The FDA operates a voluntary inspection service for oyster and shrimp processors. Under the regulation for administering this program for each of the processors they call for the equipment and conveyance vehicles to be washed with "clean unpolluted water". In addition, they call for the oysters or shrimp prior to processing to be adequately washed with "clean unpolluted water".

U.S. Environmental Protection Agency

The EPA has issued Primary and Secondary Drinking Water Regulations in response to the Safe Drinking Water Act. Within these regulations, they have defined "contaminant" as any physical, chemical, biological or radiological substance or matter in water and maximum contaminant level as the maximum permissible level of a contaminant in water which is delivered to the free flowing outlet of the ultimate user of a public water system. An exception is made in the case of turbidity where the MCL is measured at the point of entry to the distribution system. Contaminants added to the water under circumstances controlled by the user, except those resulting from corrosion of piping and plumbing caused by the water quality are excluded from the definition. The combined listing of primary drinking water standards and secondary drinking water standards as promulgated by the EPA are similar to the FDA bottled water standards. The EPA standards are discussed in more detail elsewhere in this book.

Agency Responsibility for Reuse of Wastewater in the Food Processing Industry

It is clear that the two principal agencies with responsibility for water quality for use in the food processing industry are the Food and Drug Administration under the Food, Drug and Cosmetics Act and the U.S. Environmental Protection Agency under the Safe Drinking Water Act. In order to delineate the areas of responsibility, these two agencies have executed a Memorandum of Understanding (MOU) with regard to the control of direct and indirect additives to and substances in drinking water. Under the MOU the EPA will retain primary responsibility over direct and indirect additives and other substances under the Safe Drinking Water Act, the Toxic Substances Control Act and the Federal Insecticide, Fungicide and Rodenticide Act. The FDA will have responsibility for water and substances in water used in food or for food processing and for bottled water under the Food, Drug and Cosmetic Act. All water used in food remains a food and subject to the provisions of the FFDCA. Water used for food processing is also subject to applicable provisions of the act. All substances in the water used in food are to be considered added substances except that no substances added to a public drinking water system before the water enters a food processing establishment will be considered a food additive.

The enabling legislation and the resulting regulations can be summarized as essentially requiring that water used in food processing must be of a safe and adequate quality. There is a definition of the quality of bottled waters in the Food and Drug Administration Regulations and a definition of maximum contaminant levels for public water supplies in EPA's Primary Drinking Water Standards. Although the FDA requires that "potable" water must be used for processing low acid foods, no definition of "potable" appears in any of the regulations that have been promulgated by these Federal agencies. Webster defines "potable" simply as "drinkable".

Reusing food processing water must not result in contamination of the product. The FFDCA clearly establishes that filth entering the processed food would be intolerable, that microbiological contamination is unacceptable and that the introduction of toxicants such as heavy metals or toxic organics in sufficient quantities to affect the consumer is not allowed.

Representatives of FDA and EPA have discussed wastewater reuse in the food processing industry [5,8-10]. They agree that the principal concerns in food processing operations are sanitation and microbiological quality, and if water reuse is implemented, the hazard of microbiological contamination of food would be the primary risk. Thus, the possibility of direct pathogen transfer must be eliminated by completely segregating sanitary wastewater from the food processing wastewaters to be reclaimed and reused. If the reused water may directly contact the food product, its microbiological quality must meet the requirements of the FDA Bottled Water Regulations and the EPA maximum contaminant levels for public water supplies. The reclamation treatment system for the water to be reused must be highly reliable to prevent breakdown and it must be monitored closely to allow the reclaimed water to immediately be replaced by water of adequate quality in case of system failure.

For food processing where a "potable" water is required, a reclaimed effluent meeting the regulations for bottled water or the MCL for public water supplies may be adequate for use. The MCL for organic and inorganic contaminants listed in the National Primary Drinking Water Regulations are intended to protect public health during full-time, long-term exposure. Such limitations may be more stringent than necessary for intermittent short-term uses such as contact with foods where a small amount of the water may remain in the final food product. The regulations limiting the introduction of "filth" into food products are always applicable and the physical and chemical quality of reclaimed waters must be adequate to prevent such contamination with their reuse.

Neither the EPA nor the FDA has generally declared that reclamation and reuse of food processing effluents is an acceptable practice. Both, however, recognize the need for considering this source of water supply and have embarked on a cautious and deliberate appraisal of the opportunities for reclamation and reuse of processing effluents. Their concerns, in addition to meeting the MCL or Bottled Water Regulations, include the fate of heavy metals and pesticides during reclamation and multiple use and the generation of potential toxicants during the disinfection.

REUSE OF RECLAIMED PROCESSING EFFLUENT
IN FOOD PROCESSING PLANTS

Within food processing plants there are three general areas for potential use of reclaimed effluent, listed here in order of increasing risk of product contamination:

1. areas where the water would not contact the finished or unprotected product but which may contact the product in initial processing stages or may contact the containerized product with little likelihood of entering the processed food package,
2. areas where water directly contacts the product before, during and after processing and where the water could be incorporated into the finished product package in small amounts, and
3. areas where water would be directly incorporated into the product and the product containers.

Wastewaters reclaimed by conventional reclamation treatment processes with good quality control could be reused in the first general area with little risk of product contamination. The second area of reuse would increase the contamination risk if the reused water were of less than adequate quality. If in the third area the reused water were to directly enter the food chain in greater than miniscule amounts there could be a real risk of contamination and a reduction in the wholesomeness and quality of the product if the water were of less than ideal quality.

High-acid foods, such as fruit, would generally be less suceptible to increased public health risk from reclamation and reuse of water than low-acid foods because harmful microorganisms are less likely to survive in high-acid foods. Disinfection of cooling waters is required for low-acid foods.

Within the low-acid foods category, meat processing regulations require "potable" water for all washing operations and use in the meat processing plant. It appears that a ranking of lowest to highest risk among processed products would be (1) high-acid foods; (2) low-acid foods for which "potable" water is not required and (3) low-acid foods for which "potable" water is required for processing.

Reclaimed water limited to reuse in noncontact areas or to contact with foods only prior to further processing might be used for condensors, for initial fluming, for initial washing to remove dirt and debris, for floor and gutter wash and for direct contact container cooling. The second category of use with a slightly higher risk could see reclaimed water reused for blanching, for product washing, for equipment washing, for direct cooling of the product and for spraying the equipment and product prior to package filling. In the first category there would be very little likelihood of any reclaimed water being incorporated into the final finished product for public distribution. In the second category there would be some probability of small amounts of reclaimed water incorporation into the packaged product. The third category of water use would include water incorporation directly into the packaged product in substantial quantities.

For the lowest category, where reused water would not likely enter finished product nor contaminate it, the quality of the reclaimed water must still be of an adequate level from a bacteriological standpoint to prevent entrance of pathogen or toxin-producing organisms into the finished containers. Entry could be made into the sealed cans during cooling, or by retention on product through its freezing for frozen materials. The survival of such organisms is unlikely in high-acid foods, thereby making them the lowest risk product. However, the possibility of product spoilage and loss, or the introduction of "filth", requires that treatment prior to reuse must be of an adequate nature to preclude suspended solids and high bacterial counts. The second category of reuse would require disinfection to very low levels of microorganisms, adequate to reduce the risk of any pathogens being present in the water to an acceptable level. A low level of suspended solids must be maintained in the reclaimed wastewater and levels of toxicants such as heavy metals or organics which could absorb onto the product must be maintained at or below acceptable levels following the treatment. For meat products the reclaimed water must be of "potable" quality. If a reclaimed wastewater is ever to be determined "potable" it will undoubtedly as a minimum be required to meet the FDA requirements for bottled water.

Since reclamation and reuse of the processing effluent has not been endorsed by either the EPA or FDA, it still must be demonstrated that treatment works can provide reclaimed water of adequate quality for reuse in the processing plant. This quality standard demonstration could be accomplished by first introducing reuse into the least critical processing areas such as initial product washing and conveying and container cooling, before reclamation and reuse are instituted in the direct washdown of the plant and equipment and in the processing operation. The overall concept of reclamation and reuse must be acceptable to the FDA and EPA prior to each plant instituting its own demonstration procedures with the reclamation system.

RECLAMATION OF PROCESSING EFFLUENT
FOR REUSE

Reclamation of a food processing effluent for reuse will entail some treatment of the effluent outside of the plant processing area. Food processing wastewaters are characterized by high concentrations of organic constituents washed from or leached from the food being processed and include quantities of the product itself. In order to make the effluent acceptably unlikely to introduce spoiled or putrid material into the product the effluent would necessarily be treated to reduce the organic content. This is usually accomplished by biological treatment, such as activated sludge or aerated lagoon treatment. A minimum level of biological treatment effectiveness would appear to be the best practicable treatment levels (BPT) defined by the EPA for use in the national pollutant discharge elimination system (NPDES) for establishing discharge permit requirements [11]. Reclamation of the biologically treated effluent must ensure that carryover biological solids are removed from the reclaimed effluent stream and that disinfection achieves the necessary microbiological quality. Conventional treatment processes such as filtration through granular media, possibly accompanied by chemical coagulation, can be used to remove excess suspended solids from the biologically treated effluent. Disinfection with chlorine, chlorine dioxide, ozone or other agent would be necessary for removal of microbiological organisms. Disinfection with chlorine has the advantage of a residual being present in the reclaimed water to ensure that recontamination could not occur. It can be used as a quick reference check for adequate treatment.

"Conventional reclamation treatment" may be considered to consist of biological treatment, physical treatment for reduction of suspended material (possibly including chemical coagulation) and disinfection. Adequate conventional reclamation treatment would result in a reclaimed water slightly higher in salinity, slightly higher in organic constituents (primarily those organics less removable by biological treatment, such as refractory organics) and potentially higher in heavy metals, pesticides and other priority pollutants than the original water supply. A processing effluent for reuse in food processing would not contain any sanitary wastewaters.

"Advanced reclamation treatment" would consist of more sophisticated processes for the reduction of salts or refractory organics and would add considerable expense to the reclamation process.

The conventional processes are all in common use in water and wastewater treatment systems and their construction and operation and maintenance costs are reasonably well documented. If conventional reclamation treatment processes are shown to produce a reclaimed effluent suitable for reuse in food processing plants, food processing wastewater reclamation and reuse may be implemented on a wide scale in the foreseeable future. If advanced processes such as activated carbon adsorption, ultrafiltration or reverse osmosis are required to achieve a satisfactory quality for reclamation and reuse, implementation on a commercial scale will be less likely in the near future.

DEMONSTRATION OF WASTEWATER RECLAMATION AND REUSE IN THE FOOD PROCESSING INDUSTRY

Two demonstration projects of wastewater reclamation and reuse are currently underway in the food processing industry. Each is a continuation of an earlier feasibility research and development project which established that certain objective reclaimed effluent quality parameters could be achieved. The current demonstration phase of each project has been initiated to assess the long-term reliability of the treatment processes employed and the level of priority constituents present in the reclaimed waters. The two projects are being conducted and evaluated under cooperative agreement between the processor, the EPA Health Effects Research Laboratory, the Food and Drug Administration and others, including industry representatives.

Demonstration of Reclamation and Reuse of Fruit Processing Effluent

Snokist Growers, a cooperative in Yakima, Washington, processes apples, pears, peaches, plums, crabapples and cherries at their cannery facility. In 1967 and 1968 they installed a biological treatment system (Figures 1-4) to demonstrate that aerobic treatment of fruit processing wastes was feasible for reducing their BOD loading to surface waters [12,13]. The project showed that aerobic treatment of food processing wastes was feasible. At activated sludge rates below 0.4 mg COD removed/day/mg MLVSS in excess of 95% removal of the organic load was achieved, and at clarification rates of less than 400 gpd/sq ft^2 (16 m/day) the flocculant suspended solids were effectively removed from the effluent. Sludge return rates of 1-2 times the influent flow rate were shown to be necessary for proper operation.

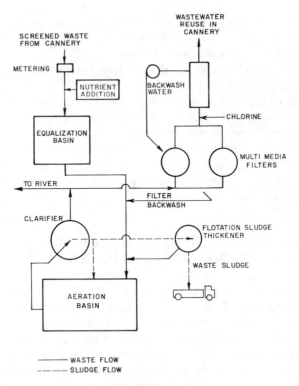

Figure 1. Wastewater treatment and reclamation system schematic flow diagram for Snokist Growers Cannery, Yakima, Washington.

Figure 2. Activated sludge aeration basin at Snokist Growers Cannery.

Figure 3. Secondary clarifier at Snokist Growers Cannery.

In 1975 multimedia filters and a chlorination disinfection system were installed and a research and development (R&D) project was initiated to determine the feasibility of reusing the treated processing effluent in the cannery [14,15]. This project showed that the biologically treated wastewater could be polished by filtration and disinfection with chlorine to a quality suitable for reuse within their cannery except during periods of high suspended solids discharge from the biological treatment system. It also showed

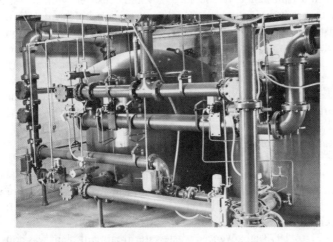

Figure 4. Granular media pressure filter system at Snokist Growers Cannery wastewater treatment and reclamation plant.

that an early warning system for deterioration of effluent quality, including continuous monitoring of reclaimed water turbidity and chlorine residual, was necessary to preclude the reuse of water of unsuitable quality. The project concluded that the effluent quality obtainable on a consistent basis (suspended solids $\leqslant$ 30 mg/l, turbidity $\leqslant$ 20 NTU, total coliform $\leqslant$ 1 organism/100 ml and total plate count $\leqslant$ 500/ml) was satisfactory for use in the initial processing area of the cannery, excluding peelers and pealed product conveyors, for floor and gutter wash down and for direct contact container cooling. It concluded that a measurable chlorine residual on a continuous basis was necessary at the point of use to ensure adequate bacteriological quality.

The reuse research and development project results indicated that the 1977 cost of reclaiming wastewater for reuse at the Snokist Growers Cannery was about $0.20/m^3$ above the cost of providing biological treatment adequate for discharge to the receiving water. The study also indicated that up to 50% of the cannery water demand could be met through effluent reclamation and reuse.

At the conclusion of the R & D project a review committee consisting of representatives of the Food and Drug Administration, the EPA and the industry recommended that a demonstration project be conducted. During the 1980 processing season demonstration of the reclamation and reuse of wastewaters at the Snokist Growers Cannery were initiated. Reuse areas include direct contact container cooling, floor and gutter wash and initial fruit conveying and washing.

Objectives of the demonstration project are documentation of the health effect potential related to commercial-scale reclamation and reuse of treated effluent in a fruit cannery. The adequacy and consistency of the reclamation system to produce a suitable low solids, disinfected effluent will be shown. Demonstration of the functioning and reliability of the quality controls for the reclaimed water, including continuous turbidity and chlorine residual monitoring, will be made. The level of priority pollutants and their increase in the reclaimed effluent will be determined. Monitoring for heavy metals, pesticides and halogenated organics will consume a major share of the analytical efforts during this project.

Reclamation of Poultry Processing Plant Effluent

In 1965 the Sterling Poultry Processing Company in Oakland, Maryland, installed a two-stage aerated lagoon treatment system for its processing wastewaters prior to their discharge into the little Youghiogheny River following chlorination.

In 1971 to 1973 an advanced wastewater treatment plant was constructed to treat the biological wastewater effluent for reuse as a supplementary water supply. The advanced treatment plant consisted of a microstrainer, chemical coagulation, flocculation and sedimentation, sand filtration, gas chlorination and storage. The intended point of use of the reclaimed effluent was mixing with the raw plant water supply. The plant water is given conventional water treatment of chemical coagulation and sedimentation followed by filtration and chlorination [16].

A two-year R & D project to determine the feasibility of reclaiming the poultry processing effluent for reuse found that the inorganic and physico-chemical characteristics of the renovated water consistently met EPA primary drinking water standards [16,17]. The study concluded that it would be highly unlikely that the contemplated reuse of water at the processing plant would pose a risk of disease to consumers of the poultry. The capital and operating costs of reclaiming the effluent following biological treatment were estimated at $0.071/m^3$.

The study recommended a trial period of full-scale reuse with monitoring of the renovated water and carcasses followed by a comprehensive evaluation prior to decision on permanent reclamation and reuse at the site. The demonstration phase of this project is being initiated for the 1980-1981 period. It will consist of operation of the reclamation system and recirculation of the reclaimed effluent to the biological treatment system inlet, but not reuse in the processing plant. It also calls for thorough evaluation of the levels of other constituents such as pesticides and halogenated organics for use by the EPA and FDA in determining suitability of this water for reuse.

The Sterling Poultry Plant Project seeks to approach wastewater reclamation and reuse from a different standpoint than Snokist Growers. Sterling intends that the entire water supply be a combined mixture of reclaimed effluent and water from their present supply source which needs augmentation. It intends to utilize the water in all processing areas. Regulations call for the water to be "potable" for use in poultry processing plants. To date reuse of the reclaimed water is not permitted since it has not been declared "potable" by the proper authority.

In 1977 the Maryland Department of Health and Mental Hygiene certified that the reclaimed water at the Sterling Poultry Processing Plant met potable water criteria according to its standards; however, the EPA Office of Water Supply (OWS) did not agree with this determination since the water did not come from a "natural" source and it (the OWS) "did not know everything it needs to know about the reclaimed water's chemistry as far as drinking water was concerned" [18]. The term "potability" needs definition and agreement among the various entities charged with the responsibility for water supplies for food processing. The 1979 memorandum of understanding (MOU) between the EPA and the FDA seems to shift the primary responsibility for such determinations to the FDA [10].

Other Wastewater Reuse Projects in the
Food Processing Industry

The reuse of treated effluents for food processing operations have been reported in other areas, but none have been documented to the extent that the Snokist Growers and Sterling Poultry Processing projects have. There are reports of treated effluent being used for condenser cooling legs in citrus processing plants and uses in other plants which would definitely not involve contact between the water and any product.

SUMMARY

Reuse of reclaimed wastewater in the food processing industry has not been implemented to any great extent in the United States to date.

Food and Drug Administration regulations promulgated under the Food, Drug and Cosmetic Act provide the principal controls for food quality and safety in the United States. These regulations call for a water supply of adequate quantity and quality in all food processing plants. The Good Management Practices codified in the Code of Federal Regulations provide guidelines for water quality for food processing. They recognize the greater inherent risk of contamination of low-acid preserved and processed products as opposed to high-acid foods. The U.S. Department of Agriculture Food Safety and Quality Service (FSQS) regulations for meat and poultry products require that "potable" water be used for processing these items. The EPA promulgates water quality requirements for public water supplies under the Safe Drinking Water Act which control the outside water supply quality. Once inside the plant, however, the water and any additives thereto are the responsibility of the Food and Drug Administration under a recent memorandum of understanding. The quality requirements for reclaimed and reused water is apparently the responsibility of FDA under this agreement.

The two wastewater reclamation and reuse projects in demonstration phases in the United States are being monitored closely by the EPA Health Effects Research Laboratory and the FDA and USDA. These projects, one at a high-acid fruit processing plant and the other at a poultry plant, are extensively monitoring reclaimed water quality on a long-term basis for priority pollutants such as heavy metals, pesticides and halogenated organics as well as microbiological quality. Each of the projects is operating on a full-scale basis reclaiming a substantial quantity of the total effluent.

The fruit processing plant is using the reclaimed water to replace a portion of its normal water supply for individual use areas, such as final contact container cooling, initial fruit conveying and washing and floor and gutter wash.

The poultry processing plant is attempting to gain long-term approval of its reclaimed effluent as a portion of its raw water supply to be given treatment in conjunction with its normal water supply and circulated back to the plant in a complete mixed combined reclaimed water and freshwater supply system, but reuse in the plant has not yet been approved and the reclaimed effluent is merely recycled to the wastewater treatment works. The reclaimed effluent has not yet been certified "potable" by all concerned agencies.

The treatment technology for reclaiming food processing plant effluents for reuse will probably consist of conventional processes. It is anticipated that biological treatment such as activated sludge or aerated lagoons as used at the fruit processing and poultry processing plants will be a prerequisite to further reclamation treatment. Treatment by granular media filtration, possibly following coagulation and sedimentation, and followed by thorough and effective disinfection appear to be satisfactory for reclamation of the biological effluent for reuse. If this level of conventional treatment technology for reclamation is acceptable, the likelihood of implementation on an industry-wide basis is much higher than if advanced reclamation technology such as reverse osmosis, activated carbon or other more expensive, processes are required. Both plants now demonstrating reclamation and reuse use conventional reclamation treatment processes.

Since the regulatory agencies are only now beginning to assess the potential for reclamation and reuse in the food processing industry it is unlikely that widespread food processing effluent reuse in the food processing area will be implemented for a few more years. The results of the current demonstration projects will not be available until approximately 1982, at which time an assessment of the level of reuse attainable within the processing plants can be made by the Food and Drug Administration and other concerned regulatory agencies. Acceptability for reuse in the lowest level of use areas, that of initial floor and gutter wash, initial food conveying and washing prior to peeling and processing and for final contact container cooling will provide incentives for many food processors to reclaim and reuse some effluents. Acceptbility in the areas of actual contact with product, but without incorporation into the product itself, would affect a wider range of potential applications and greater potential number of plants which might invest in the technology required for the reclamation and reuse of treated effluents.

The question of potability of a water supply must be addressed in order for reclamation and reuse to be a viable alternative in the meat industry and before reclaimed effluent could be reused in conjunction with the normal plant water supply in areas of container filling or product preparation in other plants. The Food and Drug Administration and the EPA must determine whether waters can be declared "potable" if they meet the current standards under the Safe Drinking Water Act and the criteria for bottled water even though they emanate from a reclaimed effluent source, not including sanitary waste. Recognition that only small quantities of this water will actually be consumed should be a consideration when this determination is made.

Perhaps the most encouraging aspects of the demonstration projects now underway are that the regulatory agencies and the industry are jointly and seriously considering the criteria under which reclaimed effluents may be reused in food processing plants. It is hoped that the culmination of these two demonstration projects will be the delineation of requirements for reclamation and reuse which can be applied on an industry-wide scale. Careful consideration and good efficient design of reclamation facilities will still be a requisite as will an effective monitoring and control system to ensure that treatment plant or reclamation facility breakdown does not affect the quality of water being used for producing processed food products for distribution to the consumer.

REFERENCES

1. "Pollution Abatement in the Fruit and Vegetable Industry," U.S. Environmental Protection Agency, EPA 625/3-77-0007, 3 volumes (July 1977).
2. "Upgrading Meat Packing Facilities to Reduce Pollution," U.S. Environmental Protection Agency Technology Transfer, 3 volumes (October 1973).
3. "Upgrading Poultry Processing Facilities to Reduce Pollution," U.S. Environmental Protection Agency Technology Transfer, 3 volumes, (July 1973).
4. Katsuyama, A. M. *A Guide for Waste Management in the Food Processing Industry* (Washington, D.C.: The Food Processing Institute, 1979).
5. Pahren, H. R. "Water Quality for the Food Processing Industry," paper presented at the Sixth Engineering Foundation Conference on Environmental Engineering in the Food Processing Industry, February 18, 1976.
6. Stanley Associates Engineering Ltd. "Land Application of Food Processing Wastewater," Environment Canada, EPS 3-WP-78-5 (June 1978).
7. "Process Design Manual for Land Treatment of Municipal Wastewater," U.S. Environmental Protection Agency, EPA 625/1-77-008 (October 1977).
8. Schaffner, R. M. "FDA Regulations on Water," paper presented at the Sixth Engineering Research Foundation Conference on Environmental Engineering in the Food Processing Industry, February 18, 1976.
9. Handwerk, R. L. "FDA Viewpoint on Water Reuse in Food Processing," paper presented at the Seventh Engineering Research Foundation Conference on Environmental Engineering in the Food Processing Industry, February 14, 1977.

10. Braude, G. L. "Quality Criteria for Water Reuse—FDA Considerations," paper presented at National Research Council, National Academy of Sciences, Washington, DC, October 2, 1979.
11. Development Documents for Effluent Limitations, Guidelines and New Source Performance Standards. Various categories, U.S. Environmental Protection Agency.
12. Esvelt, L. A. "Aerobic Treatment of Fruit Processing Wastes," Federal Water Pollution Control Administration, 12060 FAD (October 1969).
13. Esvelt, L. A., and H. H. Hart. "Treatment of Fruit Processing Waste by Aeration," *J. Water Poll. Control Fed.* (July 1970).
14. Esvelt, L. A. "Reuse of Treated Fruit Processing Waste Water in a Cannery," U.S. Environmental Protection Agency, EPA-600/2-78-203 (September 1978).
15. Esvelt, L. A., H. W. Thompson and H. H. Hart. "Reuse of Reclaimed Fruit Processing Wastewater," *Proceedings of the First International Water Reuse Symposium,* Washington, DC (March 1979).
16. Clise, J. D. "Poultry Processing Wastewater Treatment and Reuse," U.S. Environmental Protection Agency, EPA-660/2-74-060 (March 1974).
17. Andelman, J. B. "Safety Evaluation of Renovated Wastewater from a Poultry Processing Plant," U.S. Environmental Protection Agency, EPA-600/1-79-030 (August 1979).
18. Clise, J. D. "Water Recycling in a Poultry Processing Plant—Catch Twenty-Two," presented at the American Society of Agricultural Engineers' 1978 Summer Meeting, Logan, Utah, June 1978.

SECTION 4

VIRAL AND BACTERIAL REMOVAL
AND MONITORING

CHAPTER 26

VIRUS UPTAKE BY MINERALS AND SOILS

G. Wolfgang Fuhs and Dene H. Taylor
Environmental Health Institute
Division of Laboratories and Research
New York State Department of Health
Albany, New York

INTRODUCTION

Adsorption of human viruses to minerals and soils is an important consideration in the reuse of treated wastewater through groundwater recharge. At first adsorption might appear to be only one of several aspects of mechanisms of virus removal. In wastewater and wastewater treatment systems, virus can be adsorbed to inorganic or organic particulates or bacterial floc prior to recharge. Virus adsorbed onto or incorporated into microscopic or larger particulates can be expected to be retained by soils mostly by filtration and settling in combination with drying onto soil surfaces, but it may become liberated upon dissolution or decomposition of the carrier particle. Free virus, however, particularly in monodisperse form, is removed principally by adsorption.

Although adsorption is in principle reversible and in many instances does not involve sufficient damage to the virus particle to cause loss of infectivity, recovery of adsorbed virus from minerals and soil is generally incomplete. This emphasizes the practical utility of attempting to understand and to optimize the adsorption of virus to soil surfaces.

Although some limitations of laboratory studies in this field have been overcome, others remain. Naturally occurring minerals and soils do not exhibit clean surfaces but may already carry other substances adsorbed to them, or surfaces may be overgrown by microorganisms. Experimental evidence shows that naturally occurring organic matter and bacterial overgrowth tends to reduce markedly the adsorption of virus. Remaining problems are the contamination, despite elaborate purification, of laboratory-grown virus with cellular debris of tissue-culture origin. Virus in human excreta may carry fecal antibodies, which may alter to some extent the surface characteristics of the virus particles.

Laboratory studies of virus adsorption are effectively supplemented by field studies, which are dealt with in the next chapter in this volume. Field studies often do not include all of the measurements and characterizations of soil properties that would be desirable to explain the results of a specific field experiment. Laboratory studies are therefore necessary to identify the variables which, when properly adjusted, will optimize the adsorption and retention of virus in recharge beds.

Earlier reviews on the subject were written by Drewry and Eliassen [1], Bitton [2], Gerba et al. [3] and Duboise [4]; the interaction of viruses with model membranes was reviewed by Tiffany [5]. This chapter describes the more recent literature and the results of a four-year study on the adsorption of three viruses to a large number of soils and minerals, which was conducted by the Environmental Health Institute, New York State Department of Health, with grant support from the U.S. Environmental Protection Agency.

VIRUS SPECIES OF INTEREST

The group of viruses most commonly encountered in wastewaters are the enteroviruses, small icosahedral particles of 21-30 nm diam which contain a genome of single-stranded ribonucleic acid (RNA). They include the poliomyelitis viruses (poliovirus types 1, 2, and 3), Coxsackie viruses (types A1-A24, B1-B-6), the echo viruses (enteric cytopathogenic human orphan viruses, types 1-34) and the new enteroviruses types 68-71. Coxsackie and echo strains usually cause mild respiratory infections but occasionally also severe forms of illness such as encephalitis, meningitis, paralytic disease and peri/myocarditis. Vaccine strains of the poliomyelitis virus are common in sewage throughout the year; the other enteroviruses occur with seasonal maxima in summer and fall.

Reovirus is a larger icosahedral particle (70-80 nm), which contains a genome of double-stranded RNA. The three known types have been associated with fever and diarrhea but have also been isolated from healthy children.

Rotaviruses are recently discovered members of the Reoviridae. Human strains so far have not been cultured, and an animal strain (simian rotavirus) has been used for model studies. Rotaviruses cause serious infant diarrhea and are presently thought to be responsible for waterborne gastroenteritis of unknown etiology.

Infectious hepatitis (hepatitis A) virus is another important waterborne virus. It still is poorly characterized, but its morphology and size suggest that it may belong to the enterovirus group. It has not been propagated in a usable form for routine laboratory investigations.

Adenovirus is a DNA-containing icosahedral virus of 80 nm diam. There are 35 different serotypes. Adenoviruses mainly cause conjunctivitis and respiratory-tract infections, although occasionally they have caused water-borne outbreaks (swimming pool conjunctivitis). Their lesser importance may be due to their relative instability outside a pH range of 6-9.

Bacteriophages have often been used in lieu of human viruses in adsorption studies. However, not only are many phages organized quite differently from human viruses, but even those of similar size and appearance differ in their adsorptive behavior and stability. This is presumably due to differences in the chemical compositions, and hence the ionic states, of their capsid proteins (e.g., phages f2, Qβ [6], and MS2 [7,8] which resemble the enteroviruses).

There is no agreement at present on a single virus for use in model studies or to serve as an indicator species and a basis for environmental regulation.

THEORY

In earlier discussions of virus-soil interactions it was assumed that electrostatic forces between virus and adsorbent played important roles [2,3]. However, there has been almost no direct evidence on the nature of virus charges during these interactions from which to justify this assumption. Recent detailed studies of the surface electrochemical properties of viruses in dilute aqueous media [9,10] have shown directly the role of charge in virus interactions with surfaces.

This section discusses the origin and nature of charge on viruses and soil minerals, the distribution of charge and variation of electrical potential at the virus-aqueous and mineral-aqueous interfaces, and the nature of the interaction on close approach of a virus to a mineral surface.

Origin of Charge on Soil Minerals

Soil minerals derive charge from structural irregularities within the matrix, from surface group ionization or from preferential adsorption of ions at the surface. Structural irregularities include site vacancies, where an atom is missing in the crystal structure, and isomorphous replacements, where one ion replaces another ion of different charge in the matrix. Replacement is particularly common in the aluminosilicate clay minerals; for example, in montmorillonite Al^{3+} can substitute for Si^{4+} and Mg^{2+} for Al^{3+}. The result is a net permanent negative charge neutralized largely by cation adsorption at the particle surface. Minerals deriving their charge in this way are known as constant-charge materials.

The surfaces of many soil colloids, notably the oxides and hydroxides, are extensively hydroxylated and can develop charge in aqueous media by ionization or by adsorption of H^+ or OH^- [11,12] ions. These reactions may be presented simply as:

$$R{-}OH \quad \rightleftharpoons R{-}O^- + H^+ \tag{1}$$

$$R{-}OH_2 \quad \rightleftharpoons R{-}OH + H^+ \tag{2}$$

where R represents the solid. As reactions of this type are obviously pH-dependent, the mineral surfaces will have pH-dependent charge. The surface charge can be expressed as:

$$\sigma_0 = F\,(\Gamma^+ - \Gamma^-) \tag{3}$$

where   F   =   the Faraday constant and
           Γ^+ and Γ^- =   the surface densities of cationic and anionic sites (in mol m^{-2}).

Γ^+ and Γ^- can be determined by potentiometric titration with acid and base if the surface area is known. At some pH there will be no net surface charge and $\sigma_0 = 0$. This is the point of zero charge, or PZC. With increasing pH above the PZC, negative surface charge develops, while on acidification the surface becomes positively charged. For minerals such as the oxides and hydroxides H^+ and OH^- are termed the potential-determining ions, as they would establish the surface potential if the solid were a reversible electrode. If the surface is treated as ideal and if H^+ is the potential-determining ion, the surface potential ψ_0 is given by the Nernst equation:

$$\psi_0 = (kT/e)\,(pH_z - pH) \tag{4}$$

where pH_z =   the pH at the PZC,
           k   =   the Boltzmann constant,
           e   =   the electron charge, and
           T   =   the absolute temperature.

This equation is approximately valid for low potentials. Materials displaying this property are known as constant-potential minerals. A number of minerals, in particular the clay minerals kaolinite, halloysite and allophane, possess charge from both isomorphous replacement and surface group reactions.

Origin of Charge on Virus

Most viruses have coats or capsids either totally or partially composed of polypeptides. The presence of amino acids such as glutamic and aspartic acids, histidine, tyrosine and guanidine in the polypeptide introduces weakly acidic and basic groups, which on ionization give the capsid electrical charge. Typical charging reactions are:

$$-CO_2H \quad \rightleftharpoons -CO_2^- \quad + \quad H^+ \tag{5}$$

$$-NH_3^+ \quad \rightleftharpoons -NH_2 \quad + \quad H^+ \tag{6}$$

$$-C_6H_4OH \rightleftharpoons -C_6H_4O^- + \quad H^+ \tag{7}$$

Each ionizing group in the polypeptide has a characteristic dissociation constant K_i. The spread of K_i values ensures that most viruses have net charges that vary continuously with pH. Again the net surface charge σ_0 is given by Equation 3. Titration experiments have shown that not all secondary groups in proteins are able to ionize—some are buried deeply within the protein [13], and others may be blocked by reaction with non-ionizing groups. Nevertheless preliminary studies have shown close correlation between the pH-dependent charge of reovirus measured by electrokinetics (see below) and that calculated from the composition of the outermost polypeptide of its capsid [9,10]. The specific positions of ionizable groups in the polypeptide give each virus a characteristic charge mosaic. Viruses with different surface structures, including capsid mutants, will have different mosaics and will show different surface properties.

Distribution of Charge at the Solid/Aqueous Electrolyte Interface—the Electrochemical Double Layer

The charge of a surface immersed in an electrolyte solution is compensated by an equivalent but opposite charge adsorbed at the aqueous side of the interface. This separation of charge into two planes produces the so-called double layer. The neutralizing charge includes both counterions (ions of opposite charge) attracted to and coions repelled from the solid. If the electrolyte is indifferent (i.e., if its components do not form bonds directly with the surface), and the ions do not approach more closely than their hydrated radii, they will be distributed by thermal motion in a relatively thick zone. This region is the Guoy-Chapman diffuse layer, in which ion concentrations are described by a Boltzmann relationship (Figure 1). For low potentials ψ_x the potential in the diffuse region can be approximated by:

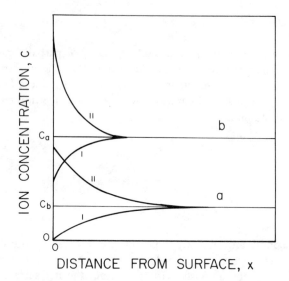

Figure 1. Concentrations of coions and counterions in Gouy-Chapman diffuse layer at the charged surface/aqueous interface. a = low electrolyte concentration; b = high electrolyte concentration; i = coion; ii = counterion.

$$\psi_x = \psi_\delta \exp(-\kappa x) \qquad (8)$$

where ψ_δ =   the potential at δ, the distance of closest approach of the ions, and
        κ =   the Debye parameter, defined by:

$$\kappa = [8\pi z^2 e^2 c/kT]^{-\frac{1}{2}} \qquad (9)$$

where   z   =   the ion valence and
        e   =   electron charge
        c   =   the ionic concentration in the bulk solution.

Thus in the diffuse layer the potential drops off by a factor of $1/e$ (0.368) over a distance of the order of $1/\kappa$, the double-layer thickness (Figure 2). The diffuse layer will be narrower for higher ionic strengths and for electrolytes of higher valence.

Many ions display specific adsorption at some surfaces, i.e., they not only have ionic interactions with the surface but also other attractions such as hydrogen bonding, hydrophobic effects, or some degree of covalent bonding. These ions are bound close to the surface within the Stern layer, i.e., a layer where they lie closer than their hydrated radius, δ (Figure 3). If these bonds are sufficiently strong, and there is an excess of counterions bound, the total charge of ions in the Stern layer may exceed the surface charge and so reverse ψ_δ. This phenomenon is called charge reversal (actually an electrokinetic term). Specific adsorption of surface-charge coions, however, will enhance ψ_δ.

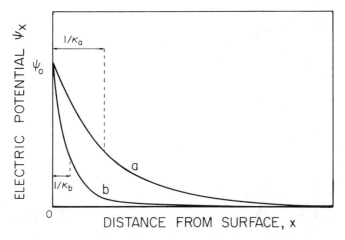

Figure 2. Variation of potential with distance in the Gouy-Chapman diffuse layer at the charged surface/aqueous interface. a = low electrolyte concentration; b = high electrolyte concentration; $1/\kappa$= double-layer thickness.

At the solid/aqueous interface there is a layer of immobilized fluid immediately adjacent to the solid. The boundary between this stationary layer and the bulk mobile liquid, the plane of shear or the hydrodynamic boundary, is presumed to be within a few molecular layers of the surface, close to but outside of the Stern layer. The potential at this plane is the zeta potential ζ. It is probably similar in value to ψ_δ (Figure 3) and reflects ion adsorption within the Stern layer as well as electrolyte changes that alter the double-layer thickness. Any set of conditions where $\zeta = 0$ is of considerable interest. This is an isoelectric point (IEP), not to be confused with the PZC ($\sigma_0 = 0$). Although the two are usually similar in indifferent electrolytes, specific interactions cause considerable differences. For example, in the presence of Ca^{2+} the IEP of Al_2O_3 increases but the PZC decreases [14].

Electrokinetics, the study of the tangential motions of charged surfaces past electrolyte solutions, is of great interest in many areas of colloid and surface chemistry. Techniques such as electrophoresis and the measurement of streaming potential allow the potential difference, ζ, between the hydrodynamic boundary of the solid and the bulk solution to be obtained. A number of assumptions must be made in calculating ζ from electrophoretic studies: unless a particle is ideal (e.g., spherical) and the potential is low, it is becoming customary to report the measured property, i.e., the electrophoretic mobility. This is the velocity of the particle (usually reported in $\mu m \cdot s^{-1} \cdot V^{-1} \cdot cm$) in an electric potential gradient. For ideal small spherical colloids in dilute electrolytes ζ is directly proportional to the mobility.

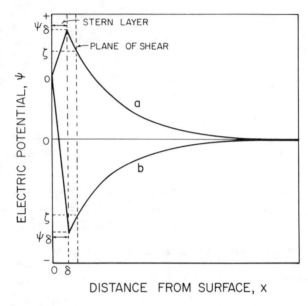

Figure 3. Variation of potential in the Stern and diffuse layers at the charged interface/ aqueous electrolyte interface. a = specifically adsorbed coions; b = specifically adsorbed counterions with charge reversal.

Electrokinetic Properties of Soil Minerals

The electrokinetic properties of soil minerals have been widely studied, mostly by microelectrophoresis, but the reports are scattered through the literature. The electrophoretic properties of some of the more important soils, hydrous oxides and clay minerals are discussed here. The most reactive hydrous oxides in soil are generally those of aluminum and iron (III). (Silica, although very common, is less reactive; it is protonated only with difficulty.) The electrophoretic mobilities of iron and aluminum oxides are characterized by relatively high IEPs and strong pH dependence in the region of the IEP (Figure 4). The mobility thus reflects surface-charge development; it, and hence ζ, decreases with increasing indifferent electrolyte concentration. Multivalent surface counterions are more effective than monovalent ions.

As described above, the clay minerals are dominantly aluminosilicates with substantial negative charge of structural origin. The presence of surface ionizable hydroxyl groups varies between minerals. This is reflected in their pH-dependent charge development and electrophoretic mobility (Figure 5). For instance, montmorillonite has only a few hydroxyl groups at plate ends; over the pH range of most natural systems it remains strongly negatively charged [11]. Halloysite, like kaolinite, has a much greater surface concentration of

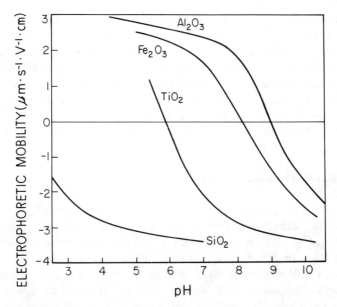

Figure 4. pH-dependent electrophoretic mobility of hydrous oxides in 10^{-3} M indifferent electrolytes.

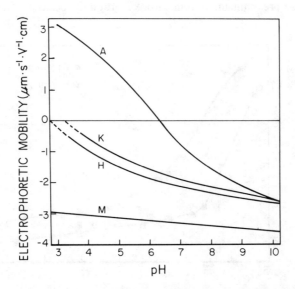

Figure 5. Electrophoretic mobilities of clay minerals in dilute electrolytes. A = Ohaupo allophane; K = Mt. Egerton kaolinite; H = Mangawara halloysite; M = Wyoming montmorillonite. A, H, M in 10^{-3} M NaCl (adopted from Ref. 15); K in 10^{-4} M KCl (adopted from Ref. 16).

hydroxyl groups, some of which are amphoteric, while others are acidic. It thus displays pH-dependent electrophoretic mobility. Allophane is an amorphous aluminosilicate coated to a variable degree with aluminum hydroxide. The aluminosilicate part has charge properties very much like kaolinite (i.e., pH-dependent with a low IEP), while the surface hydroxide gives a high IEP component. The resultant electrokinetic properties are intermediate between aluminum hydroxide and acidic aluminosilicate. In general the electrokinetic properties of clay minerals can be readily interpreted in terms of the minerals' composition and structure.

Electrokinetic Properties of Viruses

The electrophoretic properties of viruses have not been widely studied. Because they were considered too small to be seen microscopically, they were not thought to be amenable to study by microelectrophoresis. However, by using a laser for dark-field illumination in the ultramicroscope, Taylor and Bosmann [9,10] determined the electrokinetic properties in dilute electrolytes of vaccinia virus and reovirus. Vaccinia virus is 200 X 250 nm in size, while reovirus has a hydrodynamic diameter of 96 nm [17]. The mobilities of both viruses varied continuously with pH, and both had moderately low IEPs (Figure 6). At neutrality both viruses were negatively charged; the surface charge arose predominantly from carboxyl-group dissociation. The effects of

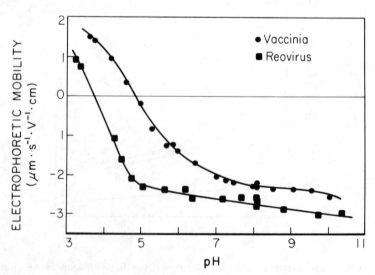

Figure 6. Electrophoretic mobilities of reovirus type 3 and vaccinia virus in 10^{-2} M NACl [9]

electrolytes on the two viruses were also similar. The mobility was reduced in magnitude at all pH values by increases in NaCl concentration. Ca^{2+} reduced the mobility of negatively charged virus much more strongly than Na^+; the relative concentrations of Ca^{2+} to Na^+ to give the same mobility at pH 8.1 were 1:50 and 1:100 for vaccinia and reovirus respectively. Both viruses can be considered as small colloids, with H^+ and OH^- as their potential-determining ions.

Specific interactions with reovirus were seen with calcium, but only as pH approached the IEP, and with Al^{3+} in the pH range 3.5 to 10. In the presence of 50 mg l^{-1} $Al_2(SO_4)_3 \cdot 16H_2O$ reovirus had pH-dependent mobility between that of the naked virus and fresh $Al(OH)_3$ precipitate with an IEP near 8. Charge reversal was seen not only with aluminum, but also in the presence of a small amount of polydiallyldimethylammonium chloride (PDADMA), a cationic polymeric flocculant. Therefore reovirus should be chemically amenable to removal from water by conventional water treatment processes with cationic flocculants.

Virus IEPs have frequently been determined by isoelectric focusing (IF). The IEP of reovirus determined this way, 3.9 [18,19], agrees closely with that from microelectrophoresis, 3.8. In the absence of suitable electrokinetic data the IEP from IF should be a useful indicator of virus charge in various environments, notwithstanding the differences in suspending media. The IEPs of a number of viruses (Table I) are seen to differ not only between viruses but also between types of the same virus.

Many viruses have been studied by electrophoresis in buffered gels or sucrose density gradients. Although absolute mobilities are not obtained by these techniques—the movement is not always reported with respect to a standard, and specific interaction with the buffer cannot always be discounted—some relative mobilities have been given (see Figure 7). Mobilities of viruses at pH 8.6 in borate buffer varied considerably. Among the most highly charged mammalian viruses are herpes, rubella, and Echo 1, while polio, Echo 7, and many Coxsackie A viruses were comparatively low.

Due to the wide variation in electrokinetic properties of viruses, their behavior in dilute electrolytes can be expected to show large differences. Electrokinetic characterization of additional virus strains and species should bear this out.

The Interaction of a Charged Sphere and a Flat Plate

The relative size differences between viruses and many soil particles suggest that the interaction may be considered as that of a sphere with a planar surface. The forces effective beyond the hydration layers on approach of a colloid to a surface are London-van der Waals forces from dipole and polarization

Table I. Isoelectric Points of Viruses

Virus	Strain	IEP	Method[a]	References
Mammalian				
Reovirus Type 3	Dearing	3.8	MEP	9,10
Reovirus Type 3	Dearing	3.9	IF	18
Reovirus Type 3	Dearing	4.0	IF	19
Vaccinia	WR	4.8	MEP	9,10
Vaccinia	Lister	3.9[b]	MEP	20
Cowpox	Brighton	4.3[b]	MEP	20
Rabbit pox	Utrecht	2.3[b]	MEP	20
Polio Type 1	Brunhilde	4.5,7.1	IF	21
Polio Type 2	P712-ch-2ab	7.8	IF	19
Plant				
Tobacco mosaic (TMV)	8 strains	3.83-4.18	Turb	22
Turnip yellow mosaic (TYMV)		4.0	IF	23
Southern bean mosaic (SBMV)	5 strains	3.62-6.03[c]	MB	24
Brome mosaic (BMV)	5 strains	6.8	IF	23
Brome mosaic (BMV)	5 strains	7.9[b]	MB	25
Satellite of tobacco necrosis (STNV)		4.3	IF	23
Satellite of tobacco necrosis (STNV))		7.0[b]	MB	26
Cowpea cholrotic mottle (CCMV)		4.1	IF	23
Cowpea chlorotic mottle (CCMV)		3.65[b]	MB	27
Bacteria				
Qβ		4.1	IF	23
Qβ		5.3[b]	MB	28
MS2		3.9[b]	MB	28
T4		4-5	Gel	29
R17		<4	Coag	30

[a]MEP = microelectrophoresis; IF = isoelectric focusing; MB = moving boundary; Turb = turbidimetry; Gel = gel electrophoresis; Coag = coagulation behavior.
[b]In buffer solutions, which may influence value.
[c]IEP varied with strain and buffer composition.

interactions and double-layer or Coulombic forces from electrical-potential interactions.

The contribution by unretarded van der Waals forces to the potential energy of interaction for a sphere of radius a at a distance x from a flat plate has been estimated [32,33] as:

$$\phi_{VDW}(x) = A_{132}[\ln (x + 2a)/x \ - 2a(a + x)/x(x + 2a)]/6 \qquad (10)$$

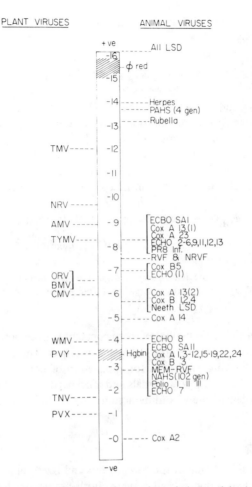

Figure 7. Composite drawing indicating the maxima of virus activity in relation to the distance migrated by rabbit hemoglobin and phenol red (Hgbin and ϕ red) in borate buffer at pH 8.6, Animal viruses: All. LSD = Allerton lumpy skin disease of cattle; PAHS = pantropic African horsesickness virus; RVF and NRVF = pantropic and neurotropic (105 mouse generations) Rift Valley fever virus, respectively; Neeth = LSD neethling lumpy skin disease of cattle; MEM-RVF = Rift Valley fever virus that had 105 generations in mouse brain followed by 56 generations in embryonated eggs and 15 generations in mouse brain; ECBO SA I and ECBO SA II = two bovine orphan viruses; NAHS = is neurotropic African horsesickness virus ($>$ 100 generations in mouse brain). The human enteroviruses are indicated by their commonly known denotations. Plant viruses: TMV = tobacco mosaic virus; TYMV = turnip yellow mosaic virus; NRV = necrotic ring-spot virus; AMV = alfalfa mosaic virus; ORV, BMV, and CMV = odontoglossum ringspot, bean mosaic, and cucumber mosaic viruses, respectively; WMV = watermelon mosaic virus; PVY and PVX = potato viruses Y and X, respectively; TNV = tobacco necrosis virus (from Ref. 31 with permission).

where A_{132} is the overall Hamaker constant for the three-component system, approximated by:

$$A_{132} \approx (A_{11}^{\frac{1}{2}} - A_{33}^{\frac{1}{2}})(A_{22}^{\frac{1}{2}} - A_{33}^{\frac{1}{2}}) \tag{11}$$

where A_{11}, A_{22} and A_{33} are the individual Hamaker constants for the materials of the sphere, plate and solution, respectively. The individual Hamaker constant, which is large for metals and metal oxides and small for hydrocarbons and water, may be calculated from surface-tension measurements.

The potential energy from interaction of the double layers of the sphere and plate is calculated from the linearized Poisson-Boltzman equation [34]:

$$\phi_{DL}(x) = \pm(\epsilon a/4)[(\psi_1 + \psi_2)^2 \ln\{1 \pm \exp(-\kappa a)\}$$

$$+ (\psi_1 - \psi_2)^2 \ln\{1 \pm \exp(-\kappa a)\}] \tag{12}$$

where ϵ = the dielectric constant of the medium and

ψ_1 and ψ_2 = the surface potentials (often approximated by ζ) of the isolated bodies.

The upper sign in this equation refers to constant-surface-potential materials, and the lower to constant-surface-charge substances. The two may be quite different. The limiting conditions for application of Equation 12 are $|\psi_i| <$ 25 mV and $\kappa a < 5$; i.e., low potentials and small particles in dilute electrolytes.

The total potential energy of interaction $\phi_T(x)$ is the sum of the two parts:

$$\phi_T(x) = \phi_{DL}(x) + \phi_{VDW}(x) \tag{13}$$

Figure 8 gives variation of the components and the total potential energies with separation of a sphere and a plate. Both are assumed to be constant-potential substances, as this approximates the natures of proteinaceous viruses and hydrous oxides. $\phi_{VDW}(x)$ has been made negative at all separations (van der Waals forces are attractive); i.e., it is assumed that A_{11} (virus) and A_{22} (surface) are both either greater or both less than A_{33} (water). This has not been proved; in fact, if $A_{11} > A_{33} > A_{22}$ or $A_{11} < A_{33} < A_{22}$, van der Waals forces will be repulsive (see Equation 11). Because, however, virus adsorption suggests a strong inherent attraction, our assumption of $\phi_{VDW} < 0$ is reasonable for this discussion.

The total potential energy favors adsorption when the surfaces are oppositely charged, when one or the other has no charge, or when the surface potentials are very different [35]. Adsorption is not favored when the surface potentials have the same sign and are either moderate or strong.

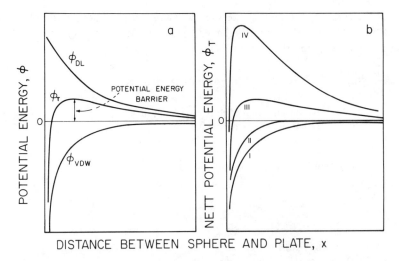

Figure 8. Potential energy of interaction on approach of a sphere to a plate of like-charge. (a) individual components; (b) net potential energy in indifferent electrolytes of indifferent electrolytes of increasing electrolyte concentration i $>$ ii $>$ iii $>$ iv [30].

Rate of Virus Adsorption to Solid Surfaces

When virus particles are adsorbed onto a solid surface from a stagnant or flowing aqueous medium, they must pass through the layer of fluid, which for all practical purposes must be considered fixed or stagnant, that surrounds the adsorbing surfaces. The rate of virus adsorption through this layer is a function of Brownian and van der Waals forces, modified by electrostatic effects.

The Brownian or diffusion character of the adsorption processes has been recognized as a correlation between the amount of virus adsorbed and the square root of time, $\sqrt{t}$ [36,37].

Valentine and Allison [38] considered the rate of adsorption of virus particles to be related to Brownian motion. The maximum rate of adsorption was defined as the rate observed when every collision event leads to adsorption. It can be measured when the effect of the electrochemical double layer is minimized by increasing the salt concentration of the medium to 0.1% NaCl. Valentine and Allison found that the rate was independent of turbulence in the medium, that the square of the number of particles adsorbed increased linearly with elapsed time and that the process was therefore consistent with Brownian theory, i.e., it was diffusion controlled.

If we consider the part of the theory dealing with the adsorption of virus-size particles to other, mostly larger, spherical elements such as sand and smaller soil particles, the appropriate model is:

$$C/C_0 = \exp\left[-4\pi n\, RD(t + R\sqrt{t/\pi D}\,)\right] \tag{14}$$

where C = the concentration of virus particles at time t,
        C_0 = the concentration at time 0,
        n = the number of adsorbing spherical particles,
        R = the radius of the adsorbing spheres, and
        D = the diffusion coefficient, given by:

$$D = \frac{kT}{6\pi \eta\, a} \tag{15}$$

where η = the viscosity, and
        a = the radius of the virus particle.

Observations on the time-dependence of virus adsorption to sands and soils [39] were consistent with this theory. In systems containing 0.5 g of sand or soil in 2 ml of suspending medium at 4°C, the rate of virus adsorption should depend on the size of both the virus and the sand or soil particles. It was found that the smaller enteroviruses, such as poliovirus (23 nm), were adsorbed to the coarser sands, such as Ottawa sand (specific surface area 0.018 $m^2 \cdot g^{-1}$), so that a three-log reduction occurred in one hour. For the larger reovirus (70 nm), removal was only 40% in 1 hour, and 24 hours were required for a three-log reduction. With finer sands (0.1 $m^2 \cdot g^{-1}$ specific surface area), a three-log reduction can be achieved in 1 hour. Such conditions can occur in a fluidized bed reactor or in a sediment suspension. Somewhat more favorable conditions occur in sand filters or soil formations, as the density of particles is greater. Calculations suggest that in a packed bed of uniform 100 mesh (0.15 mm diam) sand, a three-log reduction of poliovirus is possible in 20 minutes of contact time. For reovirus the same reduction requires one hour. (The number of logs of reduction in virus concentration is nearly proportional in contact time.)

Rates of adsorption to electronegative surfaces increase linearly with salt concentration up to 0.1% NaCl, at which maximum adsorption is observed [38]. The Schultze-Hardy rule is observed for a change of monovalent to divalent cations, but trivalent cations were no more effective than divalent ones [38].

The deposition of virus on flat surfaces, discussed earlier, is also a slow process regardless of turbulence of the suspension medium. This may explain the time of exposure on the order of hours required for tissue infection by virus [38]. Adsorption in the initial phase follows the relationship:

$$f = 1.13\sqrt{Dt/d} \qquad (16)$$

where f = the fraction of particles adsorbed,
       d = the thickness of the fluid layer, and
       D = particle diameter.

The proportionality between amount of virus adsorbed and $\sqrt{t}$ could be reinforced by, and therefore could be due in part to, heterogeneity of surface charge of the adsorbing material [40]. The particles adsorbing to the surface would not be those that pass through a stagnant layer of solvent from a larger distance by Brownian motion but rather those that attach to the surface only after a greater number of collisions with poorly adsorbing portions of the adsorbent. The same effect may also be caused by heterogeneity of the virus population (see the discussion of adsorption kinetics).

EXPERIMENTALLY DETERMINED FACTORS AFFECTING VIRUS ADSORPTION

pH Effects

The dependence of virus adsorption by soils and minerals upon pH, which has been demonstrated many times [2,3], has been assumed to indicate the effects of surface-charge development. Although pH-dependent adsorption and charge variation of both virus and adsorbent have not been investigated simultaneously through combining independent studies, the role of surface charges has been verified [41]. (Because clay minerals prepared similarly but separately were used, estimates of surface potentials would have been approximate. Thus calculations of ϕ_T would have been subject to excessive error, and so were not attempted.)

Taylor et al. [30] studied the pH-dependent interaction of two low-IEP viruses, reovirus type 3 and bacteriophage R17, with two clay minerals, Wyoming montmorillonite and Ohaupo allophane. Adsorption of the two viruses was similar. In base uptake by both clay minerals was weak; on pH reduction below 7 allophane became a strong absorbent, while montmorillonite showed adsorption only at low pH (Figure 9). Combination of these results with data on electrokinetic properties of reovirus (Figure 6) and clays (Figure 5) corroborates the qualitative description of double-layer effects in virus adsorption [41]. Adsorption is inhibited when the electrokinetic potentials of both absorbate and absorbent are high and of the same sign ($\psi_1, \psi_2 \ll 0$ or $\psi_1, \psi_2 \gg 0$). On charge reduction of either material (i.e., as pH approaches either IEP), the double-layer repulsion is reduced ($\psi_1 \to 0$ or $\psi_2 \to 0$). At the IEP of either, or when they are of opposite charge, adsorption

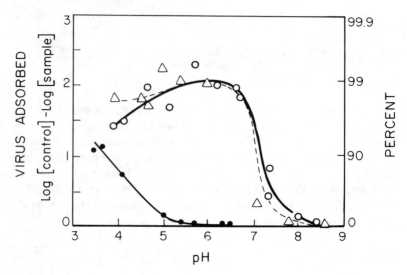

Figure 9. pH-dependent adsorption of bacteriophage R17 (●, △) and reovirus (○) by allophane (△, ○) and montmorillonite (●) [redrawn from Ref. 30, with permission].

is promoted by the inherent attractive forces and, if applicable, the double-layer interactions.

Related studies in this laboratory showed that interaction of poliovirus type 2 with soils and clays over a wide pH range also displayed a narrow region above which adsorption was weak and below which it was strong [42]. This region fell in the range 7.2-9.3 (Figure 10), close to or above the virus IEP (7.8). The electrophoretic mobilities of the solid, determined in the suspending medium, showed differences which, although substantial, were consistent with the variation of adsorption (Figure 11). For example, poliovirus adsorbed strongly to Riverhead sandy loam below pH 8.0 and to Genesee silt loam below pH 9.2. In each case at these pH levels the soil mobility indicated reduction of negative charge. (Poliovirus charge has been shown by zone electrophoresis to become more negative as pH increases above the IEP [21].) For virus and soil both negatively charged above the critical pH region, the same qualitative picture of double-layer interaction emerges. Only when virus and adsorbent are similarly strongly charged is adsorption prevented; double-layer effects are otherwise of secondary importance.

Data on other viruses support these results and the presumed underlying mechanisms. MS2 coliphage, which resembles enterovirus in its shape and size (25 nm) and which has an IEP of 3.9, shows markedly improved adsorption to sand when pH decreases from 8.4 to 6.0 [7]. The plant pathogenic virus of cytoplasmic hedrosis adsorbs well to acid soils [43].

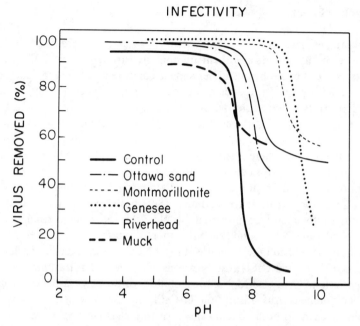

Figure 10. pH-dependent adsorption of poliovirus from a synthetic freshwater by 3 soils, a sand and a clay mineral [42].

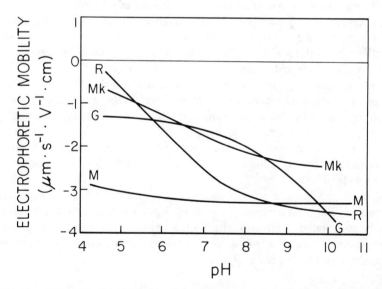

Figure 11. pH-dependent electrophoretic mobility of Arizona montmorillonite (M), Riverhead sandy loam (R), Genesee silt (G), and Wayne County muck (Mk) after equilibration in a synthetic freshwater (1:4 w/v ratio) [42].

Electrolyte Effects

pH studies have shown that double-layer interactions are clearly evident only when both virus and adsorbent are similarly, strongly charged. Electrolyte effects will be most obvious under similar conditions, whereupon they may promote adsorption by reducing surface potentials. When the influence of simple electrolyte concentration on poliovirus adsorption by soils and a clay (Figure 9) was studied at the equilibration pH of the absorbent, uptake varied only for montmorillonite [42]. For the other substrates the equilibration pH was in the region where there was no pH-induced double-layer repulsion. Thus simple electrolytes could not show profound influence. However, with montmorillonite adsorption increased with concentration of NaCl, $CaCl_2$ and Na_2SO_4; at the pH of the experiment (8-9) both colloids were negatively charged, so their surface potentials would decrease as ionic strength rose.

Carlson et al. [44] found that adsorption of phage T2, a low-IEP virus, on montmorillonite, kaolinite and illite at pH 7 increased with increasing concentrations of NaCl and $CaCl_2$, $CaCl_2$ was as effective as NaCl at one-tenth the concentration. When electrophoretic mobilities of the clays in the electrolytes were determined virus adsorption in $CaCl_2$ increased with decreasing clay mobility. Again, as both colloids are negatively charged, divalent Ca^{2+} would be particularly effective in reducing surface potentials and consequently promoting adsorption. However, the clays did not display mobility changes as NaCl concentration was increased, even though adsorption was enhanced. This suggests that double-layer interactions were of secondary importance. Nevertheless if the electrokinetic properties of T2 were determined in NaCl solutions, reductions in its surface potentials similar to those for vaccinia and reovirus [9,10] would probably be seen, and the role of the double-layer would be more fully appreciated. Kaolinite, to which $Al_2(SO_4)_3$ was added to give either a weak negative or weak positive charge, was found to be the strongest T2 absorbent of all, again reinforcing the importance of double-layer interactions. Durham and Bancroft [45] found three cationic binding sites in the protein of papaya mosaic virus and predicted that in the presence of sufficient divalent cations the particle would exhibit almost zero charge over a wide pH range.

Overall, the normal results of changes in the concentrations and the natures of simple, dilute electrolytes on virus adsorption described in the literature can be interpreted in terms of the electrochemical double-layer interactions. Observations at high electrolyte concentrations (e.g., in seawater [46]) require an analysis outside the realm of this article.

Effects of Organic Matter

Dissolved organic matter tends to compete with virus for adsorption sites and thereby reduce virus adsorption. Wastewater organics interfere with adsorption [47,48], as do proteins such as albumin [44]. Media high in protein are used to elute virus from membrane filters [49]. The effect of these organics is lessened if the divalent cation concentration is increased [50,51].

Naturally occurring humic substances have important effects on virus adsorption. Bixby and O'Brien [52] found bacteriophage adsorption to be reduced in the presence of fulvic acid. Soluble humic matter with a molecular weight of less than 50,000 interferes with virus adsorption [40,53]. This fraction can be eluted from soil, and adsorption can be improved by treatment of the soil with 0.01 N $CaCl_2$ solution. Humic matter also reduces poliovirus adsorption to magnetite [50].

DeSilva and Toth [54] studied the pH-dependent electrokinetic mobility of humic materials in the free state and when adsorbed on kaolinite and bentonite. In all cases the humic matter had an IEP below pH 3 with continuing negative charge development with increasing pH. These charge properties were conferred to the clay minerals after adsorption of the organic matter, i.e., when the humus coated much of the clay surface. Wilson and Kinney [55] confirmed that freshwater humic polymer was increasingly ionized from pH 3 to 7 due to its carboxylic acid groups. Phenolic groups dissociate in the pH range 8.5 to 12; this dissociation is greatly affected by the presence of the ionized carboxylic acid groups.

Organic matter as part of the soil matrix tends to reduce virus adsorption, often dramatically. Burge and Enkiri [37] found that poliovirus adsorbed poorly to an organic cypress-dome soil. In our studies the most highly organic soil, a muck, was also the poorest adsorbent for poliovirus and reovirus. Although humic matter tends to be acidic and hence negatively charged, our electrophoretic studies showed that the charge is not great. The mobility of the highly organic muck soil in neutral suspension was comparatively low (Figure 10); this soil did not bind poliovirus at all strongly (Figure 9).

A striking reduction in adsorption also occurs when soil particles are overgrown with a layer of bacteria, particularly of the Zoogloea or Pseudomonas type, which produce an exopolymer. Such bacteria are common in activated sludge floc and can cause the coating of soils and other adsorbent surfaces in contact with organic-rich effluents, including primary effluent. These exopolymers have been characterized as polyglucose [56] or as mucopolysaccharides [57] consisting mostly of N-acetylated amino sugars. In batch equilibration experiments, the effectiveness of magnetite sand, which in pure form removes over 99.99% of poliovirus from suspensions, was reduced to 40% when overgrown with Zoogloea [58].

Certain antibodies produced in the human intestinal tract may react with viruses and affect their surface properties.

Influence of the Substrate

Adsorption to Soils and Minerals from
a Synthetic Fresh Water

Most studies of soil-virus interactions have concentrated either on a single virus and a single adsorbent or on only a limited number of types of each. Berg [59], in his summary of the International Conference on Virus in Water held in Mexico City in 1974, recommended that tests for virus retention should be made with many soils in many areas to determine whether generalizations can be drawn.

In our laboratory an attempt was made to study the interactions of several very different viruses with a wide range of adsorbents. Results with three viruses and the 30 most important substrates shown in Table II were obtained by batch equilibration with the test virus in a synthetic freshwater of medium hardness and alkalinity (1.0 mM Ca^{2+}, 1.25 mM HCO_3^-, 4.1 mM total electrolyte content, see footnote a of Table II). The pH value, initially adjusted to 7.0, was affected by the mineral or soil substrate. Aggregation of virus, loss of viability of the virus in controls, and adsorption to the container were either controlled or taken into account.

The three viruses, although taken up to different extents by the 30 substrates (an average of 99.6% for Reovirus 3, 96% for Poliovirus 2 and 91% for Echovirus 1), showed a number of similarities. For instance, the highly negatively charged montmorillonites adsorbed all of the viruses comparatively poorly, as did glauconite, Genesee silt (a topsoil), and Colonie sandy loam. Substrates which were relatively poor adsorbents for two of the three viruses included the adobe and muck soils and the Lake Ontario dune and beach sands. Other relatively poor adsorbents include the humus-containing soils and sands.

Dickite, a neutral aluminosilicate clay mineral, was the most consistently strong adsorbent; others were magnetite (an oxide of iron), a sandstone rock and two acidic, finely divided clay minerals, halloysite and kaolinite. The shales (sedimentary rock) were also effective adsorbents. Overall, freshly ground, unweathered rocks and minerals were stronger adsorbents than soils and sands. Topsoils were particularly weak binders.

Linear correlation coefficients were calculated for uptake of the three viruses against each other and against each of a wide variety of properties of the 30 adsorbents. The correlation coefficients for poliovirus uptake against reovirus and echovirus uptake and for reovirus against echovirus uptake were

Table II. Interaction of Poliovirus 2, Echovirus 1 and Reovirus 3 with 30 Soils, Minerals, and Other Materials Suspended in a Synthetic Fresh Water[a,b]

Sample	pH	Surface Area (m^2/g)	Organic Carbon $(\%)$	Nonadsorbed Virus $(\%)$[c]		
				Echo	Polio	Reo
Dolomite	8.7	0.85	1.1	0.50	0.83	0.15
Arizona montmorillonite	9.1	41	0.06	37	8.5	0.86
Wyoming montmorillonite	9.3	32	0.16	11	5.9	0.24
Quartz	8.3	0.54	0.01	5.8	0.25	0.59
Bituminous shale	8.4	3.8	0.42	0.14	5.2	0.012
Calcite	8.5	4.0	0.52	0.13	1.8	0.29
Fossil-bearing shale	8.8	4.5	0.14	0.12	0.60	0.02
Illitic shale	7.3	55	1.36	0.54	0.81	0.035
Argillaceous shale	8.4	29	0.09	0.82	0.44	0.045
Arenaceous shale	8.7	3.5	0.11	0.76	0.55	0.05
Calcareous shale	8.4	9.4	0.24	0.07	2.1	0.04
South Carolina kaolinite	5.7	16	0.02	0.04	0.47	0.21
Utah halloysite	6.6	42	0.15	0.04	1.4	0.002
Utah dickite	8.6	2.7	0.09	0.04	0.19	0.014
Georgia attapulgite	8.2	118	0.06	0.29	2.1	0.006
Lake Ontario dune sand	8.6	4.1	0.20	29	0.44	0.65
Lake Superior dune sand	8.4	1.2	0.08	4.8	2.0	1.2
Lake Ontario beach sand	8.7	0.73	0.28	31	0.10	1.2
Magnetite sand	8.7	5.7	0.04	0.50	0.006	0.02
Colonie sandy loam	8.3	5.3	0.36	6.9	7.5	0.13
Genesee silt loam	8.1	4.2	1.65	61	25	1.1
Brown sandstone	8.3	5.6	0.12	0.03	0.52	0.01
Quartz conglomerate	8.9	3.2	0.12	3.2	1.8	0.73
Glauconite sand	7.7	54	0.10	15	5.9	0.44
Deming adobe	8.3	34	0.57	43	6.1	0.09
Genesee silt	8.0	11	0.15	2.0	2.1	0.17
Wayne county muck	7.6	1.2	20	3.2	21	2.2
Georgia kaolinite	8.1	12	0.007	0.55	1.3	0.04
Kame conglomerate	9.6	3.1	0.66	0.34	0.45	0.48
Kansas City loess	7.8	25	0.13	1.7	1.7	0.11

[a] 0.5 g of solid was shaken at 4°C with approximately 10^8-10^9 pfu of virus in 2 ml of a solution of 1.25 mM $NaHCO_3$, 1.0 mM $CaCl_2$, 0.375 mM $MgSO_4$ and 0.1 mM KCl initially at pH 7.0. After 1 hr the soil was removed by centrifugation and the supernatant assayed for infective virus by plaque assay.

[b] Poliovirus type 2 vaccine strain P712-CL-2ab, echovirus 1 strain 5199, and reovirus type 3 Dearing strain.

[c] $\%$ = Assay of sample supernatant/Assay of control supernatant x 100.

all significant, supporting the concept that substrates which adsorb one virus will behave similarly with other viruses, although the extent of adsorption may be radically different. Uptake of poliovirus and reovirus correlated strongly (inversely) with organic content by whatever means this was estimated; for echovirus uptake similar significant ($P < 0.05$) correlation was found with organic content only when muck (a highly organic soil which adsorbed echovirus surprisingly strongly) was omitted from the calculation.

Of the other properties examined (e.g., surface area, pH or content of extractable cationic components) only one was found to correlate consistently significantly with virus uptake. The capacity of the substrates to bind a cationic polyelectrolyte was inversely related to uptake of all three viruses. This capacity (determined in amount of polyelectrolyte bound per unit surface area) serves as an estimate of the negative surface charge available or exposed to a relatively large approaching body and in turn correlates strongly with content of negatively charged aluminosilicates and of humic matter. It is also related to cation exchange capacity.

Virus Uptake in Soil Series

Adsorption experiments with poliovirus and echovirus and soil series from the same site but different horizons showed that virus adsorption generally increased with increasing soil depth (Table III). Correlation of virus uptake with the soil properties of pH, surface area and content of extractable compounds was again absent, although uptake appeared to be inversely associated with organic content or capacity for the cationic polyelectrolyte.

Included in this study were two soil series: one from a field which had been irrigated with secondary treated sewage effluent for 30 years, and one from a neighboring untreated farm. Although soil from the irrigated field had a higher organic content, it was not substantially less effective as a virus adsorbent. This suggests that long-term use of an effluent disposal site need not lead to its eventual ineffectiveness for viral removal.

Comparative Studies of Different Viruses

Goyal and Gerba [60], in one of the widest studies yet performed, confirmed that virus removal by soils varies greatly, not only between virus species but also between different strains of the same species and between different isolates of the same strain. In a study of the interaction of 15 viruses (including enteroviruses, bacteriophages and a rotavirus) with nine soils they found that some soils and viruses had consistent adsorption behavior. For instance poliovirus 1 (LSc), rotavirus SA-11 and bacteriophage T4 were generally strongly adsorbed, while echovirus 1 and phages ϕX174 and f2 typically resisted removal from suspension. One soil, an acidic clayey sand of low organic content, could consistently take up 99% or more of all added viruses. A neutral siliceous sandy soil, in contrast, was a generally poor adsorbent.

Correlation coefficients between soil properties and virus uptake indicated that soil pH (the experimental pH was not reported) was the most common factor. It was significant for echovirus 1 (the reference strain and two isolates),

Table III. Adsorption of Poliovirus 2 and Echovirus 1 by Four Soil Series[a,b]

Soil	Horizon Depth (cm)	pH	Surface Area (m²/g)	Organic Carbon (%)	Nonadsorbed Virus (%)[c]	
					Polio	Echo
Genesee silt loam						
24	15-30	8.1	4.2	1.65	25	61
29	30-60	8.0	11	0.15	2.1	2.0
Riverhead sandy loam						
52	2-12	5.3	5.3	1.0	13	3.1
53	12-40	5.5	7.3	0.45	5.2	0.72
54	40-60	5.6	6.2	0.25	2.8	0.57
55	60-90	6.3	3.1	0.20	3.0	0.32
56a	90-150	6.9	3.3	0.20	0.70	0.48
56	150-180	7.4	1.25	0.10	0.24	4.7
Acuff loam I[d]						
101	0-15	7.4	6.0	1.65	3.8	
102	15-30	7.3	10.2	0.60	0.5	
103	30-60	7.5	18	0.35	0.08	
Acuff loam II[e]						
104	0-15	7.7	10.4	0.45	2.4	
105	15-30	7.9	13	0.40	0.23	
106	30-60	8.0	22	0.45	0.12	

[a]0.5 g of solid was shaken at 4°C with approximately 10^8-10^9 pfu of virus in 2 ml of a solution of 1.25 mM $NaHCO_3$, 1.0 mM $CaCL_2$, 0.375 mM $MgSO_4$ and 0.1 mM KCl initially at pH 7.0. After 1 hr the soil was removed by centrifugation and the supernatant assayed for infective virus by plaque assay.

[b]Poliovirus type 2 vaccine strain P712-Cl-2ab, echovirus 1 strain 5199.

[c]% = Assay of sample supernatant/Assay of control supernatant x 100.

[d]Site irrigated with secondary-treated sewage effluent.

[e]Site not irrigated.

coxsackie B4 and phage ϕX174. Other properties such as surface area; clay, silt or sand content; and organic content were of less general importance.

For 9 of 12 viruses uptake by the sandy loam was greater from suspension in 10^{-2} M $CaCl_2$ than from either deionized water or 10^{-3} M $CaCl_2$. The three viruses relatively unaffected were poliovirus 1, echovirus 7 and coxsackie B3 reference strains.

How do these results relate to the concept discussed earlier that, while virus/soil interactions are inherently attractive, adsorption can be inhibited by electrostatic repulsion if virus and surface are similarly strongly charged? Since poliovirus 1 has a IEP near neutrality (Table I), it will have a low charge or positive charge in more simple electrolytes and is not readily prevented from adsorbing. Echovirus 7 and coxsackie B3 are similarly relatively low in charge [31], which would facilitate their adsorption. As none of these three viruses would be strongly negatively charged, divalent cation (Ca^{2+}) would have little effect on double-layer interactions with negatively charged surfaces. On the other hand, echovirus 1 is relatively highly charged [31], making its uptake by most soils difficult (although susceptible to enhancement by addition of Ca^{2+}). As observed, acid soils would be effective adsorbents, since they generally have low charges themselves. Viruses in low-pH media would also be low in charge.

Interactions of phages T1, T2 and f2 with five soils were also not inferable from any single soil parameter, including surface area, pH, cation exchange capacity and organic carbon content [1]. The coliphage MS2, which resembles poliovirus but has a lower IEP, was adsorbed to three soils in accordance with the pH value of soil solution and suspension medium as expected, but adsorption was much less than is found with poliovirus [8].

Adsorption Isotherms

Adsorption of virus to soil surfaces follows the Freundlich isotherm at low virus concentrations [30,37,50,61-63]. Exponents of the Freundlich isotherm of 0.2 to 1.2 have been reported. Near surface saturation, however, the Langmuir representation is superior [64-66], and adsorption can be described in terms of second-order kinetics (the adsorption is first-order with respect to both virus concentration and concentration of absorbing surface)[65,66]. Langmuir representations suggest a monolayer adsorption capacity for coliphage MS2 in the range of 10^7 to 10^9 plaque forming units (PFU) per gram in two soils of approximately equal sand, silt and clay contents [8]. In our study poliovirus coverage on Ottawa sand reached 10-20% of the available surface. T4 coliphage adsorbed on coal may cover up to 0.1% of the open surface [65].

The Freundlich isotherm is an empirical formulation; attempts to develop a theoretical basis involve assumptions of a heterogeneity of either the adsorbing surface or the particle population [40,67]. The existence of two particle populations of different charge properties is suggested by soil adsorption experiments with bacteriophage ϕX174 [40], as well as by early experiments by Delbrück [68]. Deeper penetration of soil columns by a portion of added poliovirus indicated population heterogeneity [69]. Poliovirus can adopt either of two configurations, which have different IEPs [21]. The conditions under which this change occurs are related to pH but are otherwise not well characterized. Virulent poliovirus type 1 elutes much more readily from aluminum hydroxide gel at pH 7 or 8 than does the attenuated form. Virulent poliovirus types 2 and 3 are much less readily eluted, and their corresponding attenuated forms show somewhat different behavior [70]. Virus mutants and particles with deformed capsids may show altered charges [71] (Table I).

Reversibility of Adsorption

Although adsorption is in principle reversible, results of laboratory adsorption-desorption experiments suggest that the degree of reversibility is neither qualitatively nor quantitatively predictable from the adsorption models. In experiments in this laboratory with poliovirus and Ottawa sand recovery was always less than expected. Partial desorption was achieved with the same synthetic fresh water as was used in the adsorption step, and also with distilled water. In an extraction with synthetic fresh water and two with distilled water 2.5-6.8% of adsorbed virus infectivity was recovered. Very effective in recovering poliovirus from sand were solutions of sodium hexametaphosphate, a surface-active compound able to adsorb to the sand and make its effective surface charge highly negative. These results also suggest decreased infectivity of the desorbed virus material. These results essentially confirm those of earlier reports [30,37,46,72,73]. Poliovirus can be desorbed from soil columns with deionized water [63,74,75]. A period of drying reduced the recovery of virus from the soil. Drying also very effectively causes the irreversible loss of several types of enterovirus [76]. As would be expected, desorption from moist soil by media of low electrolyte content is strain-dependent [77]. These results are pertinent to the issue of mobilization of virus in recharge beds by rainwater [77,78].

Elution of virus at high pH or in media with high phosphate or organic content is common practice in the isolation of viruses from water or sediments. Each of the reagents able to desorb virus operates consistently with concepts of the binding interactions; i.e., each in some way allows or causes development of repulsive interactions between the two components. For example, distilled water lowers the ionic concentration within the diffuse and Stern layers at the virus/solution interfaces. This in turn permits establishment of repulsive double layer interactions, which will cause desorption if virus and soil have charge of the same sign and sufficient magnitude. The inability of rainwater to elute poliovirus (LSc strain) bound to a Long Island soil [77] probably reflects the fact that this virus, due to its high IEP [21] will not be strongly negatively charged in near-neutral solutions and so is not electrostatically repelled from most soil surfaces.

In alkaline media, deprotonation causes an increase of negative charge, and hence coulombic repulsion, on both materials. It may in addition dissolve binding agents such as hydrolyzed Al^{3+} species [79]. Phosphates, which are also strongly anionic, increase the negative charge of surfaces by adsorption.

Unlike the simpler reagents organic matter such as protein displaces viruses through strong adsorption and competition for available surface. Furthermore virus does not readily bind to surfaces of protein or similar material.

Stagg [80] discovered that phage MS2 adsorbed on cellulose nitrate filters from neutral 0.05 M $MgCl_2$ was not eluted in base unless EDTA, a chelating agent, was added. The chelating agent apparently competed for Mg^{2+} and so extracted it from the virus-filter interface, where it was promoting the adhesion of the virus.

Aggregation of Virus

To judge the effectiveness of soil adsorption of virus we must briefly consider the occurrence of virus in association with itself and with particulate matter. Virus, when liberated from infected tissue by laboratory techniques involving cell breakage through freezing and thawing, is always in the form of aggregates [81], and these are likely to persist in water to some extent. Aggregates also form in wastewater coagulation processes involving use of cationic agents and formation of hydroxide floc [82]. Association with these and organic sludge particles [83] does not reduce infectivity [81,84], but virus occluded in large aggregates or organic floc resists chemical disinfection to a considerable extent [81,85,86].

Monodisperse virus also tends to aggregate considerably in dilute media, particularly at a pH below neutral, although this tendency is of course variable with virus and ion species [87-90]. Aggregation of reovirus and poliovirus is dominated by the ionic composition of the medium [16]. Neutral suspensions dispersed in buffers of near-physiologic strength (0.14 M NaCl) will aggregate on dilution into distilled water or on pH reduction. At low pH, when the virus is positively charged, aggregation in the buffered solution is influenced by addition of salts. Cations, particularly multivalent ones, have a much greater effect than anions. This suggests that virus can interact specifically with cations such as Mg^{2+} and Ca^{2+}. Poliovirus, although it tends to form aggregates more easily than reovirus, did not aggregate at its IEP. The observed deviations from typical colloid behavior have not been entirely resolved; determination of zeta potentials concurrently with aggregation measurements would facilitate interpretation of these observations.

Virus aggregation may not occur during effluent recharge to soils if the concentration of virus is low or the availability for adsorption of a soil surface is very large relative to the virus particle surface. The existence of preformed virus aggregates in effluents is nevertheless of practical importance, not only because the larger aggregates may be retained in soil by mechanisms other than those affecting monodisperse virus (i.e., mechanical trapping), but also because virus aggregates may be—and upon disruption certainly are—more infectious than single virions, both in tissue culture and in the host organism. The increase in infectivity upon disruption of an aggregate is not directly proportional to the increase in the number of physical units, however, as not every particle is infective [91].

Particle-Associated Virus

Numerous publications emphasize that virus in wastewaters and treated effluents is largely associated with suspended solids. This means that in the soil the properties of the larger particles to which the virus is attached, rather than the properties of the virus itself, will determine the fate of the virus. Larger particles are retained in soil primarily by filtration and sedimentation rather than by adsorption. Virus desorbed from such particles will, however, travel in soil and become adsorbed according to the mechanisms described in this chapter. Virus occluded in organic particulate matter will be protected until the organic matter is degraded. The biologically active layer that filters and degrades particulate organic matter, may extend one centimeter to several decimeters below the soil surface. Virus liberated from the organic particles and not retained on the poorly adsorbing surfaces of the biologically active layer will penetrate the soil further and will become subject to adsorption to soil minerals. The fate of solids-associated virus will thus ultimately be determined by the mechanisms described above.

Virus-Inactivation Mechanisms

All observations of changes in virus infectivity in the presence of adsorbents must take into account the experimental setting. They may indicate irreversible adsorption of virus and/or adsorption followed by release of virus sufficiently altered to be incapable of infecting host tissue. Loss of infectivity of virus can have been caused already by small changes in the conformation of a coat protein or even by the adsorption of a suitable substance to the coat protein [92,93]. On the other hand, disintegration of the virus particle can yield free nucleic acid, which under suitable conditions can be infective (see Reference 4 for a detailed review). Laboratory techniques are available to discriminate between the various events that lead to apparent virus loss [94]. Certain viruses are sensitive to several surfactants as a function of pH [6,95], as well as to the very high pH values (11-12) in wastewater and water-treatment processes [96]. Poliovirus is sensitive to free ammonia [97]. Experiments with radiolabeled viruses indicate that only a fraction of the particles adsorbed to soils and minerals can be removed. Furthermore, Yeager and O'Brien [76,98] found two mechanisms of inactivation of poliovirus in soil: dissociation of the ribonucleic acid genome from the capsids in dry soil, and nucleic acid damage in moist soil. Taylor et al. [30] found that reovirus eluted from clay with sodium hexametaphosphate was not disintegrated but was of lower infectivity. However, aggregates of the virus may have formed, accounting for this result.

The nature of the particles can affect virus infectivity. Schaub and Sagik [99] and Gerba et al. [100] suggested that the adsorbent can have a protective effect on the adsorbed virus in wastewater-soil systems. Pure silica (SiO_2) has little effect on poliovirus infectivity, but CuO causes infectivity loss by particle disruption, Al_2O_3 causes some inactivation, MnO_2 less, and Fe_2O_3 no inactivation [101].

Finally, biological mechanisms of virus inactivation are as yet little understood. Proteolytic action [102-104] and grazing by protozoa in complex biological systems [105] have been shown to be important. An unidentified factor in certain California soils inactivates a plant pathogenic virus [106].

CONCLUSIONS AND RECOMMENDATIONS

In general viruses behave as small proteinaceous colloids that develop charge by protonation and deprotonation of exposed ionizable groups in their capsids. Their surface properties vary from species to species and strain to strain. If their IEPs are known, however, much of their behavior in dilute electrolytes can be predicted.

Most viruses appear to have an inherent tendency to adsorb firmly to inorganic surfaces. Adsorption is inhibited, however, by electrostatic repulsion if virus and surface are similarly and strongly charged or if the surface is coated with organic matter. As many viruses have low IEPs, they, like most soils, are negatively charged at pH values around neutral. Adsorption can be enhanced by acidification or by addition of simple electrolytes, particularly Ca^{2+} and Mg^{2+}, all of which effectively reduce the repulsive double layer.

At present no general formula is available to predict uptake of any virus by any soil. However, the following chemical parameters should be considered if a soil is intended for use in recharge beds. First, the soil should not be basic; although neutral soils may be suitable, acidic soils have been consistently found most effective. Second, if the soil has a low buffering capacity, it should not be exposed to basic waste. Third, soils with a significant organic content will be most likely to permit passage of viruses. Fourth, soils formed predominantly of montmorillonite clays will probably be less suitable than those containing kaolinite, halloysite and other less well-ordered clay minerals. In general substantial virus removal will be effected quickly and within short distances if pH is maintained near or below neutrality and if divalent cationic species are present.

For practical application of these theoretical and laboratory findings we recommend adequate testing, including physical and mineralogic characterization of the recharge area, when a soil-recharge system is being planned. Our studies [9,39] have shown that surface area, equilibrium pH, organic content and capacity for binding synthetic polyelectrolytes are all useful parameters for identifying a soil's ability to adsorb proteinaceous colloids.

Unfortunately an indicator virus suitable for both laboratory and pilot studies has not yet been identified, but comparative tests with poliovirus should be valuable, as a substantial body of data is already available. Less readily adsorbed strains such as echo 1 or coliphage MS2, however, might be very useful, as might rotaviruses in view of their possible public health significance. Field measurements will often provide ultimate proof of the functioning of a treatment-and-recharge system.

If virus concentration in municipal wastewater is 10^3-10^4 per liter and an acceptable standard for drinking water is 1 pfu in 100-1000 liters [107] the virus removal rate required to occur in the treatment and recharge of municipal wastewater is 10^6-10^7. It would be desirable to recharge into drinking water aquifers so that this standard is met at the perimeter of the recharge area. Then: (1) virus would be controlled at the source and breakthrough could be clearly traced to a particular recharge system and (2) water drawn practically anywhere in the aquifer would be potable, with disinfection required only as an additional barrier.

In theory, proportions of the total removal efficiency can be allocated among the wastewater treatment and recharge components of the system, but uncertainties in the reliability of each of the components dictate a conservative approach to such allocations [108].

In this situation we suggest that treatment of wastewater be optimized to enhance the effectiveness of recharge beds to the point where the necessary removals can be achieved even without chlorination or other chemical disinfection. This would be desirable, as chlorination tends to produce persistent by-products.

Several elements of advanced treatment, flocculation-precipitation reactions, followed by sedimentation and dual-media filtration, benefit virus removal during treatment and improve the performance of recharge beds. These processes: (1) remove virus; (2) remove organic matter that reduces the soil's adsorption efficiency either directly or indirectly by causing the buildup of bacterial films; (3) remove organic particulates which, although mechanically retained in soil, may release encapsulated virus upon decomposition; (4) add cations such as Ca^{2+}, Fe^{3+} and Al^{3+}, all of which are strong promoters of virus adsorption to the effluent and (5) may include high-pH lime treatment (pH 11.5 or higher), which destroys virus either directly or assisted by high levels of free ammonia [97,109,110]. High-pH treatment must be followed by adjustment of pH to a value of 7 or lower to facilitate adsorption. Mixed- or dual-media filtration is generally recommended for effluent polishing.

Nonbasic effluents with a naturally high Ca^{++} content are suitable for recharge with careful removal of organics and particulate matter. If dissolved solids must be reduced by reverse osmosis, substantial virus removal is of course obtained, so the demands on the quality of the recharge bed are minimal.

The relationship of virus density in effluents to available soil surface suggests that saturation of the soil with virus will hardly, if ever, occur even in a long-term operation. Predictions based on these factors alone do not consider either the possible existence of competing organic materials or the possible ultimate inactivation and physical destruction of the virus. Where properties of local soils are unfavorable, preparation of the recharge sites with more suitable soils is possible if such materials can be found within a reasonable distance. Use of magnetite sand in adsorption beds or in advanced treatment [111-113] should be considered.

Recharge beds will be most effective if the hydraulic loading is kept high and constant to minimize the virus-mobilizing effect of rains. Periodic and thorough drying of the beds is an alternative in arid climates.

Careful design of both wastewater treatment and recharge operations should permit a reduction of virus concentration in excess of 10^7 at or near the perimeter of the recharge area. (This should be substantiated by sampling of monitoring wells.) In actual applications, additional residence in ground water on the order of weeks or months, combined with distances over 1 km to point of withdrawal, can often be provided, thus affording a substantial additional margin of safety in accordance with the multiple-barrier concept.

The drawbacks of incomplete treatment and reduced efficiency of a recharge system can combine to result in a much inferior performance of the overall process. Shortcuts to advanced wastewater treatment are acceptable only in situations where the aquifer is not intended to provide potable water.

REFERENCES

1. Drewry, W. A., and R. Eliassen. "Virus Movement in Groundwater," *J. Wat. Poll. Control Fed.* 40:R257-271 (1964).
2. Bitton, G. "Adsorption of Viruses onto Surfaces in Soil and Water," *Water Res.* 9:473-484 (1975).
3. Gerba, C. P., C. Wallis and J. L. Melnick. "Fate of Wastewater Bacteria Viruses in Soil," *J. Irrig. Drain. Div., Am. Soc. Civil Eng.* 101:157-174 (1975).
4. Duboise, S. M. "Poliovirus Survival and Movement in Soils Following Application of Wastewater Effluent," M.A. Thesis (Austin, TX: University of Texas, 1977).
5. Tiffany, J. M. "The Interaction of Viruses with Model Membranes," in *Virus Infection and the Cell Surface,* G. Poste and G. L. Nicholson, Eds. (Amsterdam: Elsevier North Holland Biomedical Press, 1977), pp. 157-194.
6. Ward, R. L., and C. S. Ashley. "pH Modification of the Effects of Detergents on the Stability of Enteric Viruses," *Appl. Envir. Microbiol.* 38:314-322 (1979).

7. Chaudhuri, M., K. V. A. Koya and N. Sriramulu. "Some Notes on Virus Retention by Sand," *J. Gen. Appl. Microbiol.* 23:337-344 (1977).
8. Koya, K. V. A., and M. Chaudhuri. "Virus Retention by Soil," *Progr. Wat. Technol.* 9:43-52 (1977).
9. Taylor, K. H., and H. B. Bosmann. "Measurement of the Electrokinetic Properties of Vacinnia and Reovirus by Laser-Illuminated Whole-Particle Electrophoresis," *J. Virol. Methods* 2:251-260 (1981).
10. Taylor, D. H., and H. B. Bosmann. "The Electrokinetic Properties of Reovirus Type 3: Electrophoretic Mobility and Zeta Potential in Dilute Electrolyte," *J. Colloid Interface Sci.* (in press).
11. van Olphen, H. *Introduction to Clay Colloid Chemistry,* 2nd Ed. (New York: John Wiley and Sons, 1977).
12. Parks, G. A., and P. L. deBruyn. "The Zero Point of Charge of Oxides," *J. Phys. Chem.* 66:967-973 (1962).
13. Tanford, C. *Physical Chemistry of Macromolecules* (New York: John Wiley and Sons, 1961).
14. Huang, C. P., and W. Stumm. "Specific Adsorption of Cations on Hydrous γ-Alumina," *J. Colloid Interface Sci.* 43:409-420 (1973).
15. Taylor, D. H. "The Colloid and Surface Chemistry of Allophane," D. Phil. Thesis (Hamilton, New Zealand: Univ. of Waikato, 1977).
16. Buchanan, A. S., and R. C. Oppenheim. "The Surface Chemistry of Kaolinite," *Aust. J. Chem.* 21:2367-2371 (1968).
17. Harvey, J. D., J. A. Farrell and A. R. Bellamy. "Biophysical Studies of Reovirus Type 3, II. Properties of the Hydrated Particle," *Virology* 62:154-160 (1974).
18. Floyd, R., and D. G. Sharp. "Viral Aggregation: Effects of Salts on the Aggregation of Poliovirus and Reovirus at Low pH," *Appl. Envir. Microbiol.* 35:1084-1094 (1978).
19. Moore, R. S., D. H. Taylor, M. Chen and L. S. Sturman. "Adsorption of Reovirus by Minerals and Soils" (in preparation).
20. Douglas, H. W., B. L. Williams and C. J. M. Rondle. "Microelectrophoresis of Pox Viruses in Molar Sucrose," *J. Gen. Virol.* 5:391-396 (1969).
21. Mandel, B. "Characterization of Type 1 Poliovirus by Electrophoretic Analysis," *Virology* 41:554-568 (1971).
22. Ginoza, W., and D. E. Atkinson. "Comparison of Some Physical and Chemical Properties of Eight Strains of Tobacco Mosaic Virus," *Virology* 1:253-260 (1955).
23. Rice, R. H., and J. Horst. "Isoelectric Focussing of Viruses in Polyacrylamide Gels," *Virology* 49:602-604 (1972).
24. Magdoff-Fairchild, B. S. "Electrophoretic and Buoyant Density Variants of Southern Bean Mosaic Virus," *Virology* 31:142-153 (1967).
25. Bockstahler, L. E., and P. Kaesberg. "The Molecular Weight and other Biophysical Properties of Bromegrass Mosaic Virus," *Biophys. J.* 2:1-9 (1962).
26. Kassanis, B., and A. Kleczkowski. "Mutual Precipitation of Two Viruses," *Nature, London* 205:310 (1965).
27. Bancroft, J. B., E. Herbert, M. W. Rees and R. Markham. "Properties of Cowpea Chlorotic Mottle Virus, its Protein and Nucleic Acid," *Virology* 34:224-239 (1968).
28. Overby, L. R., et al. "Comparison of Two Serologically Distinct Ribonucleic Acid Bacteriophages," *J. Bact.* 91:442-448 (1966).

29. Childs, J. D., and H. C. Birnboim. "Polyacrylamide Gel Electrophoresis of Intact Bacteriophage T4D Particles," *J. Virol.* 16:652-661 (1975).
30. Taylor, D. H., A. R. Bellamy and A. T. Wilson. "The Interaction of Bacteriophage R17 and Reovirus Type 3 with the Clay Mineral Allophane," *Water Res.* 14:339-346 (1980).
31. Polson, A., and B. Russell. "Electrophoresis of Viruses," in *Methods in Virology*, Vol. II. K. Maramorosch and H. Koprowski, Eds. (New York: Academic Press, 1967), pp. 391-426.
32. Hamaker, H. C. "The London-van der Waals Attraction Between Spherical Particles," *Physica* 4:1058-1072 (1937).
33. Visser, J. "Adhesion of Colloidal Particles," *Surface Colloid Sci.* 8:3-84 (1976).
34. Prieve, D. C., and E. Ruckenstein. "Role of Surface Chemistry in Particle Deposition," *J. Colloid. Interface Sci.* 60:337-348 (1977).
35. Bleier, A., and E. Matijević. "Heterocoagulation. I. Interactions of Monodispersed Chromium Hydroxide with Polyvinyl Chloride Latex," *J. Colloid Interface Sci.* 55:510-524 (1976).
36. Pollard, E. C. *The Physics of Viruses* (New York: Academic Press, Inc., 1953), pp. 122-133.
37. Burge, W. D., and N. K. Enkiri. "Virus Adsorption by Five Soils," *J. Envir. Qual.* 7:73-76 (1978).
38. Valentine, R. C., and A. C. Allison. "Virus Particle Adsorption. I. Theory of Adsorption and Experiments on the Attachment of Particles to Non-Biological Surfaces," *Biochim. Biophys. Acta* 34:10-23 (1959).
39. Fuhs, G. W., R. S. Moore, M. M. Reddy, L. S. Sturman and D. H. Taylor. "A Laboratory Study of Virus Uptake by Minerals and Soils," *Proceedings of the Water Reuse Symposium*, Vol. 3 (Denver, CO: American Water Works Association Research Foundation, 1979), pp. 2274-2281.
40. Burge, W. D., and N. K. Enkiri. "Adsorption Kinetics of Bacteriophage ∅X-174 on Soil," *J. Envir. Qual.* 7:536-541 (1978).
41. Taylor, D. H. "Interpretation of the Adsorption of Viruses by Clays from their Electrokinetic Properties," in *Chemistry in Water Reuse*, W. J. Cooper, Ed. (Ann Arbor, MI: Ann Arbor Science Publishers, Inc., 1981).
42. Taylor, D. H., R. S. Moore and L. S. Sturman. "Influence of pH and Electrolyte Composition on the Adsorption of Poliovirus by Soils and Minerals," *Appl. Environ. Microbiol.* (in press).
43. Hukuhara, T., and H. Wada. "Adsorption of Polyhedra of a Cytoplasmic Polyhedrosis Virus by Soil Particles," *J. Invert. Pathol.* 20:309-316 (1972).
44. Carlson, G. F., Jr., F. E. Woodward, D. F. Wentworth and O. J. Sproul. "Virus Inactivation on Clay Particles in Natural Waters," *J. Water Poll. Control Fed.* 40:R89-R106 (1968).
45. Durham, A. C. H., and J. B. Bancroft. "Cation Binding by Papaya Mosaic Virus and its Protein," *Virology* 93:246-252 (1979).
46. Fildes, P., and D. Kay. "The Conditions which Govern the Adsorption of a Tryptophan-Dependent Bacteriophage to Kaolin and Bacteria," *J. Gen. Microbiol.* 30:183-191 (1963).
47. Lo, S. H., and O. J. Sproul. "Polio-Virus Adsorption from Water onto Silicate Minerals," *Water Res.* 11:653-658 (1977).

48. Bitton, G., N. Masterson and G. E. Gifford. "Effect of a Secondary Treated Effluent on the Movement of Viruses Through a Cypress Dome Soil," *J. Envir. Qual.* 5:370-375 (1976).
49. Sobsey, M. D. "Methods for Detecting Enteric Viruses in Water and Wastewater," in *Viruses in Water,* G. Berg et al., Eds. (Washington, DC: American Public Health Association, 1976), pp. 89-127.
50. Bitton, G., O. Pancorbo and G. E. Gifford. "Factors Affecting the Adsorption of Virus to Magnetite in Water and Wastewater," *Water Res.* 10:973-980 (1976).
51. LaBelle, R. L., and C. P. Gerba. "Influence of pH, Salinity, and Organic Matter on the Adsorption of Enteric Viruses to Estuarine Sediment," *Appl. Envir. Microbiol.* 38:93-101 (1979).
52. Bixby, R. L., and D. J. O'Brien. "Influence of Fulvic Acid on Bacteriophage Adsorption and Complexation in Soil," *Appl. Envir. Microbiol.* 38:840-845 (1979).
53. Scheuerman, P. R., G. Bitton, A. R. Overman and G. E. Gifford. "Transport of Viruses Through Organic Soils and Sediments," *J. Envir. Eng. Div., Am. Soc. Civil Eng.* 105:629-640 (1979).
54. DeSilva, J. A., and S. J. Toth. "Cation-Exchange Reactions, Electrokinetic and Viscometric Behavior of Clay-Organic Complexes," *Soil Sci.* 97:63-73 (1964).
55. Wilson, D. E., and P. Kinney. "Effects of Polymeric Charge Variations on the Proton-Metal Ion Equilibria of Humic Materials," *Limnol. Oceanogr.* 22:281-289 (1976).
56. Friedman, B. A., P. R. Dugan, R. M. Pfister and C. C. Remson. "Fine Structure and Composition of the Zoogloeal Matrix Surrounding *Zoogloea ramigera*," *J. Bact.* 96:2144-2153 (1968).
57. Tezuka, Y. "A *Zoogloea* Bacterium with Gelatinous Mucopolysaccharide Matrix," *J. Water Poll. Control Fed.* 45:531-536 (1973).
58. Moore, R. S., M. Chen and G. W. Fuhs. Unpublished results.
59. Berg, G. "Perspectives—Reclamation and Disposal," in *Viruses in Water,* G. Berg et al., Eds. (Washington, DC: American Public Health Association, 1976), p. 255.
60. Goyal, S. M., and C. P. Gerba. "Comparative Adsorption of Human Enteroviruses, Simian Rotavirus, and Selected Bacteriophages to Soil," *Appl. Envir. Microbiol.* 38:241-247 (1979).
61. Drescher, J. "Adsorptionsvorgänge bei Virusarten, I. Mitt. Aufnahme der Adsorptionsisotherme und Prüfung des pH-Wertes und der Molarität für das System Pre-Influenzavirus-γ-Aluminiumoxyd," *Zbl. Bakt. I. Orig.* 168:217-234 (1957).
62. Drescher, J. "Comparison of the Adsorption of Influenza Virus Strain B/Berlin/2/55 on Aluminum Oxide and on Aluminum Oxide and on Aluminum Hydroxide," *Am. J. Hyg.* 74:104-118 (1961).
63. Gerba, C. P., and J. C. Lance. "Poliovirus Removal from Primary and Secondary Sewage Effluent by Soil Filtration," *Appl. Envir. Microbiol.* 36:247-251 (1978).
64. Cookson, J. T., and W. J. North. "Adsorption of Viruses on Activated Carbon: Equilibria and Kinetics of Attachment of *E. coli* Bacteriophage T4 on Activated Carbon," *Environ. Sci. Technol.* 1:46-52 (1967).
65. Oza, P. P., and M. Chaudhuri. "Removal of Viruses from Water by Sorption on Coal," *Water Res.* 9:707-712 (1975).

66. Oza, P. P., and M. Chaudhuri. "Virus-Coal Adsorption Interaction," *J. Envir. Engin. Div., ASCE* 102:1255-1262 (1976).
67. Cerofolini, G. F., M. Jaroniec and S. Sokolowski. "A Theoretical Isotherm for Adsorption on Heterogeneous Surface," *Colloid Interface Sci.* 256:471-477 (1978).
68. Delbrück, M. "Adsorption of Bacteriophage Under Various Physiological Conditions of the Host," *J. Gen. Physiol.* 23:631-642 (1940).
69. Lance, J. C., and C. P. Gerba. "Poliovirus Movement during High-Rate Land Filtration of Sewage Water," *J. Envir. Qual.* 9:31-34 (1980).
70. Woods, W. A., and F. C. Robbins. "The Elution Properties of Type 1 Poliovirus from Al(OH)$_3$ Gel. A Possible Genetic Attribute," *Proc. Natl. Acad. Sci. USA* 47:1501-1507 (1961).
71. Brinton, C. C., Jr., and M. A. Lauffer. "The Electrophoresis of Viruses, Bacteria and Cells and the Microscope Method of Electrophoresis," in *Electrophoresis, Theory, Methods and Applications,* M. Bier, Ed. (New York: Academic Press, Inc., 1959), pp. 427-492.
72. van Regenmortel, M. H. V. "Electrophoresis," in *Principles and Techniques in Plant Virology,* C. I. Kado and H. O. Agrawal, Eds. (New York: Van Nostrand Reinhold Co., 1972), pp. 390-412.
73. Moore, B. E., B. P. Sagik and J. F. Malina, Jr. "Viral Association with Suspended Solids," *Water Res.* 9:197-203 (1975).
74. Lance, J. C., C. P. Gerba and J. L. Melnick. "Virus Movement in Soil Columns Flooded with Secondary Sewage Effluent," *Appl. Envir. Microbiol.* 32:520-526 (1976).
75. Duboise, S. M., B. E. Moore and B. P. Sagik. "Poliovirus Survival and Movement in a Sandy Forest Soil," *Appl. Envir. Microbiol.* 31:536-543 (1976).
76. Yeager, J. G., and R. T. O'Brien. "Enterovirus Inactivation in Soil," *Appl. Envir. Microbiol.* 38:694-701 (1979).
77. Landry, E. F., J. M. Vaughn, McH. Z. Thomas and C. A. Beckwith. "Adsorption of Enteroviruses to Soil Cores and Their Subsequent Elution by Artificial Rainwater," *Appl. Envir. Microbiol.* 38:680-687 (1979).
78. Wellings, F. M., A. L. Lewis, C. M. Mountain and L. V. Pierce. "Demonstration of Virus in Groundwater after Effluent Discharge into Soil," *Appl. Envir. Microbiol.* 29:751-757 (1975).
79. Kessick, W. A., and R. A. Wagner. "Electrophoretic Mobilities of Virus Absorbing Filter Materials," *Water Res.* 12:263-268 (1978).
80. Stagg, C. H. "Inactivation of Solids-Associated Virus by Hypochlorous Acid," PhD Thesis (Houston, TX: Rice University, 1976).
81. Young, D. C., and D. G. Sharp. "Poliovirus Aggregates and Their Survival in Water," *Appl. Envir. Microbiol.* 33:168-177 (1977).
82. Farrah, S. R., S. M. Goyal, C. P. Gerba, C. Wallis and J. L. Melnick. "Concentration of Poliovirus from Tap Water onto Membrane Filters with Aluminum Chloride at Ambient pH Levels," *Appl. Envir. Microbiol.* 35:624-626 (1978).
83. Farrah, S. R., et al. "Comparison between Adsorption of Poliovirus and Rotavirus by Aluminum Hydroxide and Activated Sludge Flocs," *Appl. Envir. Microbiol.* 35:360-363 (1978).
84. Ward, R. L., and C. S. Ashley. "Mode of Initiation of Cell Infection with Sludge-Associated Poliovirus," *Appl. Envir. Microbiol.* 38:329-331 (1979).

85. Boardman, G. D., and O. J. Sproul. "Protection of Viruses During Disinfection by Adsorption to Particulate Matter," *J. Water Poll. Control Fed.* 49:1857-1861 (1977).

86. Sharp, D. G., R. Floyd and J. D. Johnson. "Nature of the Surviving Plaque-Forming Unit of Reovirus in Water Containing Bromine," *Appl. Envir. Microbiol.* 29:94-101 (1975).

87. Floyd, R. "Viral Aggregation: Mixed Suspensions of Poliovirus and Reovirus," *Appl. Envir. Microbiol.* 38:980-986 (1979).

88. Floyd, R., and D. G. Sharp. "Aggregation of Poliovirus and Reovirus by Dilution in Water," *Appl. Envir. Microbiol.* 33:159-167 (1977).

89. Floyd, R., and D. G. Sharp. "Viral Aggregation: Quantitation and Kinetics of the Aggregation of Poliovirus and Reovirus," *Appl. Envir. Microbiol.* 35:1079-1083 (1978).

90. Floyd, R., and D. G. Sharp. "Viral Aggregation: Buffer Effects in the Aggregation of Poliovirus and Reovirus at Low and High pH," *Appl. Envir. Microbiol.* 39:395-401 (1979).

91. Galasso, G. J., and D. G. Sharp. "Virus Particle Aggregation and Plaque-Forming Unit," *J. Immunol.* 88:339-347 (1962).

92. Fuchs, P., and A. Levanon. "Inhibition of Adsorption of West-Nile and Herpes Simplex Viruses by Procaine," *Arch. Virol.* 56:163-168 (1978).

93. Ostle, A. G., and J. G. Holt. "Elution and Inactivation of Bacteriophages on Soil and Cation-Exchange Resin," *Appl. Envir. Microbiol.* 38:59-65 (1979).

94. Salo, R. J., and D. O. Cliver. "Inactivation of Enteroviruses by Ascorbic Acid and Sodium Bisulfite," *Appl. Envir. Microbiol.* 36:68-75 (1979).

95. Salo, R. J., and D. O. Cliver. "Effect of Acid pH, Salts, and Temperature on the Infectivity and Physical Integrity of Enteroviruses," *Arch. Virol.* 52:269-282 (1976).

96. Thayer, S. E., and O. J. Sproul. "Virus Inactivation in Water-Softening Precipitation Processes," *J. Am. Water Works Assoc.* 58:1063-1074 (1966).

97. Ward, R. L. "Mechanism of Poliovirus Inactivation by Ammonia," *J. Virol.* 26:299-305 (1978).

98. Yeager, J. G., and R. T. O'Brien. "Structural Changes Associated with Poliovirus Inactivation in Soil," *Appl. Envir. Microbiol.* 38:702-709 (1979).

99. Schaub, S. A., and B. P. Sagik. "Association of Enteroviruses with Natural and Artificially Introduced Colloidal Solids in Water and Infectivity of Solids-Associated Virions," *Appl. Envir. Microbiol.* 30:212-222 (1975).

100. Gerba, C. P., C. H. Stagg and M. G. Abadie. "Characterization of Sewage Solid-Associated Viruses and Behavior in Natural Waters," *Water Res.* 12:805-812 (1978).

101. Murray, J. P., and S. J. Laband. "Degradation of Poliovirus by Adsorption to Inorganic Surfaces," *Appl. Envir. Microbiol.* 37:480-486 (1979).

102. Cliver, D. O., and J. E. Herrmann. "Proteolytic and Microbial Inactivation of Enteroviruses," *Water Res.* 6:797-805 (1972).

103. Herrmann, J., and D. O. Cliver. "Degradation of Coxsackievirus Type A9 by Proteolytic Enzymes," *Infect. Immunity* 7:513-517 (1973).

104. Herrmann, J. E., J. E. Kostenbader and D. O. Cliver. "Persistence of Enteroviruses in Lake Water," *Appl. Microbiol.* 28:895-896 (1974).

105. Groupé, V., E. C. Herrmann, Jr. and F. J. Rauscher. "Ingestion and Destructions of Influenza Virus by Free-Living Ciliate *Tetrahymena pyriformis*," *Proc. Soc. Exp. Biol.* 88:479-482 (1955).
106. Cheo, P. C. "Antiviral Factors in Soil," *Soil Sci. Soc. Amer. J.* 44:62-67 (1980).
107. WHO Scientific Group. "Human Viruses in Water, Wastewater and Soil," *Technical Report Series* No. 639 (Geneva: World Health Organization, 1979), 50 pp.
108. Malina, J. E. "Effect of Unit Processes of Water and Wastewater Treatment on Virus Removal," in *Viruses and Trace Contaminants in Water and Wastewater,* J. A. Borchardt, et al., Eds. (Ann Arbor, MI: Ann Arbor Science Publishers, Inc., 1977), pp. 33-54.
109. Sattar, S. A., S. Ramia and J. C. N. Westwood. "Calcium Hydroxide (Lime) and the Elimination of Human Pathogenic Viruses from Sewage: Studies with Experimentally-Contaminated (Poliovirus Type 1, Sabin) and Pilot Plant Samples," *Can. J. Public Health* 67:221-225 (1976).
110. Grabow, W. O. K., I. G. Middendorff and N. C. Basson. "Role of Lime Treatment in the Removal of Bacteria, Enteric Viruses, and Coliphages in a Wastewater Reclamation Plant," *Appl. Envir. Microbiol.* 35:663-669 (1978).
111. Warren, J., A. Neal and D. Rennels. "Adsorption of Myxoviruses on Magnetic Iron Oxides," *Proc. Soc. Exp. Biol. Med.* 121:1250-1253 (1966).
112. Rao, V. C., R. Sullivan, R. B. Read and N. A. Clarke. "A Simple Method for Concentrating and Detecting Viruses in Water," *J. Am. Water Works Assoc.* 60:1288-1294 (1968).
113. Bitton, G., and R. Mitchell. "The Removal of Eschericha coli Bacteriophage T7 by Magnetic Filtration," *Water Res.* 8:547-551 (1974).

CHAPTER 27

VIRUS REMOVAL WITH LAND FILTRATION

J. C. Lance and C. P. Gerba
U.S. Water Conservation Laboratory
U.S. Department of Agriculture
Phoenix, Arizona

Since more than 100 types of viruses are found in sewage water, virus removal is an important consideration in determining the best use for wastewater and the amount of treatment needed before various kinds of wastewater reuse.

Viruses are ultramicroscopic intracellular parasites, incapable of replication outside a host organism. They consist of a nucleic acid genome enclosed in a protective protein coat. Viruses that are shed in fecal matter are referred to as enteric viruses; they are characterized by their ability to infect tissues in the throat and gastrointestinal tract, but are capable of replicating in other organs of the body as well. Enteric viruses are excreted in concentrations as high as one million viruses per gram of feces, and concentrations as high as 463,500 infectious virus particles per liter have been detected in raw sewage [1].

A number of studies have shown that many soils readily remove viruses from wastewater moving through the soil [2-4]. However, much of the data are from laboratory studies and only a few scattered field sites have been studied. This discussion is an attempt to summarize the available data in order to determine how much of the laboratory data can be extrapolated to field conditions and to describe factors controlling virus movement through various soils and land treatment systems.

LABORATORY BATCH STUDIES

When small quantities of soil were mixed with viruses suspended in solutions and adsorption was determined after a given period of time, virus adsorption increased with increasing cation exchange capacity and specific surface area, and decreased with increasing pH [5,6]. However, in each study, adsorption to at least one kind of soil did not fit this pattern. For example, Drewry and Eliassen [6] found that a soil with a low cation exchange was one of the highest in the test group for virus adsorption.

Batch tests have also indicated that virus adsorption by clays increased as the salt content of the suspending solution increased; adsorption could be reversed by suspending the clay in distilled water; adsorption varied for different kinds of clays; and proteinaceous materials like egg albumin interfered with virus adsorption [7,8]. Similarly, fulvic acid complexed bacteriophage, which prevented its adsorption to soil [9]. In another batch experiment, the adsorption of virus types and strains to nine different soils was compared using suspensions in deionized water [10]. A great deal of variability among types and strains was noted, along with considerable variability among different soils. In general some viruses, such as polio, adsorbed to all of the soils, whereas soils with a pH below the isoelectric point of the viruses had a high adsorption rate for all viruses. Soil pH was the only soil characteristic which consistently affected virus adsorption. Much less variability among virus strains was noted when the viruses were suspended in 0.01 M $CaCl_2$ instead of deionized water. One unique feature of these batch experiments was that the soil from the same site also was used in soil column experiments and in field studies. The adsorption of Echo 1 and Echo 29, two reference strains that adsorbed poorly to the loamy sand taken from basins used for groundwater recharge with sewage water, was compared to adsorption of poliovirus type 1 in a 250-cm-long column packed with the same soil [11]. The adsorption patterns of Echo 29 and Polio 1 were almost identical with 90% adsorption in the top 2 cm and no viruses detected below 160 cm. Adsorption of Echo 1 was 77% for the top 2 cm but the adsorption patterns were similar below the 40-cm depth. Thus, more Echo 1 than Polio 1 viruses moved to the 40-cm depth but numbers moving to lower depths were similar. Therefore the potential for movement through a soil to groundwater would be similar for the two viruses if the water table was deeper than 40 cm below the soil surface. The desorption and movement of the polio and echo virus were similar when the column was flooded with deionized water.

When adsorption data from the batch studies are compared with soil column adsorption data, the 0.01 M $CaCl_2$ solutions were the only batch tests that accurately assessed the potential for virus adsorption by the soil column. This test showed that adsorption of Echo 29 and Polio 1 were similar and slightly more than adsorption of Echo 1, as the column experiments indicated. Batch tests with other solutions such as deionized water, secondary effluent, soil extract, and 0.001 M $CaCl_2$ showed poor adsorption of the two echo virus strains. However, virus adsorption by soil columns was tested for only one soil, and the 0.01 $CaCl_2$ solution batch might not accurately assess the adsorption potential of columns packed with other soils. The calcium content of the water increased as the water moved through the calcareous recharge basin soil as the calcium carbonate in the soil dissolved. This may at least partly explain why virus adsorption from suspensions of virus in 0.01 $CaCl_2$ was comparable with virus adsorption by the soil column.

Thus, it appears that the ionic composition of the suspending solutions for the viruses must approximate that of the soil solution for batch studies to approximate adsorption by soil columns. Even then, important factors such as flow velocity cannot be taken into account with batch studies.

In summary, batch studies showed that increasing the salt concentration increased adsorption, some organic compounds interfered with adsorption, and distilled water desorbed viruses. These findings have been verified by column studies [3,4]. Differences in adsorption rates of different viruses and the effect of various soil characteristics on virus adsorption have not been confirmed. Thus, batch tests give information on some factors affecting virus adsorption but cannot yield quantitative estimates of virus adsorption by different field soils.

COLUMN STUDIES

Virus Adsorption by Short Columns

A number of studies with various soils have shown that many soils have a capacity to adsorb viruses. Some of the earliest work on virus migration through porous media centered on sand filters used in the treatment of drinking water. Robeck et al. [12] allowed poliovirus suspended in dechlorinated tap water to percolate through 60 cm of California dune sand. With virus doses of 10,000-60,000 PFU/mℓ (plaque-forming units/milliliter) the sand was capable of removing 99% of the viruses for 98 days of operation. Virus removal was found to depend on flow rate through coarse sand; removal greater than 90% occurred when flow rates were below 1.2 m/day. Higher flow rates resulted in considerably more virus breakthrough. Drewry and Eliassen [6] found that when virus suspended in distilled water was passed through columns of 40-50 cm of sterile soil, over 99% of the virus was removed. Radioactivity-tagging experiments indicated that most of the virus was retained in the top 2 cm of the column. In Hawaii 10-cm columns of two acid soils from the island of Oahu, Wahiawa (Tropeptic Eutrustox) and Lahaina (Tropeptic Haplustox), removed 99.3% of the poliovirus applied in distilled water, whereas Tantalus cinder (Typic Dystrandepts) removed 78% of the applied virus [13].

In Israel, poliovirus and bacteriophage f2 were passed through 20-cm columns of sand sterilized with HCl and rinsed with distilled water [14]. Most of the virus was found in the top sand fractions and virus removal was highly dependent on cation concentration. About 37% was removed when the divalent ion concentration was 10^{-3} N but more than 99% was removed when the Ca^{++} and Mg^{++} concentrations were 10^{-2} N. The Na^+ concentrations as high as 0.5 N had little effect on virus retention.

Bitton et al. [15] and Scheuerman et al. [16] showed that humic substances reduced virus retention by a sandy soil taken from a cypress dome in Florida and packed in soil column lengths of 13-50 cm. Duboise et al. [17] showed that viruses could be desorbed from 19.5-cm soil columns by flooding with distilled water to simulate rain.

Landry et al. [18] reported that all polioviruses tested, including both reference and field strains, adsorbed well to 12.5-cm soil cores; however, elution of the viruses by sewage effluent or rain water rinses appeared to be strain dependent.

These studies with relatively short columns give information on factors affecting virus adsorption by relatively thin layers of soil. However, predicting the potential for groundwater pollution from these data or making quantitative predictions of virus movement through various soils is difficult because short columns represent layers of soil rather an entire soil profile extending to the water table. Also, short columns do not give an accurate assessment of soil virus interactions during wetting and drying cycles because they do not drain during the drying cycle due to the capillary action of water in soil pores.

Virus Adsorption by Long Columns

Lance et al. [19] studied virus movement through 250-cm columns of fine sand taken from basis in the dry Salt River bed near Phoenix, Arizona. The columns were used for groundwater recharge of secondary sewage effluent. Previous studies on infiltration, nutrient removal, and coliform removal had demonstrated that the columns were good models of the field groundwater recharge system. Ceramic soil water samplers were used to monitor virus concentrations at various depths. Wang et al. [20] showed that the samplers constructed from ceramic with a pore size of 20 μm gave high virus recoveries (82-100%) from sewage water or tap water while two commercially available samplers were unsuitable due to low virus recoveries.

Most of the poliovirus suspended in secondary sewage effluent was adsorbed in the top 5 cm of the soil columns [19]. Viruses were detected in only three of the forty-three 1-mℓ samples from the 160-cm depth, and none were detected in 1-mℓ samples from the 240- and 250-cm depths. Viruses were detected in five of forty-three 100-mℓ samples of the daily cumulative drainage from the columns. The virus concentration in the sewage water was reduced by about 2 logs (from a starting concentration of 3×10^4 PFU/mℓ) during the first 2 cm of travel but an additional 158 cm of travel was required for another 2 log reduction. Flooding a column with the sewage water-virus mixture for 27 days did not saturate the surface layer of soil with virus. The virus adsorption profile at the end of the 27-day period was almost identical to the pattern observed on the first day. Thus, a soil that had been intermittently flooded with sewage for 10 years was not saturated by applying heavy loads of virus. When viruses were eluted from soil samples taken from groundwater recharge basins, the greatest amount of virus was found in the upper 2.5 cm of soil [21]. The virus concentrations in the top 2.5-cm layer were about 10 to 15 times as high as those found in the 2.5- to 10-cm zone. This indicates that the virus adsorption pattern of the soil columns accurately represented the field adsorption pattern.

When a column packed with soil that had not previously been flooded with sewage water was flooded with virus enriched sewage, most of the viruses still were adsorbed near the soil surface [22]. This showed that the concentration of viruses near the surface of the soil columns was not due to a build up of organic material or salts near the soil surface.

Flooding with deionized water immediately after application of the sewage water virus mixture caused considerable virus desorption and movement [19]. Most of the viruses were adsorbed again lower in the column but a few were detected in the column outflow after 2 days of flooding with deionized water. Only a few viruses were detected in profile samples when 1 or 3 mM CaCl$_2$ solution was applied instead of deionized water and none were detected below the 80-cm depth. Desorption of virus by deionized water was greatly reduced when the soil was dried for 1 day between the application of virus-enriched sewage and deionized water; no viruses were desorbed when the drying time was increased to 5 days. No viruses were detected below the 80-cm depth when at least 1 day of drying was allowed between the application of virus-enriched sewage and of deionized water.

When a soil column was flooded with the sewage water-virus mixture followed by 10 cm of deionized water and then by sewage effluent with no added virus, a peak concentration of 1000 PFU/mℓ was detected at the 10-cm depth in samples extracted 1 hour after the sewage without virus was applied (or 1 hour after the 10 cm of deionized water infiltrated). No viruses were detected in 100-mℓ samples of the cumulative outflow taken 1 day and 2 days after application of the deionized water.

Thus, these experiments with deionized water show that virus movement due to rainfall would be slight unless the rainfall occurred during the first day of a drying period following application of sewage. Heavy rains during the first few hours of the drying period could cause some movement but would not move viruses through 250 cm of the Flushing Meadows sand. Virus movement probably could be almost completely eliminated by resuming the application of sewage immediately if rain fell during the first few hours of the drying period or by applying salts to the basins.

Virus adsorption was greater when poliovirus was suspended in tap water than it was when poliovirus was added to secondary sewage effluent, indicating that organic compounds reduced virus adsorption. However, when a soil column was flooded with virus-enriched secondary effluent and later with virus-enriched primary effluent, the adsorption patterns were quite similar [23]. Thus, the additional organic carbon in the primary effluent (69.5 mg/ℓ for the primary vs 10 mg/ℓ for the secondary) had no added effect on virus adsorption.

When the concentration of poliovirus added to the sewage water was increased from 0.9×10^2 to 2.6×10^4 PFU/mℓ, the number of viruses found at each depth in the soil increased as the concentration of applied virus increased, but the percentage of the applied virus found at each depth was the same for all concentrations [22]. Thus, the maximum depth of penetration by the virus was independent of the concentration of applied virus.

This suggests that the adsorption reaction rate between viruses and soil particles is not concentration dependent; the number of virus adsorption sites in the soil may be so large that adsorption did not effectively reduce the quantity of sites available. Thus, reducing virus concentrations in the sewage effluent by some pretreatment like chlorination could not prevent penetration of some viruses to a particular depth. Pretreatment could reduce the number of viruses penetrating to that depth, however. It is possible that virus adsorption is affected by the strength of the negative change on virus particles. Perhaps viruses with a change below a certain level are immediately adsorbed and other viruses move farther down the column. When a virus population is diluted to provide various concentrations, the percentage in each virus concentration having a particular charge strength would not change, and therefore, the percent adsorption at each depth would not change either. This would also explain why a 2-log reduction in virus concentration occurred in the top 5 cm of the soil and the remaining viruses were adsorbed at a lower rate as the water moved down the column.

Viruses were not detected below 160 cm when a column packed with coarse sand was flooded with virus enriched sewage at a flow rate of 0.6 cm/ day [22]. However, some viruses broke through the column when the flow rate was increased to 1.2 m/day. Virus adsorption was not greatly affected by further increases in the flow rate. Apparently virus adsorption is not affected by increases in the flow rate up to some breakthrough point. After that breakthrough point was reached, some viruses moved through, but 99% were still removed at flow rates of 12 m/day. The breakthrough velocity probably corresponds to the velocity where some water begins to move only through the large soil pores, allowing little or no contact between viruses in the water and adsorptive surfaces. The breakpoint apparently occurred between 0.6 and 1.2 m/day for the Salt River bed sand columns, and at about 1.2 m/day for the sand columns used by Robeck et al. [12], but it might be different for other soils. Flow rates about 1.2 m/day could not be achieved for many soils.

In adsorption studies with different soils packed into 87-cm columns, poliovirus and echovirus did not move below 67 cm in three columns but moved completely through another column [24]. Virus adsorption was inversely proportional to the permeability of the different soils. These experiments on flow velocity suggested that permeability may be the most important factor affecting virus movement. Therefore, the experiments with long soil columns confirm that most of the viruses are adsorbed near the soil surface and showed that few viruses move below the 160-cm depth. The depth of virus penetration was not affected by the concentration of applied virus. Organic compounds interfered with virus adsorption, and salts promoted adsorption as batch studies and short columns indicated. Extensive leaching with deionized water desorbed virus, as was reported for short columns, but the viruses were readsorbed lower in the column. Also, very few viruses were desorbed when the free water drained from the soil columns between additions of virus-enriched sewage and deionized water. Thus far, soil permeability seems to be the most important soil characteristic affecting virus adsorption.

FIELD STUDIES

At the Santee Project [25] viruses were never isolated from wells located distances of 60 and 120 m from groundwater basins after poliovirus was seeded into the sewage water. However, methods for concentrating viruses were not available when that work was done. Only since the early 1970s have methods of concentrating viruses from large sample volumes been available for use in field studies of virus movement. Viruses have been detected in the groundwater below several field sites (Table I) located on sandy or gravelly soils with high infiltration rates [26-33]. The Fort Devens, Vineland, East Meadows, and Holbrook sites have mixtures of coarse sands and gravels. Vaughn and Landry [31] detected virus movement through a coarse sand and gravel test basin at infiltration rates of 144 cm/day at another Long Island site similar to the East Meadows and Holbrook sites. They found little virus retention at flow rates of 18-24 m/day. These results agree with soil column data showing virus breakthroughs at flow rates around 1.2 m/day [12,22].

The site at St. Petersburg, Florida, was located on a sand which probably had a high infiltration rate, and viruses were detected after a 3-month period

Table I. Reported Isolations of Virus Beneath Land Treatment Sites

| Site Location | Distance of Virus Migration (m) | | Reference |
	Vertical	Horizontal	
St. Petersburg, FL	6	–	29
Cypress Dome, FL	3	7	30
Fort Devens, MA	18.3	182	26
Vineland, NJ	16.8	250	27
East Meadows, NY	11.3	3	28
Holbrook, NY [a]	6.1	45.7	28

[a] Viruses were isolated from wells at three other similar locations on Long Island: Sayville [33], Twelve Pines [31] and North Masapequa [32].

when 70 cm of rainfall was recorded [29]. Bitton et al. [15] showed that organic compounds interfered with virus adsorption at the Cypress Dome site.

In studies at field projects with finer textured soils and lower infiltration rates [34-38], researchers have not detected viruses in samples from wells or lysimeter leachates (Table II). Concentrated duplicate samples ranging from 174 to 454 ℓ from each of four wells at about 2-month intervals during a year of operation at the Flushing Meadows Project in Arizona did not yield any positive samples for viruses [34]. The Flushing Meadows Project was a groundwater recharge system for renovation of secondary sewage effluent where infiltration rates ranged from 30 to 70 cm/day. The rainfall in the area is low, but soil column studies mentioned previously indicated that rainfall would have little effect of virus movement at the site [19]. With a single questionable exception, viruses were not isolated from leachate samples collected from 150-cm-deep lysimeters located in grass sod and sugarcane fields in Hawaii that were regularly irrigated with secondary effluent [35]. Studies at Roswell, New Mexico, also failed to yield viruses in 151- to 333-ℓ samples taken from shallow wells (3-37 m) beneath farmland that was intermittently flooded with secondary sewage effluent [36]. Concentrations of viruses in the Roswell wastewater before land application ranged from 960 to 5070 PFU/ℓ.

Viruses added to sludge placed on lysimeters in Denmark did not move through the soil even though they survived for 23 weeks [37]. Water movement was slow since no water was added to supplement the 30 cm of rainfall during the test period.

Viruses were not detected in groundwater samples of 189-378 ℓ from a site in Florida where sludge was spread on agricultural land [38]. Probably viruses applied in sludge are less mobile than those applied in sewage water because of adsorption to sludge solids.

Schaub et al. [39] measured the removal of enteric viruses and tracer bacteriophages from raw, primary, and secondary treated wastewaters by an

Table II. Field Sites Where Viruses Have Not Been Detected
in Wells or Lysimeter Leachates

Locations	Sampling Method	Soil Texture	Reference
Arizona	wells	fine sand	34
Hawaii	lysimeters	silty clay	35
New Mexico	wells	silty clay loam	36
Florida[a]	wells	sandy loam	38
Denmark[a]	lysimeter	two clays, two sands	37

[a]Virus applied in sludge.

overland flow treatment system. Tracer bacteriophage f2 seeded in waste-water was reduced by 30-60% while enteric viruses naturally occurring in sewage were reduced 68-85% during travel down a 36-m slope.

Thus, a summary of results from field studies indicates that virus movement to the groundwater is a problem primarily in soils composed of coarse sands and gravels. Field data agree with the column studies that pointed to water flow velocity as one of the most important and possibly the most important factor affecting virus movement. Column studies suggested that virus movement through a coarse sand could be prevented by reducing the infiltration rate [12,22]. Complete virus removal would not be expected for overland flow systems, but groundwater contamination would not occur below those systems since they are placed on soils with a very low permeability.

VIRUS SURVIVAL

Survival in Soil-Laboratory Studies

Viruses removed from sewage water can remain in the soil for some time. Schaub et al. [40] have shown that viruses adsorbed to clay are still infectious. In laboratory studies Leffler and Kott [14] and Duboise et al. [17] found that poliovirus added to sand columns survived about 90 days stored at 18-20°C. However, 69-90% of the virus was inactivated after 7 days. Survival time was greatly prolonged at 4°C with viruses still surviving after 175 days.

Hurst et al. [41] showed that viruses survived much longer under sterile conditions than nonsterile conditions in an aerobic environment at 23°C. Survival times were similar for sterile and nonsterile soils under anaerobic conditions, which indicated that aerobic soil microorganisms are antagonistic to viruses. The survival curves for Poliovirus 1, Coxsackie B3, Echovirus 1 and Rotavirus SA11 were similar, while Coxsackie A9 was inactivated much faster. Virus survival time increased as temperature decreased, as had been reported by others. When sewage water was mixed with distilled water in different ratios, virus survival was not related to the sewage effluent concentration. In a study of the effect of soil properties on virus survival, increases in virus adsorption and in exchangeable aluminum prolonged virus survival. Virus survival decreased with increases in pH and resin extractable phosphorus. It is possible that the last three variables (Al, pH, and P) affect virus survival indirectly through their effect on virus adsorption in batch studies with small soil samples.

Yeager and O'Brien [42], using radioactively labeled poliovirus and coxsackie virus B1, recorded no infectivity from saturated soil after 12 days at 37°C, whereas virus persisted for 180 days at 4°C. Viruses persisted longer in soil saturated with septic tank liquor than in soils saturated with groundwater or Rio Grande water. Infectivity of soil samples dried in a desiccator decreased only slightly as the moisture content decreased from 18 to 2% over a 4-hr period. However, infectivity was considerably reduced at moisture contents of 1.0 and 0.6%. This shows that drying to very low moisture contents inactivated viruses directly in contrast to the indirect effect of soil drying, which produces aerobic conditions and allow aerobic microorganisms to increase virus inactivation as noted by Hurst et al. [21]. However, the direct drying effect would seldom occur under field conditions because soils in the natural state seldom dry to a 2% moisture content particularly in land treatment systems where drying periods are usually limited to a few days. The soil surface can become very dry but even at the surface the viruses are probably inactivated before the low moisture content is reached. Hurst et al. [21] found that after 5 days drying viruses could not be recovered from the surface of a basin used to recharge primary sewage effluent even though the moisture content was 8%.

Ostle and Holt [43] adsorbed bacteriophages to soils, clays and Dowex-50 resin, eluted them and measured inactivation. Examination by electron microscopy indicated that physical damage to bacteriophage tails during elution caused inactivation. The authors suggested that bacteriophages could not be used as models of vertebrate viruses in survival studies since no vertebrate viruses have tail structures.

Survival in Soil-Field Studies

Tierney [44] reported that, in soils irrigated with sewage, seeded poliovirus survived for 11 days during the summer and 96 days during the winter. Sadovski et al. [45] found a loss in virus titer of 0.25 $\log_{10}$ units/day in 10 days in plots irrigated with sewage water.

Hurst et al. [21] found that most of the viruses found in groundwater recharge basins that had been extensively flooded with sewage water were concentrated near the soil surface. This is in agreement in soil column studies mentioned earlier [19].

Viruses were not detected in soil samples after the fifth day of the drying cycle. Poliovirus and echovirus were seeded in open, 15-cm diameter pipes that were filled with soil and buried in a vertical position. The basins were flooded with primary sewage effluent on a schedule of 5 days flooding alternated with 9 days drying. The survival times were similar for seeded polio and echo viruses in the buried pipe sections and for viruses naturally occurring in sewage water that were eluted from soil samples in other areas of the same basin where the pipes were buried. The similar survival times for the different viruses indicates that poliovirus may be a good indicator virus for use in virus survival studies. The rate of virus titer reduction was about 0.5 $\log_{10}$ units/day in laboratory soil samples at a constant 15% moisture content and about the same temperature (25°C) as the basin soil. Viruses survived longer during flooding cycles than during drying cycles and longer in closed tubes than in open tubes. This confirmed laboratory results showing longer survival times under anaerobic conditions than under aerobic conditions and showed that drying cycles are important in preventing an accumulation of viruses in soils treated with sewage water. The effect of drying cycles in the field probably was due to aeration of the soil rather than the soil drying per se because the soil moisture content usually exceeded 10% at the end of the drying period.

Virus Survival in Water

It is important to minimize virus movement to groundwater because it has been shown that under certain conditions poliovirus can travel at least 90 m underground [46]. Viruses could be adsorbed in aquifers if water is not moving through gravel layers or fractured rock and survival time depends upon the water temperature. Clarke et al. [47] reported that enteroviruses survived 5-20 days in Ohio River water samples held in the laboratory at 20°C and survived longer in clean water than in heavily polluted waters. Joyce and Weiser [48] found the enteroviruses survived in Ohio farm pond water as long as 84 days at 20°C and were still present after 91 days at 4°C.

Lefler and Kott [14] studied the survival of f2 bacteriophage and poliovirus in sand saturated with distilled water, distilled water containing cations, tapwater, and oxidation pond effluents. The poliovirus titer was lost between 63 days and 91 days in distilled water, while f2 bacteriophage survived longer than 175 days. The viruses survived even longer in distilled water containing cations. When tapwater or oxidation pond effluent was used, the researchers noted a very high initial kill of virus, but poliovirus particles could still be detected after 91 days. In these media, f2 particles again survived longer than 175 days. Thus viruses that reach surface water could be a hazard for a long time, particularly in cold water.

Virus Survival on Crops

Larkin et al. [49] reported that poliovirus survived on lettuce or radishes in field plots for 14-36 days, although detectable viruses decreased 99% during the first 5-6 days. The survival time was longer for plants grown during the fall than for those grown during the summer.

Kott and Fishelson [50] found that solar radiation accelerated the inactivation of polioviruses on vegetables dipped into oxidation pond effluent seeded with attenuated poliovirus. The minimum radiation for mass inactivation was about 0.35 cal/cm^2/min. No active viruses were detected after 28 hours of exposure to sunlight. Chlorination slightly accelerated the rate of inactivation.

Virus Survival in Aerosols

Sorber [51] concluded that aerosol production during sprinkler irrigation of wastewater seemed to range from 0.2 to 0.4%, depending on sprinkler equipment and the prevailing meteorological conditions. Sunlight, temperature, and relative humidity affect virus survival in aerosols. Atmospheric stability is the most important meteorological variable affecting virus survival. Sorber indicated that under the least desirable meteorological conditions, less than 200 m of travel would be required to reduce aerosolized virus concentrations by three orders of magnitude. This suggests that low levels of enteroviruses would be present 200 m from irrigated fields if secondary sewage effluent was sprayed, but considerably more would be present if raw sewage was sprayed.

COMPARISON OF VIRUS REMOVED IN LAND TREATMENT SYSTEMS AND IN SEWAGE TREATMENT PLANTS

Although concentrations of enteric viruses near 500,000 PFU/ℓ have been reported for wastewater in some parts of the world [1], the expected average concentration in the United States has been estimated at about 7000 PFU/ℓ in raw sewage [52]. Virus concentrations may vary in different locations depending upon the time of year, the hygienic conditions in the community, the per capita consumption of water, etc. The virus concentrations mentioned above and estimated removal efficiencies were used [52-55] to calculate the viruses that would be left after various treatment sequences (Table III).

Sample from test wells below the basins at Vineland, New Jersey, were positive for viruses in 18 of 38 samples with virus concentrations ranging from 0 to 8 PFU/ℓ [27]. Samples from wells downslope from the basins were

Table III. Estimated Enterovirus Reduction by Various Sewage Treatment Methods

Treatment	Percent of Virus Removal Expected	Virus Remaining in Effluent (PFU/ℓ)	
		Sproul[a] [52]	Buras[b] [1]
1. Primary treatment			
Sedimentation	0	7000	500,000
Chlorination	50	3500	250,000
2. Secondary treatment			
Stabilization ponds	90	700	50,000
Trickling filters	50	3500	250,000
Chlorination (after trickling filters)	50	1750	125,000
Activated sludge	90	700	50,000
Chlorination (after activated sludge)	50	350	25,000
3. Tertiary treatment (after sedimentation) Excess lime precipitation	90-99.99	700-0.7	50,000-50
Alum precipitation (after lime)	90	7-0.07	5,000-5
Chlorination (after lime and alum	99-99.99	0.07-0.000007	50-0.0005
Activated carbon adsorption	0-50	--	--

[a] Based on an estimate of 7000 PFU/ℓ in wastewater in U.S.
[b] Based on a finding of 500,000 PFU/ℓ in wastewater in Israel.

positive for 2 of 10 samples, with concentrations ranging from 0 to 2.2 PFU/ℓ. Schaub et al. [26] found viruses in about half of the well samples from the Fort Devens site. The average virus concentration in the well downslope from the basin was 4.2 PFU/ℓ. These systems, where primary effluent is applied to coarse sands and gravels, should represent land treatment systems that are least effective in removing virus. A comparison of the virus concentrations found in these wells and the concentrations shown in Table III for treatment plants indicates that these systems are more effective than secondary treatment plus disinfection. Dryden et al. [56] determined virus removal by several tertiary treatment methods which included disinfection but not excess lime precipitation. They reported removal rates similar to those calculated in Table III when excess lime precipitation is omitted. Naturally occurring viruses were isolated in final tertiary effluent in 9 of 56 tests with concentrations varying from 0 to 1.1 PFU/ℓ. Thus, the least effective land treatment systems described above approach the effectiveness of the tertiary treatment systems described by Dryden et al. [56].

A more effective land treatment system, the Flushing Meadows Project in Phoenix, was located on a fine sand with an infiltration rate of about 0.6 m/day [34]. Viruses were not detected in two wells within the recharge area and two wells adjacent to the area that were sampled every 2 months during 1 year. Column studies described previously showed that this soil has a very high adsorption capacity for viruses and that maintaining infiltration rates below 1 m/day would prevent virus movement through 2.5 m of this sand [22]. Thus, these data suggest that a land treatment system placed on soils suitable for virus removal and designed and managed for optimum sewage treatment would be more effective than tertiary treatment and disinfection. Virus removal has not been considered previously in the development of design and management criteria for land treatment systems. However, many practices described above and summarized in Table IV potentially could enhance virus removal.

SUMMARY AND CONCLUSIONS

Laboratory studies have shown that the ionic composition of the suspending solution greatly affects virus adsorption. Virus adsorption appears to increase as the ionic strength of the suspending solution increases and viruses can be desorbed by distilled water. This was shown by laboratory batch studies, column studies, and some limited field data [7,17-19,30]. The effect of salts can be explained using double layer theory. In most soils except very acid soils ($<$ pH 5), both viruses and soil particles have a net negative charge that causes the two kinds of charged particles to repel each other. Increasing

Table IV. Potential Treatment Practices That May Enhance Virus Removal

Practice	Comment	Reference
Alternating infiltration and drying periods	Drying periods reduce virus movement by enhancing inactivation	19,21,57
Addition of cations to sewage	Cations enhance virus adsorption to soil; effects of addition to sewage have not been tested	10,11,14, 19
Flooding with wastewater after rainfall	Reduces virus movement by reducing the number of viruses desorbed by rainwater (tested in long soil columns)	19
Infiltration rate	Slower rates promote virus adsorption	12,22,29
Soil modification	Addition of clays or other substances to soils might enhance virus adsorption (not yet tested)	3,4,57
Site evaluation	Depth to groundwater, lateral movement to collection points, nature of soil and substructure	54,57,58

the ionic strength of the solution decreases the thickness of the double layers (clouds of ions) around the viruses and soil particles and allows the viruses to move close enough to the soil particle surface to be bound by London Van der Waals forces. Adding distilled water decreases the ionic strength and causes the double layer to expand and the negatively charged viruses and soil particles to repel each other. Draining the free water from a saturated soil may also allow viruses and soil particles to move close enough together to form a bond strong enough to resist desorption when distilled water (or rain) is added. This would explain why distilled water eluted viruses from saturated soil columns but not from soil columns at field capacity [19].

Batch studies have been important in showing how some factors, such as the ionic strength of the solution and interference by organic compounds, affect virus adsorption. However, they have not been effective in predicting quantitatively the effect of soil characteristics on virus movement and of course cannot show the effects of flow rates and length of the flow path on virus movement. One problem with batch studies is that using high virus concentrations and high water-to-soil ratios saturates some soil samples with viruses. This makes surface area and other characteristics that are correlated with surface area, like cation exchange capacity, seem to be limiting factors. In field soils even thin layers of soil probably are not saturated with viruses as flooding of a soil column for 27 days with sewage water seeded with poliovirus showed [19]. Even the surface layer where most of the viruses were adsorbed was not saturated.

Soil columns have been useful in further describing the effect of various factors on virus adsorption. However, to accurately represent a field system, the column length must be matched with the depth of the soil layer to be studied and the infiltration rate must be matched with the field infiltration rate. Also, the sewage water used in column studies should be the same as that applied in the field. Column studies have indicated that flow rates below about 1 m/day do not affect virus adsorption but that a virus breakthrough occurs at rates above 1 m/day [12,22].

The soil characteristic which seemed to be the most important in laboratory experiments with long soil columns and in field studies was flow velocity [22,24,31]. Flow velocity seemed to overshadow the effect of other soil characteristics such as surface area and pH, although more work is needed to definitely establish this point. Exceptions to this rule would include low pH soils (< pH 5.0), where virus adsorption should be very high, and organic soils that are poor virus adsorbers. Therefore, soil texture is an important characteristic because it is the most important factor limiting soil permeability. Also, if surface area is shown to be a limiting factor for some soils, texture will be the most important indicator of surface area.

Studies with soil columns indicate that about 1.0 m of travel through a soil with an infiltration rate below 1 m/day is needed for a 2-log reduction in virus concentration and about 2.50 m travel is needed for a 4- or 5-log reduction [12,19,22]. Even columns with infiltration rates of 12 m/day reduced virus concentrations by about 2 logs during 2.50 m of travel [22]. In a field system additional adsorption occurs after the water reaches the water table unless the water moves through fractures or coarse gravel layers. Field studies showed that virus movement does occur in land treatment systems where the soil is primarily coarse sand and gravel, but even in these less effective systems virus removal may approach the removal level of some tertiary treatment plants [26,27,56]. Virus removal in systems with infiltration rates below 1 m/day was better than tertiary treatment including disinfection; viruses generally were not detected in well samples or lysimeter samples from such systems.

Field and laboratory studies have shown that virus survival is very much affected by temperature. Most viruses survived only a few days at temperatures of 35°C or more but they may survive for months at temperatures near freezing or in frozen soils.

Temperature has both direct and indirect effects on virus survival. Virus inactivation rates under sterile and nonsterile conditions were similar at temperature extremes of 1° and 37°C [41]. Inactivation rates were very slow at 1°C and high at 37°C. At an intermediate temperature of 23°C the inactivation rate under nonsterile conditions was about five times that under sterile conditions. Inactivation rates under anaerobic conditions (nonsterile) were similar to those for sterile soils, indicating that aerobic bacteria inactivate viruses. Thus temperature has an indirect effect on virus inactivation due to its effect on the growth of aerobic bacteria.

Similarly, soil drying has an indirect effect on virus inactivation because drying the soil allows air to enter and promote the growth of aerobic bacteria. Thus, drying cycles are important in increasing the inactivation rate of viruses in land treatment systems. Drying also has a direct effect but only at very low moisture contents, which are not usually reached in field soils [42]. Various factors that affect virus movement and survival are summarized in Table V.

Since aerobic conditions are necessary for high virus inactivation rates, virus survival is prolonged if viruses penetrate deep into the soil or into groundwater where anaerobic conditions are prevalent.

In general, if land treatment systems are carefully designed and managed, they will produce high quality renovated water that can be reused for a variety of purposes.

Table V. Factors Affecting Virus Removal by Land Filtration

Factor	Survival [a]	Adsorption [a]
Hydrogeological		
Soil texture	+	+
Humic acids	?	+
Cations	+	+
pH	+	+
Ionic strength	+	+
Permeability	+	+
Biological		
Virus type	+	+
Microbial antagonism	+	−
Meteorological		
Rainfall	+	+
Temperature	+	−
Desiccation	+	−
Sunlight	+	−

[a] + = effect; − = no effect.

REFERENCES

1. Buras, N. "Recovery of Viruses from Wastewater and Effluent by the Direct Inoculation Method," *Water Res.* 8:19-22 (1974).
2. Bitton, G. "Adsorption of Viruses onto Surfaces in Soil and Water," *Water Res.* 9:473-484 (1975).
3. Gerba, C. P., C. Wallis and J. L. Melnick. "Fate of Wastewater Bacteria and Viruses in Soil," *J. Irrig. Drain. Div. Am. Soc. Civil Eng.* 101:157-174 (1975).
4. Lance, J. C. "Fate of Bacteria and Viruses in Sewage Applied to Soil," *Trans. Am. Soc. Agric. Eng.* 21(6):1114-1122 (1978).
5. Burge, W. D., and N. K. Enkiri. "Virus Adsorption by Five Soils," *J. Environ. Qual.* 7:73-76 (1978).
6. Drewry, W. A., and R. Eliassen. "Virus Movement in Groundwater," *J. Water Poll. Control Fed.* 40:R257-R271 (1968).
7. Carlson, G. F., Jr., F. E. Woodward, D. F. Wentworth and O. J. Sproul. "Virus Inactivation on Clay Particle in Natural Waters," *J. Water Poll. Control Fed.* 40:R89-R106 (1968).
8. Lo, S. H., and O. J. Sproul. "Poliovirus Adsorption from Water onto Silicate Minerals," *Water Res.* 11:653-658 (1977).
9. Bixby, R. L., and D. J. O'Brien. "Influence of Fulvic Acid on Bacteriophage Adsorption and Complexation in Soil," *Appl. Environ. Microbiol.* 38:840-845 (1979).
10. Goyal, S. M., and C. P. Gerba. "Comparative Adsorption of Human Enteroviruses, Simian Rotavirus, and Selected Bacteriophages to Soils," *Appl. Environ. Microbiol.* 38:241-247 (1979).

11. Lance, J. C., and C. P. Gerba. "Virus Removal from Sewage During High Rate Land Filtration," *Proceedings of the 1979 AWWA Water Reuse Symposium*, Vol. 3, (Denver, CO: American Water Works Association, 1979), pp. 2282-2290.

12. Robeck, G. G., W. A. Clarke and K. A. Dostal. "Effectiveness of Treatment Processes," *J. Am. Water Works Assoc.* 54:1275-1288 (1962).

13. Young, R. H. F., and N. C. Burbank, Jr. "Virus Removal in Hawaiian Soils," *J. Am. Water Works Assoc.* 65:598-604 (1973).

14. Leffler, E., and Y. Kott. "Virus Retention and Survival in Sand," in *Virus Survival in Water and Wastewater Systems*, J. F. Malina and B. P. Sagik, Eds. (Austin, TX: Center for Research in Water Resources, 1974), pp. 85-91.

15. Bitton, G., H. Masterson and G. E. Gifford. "Effect of a Secondary Effluent on the Movement of Viruses through a Cypress Dome Soil," *J. Environ. Qual.* 5:370-375 (1976).

16. Scheuerman, P. R., G. Bitton, A. R. Overman and G. E. Gifford. "Transport of Viruses Through Organic Soils and Sediments," *J. Environ. Eng. Div., Am. Soc. Civil Eng.* 105(EE 4):629-640 (1979).

17. Duboise, S. M., B. E. Moore and B. P. Sagik. "Poliovirus Survival and Movement in a Sandy Forest Soil," *Appl. Environ. Microbiol.* 31:536-543 (1976).

18. Landry, E. F., J. M. Vaugh, McH. Z. Thomas and C. A. Beckwith. "Adsorption of Enterovirus to Soil Cores and Their Subsequent Elution by Artificial Rainwater," *Appl. Environ. Microbiol.* 38:680-687 (1979).

19. Lance, J. C., C. P. Gerba and J. L. Melnick. "Virus Movement in Soil Columns Flooded with Secondary Sewage Effluent," *Appl. Environ. Microbiol.* 32:520-526 (1976).

20. Wang, De-S., J. C. Lance and C. P. Gerba. "Evaluation of Various Soil Water Samplers for Virological Sampling," *Appl. Environ. Microbiol.* 39:662-664 (1980).

21. Hurst, C. J., C. P. Gerba, J. C. Lance and R. C. Rice. "Survival of Enteroviruses in Rapid Infiltration Basins During the Land Application of Wastewater," *Appl. Environ. Microbiol.* (in press).

22. Lance, J. C., and C. P. Gerba. "Poliovirus Movement During High Rate Land Filtration of Sewage Water," *J. Environ. Qual.* 9:31-34 (1980).

23. Gerba, C. P., and J. C. Lance. "Virus Removal from Primary and Secondary Sewage Effluent by Soil Filtration," *Appl. Environ. Microbiol.* 36:247-251 (1978).

24. Wang, De-S, J. C. Lance and C. P. Gerba. "Removal of Virus from Wastewater During Land Disposal," *Abstracts of the National Meeting of the American Chemical Society* 20:170-171 (1980).

25. Merrell, J. C. et al. "The Santee Recreation Project, Santee, California," Final Report, Pub. No. WP-20-7. Federal Water Pollution Control Administration, Washington, D.C. (1967).

26. Schaub, S. A., and C. A. Sorber. "Virus and Bacteria Removal from Wastewater by Rapid Infiltration Through Soil," *Appl. Environ. Microbiol.* 33:609-619 (1977).

27. Koerner, E. L., and D. A. Haws. "Long Term Effects of Land Application of Domestic Wastewater: Vineland, New Jersey Rapid Infiltration Site," U.S. EPA, Ada, OK (1979).

28. Vaughn, J. M., et al. "Survey of Human Virus Occurrence in Wastewater Recharged Groundwater on Long Island," *Appl. Environ. Microbiol.* 36: 47-53 (1978).
29. Wellings, F. M., A. L. Lewis and C. W. Mountain. "Virus Survival Following Wastewater Spray Irrigation of Sandy Soils," in *Virus Survival in Water and Wastewater Systems,* J. F. Malina and B. P. Sagik, Eds. (Austin, TX: Center for Research in Water Resources, 1974), pp. 253-260.
30. Wellings, F. M., A. L. Lewis and C. W. Mountain. "Demonstration of Virus in Ground Water After Discharge onto Soil," *Appl. Microbiol.* 29: 751-757 (1975).
31. Vaughn, J. M., and E. F. Landry. "The Occurrence of Human Enteroviruses in a Long Island Groundwater Aquifer Recharged with Tertiary Wastewater Effluent," in *State of Knowledge in Land Treatment of Wastewater,* H. L. McKim, Ed. (Hanover, NH: U.S. Army Corps of Engineers, Cold Region Research and Engineering Laboratory, 1978), pp. 233-243.
32. Vaughn, J., and E. F. Landry. "Nassau-Suffolk Regional Planning Board Virus Study," Hauppage, NY (1979).
33. Vaughn, J. M., and E. F. Landry. "An Assessment of the Occurrence of Human Viruses in Long Island Aquatic Systems," Brookhaven National Laboratory, Upton, NY (1977).
34. Gilbert, R. G., et al. "Wastewater Renovation and Reuse: Virus Removal by Soil Filtration," *Science* 192:1004-1005 (1976).
35. Dugan, G. L., et al. "Land Disposal of Wastewater in Hawaii," *J. Water Poll. Control Admin.* 47(8):2067-2085 (1975).
36. Koerner, E. L., and D. A. Haws. "Long-Term Effects of Land Application of Domestic Wastewater: Roswell, New Mexico, Slow Rate Irrigation Site," U.S. EPA 600-279047 (1979).
37. Damgaard-Larsen, S., K. O. Jensen, E. Lund and B. Nissen. "Survival and Movement of Enterovirus in Connection with Land Disposal of Sludges," *Water Res.* 11:503-508 (1977).
38. Bitton, G., J. M. Davidson and S. R. Farrah. "On the Value of Soil Columns for Assessing the Transport Pattern of Viruses through Soils: a Critical Outlook," *Water Air Soil Poll.* 12:449-457 (1979).
39. Schaub, S. A., K. F. Kenyon, B. Bledsoe and R. E. Thomas. "Evaluation of the Overland Runoff Mode of Land Wastewater Treatment for Virus Removal," *Appl. Environ. Microbiol.* 39:127-134 (1980).
40. Schaub, S. A., and B. P. Sagik. "Association of Enteroviruses with Natural and Artifically Induced Colloidal Solids in Water and Infectivity of Solids-Associated Virions," *Appl. Environ. Microbiol.* 30:212-222 (1975).
41. Hurst, C. J., C. P. Gerba and I. Cech. "The Effects of Environmental Variables and Soil Characteristics on Virus Survival in Soil," *Appl. Environ. Microbiol.* 40:1067-1079 (1980).
42. Yeager, J. G., and R. T. O'Brien. "Enteroviruses Inactivation in Soil," *Appl. Environ. Microbiol.* 38:694-701 (1979).
43. Ostle, A. G., and J. G. Holt. "Elution and Inactivation of Bacteriophages on Soil and Cation Exchange Resin," *Appl. Environ. Microbiol.* 38:59-65 (1979).
44. Tierney, J. T., R. Sullivan and E. P. Larkin. "Persistence of Poliovirus 1 in Soil and on Vegetables Grown in Soil Previously Flooded with Inocu-

lated Sewage Sludge or Effluent," *Appl. Environ. Microbiol.* 33:109-113 (1979).

45. Sadovski, A. Y., et al. "High Levels of Microbial Contamination of Vegetables Irrigated with Wastewater by the Drip Method," *Appl. Environ. Microbiol.* 36:824-830 (1978).

46. Mack, W. N., Y.-S. Lu and D. B. Coohan. "Isolation of Poliomyelitis Virus from a Contaminated Well," *Health Services Reports* 87:271-274 (1972).

47. Clarke, N. A., G. Berg, P. W. Kabler and S. L. Chang. "Human Enteric Viruses in Water: Source, Survival and Removability," *in Proceedings of the First International Conference on Water Pollution Research* (1962), p. 523.

48. Joyce, G., and H. H. Weiser. "Survival of Enteroviruses and Bacteriophage in Farm Pond Waters," *J. Am. Water Works Assoc.* 59:491-501 (1967).

49. Larkin, E. P., J. L. Tierney and R. Sullivan. "Persistence of Virus on Sewage-Irrigated Vegetables," *J. Environ. Eng. Div., Am. Soc. Civil Eng.* 102:29-35 (1976).

50. Kott, H., and L. Fishelson. "Survival of Enteroviruses on Vegetables Irrigated with Chlorinated Oxidation Pond Effluents," *Israel J. Technol.* 12:290-297 (1974).

51. Sorber, C. A. "Virus in Aerosolized Wastewater," in *Virus Aspects of Applying Municipal Wastes to Land,* L. R. Baldwin, J. M. Davidson and J. F. Gerber, Eds. (Gainesville, FL: University of Florida, 1976), pp. 83-86.

52. Sproul, O. J. "Removal of Viruses by Treatment Processes," in *Virus in Water,* G. Berg, H. L. Bodily, E. H. Lennette, J. L. Melnick and T. G. Metcalf, Eds. (Washington, D.C.: Amer. Publ. Health Assoc., 1976), pp. 167-179.

53. Gerba, C. P., C. Wallis and J. L. Melnick. "Viruses in Water: A Definition of the Problem and Approaches to its Solution," *Environ. Sci. Technol.* 9:1122-1126 (1975).

54. Lance, J. C., and C. P. Gerba. "Pretreatment Requirements Before Land Application of Municipal Wastewater," in *State of Knowledge of Land Treatment of Wastewater,* H. L. McKim, Ed. (Hanover, NH: U.S. Army Corps of Engineers Cold Region Research and Engineering Laboratory, 1978), pp. 293-304.

55. Clarke, N. A., R. E. Stevenson, S. L. Chang and P. W. Kabler. "Removal of Enteric Viruses from Sewage by Activated Sludge Treatment," *Am. J. Pub. Health* 51:1118-1129 (1961).

56. Dryden, F. D., C. Ching-lin and M. W. Selna. "Virus Removal in Advanced Wastewater Treatment Systems," *J. Fed. Water Poll. Control Admin.* 51:2098-2109 (1979).

57. Bouwer, H. "The Use of the Earth's Crust for Treatment or Storage of Sewage Effluent and Other Waste Fluids," in *Critical Review Environ. Control* 6:111-130 (1976).

58. Duboise, S. M., B. E. Moore, C. A. Sorber and B. P. Sagik. "Viruses in Soil Systems," in *Critical Reviews in Microbiol.* 7:245-285 (1979).

CHAPTER 28

AUTOMATED WATER-QUALITY MONITORING

Kenji Nishioka
Reuse Water Monitoring and Control System Program
Ames Research Center, NASA
Moffett Field, California

Automated monitoring, which can provide continuous real-time or near real-time measurement of various water-quality parameters, is necessary if the supply of high-quality (potable) water at a reasonable cost is to be maintained. This chapter provides some insight into automated water-quality monitoring using state-of-the-art technology. An overall experimental system and water-quality parameters amenable to automated measurement are described. Problems that may be encountered, their solutions, and typical measurement results that can be expected from an automated system are discussed. Because the benefits of automated water-quality monitoring can only be fully realized when the data are used in real-time control of a water-processing plant, this chapter closes with a description of how this might be accomplished.

Commercial companies are cited in this chapter for illustrative purposes only; no endorsement of their products or services is intended. Comments, opinions and conclusions expressed in the chapter are those of the author, and they should not be interpreted to reflect those of his employer.

BACKGROUND

Automated water-quality monitoring is defined here as the capability to obtain desired water-quality data without human intervention, and to do so continuously, sample after sample, from sample acquisition through sample preparation, sample analysis, data acquisition, data analysis, data display, and data storage. The term "automated" applies to a system capable of unattended operation as its sensors continuously analyze a flowing sample stream and automatically perform other functions necessary to accomplish the analysis, such as switching sample streams.

Examples of water-quality parameters that must be monitored to meet U.S. Environmental Protection Agency (EPA) potable water specifications [1] are summarized in Table I. They include physical, chemical, and biological parameters. All these water-quality parameters can be detected using state-of-the-art instrumentation and standard biological and chemical laboratory analysis techniques. Because of manpower requirements, however, these standard methods can be expensive, especially if a large number of samples must be collected and analyzed. In dealing with parameters that require only periodic sampling, it is common practice to ship samples to distant laboratories for analysis. In the meantime, the composition of the samples can change substantially (e.g., when finally analyzed, the amount of volatiles in the samples could be substantially lower than at the source). Moreover, the results

Table I. Examples of Some EPA Potable Water-Quality Specifications[a]

Contaminants	Maximum Contaminant (mg/l)
Inorganic chemicals	
Arsenic	0.05
Barium	1
Cadmium	0.01
Chromium	0.05
Lead	0.05
Mercury	0.002
Nitrate (as N)	10
Selenium	0.01
Silver	0.05
Organic chemicals	
Chlorinated hydrocarbons	
Endrin	0.0002
Lindane	0.004
Methoxychlor	0.1
Toxaphene	0.005
Chlorophenoxys	
2, 4-D	0.1
2, 4, 5-T silvex	0.01
Microbiological	
Coliform bacteria (membrane filter, avg of all samples for 1 month)	1/100 ml
Turbidity (monthly average)	1 TU

[a]For specific details concerning sampling and analysis techniques, frequency of samples, contaminant levels, etc., refer to Federal Register 40 (248): Part IV [1].

from these analyses may not be made known for several weeks. In contrast, it is possible to automate the sampling and analysis for many of the water-quality parameters listed in Table I. Parameters amenable to automated monitoring are shown in Table II, and that list will continue to grow as new instruments and supporting technology are developed.

Some typical examples of prototype instruments currently being tested are summarized in Table III. Of the four instruments shown, the heavy metal (cation), biomass and coliform sensors were specifically developed for automated operation. The viral sensor was not designed for automated operation; therefore, it will require further development after its current feasibility testing is completed before it can be integrated into an automated system.

Several companies are working on various instruments that are automated or semiautomated for monitoring some specific facet of water quality. An example is the automatic dissolved metals analyzer developed by the Magnavox Company, Fort Wayne, Indiana [2], for the EPA Laboratory in Cincinnati, Ohio. This device can detect 16 metals simultaneously in the parts-per-billion range in 20 minutes; it is being evaluated by the EPA. A second example is an instrument capable of detecting chemical oxygen demand (COD) and total organic carbon (TOC) automatically; it was developed by Life Systems Incorporated, Cleveland, Ohio [3], for the U.S. Army Medical Bioengineering Research and Development Laboratory, Fort Detrick, Maryland. The instrument can measure 0.1-500 ppm TOC and 100-1500 ppm COD, but the manufacturer claims that an instrument capable of detecting COD even at 0.5 ppm is possible.

Basically, the concept of automated water-quality monitoring is a feasible one. The instrumentation, sensors, computer hardware and software, and technology are, in general, available. The following discussion is based primarily on experience [4] with the automated water-quality monitoring system designed and assembled to National Aeronautics and Space Administration (NASA) specifications by the Boeing Company. This systems has operated successfully for the last 2.5 years at an experimental test site in Palo Alto, California. The data used in this chapter are derived from that experiment. The results show why continuous water-quality monitoring is desirable and, further, that valid results can be obtained from an automated system.

Table II. Typical Water-Quality Parameters Measurable by Automatic Monitoring

Ammonia, NH_3	Dissolved Oxygen, DO	Residual Chlorine
Biomass	Halocarbons	Sodium, Na
Chlorides	Hardness	Temperature
Coliform	Nitrate/Nitrite	Total Organic Carbon, TOC
Conductivity	pH	Turbidity

Table III. Examples of Prototype Automated Water-Quality Instruments Being Tested

Measured Parameter	Sensor Principle	Organization
Heavy Metals	X-ray fluorescence	Magnavox
Biomass, total/viable	Chemiluminescence	Ames Research Center
Coliform	Electrochemical (hydrogen)	Ames Research Center
Virus	Laser fluorescence	Ames Research Center

DESCRIPTION OF AN AUTOMATED WATER-QUALITY MONITORING SYSTEM

Before describing an automated water-quality monitoring system, it is appropriate to discuss the key reasons for using automation. The whole area of water quality from regulations, to measurement techniques, to understanding biological effects of contaminants, to an increase in the number of artificial contaminants, is undergoing rapid change. All these factors indicate the additional tasks and thus costs that will be necessary to provide quality-assured water in the future [5]. A historic example is the use of chlorine or its compounds to disinfect drinking water. The practice started in 1904 when the London Metropolitan Water System began using sodium hypochlorite; the Chicago Water Supply, which was the first in the United States to use chlorine compounds, began using calcium hypochlorite in 1908 [6]. Chlorine is an effective disinfectant, but recent findings show that it may be reacting with substances in water to form known carcinogens, such as chloroform, trichloroethylene, and other compounds.

As the requirements for water quality and quality assurance are further increased, the cost could become prohibitive if present analytical practices are continued. A good candidate for change is the current monitoring requirements for heavy metals, pesticides, and organochlorides, for which samples need only be analyzed periodically. In many cases, the results of these analyses may not be known for weeks after the samples are taken. If safe limits are exceeded, the water companies are only required to publicly announce that this has occurred. This problem could be avoided or minimized with automated water-quality monitoring. Automation can provide rapid, on-line, reliable and cost-effective water-quality analysis.

The overall monitoring system is shown in Figure 1. There are three primary subsystems: sample collector, sensor and computer [7]. The computer is the key subsystem because it makes automated operation possible. Each of these subsystems is described in detail in the following sections.

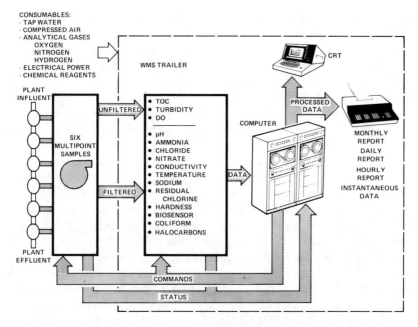

Figure 1. Schematic of an automated water-quality monitoring system.

Computer Subsystem

Basically, the computer subsystem is composed of the computer with input-output devices and an on-line disc for data storage [8]. The core requirements are dictated by the data acquisition (the number of sample streams to be monitored, the frequency of monitoring, the number of sensors, and the redundance built into the overall monitoring system), and overall system control requirements. For a nominal system, a minicomputer with a working memory (RAM) of 128,000 words and disc storage of 10 megabytes is sufficient. A CRT terminal with command keyboard for instantaneous display of current and stored information is required, and an additional CRT-keyboard unit for simultaneous program development would be desirable. Also, a hardcopy printer is necessary for printing current and historic data such as monthly data summary reports. A data logger, with an analog-to-digital converter and interface device, is located between the sensors. The computer is required to scan the sensors and convert the data and status signals from the sensors, whose signals generally are in analog form. The type of data logger used depends on the number of sensors being monitored, and the interface device must be matched to the computer.

In addition, a digital clock is required to provide the system with a time base. An accurate and reliable clock is necessary if the computer is to keep track of system status and hardware tasks, such as valve switching and data acquisition at preselected times. It is also important that the data be annotated with the time of acquisition for archiving and future use.

The last and most important part of the computer system is the software. It provides the logical instructions to the computer so that all the required tasks related to system management are accomplished, including acquisition of desired water-quality information. The block diagram shown in Figure 2 illustrates the main elements of the required software. The executive routine maintains and schedules the flow of information in and out of the computer and sends commands from the computer to the various sensors and other sub-systems in the water-quality monitoring system. The executive routine keeps track of what tasks are required and calls on the appropriate subroutines to execute those tasks. Each subroutine should accomplish specific tasks; for example, the data acquisition subroutine should be written so that it is capable of scanning each sensor at set times and allowing those set times to be individually changed. It must delete data for a given period, while a sensor is stabilizing and when sample streams are changed, and average data over a set time period. The bounds for each subroutine are flexible. With an automated

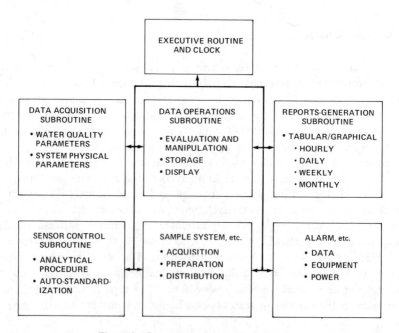

Figure 2. Computer software block diagram.

monitoring system, an alarm system is required to alert plant operators of malfunctions in the system as well as out-of-tolerance conditions. The alarm-control function software should be able to identify allowable values for water-quality parameters, system physical parameters, etc. The alarm subroutine could be wired to sound a bell in some manned area and display the problem on a CRT.

All the useful water-quality data must be readily accessible and be conveniently displayed. Table IV shows a sample printout of data that might be displayed on-line using a CRT. As seen, it can instantly provide a system status preview ranging from which sample line is being monitored to which sensors are operational. For summary type reports, the data can often be shown

Table IV. Printout of Instantaneous CRT Display of Sensor System Status

Time − 190:15:00:00   Sampling point: sand filtered/ozonated 4

Cha No.	Sensor	Units[a]	Status[b]	Inst Value	Averages 15 min	30 min	1 hr	Running
1.	Total biomass	mil c/ml	NDTA	−8.919	0.000	0.757	0.757	0.599
2.	Viable biomass	mil c/ml	NDTA	−8.919	0.333	0.333	0.333	0.172
5.	Res chlorine	mg/l		1.1	1.1	1.1	1.1	1.1
6.	Turbidity-SiO$_2$	mg/l		12.8	14.5	14.6	13.7	11.9
7.	Dis oxygen	mg/l		7.0	7.0	7.1	7.1	7.0
10.	Ammonia	mg/l		12.9	12.9	12.9	12.9	12.9
11.	Nitrate	mg/l		8.8	8.8	8.8	8.8	8.8
12.	pH	pH		7.08	7.08	7.08	7.08	7.08
13.	Tot org carbon	mg/l		11.7	11.4	11.5	11.5	11.5
14.	Conductivity	mmmho/cm		1392.0	1398.9	1397.3	1392.4	1189.8
15.	Temperature #1	deg F		72.4	72.0	71.9	71.8	69.4
16.	Hardness	mg/l		250.0	254.0	257.0	258.0	258.0
17.	Sodium	mg/l		155.0	160.0	160.0	159.0	159.0
20.	Ambient temp	deg F		81.6	80.0	79.9	80.3	75.3

Time − 190:15:00:00   Sampling point: reclamation facility effluent

Cha No.	Sensor	Units[a]	Status[b]	Inst Value	Averages 15 min	30 min	1 hr	Running
3.	Air comp	psia		84.8	83.6	83.3	83.4	83.4
18.	Turbidity-hw	ftu	Out					
23.	Effluent	psia		28.5	28.5	28.5	28.5	28.4
31.	Flash mix pH	pH		11.3	11.3	11.3	11.3	11.4
32.	Plant flow	mgd		1.5	1.5	1.5	1.5	1.1
33.	Sludge density	% solids		0.4	0.4	0.4	0.5	2.4
34.	Sludge pump	rpm		664.8	657.6	661.0	649.3	679.1
35.	Lime feed vl	% open		52.2	52.2	53.1	52.0	41.8
38.	Total biomass	mil c/ml	NDTA	−8.919	0.000	0.000	0.000	0.000
39.	Viable biomass	mil c/ml	NDTA	−8.919	0.000	0.000	0.000	0.000

[a] mil c/ml = million cells per milliliter; mmmho = μmho.
[b] NDTA = no data; out = sensor is not operational.

graphically (Figure 3). The daily average curve might indicate water-quality trends that may be developing during a month; the daily data shown as plus or minus 1-sigma variation curves surrounding the daily average curve give an indication of the water-quality variability for each day. At times, tabular data are required to show the actual values of the water-quality parameters; Table V illustrates one way of displaying the data.

Sensor Subsystem

Solid-state electronics and advances in technology have enabled the instrumentation industry to manufacture sensors that can operate reliably for long periods of time. Required sensors can be integrated into packages for automated operations to meet the goals for most applications. This is true for a simple pH meter or a complex gas chromatograph with mass spectroscopy. In the past, for example, TOC sensors have required a high-temperature (850°C) reactor utilizing nitric acid, which caused a longevity problem; but recent technology developments have led to TOC sensors that use a low-temperature (60°C) reactor.

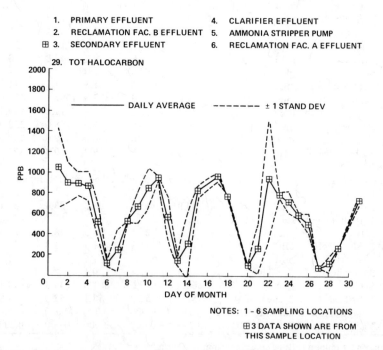

Figure 3. Monthly summary data—total halocarbon (NASA/WMS–SCVWD Palo Alto Water Reclamation Facility).

To properly measure water quality, many different types of sensors are necessary [8]. These can be categorized by the detection principle used, as shown in Table VI. In general, sensors using thermoelectrical or electrochemical-impedance are the easiest to automate. These are followed, in order, by those using chemical reaction, optical (light), and wet chemistry. Because of their nature, biological sensors are the most difficult to use. That is, the parameters to be measured are living organisms and they have a tendency to aggregate, thereby leading to false readings. Generally, the concentrations of organisms of interest are very low; for example, the concentration level of interest for coliform bacteria in potable water is 1.0 organism per 100 ml (Ref. 1). It is currently difficult to directly detect the bacteria; therefore, they must be cultured and grown for manual or automated detection. For automated detection, its by-product, hydrogen gas, can be detected, but this requires an incubation chamber as part of the sensor package. In general, the key to successfully detecting any contaminant that affects water quality is closely linked to the concentration level of the contaminant.

Table V. Sample Data–Daily Averages for Typical Week

Parameter	Sample Source[a]	Concentrations, 22-28 January 1980						
		T	W	T	F	S	S	M
TOC (mg/l)	2	21.4	22.1	24.3	24.3	19.6	17.0	19.7
	6	12.5	13.1	14.9	13.3	11.8	7.8	9.0
Turbidity (mg/l)	2	24.1	35.5	75.7	14.7	16.7	13.4	16.2
	6	4.3	4.5	5.7	8.1	5.9	3.4	3.8
Viable biomass	2	2.7	4.86	4.46	5.44	3.50	2.66	2.94
(10^6 cells/ml)	6	0.14	0.17	0.51	0.76	0.21	0.55	0.37
Total biomass	2	7.86	11.08	11.42	15.34	11.67	9.35	12.54
(10^6 cells/ml)	6	0.49	0.58	1.39	1.96	1.04	0.85	0.74
Ammonia (mg/l)	2	15.3	13.7	12.8	13.9	12.0	–	15.2
	6	14.3	14.5	13.1	10.6	12.2	–	12.3
Conductivity	2	1484	1371	1416	1388	1370	1387	1377
(μmho/cm)	6	1356	1351	1396	1381	1322	1357	1316
Temperature ($^\circ$F)	2	72.1	70.3	69.2	70.8	71.7	71.7	71.5
	6	72.0	70.6	70.2	69.5	71.6	71.7	70.8
DO (mg/l)	2	6.4	6.2	8.6	5.8	4.0	4.1	4.0
	6	7.2	6.7	5.9	5.4	5.7	6.1	6.3
Nitrate (mg/l)	2	0.4	–	1.7	1.5	1.5	–	0.2
	6	1.5	3.8	4.6	1.7	–	–	8.4
pH	2	6.71	6.50	6.34	6.33	6.25	6.26	6.28
	6	7.93	8.07	7.28	5.84	7.38	7.75	8.07
Residual chlorine	2	5.4	6.4	5.8	5.8	6.7	6.8	7.0
(mg/l)	6	3.6	6.9	5.9	45.7	8.5	4.3	3.0

[a]Source 2 is the influent into the reclamation plant; Source 6 is the final effluent from the reclamation plant.

Table VI. Some Characteristics of Sensors Used By NASA for Automated Water-Quality Monitoring Systems[a]

	Principle	Range	Manufacturer
1. Turbidity/ Photometer	Nephelometric– light scatter by particles in suspension	0.1-5000 mg/l	Sigrist
2. Dissolved Oxygen/Electrode	Electrochemical oxidation-reduction	0-20 mg/l	Delta Scientific
3. Total Organic Carbon/Infrared Detector	$C+O_2 \xrightarrow[pH=1.5]{60°C}$ (IR absorption)	0.1-1000 mg/l	Astro-Ecology
4. Conductivity/ Impedance Bridge	Ion activity $\approx$ 1/resistance	0-2000 μmho/cm	Beckman
5. Total Residual Chlorine/Electrodes	Wet chemistry- electrochemical	0.0-1000 ppm	Orion
6. Hardness/Electrodes	Wet chemistry- electrochemical	1-1000 mg/l	Orion
7. Ammonia/ Photometer	Wet chemistry- spectrophotometry	0-80 ppm	Delta Scientific
8. Sodium Electrode	Electrochemical	10-1000 ppm	Beckman
9. pH/Electrode	Electrochemical	2-12 pH units	Orion-Great Lakes
10. Chloride/Electrode	Electrochemical	10-1000 ppm	Great Lakes
11. Temperature/RTD	Thermoelectrical	0-200°F	Action PAC
12. Coliform	Electrochemical	20-50,000 units/ 100 ml	NASA/Boeing
13. Twin-Column Gas Chromatograph	Electron capture detector	0.1-1000 ppb	NASA/TRW
14. Chemiluminescence Biosensor	Chemiluminescence	10^6-10^8 units/ml	NASA/Boeing

[a]The coliform sensor does not have potable-water capability and the potable-water capability of the chemiluminescence biosensor is unknown. All other sensors listed have potable-water capability.

Another problem that needs to be considered is the interference caused by high concentrations of foreign particles. For example, the presence of chlorides, phosphates, and carbonates will interfere with the detection of chromium anions (a water-quality parameter) when selective anion-exchange membrane-filtration for x-ray fluorescence detection of the chromium ion is used.

All of the factors touched on above are important to successful operation of automated sensors. Several new types of automated sensors are being developed by governmental agencies and commercial companies. In addition to the examples cited earlier, prototype sensors for detecting coliform, halocarbons (gas chromatograph), and total and viable biomass have been developed by NASA. These unique sensors are described in detail below.

Coliform Sensor

The automated coliform sensor [8] is based on a method similar to the *Standard Methods* Most Probable Number (MPN) Test. The sensor employs redox electrodes to monitor changes in the coliform specific media when bacterial growth takes place. The hydrogen gas evolved by the coliform bacteria as they metabolize lactose in a broth solution is first measured by a redox electrode in the broth and is then vented to a second set of redox electrodes to positively identify the bacterial presence as coliform. [During experimentation, it was discovered that bacteria other than coliform can affect the redox electrodes placed in the incubator media, thus causing false positive identifications. By venting the hydrogen gas (which is only evolved by coliform bacteria) to a second set of electrodes, the false positive readings were eliminated.] The time required for the electrodes to detect a change in the media is directly related to the initial concentration of coliform bacteria in the sample. In addition, the differentiation of total and fecal coliform can be achieved by using different incubation temperatures.

The current NASA coliform sensor consists of four sets of paired thermostatically controlled incubation cells. The temperatures in the cells are maintained at 37°C for monitoring total coliforms and at 44°C for monitoring fecal coliforms. The cycle is initiated when 14-ml samples are automatically injected into each of the first four cells, each containing 14 ml of lauryl sulfate broth as the media. The redox potential of the media is continuously monitored as the coliform bacteria grow and multiply. Calculations of the initial coliform concentration are based on the time required to achieve a 200-mV change in the electrode output. Hydrogen gas from each of the first cells in the pairs of incubation cells is vented to the bottom of the second cells in the pairs which contain 14 ml of 1.0% potassium phosphate buffer solution. The hydrogen gas dissolves in the buffer solution and causes a redox potential change in the solution. A 200-mV change in the electrode output is monitored to verify the readings from the first four cells. After each cell as reacted, or the maximum reaction time of 14 hr has elapsed, the sensors are automatically prepared for another analysis. The first four cells are drained of spent media and then flushed with 0.1 N nitric acid to sterilize and clean the platinum electrodes. These cells are then flushed with 95°C water and drained, then finally filled with 14 ml of lauryl sulfate broth and heated to 85°C for a final sterilization. The media are allowed to cool to the appropriate temperature for the next sample. In the meantime, compressed air is bubbled through the second set of four cells to purge the hydrogen dissolved in the buffer solution to make it ready for the next sample.

Biosensor, Luminol-Carbon Monoxide

A chemiluminescent method for monitoring total and viable (live) biomass (bacteria) which employs an alkaline luminol-hydrogen peroxide reaction with bacterial iron porphyrins to quantify bacteria has been developed [8]. The sensor uses a combination of techniques, including hydrogen peroxide pretreatment, reaction rate differentiation, and ethylene diamine tetracetic acid (EDTA) to provide the specificity for monitoring bacteria in water and wastewater. In addition, the sensor uses a new technique whereby carbon monoxide gas reacts with the iron porphyrins in the viable biomass only, which permits differentiation between viable and dead biomass. The CO reaction with the iron porphyrins in the viable biomass results in the suppression of the chemiluminescence action in the second measurement and thus the luminescence measured will be from the dead bacteria only. This second luminescence measurement subtracted from the first luminescence measurement provides the viable biomass present. Typical analysis times required for both luminescence measurements is about 60 min.

Gas Chromatograph

A computer-controlled automated twin-gas chromatograph (GC) capable of monitoring 10 volatile halocarbons within an hour was developed [8]. The GC contains an automatic sampler that injects a 120-μl water sample into a preparative GC for separating the organic fraction from the water. The organics are first collected in a Tenax trap; then the trap is heated to introduce the compounds into an analytical GC for detection by an electron capture detector (ECD). As with any GC, it is necessary to periodically run standards for each halocarbon monitored in order to be able to interpret the chromatogram. The GC-derived instrument developed by NASA is capable of quantitative analysis at the parts-per-billion level.

Commercial Sensors

Included in Table VI are some typical commercial sensors that are either ready for use in automated operations or could be readily converted for such use. The table identifies the manufacturer, the sensitivity range, and the operating principle for the sensors.

Comments

Probably the most important sensor-related considerations are: (1) the frequency of required calibrations, (2) the stability of the standards used and (3) the amount and stability of reagent(s) required. Together, these considerations will usually determine the level of repetitive maintenance necessary and this in turn will have a strong effect on the operating cost. Other sensor considerations include sample filtration needs, flow rates required, pressurization, chemical treatment requirements, and residence time required for detection. These factors will also affect the amount of routine maintenance necessary and thus the operating cost of the water monitoring system.

Although a good system design for automated water-quality monitoring, based on the considerations stated above, would seem to ensure the success of the system, such is not the case. Unfortunately, many sensor packages present problems with reliability, sensitivity and maintainability. Often, these problems can be resolved by replacing components or making other fixes, but they nevertheless can be frustrating when schedules must be met. Some further comments are found in the Problems section.

Sample and Sample Preparation Subsystem

Obviously, before the sensors can detect and determine water quality, a sample of the water must be made available to the sensor. Since the discussion here is of automated water-quality monitoring, the requirement is that the sample must be obtained and presented to the sensor automatically and continuously.

The sample subsystem can be divided into four parts: (1) the sample source pump and sample line, (2) the electronic valves and manifold; (3) the filter mechanisms, and (4) the sample distribution network to the sensors.

Because the system is to be automated, operations should be as simple as possible in the interest of reliability of operation. Accordingly, the sample source pump should supply a continuous stream to the laboratory, and the supply must be diverted to an outfall when the sample is not needed. A continuous flow in the sample line will ensure that a fresh sample is always present for testing. A fluid-flow sensor will be required in the sample line to shut down the electric pump when the plant stops. This can be programmed into the computer, so that the status of the pump is simultaneously displayed on the CRT, and an alarm would be provided.

The sample lines should all be equipped with two-way pneumatic or electronic valves, operable either manually or automatically for complete control of sample-line selection for monitoring. Actuation of the valves will normally be via computer control, and the valves should always fail in the open position to the outfall manifold.

The sample flow is split into two streams, filtered and unfiltered. The unfiltered stream is distributed through the laboratory to sensors that require unfiltered samples, such as TOC, turbidity and DO (dissolved oxygen) sensors. The other sample stream is filtered through 50-μm stainless steel mesh screens and ends in a second sample distribution network in the laboratory for distribution to the appropriate sensors. Some means as backwashing should be provided to keep the filter screens clear as long as possible. Bypass filters, in which only a small portion of the flow actually passes through the filter while the major portion of the sample flows across the surface of the filter, work well for this type of application.

Both sample distribution streams in the laboratory should flow continuously, and the lines to the individual sensors should be allowed to flush themselves and, where appropriate, the sensor probes whenever the sample source being analyzed is changed. The time required for a sensor to equilibrate following a sample source change will vary from sensor to sensor.

Support Subsystem

Support functions required for the water-monitor system are electrical power, water, air conditioning, analytical reagents and analytical gases. The power that is provided should be conditioned so that the computer and instruments are protected from power surges. Some means should be provided to supply the instruments with either distilled or deionized sterile water for analysis and cleanup. A distribution network for the distilled or deionized water to each instrument (ammonia, nitrate, TOC, etc.) that requires such water must be provided. Air conditioning is necessary for reliable computer operation, and the performance of the sensors will also be improved if the ambient temperature can be kept constant. The specific reagents required to support the analytical needs of the sensors would naturally be dictated by the types of sensors chosen. The choice of sensors can thus play a significant role in minimizing the need for reagents that are difficult to handle and have short half lives, important criteria for automated operation. There will probably be some need for gases that meet laboratory analytical purity requirements, such as argon-methane mixture for gas chromatographs, nitrogen as a general purging gas, carbon monoxide for viable biomass detection, and, possibly, hydrogen and oxygen for general use.

Generally, the support requirements for automation are no different from what would be required for a standard water chemistry laboratory.

Other Desirable Water-Quality Sensors

Existing sensors that are already automated or capable of being easily modified for automated operation have been discussed; however, in order to meet EPA criteria for potable water, additional sensors must be developed. Some reliable, easy-to-use sensors for detecting heavy metals (anions and cations), organic compounds (herbicide, pesticide and others), total nitrogen, and toxic viruses are of immediate interest. As stated earlier, some work on detecting heavy metal cations has been done by the Magnavox Corporation for the EPA, but that unit is still a long way from being operationally usable. Some work on a similar method for detecting anions has been done by Dr. Alan Ling at San Jose State University, San Jose, California, but interference problems caused by carbonate and other anions have not been solved. Some techniques of selectively eliminating the major interfering anions are currently being investigated.

The state of the art in detecting organics is the gas chromatograph with a mass spectrometer, or high-pressure liquid chromatography, or both. Prospects of automating the operation of these instruments appear bright, based on NASA experience with automating a twin-column GC for detecting halocarbons.

At the present time, there are several total nitrogen sensors available on the market, but, unfortunately, none of them is designed to be operated continuously in an automated mode. Perhaps a totally new technique is necessary to solve this seemingly straightforward problem of automatically measuring total nitrogen.

NASA has sponsored research in developing an instrument for detecting viruses in a flowing sample stream. The instrument was developed using a laser illumination scattering technique to detect individual viruses, but some questions remain whether the instrument is sensitive enough to detect viral presence lower than 10^5 units/100 ml. The minimum required sensitivity would be about 10^3 units/100 ml for most practical applications. The instrument is currently undergoing final evaluation at the Desert Research Institute of the University of Nevada, Stead, Nevada, by Dr. Vern Smiley.

Problems

The difficulties that can be expected in implementing automated water-quality monitoring can be categorized as equipment-related (hardware and software) and people-related.

Equipment problems can be expected to range from simple mechanical and electrical failures to complex control-logic difficulties with computer software (programming "bugs") and may even include cross wiring of sensors and incorrect plumping connections. Probably the two most frustrating problems that must be coped with are the unexplained intermittent computer stoppages, which are certain to occur, and the elusive gremlins in the computer sensor link that cause either occasional data bits to be dropped or erroneous data readings.

The people-related problems can be minor or overwhelming. In general, good communication with and early training of individuals who will be involved in the operation of automated water-quality monitoring systems will significantly reduce the extent of this problem. An additional problem is the limited number of qualified persons who can design and implement an automated system. Experts in specific areas—such as computer programming, instrumentation, water chemistry, and electronics—will have to be chosen carefully and used to design and implement an automated system.

The problems alluded to above are typical examples; other problems may develop, but the expectation is that only a few of the problems discussed will be encountered.

FUTURE APPLICATION OF AUTOMATED WATER-QUALITY MONITORING

The successful application of automated water-quality monitoring requires that the data generated be used as effectively as possible. For example, an obvious application would be to use the data from the automated water monitoring system to control the water treatment plant, thus having a closed loop between the monitoring function and water treatment. A simplified block diagram of this application is shown in Figure 4. Before this can be a reality, however, control algorithms will have to be developed to effectively tie the data-gathering function to the treatment processes. When this is accomplished, the water treatment plant will be automated, producing quality-assured water continuously. The use of chemicals for water treatment will be

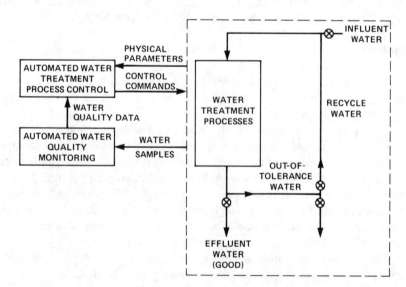

Figure 4. Block diagram of closed-loop automated water treatment plant.

optimized, which means that the water will contain fewer chemicals and that the cost of treatment chemicals will be reduced. With the treatment process optimized, the amount of energy required should also be minimized.

REFERENCES

1. U.S. Environmental Protection Agency. "National Interim Primary Drinking Water Regulations," *Federal Register* 40(248):59570-59573 (1975).
2. The Magnavox Company. "Operation and Service Manual for the Automatic Dissolved Metals Analyzer," Report No. FWD78-1816 for U.S. EPA Environmental Monitoring and Support Laboratories, Cincinnati, OH (1978).
3. Life Systems, Inc. "Evaluation of a Breadboard Electrochemical TOC/TOD Analyzer," Report No. LSI ER-310-4-2 for U.S. Army Medical Research and Development Command, Ft. Detrick, MD (1978).
4. Nishioka, K., and R. Brooks. "Automated In-Situ Water Quality Monitoring," *Intech* 26(10):47-51 (1979).
5. Robertson, J. H., W. F. Cowen and J. Y. Longfield. "Water Pollution Control," *Chem. Eng.* 87(13):103-119 (1980).
6. Sawyer, C. N., and P. L. McCarty. *Chemistry for Sanitary Engineers,* 2nd ed. (New York: McGraw-Hill, 1967), p. 365.
7. Nishioka, K., et al. "On-Line Automated Water Quality Monitoring," in *Proceedings of the 1979 AWWA Water Reuse Symposium,* Vol. 3 (Denver, CO: American Water Works Association, 1979), pp. 1779-1799.
8. Jeffers, E. L., et al. *Water Monitoring System Technical Summary Report,* Boeing Aerospace Company, NASA Contract NAS2-9885 (1979).

CHAPTER 29

CONCENTRATION OF VIRUSES FROM AQUEOUS ENVIRONMENTS: A REVIEW OF THE METHODOLOGY

Georges Belfort and David M. Dziewulski
Department of Chemical and Environmental Engineering
Rensselaer Polytechnic Institute
Troy, New York

INTRODUCTION

The Need for Detection and Concentration

The ever-increasing demand for additional sources of water has resulted in the direct and indirect reuse of wastewater. This, coupled with the ability of potentially pathogenic viruses to survive in finished and unfinished waters [1,2], emphasizes the need for adequate detection methodology. The increasing attractiveness of the direct land disposal of sewage and the persistence of viruses in the soil and on crops at such sites [3] is a specific instance in which efficient virus monitoring methods would help to minimize potential health hazards associated with direct reuse of wastewater. Indirect reuse of wastewater for urban, industrial and agricultural purposes is presently occurring; in fact, it has been estimated that 100 million people throughout the world are currently being supplied with drinking water via indirect wastewater reuse [4].

There is increasing evidence that conventional water treatment processes are unable to completely remove viruses from domestic wastewater before their discharge into the environment [5]. For example, human enteroviruses and other enteric viruses have been detected in treated waters in spite of the presence of total residual chlorine levels at, or in excess of, 0.5 mg/l [6]. The effectiveness of chlorine treatment is dependent on several factors: (1) the pH of the suspending medium strongly influences the relative concentration of HOCl and OCl⁻ and therefore determines the rate of virus inactivation in chlorine-treated waters [7] ; (2) the presence of organic solids protects viruses from the effects of chlorine treatment [8] ; and (3) the reaction of chlorine with ammonia commonly found in wastewater results in the formation of the less effective chloramines and, hence, reduces the actual level of free residual chlorine for disinfection.

Coliform indicators have been found generally to be much more sensitive than poliovirus [9,10] and other enteroviruses [11] to chlorine treatment. The difference in the effectiveness of chlorine inactivation of enteroviruses, other enteric viruses and the coliforms strongly suggest that the absence of indicator bacteria does not indicate that the waters are virus-free [6]. Nevertheless, the coliform indicator system remains the simplest and most rapid means of determining the microbiological quality of water.

Anaerobic digestion of sewage sludge was shown to be effective in the inactivation of polioviruses and other enteroviruses [12,13] but reovirus was found to be quite resistant to these same conditions [14]. The component responsible for this viricidal activity was determined to be ammonia in its uncharged state [14]. Berg and Berman [15] found that anaerobic mesophilic and thermophilic digestion of sludge were both useful in removing fecal and total coliforms but fecal streptococci and viruses remained. Because the removal rate of the fecal streptococci was similar to the reduction of virus numbers in the sample, fecal streptococci were suggested as potentially useful indicators of viruses in digested sewage sludge. In most instances, however, enteric viruses could be expected to survive beyond the point at which bacterial indicators are not longer recoverable [10].

Attempts to use coliphage as an indicator of human and animal virus pollution in various surface waters have resulted in inconclusive results. Kott et al. [16] reported that enteric viruses and coliphage could be shown to occur in a given sample of sewage or contaminated seawater. A subsequent study of estuarine waters [17] indicated that the coliphage and enteric viruses did not always occur jointly in the same sample. Therefore, the presence (or absence) of bacteriophage would seem to be a questionable method of monitoring human- or animal-associated viruses in water.

Due to the persistence of enteric viruses in various waters and the apparent disparity in the correlation between indicator systems and the presence of viruses, the general trend has been toward development of methods for direct concentration and isolation of animal-associated viruses from the environment.

Viruses in Water

At present the standard method of determining the microbiological quality of drinking water requires measurement of the coliform group [18]. Realizing that enteric viruses may pass through water treatment plants [19] and exhibit more resistance than coliform bacteria to disinfection [10,20], the World Health Organization has suggested that, whenever possible, drinking water should be free from enteric viruses [5].

Because the natural reservoir for the enteric viruses is the mammalian gut, the fecal wastes associated with populated areas serve as a constant source for the introduction of these agents into the environment. Thus, there exists the possibility of finding in water human enteric viruses of over 100 serologically recognized types from three major groups [21]. These three major groups are: culturable enteroviruses (poliovirus, echovirus, coxsackieviruses A and B and enteroviruses intermediate in properties between the polio- and echo-virus groups), reovirus and adenoviruses (Table I).

Table I. Viruses Occurring in the Environment[a]

Virus Group	Number of Virus Types in Group	Size (nm)	Nucleic acid[b]	Associated Disease States
Enterovirus				
Poliovirus	3	20-30	ssRNA	Aseptic meningitis, paralysis
Echovirus	34	20-30	ssRNA	Upper respiratory tract infections, aseptic meningitis, pericarditis, myocarditis
Coxsackievirus A	24	20-30	ssRNA	Upper respiratory tract infections, aseptic meningitis, paralysis, herpangina
B	6	20-30	ssRNA	Pericarditis, pancreatitis, paralysis, aseptic meningitis
Reovirus	3	70-80	dsRNA	Respiratory involvement; gastroenteritis
Rotavirus	?	70-80	dsRNA	Infantile gastroenteritis
Adenovirus	33	70-80	dsDNA	Acute respiratory disease, general upper respiratory tract infections, conjunctivitis
Norwalk-type Agent	1	∿27	DNA?	Acute nonbacterial gastroenteritis
Hepatitis A	--	27	RNA?	Infectious hepatitis
B	--	42	dsDNA	Serum hepatitis
non-A, non-B	--	--	--	--

[a] Compiled from References 22-24.
[b] ss = single stranded; ds = double stranded.

A member of the reovirus group, the rotavirus, has been established as a gastroenteritis agent of significant importance. The rotaviruses have been implicated in episodes of nonbacterial gastroenteritis in human infants [25,26], juveniles [27] and adults [28] and are considered to be the major etiologic agent of nonbacterial infantile diarrhea. Although the virus has been detected in 40% of infants with diarrhea [5] no method is currently available for the routine isolation of human-associated rotaviruses from the environment. Viruses of the Norwalk type have also been identified as the cause of self-limiting episodes of acute gastroenteritis [29,30]. Other groups of viruses infecting man, such as cytomegaloviruses, herpes viruses and rhinoviruses are less likely to be present in water due to their sensitivity to environmental conditions [21].

In spite of the potential presence of any or all of the above viral agents in water, only for hepatitis A is there epidemiological evidence of transmission via the water route. Outbreaks of hepatitis A have usually been due to the ingestion of raw [31-33] or incompletely cooked shellfish [34] but waterborne cases have been reported [35]. The incidence of hepatitis virus in shellfish and shellfish-growing waters remains a problem since outbreaks of hepatitis have been linked to areas designated for the growth of virus-free shellfish [36].

The frequency and concentration of these viral agents are largely dependent on the number of individuals infected at any given time. Unlike the bacterial indicator organisms which are part of the normal flora of the human population and are shed continuously in large numbers, enteroviruses and other enteric viruses are shed in comparatively low numbers and generally show annual peak concentrations in the late summer and early fall. The concentration of viruses detected in water samples varies not only according to seasonal fluctuations but also with the type of sample, the collection procedure used, the area of collection and even the time of day of the sampling. Virus concentrations in sewage in the United States determined by a variety of techniques, were estimated to be in the range of 10^2-10^3 virus units/l, whereas researchers in other countries generally report higher recoveries [6]. Human enteric viruses also have been isolated from drinking water treatment plant intakes [21], from groundwater over which sewage effluents were applied [37] and in oysters from estuarine waters [38].

The apparent frequency with which a considerable variety of potentially pathogenic viruses might occur in the environment brings up the question of what dose is required for an individual to become involved in a disease episode. Theoretically, one plaque-forming unit (PFU) is sufficient to cause infection in tissue culture [39]. Plotkin and Katz [40] suggested that just one tissue culture infectious dose of poliovirus could infect man; similar results were found by other workers when man and various animals were used as the hosts [41]. The chances of acquiring an infection from an environmental source remain unclear (except possibly in the case of hepatitis) [41]. However, viruses in the environment may be responsible for some increase in the incidence of viral disease beyond the normal person-to-person transmission.

Virus Concentration Methodology

The techniques employed for concentration of viruses from aqueous environments rely on the surface chemical and physical characteristics of the viruses and on the properties of the suspending medium. By virtue of their proteinaceous nature, viruses behave as amphoteric hydrophilic colloids in aqueous suspension. Viruses have a characteristic isoelectric point (pI) which is determined by the number and type of ionizable groups existing at the surface which, in turn, is dependent on the amino acid composition of the protein coat. The pH and ionic composition of the suspending medium, along with the degree of dissolved organic material and particulate matter present, have a direct effect on the fate of the virus in the concentration procedure, its recovery from a given sample and the performance of the components in the concentrator system.

Current methods for isolation of viruses from large volumes of water involve either: (1) the adsorption of virus to some filter material with a subsequent elution step or (2) retention of virus (solute) by a membrane of specified porosity which allows the solvent, small molecules and ions to permeate the membrane, thus dehydrating the sample. The first method has been extensively examined, the second is in developmental stages with regard to application for virus concentration from natural waters (Table II).

Briefly, adsorption-elution involves water conditioning with acid and/or salt to optimize the conditions for virus adsorption to filters. After filtration, virus is recovered by elution with highly alkaline buffers (e.g., glycine, pH 10.0-11.5) or moderately alkaline protein solutions which, depending on sample size and method of elution, are reconcentrated by various procedures. Reconcentration by readsorption and elution to smaller filters results in reduced eluate volumes and is one of the recommended second-step concentration methods. Use of highly alkaline glycine buffers is necessary in instances where $AlCl_3$ is used for enhancement of virus adsorption so that aluminum hydroxide floc formation is minimized (or solubilized in cases where it forms) and virus can be released. Protein solutions and flow conditions may contribute to elution by one or a combination of several mechanisms including shear forces, chelation of cations, competition for adsorptive sites and electrostatic charge reversal due to pH changes.

Recovery of virus from ultrafiltration membranes also involves an "elution" step. Where crossflow is used to introduce shear forces at hollow fiber membrane surfaces to maintain permeation fluxes, virus is polarized by a combi-

Table II. Summary of Applications and Limitations of Concentration Methods of Waterborne Viruses

Method	Initial Volume of Water	Quality of Water	Concentration of Virus in Water	Suggested Application	Limitations or Disadvantages
1. Filter adsorption-elution procedures	large	Low concentration of competitive adsorbers present	low	Reclaimed wastewater and finished waters	Additional development needed for application to wastewaters and polluted natural waters
2. Precipitation by or adsorption to various charged species	small	High organics concentration acceptable	high	Raw or partially treated sewage	Yet to be shown efficient with large volumes at low virus concentration
3. Polymer two-phase separation	small	High organics concentration acceptable	high	Sewage, treated effluents and other raw materials	Processing is slow and virus recovery is inaccurate and of low precision
4. Membrane Filtration Methods					
a. ultrafiltration					
1. soluble filters	small	Good	–	Clean waters	Poor performance with unclarified effluents
2. tangential flow	large	Unknown	low	Treated effluents & finished waters	Looks promising but needs additional testing with waste and polluted waters
b. reverse osmosis (hyperfiltration)	small	Unknown	low	Treated effluents & finished waters	Also concentrates cytotoxic compounds which adversely affect assay methods
c. hydroextraction	small	High organics	large	Raw and treated sewage	High virus loss with wastewaters
d. electro-osmosis	small	Low concentration pressure	low	Clear waters	Equipment expensive and not portable

5. Other methods

a. ultra-centrifuge	small	Good	high	Clean waters	Equipment expensive but not portable
b. electrophoresis	small	Unknown	high	Clean waters	Equipment expensive but not portable
c. freeze concentration	small	Good	high	Clean waters	Not practical yet
d. gauze pad	large	Unknown	unknown	Treated waters	Not quantitative or reproducible
e. vegetable floc	small	Good	high	Clean waters	Not tested extensively
f. affinity chromatography	small	Good	high	Clean waters	Not tested extensively

nation of axial and radial flows and is retained on the membrane surface. A backwash and/or flushing step is used to "desorb" virus from the membrane. Use of ultrafiltration membranes without crossflow results in extensive concentration polarization of solutes with significantly reduced permeation fluxes and increased difficulty in recovering surface-associated virus particles.

Other methods of virus concentration include: (1) organic flocculation (using beef extract) [42], (2) precipitation with protamine sulfate and bovine serum albumin [43], (3) aqueous two-phase polymer separation [44-51], (4) adsorption to and flocculation of insoluble polyelectrolytes [52-54] and cation salts [55-57] and (5) hydroextraction with polyethylene glycol [58, 59]. These methods are capable of efficiently processing several liters of water (or less) and their usefulness has been extensively demonstrated when employed for highly contaminated waters where small sample size will suffice for virus detection or as second-step concentration (reconcentration) methods.

In the following sections, adsorption-elution and ultrafiltration are presented as methods capable of handling large volumes of water. The final section reviews the various reconcentration methods generally used for samples of limited volume.

THEORY

Capture and Attachment in Adsorption-Elution Depth Filters: A Theoretical Analysis

The capture of fine particles such as viruses in suspension by filtration through a porous medium may be divided into two steps: attachment and transport. Another step not well understood involves reentrainment or detachment.

Attachment

The surface characteristics of the virus and the grain (or filter surface) determine directly whether attachment will occur. This is independent of the mechanism of transport to the grain, but can be influenced by the ionic strength of the aqueous solution. For the virus to attach itself to the clean grain surface or previously adsorbed deposits, it has to be colloidally unstable. The classical theory of Derjagiun-Landau-Verwey-Overbeek (DLVO) describing colloidal destabilization can be used to study virus attachment [60,61]. Electric repulsion forces are counterbalanced by London (van der Waal's) dispersion attractive forces. A detailed analysis based on the DLVO theory for virus attachment will soon be published [62]. However, other mechanisms besides the purely electrostatic interactions of the DLVO theory could also effect attachment. These include charge-neutralization, adsorption of charged polymers and bridging [63]. Recently, Fuhs et al. [64], using a laser electrophoresis unit, were able to study the zeta-potential of reovirus and various flocculants and clays as a function of pH. They explained virus attachment via the electrostatic model referred to above. Murray and co-workers have also used the DLVO theory to explain virus association with soil particles [65]. Additional details will be presented in a forthcoming book by Belfort and Dziewulski [62].

Transport [66]

For aqueous filtration the flow (rate) through porous media is usually laminar and proportional to pressure drop according to Darcy's law.

The dominant capture mechanisms are shown in Figure 1. Although straining has been omitted there is always the possibility that at the beginning of the run or as the filter nears exhaustion a cake may form at the top of the filter surface due to particle straining. This phenomenon of cake-formation is most likely to occur on membrane-filter surfaces during the reverse osmosis and the ultrafiltration process [67]. A dimensionless ratio is used to characterize each capture mechanism and to identify the important parameters.

1. Diffusion. Very small particles such as viruses move randomly about a fluid as a result of thermal energy, kT. The Stokes-Einstein diffusion coefficient D is given by

$$D = kT/3\pi\mu d_v \tag{1}$$

where k = Boltzmann's constant,
         T = absolute temperature,
         μ = fluid viscosity
         d_v = virus diameter.

A convenient dimensionless ratio characterizing diffusion relative to convection is given by the reciprocal Peclét number:

$$N_{diff} = Pe^{-1} = \frac{D}{dv} = \frac{kT}{3\pi\mu dvd_v} \tag{2}$$

where d = grain (or pore) diameter
         v = approach convective velocity.

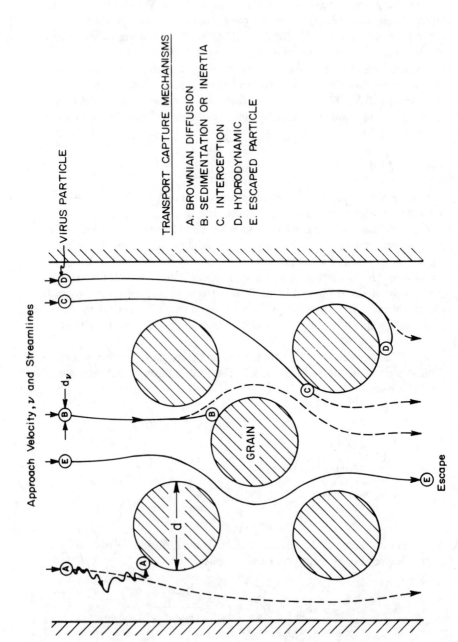

Figure 1. Transport capture mechanisms for virus and virus-associated particles in passing through a depth filter.

2. Sedimentation. The settling of particles due to gravity has been an important mechanism for relatively large particles (such as virus associated macrocolloids) in filtration. Stokes' law describes the settling of particles of density ρ_v in a fluid density ρ:

$$v_s = \frac{g(\rho_v - \rho)d^2}{18\mu} \tag{3}$$

where g is the acceleration due to gravity. By relating v_s to v, a characteristic dimensionless number of sedimentation is obtained

$$N_{sed} = \frac{v_s}{v} = \frac{g(\rho_v - \rho)d_v^2}{18\mu v} \tag{4}$$

3. Inertia. If a particle possessess sufficient inertia it could conceivably veer-off its fluid streamline allowing capture. The inertial efficiency group is given by

$$N_{iner} = \frac{\rho_s d_v^2 U}{18\mu d} \tag{5}$$

where U is the velocity of the fluid at an infinite distance relative to the spherical grain.

4. Interception. If the center of a particle is moving in a streamline that is closer to the grain than the radius of the particle it will contact the grain. This mechanism of capture is called interception and is characterized by

$$N_{int} = d_v/d \tag{6}$$

5. Hydrodynamic. Because of velocity gradients across a moving particle, lateral migration across streamlines could occur due to pressure gradients across the rotating particle and due to inertial effects [68]. For want of a better number, inverse Reynolds number N_{Re} has been used to characterize this lift phenomenon. We suggest a new characteristic number directly related to lift velocity [69]

$$v_L = \frac{K_1 v N_{Re}}{\epsilon} (d_v/d)^3 \, f(\acute{\eta}) \tag{7}$$

where K_1 = a constant,

N_{Re} = $\rho dv/\epsilon\mu$,

$f(\acute{\eta})$ = a function of $\acute{\eta} = d_v/d$ explicitly given by Ho and Leal [70].

The new characteristic number is given by

$$N_{hydro} = v_L/v = \left[\frac{K_1 N_{Re}}{\epsilon} \left(\frac{d_v}{d}\right)^3 f(\acute{\eta}) \right]^{-1} \tag{8}$$

The efficiency of particle retention has been expressed through the so-called filter coefficient defined by the Iwasaki rate equation

$$\frac{\partial c}{\partial z} = -\lambda c \tag{9}$$

where c = the particle bed concentration

z = the depth of the bed.

The filter coefficient is related to the total collection efficiency η resulting from all the mechanisms of capture [71]

$$\lambda \simeq [3(1 - \epsilon)/2d] \, \eta \tag{10}$$

where η = constant $N_{diff}^{\beta} \, N_{sed}^{\gamma} \, N_{iner}^{\xi} \, N_{int}^{\alpha} \, N_{hydro}^{\delta}$

$$= \text{constant} \left[\frac{kT}{3\pi\mu dv d_v} \right]^{\beta} \left[\frac{g(\rho_v - \rho)d_v^2}{18\mu d} \right]^{\gamma} \left[\frac{\rho_s d_v^2 U}{18\mu d} \right]^{\xi} \left[\frac{d_v}{d} \right]^{\alpha} \left[\frac{\rho dv}{\mu} \left(\frac{d_v}{d}\right)^3 f(\acute{\eta}) \right]^{-\delta} \tag{11}$$

where $\alpha, \beta, \gamma, \delta$ and ξ are positive exponents. Collecting items together,

$$\eta = \text{constant} \frac{d_v^{a}(U\rho_s)^{\xi} [\rho f(\acute{\eta})]^{-\delta} (kT)^{\beta} (\rho_v - \rho)^{\gamma}}{d^{b} \mu^{c} v^{d}} \tag{12}$$

where the exponents are given by

$$
\begin{aligned}
a &= \alpha - \beta + 2\,(\gamma + \xi) - 3\,\delta \\
b &= \alpha + \beta - 2\,\delta + \xi \\
c &= \beta + \delta + \gamma - \xi \\
d &= \beta + \gamma + \delta
\end{aligned}
\tag{13}
$$

Equations 12 and 13 indicate that η increases with: (1) virus diameter d_v provided a is positive; i.e., $\alpha + 2\,(\gamma + \xi) > 3\delta + \beta$ (for very small particles a may be negative and diffusion would dominate); (2) decrease in grain diameter d, provided life is negligible relative to the other forces; i.e., $\alpha + \beta + \xi > 2\,\delta$; (3) decrease in viscosity μ; or (4) decrease in approach velocity v.

In fact, experimental results show that for particles smaller than $\sim 1\ \mu m$, Brownian diffusion (large β) is the dominant mechanism of removal, and for particles larger than $\sim 5\ \mu m$, interception (large α) and sedimentation (large γ) are the dominant removal mechanisms [72]. Inertial effects are generally very small in aqueous systems and usually can be ignored (small ξ).

In designing an adsorption-elution system for the capture of viruses of diameter d_v, the maximum virus capture efficiency should occur at the highest fluid temperature tolerable by the virus [high $(kT)^\beta$ and low μ^c], at the lowest fluid approach velocity (low v^d) and with the smallest diameter capture surface or grain size (small d^b). Clearly, for completeness, a detailed attachment analysis incorporating the DLVO electrostatic theory is also necessary. This will soon be published [62].

Transport and Hydrodynamics in Hollow Fiber Modules

A theoretical basis for comparing and analyzing the performance of different hollow fiber modules operating under different conditions is presented here [73].

Solute Removal

As described previously [73], a convenient method of comparing the removal rates of solutes (number of particles N removed) and fluid (volume V_f removed) from the feed solution during an ultrafiltration experiment uses the following equations:

$$
N = (\overline{V}_f)^{1-R}
\tag{14}
$$

or

$$
\overline{C} = (\overline{V}_f)^{-R}
\tag{15}
$$

The reduced number $\overline{N}$ or reduced concentration $\overline{C}$ represents quality parameters such as virus number (PFU), turbidity (FTU) or dissolved organic carbon (DOC in mg/l). From a plot of log $\overline{N}$ (or log $\overline{C}$) versus log $\overline{V}_f$, the retention efficiency or average rejection R is obtained. $R \to 1$ implies $\overline{N} \to N_{f_{i-1}}$ which means there is no removal of N_f during the experiment and $\overline{C} \to \overline{V}_f$. However, $R \to 0$ implies that N_f is removed from the solution at approximately the same rate that $\overline{V}_f$ is reduced, resulting in $\overline{C} \to 1$ or a constant concentration during the dehydration experiment.

These conclusions can best be explained from the usual definition of retention, viz.,

$$R = 1 - v_s/(v_w C_f) \tag{16}$$

Thus for $R = 0$, the solute flux $v_s = v_w C_f$ and is simply due to permeation. For $R = 1$, the solute flux $v_s = 0$ since v_w and C_f are observable nonzero quantities here.

During the experiment the solute could be adsorbed either onto or into the membrane or it could pass through the membrane with the permeate. With the asymmetric membranes of low molecular-weight cut-off used in this study, the viruses apparently do not permeate the membrane but are "adsorbed" for later recovery by backwashing [74,75].

Modified Filtration Equation

When comparing the performances of different hollow fiber modules or different runs for the same module, it is preferable to analyze the results as a function of reduced and accumulated permeation volume, $\overline{V}_p$. Belfort and Marx [67] have adapted the standard filtration approach to study fouling of reverse osmosis membranes. Their modified filtration theory is described below for analysis of the hollow fiber ultrafiltration experiments reported later.

The modified-filtration equation ignoring osmotic effects ($\Delta\pi \approx 0$) and imperfections in the membrane such as micro-pin-holes, can be written as follows:

$$\overline{v}_w^{-1} = \frac{\mu\alpha' D_w W}{R\Delta P_T S} = \frac{\mu r' D_w}{R\Delta P_T} \tag{17}$$

where the total accumulated weight of suspended solids deposited on the membrane surface is given by

$$W = \int_{0}^{V_P} w \, dV_P = - \int_{V_{f_i}}^{V_f} w \, dV_f \tag{18}$$

Substituting Equation 15 for $\overline{C} = \overline{w} = w/w_i$ into Equation 18 and integrating, and then into Equation 17, the following general result is obtained

$$\overline{v}_w^{-1} = -k_1(1 - \overline{V}_p)^{1-R} + k_2 \tag{19}$$

where

$$\overline{V}_f = 1 - \overline{V}_p$$

$$k_1 = \frac{\mu \alpha' D_w w_i V_{f_i}}{R \Delta P_T S(1-R)}$$

$$k_2 = \frac{\mu r' D_w}{R \Delta P_T} + k_1$$

Expanding the brackets in the first term on the right-hand side of Equation 19 for $\overline{V}_p^2 < 1$ and $R \leqslant 1$,

$$\overline{v}_w^{-1} = -k_1 [1-(1-R)\, \overline{V}_p + \frac{(1-R)(-R)}{2}\overline{V}_p^2 + \ldots] + k_2 \tag{20}$$

For small R (e.g., < 0.6) the terms of $O(\overline{V}_p^2)$ and greater can be neglected in the square brackets. Thus,

$$\overline{v}_w^{-1} = k_1' \, \overline{V}_p + k_2' \tag{21}$$

where

$$k_1' = k_1(1-R)$$
$$k_2' = k_2 - k_1.$$

For constant suspended solids concentration w_i in the feed, such as in cake filtration or for ultrafiltration with no feed recycle, $W = w_i V_p$ from Equation 18, and Equation 21 is obtained directly on substitution into Equation 17.

For batch recycle of the feed in an ultrafiltration system, such as that used in these studies, the predominant radial force acting on the suspended particles is due to permeation (see the discussion in Ref. 73 comparing permeation with Brownian motion). A plot of $\overline{v}_w^{-1}$ versus $\overline{V}_p$ gives k_1' and k_2' and from these $\alpha' w_i$ and r', respectively.

Solute Hold-up Index, $\overline{H}$

As described in the previous section, under certain conditions the permeation force is the dominant force acting on the fluid and solute particles during their passage through the hollow fiber module. For a given module geometry (R,L,n), permeation characteristics (v_w) and average recycle velocity inside the tubes ($<v_z>$) it would be convenient to have a measure of the potential solute hold-up or retention during a concentration run. $(1-R)$ is such a measurement. Unfortunately, R is difficult and expensive to obtain in the field and, although easily measured in the laboratory, requires a costly and time-consuming effort to characterize a particular module and operating conditions.

A simple, easily obtained qualitative measure of the potential solute hold-up can be estimated heuristically. Assuming relative destabilization (attachment) on contact between the virus particles and the membrane surface and negligible reentrainment, it seems reasonable to propose a hold-up index $\overline{H}$ as follows:

$$\overline{H} \equiv 8\pi \times v_w P_f L / \, |\dot{\gamma}_w| \tag{22}$$

Substituting for $|\dot{\gamma}_w| = |-<v_z> 4/R|$ for laminar flow in tubes and

$$P_f = t_T/t_1 = \frac{\Delta V_p/(2\pi RLv_w n)}{L/<v_z>} \tag{23}$$

into Equation 22, the following simple expression is obtained:

$$\overline{H} \equiv |\Delta V_p/(Ln)| \tag{24}$$

Reference 73 provides additional hydrodynamic analysis on convective-induced polarization and particle capture, Taylor dispersion analysis under low seepage rates and a comparison of permeation versus Brownian motion for three different hollow fiber modules and other enteroviruses.

CURRENT STATE OF METHODOLOGY

Adsorption-Elution Methodology

Early attempts at recovering viruses from water employed grab sampling or the suspended gauze pad technique [76]. Since the gauze pad method could not correlate the amount of water tested to the number of viruses recovered, attempts at quantitation were made using a flow-through gauze pad sampler [77]. Liu et al. [78] examined such a device using poliovirus-seeded tap water and sea water and were able to recover 2% and 15-19%, respectively, of the input viruses. It was noted that addition of NaCl to tap water aided adsorption since, upon elution, 49% of the virus input was recoverable.

One of the first to recognize the ability of viruses to adsorb to membrane filters was Metcalf [79] who used membrane filters to eliminate bacterial contamination from prepared suspensions of influenza virus and clinical samples. In spite of porosities exceeding the diameter of the virus, retention of virus by the membrane still occurred; electrostatic forces were suggested as a possible means of virus retention.

In the development of methods for the isolation of viruses from food extracts Cliver [80] described the adsorptive capability of a series of membrane filters ranging in porosity from 50 to 220 μm using poliovirus- and coxsackievirus-seeded solutions. Extensive decreases in virus titer were determined after passage of the solutions through the filters. However, the presence of various proteins in the diluent or pretreatment of membranes with protein solutions effectively reduced the adsorptive capacity of the filters and allowed passage of the viruses.

Application of membrane filters for the concentration of viruses was reported by Wallis and Melnick [81]. Viruses were concentrated from cell culture harvests after removal of organic membrane coating components (MCC) by ion-exchange chromatography. Salts were found to facilitate adsorption and viruses could be eluted from filters using solutions containing MCC.

Concentration of Virus from Tap Water

After initial studies determined that adsorption of viruses to filters of various composition (see Table III) was a possible means of concentration, application of the methodology to experimentally seeded and natural tap water, wastewater effluent, sea water and sewage was implemented. Most model systems were initially examined using various volumes of tap water seeded with poliovirus or other enterovirus. An abbreviated survey of experiments using adsorption-elution methods with tap water is presented in Table IV. In the evolution of this particular virus concentration methodology, changes have been made to improve adsorption, increase elution efficiency and, in general, simplify the protocol. Some noteworthy changes or amendments to the methodology are presented below.

Wallis et al. [83] compared the efficiency of divalent and trivalent cations for the enhancement of virus adsorption to cellulose membranes and determined that $AlCl_3$ was the most efficient cation for adsorption. In large-scale experiments, 1890 liters of tap water seeded with poliovirus 1 were adjusted to pH 3.0 and a final concentration of 0.0005 M $AlCl_3$. Samples were initially passed through 293-mm-diam cellulose membranes and adsorbed virus was eluted with 0.05 M glycine, pH 11.5. Reconcentration was accomplished by adjusting the initial eluate to pH 3.5 and passing the samples through 25-mm-diam membranes. Virus was subsequently eluted with 5-ml of 0.05 M glycine, pH 11.5. The overall average recovery efficiency for the two-step process using virus inputs of 250, 2500 and 25000 per 500-gal sample was 92%. Further examination of this procedure involved the development of a portable apparatus for the concentration of virus onto cellulose membranes [95].

Table III. Various Filters Used for Virus Concentration Procedures

Material	Manufacturer
Adsorbent Filters[a]	
Nitrocellulose (HA)	Millipore Corp.
Cellulose ester (GS)	Bedford, MA
Epoxy-fiberglass	Filterite Corp.
	Timonium, MD
Epoxy-fiberglass-asbestos	Cox Instrument Corp.
(Epoxy-fiberglass)	Detroit, MI
Glass microfiber-epoxy resin	Balston, Inc.
	Lexington, MA
Fiberglass	Commercial Filter Div.
Cellulose acetate	Carborundum Co.
	Lebanon, IN
Cellulose-diatomaceous earth	AMF-Cuno
charge-modified resin	Meriden, CT
Clarifying Filters[b]	
Orlon	Commercial Filter Div.
Cotton	Carborundum Co.
Cellulose acetate	Lebanon, IN
Fiberglass (AP)	Millipore Corp.
	Bedford, MA

[a] Adapted from Ref. 82.
[b] Usually pretreated before use.

Table IV. Concentration of Virus from Tap Water Using the Adsorption-Elution Method

Filter Material	Initial Volume, v_i (liter)	Sample Treatment	Virus	Initial Virus Titer (PFU)	Eluate	Reconcentration Method	Final Volume, V_f (ml)	Recovery (%)	Flow Rate, Q (liter/min)	Ref.
Cellulose membrane	1890	pH 3.0 + 0.005 M AlCl3	Poliovirus 1	(1) 250 (2) 2500 (3) 25000	0.05 M glycine, pH 11.5	Ads-El[a]	5.0	(1) 94 (2) 88 (3) 94	NR[b]	83
Cellulose nitrate membrane	1	0.05 M[c] Na2HPO4 + citric acid to pH 7.0	(1) Polio 1 (2) Reo 1 (3) Echo 7 (4) Coxsackie	(1) 72-100 (2) 32-48 (3) 58-1185 (4) 31	(1) 20-min sonication (2) 40-min sonication (3) 20-40 min sonication (4) 10-20 min sonication in beef extract	ND[d]	ND	(1) up to 100 (2) 83-94 (3) 96-100 (4) 48-97	NR	84
Wound fiberglass depth filter + epoxy fiberglass disc	378	pH 3.5	Polio 1	380-2500	0.05 M glycine, pH 11.5	Ads-El	10.0	48-95	7.6	85
Cellulose ester pleated membrane filter	378	pH 4.5 + MgCl2 (1200 µg/ml)	Polio 1	24-194	5x nutrient broth in carbonate-bi-carbonate buffer	Two-phase[e]	3.0	24-67	9.5	86
Nitrocellulose or fiberglass microfilters	378	pH 3.5-4.5 + MgCL2, AlCl3	Polio 1	16-50	NR	NR	NR	20-50	3.8-7.6	87

Table IV, continued

Filter Material	Initial Volume, V_i (liter)	Sample Treatment	Virus	Initial Virus Titer (PFU)	Eluate	Reconcentration Method	Final Volume, V_f (ml)	Recovery (%)	Flow Rate, Q (liter/min)	Ref.
Epoxy fiberglass filter tubes	378	pH 3.5	Polio 1	19-153	0.05 M glycine, pH 11.5	Ads-El	10.0	42-57	3.8	88
(1) Nitrocellulose membrane	380	pH 3.5	Polio 1	12-173	0.05 M glycine, pH 11.5	Ads-El	14.0	(1) 15-52	3.8	89
(2) Epoxy fiberglass disc								(2) 8-62		
(3) Epoxy fiberglass filter tubes								(3) 11-74		
(4) Wound fiberglass depth filter + epoxy fiberglass disc								(4) 0-52		
(1) Nitrocellulose membrane	1900	pH 3.5	Polio 1	12-22	0.05 M glycine, pH 11.5	Ads-El	14.0	(1) 25-50	9.5	89
(2) Epoxy fiberglass filter tubes								(2) 25-80		
Pleated epoxy-fiberglass	1900	pH 3.5 + 0.0005 M AlCl$_3$	Polio 1	5×10^0 - 8.5×10^6	0.05 M glycine, pH 10.5	Al(OH)$_x$ floc	30	40-67	26	90

Material	Volume	Conditions	Virus	Virus input	Elution	Reconcentration				Ref.
Pleated epoxy-fiberglass or wound fiberglass depth filter	472-1000	pH 3.5 + 0.0005 M AlCl$_3$	Polio 1	2.2×10^3-1.5×10^6	0.05 M glycine, pH 10.5	Al(OH)$_x$ floc + hydro-extraction	20	41-82	37.8	91
Epoxy-fiberglass	1000	2×10^{-5} M AlCl$_3$	Polio 1	1.76×10^2-7.3×10^6	0.05 M glycine, pH 11.0	Al(OH)x floc	28-80	59-79	26	92
Cellulose diatomaceous earth "charge-modified" resin	(1) 12 (2) 378	(1) pH 7- (2) pH 7.1	Polio 1	(1) 22-217 (2) 413-426	0.05 M glycine, pH 10.0	(1) ND (2) Ads-El	(1) 15 (2) 25.0	(1) 64 (2) 22.5	NR	93
Cellulose "charge-modified" resin	(1) 12 (2) 1000	pH 7.5	Polio 1	(1) c.10^4	(1) 0.05 M glycine, pH 11.0; or 0.3% beef extract in glycine, pH 0.5 (2) 0.3% beef extract in glycine, pH 9.5	(1) ND (2) Organic floc	(1) ND (2) 40-75	(1) 48	NR (2) 30	94

a Reconcentration by adsorption to and elution from smaller filters.
b Not reported.
c Distilled deionized water.
d Not done.
e Aqueous two-phase polymer separation.

The concentration of viruses from large volumes of clean waters using the portable virus concentrator was simplified when it was determined that acidification of water samples would allow for virus adsorption in the absence of any added salts [85]. Modifications of the original portable concentrator described by Wallis and Melnick [95] involved: (1) elimination of clarifying filters, (2) injection of HCl as the only conditioning necessary for water samples before passage through adsorbing filters and (3) changes in absorbent filters. The cellulose membranes originally used as adsorbent were replaced with a series containing a fiberglass wound-fiber depth filter and epoxy fiberglass discs. (See Table III and Ref. 83 for filter manufacturer and codes to filters.) Reconcentration in the two-step procedure required the addition of small amounts of $AlCl_3$ to overcome the interfering components present in tap water concentrated initially by the procedure so that readsorption would be possible. In 380-liter experiments using total virus inputs of 380-2500 PFU, the average virus recovery was 77% [85].

In a comparative study of four commonly used microporous filters for the concentration of viruses from drinking water, Jakubowski et al. [89] found no significant difference in virus recovery efficiency among the filters. In this study water was acidified to pH 3.5 and virus input ranged from 12 to 173 PFU/378-liter sample. After elution with 0.05 M glycine, pH 11.5, the samples were reconcentrated by readsorption to and elution from smaller filters. Microfiber-epoxy resin filter tubes resulted in the highest recovery, 45.3%. The 267-mm-diam epoxy-fiberglass-asbestos filters, 293-mm-diam nitrocellulose membranes and the fiberglass depth filter combined with a 127-mm epoxy-fiberglass-asbestos diameter filter yielded recovery percentages of 39, 30.9 and 28.8, respectively. The nitrocellulose membranes and filter tubes were further evaluated under conditions of low virus multiplicity. Virus inputs of 12-22 PFU/1900 liters resulted in recoveries of 25-50% for the nitrocellulose membrane and 25-80% for the filter tube using the methods of the comparative portion of the study. It also was determined that elution could be performed at a pH lower than 11.5. A subsequent study using various combinations of the same filter substrates indicated no significant difference among the sensitivities of the various systems [96].

In a fast-flow system (37.8 liter/min) described by Farrah et al. [90] virus was adsorbed to a fiberglass depth filter and pleated epoxy-fiberglass filter in a series in the presence of $AlCl_3$ at pH 3.5. Elution was performed with 0.05 M glycine, pH 10.5, and eluates were reconcentrated by aluminum hydroxide flocculation. Recoveries of 40-50% were reported from 1900-liter samples of tap water seeded with as low as 5 PFU and as high as 8.5×10^6 PFU. Significantly, the high flow rates did not adversely affect the efficiency of virus adsorption to filters. Treatment of the pleated epoxy-fiberglass filters with 0.1 N NaOH or autoclaving was found to successfully regenerate the filter for reuse [97].

Gerba et al. [91] examined large volumes (472-1000 liters) of tap water as part of the development of a unified scheme for the efficient concentration of viruses from waters of various quality. Virus samples were adsorbed to a series of pleated epoxy-fiberglass filters in the presence of $AlCl_3$ at pH 3.5. Elution from filters was performed with a glycine eluent at pH 10.5 or 11.5 for tapwater experiments and at the latter pH for experiments involving treated sewage or sea water. Separate experiments indicated that the majority of solids-associated virus was recoverable using the pH 11.5 eluent. Recovery efficiency of seeded poliovirus from tap water, treated sewage and sea water was approximately 50% in each instance. The major advantages of the system were the capacity to operate at high flow rates (37.8 liter/min), resist clogging and maintain recovery efficiency regardless of water type.

In the study described above a comparison of the in-line injection method of water conditioning used in the field with batch preparation of the sample before filtration commonly used in laboratory models was conducted to determine the validity of the latter procedure. No significant differences were evident.

Farrah et al. [92] examined the concentration of poliovirus from large volumes of tap water (1000 liters) in the presence of 2×10^{-5} M $AlCl_3$ at pH 7.4-7.9. At these pH values a slight turbidity due to aluminum hydroxide flocculation was evident. Enhanced virus removal was attributed to the turbidity present during adsorption. Pleated filters were used since they exhibited a good resistance to fouling. In effect, the acidification step was circumvented and virus could simply be recovered by solubilization of the captured $Al(OH)_x$ flocs with an alkaline eluent. Processing of 1000-liter tapwater samples in this manner resulted in an average recovery after aluminum hydroxide reconcentration of 71%.

Sobsey and Jones evaluated more electropositively charged microporous filter discs (PC filters) for the concentration of poliovirus from tapwater samples [93]. Using 12-liter volumes of tap water with low concentrations of poliovirus 1 LSc (22-217 PFU) an average recovery of 64% was reported when virus was adsorbed at pH 7.0-7.5 and eluted with pH 10.0 glycine-NaOH. In large-scale experiments (378 liter) employing a two-step adsorption-elution method, virus was adsorbed to PC filters at pH 7.1 and elution was performed using pH 10.0 glycine. For reconcentration virus was adsorbed at pH 6.0. In two trials, virus recoveries of 25 and 20% were reported. An epoxy-fiberglass filter disc system used under the identical conditions of the PC substrate and the conditions prescribed in the 14th edition of *Standard Methods* resulted in recoveries of 2 and 4%, respectively.

A positively charged filter substrate capable of being formed into pleated cartridges was evaluated along with positively charged filter discs and epoxy-fiberglass filters [94]. Preliminary studies showed that both PC filters could efficiently adsorb poliovirus ($\geqslant$ 80%) between pH 3.5 and 7.5. However, single layers of the PC filter material used for the pleated cartridge were less efficient than the PC filter disc under similar conditions. It was suggested that this difference was due to filter depth and hence the degree of contact between virus and filter, the original filter disc being 5-mm thick and the material for the pleated filter 1-mm thick. Double layers of the latter filter performed as well as the PC filter disc (95 vs 99%) in adsorption efficiency at pH 7.5. Recovery of virus from single- and double-layers of 47-mm diameter filters with beef extract in 0.05 M glycine, pH 9.5, was 67 and 66%, respectively.

In the same study, poliovirus was concentrated from 1000-liter tapwater samples with PC and epoxy-fiberglass pleated cartridge filters. Adsorption conditions for the conventional medium were pH 3.5 and a final concentration of 5×10^{-3} M $MgCl_2$. Water conditioning prior to filtration through the positively charged filter was simply adjustment to pH 7.5. Elution was performed using the beef extract-glycine eluent and after reconcentration by organic flocculation, the average recoveries were 33% for the conventional medium and 30% for the positively charged filter.

Concentration from Natural and Virus-Seeded Natural Waters

The isolation and concentration of viruses from various natural surface waters and sewage poses problems not evident in model systems tailored for application with finished waters. Early in the development of adsorption-elution procedures, several workers recognized the ability of soluble proteinaceous (and organic) materials to interfere with adsorption of virus to filter surfaces by competing for potential virus adsorption sites [80,81]. In addition the presence of suspended solids in samples severely limited the capacity of a given filter to perform optimally for extended periods under turbid conditions [98]. Attempts to remove turbidity and other foulants by the introduction of protective clarifying filters often will result in the removal of solids-associated virus [99] or a situation in which the continuous build-up of particulates or metal ions on the clarifying filter results in an adsorbing substrate being "deposited" over time on the filter surface. Various methods have been used to obviate these problems in field applications and seeded laboratory experiments using natural waters.

Experimental conditions and results of work dealing with methods for the concentration of viruses from natural surface waters, sewage and sewage effluents are summarized in Table V. In the discussion below, major advances obtained in the various studies presented in Table V are highlighted.

Recovery of seeded poliovirus from sewage samples was attempted by Wallis and Melnick [100] using a system employing prefilters. Samples were clarified of turbidity by passage through a fiberglass prefilter and a 0.22-μm porosity cellulose nitrate membrane. The resultant filtrate was then treated with an ion exchange column to remove membrane coating components (MCC). Samples were treated with $MgCl_2$ and virus was adsorbed to cellulose ester membranes. Elution by grinding of adsorbent membranes was performed in Melnick's B medium containing 10% fetal calf serum. Recoveries of 80% were attained with this procedure.

Rao and Labzoffsky [102] found in their system—in which seeded poliovirus was concentrated from lake and river water in the presence of added $CaCl_2$—that virus did adsorb to both their prefilter and the adsorbent membrane. Virus was added to 500-ml samples to a final concentration range of 26-37 PFU and recoveries of 53 to > 100% were reported after elution with approximately 10 ml of a 3% beef extract solution.

In a system evaluated for both tap and estuarine waters employing membrane cartridge filters, an average of 70% of the seeded poliovirus was recoverable from artificial and natural estuarine waters [86]. The highest recovery of 97% was obtained with poliovirus-seeded natural estuarine water adsorbed at pH 4.5 in the presence of $MgCl_2$ and eluted with 5x nutrient broth in carbonate-bicarbonate buffer, pH 9.0. In all but one experiment virus penetrated membrane filters and it was suggested that this was an effect of MCC. The adverse effect of MCC on adsorption was made evident by the use of stored estuarine waters. Although these waters were prefiltered before use in seeded control experiments, overall recoveries were reduced to 43%.

Rao et al. [103] demonstrated that small volumes of highly contaminated sewage samples were sufficient for the isolation of enteroviruses and that processing could be fast and efficient. Initial laboratory studies involved the clarification of samples by centrifuging poliovirus-infected raw sewage at 1800 $\times g$ to remove gross particulates, blending the supernatant fluid at pH 3.0 and an additional centrifugation at 9230 $\times g$ after floc formation to further clarify the sample. The final supernatant was collected with no reported loss of virus. Samples were then adsorbed to cellulose ester filters at pH 3.0 in the presence of 0.05 M $MgCl_2$. Elution was done with 3% beef extract, pH 8.0, and resulted in recovery of poliovirus with efficiencies in the range of 88-97% from 100-ml samples. Application of the technique to 40-ml raw sewage and 320-ml effluent samples was made over a one-year period. Virus recovery was made in all raw sewage samples and ranged from 1050 to 11575 PFU/l. In only one instance did an effluent sample result in no virus recovery; other samples yielded a range of 3-90 PFU/l.

In the development of a portable virus concentrator for use in the field, Wallis et al. [112] examined the use of wound-fiber depth filters for the clarification of water samples prior to contact with adsorbent filters. Filters of various synthetic and natural composition were used to protect 0.45-μm porosity cellulose ester membranes or insoluble polyelectrolyte layers (PE 60) in the processing of Houston tap water. Waters high in organic material could

Table V. Concentration of Virus from Natural Surface Waters, Sewage and Sewage Effluent Using Adsorption-Elution Methods

Water Type	Initial Volume V_i (liter)	Sample Pretreatment	Adsorption Conditions	Filter Materials	Virus	Initial Virus Titer (PFU)	Eluate	Reconcentration Method	Final Volume V_f (ml)	Recovery (%)	Flow Rate Q (liter/min)	Ref.
Sewage	3.78	Fiberglass prefilter, cellulose nitrate membrane, ion exchange	0.05 M MgCl$_2$	Cellulose ester	Poliovirus 1	$\sim$4x10^6	Melnick's B+ 2% FCS; trituration	ND[a]	4.0	80	NR[b]	100
Pilot filter activated sludge influent and effluent	3.8-19.0	Fiberglass prefilter, cellulose nitrate membrane, protamine sulfate precipitation	0.1 M MgCl$_2$	Cellulose ester	Poliovirus 1	2 to 5.9 x10^3	Eagles basal medium + 10% FCS; trituration	Al(OH)$_x$ floc	3.8-10.0	0.6-5.0	NR	101
Pilot filter sludge influent and effluent	3.8	Serum treated fiberglass prefilter, ion exchange[c]	0.05 M MgCl$_2$	Cellulose ester	Poliovirus 1	$\sim$7.6x10^6	Eagles basal medium + 10% FCS; trituration	ND	4.0	26.0-31.0	NR	101
Lake and river water	0.5	Fiberglass prefilter	1200 mg/ℓ CaCl$_2$	Cellulose ester	Poliovirus 1	26-37	3% beef extract	ND	10.0	53-100	NR	102
Sea water	37.8-340	ND	ND	Gauze pad (cotton)	Poliovirus 1	1.9x10^6-1.7x10^7	Sample expressed; adjusted to pH 8.0+5% FCS	ND	570-675	15.0-19.4	3.78	78

Estuarine (artificial and natural)	380-420	Fiberglass cartridge filter	1200 mg/ℓ $MgCl_2$ + pH 4.5	Cellulose ester	Poliovirus 1	4.75×10^6-7.12×10^6	5× Nutrient broth in carbonate-bicarbonate pH 9.0	Two-phase[d]	2-3	42-97	9.5	86
Raw sewage	0.1	Centrifugation; blending recentrifugation	0.05 M MgCl2	Cellulose ester	Poliovirus 1 (other enteroviruses)	10-84	3% beef extract, pH 8.0	ND	5.0	88-97	Suction	103
Artificial sea water	19.0	Orlon wound fiber depth filter	0.05 M $MgCl_2$	Cellulose acetate wound fiber depth filter	Poliovirus 1	1.9×10^6-1.4×10^8	Ca-Mg free PBS, pH 11.5	ND	1600	85-95	7.56	104
Channel water and sewage plant effluent	19.0-606	Orlon and tween treated cellulose acetate wound fiber depth filter	0.0005 M $AlCl_3$, + pH 3.5	Fiberglass fiber depth filter + epoxy fiberglass disc	Natural	—	0.05 M glycine, pH 11.5	Resin treatment and Ads-El.[e]	10	Natural virus	7.56	105
Raw sewage	60	Orlon and tween treated cellulose acetate wound fiber depth filter	0.0005 M $AlCl_3$, + pH 3.5	Fiberglass wound fiber depth filter	Poliovirus 1	1.6×10^9	0.05 M glycine, pH 11.5	ND	1000	81	2.87	106

Table V, continued

Water Type	Initial Volume V_i (liter)	Sample Pretreatment	Adsorption Conditions	Filter Materials	Virus	Initial Virus Titer (PFU)	Eluate	Reconcentration Method	Final Volume V_f (ml)	Recovery (%)	Flow Rate Q (liter/min)	Ref.
Estuarine	190	ND	0.0005 M AlCl$_3$ + pH 3.5	Fiberglass wound depth filter	(1) Poliovirus 1 (2) Poliovirus 1 Echovirus 7; Coxsackieviruses A9,B3	(1) 1.8×10^7–5.6×10^7 (2) 10–6.5×10^3	0.05 M glycine, pH 11.5	FeCl$_3$ ppt	15-20	(1) 22-76 (2) 23-80	7.56-15.0	107
Estuarine	400	ND	0.0015 M AlCl$_3$ + pH 3.5	Fiberglass wound fiber depth filter + pleated fiberglass paper-epoxy	Natural virus	—	0.05 M glycine, pH 11.5	Al(OH)$_x$ floc + hydroextraction	20	Natural virus	24	108
0.05%-0.1% raw sewage in tap water	(1) 1	ND	Earles BSS (1:100) + pH 6.0	Talc-Celite layer	(1) Poliovirus, Echovirus 1, Coxsackievirus B3, Reovirus	(1) 1.8×10^4–2.2×10^5	10% FCS saline, pH 9.0	Hydroextraction	10	(1) 82-93	NR	109

Estuarine (1) 177–385, (2) 189–389	0.01% Celite; fiberglass prefilter	0.0005 M $AlCl_3$, + pH 3.5	(1) Epoxy-fiberglass discs, (2) Cellulose ester	Polio-virus 1; (2) Pre-titered indigenous virus	5 × Nutrient broth in carbonate-bicarbonate pH 9.0	(2) 20	(1) ND, (2) a. ND b. two-phase	(1) 3.17×10^5–3.22×10^5, (2) a. 3.22×10^5–6.52×10^5 b. 45–692	(2) 13–22	(1) 1000, (2) a. 1000 b. 2–3	(1) 59–74, (2) a. 35–63 b. 0.4–3.9	(2) 86–87; NR	98
Estuarine 189	Tween or beef extract treated orlon and cotton wound-fiber depth filters	0.05 M $MgCl_2$ + pH 6.0[f]	Fiber-glass wound fiber depth filter + epoxy-fiberglass disc	Natural adeno-virus	3% beef extract, pH 9.0		two-phase	—		2.0	Natural virus	3.78	110
Estuarine 378	ND	0.0015 M $AlCl_3$, + pH 3.5	Pleated fiber-glass paper-epoxy	Polio-virus 1	0.05 M glycine, pH 11.5		$Al(OH)_x$ floc or $FeCl_3$ ppt	8.5×10^1–1.1×10^7		10–100	48–58	22.68	111

[a] Not done.
[b] Not reported.
[c] Virus seeded after pretreatment.
[d] Aqueous two-phase polymer separation.
[e] Reconcentration by adsorption to and elution from smaller filters.
[f] Samples with salinities of 21 °/oo or greater did not require salt.

(at the operator's option) be passed through an anion exchange column to remove anions. In a 189-liter experiment using dechlorinated tap water seeded with 1200 infectious units of poliovirus, recovery of 83% of seed virus was attained after elution with borate or Tris buffer containing fetal calf serum. In determining the fate of virus in the system using high virus concentrations, it was established that virus was present in the filtrate of each clarifying filter used but no virus was detectable in the filtrate after passage through the cellulose ester adsorbent. Upon elution 65-90% of the virus was recoverable. The inability to recover higher amounts of virus was attributed to the formation of ferric complexes on the adsorbent filters. Although virus was adsorbed to these complexes in the presence of $MgCl_2$, they could not be efficiently eluted. The evaluation of clarifiers to remove $FeCl_3$, phosphates and carbonates without severe removal of virus found that the cotton-wound filter was the choice textile filter. Tween-80 treated filters removed ferric complexes but generally allowed passage of more virus than untreated controls.

In a further development of the portable virus concentrator, Wallis and Melnick [95] used orlon and Tween-80 treated cotton depth filters to clarify waters of particulates and complexed metals, respectively. Adsorbent filters used were fiberglass or cellulose acetate. Adsorption included the use of $MgCl_2$ as an enhancing salt. Virus was eluted from both adsorbents with one liter of Ca-Mg-free phosphate buffered saline, pH 11.5. Reconcentration was by adsorption of elution and resulted in recoveries of 78%.

The use of clarifying filters to remove suspended material and protect adsorbent filters from premature fouling resulted, therefore, in the ability to process large volumes of water at relatively high flow rates with low process times (> 1100 liter/hr). Application of systems containing clarifying filters to laboratory studies using artificial sea water and actual field studies of estuarine environments constituted a critical examination of a model system. Since sea water might contain sufficient salts to allow adsorption of virus to both clarifying and adsorbent filters, examination of a variety of substrates was necessary to determine the fate of virus in filter systems before actual field application.

In the evaluation of natural and synthetic wound fiber filters for the concentration of viruses from sea water, Metcalf et al. [104,105] found that orlon, polyester and polypropylene filters removed the least amount of virus from seeded artificial seawater samples. Fiberglass, cellulose acetate and dynel allowed the least amount of virus to break through in the filtrates and, hence, were chosen as virus adsorbents. In trial runs with 19 liters of artificial sea water, orlon filters were used for clarification and fiberglass or cellulose acetate were used for adsorption. Two trials involving the cellulose acetate filters as adsorbent resulted in almost complete passage of virus through the orlon clarifier and 90% average recovery of the input virus upon elution from the cellulose acetate adsorbent. In one trial using the fiberglass filter, 97.5% recovery of input virus was reported.

In this same study, cellulose acetate and fiberglass filters were challenged in parallel with 2.5-40 gal of virus-seeded river channel water clarified with an orlon filter. Filtrates of the fiberglass filter showed little virus penetration, but cellulose acetate filters exhibited breakthrough of virus with successively larger challenge volumes. The effect of MCC adsorption to cellulose acetate filters was suggested as the reason for the efficiency decreases with increases in the volume size tested. Similar tests using 2.5-160 gal of seeded estuarine water resulted in virus breakthrough in filtrates from both adsorbents. However, no difference in adsorptive capability with larger volumes was evident.

Further study of this system [105] determined that polluted natural waters could be clarified without significant virus losses when an apparatus using the series of previously reported clarifying filters was employed. Also, natural waters containing salts still required addition of $MgCl_2$ or $AlCl_3$ for efficient virus adsorption. In a two-step procedure involving readsorption of primary eluates to smaller filters for reconcentration, passage of the primary eluate through cation resin exchangers to remove MCC allowed for increased detection of naturally occurring viruses in sea water and sewage plant effluents.

Hill et al. [98] used celite as a filter aid in the processing of poliovirus-seeded laboratory samples (200 ml) and raw estuarine waters (57-378 liter). In a preliminary study without virus inputs, filter fluxes were examined with 8.0 JTU estuarine water with and without celite added to the water sample. In a 30-min period the control without celite yielded 5.9 gal, and the treated sample allowed passage of 34.9 gal. Conditions for adsorption of virus from turbid waters to epoxy fiberglass filter discs or cellulose ester membranes were pH 3.5 with $AlCl_3$ added to 0.0005 M. Celite was added at a concentration of 0.01%. Elution was performed using 5x nutrient broth in carbonate buffer, pH 9.0. In laboratory-based studies in which sea water was made turbid with marine silt (5, 10 and 15 JTU) virus was added at a rate of 1.06 $\times 10^4$ PFU/200 ml sample. Mean virus recoveries were: cellulose ester membrane, 73%; epoxy-fiberglass filter, 78%; epoxy-fiberglass filter with fiberglass prefilter, 72%. Large volume experiments (57-378 liter) were seeded with high multiplicities of poliovirus; the turbidity tested was either 8.5 or 15 JTU and virus recoveries ranged from 35 to 74% using filters employed in the preliminary work. High turbidities, up to 35 JTU, were examined using virus inputs of 56-692 PFU in 100-gal samples. Virus recoveries were < 1 to 2.2% but virus was detected in every experiment. Filter clogging at turbidities of 35 JTU limited to 25-30 gal the amount of the original sample that could be processed. Increases in turbidity to 60-80 JTU severely limited the amount of sample filtered to 15-30 gal and resultant virus recoveries were 0.8-3.9%.

Recognizing the potential for the possible presence of viruses other than those belonging to the enterovirus group in surface waters, Fields and Metcalf [110] developed an amended methodology for isolation and concentration of adenovirus from seawater (estuarine) samples. Preliminary laboratory studies examined the current methodology to determine at which points adenovirus may be inactivated due to the conditions presented in the various procedures. Adenovirus 5 was found to be sensitive to pH < 6.0 and > 9.0 and the methods employed were then tailored to maintain optimum pH conditions for virus survival since the current methods exceeded these pH limits. Pretreatment of orlon and cotton wound-fiber depth filters with Tween-80 or beef extract allowed passage of adenovirus during clarification of water samples. Adenovirus adsorption to a series of one fiberglass wound-fiber filter followed by an epoxy-fiberglass disc were established at pH 6.0 with a final concentration of 0.05 M $MgCl_2$. Salts addition was necessary in tapwater experiments but seeded estuarine waters with salinities of 21 $^o/_{oo}$ did not require added salt for adsorption enhancement. Elution was performed with a 3% beef extract solution at pH 9.0. Reconcentration of beef extract eluates was accomplished by an aqueous two-phase polymer system modeled after Wesslen et al. [113] since the pH of the method was within acceptable limits and the procedure was considered to be selective for larger viruses. Application of the developed methodology to a 190-liter waste treatment plant effluent sample resulted in 116 isolates, 106 of which were identified as adenovirus.

A modified portable virus concentrator utilizing pleated membrane filters [90] was used with turbid estuarine waters (6-19 JTU) seeded with poliovirus and resulted in overall recoveries of 50% of the input virus from 378-liter samples [111]. Optimum adsorption of virus to adsorbent filters was found at pH 3.5 in the presence of 0.0015 M $AlCl_3$ used for processing tapwater samples [90], the increased $AlCl_3$ concentration allowed consistent and complete adsorption of virus to membranes and the combination of a 3.0- and a 0.45-μm adsorbent filter in series allowed processing of > 400 gal of estuarine water in 90 min.

Reconcentration of primary eluates from natural samples in two-step adsorption-elution methods are plagued with floc formation after acidification of the primary eluate [99]. This in effect concentrates membrane coating components and makes complete virus adsorption in the second step extremely difficult and inefficient. This problem of organic floc formation is particularly notable in the processing of sewage samples. Farrah et al. [114] used the portable virus concentrator for the first step in the concentration of seeded poliovirus from secondary sewage effluents. Elution with pH 10.5 glycine buffer was followed by modification of second-step concentration methods designed to circumvent precipitable materials from interfering with virus recovery. The organic load of the primary eluent was measured spectrophotometrically and an appropriately sized column of charcoal (BPL grade,

Calgon Corp.) was constructed. Samples were treated with ethylenediamine tetraacetic acid and passed through the column. The eluate was then passed through a resin column after which it was adjusted for reconcentration by adsorption-elution. Initial eluates were reconcentrated 120X and yielded an average recovery of 40% of their virus content.

Another study [107] examining turbid estuarine waters (4-30 JTU) processed through clarifying filters at various combinations of pH turbidity and salinity determined that 75% of virus was lost to these filters in laboratory studies. A virus adsorbent system containing an unprotected fiberglass depth filter and a series of epoxy-fiberglass filter discs was challenged with approximately 190 liters of turbid estuarine water seeded with 1.8×10^7 to 5.6×10^7 total PFU. Upon elution, glycine eluates were reconcentrated by $FeCl_3$ precipitation and yielded recoveries of 22-76%. Similar experiments with other representative enteroviruses seeded at low multiplicities (10-2000 PFU total) into 170-190 liters of estuarine water resulted in a grand mean of 41% recovery. As in the study by Metcalf et al. [105], added salt was necessary, presumably to overcome the effect of MCC.

Talc-Celite layers sandwhiched between filter paper with a fiberglass prefilter were the basis of a virus concentration technique developed by Sattar and Ramia and tested with potable waters [115]. Use of this method with tap water resulted in recovery ot 58-64% of input virus from 100 to 1000 liters containing as low as 1.2 PFU/l. Application of the technique to 20-liter water samples contaminated with 0.05% or 0.1% raw sewage was successful in recovering 86.0% of pretitered indigenous virus from the sample [109]. The system involved adjustment of water samples to pH 6.0 and a 1:100 final concentration of Earle's balanced salt solution as a source of cations. Samples were then filtered through talc-Celite layers. Elution was performed with 10% fetal calf serum in saline, pH 9.0 and reconcentration by hydroextraction with polyethylene glycol completed the procedure.

Some Field Trials

A virus profile of communities along the Texas coast was assembled by Goyal et al. [116]. Methods used were those of Wallis and Melnick [95] with modification of the portable virus concentrator described by Payment et al. [111]. In addition to virus isolation, the total coliform, fecal coliform and salmonella content of both overlying waters and sediments was monitored along with pH, salinity and turbidity. Each of the six sampling sites in the study yielded a positive virus sample at least once during the year. A total of 1078 virus isolates was made and 152 of these were identified as enteroviruses. Poliovirus 1 was isolated more frequently than all other viruses but echoviruses and coxsackieviruses were also isolated. Correlation of virus presence and most probable number (MPN) data were made: Virus was isolated from waters negative for fecal coliforms and 44 samples were positive for a virus although the fecal coliform content was at a level of acceptable quality for shellfish harvesting water ($<$ 70 fecal coliforms/100 ml). There was, however, a strong correlation between virus isolation and total coliforms in bottom sediments and this measure was suggested as a better indicator of viruses than either total coliforms in water or fecal coliforms in water or sediment.

Vaughn et al. [117] conducted a monthly examination of indigenous enterovirus and coliform bacteria from lake, creek and marine waters over a one-year period. Lake and creek waters yielded two positive samplings each. Virus was isolated from marine embayments which were open and closed to shellfish collection; coliform profiles in these areas were comparatively low and did not differ significantly. Although a variety of enteroviruses was demonstrable in wastewater effluents regularly discharged into one of the embayments tested, only one sample yielded virus. Various mechanisms of removal were suggested as the reason for the difference between virus numbers in the effluents and the lone positive isolate from the bay during the survey.

Adsorption-elution methods were used to evaluate the ability of Orange County's (CA) Water Factory #21 to remove virus [118]. The factory includes a 5-mgd reverse osmosis plant in an advanced water treatment train. Optimization of virus detection methods evolved over a period of time and resulted in a procedure similar to those suggested in *Standard Methods*. Recovery efficiency of the virus concentrator system was evaluated with 19-liter influent samples and 756-liter effluent samples and resulted in average recoveries of 37.2± 17.9% and 56.5± 91.3%, respectively. Native virus detected in influent samples resulted in a median concentration of 5.0 PFU/gal (1.32 PFU/l). Seasonal variations occurred with a peak in October. Of 123 samples of the plant's chlorinated effluent, two were virus positive and occurred on days in which plant operations were atypical. Poliovirus 2 was recovered from samples with high turbidity (2.3 TU), carbon fines and the possible absence of a free chlorine residual. A second isolate was identified as echovirus 7 and occurred on a day with exceptionally high virus concentrations in the influent samples (169 PFU/gal, or 44.64 PFU/l) and, again, a possible combined chlorine residual.

The above examples demonstrate the ability to recover viruses from waters of various quality using adsorption-elution but are not able to (or were not designed to) reflect the degree of efficiency with which viruses are isolated and determine if all viruses of interest can be recovered. Recent studies have shown that any one method involving certain prescribed conditions may not be sufficient for the isolation of all virus groups of interest or even members of the same group of virus.

La Belle and Gerba [119] studied the factors controlling the degree of enteric virus adsorption to marine sediments. Several enteroviruses and one rotavirus (simian) were examined under different conditions of pH, salinity and amount of soluble organics present. Results showed that most entero-viruses will readily adsorb to sediments (> 99%), however, instances of vari-able adsorption can occur. For example, coxsackievirus B4 differed signifi-cantly in adsorption behavior under similar conditions when compared to poliovirus type 1 since little adsorption of the coxsackievirus occurred. Alter-ation of pH or salinity determined that only one virus (echovirus) was able to desorb to any degree under the conditions tested. Gerba et al. [120] also found type and strain dependence of enterovirus adsorption to activated sludge, soils and estuarine sediments.

An evaluation of the tentative standard method using a glycine eluate for the concentration of laboratory and naturally-occurring viruses seeded into tap water determined that the method was generally inefficient [121]. Entero-virus recovery efficiencies by the two-step filtration-elution method ranged from 0.7 to 32%. A wide variety of recovery efficiencies among enteroviruses was also determined and it was suggested that good recoveries for one virus member of a particular taxonomic group did not imply the same results for other members of the group.

In this same study [121], reovirus, adenovirus and parvovirus also were examined. Adenovirus type 5 and reovirus type 3 were not recovered and the parvovirus (minute virus of mice) was recovered with only a 5% average effi-ciency. Sensitivity to the alkaline pH necessary for elution was suggested.

In a subsequent paper, modifications of this tentative standard method were proposed including moderation of elution pH by use of beef extract and examination of alternative adsorbent filters to circumvent the difficulty of eluting certain viruses from filters after use of $AlCl_3$ as an enhancing salt [122]. Others have demonstrated that adenovirus [110] and reovirus [123] were extremely sensitive to high alkaline conditions and that modification of techniques or alternative methods of reconcentration were necessary for efficient virus recovery.

Sobsey and Jones [93] investigated the use of a more electropositively charged filter composed of a diatomaceous earth-cellulose-charge modified resin mixture as an alternative method (see Refs. 93,94 and text). The filters were reported to have a net positive charge up to a pH of 5-6 and adsorbed poliovirus at pH 7.0-7.5 without the addition of cations. Elution was per-formed using a 3% beef extract solution, pH 9.0. Comparison with conven-tional net-negatively charged filters showed that under similar conditions the more positively charged filters recovered 64 and 22.5% of the input virus in one- and two-step procedures, respectively, whereas the tentative standard method only recovered 5% of the input virus.

Further comparison of electropositively charged filters with conventional epoxy-fiberglass filters in large-scale (1000 liter) experiments determined that the overall results with both filter types were comparable. The use of the modified filter was reported to simplify the process since conditioning of water samples with acid and cationic salts was unnecessary [94].

In conclusion, the usefulness of the adsorption-elution technique has been demonstrated over the years of its development by application to natural surface waters in laboratory and field studies. This critical test of methodology has shown that human-associated viruses can be recovered from large volumes of water samples and some degree of confidence can be drawn from the results. In light of the variable behavior of different viruses under similar conditions, whether due to the nature of the virus itself or the conditions imposed by a particular concentration technique, further study of this methodology is clearly warranted.

Membrane Filtration Methods

Various membrane filtration methods based on solute size exclusion have been evaluated as concentration methods for waterborne viruses. In four methods—reverse osmosis (also called hyperfiltration), ultrafiltration, hydro-extraction and electropolarization—the smaller solvent molecules are removed from the virus-containing feed stream through controlled pore-size polymeric membranes by various driving forces. Pressure, electrical or osmotic (chemical) potential differences are imposed across the semipermeable membrane to effect separation or dehydration.

Although synthetic membrane barriers have been studied and used for over a hundred years, two recent developments have spurred interest in their use. The first was the development of the high-flux asymmetric polymeric membrane by Loeb and Sourirajan in 1962 [124]. The second development involved the realization that by optimizing the fluid flow conditions adjacent to the membrane-solution interface high permeation fluxes could be attained by minimizing cake formation and concentration polarization [125]. Thus membrane modules with tangential flow across the membrane surface were introduced. These included plate-and-frame spiral-wound, tubular, capillary, and even hollow fine fiber modules. In spite of these advances, many investigators have chosen to ignore the significant advantages in using asymmetric polymer membranes with tangential flow hydrodynamics.

Besides Loeb and Sourirajen's asymmetric cellulose acetate membrane [124], many new membrane (materials) now are available commercially. The most attractive asymmetric membranes for virus concentration include polysulphone, polyamide (Nomex) and the new proprietary thin-film composite membrane recently described by Cadotte et al. [126]. These membranes are characterized by high water flux, excellent solute rejection, stability over a wide range of pH, a degree of chlorine resistance (except for the polyamide membrane), stability over a larger temperature range and low biodegradability. These membranes can usually be tailor-made to a predetermined specification with respect to molecular weight retention, and are commercially available at cost in flat sheets and hollow capillaries with the skin or active layer on the bore-side. A cross section of a typical ultrafiltration hollow fiber is shown in Figure 2.

Simultaneous to the appearance of asymmetric membranes, soluble alginate ultrafilters were used for virus concentration from waters that normally would cause membranes to clog. After the concentration step the alginate ultrafilters are dissolved along with the retained viruses in small volumes of nontoxic (to the virus and assay cultures) solvent for subsequent virus assay. Since the use of sodium alginate filters has been limited to rather small initial volumes (< 10 liter), and since not much new work has been reported using these membranes, the reader is referred to prior review chapters for additional details on performance and limitations of the method [127,128].

Results obtained in concentrating virus (phage and enteroviruses) with each of the methods referred to above are summarized and discussed below.

Reverse Osmosis

It was recognized soon after Loeb and Sourirajan developed the anisotropic cellulose acetate desalination membrane that other solutes besides ions could effectively be rejected [125]. This included viruses, since several workers were interested in using the reverse osmosis process in advanced waste treatment to polish the water prior to reuse. In fact, 5.0 mgd of treated secondary effluent after passing through a battery of pretreatment steps (chemical coagulation, sand and activated carbon filtration) is currently being processed by a spiral wound reverse osmosis (RO) system. No significant virus titers have been detected in the permeate [118].

One of the first direct attempts to evaluate the RO process for concentrating viruses from a phosphate-buffered saline (PBS) and a lake water was conducted by Sorber et al. [129]. Three grades of asymmetric cellulose acetate membranes were used, ranging from a low flux/high salt rejecting membrane to a high flux/non-salt rejecting membrane. Their feed stream was operated at 600 psi and 500 ml/min recirculating flow. Unfortunately, the test cell (2 in^2 membrane area) was of the impingment-type where the fluid containing the virus impacted directly normal onto the membrane. This type of test cell has lost favor since it has a nonlinear pressure distribution along the flow path. One of the main conclusions of their study in 1972 was that even the so-called ion-rejecting membranes used in reverse

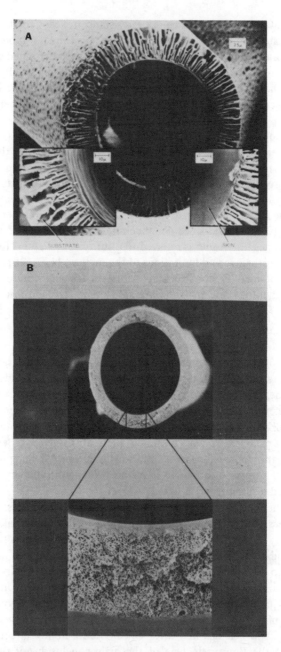

Figure 2. Hollow fibers with dense fibers with dense layers (skin) on the pore-side (a) polysulphone (courtesy of Amicon Corp.; Danvers, Mass., USA) and (b) polyamide (courtesy of Berghof Forschungs-institut, Tubingen, Germany). The inside diameter of the polyamide fiber is 1.5 mm.

osmosis allowed poliovirus (300Å) to penetrate randomly through the membrane. This conclusion is misleading, since today it is known that most of the commercial reverse osmosis membranes available then were covered with surface imperfections such as micro-pin-holes. Large subsurface voids were also found randomly all over the membrane area. Scanning micrographs by Strathman and Kock [130] and Belfort (unpublished) have shown that the surface micro-pin-holes could connect to the subsurface voids when exposed to relatively high pressures (30-40 bar) resulting in convective short-circuit passages through the membrane. Membrane manufacturers today have learned how to eliminate the subsurface voids, thereby reducing the possibility of virus penetration significantly.

In conclusion, virus transport and penetration would be minimized with tangential flow hydrodynamics and the new void-free commercial reverse osmosis membranes. The use of RO membranes, because of their excellent total solute rejection, poses another and more serious problem for virus concentration and subsequent assaying. The co-concentration, with the desired virus, of solutes that either inactivate the virus, aggregate the virus and/or are toxic to the cells during the assay procedure, could seriously affect virus recovery and quantification. These effects have been observed by Ellender and Sweet [131] with reverse osmosis cellulose acetate membranes and osmotically induced water flow.

Hydroextraction

Ellender and Sweet [131] have evaluated the much slower and usually less efficient osmotic driving through RO membranes to dehydrate virus containing distilled and tap water. Loeb [132] discovered that the theoretical osmotic pressure across a desalination membrane was not approached experimentally due to losses from the long path-length through the whole membrane (skin and porous understructure). The hydroextraction process, because it depends on diffusion, is much slower than the convective pressure-driven reverse osmosis and ultrafiltration processes. Some results from Ellender and Sweet's paper [131] are summarized in Table VI. They were able to dehydrate 10-20 ℓ of distilled or dechlorinated activated carbon filtered tap water 40-100 fold in about 2-4 hr. Above ∿ 1000 PFU/ℓ they obtained virus recoveries > 50%. However at < 1000 PFU/ℓ with tap water, recoveries for poliovirus dropped to 0-35% and reovirus was not recovered at all. They intimated that co-concentrates in the tap water were responsible for the inefficient virus recoveries. Prefiltered tap water resulted in increased virus recoveries.

Any process requiring prefiltration of particulates could lose substantial virus counts in natural waters due to the well known virus-associated-solids

Table VI. Summary of Poliovirus Concentration Experiments Using Salt Rejecting Asymmetric Cellulose Acetate Membranes

Concentration Method[a]	Water Type	Configuration	Retention (MW cutoff) or Rejection (%)	Water Flux (ml cm^{-2} min^{-1})	Operating Pressure (psi)	Total Virus (PFU)	Initial Volume, V_i (ℓ)	Final Volume, V_f (ml)	Recovery (%)	Ref.
RO	Lake water	flat membrane impinged flow	82-93% conductivity rejection	$5.09^{10^{-2}}$-8.49×10^{-2}	600	3.2×10^2-3.2×10^3	0.8-8.0	--	98.81-100	127
RO-UF	PBS[b]	same	50% conductivity	1.41×10^{-1}	600	8×10^8	11.4	--	99.999	127
RO-UF	Lake water	same	10-80% conductivity	8.49×10^{-2}-3.68×10^{-1}	600	3.2×10^3-9.1×10^8	5.6-11.4	--	99.77-99.999	127
UF	Lake water	same	40% sucrose	1.7×10^{-1}	600	63×10^8	8	--	>99.999	127
OU	Distilled	cross-flow	90% conductivity	2.00×10^{-2}-2.50×10^{-2}	590-1180	1×10^2-1×10^5	10-20	~100	50-100	131
OU	Tap	cross-flow	same	same	same	~3×10^3	10	~100	0-30	131

[a]RO = reverse osmosis, UF = ultrafiltration and OU = osmotic ultrafiltration.
[b]Phosphate buffered saline.

problem. This process suffers further by virtue of its slow dehydration rate and the co-concentration of potential cytotoxic and other flocculant inducing solutes.

Electro-Polarization (With Reverse Osmosis Membranes)

In the late 1960s Bier and coworkers built an electrophoresis module to concentrate bacteriophage Tl [133]. They obtained nearly complete phage recovery with 100-fold concentration from water. Sweet et al. [134] and Ellender et al. [135] concentrated poliovirus 1 from initial feed volumes of 0.5-1 liter of buffered distilled water with an average recovery of 66%. The major problem with their experiments were conceptual. For example, in one experimental setup, Sweet et al. [134] tried to electrically polarize the virus *counter* to the permeation stream and found (not surprisingly) that extremely high impractical voltages were probably necessary to do this successfully.

In another experiment, they relied on an electric field to draw poliovirus through a (bacterial) depth filter from a flowing stream in a cell only $\sim$ 6.5 in. long. Adsorption on the filter and anode severely limited virus recoveries, while the expected virus carryover also occurred. Belfort and coworkers are currently studying the potential of moving viruses by cross-polarization fields (convective and electric) out of the flow field and into the channel wall-fluid interfacial region. Later depolarization will recover the viruses [73]. (Convective polarization of viruses is discussed in the section on ultrafiltration.)

Sweet and coworkers [134] and Bier and coworkers [133] have shown the potential of electropolarization for virus concentration; however, the process has yet to be developed into a viable, practical and efficient method.

Sweet and Ellender also evaluated a combination process, electroosmosis, in which both osmotic and electrical potentials are imposed on the virus containing fluid simultaneously [136]. The electrical potential increases the water dehydration rate by electroosmosis. The viruses are drawn to the membrane by the potential field and later collected by polarity reversal. The results were disappointing when poliovirus and reovirus were concentrated from activated carbon filtered tap water. Volume reductions from 5 liter to $\sim$ 100-200 ml, with seeded virus at 242-422 PFU (polio) and 700-2000 PFU (reo) resulted in 33-100% poliovirus and 3-44% reovirus recoveries. Detecting low viral inputs proved to be difficult. Again co-concentrates were thought to be responsible for relatively poor results.

It appears that processes such as reverse osmosis (hyperfiltration), osmotic ultrafiltration and electropolarization (with RO membranes) all have the major disadvantage of co-concentrating, with the virus, solutes that could aggregate and inactivate the virus or prove cytotoxic to the cells during assaying. Clearly attractive are processes such as adsorption-elution and tangential-flow-ultrafiltration which allow some of these undesirable solutes to pass through the filter, thus reducing virus aggregation and inactivation and possible cytotoxic effects.

Ultrafiltration

The term ultrafiltration generally has been used for filtration with membranes that retain solutes larger than ionic species but smaller than submicron-size colloidal species (e.g., < 0.1 μm). Filtration with membranes such as the typical cellulose acetate/nitrate mixed ester membranes with pore sizes greater than 0.2 μm is commonly called microfiltration. These membranes differ from the highly anistropic skinned ultrafiltration and hyperfiltration (reverse osmosis) membranes in that they are relatively isotropic, have a wide-pore-size distribution, often act as depth rather than surface filters and have a spaghetti-type open porous structure. In contrast, ultrafiltration membranes remove solutes by surface sieving due to their dense skin (see Figure 2). Unfortunately some confusion exists in the literature because the term ultrafiltration was used when microfiltration was meant [137] and hyperfiltration was incorrectly used synonomously with ultrafiltration [131].

Since ultrafiltration removes solutes such as polymers and macrosolutes and passes ionic species, relatively low osmotic pressures as compared to hyperfiltration need to be overcome. Thus, the latter process operates at applied driving pressures (> 300 psi or 2.068×10^6 Nm^{-2}) much higher than the former (< 100 psi or 6.895×10^5 Nm^{-2}).

Two reasons favor ultrafiltration over reverse osmosis for virus concentration: (1) most important is that the more open pores would possibly allow the low molecular weight solutes to permeate the membrane thereby possibly reducing virus aggregation, inactivation and cytotoxic cell effects; and (2) much higher permeation fluxes can be obtained with ultrafiltration reducing the time to concentrate a given volume.

Some of the first studies to concentrate biological fluids and wastewater for virus enumeration were conducted in the late 1960s and early 1970s [127, 138]. Most of these studies treated < 100 ml feed volume in small magnetically stirred batch test cells. Nupen and Stander [139] used flat polysulphone membranes of 30,000 MW cut-off in batch cells to concentrate 10 liter of Windhoek wastewater reclamation effluent and showed that activated carbon and ion-exchange filters must not be used for the removal of viruses and pathogens. In one field study [140], a flat filter press module was used to detect seeded coliphage from large volumes of processed water (1860 gpd). Foliguet and coworkers [141] appear to have investigated ultrafiltration as early as 1971, and some results of their early studies are summarized in Table

VII. They were able to recover viruses from river and sewage using hollow fiber polymeric membranes from 20 liter of feed at a 90-95% recovery efficiency.

Sweet et al. [127] conducted a comprehensive evaluation of ultrafiltration membranes and configurations. They varied initial virus titer (low and high), molecular weight cut-off (10,000-300,000), membrane type (mixed polyelectrolyte polymers, polysulphone and cellulose acetate) and configuration (flat and hollow fiber). Their results are reviewed in Table VII (2a-6b). The main conclusion from their study was that excellent virus recoveries from distilled water are possible with high initial virus feed concentrations. However for low concentrations, virus loss perhaps due to virus-membrane surface interactions (termed adsorption) is a problem. They suggested that membrane pretreatment with 0.1% bovine albumin solution (Fraction V) enhanced recoveries of poliovirus ($>$ 90%). In a small system (114 liter) they dehydrated to 50 ml in about 2 hr using 5 ft^2 (2.32 m^2) of membrane area. This equals an excellent permeation flux of about 0.41 liter/m^2 of membrane area/min.

Although the results of the virus concentration studies using ultrafiltration of Sweet et al. described above were very encouraging, two major problems remained: (1) the virus-loss during concentration especially with the high-flux noncellulosic hollow fiber modules and (2) the membrane plugging or fouling problem with water other than the distilled water used by Sweet et al. [127]. In addition, virus concentration methodology appeared to be an art with little or no mathematical formalism to help model the process and characterize its dynamic behavior. With these problems in mind and the objective of describing the process from an engineering viewpoint, Belfort embarked on a comprehensive series of studies to evaluate, optimize and model tangential flow hollow fiber ultrafiltration (TFHFU) for clean (tap) and turbid feed waters.

1. Clean Water Studies. The initial objective of these studies was to repeat the results of Sweet et al. with deionized water and several different asymmetric hollow fiber membrane modules while keeping in mind the virus-loss problem. The studies were then extended to tap water increasing the initial feed volumes from 5 to 100 liters and decreasing the virus titer from 3000 to 39×10^{-5} PFU ml^{-1}. Results were analyzed according to the mathematical formalism presented earlier in this chapter. During these studies, techniques were continually being sought either to reduce virus-loss or recover virus from the system after dehydration with various backwash solutions. It soon became evident that hollow fiber membrane technology could be automated for relatively clean surface waters and modified for more turbid waters.

A summary of the results from three publications is presented in Table VIII [74,75,142]. The experimental set-up was similar for all three sets of

Table VII. Summary of Early Poliovirus Concentration Experiments Using Different Ultrafiltration Membranes

Water Type	Configuration	Membrane	Retention (MW cut-off)	Initial Volume, V_i (ℓ)	Final Volume, V_f (ml)	Water Flux (ml cm^{-2} sec^{-1})	Operating Pressure (psi)	Total Virus (PFU)	Virus Recovery Range, avg. (%)	Reference
1a Tap	flat	cellulosic	--	20	350	--	--	2000[a]	50-80	139 from 128
1b Filtered river	flat	cellulosic	--	50	350	--	--	2000[a]	25-50	139 from 128
1c Unfiltered river	flat	cellulosic	--	50	350	--	--	2000[a]	60-100	139 from 128
1d River and sewage	hollow fiber	polymeric(?)	10,000	20	50	--	--	--	90-95	Foliguet from 128
2a Distilled	flat	mixed polyelectrolyte (XM-300)	300,000	30	380	2.5	20-26	6×10^9	15	127
2b Distilled	flat	mixed polyelectrolyte (XM-300)	300,000	30	295	2.5	20-25	4×10^4	0.4-16 (4.6)	127
3a Distilled	flat	polysulphone (PM-30)	30,000	30	295	0.8	20-26	8.6×10^8	99.98	127
3b Distilled	flat	polysulphone (PM-30)	30,000	30	74	0.8	20-25	296	50-60 (56)	127
4a Distilled	hollow fiber	mixed polyelectrolyte (X-50HF)	50,000	4.2	33	0.1	15	8×10^7	>41	127
4b Distilled	hollow fiber	mixed polyelectrolyte (X-50HF)	50,000	40-114	<50	0.1	15	6.6×10^3	40-98 (66)	127
5a Distilled	hollow fiber	polysulphone (CP-10HF)	10,000	5	28	0.1	15	2.3×10^6	--	127
5b Distilled	hollow fiber	polysulphone (CP-10HF)	10,000	7 (avg.)	<35	0.1	15	180-1620	30-100 (46)	127

| 6a Distilled hollow fiber | cellulose triacetate[b] | 80,000 | 1.14 | 26 | 0.009 | 15 | 3.6×10^6 | 81 | 127 |
| 6b Distilled hollow fiber | cellulose triacetate[b] | 80,000 | 1.14 | <28 | 0.008 | 15 | 90-300 | 59-100 (89) | 127 |

[a]Most probable number of cytopathogenic units (MPNCPU).
[b]Symmetric membranes.

Table VIII. Summary of Poliovirus 1 Concentration by Symmetric Semidense Cellulose Acetate, Asymmetric Polysulphone and Asymmetric Polyamide Hollow Fiber-Capillary Membranes with Core Feed[a]

Membrane Type	Initial Feed Volume, V_f (liter)	Feed-water Type	Average Total Recovery, $\bar{R}$ (%)	Average Rejection, R	Average Permeability Coefficient $\bar{L}_p$ (ml cm^{-2}h^{-1}atm^{-1})	Average Backwash Recovery, $\bar{B}$ (%)	Backwash Solution	No. of Exps.
Symmetric Semidense Cellulose Acetate[b]	5	deionized	84±9	0.759	0.599	36±17	deionized	11
Asymmetric Polysulphone[c]	50	deionized	77±33	0.959	11.31	26±14	deionized	12
	5	deionized	61±22	0.963	13.25	46±22	deionized	4
	50	tap	52±14	0.846	7.61	38±19	deionized	4
Asymmetric Polyamide[d]	100	prefiltered tap	106±42	~0.330	16.76	101±42	0.05 M glycine at pH ±10.5	5
	100	tap	67±35	~0.270	13.61	65±34	0.05 M glycine at pH = 10.5	5
	100	tap	67±19	--	13.34	48±24	1% beef extract at pH = 9.0	4
	100	tap	80±20	--	13.59	--	1% beef extract at pH = 9.0 with organic flocculation	5

[a] The reported errors are for 1 standard deviation.
[b] Results from Ref. 74 with initial virus concentrations of 0.047 $<C_{f_o}<$ 3000 PFU ml^{-1}.
[c] Results from Ref. 75 with initial virus concentrations of 0.0066 $<C_{f_o}<$ 934 PFU ml^{-1}.
[d] Results from Ref. 142 with initial virus concentrations of 39x10^{-5} $<C_{f_o}<$ 71 PFU ml^{-1}.

results, except that a different module was used in each. The feed water was seeded with poliovirus 1 and recycled through the hollow fiber modules for multiple passages with a positive displacement pump. A vacuum system was attached to the permeate side of the hollow-fiber ultrafiltration modules to increase the permeation flux (dehydration) rate. To increase virus recovery at the end of the dehydration step, backwashing without or with glycine or beef extract solution at pH 9-10.5 was performed. In one set of results (see the last row in Table VIII), a second step concentration by organic-flocculation was performed on the final virus concentrate from the hollow-fiber ultrafilter. Additional details on the materials and methods used in each experimental set are available in the literature [142]. The symmetric semidense cellulose acetate and asymmetric polysulphone membranes with < 1.0-mm i.d. are arbitrarily called hollow fibers, while the asymmetric polyamide membranes with > 1.0-mm i.d. are arbitrarily called capillaries.

The results presented in Table VIII for tapwater virus concentration with the asymmetric polyamide capillaries represent a significant improvement over the cellulose acetate and polysulphone hollow fibers. Higher total average virus recoveries $\overline{R}$, higher average permeation coefficients $\overline{L}_p$, larger initial volumes V_f and lower initial virus concentrations C_{f_o}, were obtained for the latter capillary study. The values for the average virus rejection R for the capillary membranes were much lower than previously measured and, consequently, average virus recovery $\overline{B}$ from backwashing was much higher. Because of the low initial virus concentration for the capillary polyamide membranes, it was not expected that the permeability L_p (or flux v_w, cm sec^{-1}) would be reduced during dehydration due to the presence of the virus particles alone. Thus fouling due to the virus itself is not expected. (See the discussion on hydrodynamics, where the reduction in v_w ($=L_p\Delta P_{ln}$ mm/ 3600) is attributed to suspended solids in the Jerusalem tap water rather than the seeded virus.) Buffered backwash solutions containing glycine or beef extract were superior in recovering missing (i.e., adsorbed) virus after concentration by ultrafiltration than previously used deionized water.

In order to describe, optimize and compare hollow fiber performance, Belfort recently presented a detailed hydrodynamic analysis for the various flow conditions usually observed in virus concentration studies [73]. Some of this analysis, presented earlier in the theoretical section, will be applied to actual laboratory virus concentration studies with different hollow fiber modules.

a. Laminar flow in circular tube. Based on the manufacturer's supplied information and on the chosen operating conditions, the specifications for hollow fiber and capillary membrane modules are compared in Table VIII. Using these specifications several hydrodynamic parameters are calculated and presented in Table IX for laminar tubular flow [73].

Table IX. Calculated Hydrodynamic Parameters for Laminar Tubular Flow[a] Pressure Considerations[b]

Module	Feed Entrance Pressure, P_{f_o} (atm)	Internal Pressure Loss, ΔP_f (atm)	Log Mean Differential Pressure, $\Delta P \ell n\ mn$ (atm)	Shear $\dot{\gamma}_w$ Rate[c] at Membrane (sec^{-1})	Number of Fluid Passes,[d] P_f	Reynolds Number, N_{Re}
Miniplant 80	2.0	0.07185	0.964	1820	656	39-60
H10P10	2.7	0.3357	1.526	8488	14900	27-423
BMR 500515	2.0	0.011	0.995	838	1600	263

[a] See Ref. 143 for analysis and assumptions of laminar flow through a circular tube. Input parameters such as n, Q, L and d are obtained from Ref. 73.

[b] Internal pressure loss is given by $\Delta P_f = (128\ LQ)/(\pi\ nd^4)$ where maximum Q was used. The log mean differential pressure $\Delta P \ell n\ mn = (\Delta P_o - \Delta P_L)/\ln(\Delta P_o/\Delta P_L)$ where $\Delta P_o = P_{f_o} - 1$ and $\Delta P_L = P_{f_o} - \Delta P_f^{-1}$.

[c] For a Newtonian fluid, the laminar flow shear rate at the membrane-solution interface $\dot{\gamma}_w = -32Q/(\pi nd^3)$.

[d] $P_f = \bar{t}_T/t_1$ where t_1 = single pass time, and $\bar{t}_T$ = average time for the experiment.

All the Reynolds numbers N_{Re} in Table IX are clearly < 2100 and are well within the laminar regime, although N_{Re} for the capillary unit is more than four times that of the other units [73]. This high N_{Re} value is misleading as it cannot be used to explain the higher $\overline{L}_p$ for the capillary unit presented in Table VIII. Because of the large capillary inside diameter, a low value of the shear rate at the membrane-solution interface $\dot{\gamma}_w$ results (see Table IX). The higher $\overline{L}_p$ can rather be explained by the higher molecular weight cut-off and relatively high intrinsic water permeability for polyamide membranes as compared to most other (not cellulose acetate) membranes [144,145].

b. Virus removal during concentration. Table VIII shows for the polyamide capillaries that most of the virus is held up in the system and finally recovered by backwashing. Thus $\overline{B}$ values are close to $\overline{R}$, and R values are expectedly low. Since no virus was detected in the permeate, it is assumed that the well-known concentration polarization phenomenon due to permeation is responsible for this high virus loss. The low-shear rate, long flow path, and high average number of fluid passes P_f, for the BMR 500515 unit may explain the high tendency to adsorb or hold up virus particles [73]. It should be noted, however, that the average number of virus passes P_s is probably not equal to the average number of fluid passes P_f due to the polarization and retardation (R_s) of the viruses adjacent to the membrane-solution interface.

In Table X, the hold-up index $\overline{H}$ is compared with $(1-R)$ as a measure of virus hold-up during concentration. Wide possible variations due to experimental errors in both $\overline{B}$ and $\overline{R}$ could explain the relatively poor linear corre-

Table X. Virus Hold-Up During Concentration[a]

Module	Feedwater Type	Fraction Recovered by Backwashing, $\overline{B}/\overline{R}$	Hold-Up Index, H (cm^2)	$1-R$
Miniplant 80	deionized	0.429	0.04	0.24
H10P10	deionized	0.338	2.50	0.04
	deionized	0.754	0.25	0.03
	tap	0.731	2.50	0.15
BMR 500515	prefilterd tap	0.953	9.52	0.67
	tap	0.970	9.52	0.73
	tap	0.716	9.52	--

[a]Data are from Tables VIII, IX and Ref. 73 where $H = [\Delta V_p/(Ln)]$. A linear correlation of $\overline{B}/\overline{R}$ versus $\overline{H}$ and $1-R$, and H versus $1-R$ gives the following ($y = mx + b'$):

x	y	intercept b'	slope m	corr.coef. $\overline{r}$
H	$\overline{B}/\overline{R}$	0.525	0.036	.673
$1-R$	$\overline{B}/\overline{R}$	0.504	0.617	.734
$1-R$	$\overline{H}$	0.065	12.870	.921

lations of $\overline{B}/\overline{R}$ with $\overline{H}$ and $(1-R)$, i.e., $\overline{r} = 0.673$ and 0.734, respectively. The linear correlation between $(1-R)$ and $\overline{H}$, however, is good ($\overline{r} = 0.921$). This suggests the use of $\overline{H}$ as a good qualitative estimate of virus hold-up during concentration.

We have previously shown that during concentration poliovirus 1 does not noticeably aggregate as a result of orthokinetic floccuation, but is removed from the feed water onto the membrane at about the same rate as the water permeation flux [146]. Equation 15 was used to show that $R\approx0$ and $\overline{C} \to 1$ or a constant virus concentration is maintained during dehydration.

An analysis similar to that presented for virus removal has been conducted for suspended and dissolved solids removal [73]. Virus adsorption onto and protection by suspended and dissolved solids is well-known. The presence of these solids can also clog the membrane and reduce the permeation flux v_w during the concentration process [73], although recent design and operational improvements have diminished this fouling effect even with turbidities as high as 975 NTU for hollow fiber ultrafiltration [147].

c. Solvent removal during concentration (temperature effects). Because the pump imparts energy into the fluid during recycle, the temperature increases from ~ 20 to $36°C$ when concentrating 100 liter. This is very fortunate as long as virus viability is maintained, as was the case in the above experiments, since the relative flux increases near the end of the experiment due to this temperature-increase. This compensates the expected flux-decline due to membrane clogging especially near the end of the experiment and results in better performance than expected. Without temperature compensation, the inverse reduced flux $\overline{v}_w$ $(T)^{-1}$ shows a maximum versus $\overline{V}_p$ in Figure 3. With temperature compensation, $\overline{v}_w^{-1}$ shows a linear correlation fit with $\overline{V}_p$ of $r = 0.9942$ as predicted by Equation 21.

From the dispersion analysis with the BMR 500515 modules, about 98.6% of the virus is removed from the feed fluid during one fluid pass [73]. The final dehydration rates based on the data in Tables VIII and IX are calculated in Table XI. For the tap water feed, the BMR 500515 module takes about 90 min to treat 100 liter in a 0.5 m^2 module.

The total fraction $f_T = \Delta N/N$ removed from the feed solution after P_f integer passes can be calculated from

$$f_T = [1 - (1-f_1)^{P_f}] \tag{25}$$

where $f_1 = 0.986$ is the assumed constant removal fraction for each pass.

To get $f_T = 0.999(9)$ only three fluid passes are needed. For similar total removals, Equation 25 shows that if $f_1 = 0.9, 0.8, 0.7, 0.5, 0.3$, and 0.1, then $P_f = 5, 7, 9, 15, 28$ and 94, respectively.

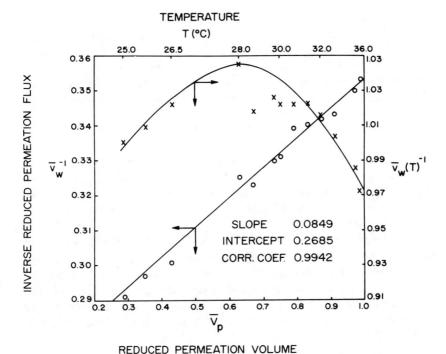

Figure 3. Variation of permeation flux with and without temperature compensation during dehydration with the Berghof module.

Table XI. Dehydration Rates for Ultrafiltration

Module[a] $(2\pi RLn)10^{-3}$ (cm^2)	Volume Permeated,[b] ΔV_p (ℓ)	Permeation Flux,[c] $v_w \times 10^3$ $(cm\ s^{-1})$	Volumetric Permeation Rate,[d] Q_p $(ml\ s^{-1})$	Duration of Permeation,[e] θ_T (min)
Miniplant 80 (8.000)	5	0.160	1.28	65
H10P10 (1.257)	50	4.795	6.03	138
	5	5.617	7.06	12
	50	3.226	4.06	206
BMR 500515 (4.948)	100	4.632	22.92	73
	100	3.762	18.61	90
	100	3.687	18.24	91
	100	3.756	18.59	90

[a] Active surface area = $2\pi RLn$, cm^2 (see Table II in Ref. 73 for data).
[b] Assume $\Delta V_p \approx V_{fi}$.
[c] Obtained from Tables VIII and IX.
[d] Q_p = Volumetric permeation rate per pass = $v_w(2\pi RLn)$, $cm^3\ s^{-1}$.
[e] $\theta_T = \Delta V_p \times 10^3/(Q_p\ 60)$, min.

From a dispersion experiment and subsequent analysis [73], it appears that only three fluid passes of each initial volume are necessary for $f_T = 0.9999(9)$. The actual number $P_f = 1600$ is shown in Table IX. This amazing difference is due to the fact that the module is surface-area limited. If, for example, the largest commercially available BMR 505015 module is used (i.e., 5 m^2), then $\theta_T \approx 9$ min with the actual P_f significantly lower to treat 100 liter and 1900 liter could conceivably be treated in about 2.85 hr.

In summary, the virus-loss problem has been solved with appropriate backwash and recovery methods; the flux-decline problem due to membrane fouling did not materialize with recycle systems of 100 liters feed volume due to temperature compensation effects; and a simple engineering formalism has been used to analyze the data, understand the dynamic phenomena and compare the performance of different hollow fiber modules. (Additional details are given in Ref 73.)

Virus concentration performance for the BMR 500515 module was found to be the best so far. Higher final average virus recoveries with higher average fluxes were obtained with this module than with the other two. It should be pointed out, however, that the HIDPIO module runs a close second and could, with some modifications in operational procedures (such as using organic additives during backwashing), attain higher recoveries. Its lower flux may still be a limitation.

Recently several investigators from the U.S. EPA in Cincinnati using 10,000 molecular weight cut-off flat sheet asymmetric (polysulphone?) membranes, which were pretreated with a flocculated 3% beef extract, yielded a poliovirus in 2-liter distilled water recovery of 66.6% [148]. No dynamic or geometric data were given precluding any comparison with the above results except to say that using the polyamide hollow fiber module, Belfort et al. obtained equivalent or better recoveries with 100 liters of tap water [73]. (The results are presented in Table VIII.)

2. Turbid Water Studies. Virus recovery from highly turbid wastewaters such as sewage is discussed under Other Methods. In this section virus concentration from relatively turbid (compared to tap and laboratory-prepared water) wastewaters is considered.

In a recent report Nupen and co-workers evaluated the efficiency of nonflow magnetically stirred flat ultrafiltration membranes [149]. Although they chose to use a nonflow system resulting in extensive polarization of solutes and extremely low permeation rates (40-72 hr to filter 50 liter of distilled, tap or final reclaimed effluent), their recovery efficiencies for polioviruses 1, 2 and 3, reovirus, a rotavirus and seven different morphological types of coliphages were reported at an average of 94%.

In recent studies, Belfort and Dziewulski (unpublished) have developed membrane flux-maintenance techniques resulting in a net decrease in flux decline of 50%. In preliminary fast axial flow experiments, they recovered 49% poliovirus from 50 liters of tap water initially containing kaolin turbidity of 95 NTU. It took 50 min to filter 50 liters of water increasing in turbidity from 75 to 975 NTU. The "lost virus" was not detected in the permeate and probably was associated with the 975 NTU kaolin turbidity.

3. Automated Hollow Fiber Ultrafiltration. By closing the permeate and exit filter ports during hollow fiber filtration, the filtration and permeate pressures will rise and permeation flux will decrease to zero as the two pressures equilibrate. If the filtration pump is halted and the filtration fluid opened and drained to atmospheric pressure, the accumulated pressure on the permeate side will backflush permeate through the membrane removing any adsorbed viruses (and other solutes) off the membrane surface and concentrating it in a relatively small volume. This relatively simple procedure is presented diagrammatically in Figure 4. A photograph of automated system is shown in Figure 5.

Results using this automated hollow fiber system for pyrogen removal and phage recovery from 20-liter tap and 100-liter Lake Michigan water have recently appeared in the literature [150]. A mathematical model to describe backflush recovery also has been developed and used to estimate the number of phage initially held onto the membrane. For bacteriophage f^2 concentration studies, initial phage concentrations ranged from 40 to 4900 PFU/ml, with concentration ratios from 14 to 450, and recovery efficiencies from 39 to 252% (the mean value for 9 experiments was 127%).

Belfort and Palusczek (unpublished results) have very recently used this dead-ended automatic hollow fiber ultrafiltration system to recover about $100\pm30\%$ poliovirus 2 in 192 ml from 3 liters of tap water in 30 min. These results compare favorably with the results of Nupen et al. [149] quoted above with a *significant* reduction in the duration of the experiment and with the advantages of automation.

In summary, the concentration of enteroviruses and phages from distilled, tap and lake waters, using hollow fiber ultrafiltration has progressed a long way since Sweet and Ellender first began their studies in the early 1970s. Virus-loss has been minimized, flux-decline problems can be controlled and do not appear to be at all significant for tap waters. A mathematical formulism has been developed to describe and understand the process. Finally, the process has been automated for relatively clean waters with push-button technology.

Future efforts will surely be in the direction of organic loaded and other turbid waters.

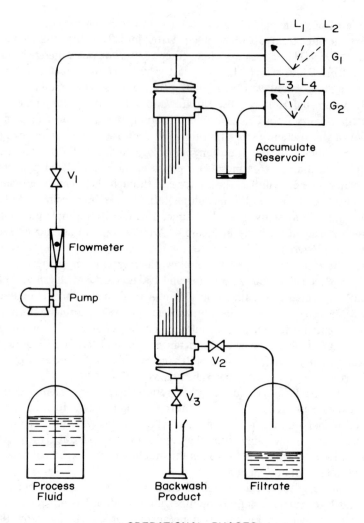

OPERATIONAL PHASES

	Filtrate	Accumulate	Backwash
Pump	Running	Running	Off
G_1	Rising	Rising	Falling
G_2	Stable	Rising	Falling
V_1	Open	Open	Closed
V_2	Open	Closed	Closed
V_3	Closed	Closed	Open

Figure 4. Flow sheet for automated hollow fiber ultrafiltration system.

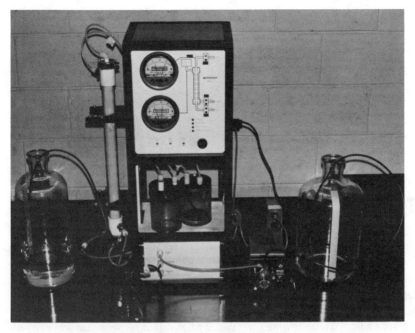

Figure 5. Photograph of automated hollow fiber ultrafiltration system.

Other Methods

A variety of chemical methods has been developed for the concentration of viruses from aqueous solutions. In general, these methods are most effective when used with small volumes and, therefore, have been primarily used as second-step concentration methods of eluents or for samples which have relatively high concentrations of virus.

A brief summary is presented here of methods used either as reconcentration procedures (suggested in *Standard Methods* [151]) or as a concentration method in their own right. Reviews containing additional details of these methods and closely related ones are available [128,152,153].

Aqueous Two-Phase Polymers Separation

Variations of this method involve dissolving two polymers in an aqueous solution under specified conditions of pH and ionic strength which will ultimately allow them to separate into two phases [44]. The mechanism of virus partition has been explained on purely electrostatic grounds [44], and the phase separation appears to be the result of the well-known "salting-out" effect. The ionic strength influences the partitioning by producing a chemical potential between the two polymer phases and, hence, determines movement of charged virus particles into one of the two phases. Concentration is accomplished by making the phase containing the virus small in comparison to the original sample volume [45].

The most common polymers used in the early application of two-phase systems were dextran sulfate and polyethylene glycol (PEG) with NaCl added to change the ionic strength and hence the chemical potential between the phases. Lund and Hedstrom [46] used this method to recover poliovirus from artificially contaminated sewage with a resultant concentration factor of 100 and a 99% recovery efficiency. Notably, results indicated that three times as many virus isolates were possible with the two-phase process when compared to simple direct plating of samples.

Shuval et al. [47] used a single-step method in which the lower dextran-rich phase was collected and KCl was added to a final concentration of 1M. The suspension was then centrifuged and the supernatant was assayed for virus content. Concentration factors of 52-200 were achieved and recovery ranged from 37 to 98%. In this study a trial of a two-step method was attempted in which NaCl rather than KCl was added to the collected dextran sulfate phase. After allowing for phase separation the upper phase was assayed for virus. Recovery of 51% was reported with a 274-fold concentration of the original sample.

Further study by Shuval et al. [48] of the two-step procedure resulted in concentration factors of 520 and recoveries > 99%. Both lower and interphases were collected and parallel examination of a one-step method was undertaken. Recovery efficiency for the one-step method was 87% and resulted in concentration of samples 173-fold.

Field application of a method similar to that developed by Shuval et al. [48] was made by Nupen [49]. All virus isolations during the study were made prior to final treatment of sewage. Control experiments using both one- and two-step procedures resulted in 22 and 44% recoveries, respectively.

Grindrod and Cliver [50] determined that dextran sulfate interfered with the detectability of some viruses in cell culture. An alternative method using dextran rather than sulphonated polysaccharide caused little or no inhibition of virus [51]. Recoveries with dextran ranged from 59 to 100% compared to a 0.001-100% recovery range when dextran sulfate was used.

Application of polymer two-phase systems has been limited as a result of several disadvantages [128]. One major disadvantage is the slow processing of the sample due to the time required for phase development (overnight) and the additional time needed for further concentration if a two-step procedure is used. Also, the system is not designed to recover virus from large volumes of water; its application is limited to the concentration of several liters. Nevertheless, its usefulness as a reconcentration method of beef extract filter eluates for viruses which are sensitive to pH extremes has been demonstrated [110].

In comparing the effectiveness of eight different concentration methods using 5-liter samples, Shuval and Katzenelson [154] found the phase separation was capable of consistently high recovery (about 100%) of poliovirus and echovirus. However, when wild-type enteroviruses were used, the two-phase separation method showed reduced recovery and indications of selectivity among the viruses used. Clearly the efficiency of the method is also directly dependent on the ionic strength and composition of the original water sample.

Adsorption to Precipitable Salts, Insoluble Polyelectrolytes and Minerals

Adsorption of viruses to precipitable salts, notably aluminum hydroxide flocs, has been used extensively for reconcentration of virus from filter eluates and is suggested as a second-step procedure in *Standard Methods* [151].

Aluminum [Al(III)] exists as the hydrated ion in an aqueous medium. With the coordination of hydroxyl groups, the positive charge of the ion decreases and the degree of repulsion between adjacent ions also decreases. The net result is polymerization of simple hydroxocomplexes and the formation of insoluble metal hydroxide precipitates [153]. Cookson [153] suggests several possible mechanisms for virus adsorption to aluminum hydroxides: (1) adsorption of viruses by coordination of OH^- groups on the virus and hydroxo-metal complexes, (2) electrostatic attraction between negatively charged virus and the positively charged aluminum hydroxide complex behaving as an amphoteric polyelectrolyte and (3) covalent bonding between aluminum atoms and sites on the virus surface. The last mechanism is the least likely since it should also occur with $AlPO_4$ precipitates which has been shown, however, to have a limited adsorptive capacity [56].

Early in the development of this methodology Wallis and Melnick [55,56] reported the ability of various viruses to adsorb to aluminum phosphate, aluminum hydroxide and calcium phosphate precipitates. Enteroviruses, including poliovirus, coxsackievirus and echovirus adsorbed well to aluminum hydroxide and calcium phosphate precipitates. Acid sensitive viruses such as herpesvirus, influenza and parainfluenza adsorbed only to aluminum phosphate. Adenovirus adsorbed only to aluminum hydroxide and reovirus did not adsorb to any of the salts tested. Wallis and Melnick also found that low multiplicities of poliovirus [< 1 PFU/ml] could be readily detected by this procedure.

In a review of chemical methods for virus recovery, England [152] reported the work of Palfi [155] in which calcium phosphate was used as the adsorbent for the isolation of virus from sewage. A total of 489 virus isolates were made during the study, 31% of which were reovirus. This is in contrast to the results of Wallis and Melnick [55] which indicated that reovirus could *not* be isolated by this method.

In a comparison of several isolation methods, Fattal et al. [154,156] examined the aluminum hydroxide precipitation method using 5 liters of filtered tap water with or without sewage added to a concentration of 0.01-0.025%. Samples of tap water seeded with low (8-64 PFU/5 liter) and high 391-792 PFU/5 liter) multiplicities of poliovirus 1 were concentrated 1000-fold with resultant recoveries of > 99%. Echovirus of recoveries from tap water averaged 54% while natural enteroviruses were recovered with a 34% efficiency from sewage-contaminated samples.

Application of the aluminum hydroxide flocculation method as a virus reconcentration procedure after adsorption and elution from filters has been extensive (examples were presented in previous sections and in Tables IV and V). Other precipitable salts such as $FeCl_3$ [107,111,157] and aluminum phosphate [153,157-159] have been examined by various workers but have received less attention as a routine concentration procedure for enteroviruses because of the degree of success using $AlCl_3$ compared to other methods [108]. However, it has been reported that the procedure has not been successful in concentrating both reovirus [55] and rotavirus [57]. Similar concentration difficulties may also occur among various types and strains of enteroviruses.

Insoluble polyelectrolytes have also been used for the adsorption and removal of viruses from water samples. Virus adsorption by this method is thought to occur between sites on the virus surface and carboxyl groups and ammonium radicals of the polymer [153].

A divinylbenzene-crosslinked styrene/maleic anhydride copolymer was found to adsorb tobacco mosaic virus with an efficiency of 100% [52]. The same study indicated that poliovirus was removed from water samples in excess of 99%.

In a comparison of aluminum hydroxide precipitation, depth (membrane) filtration and adsorption to an insoluble, cross-linked copolymer of isobutylene maleic anhydride (PE 60, Monsanto), results indicated that concentration with PE-60 was more rapid and efficient than the other methods tested [53]. Relative recovery rates for the three methods were: aluminum hydroxide, 56%; cellulose membrane, 64%; PE-60, 93%. Elution of virus from the insoluble polyelectrolyte was accomplished by suspension in phosphate buffer pH 8.0 or borate buffer at pH 9.0. In this study the polyelectrolyte was also sandwiched between fiberglass filters for use as a flow-through sampler. Eluates obtained were readsorbed to another polyelectrolyte (PE-52) and again eluted with physiological saline. Reduction of sample volumes greater than 1000-fold from 3.78 liters was achieved while efficiency was maintained.

A subsequent study by Wallis et al. [54] determined that thin layers of insoluble polyelectrolyte prepared on 90 or 293 mm fiberglass filter pads could efficiently adsorb poliovirus from large volumes of artificially contaminated water. Virus added to 95-380 liters of water were recovered upon elution with borate buffer with efficiencies up to 80%. An 1134-liter sampled showed a decreased recovery of 40%.

Although results with insoluble polyelectrolytes as virus adsorbents indicated reasonable efficiency, the method was not recommended for quantitative work because of several difficulties [128]: instability of the polymer [128], variation in chemical characteristics and inability to adsorb a variety of enteric viruses under similar experimental conditions [53].

Minerals such as iron oxide [160], bentonite (aluminum silicate) [161] and talc (magnesium silicate) [162] have been used as virus adsorbents but are no longer routinely used for virus concentration. Recent applications of a talc-celite (diatomaceous earth) layer in the form of a filter "sandwich" was presented in a previous section and in Table V (see References 111 and 115).

Various proteins have been employed as flocculation/precipitation agents for the concentration of virus from aqueous media. England [43] developed a method involving adsorption of viruses to the sulfate salt or protamine. Protamines are low molecular weight proteins which contain a large number of arginine residues. This predominance of arginine confers on the protein a high pI value ($\sim$ 12.0) and hence results in a positively charged protein in most aqueous suspensions. The positive charge of the protamines at or near neutral pH would readily allow adsorption of net negatively charged virus particles. England's method involved the addition of bovine serum albumin to a final concentration of 0.25%. The sample was then adjusted to pH 7.5 and protamine sulfate was added for a final concentration of 0.025%-0.05% to the albumin supplemented sample. After an appropriate mixing time the precipitate was collected by Tween-treated fiberglass filter discs, dissolved with 1M NaCl and evacuated from the filter into a receiving tube. This method was able to recover 80-100% of reovirus and adenovirus from artificially contaminated sewage. Enterovirus recoveries varied but the method was primarily designed to selectively concentrate the reoviruses and adenovirus so that these viruses could have an increased possibility of being detected in tissue culture free from interference by the faster replicating enteroviruses.

The use of alkaline beef extract solutions as eluents precluded the readsorption of virus onto smaller filters for reconcentration since adjustment to acidic pH caused considerable flocculation. Katzenelson et al. [42] examined the fate of viruses in these precipitated solids. Poliovirus was seeded into 3% beef extract solutions followed by precipitate formation at pH 2.5 or 4.0. After mixing at room temperature the precipitate was collected by centrifugation and the pellet was resuspended with 0.15 M Na_2HPO_4, pH 9.0. Essentially 100% of the poliovirus added to the sample remained in the pellet. In a comparison of organic flocculation with glycine elution followed by readsorption to smaller filters, mean recovery efficiencies were 74 and 35%, respectively.

Landry et al. [163] evaluated beef extract for the recovery of poliovirus from wastewater treatments. Use of alkaline beef extract solutions (pH 9.0) followed by flocculation yielded a mean recovery efficiency of 85%; elution with 0.1 M glycine, pH 11.5, followed by aluminum hydroxide flocculation resulted in an average recovery of 36%. In addition, the use of beef extract solutions at concentrations < 3% were found to be sufficient for poliovirus recovery from renovated wastewater.

Bitton and co-workers examined the flocculations of nonfat dry milk (NFDM) casein as an organic concentration method [58]. Virus was eluted from filters using 1% NFDM in 0.05 M glycine, pH 9.0. Reconcentration was performed by acidification to pH 4.6-4.7, centrifugation to collect the precipitate protein and dissolution of the pellet in 0.15 M Na_2HPO_4. Poliovirus 1 was recovered with an overall efficiency of 70% or more. Other representative enteroviruses resulted in reasonable recoveries but both echoviruses studied resulted in reasonable recoveries but both echoviruses studied gave the lowest overall recoveries (8 and 31%) of all the viruses examined.

Other methods of concentration attempted by various workers include hydroextraction of samples with polyethylene glycol [58,59], flocculation with lettuce extract [164,165], freeze concentration [166] and ultracentrifugation [167,168]. Of these methods only hydroextraction with polyethylene glycol is frequently used in procedures for isolation and concentration of virus from natural samples. However, it is usually used to further reduce the volume of a product from some other second step concentration method.

SUMMARY AND CONCLUSIONS

The ideal virus concentration technique has been described as: (1) relatively simple, (2) rapid, (3) inexpensive, (4) efficient, (5) able to handle large volumes of water without regard to the quality of water and (6) consistently able to detect a wide variety of viruses at low concentrations.

Variations of filter adsorption-elution techniques examined by numerous workers have been able to fulfill several of the above requirements. This method has been extensively tested in the laboratory and field studies have lent a degree of confidence to its ability to recover viruses from different types of water. However, evaluation of adsorption-elution has shown that the technique has drawbacks when applied to viruses other than those of the enterovirus group; presumably elution and reconcentration may be responsible for severe reductions in recovery of enteric viruses sensitive to pH extremes or with surface properties which are considerably different than the entero-viruses. Reliance on the removal of a solute (virus) from a flowing stream via transport and capture onto a solid substrate clearly depends on the hydro-dynamics of the system which, in turn, can be affected by water quality, i.e., the presence of competing particles or colloids, dissolved organics, ionic strength, etc. The co-concentration of virus and interfering substances can reduce virus adsorption efficiency and ultimately reduce the efficiency of the system upon elution since virus and other co-concentrates will occur concom-itantly in the sample. With the development of a more positively charged fil-ter substrate, adsorption-elution methods have been simplified and some of the difficulties associated with variable virus adsorption due to type- and strain-dependence have been reduced.

A host of membrane processes have been used in attempts to concentrate virus. Most have suffered from problems of slow processing times and extreme sensitivity to foulants. Ultrafiltration, however, is a promising technique that warrants further examination, especially in cases where crossflow is used to reduce membrane fouling and maintain flux. Since the method does not rely exclusively on adsorption, difficulties associated with adsorption of virus to substrate are avoided. Since virus particles are essentially retained by size-exclusion, solvent and solutes including toxic materials smaller than the cut-off retention size permeate the membrane during dehydration. Operating time becomes an overall factor of cost when field studies are performed and emphasis has been placed on rapidity. Although process times for ultrafil-tration are longer than for adsorption-elution, advances in tangential flow hol-low fiber ultrafiltration have made it possible to produce permeate at fairly high rates (1-2 liter/min).

All methods suffer when samples containing high organic loads are pro-cessed but with the use of various reconcentration methods, co-concentration of toxic solutes can be reduced. Alternate means of reconcentration which reduce extremes of pH should be critically examined since these can only serve to increase the detection of a wider variety of virus types.

Most important to the development of virus concentration techniques is a reassessment of currently available methods and further examination of recent improvements in sample processing that emphasize both the ease of application to turbid natural waters and the recovery of a diversity of natu-rally occurring animal viruses.

REFERENCES

1. Akin, E. W., W. H. Benton and W. F. Hill. "Enteric Viruses in Ground and Surface Waters: A Review of Their Occurrence and Survival," in *Viruses and Water Quality: Occurrence and Control,* V. Snoeyink, Ed. (Urbana, IL: University of Illinois, 1971), pp. 59-74.
2. Akin, E. W., D. A. Brashear, E. C. Lippy and N. A. Clarke. "A Virus-in-Water Study of Finished Water from Six Communities," Environmental Health Effects Research Series, U.S. EPA Report-600/1-75-003, (1975).
3. Tierney, J. T., R. Sullivan and E. P. Larkin. "Persistence of Poliovirus 1 in Soil and on Vegetables Grown in Soil Previously Flooded with Inoculated Sewage Sludge or Effluent," *Appl. Environ. Microbiol.* 33:109-113 (1977).
4. Shuval, H. I. "Health Considerations in Water Renovation and Reuse," in *Water Renovation and Reuse* (New York: Academic Press, 1977), pp. 23-72.
5. World Health Organization. "Human Viruses in Water, Wastewater and Soil," WHO Scientific Group Technical Report Series, 639 (Geneva, Switzerland: World Health Organization, 1979).
6. "Human Viruses in the Aquatic Environment. A Status Report with Emphasis on the EPA Research Program," U.S. EPA Report-570/9-78-006 (1978).
7. Sharp, D. G., D. C. Young, R. Floyd and J. D. Johnson. "Effect of Ionic Environment on the Inactivation of Poliovirus in Water by Chlorrine," *Appl. Environ. Microbiol.* 39:530-534 (1980).
8. Hejkal, T. W., F. M. Wellings, P. A. LaRock and A. L. Lewis. "Survival of Poliovirus Within Organic Solids During Chlorination," *Appl. Environ. Microbiol.* 38:114-118 (1979).
9. Scarpino, P. V., G. Berg, S. L. Chang, D. Dahling and M. Lucas. "A Comparative Study of the Inactivation of Viruses in Water by Chlorine," *Water Res.* 6:959-965 (1972).
10. Berg, G., D. R. Dahling, G. A. Brown and D. Berman. "Validity of Fecal Coliforms, Total Coliforms and Fecal Streptococci as Indicators of Viruses in Chlorinated Primary Sewage Effluent," *Appl. Environ. Microbiol.* 36:880-884 (1978).
11. Englebrecht, R. S., and E. O. Greening. "Chlorine Resistant Indicators," in *Indicators of Viruses in Water and Food,* G. Berg, Ed. (Ann Arbor, MI: Ann Arbor Science Publishers, Inc., 1978).
12. Ward, R. L. and C. S. Ashley. "Inactivation of Poliovirus in Digested Sludge," *Appl. Environ. Microbiol.* 31:921-930 (1976).
13. Fenters, J., J. Reed, C. Lue-Hing and J. Bertucci. "Inactivation of Viruses by Digested Sludge Components," *J. Water Poll. Control Fed.* 51:689-694 (1979).
14. Ward, R. L., and C. S. Ashley. "Identification of the Virucidal Agent in Wastewater Sludge," *Appl. Environ. Microbiol.* 33:860-864 (1977).
15. Berg, G., and D. Berman. "Destruction by Anaerobic Mesophilic and Thermophilic Digestion of Viruses and Indicator Bacteria Indigenous to Domestic Sludges," *Appl. Environ. Microbiol.* 39:361-368 (1980).

16. Kott, Y., N. Roze, S. Sperber and N. Betzlr. "Bacteriophages as Viral Pollution Indicators," *Water Res.* 8:165-171 (1974).
17. Vaughn, J. M., and T. G. Metcalf. "Coliphages as Indicators of Enteric Viruses in Shellfish and Shellfish Raising Estuarine Waters," *Water Res.* 9:613-616 (1975).
18. *Standard Methods for the examination of water and wastewater,* 13th ed. (New York: American Public Health Association, 1975).
19. Sproul, O. J. "Removal of Viruses by Treatment Processes," in *Viruses in Water,* G. Berg, et al., eds. (Washington, D.C.: American Public Health Association, 1976).
20. Shuval, H. I. "Disinfection of Wastewater for Agricultural Utilization," *Prog. Water Technol.* 1:857-867 (1975).
21. Jakubowski, W., N. A. Clarke, W. F. Hill and E. W. Akin. "Viruses and Drinking Water Quality: Observations and Reflections in Methods and Standards," paper presented at the Water Research Center Colloquim, HTS Management Center, Lane End High Wycombe, Buckinghamshire, England.
22. Bitton, G. *Introduction to Environmental Virology* (New York: John Wiley and Sons, Inc., 1980).
23. Fenner, F., B. R. McAuslan, C. A. Mims, J. Sambrook and D. O. White. *Animal Viruses,* student edition (New York: Academic Press, Inc., 1974).
24. Fenner, F. J., and D. O. White. *Medical Virology,* 2nd ed. (New York: Academic Press, 1976).
25. Kapikian, A. Z., et al. "Human Reovirus-Like Agent as the Major Pathogen Associated with "Winter" Gastroenteritis in Hospitalized Infants and Young Children," *New England J. Med.* 294:965-972 (1976).
26. Murphy, A. M., M. B. Albrey and E. B. Crewe. "Rotavirus Infections in Neonates," *Lancet* 2:1149-1150 (1977).
27. Elias, M. M. "Distribution and Titers of Rotavirus Antibodies in Different Age Groups," *Brit. J. Med.* 24:2-9 (1977).
28. Bolivar, R., et al. "Rotavirus in Travelers' Diarrhea: Study of an Adult Student Population in Mexico," *J. Infect. Dis.* 137:324-327 (1978).
29. Alder, J. L., and R. Zickl. "Winter Vomiting Disease," *J. Infect. Dis.* 119:668-673 (1969).
30. Dolin, R., et al. "Transmission of Acute Infectious Nonbacterial Gastroenteritis to Volunteers by Oral Administration of Stool Filtrates," *J. Infect. Dis.* 123:307-312 (1971).
31. Mason, J. O., and W. R. McLean. "Infectious Hepatitis Traced to the Consumption of Raw Oysters: an Epidemiologic Study," *Am. J. Hyg.* 75: 90-111 (1962).
32. Dismukes, W. E., A. L. Bisno, S. Katz and R. F. Johnson. "An Outbreak of Gastroenteritis and Infectious Hepatitis Attributed to Raw Clams," *Am. J. Epidemiol.* 89:555-561 (1969).
33. Ruddy, S. J., et al. "An Epidemic of Clam-Associated Hepatitis," *J. Am. Med. Assoc.* 208:649-655 (1969).
34. Dienstag, J. L., I. D. Gust, C. R. Lucas, D. Wong and R. H. Purcell. "Mussel-Associated Viral Hepatitis Type A: Serological Confirmation," *Lancet* 1:561-564 (1976).
35. Mosley, J. W. "Water-Borne Infectious Hepatitis," *New England J. Med.* 261:703-708, 748-753 (1959).

36. Portnoy, B. L., et al. "Oyster-Associated Hepatitis: Failure of Shellfish Certification Programs to Prevent Outbreaks," *J. Am. Med. Assoc.* 233: 1065-1068 (1975).

37. Wellings, F. M., A. L. Lewis, C. W. Mountain and L. V. Pierce. "Demonstration of Virus in Ground Water After Effluent Discharge into Soil," *Appl. Microbiol.* 29:751-757 (1975).

38. Metcalf, T. G., and W. C. Stiles. "The Accumulation of Enteric Viruses by the Oyster, *Crassostea virginica*," *J. Infect. Dis.* 115:68-76 (1965).

39. Beard, J. W. "Host Virus Interaction in the Initiation of Infection," in *Transmission of Viruses by the Water Route*, G. Berg, Ed. (New York: J. Wiley & Sons, Inc., 1967), pp. 167-192.

40. Plotkin, S. A., and M. Katz. "Minimal Infective Doses of Viruses for Man by the Oral Route," in *Transmission of Viruses by the Water Route*, G. Berg, Ed. (New York: J. Wiley & Sons, Inc., 1967), pp. 151-166.

41. Westwood, J. C. N., and S. A. Sattar. "The Minimal Infective Dose," in *Viruses in Water*, G. Berg, Ed. (Washington, D.C.: American Public Health Association, Inc., 1976).

42. Katzenelson, E., B. Fattal, and T. Hostovesky. "Organic Flocculation: an Efficient Second-Step Concentration Method for the Detection of Viruses in Tap Water," *Appl. Environ. Microbiol.* 32:638-639 (1976).

43. England, B. "Concentration of Reovirus and Adenovirus from Sewage and Effluents by Protamine Sulfate (Salmine) Treatment," *Appl. Microbiol.* 24:510-512 (1972).

44. Albertsson, P. A. "Concentration of Viruses by Two Phase Separation," in *Virus Survival in Water and Wastewater Systems, Water Resources Symposium No. 7*, J. F. Malina and B. P. Sagik, Eds., (Austin, TX: University of Texas, 1974), pp. 16-18.

45. Philipson, L., P. A. Albertsson and G. Frick. "The Purification and Concentration of Viruses by Aqueous Polymer Phase System," *Virology* 11:553-571 (1960).

46. Lund, E., and C. E. Hedstrom. "The Use of an Aqueous Polymer Phase System for Enterovirus Isolation from Sewage," *Am. J. Epidemiol.* 81: 287-294 (1966).

47. Shuval, H. I., S. Cymbalista, B. Fattal and N. Goldblum. "Concentration of Enteric Viruses in Water by Hydroextraction and Two Phase Separation," in *Transmission of Viruses by the Water Route*, G. Berg, Ed. (New York: John Wiley and Sons, Inc., 1967), pp. 45-55.

48. Shuval, H. I., B. Fattal, S. Cymbalista and N. Goldblum. "The Phase Separation Method for the Concentration and Detection of Viruses in Water," *Water Res.* 3:225-240 (1969).

49. Nupen, E. M. "Virus Studies in the Windhoek Waste-Water Reclamation Plant (South-West Africa)," *Water Res.* 1:661-672 (1970).

50. Grindrod, J., and D. O. Cliver. "Limitations of the Polymer Two Phase System for Detection of Viruses," *Archiv. Ges. Virusforsch.* 28:337-347 (1969).

51. Grindrod, J., and D. O. Cliver. "A Polymer Two Phase Adapted to Virus Detection," *Archiv. Ges. Virusforsch.* 31:365-372 (1970).

52. Johnson, J. H., J. E. Fields and W. A. Darlington. "Removing Virus from Water by Polyelectrolytes," *Nature* 213:645-667 (1967).

53. Wallis, C., S. Grinstein, J. L. Melnick and J. E. Fields. "Concentration of Viruses from Sewage and Excreta on Insoluble Polyelectrolytes," *Appl. Microbiol.* 18:1007-1014 (1969).
54. Wallis, C., J. L. Melnick and J. E. Fields. "Detection of Viruses from Large Volumes of Natural Waters by Concentration on Insoluble Polyelectrolytes," *Water Res.* 1:787-796 (1970).
55. Wallis, C., and J. L. Melnick. "Concentration of Viruses on Aluminum Phosphate and Aluminum Hydroxide Precipitates," in *Transmission of Viruses by the Water Route,* G. Berg, Ed. (New York: John Wiley & Sons, Inc., 1967), pp. 129-139.
56. Wallis, C., and J. L. Melnick. "Concentration of Viruses on Aluminum and Calcium Salts," *Am. J. Epidemiol.* 85:459-468 (1967).
57. Farrah, S. R., S. M. Goyal, C. P. Gerba, R. H. Conklin and E. C. Smith. "Comparison Between Adsorption of Poliovirus and Rotavirus by Aluminum Hydroxide and Activated Sludge Flocs," *Appl. Environ. Microbiol.* 35:360-363 (1978).
58. Cliver, D. O. "Detection of Enteric Viruses by Concentration with Polyethylene Glycol," in *Transmission of Viruses by the Water Route,* G. Berg, Ed. (New York: John Wiley & Sons, Inc., 1967), pp. 109-120.
59. Shuval, H. I., S. Cymbalista, B. Fattal and N. Goldblum. "Concentration of Enteric Viruses in Water by Hydroextraction and Two Phase Separation," in *Transmission of Viruses by the Water Route,* G. Berg, Ed. (New York: John Wiley & Sons, Inc., 1967), pp. 45-54.
60. Derjaguin, B. V., and L. D. Landau. "Theory of Stability of Strongly Charged Lyophobic Sols and of the Adhesion of Strongly Charged Particles in Solutions of Electrolytes," *Acta Physiochim URSS* 14:633 (1941).
61. Verwey, E. J. W., and J. T. C. Overbeek. *Theory of the Stability of Lyophobic Colloids* (Amsterdam: Elsevier, 1948).
62. Belfort, G., and D. Dziewulski. "Transport of and Adsorption of Viruses in Porous Media," in a book soon to be published by CRC Press, Inc., Cleveland, OH, G. Berg, Ed.
63. Weber, J., Jr. *Physicochemical Processes for Water Quality Control* (New York: Wiley-Interscience, 1972).
64. Fuhs, G. W., et al. "Virus Uptake by Minerals and Soils," paper presented at the 53rd Annual Conference of the Water Pollution Control Federation, Las Vegas, NV, Sept. 28-Oct. 3, 1980.
65. Murray, J. P. "Physical Chemistry of Virus Adsorption and Degradation on Inorganic Surfaces," U.S. EPA Report-600/2-80-134, Municipal Environmental Research Laboratory (1980).
66. Ives, K. J. "Capture Mechanisms in Filtration," in *The Scientific Basis of Filtration,* K. J. Ives, Ed. (Leyden: Noordhoof International Publishing Co., 1975), pp. 183-201.
67. Belfort, G., and B. Marx. "Artificial Particulate Fouling of Hyperfiltration Membranes—II, Analysis and Protection from Fouling," *Desalination* 28:13-30 (1979).
68. Happel, J., and H. Brenner. *Low Reynolds Number Hydrodynamics* (Leyden: Noordhoff International Publishing Co., 1973), p. 298.
69. Green, G., and G. Belfort. "Fouling of Ultrafiltration Membranes: Lateral Migration and the Particle Trajectory Model," paper presented at the

Conference on Membrane Technology in the 80's, Ystad, Sweden, Sept. 29-Oct. 1, 1980.

70. Ho, B. P., and L. G. Leal. "Inertial Migration of Rigid Spheres in Two-Dimensional Unidirectional Flows," *J. Fluid Mech.* 65:365 (1974).

71. Rajagopalan, R., and C. Tien. "The Theory of Deep Bed Filtration," in *Progress in Filtration and Separation I*, R. J. Wakeman, Ed. (Amsterdam: 1979).

72. Yao, K-M. "Influence of Suspended Particle Size in the Transport Aspect of Filtration," Ph.D. Dissertation, University of North Carolina, Chapel Hill, NC (1968).

73. Belfort, G., Y. Rotem-Borensztajn and E. Katzenelson. "Virus Concentration from Waters by Tangential-Flow Hollow Fiber Ultrafiltration: the Importance of Hydrodynamics," to be published by ASCE, J. Environ. Eng.

74. Belfort, G., Y. Rotem-Borensztajn and E. Katzenelson. "Virus Concentration Using Hollow Fiber Membranes," *Water Res.* 9:79 (1975).

75. Belfort, G., Y. Rotem-Borensztajn and E. Katzenelson. "Virus Concentration Using Hollow Fiber Membranes—II," *Water Res.* 10:279 (1976).

76. Gravelle, C. R., and T. D. Y. Chin. "Enterovirus Isolations from Sewage: a Comparison of Three Methods," *J. Infect. Dis.* 109:205-209 (1961).

77. Coin, L., M. L. Menetrier, J. Labonde and M. C. Hannoun. "Modern Microbiological and Virological Aspects of Water Pollution," *2nd International Conference on Water Pollution Research* (New York: Pergamon Press, 1966).

78. Liu, D. C., D. A. Brashear, H. R. Sheraichekas, J. A. Barnick and T. G. Metcalf. "Virus in Water. I. A Preliminary Study on a Flow-Through Gauze Sampler for Recovering Virus from Waters," *Appl. Microbiol.* 21:405-410 (1971).

79. Metcalf, T. G. "Use of Membrane Filters to Facilitate the Recovery of Virus from Aqueous Suspensions," *Appl. Microbiol.* 9:376-379 (1961).

80. Cliver, D. O. "Factors in the Membrane Filtration of Enteroviruses," *Appl. Microbiol.* 13:417-425 (1965).

81. Wallis, C., and J. L. Melnick. "Concentration of Enteroviruses on Membrane Filters," *J. Virol.* 1:472-477 (1967).

82. Wallis, C., J. L. Melnick and C. P. Gerba. "Concentration of Viruses from Water by Membrane Chromatography," *Ann. Rev. Microbiol.* 33:413-437 (1979).

83. Wallis, C., M. Henderson and J. L. Melnick. "Enterovirus Concentration on Cellulose Membranes," *Appl. Microbiol.* 23:476-480 (1972).

84. Berg, G., D. R. Dahling and D. Berman. "Recovery of Small Quantities of Viruses from Clean Waters on Cellulose Nitrate Filters," *Appl. Microbiol.* 22:608-614 (1971).

85. Sobsey, M. D., C. Wallis, M. Henderson and J. L. Melnick. "Concentration of Enteroviruses from Large Volumes of Water," *Appl. Microbiol.* 26:529-534 (1973).

86. Hill, W. F., Jr., E. W. Akin, W. H. Benton and T. G. Metcalf. "Virus in Water. II. Evaluation of Membrane Cartridge Filters for Recovering Low Multiplicities of Poliovirus from Water," *Appl. Microbiol.* 23:880-888 (1972).

87. Hill, W. F., Jr., E. W. Akin, W. H. Benton, C. J. Mayhew and W. Jaku-bowski. "Apparatus for Conditioning Unlimited Quantities of Finished Waters for Enterovirus Detection," *Appl. Microbiol.* 27:1177-1178 (1974).

88. Jakubowski, W., J. C. Hoff, N. C. Anthony and W. F. Hill, Jr. "Epoxy Fiberglass Adsorbent for Concentating Viruses from Large Volumes of Potable Water," *Appl. Microbiol.* 28:501-502 (1974).

89. Jakubowski, W., W. F. Hill, Jr. and N. A. Clarke. "Comparative Study of Four Microporous Filters for Concentrating Viruses from Drinking Water," *Appl. Microbiol.* 30:58-65 (1975).

90. Farrah, S. R., C. P. Gerba, C. Wallis and J. L. Melnick. "Concentration of Viruses from Large Volumes of Tap Water Using Pleated Membrane Filters," *Appl. Environ. Microbiol.* 31:221-226 (1976).

91. Gerba, C. P., S. R. Farrah, S. M. Goyal, C. Wallis and J. L. Melnick. "Concentration of Enteroviruses from Large Volumes of Tap Water, Treated Sewage and Seawater," *Appl. Environ. Microbiol.* 35:540-548 (1978).

92. Farrah, S. R., S. M. Goyal, C. P. Gerba, C. Wallis and J. L. Melnick. "Concentration of Poliovirus from Tap Water onto Membrane Filters with Aluminum Chloride at Ambient pH Levels," *Appl. Environ. Microbiol.* 35:624-626 (1978).

93. Sobsey, M. D., and B. C. Jones. "Concentration of Poliovirus from Tap Water Using Positively Charged Microporous Filters," *Appl. Environ. Microbiol.* 37:588-595 (1979).

94. Sobsey, M. D., and J. S. Glass. "Poliovirus Concentration from Tap Water with Electropositive Adsorbent Filters," *Appl. Environ. Microbiol.* 40:201-210 (1980).

95. Wallis, C., and J. L. Melnick. "A Portable Virus Concentrator for Use in the Field," *Water Res.* 6:1249-1256 (1972).

96. Hill, W. F., Jr., W. Jakubowski, E. W. Akin and N. A. Clarke. "Detection of Virus in Water: Sensitivity of the Tentative Standard Method for Drinking Water," *Appl. Environ. Microbiol.* 31:254-261 (1976).

97. Farrah, S. R., C. P. Gerba, S. M. Goyal, C. Wallis and J. L. Melnick. "Regeneration of Pleated Filters Used to Concentrate Enteroviruses from Large Volumes of Tap Water," *Appl. Environ. Microbiol.* 33:308-311 (1977).

98. Hill, W. F., Jr., E. W. Akin, W. H. Benton, C. J. Mayhew and T. G. Metcalf. "Recovery of Poliovirus from Turbid Estuarine Water on Microporous Filters by the Use of Celite," *Appl. Microbiol.* 27:506-512 (1974).

99. Farrah, S. H., S. M. Goyal, C. P. Gerba, C. Wallis and P. T. B. Shaffer. "Characteristics of Humic Acid and Organic Compounds Concentrated from Tap Water Using the Aquella Virus Concentrator," *Water Res.* 10: 897-901 (1976).

100. Wallis, C., and J. L. Melnick. "Concentration of Viruses from Sewage by Adsorption on Millipore Membranes," *Bull. World Health Org.* 36: 219-225 (1967).

101. Moore, M. L., P. P. Ludovici and W. S. Jeter. "Quantitative Methods for the Concentration of Viruses in Wastewater," *J. Water Poll. Control Fed.* 42:R21-R28 (1970).

102. Rao, N. U., and N. A. Labzoffsky. "A Simple Method for the Detection of Low Concentrations of Viruses in Large Volumes of Water by the Membrane Filtration Technique," *Can. J. Microbiol.* 15:399-403 (1969).

103. Rao, C. U., U. Chandorkar, N. U. Rao, P. Kumaran and S. B. Lahke. "A Simple Method for Concentrating and Detecting Viruses in Wastewater," *Water Res.* 6:1565-1576 (1972).

104. Metcalf, T. G., C. Wallis and J. L. Melnick. "Concentration of Viruses from Seawater," in *Advances in Water Pollution Research. Proceedings of the 6th International Conference,* S. M. Jenkins, Ed. (Oxford: Pergamon Press, 1973).

105. Metcalf, T. G., C. Wallis and J. L. Melnick. "Environmental Factors Influencing Isolation of Enteroviruses from Polluted Surface Waters," *Appl. Microbiol.* 27:920-926 (1974).

106. Homma, A., M. D. Sobsey, C. Wallis and J. L. Melnick. "Virus Concentration from Sewage," *Water Res.* 7:945-950 (1973).

107. Sobsey, M. D., C. P. Gerba, C. Wallis and J. L. Melnick. "Concentration of Enteroviruses from Large Volumes of Turbid Estuary Water," *Can J. Microbiol.* 23:770-778 (1977).

108. Farrah, S. R., S. M. Goyal, C. P. Gerba, C. Wallis and J. L. Melnick. "Concentration of Enteroviruses from Estuarine Water," *Appl. Environ. Microbiol.* 33:1192-1196 (1977).

109. Sattar, S. A., and S. Ramia. "Talc-Celite Layer in Virus Recovery from Potable Waters Experimentally Contaminated with Field Isolates and Sewage," *Water Res.* 13:1351-1353 (1979).

110. Fields, H. A., and T. G. Metcalf. "Concentration of Adenovirus from Seawater," *Water Res.* 9:357-364 (1975).

111. Payment, P., C. P. Gerba, C. Wallis and J. L. Melnick. "Methods for Concentrating Viruses from Large Volumes of Estuarine Water on Pleated Membranes," *Water Res.* 10:893-896 (1976).

112. Wallis, C., A. Homma and J. L. Melnick. "Apparatus for Concentrating Viruses from Large Volumes," *J. Am. Water Works Assoc.* 64:189-196 (1972).

113. Wesslen, T., P. A. Albertsson and L. Philipson. "Concentration of Animal Viruses Using Two-Phase Systems of Aqueous Polymer Solutions," *Arch. Ges. Virusforsch.* 9:510-520 (1959).

114. Farrah, S. H., C. Wallis, P. T. B. Schaffer and J. L. Melnick. "Reconcentration of Poliovirus from Sewage," *Appl. Environ. Microbiol.* 32:653-658 (1976).

115. Sattar, S. A., and S. Ramia. "Use of Talc-Celite Layers in the Concentration of Enterovirus from Large Volumes of Potable Waters," *Water Res.* 13:637-643 (1979).

116. Goyal, S. M., C. P. Gerba and J. L. Melnick. "Prevalence of Human Enteric Viruses in Coastal Canal Communities," *J. Water Poll. Control Fed.* 50:2247-2256 (1978).

117. Vaughn, J. M., E. F. Landry, M. Z. Thomas, T. J. Vicale and W. F. Penello. "Survey of Human Enterovirus Occurrence in Fresh and Marine Surface Waters on Long Island," *Appl. Environ. Microbiol.* 38:290-296 (1979).

118. Montgomery, James M., Consulting Engineers, Inc. "Water Factory 21 Virus Study, Orange County Water District" (1979).

119. LaBelle, R. L., and C. P. Gerba. "Influence of pH, Salinity, and Organic Matter on the Adsorption of Enteric Viruses to Estuarine Sediments," *Appl. Environ. Microbiol.* 38:93-101 (1979).
120. Gerba, C. P., S. M. Goyal, C. J. Hurst and R. L. LaBelle. "Type and Strain Dependence of Enterovirus Adsorption to Activated Sludge, Soils and Estuarine Sediments," *Water Res.* 14:1197-1198 (1980).
121. Sobsey, M. D., J. S. Glass, R. J. Carrick, R. R. Jacobs and W. A. Rutala. "Evaluation of the Tentative Standard Method of Enteric Virus Concentration from Large Volumes of Tap Water," *J. Am. Water Works Assoc.* 75:292-299 (1980).
122. Sobsey, M. D., J. S. Glass, R. R. Jacobs and W. A. Rutala. "Modifications of the Tentative Standard Method for Improved Virus Recovery Efficiency," *J. Am. Water Works Assoc.* 76:350-355 (1980).
123. Dziewulski, D. M. "A Model System for the Concentration of Reovirus from Aqueous Environments," Ph.D. Thesis, University of New Hampshire, University Microfilms (1980).
124. Loeb, S., and S. Sourirajan. "Sea Water Demurization by Means of an Osmotic Membrane," *Adv. Chem. Ser.* 38:117 (1962).
125. Belfort, G. "Pressure-Driven Membrane Processes and Wastewater Renovation," in *Water Renovation and Reuse* (New York: Academic Press, 1977), pp. 129-189.
126. Cadotte, J. E., R. J. Petersen, R. E. Larson and E. E. Erickson. "A New Thin-Film Composite Seawater Reverse Osmosis Membrane," paper presented at the 5th Seminar on Membrane Separation Technology, Clemson University, SC, May 12-14, 1980.
127. Sweet, B. H., R. D. Ellender and J. K. L. Leong. "Recovery and Renewal of Viruses from Water-Utilizing Membrane Techniques," in *Developments in Industrial Microbiology*, E. D. Murray and A. W. Bourquin, Eds. (Arlington, VA: American Institute of Biological Sciences, 1973), pp. 143-159.
128. Sobsey, M. D. "Methods for Detecting Enteric Viruses in Water and Wastewater," in *Viruses in Water*, G. Berg et al., Eds. (Washington, DC: American Public Health Assoc., Inc., 1976), pp. 89-127.
129. Sorber, C. A., J. F. Malina and B. F. Sagik. "Virus Rejection by the Reverse Osmosis-Ultrafiltration Process," *Water Res.* 6:1377-1388 (1972).
130. Strathman, H., and K. Kock. "The Formation Mechanism of Phase Inversion Membranes," *Desalination* 21:241-255 (1977).
131. Ellender, R. D., and B. H. Sweet. "Newer Membrane Concentration Processes and Their Application to the Detection of Viral Pollution of Waters," *Water Res.* 6:741-746 (1972).
132. Loeb, S. Private communication.
133. Bier, M., G. C. Bruckner, F. C. Cooper and H. E. Roy. "Concentration of Bacteriophage by Electrophoresis," in *Transmission of Viruses by the Water Route*, G. Berg, Ed. (New York: John Wiley & Sons, Inc., 1967), pp. 51-75.
134. Sweet, B. H., J. S. McHale, K. J. Hardy and E. Klein. "Concentration of Virus from Water by Electro-Osmosis and Forced-Flow Electrophoresis," *Prep. Biochem.* 1:77-89 (1971).

135. Ellender, R. D., F. Morton, J. Whelan and B. H. Sweet. "Concentration of Virus from Water by Electro-Osmosis and Forced-Flow Electrophoresis. Improvement of Methodology and Application to Tap Water," *Prep. Biochem.* 2:215-228 (1972).

136. Sweet, B. H., and R. D. Ellender. "Electro-Osmosis: a New Technique for Concentrating Viruses from Water," *Water Res.* 6:775-779 (1972).

137. Strohmaier, K. "Virus Concentration by Ultrafiltration," in *Methods in Virology, Vol. II,* K. Maramcrosh and H. Koprowski, Eds. (New York: Academic Press, 1967), pp. 245-274.

138. Foliguet, J. M., J. Lavillaureix and L. Schwartzbrod. "Viruses in Water II. General Review of the Methods Available to Detect Viruses in Water," *Rev. Epid. Med. Soc. et Santze Publ.* 21:185-259 (1973).

139. Nupen, E. M., and G. J. Stander. "The Virus Problem in the Windhoek Wastewater Reclamation Plant," in *Advances in Water Pollution Research,* S. H. Jenkins, Ed. (New York: Pergamon Press, 1972), pp. 133-142.

140. Smith, C. V., and D. DiGregorio. "Ultrafiltration Water Treatment," in *Membrane Science and Technology: Industrial, Biological and Waste Treatment Processes,* J. E. Flinn, Ed. (New York: Plenum Press, 1970), pp. 209-220.

141. Foliguet, J. M. "Public Health Aspects of Viruses in Water," N. A. Clarke, Ed., *U.S. EPA J.* 3:8-9 (1971).

142. Belfort, G., Y. Rotem and E. Katzenelson. "Virus Concentration by Hollow Fiber Membranes: Where to Now?," *Prog. Water Tech.* 10:357 (1978).

143. Bird, R. B., W. E. Stewart and E. N. Lightfoot. *Transport Phenomena* (New York: John Wiley & Sons, Inc., 1960).

144. Dresner, L., and J. S. Johnson, Jr. "Hyperfiltration (Reverse Osmosis)," in *Principles of Desalination,* 2nd ed. K. S. Spiegler and A. D. K. Laird, Eds. (New York: Academic Press, 1980).

145. Frommer, M. A., J. S. Murday and R. M. Messalem. *European Polym. J.* 9:367 (1973).

146. Rotem-Borensztajn, Y., E. Katzenelson and G. Belfort. "Virus Concentration by Capillary Ultrafiltration," *J. Environ. Eng. Div.,* ASCE, 105 (EE2):401-407 (1979).

147. Belfort, G., and D. Dziewulski, unpublished.

148. Berman, D., M. E. Rohr and R. S. Safferman. "Concentration of Poliovirus in Water by Molecular Filtration," *Appl. Environ. Microbiol.* 40: 426-428 (1980).

149. Nupen, E. M., N. C. Basson and W. O. K. Grabow. "Efficiency of UF for the Isolation of Enteric Viruses and Coliphages from Large Volumes of Water in Studies on Wastewater Reclamation," *Prog. Water Technol.* 12:851-863 (1980).

150. Belfort, G., T. F. Baltutis and W. F. Blatt. "Automatic Hollow Fiber Ultrafiltration: Pyrogen Removal and Phage Recovery from Water," in *Ultrafiltration Membranes and Applications,* A. R. Cooper, Ed. (New York: Plenum Publishing Corporation, 1980), pp. 439-474.

151. Greenberg, A. E., J. J. Connors and D. Jenkins (Editorial Board), *Standard Methods for the Examination of Water and Wastewater,* 15th ed. published jointly by APHA, AWWA and WPCF (1980).

152. England, B. "Recovery of Viruses from Waste and Other Waters by Chemical Methods," in *Developments in Industrial Microbiology. Vol. 15*, E. D. Murray and A. W. Bourquin, Eds. (Arlington, VT: American Institute of Biological Sciences, 1973), pp. 174-183.

153. Cookson, J. T. "The Chemistry of Virus Concentration by Chemical Methods," in *Developments in Industrial Microbiology. Vol. 15.*, E. D. Murray and A. W. Bourquin, Eds. (Arlington, VT: American Institute of Biological Sciences, 1973).

154. Shuval, H. I., and E. Katzenelson. "Detection and Inactivation of Enteric Viruses in Wastewater," EMSL-CIN-009 U.S. EPA (1976).

155. Palfi, A. B. "Examination of Sewage in Budapest," paper presented at the Second International Congress for Virology, Budapest, Hungary, June 22-July 2, 1971.

156. Fattal, B., E. Katzenelson and H. I. Shuval. "Comparison of Methods for Isolation of Viruses in Water," in *Virus Survival in Water and Wastewater Systems, Water Resources Symposium No. 7*, J. F. Malina, Jr. and B. P. Sagik, Eds. (Austin, TX: University of Texas, 1974), pp. 19-30.

157. Lal, S. M., and E. Lund. "Recovery of Virus by Chemical Precipitation Followed by Elution," *Prog. Water Technol.* 7:687-693 (1975).

158. Lydholm, B., and A. L. Nielsen. "Methods for Detection of Virus in Wastewater, Applied to Samples from Small Scale Treatment Systems," *Water Res.* 14:169-173 (1980).

159. Nielsen, A. L., and B. Lydholm. "Methods for the Isolation of Virus from Raw and Digested Wastewater Sludge," *Water Res.* 14:175-178 (1980).

160. Rao, V. C., R. Sullivan, R. B. Read and N. A. Clarke. "A Simple Method for Concentrating and Detecting Viruses in Water," *J. Am. Water Works Assoc.* 60:1288-1294 (1968).

161. Moore, B. E. D., L. Funderberg and B. P. Sagik. "Application of Viral Concentration Techniques to Field Sampling," in *Virus Survival in Water and Wastewater Systems, Water Resources Symposium No. 7*, J. F. Malina and B. P. Sagik, Eds. (Austin, TX: University of Texas, 1974), pp. 3-15.

162. Kalter, S. S., and C. H. Millstein. "Efficacy of Methods for the Detection of Viruses in Treated and Untreated Sewage," in *Virus Survival in Water and Wastewater Systems, Water Resources Symposium No. 7*, J. F. Malina and B. P. Sagik, Eds. (Austin, TX: University of Texas, 1974), pp. 33-44.

163. Landry, E. F., J. M. Vaughn, M. Z. Thomas and T. J. Vicale. "Efficacy of Beef Extract for the Recovery of Poliovirus from Wastewater Effluents," *Appl. Environ. Microbiol.* 36:544-548 (1978).

164. Konowalchuck, J., and J. I. Speirs. "Enterovirus Recovery with Vegetable Floc," *Appl. Microbiol.* 26:505-507 (1973).

165. Konowalchuck, J., J. I. Speirs, R. D. Pontefract and G. Bergeron. "Concentration of Enteric Viruses from Water with Lettuce Extract," *Appl. Microbiol.* 28:717-719 (1974).

166. Rubenstein, S. H., et al. "Viruses in Metropolitan Waters: Concentration by Polyelectrolytes, Freeze Concentration and Ultrafiltration," *J. Am. Water Works Assoc.* 65:200-202 (1973).

167. Cliver, D. O., and J. Yeatman. "Ultracentrifugation in the Concentration and Detection of Enteroviruses," *Appl. Microbiol.* 13:387-392 (1965).
168. Anderson, N. G., G. B. Cline, W. W. Harris and J. G. Green. "Isolation of Viral Particles from Large Fluid Volumes," in *Transmission of Viruses by the Water Route,* G. Berg, Ed. (New York: John Wiley & Sons, Inc., 1967), pp. 75-88.

SECTION 5

HEALTH EFFECTS AND
PERSISTENT COMPOUNDS

CHAPTER 30

HEALTH EFFECTS AND LAND APPLICATION
OF WASTEWATER

Sherwood C. Reed
    Cold Regions Research and Engineering Laboratory
    U.S. Army Corps of Engineers
    Hanover, New Hampshire

INTRODUCTION

Protection of public health is the fundamental purpose of waste treatment. It is the responsibility of the engineers, scientists and public officials involved to ensure that waste treatment systems achieve this goal. It is unacceptable to build a system that does not address the problem but rather shifts the impact to some other sector. This is a basic premise for land treatment as well as any other waste treatment technology. Land application of wastewater was a common treatment/disposal option in the late nineteenth century. The treatment responses that occurred were recognized but not clearly understood. The advent in the early twentieth century of chlorine disinfection and high-rate mechanical systems made it more expedient to allow discharge of partially treated wastewaters to rivers rather than insist on the higher level of treatment and protection provided by land application.

Since the early 1970s there has been a trend back to the use of natural biological systems, including land application, for wastewater treatment. This has happened because of the growing recognition that the high-rate mechanical systems were very expensive in terms of both costs and energy and that they did not always satisfy the goals of environmental and health protection. Research has resulted in greater insight into process kinetics and the internal reactions and responses that occur in land treatment. It is now possible to design a system with the assurance that if properly managed it will perform in a consistent and reliable manner. However, land treatment is still considered by many as a new or unproven technology. There are lingering concerns on the part of some regulatory officials, engineers and scientists, and lay persons. These concerns are essentially based on the potential for accumulation and/or translocation of pollutants in land treatment systems. Questions often heard at public hearings and planning meetings include:

- Is it safe to live near a site?
- Can the crops or grazing animals be used in the human food chain?
- Is it safe to use wastewater in a park or on a golf course?
- Will pollutants accumulate and poison the soil or groundwaters?
- Will long term operation damage the environment or endanger public health?
- Can the site be safely used for other purposes after the land treatment operation ceases?

This chapter describes the current state-of-the-art of land treatment technology and considers whether or not potential health risks exist and, if so, what mitigating approaches are possible.

LAND TREATMENT CONCEPTS

The land application systems considered here are those specifically designed and properly managed to achieve the treatment and renovation of wastewaters. The indiscriminate discharge and disposal of wastewaters or other system by-products are not included. A detailed discussion of the various concepts and the criteria for their design can be found in the Process Design Manual for Land Treatment of Municipal Wastewater [1]. A brief description of the three major land treatment concepts is given below.

The slow-rate (SR) system, the most commonly used, is similar to conventional irrigation practice, with application via sprinklers, surface flooding, etc. The application rate can vary from 2 to 20 ft per year, and the wastewater applied can vary from raw sewage to tertiary effluent. A critical component in the renovative process is surface vegetation, which has included all types of crops, pastures, forests, greenbelts, golf courses, etc. Soils most often used are medium to fine textured with moderate permeability, through which the applied water percolates downward to underdrains and/or groundwater table. Climate constraints on crops may require some winter storage of wastewater.

The rapid infiltration (RI) system usually involves intermittent flooding of shallow basins in relatively coarse-textured soils of rapid permeability. The wastewater can vary from raw sewage to tertiary effluent with application rates varying from 20 to 600 ft per year. Water movement is by percolation to underdrains, recovery wells or groundwater. In this system vegetation is not a critical factor in the renovative process, nor is climate usually a constraint on application.

The overland flow (OF) system has application techniques similar to SR but on gently sloping fields with essentially impermeable soils, so that water moves via sheet flow down the slope to collection ditches at the toe. The wastewater applied varies from raw sewage to secondary effluent with rates varying from 10 to 70 ft of water per year. Vegetation comprised of water-tolerant grasses is an essential component in this sytem, and some winter wastewater storage may be needed in cold climates. Overland flow is a relatively new concept compared to SR and RI. Successful performance with industrial wastewaters, such as food processing wastes, and with municipal wastewaters has been demonstrated.

Table I summarizes current U.S. EPA guidance on the level of preapplication treatment required for these three concepts. The guidance is directly based on concern for public health. As the degree of public access increases or when any crop enters the food chain at a higher level the requirements become more stringent.

MAJOR CONCERNS

The major health concern is possible pollution by nitrogen, phosphorus, metals, pathogens, and organics. These issues and their potential pathways of greatest concern are summarized in Table II and considered below in more detail.

Table I. Guidance for Assessing Level of Preapplication Treatment[a]

I. Slow-Rate Systems
   A. Primary treatment. Acceptable for isolated locations with restricted public access and when limited to crops not for direct human consumption.
   B. Biological treatment by lagoons or inplant processes plus control of fecal coliform count to less than 1000 MPN/100 ml. Acceptable for controlled agricultural irrigation except for human food crops to be eaten raw.
   C. Biological treatment by lagoons or inplant processes with additional BOD or SS control as needed for aesthetics plus disinfection to log mean of 200/100 ml (EPA fecal coliform criteria for bathing waters). Acceptable for application in public access areas such as parks and golf courses.

II. Rapid-Infiltration Systems
   A. Primary treatment. Acceptable for isolated locations with restricted public access.
   B. Biological treatment by lagoons or inplant processes. Acceptable for urban locations with controlled public access.

III. Overland-Flow Systems
   A. Screening or comminution. Acceptable for isolated sites with no public access.
   B. Screening or comminution plus aeration to control odors during storage or application. Acceptable for urban locations with no public access.

[a]Reference sources include Water Quality Criteria 1972, EPA-R3-78-003, Water Quality Criteria EPA 1976, and various state guidelines.

Table II. Pollutants and Pathways of Concern

Pollutant	Pathway
Nitrogen	
Health	Infant water supply
Environmental	Eutrophication
Phosphorus	
Health	No direct impact
Environmental	Eutrophication
Metals	
Health	Water supply, to crops, or animals in human food chain
Environmental	Longterm soil damage, toxic to plants or wildlife
Pathogens	
Health	Water supplies, crops, aerosols
Environmental	Soil accumulation, infect wildlife
Trace Organics	
Health	Water supplies, food chain, crops or animals
Environmental	Soil accumulation

Nitrogen

The major health concern with nitrogen is excess concentrations of nitrate in drinking waters for infants under 6 months of age. The primary drinking water standards limit nitrate to 10 mg/l (as N) for this reason. The major pathway of concern in land treatment is conversion of the wastewater nitrogen to nitrate and percolation to drinking water aquifers.

All three land treatment concepts are quite efficient in nitrification of the wastewater nitrogen. It is considered prudent in design to assume that all of the applied nitrogen is converted to nitrate. Since overland flow is a surface discharge system it is only the slow-rate and rapid-infiltration systems that are of concern for groundwater impacts.

Nitrogen is often the limiting factor for design of SR land treatment systems. As indicated in the design manual and numerous other sources it is possible to design an SR system to achieve almost any level of percolate nitrogen desired. This is done through proper selection of crop, application rate and season. Achieving 10 mg/l nitrogen in the percolate is usually no problem with SR land treatment, although winter storage may be required.

The very high loading rates inherent in RI systems result in the greatest potential for nitrate contamination of drinking waters. The principal mechanism for nitrogen removal in these systems is denitrification following in situ nitrification of the wastewater nitrogen. The effectiveness is dependent on temperature, carbon source, hydraulic loading rate, loading period and physical properties of the soil.

At the Hollister, California, system [2] where all conditions were favorable, a 93% nitrogen removal was achieved within a depth of 30 ft. Nitrate concentration at this depth was approximately 1 mg/l. Experimental work in

Phoenix, Arizona [3], approached 80% nitrogen removal by careful control of application rates and schedules. For the general case nitrogen removals of 30-50% are commonly achieved [1]. Therefore, the nitrate concentration in the percolate immediately beneath the basins will probably exceed the 10 mg/l standard. At Ft. Devens, Massachusetts, for example, the total nitrogen in the applied primary effluent averaged 47 mg/l and in sampling wells near the perimeter of the site it ranged from 15 to 20 mg/l [4]. Essentially all of this nitrogen was in the nitrate form.

Alternately, where the loading is too high, or lasts too long, or when the soil permits too rapid a movement of water, nitrification may not be complete. At Milton, Wisconsin [5], where there was continuous flooding on very coarse soils, there was no nitrification and therefore no denitrification and insignificant removal of nitrogen within the basin complex.

Temperature also inhibits the biochemical reactions involved and this has been demonstrated at Boulder, Colorado [6], Lake George, New York [7], and Ft. Devens [4], where ammonia concentrations in the percolate were higher in the colder months. At Vineland, New Jersey [8], percolate nitrification is minimal regardless of the season, probably because of the short loading cycles and rapid water movement through the soil profile. Nitrogen removal still averages about 40% and is probably due to entrapment of suspended matter and subsequent reactions in the near surface layer. At Ft. Devens the nitrogen removal was about 62% within the basin complex.

It is therefore prudent to assume for the general case that the nitrogen content of the percolate beneath RI basins will probably exceed 10 mg/l. However, these relatively high nitrogen concentrations do not then necessarily represent a health risk. In the case of Ft. Devens the ground water has no potential for drinking water and it emerges in a very short distance as subflow in an adjacent river. This subflow represents a much higher quality input to the river than could be achieved by conventional mechanical treatment and surface discharge. It should be possible to utilize this experience elsewhere and locate the RI basins close to existing surface water where geohydrologic studies have documented the groundwater flow paths. The area between the basins and the surface water body could then be zoned to prohibit extraction of ground water for drinking purposes. Such an approach would provide a higher quality product at a far lower cost than treatment systems with a surface discharge to the same water body.

Mixing and dispersion in the in situ ground water can also be effective. At Milton [5] the total nitrogen concentration was 20 mg/l at the edge of the basin complex and 10.6 mg/l approximately 160 ft down gradient. Other alternatives are to locate the basins over nondrinking water aquifers or to recover the percolate with wells or underdrains for reuse elsewhere, or for surface discharge. The system in Phoenix recovers the percolate for unrestricted use in irrigation. An alternative that has been proposed is to use a two-step land treatment consisting of overland flow followed by rapid infiltration. The overland flow slope receiving screened raw sewage would reduce nitrogen to acceptable levels prior to application on the RI basins. This concept was successfully demonstrated in small scale tests in Ada, Oklahoma [9].

Phosphorus

At present there are no drinking water or irrigation water standards limiting phosphorus so the issue of concern is with respect to eutrophication of surface waters that ultimately receive the product water from land treatment.

The principal pathways for phosphorus removal are to the crop or retention in the soil. Results from typical SR and OF systems indicate that generally 20-30% will be removed by agricultural crops. Any further removal must occur within the soil matrix and this is essentially the only pathway for rapid infiltration. In OF systems where the wastewater has limited contact with the underlying soil, phosphorus removal is limited to crop activity unless pre- or post-chemical treatments are used.

The percolate quality with respect to phosphorus is listed for several land treatment systems in Table III. Slow-rate systems can typically reduce phosphorus concentrations to less than 0.1 mg/l within very short travel distances. Further reductions to background levels would be expected with additional travel distance or detention time in the soil matrix. The coarser textured soils required for RI systems are less efficient at phosphorus removal as shown by Table III. This is due to the short contact time for the liquid and the lack of

Table III. Typical Phosphorus Responses

Location	Soil Type	Travel Distance (ft)	Percolate Phosphorus (mg/l)	Reference
Slow-Rate Systems:				
Hanover, NH	sandy loam	5	0.05-0.06	10
Muskegon, MI	loamy sand	5	0.03-0.05	11
Tallahassee, FL	fine sand	4	0.1	12
Tallahassee, FL	fine sand	35	0.0	12
Penn State (forest)	silt loam	4	0.08	13
Helen, GA (forest)	sandy loam	4	0.17	14
Rapid-Infiltration Systems:				
Hollister, CA	gravelly sand	22	7.4	2
Flushing Meadows, AZ	gravelly sand	30	4.5	3
Flushing Meadows, AZ	gravelly sand	330	1.0	3
Ft. Devens, MA	gravelly sand	5	9.0	4
Ft. Devens, MA	gravelly sand	370	1.1	4
Calumet, MI	gravelly sand	30	0.1	15

material with high sorptive capacity in the soil matrix. However, even in these cases sufficient travel distance or detention time should reduce phosphorus to acceptable levels. At both Flushing Meadows and Ft. Devens [3,4] several hundred feet of travel reduced phosphorus concentrations to about 1 mg/l. Further reductions would be expected at greater distances.

The mechanisms for phosphorus removal are thought to be adsorption followed by slower precipitation and mineralization reactions. These subsequent reactions then restore the adsorption sites for additional phosphorus removal. The capacity, or design life, of a site for phosphorus removal cannot therefore be predicted by simple adsorption chemistry. The Calumet, Michigan, system has been in operation for over 90 years and is still effective in removing phosphorus. Laboratory studies were conducted with the Hollister, California, rapid infiltration system soils [2] and compared to the actual phosphorus content at the treatment site. The adsorption predictions underestimated the actual 30-year accumulation by as much as a factor of 15. The useful life of the soils at the Phoenix rapid infiltration site have been estimated to be in excess of 100 years. If these soils were used in the slow-rate treatment mode that time might extend to several centuries.

Metals

The pathways of potential concern are movement of metals to drinking water sources and translocation of metals through the food chain to man. The latter is the more critical and may involve plants grown on a wastewater site consumed directly by man or the plants used as feed by animals, with further accumulation of metals in the animal and then to man. Cadmium is the metal of most concern in all of these situations. Metals are present in all municipal wastewaters; the concentration will depend on the type of community and industrial contributions. Table IV lists the range of concentrations of metals typically found in raw municipal sewage in the United States. Most municipal wastewaters tend to be near the lower end of this range. Also shown

Table IV. Metals in Wastewaters and Allowed Concentrations
in Drinking and Irrigation Waters [1]

Element	Raw Sewage (mg/l)	Drinking Water (mg/l)	Irrigation 20 yr[a]	Continuous[b] (mg/l)
Cadmium	0.004-0.14	0.01	0.05	0.005
Chromium	0.02-0.70	0.05	20	5.0
Lead	0.05-1.27	0.05	20	5.0
Zinc	0.05-1.27	0.05	20	5.0

[a]For fine textured soils only, normal irrigation practice for 20 yr.
[b]For any soil, normal irrigation practice, no time limit.

in Table IV are the concentration limits for these same metals for drinking and irrigation waters. For the general case the concentration of metals in untreated raw sewage is already below drinking and irrigation limits. Application of untreated sewage to the land, even if no renovation occurred in the land system, should not then pose a direct health risk for drinking waters or crops grown on the site.

Renovation and removal of metals is quite effective in all three land treatment modes with that effectiveness retained for considerable periods of time [1,16,17]. Studies at the previously described RI sites [1,4] indicate no adverse impact on groundwaters. Recently completed research at Hollister, California, San Angelo, Texas, Lake George, New York, Roswell, New Mexico, and Vineland, New Jersey [2,7,18] indicate no adverse impacts on either groundwaters or crops grown on the sites. Three (Hollister, Lake George and Vineland) are RI sites, the others are slow-rate with a variety of crops produced.

The removal of metals by the soils in a land treatment site is a complex, and often sequential interaction of reactions and responses including ion exchange, adsorption, complexation and precipitation, etc. It is not possible to predict the actual renovative capacity of a site with simple ion exchange or soil adsorption theories. Although the metals are accumulated in the soil profile the accumulation does not seem to be continuously available for crop uptake. Work with sludges by Hinesley [19,20] and others demonstrates that the metal uptake by crops in a given year is more dependent on the concentration of metals in the sludge most recently applied and not on the total accumulation of metals in the soil. The data in Table V demonstrate essentially the same relationship for forage grasses on land treatment sites. At Melbourne, Australia, after 76 years of application of raw sewage the cadmium concentration in the grass on the treatment site is just slightly higher than in the grass on the control site that received no sewage. The other three locations in the table are in California and range from 9 to 66 years of wastewater application at the time the samples were taken but the concentration of

Table V. Metal Content[a] of Forage Grasses at Land Treatment Sites [22]

Parameter	Melbourne, Aust. (started 1896) 1972 Control Sample	1972 Sample	Fresno, CA (started 1907) 1973 Sample	Manteca, Ca (started 1961) 1973 Sample	Livermore, CA (started 1964) 1973 Sample
Cadmium	0.77	0.89	0.9	1.6	0.3
Copper	6.5	12	16	13	10
Nickel	2.7	4.9	5	45	2
Lead	2.5	2.5	13	15	10
Zinc	50	63	93	161	103

[a]Concentrations are given in mg/l.

cadmium in the grasses is the same order of magnitude as at Melbourne. This seems to indicate that the metals in "this year's" wastewater application are potentially available for uptake by vegetation but that the long term accumulation in the soil is not. Recent studies in California with corn grown on sludge amended soils provides additional confirmation of this relationship [21].

Metals tend to accumulate in the liver and kidney tissue of animals grazing on a land treatment site or fed harvested products. A number of studies with domestic and indigenous animals do not show adverse effects. Tests done on a mixed group of 60 Hereford and Angus steers that graze directly on the' grasses at the Melbourne, Australia, site showed that: "the concentration of zinc, cadmium and nickel found in the liver and kidney tissues of this group of cattle are within the expected normal range of mammalian tissue" [23]. Hinesly [19] has fed corn grown on sludge-treated fields to pheasants and Anthony [24] has reported on metals in bone, kidney and liver tissue in mice and rabbits indigenous to the Pennsylvania State University slow-rate land treatment site with no adverse responses noted. In the latter case animals on the wastewater site were compared to animals from a nearby control site. For rabbits, the cadmium, nickel and lead were higher in kidney tissue than in those from the control site. Only copper was accumulating in bone tissue in the rabbits on the wastewater site. The mice on the wastewater site had higher concentrations of lead and cadmium in liver and kidney tissue but zinc and nickel were higher in the mice from the control site.

The useful life for metal retention by the soils at a land treatment site should be at least the same order of magnitude as previously discussed for phosphorus. Recent experimental work with typical New England soils indicates that the useful life for copper and cadmium, based only on ion exchange capacity, could extend to several hundred years for SR systems with seasonal application.

The average metal concentrations in the shallow groundwater beneath the Hollister rapid-infiltration site are shown in Table VI. After 33 years of oper-

Table VI. Trace Metals in Groundwater Under Hollister, CA, Rapid Infiltration Site [2]

Metal	Average Concentration (mg/l)
Cadmium	0.028
Cobalt	0.010
Chromium	0.014
Copper	0.038
Iron	0.36
Manganese	0.96
Nickel	0.13
Lead	0.09
Zinc	0.081

ation the concentration of cadmium, chromium and cobalt were not significantly different from normal offsite groundwater quality. The concentrations of the other metals were higher than background. Metal concentrations in the upper foot of soil at the Hollister site are compared in Table VII to values typically found in agricultural soils and plants. After 33 years of operation the accumulation of metals in the Hollister soils are still below or near the low end of the range cited as typical. This level of metal concentration would not adversely affect the future agricultural potential of the Hollister site. If the site had been operated in the SR mode rather than RI it would take over 150 years to apply the same volume of wastewater and contained metals.

When a land treatment operation is terminated a new set of equilibrium conditions develop in the soil matrix so there is a concern regarding potential movement of the previously retained and unavailable metals to either groundwaters or crops. Recent work at Otis AFB on Cape Cod lends some insight. Sand beds operated in the RI mode had been treating wastewater for 33 years [26]. One of the beds was inactivated in 1973. The metal concentrations in the upper soil layers were measured approximately one year later. The only liquid received on the bed in the intervening period was natural precipitation. Based on past records and extrapolation of metal concentrations in the present wastewater an analysis was made of how much the 33 year metal loading was still retained in the soil. This indicated that all of the applied cadmium and lead could be accounted for and that 85% of the copper, 62% of the chromium, and 49% of the zinc were still retained in the top 20 in. of the bed profile. Over 95% of all of the metals present were contained within the top 6 in. of the sand bed. The data do not indicate any pattern of mobility or downward movement of the contained metals during the year after wastewater applications ceased. Sandy soils of the type found at Otis AFB have a relatively high potential for ion mobility when compared to most agricultural soils.

Pathogens

The pathogens of concern in land treatment systems are bacteria, parasites and virus. The major pathways of concern are to the groundwater, internal or external contamination of crops, translocation to grazing animals, and offsite

Table VII. Trace Metals in Hollister, CA, Soils Compared to Typical Levels

	Metal (ppm)			
	Zn	Cd	Cu	Ni
Hollister treatment site, DTPA extractable, [2], 0-1 ft	1.1-11	0.04-0.24	1.4-8.2	0.33-2.4
Range, typically found in soils [25]	10-300	0.01-7.0	2-100	10-1000
Range, typically found in plants grown in typical soils [25]	15-200	0.2-0.8	4-15	1.0

transmission via aerosols or runoff. All will be discussed but the major concerns are with movement of viral particles either to the groundwater or via aerosols.

Parasites

Parasites may be present in all wastewaters. Larkin [27] describes recovery of *Ascaris, E. histolytica,* helminths and other types from sewages and sludges. Under optimum conditions the eggs of these parasites, particularly *Ascaris,* can survive for many years in the soil. Because of their weight parasite cysts and eggs will settle out in preliminary treatment or storage ponds so most will be found in sludges [28]. There is no evidence available indicating transmission of parasitic disease from application of wastewater in land treatment systems. Hays [28] described several situations where the use of raw sewage in gardens was the suspected source of disease. Such a practice is outside the limits recommended by the EPA (Table I) so should not be a factor in the United States. Transmission of parasites via aerosols should not be a problem due to the weight of the cysts and eggs. There is some evidence of animal infection. Beef measles and tapeworm have been documented [28] but are due to direct contact or drinking of the sewage by animals. Schistosomiasis which is a very serious parasitic problem in many parts of the world due to direct contact by humans with polluted waters is not in the continental U.S. because the alternate host snails are not present. Dean [29] indicates that *Ascaris* eggs have occasionally been found in the soil at the Paris, France, sewage farm which received essentially raw sewage but that vegetables produced on the farm are approved by the Ministry of Health and sold on the open market in Paris. Vegetables to be eaten raw are not permitted. This is compatible with the EPA guidance in Table I.

Table VIII. Retention of Metals in a Sandy Soil [26]

Depth (in.)	Percent of Total Metal Contained			
	Cd	Cr	Zn	Pb
0-1.5	84	87	82	88
1.5-2.4	12	10	13	12
6	1	0	1	0
10	1	2	2	0
12	1	0	0.8	0
18	0.5	1	1.2	0
20	0.5	0	0.0	0

Crop Contamination

Concerns with respect to crop contamination focus mainly on surface contamination and then persistence of the pathogens until consumed by man or animals or the internal infection of the plant via the roots. Larkin [30] demonstrated the persistence of poliovirus on lettuce and radishes for up to 36 days. However, in accordance with the guidance in Table I neither crop would be grown on a land treatment system in the United States. In addition 99% of the detectable virus were gone within the first 5-6 days. Shuval [31] described internal contamination of plants with virus via transport from the roots to the leaves. However, these results were obtained with soils inoculated with high concentrations of virus and then the roots were damaged or cut. No contamination was found when roots were undamaged or when soils were not inoculated with the high concentrations.

Criteria for irrigation of pasture with primary effluent in Germany require a period of 14 days after sprinkling before animals can be allowed to graze [32]. Experiments by Bell [33] in Canada indicated that fecal coliforms on the surfaces of alfalfa hay, from sprinkling with wastewater, were killed by 10 hr of bright sunlight. He also experimented with Reed Canary grass and found 50 hr of sunlight required. The longer period required is probably due to the sheath on the grass leaf which is not present on alfalfa. He recommended a 1-week period prior to grazing to ensure sufficient sunlight, for Reed Canary, orchard and brome grasses used for forage or hay. Since fecal coliforms have similar survival characteristics to salmonella he suggests results should be applicable to both organisms.

Runoff Contamination

Runoff from a land treatment site might be a potential pathway for pathogen transport. A proper system design would eliminate runoff from adjacent lands entering the site and runoff of applied wastewater, except for the overland flow case. Runoff of renovated effluent is a design feature of overland flow. It is controlled runoff, collected in ditches prior to surface discharge or further treatment. Deemer [34] indicates that the quality of rainfall runoff from overland flow sites is equal to or better in quality than the normal renovated wastewater runoff. Runoff is not a factor of concern for rapid infiltration systems. For slow rate systems if proper measures for erosion control are taken then runoff quality, if any occurs, should be no different than expected from normal agricultural practice.

Groundwater Contamination

The risk of groundwater contamination by pathogens involves movement of bacteria or virus to aquifers that are then used for drinking purposes without further treatment. The risk is not an issue for overland flow systems but has the highest potential for rapid infiltration due to the high hydraulic loading rates and coarse textures of the receiving soils.

Bacteria removal in the finer textured agricultural soils commonly used for slow-rate systems can be quite high. Results from a five-year study in Hanover, New Hampshire [10] applying both primary and secondary effluents to two different soils indicated essentially complete removal of fecal coliforms within a 5-ft soil profile. The soils involved were a fine textured silt loam and a coarser textured loamy sand; concentrations of fecal coliforms in the applied wastewater ranged from 105 for primary effluents to 103 for secondary effluents. Similar research in Canada [33] applied undisinfected effluent [4.5 cm/wk) to grass-covered loamy sand. Most of the fecal coliforms were retained in the top 3 in. of soil; none penetrated below 27 in. Die-off occurred in two phases: an initial rapid phase within 48 hr of application during which 90% of the bacteria died, followed by a slower decline during a two-week period when the remaining 10% were eliminated. Removal of virus which is dependent on adsorption reactions should also be quite effective on these finer textured soils.

Most of the concern, and the research work on virus transmission through soils has focused on RI systems and coarse textured soils. Table IX is a summary of results from a number of studies on this topic in the U.S.

The RI basins in the Phoenix work consisted of about 30 in. of loamy sand underlain by coarse sand and gravel layers. During the study period indigenous virus were always found in the applied wastewater but none were recovered in the sampling wells as shown in Table IX.

At Santee secondary effluent was applied to percolation beds in a shallow stratum of sand and gravel [37]. Percolate moved to an interceptor trench approximately 1500 ft from the beds. Enteric virus were isolated from the treatment ponds but none ever found at observation wells at 200 ft and 400 ft from the site. The data shown in Table IX for Ft. Devens are from a well beneath one of the basins receiving primary effluent. The soils in this case are a mixture of coarse sands and gravels with some pockets of finer textured soils. The stratification and layering are irregular in these glacial deposits so that underground channels of very high lateral permeability are possible. With respect to the indigenous virus naturally present in the wastewater an average of 97% were removed in the 17 meters of travel to this sampling well.

Indigenous virus were also recovered on an irregular basis from some of the wells at the perimeter of the site. An experiment was conducted where

Table IX. Virus Transmission Through Soil at Land Application Sites

Location	Sampling Distance (m)	Virus Concentration (PFU/l)	
		At Source	At Sample Point
Phoenix, AZ [3]	3-9	8	0
(Jan-Dec 1974)		27	0
		24	0
		2	0
		75	0
		11	0
Gainesville, FL [35,36]	7	0.14 (avg)	0.005
(April-Sept 1974)		0.14 (avg)	0
		0.14 (avg)	0
		0.14 (avg)	0
		0.14 (avg)	0
		0.14 (avg)	0
		0.14 (avg)	0
		0.14 (avg)	0
Santee, CA [37]	61	Concentrated Type 3 Polio	0
(1966)			
Ft. Devens, MA [4]	17	Indigenous virus	
(1974)		276 (avg)	8.3 (avg)
		f_2 Bacteriophage seed	
		2.2×10^5	1.3×10^5
Medford, NY [38]	0.75-8.34	Indigenous virus	17 samples
(Nov 1976-Oct 1977)		1.1-81.0	negative
			6 positive, at
			0.47 (avg)
			range 0.14-0.66
		Polio virus seed	
	0.75	7×10^4 (6 cm/hr) infiltration rate)	28.25
	0.75	1.84×10^5 (100 cm/hr infiltration rate)	97.5×10^4
Vineland, NJ [8]	0.6-16.8	13 (avg)	9 of 10 positive,
(Aug 1976-May 1977)	(10 wells)		1.62 (avg)
		13 (avg)	7 to 10 positive
			1.95 (avg)
		13 (avg)	2 of 10 positive
			0.48 (avg)
		13 (avg)	0 of 10 positive

the wastewater in the basin was spiked with f2 bacteriophage and a concentration of $2.2 \times 10^5/l$ maintained during the infiltration period. This is 1000 times the concentration of virus normally present in the wastewater. As shown in Table IX samples from the well contained a phage concentration in the same order of magnitude as applied. Perimeter wells also showed high concentrations of the phage but on an inconsistent, irregular basis.

The results from Medford, New York [38] are generally similar to those described above for Ft. Devens. The indigenous virus in the tertiary effluent at Medford were about the same concentration as in undisinfected primary effluent at Ft. Devens. The majority of composite samples of percolate taken directly under the Medford basin were negative for indigenous virus. Six samples were positive and indicated an average concentration of 0.47 pfu/l. Spiking experiments were also conducted here. The first with a concentration of 7×10^4 pfu/l was conducted at an infiltration rate more than 10 times higher than previously used produced high concentrations in the composite samples.

At Vineland, the soils are finer textured sands as compared to Ft. Devens. An average of 13 pfu/l of indigenous viral particles were measured in the applied wastewater. Samples from 10 wells at various locations and depths on the site were sampled on four occasions. Most were positive in the first set, none were positive in the last as shown on Table IX. Samples from off-site wells were negative.

Comparing the results in Table IX it would appear that application of virus in very high concentrations at very high infiltration rates on very coarse soils can result in movement into the soil profile. The probability of these three factors occurring together in a properly designed and managed system seems unlikely.

Lance (see Section VI, Chapter 2) and others have examined the problem of virus desorption in the laboratory. Using soil columns it can be shown that applications of distilled water or rainwater can cause adsorbed virus to move deeper into the soil profile under certain conditions. In Lance's work viruses were not desorbed if the free water in the column drained prior to application of the distilled water. This suggests that the critical period would be the first day or two after the wastewater infiltrated. Rainfall after that period should not then cause movement of virus. Even if some movement does occur, the soil profile in nature does not necessarily have a finite bottom like a soil column. A desorbed viral particle should have further opportunities for read-sorption in the natural case.

With the exception of those at Santee and Phoenix all of the studies shown in Table IX did show positive samples for virus even though the concentration and frequency may have been very low. This raises the issue of how many viral particles represent a health risk. The issue is at present unresolved and experts on both sides seem very positive in their opinions. Some will argue that a single PFU has the capability of initiating infection in man. The opposing point of view has been summarized by Lennette [39].

> The contention that a single viral particle invariably or even frequently constitutes a minimal infecting dose for man is simplistic and misleading. It fails to take into account the considerable amount of data from oral polio vaccine studies which are non-supportive of this contention and ignores the manifold factors associated with the humoral and cellular immune responses....

The health threat of the indigenous virus levels cited in Table IX to potential users of groundwaters at some distance seems minimal or almost nonexistent when compared to the impact of household septic tank-leach field systems on groundwaters. About 26% of the population of the U.S. still depend on on-site septic tank disposal and much of that group still depend on nearby untreated groundwater sources. If transport of pathogens through the soil was really a significant problem it would seem the incidence of disease in these groups would have been observed some time ago.

Aerosols

The potential for aerosol transport of pathogens from land treatment has probably been the single most controversial health issue in recent years and has involved the broadest spectrum of participants. The issue goes beyond technical and health aspects and involves aesthetics and emotions. The lay public, and many professionals, tend to misunderstand what aerosols are and confuse them with the droplets that emerge from the sprinkler nozzles. Aerosols are almost colloidal in size ranging from 20 microns in diameter or smaller. It is prudent to design any land treatment system so that the larger droplets are contained within the sites. The public acceptance of a project will certainly be enhanced if it is understood that neither their persons nor their property will get "wet" from sprinkler droplets.

Bacterial aerosols are present in all public situations and will tend to increase with the number of people and their proximity. Sporting events, theaters, public transportation, public toilets, etc., are all potential locations for airborne infection. Thirty years ago it was common to close the theaters during a polio outbreak since airborne transport of the virus was suspected.

Table X summarizes data on aerosol bacteria concentrations at various sources, all of which involve the use or treatment of wastewaters. The cooling water for the power plant that is cited uses some disinfected effluent as make-up water. It is interesting to note that the aerosol concentration at this cooling tower is roughly the same as measured just outside the sprinkler zone

Table X. Aerosol Bacteria at Various Sources

Source	Downwind Distance (m)	Total Aerobic Bacteria[a] (no./m^3)	Total Coliform (no./m^3)
Activated Sludge Tank, Chicago [41]	10-30	14,000	6
Activated Sludge Tank, Sweden [42]	0	100,000	–
Power Plant Cooling Tower (disinfected), Calif. [40]	0	2,921	–
Mechanical Aerated Pond, Israel [43]	30	–	279
Sprinklers (disinfected) Ohio [41]	30	500	2
Sprinklers, (undisinfected) Israel [43]	30	–	116
Sprinklers, (undisinfected) Arizona [44]	46	800	6
Sprinklers, (undisinfected) Pleasanton, Calif. [41]	10-30	2,580	6

[a]Aerosol counts are per cubic meter of air.

in Pleasanton, California, where undisinfected effluent is used. It does not appear from these data that the bacterial aerosols at or near land treatment sites are worse than other sources. In fact, the opposite seems true; the aerated pond in Israel and the activated sludge systems cited have higher concentrations than the comparable land treatment sites. Adams [40] cites aerosol studies in metropolitan areas and indicates a bacterial concentration of 155/m^3 during the day time in Louisville, and an annual average of 2019/m^3 (a range of 1084-2977) for Odessa, Russia. The bacterial aerosols from land treatment systems listed in Table X fall within this range. Table XI compares the downwind aerosol concentrations of a variety of bacteria and enterovirus from an activated sludge plant and an SR land treatment system. The latter was sprinkling undisinfected effluent and the 10-30 meter sampling point would be just outside the wetted impact circle of sprinkler droplets since the distance cited is from the sprinkler nozzle. The concentrations at the distances cited are about the same for the two sources.

Table XI. Aerosol Microorganisms–Activated Sludge and Slow Rate
Land Treatment [41]

Microorganism	Distance Downwind (m)	Concentration[a] (no./m^3)	
		Activated Sludge[b]	Land Treatment Sprinklers[c]
Total Coliform	10-30	5.9	5.5
	31-80	3.3	2.4
	81-200	1.5	0.7
Fecal Coliform	10-30	0.35	1.00
	31-80	0.25	0.47
	81-200	0.12	0.23
Coliphage	10-30	0.04	0.34
	31-80	0.02	0.38
	81-200	<.02	0.21
Fecal Strep	10-300	<1	1.44
	31-80	7	0.61
	81-200	<1	0.42
Enterovirus	30-50	<0.01	0.014

[a]Aerosol counts are per cubic meter of air.
[b]John Egan activated sludge plant, M.S.D. Chicago, IL.
[c]Pleasanton, CA, sprinkling undisinfected effluent.

Epidemiological studies were originally planned at both of these locations. The work at the activated sludge plant in the Chicago area [41] documented bacteria and virus in aerosols at the plant. However, microbial and chemical content of the air, the soil and the surface waters in the surrounding area were not different than background levels and no significant illness rates due to the activated sludge plant were revealed within a 5-km radius. The epidemiological investigation at the California land treatment site was cancelled. Another effort was undertaken at an activated sludge plant in Oregon with a school playground approximately 10 meters from the aeration tanks [45]. Positive counts for aerosol bacteria were noted in the school yard but no adverse health responses, as reported in school absentee records, were observed in the children. It can be inferred from these studies that since the concentrations of virus and bacteria in aerosols from land treatment are equal to or less than the concentrations from activated sludge and since there were no adverse health effects in the latter case there then should not be any for land treatment.

Work by Katzenelson [43] in Israel also documents the comparability of land treatment versus conventional modes. Table XII summarizes downwind coliform concentrations from sprinklers and a pond with surface aerators.

Katzenelson [46] has also published results of a study that implied a connection between aerosols from wastewater sprinklers and disease incidence in agricultural communities in Israel. The study was based on an examination of

Table XII. Aerosol Coliforms from Sprinklers and Aerated Pond in Israel [43]

Source	Concentration (no./m^3) at Distance				
	30 m	100 m	150 m	200 m	250 m
Aerated Pond[a]	279	4	4	0	0
Sprinklers[b]	116	21	7	0	4

[a]Temperature: 29°C; humidity: 75%; solar rad: 7-8; wind: 4-7 m/sec; endo medium.
[b]Temperature: 24°C; humidity: 75%; solar rad: 0-6; wind: 2-5 m/sec; endo medium.

records in the Ministry of Health, not on actual site visits. The records of 77 communities using wastewater for irrigation were compared to 130 communities that did not use wastewater. The incidence of shigellosis, salmonellosis, typhoid, and hepatitis was higher in the communities practicing wastewater irrigation. It was implied that aerosols were the responsible pathway.

No data were provided on how many of the 77 communities were actually downwind of the wastewater sprinklers where aerosols might be expected. The wastewater involved is essentially raw sewage by United States standards and in addition to use for irrigation it is used in fish ponds and in some cases the fish are consumed directly by the same community. A communal life style is practiced in these settlements with central kitchen and dining facilities. Field workers may also work in the kitchen and there is direct contact at meal time in the central dining area. It seems far more likely that these factors are responsible rather than aerosols. It would be more reasonable to impose routine hygienic procedures for the workers and kitchen staff rather than require a high level of wastewater treatment and disinfection prior to sprinkling when aerosols are probably not the cause of the problem. Regardless of the vectors involved, such results would not be anticipated at any land treatment system in the United States since the communal life style is not present.

The aerosol measurements at Pleasanton [41] demonstrated that salmonella and viruses survived longer than did the traditional coliform indicators. However, as shown in Table XI, the downwind concentrations of virus were very low, 0.014 PFU/m^3. The source was undisinfected effluent from high-pressure impact sprinklers and the sampling point was about 50 m from the sprinklers. The concentration cited is equal to 1 virus particle in every 71 m^3 of air. Assuming a normal breathing intake of approximately 0.02 m^3/min it would take 59 hr of continuous exposure for the operator to inhale that much air. It is estimated that the operators might spend about an hour per work day that close to the sprinklers, which is equivalent to the time spent servicing the aeration tanks at an activated sludge plant. At that rate the operator at Pleasanton would be exposed to less than 5 virus particles per year and the risk to the adjacent population would seem to be nonexistent. The exposure at an aeration tank would be much higher.

The EPA guidance (Table I) recommends a fecal coliform count of 1000/100 ml for agricultural use and 200/100 ml for recreational-type application. These two criteria are based on preexisting standards for general irrigation water and for bathing waters and body contact sports. With respect to the aerosol risk of sprinkling such waters, Shuval [47] has reported than when the coliform concentration at the nozzle was below 1000/100 ml there was no virus detected at downwind sampling stations.

It is possible to achieve reduction, in both bacteria and virus, to these recommended levels without complex mechanical treatment and disinfection. Figure 1, taken from the land treatment design manual, illustrates the storage time required for SR and probably OF systems in the United States. Research in Texas [3], as illustrated in Figure 2, shows that the reduction of virus in ponds is quite rapid at warm temperatures. Similar results have been demonstrated for fecal coliforms in wastewater ponds in Utah. Based on this work, Bowles [49] has reported an equation, based on Chick's law, which describes the die-away of fecal coliforms in a pond system as a function of time and temperature:

$$t = \frac{ln \dfrac{C_i}{C_f}}{K \theta^{(T-20)}}$$

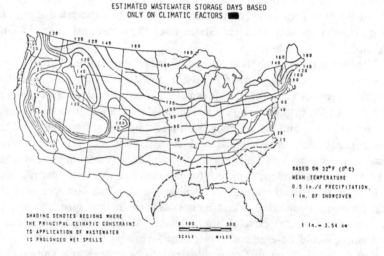

ESTIMATED WASTEWATER STORAGE DAYS BASED ONLY ON CLIMATIC FACTORS

BASED ON 32°F (0°C)
MEAN TEMPERATURE
0.5 in./d PRECIPITATION,
1 in. OF SNOWCOVER

SHADING DENOTES REGIONS WHERE
THE PRINCIPAL CLIMATIC CONSTRAINT
TO APPLICATION OF WASTEWATER
IS PROLONGED WET SPELLS

0   100   500
SCALE   MILES

1 in. = 2.54 cm

Figure 1. Estimated wastewater storage days based only on climatic factors.

where   t = "actual" detention time in days,
         C_i = entering concentration no./100 ml,
         C_f = final concentration no./100 ml,
         K = 0.5 warm months and 0.03 cold months,
         θ = 1.072 and
         T = liquid temperature (°C).

The time involved is the actual detention time since short circuiting occurs to varying degrees in almost every pond. The actual detention time in their work ranged from 25 to 89% of the design time with a geometric mean of 45%.

Using this equation it is possible to plot the curves on Figure 3 which illustrates the detention time required to achieve the EPA fecal coliform values at a particular liquid temperature. Approximately 19 days would be required to reach 1000/100 ml at 20°C. This is comparable to the time required for virus removal shown in Figure 2. Multiple cell ponds (at least 3) are recommended.

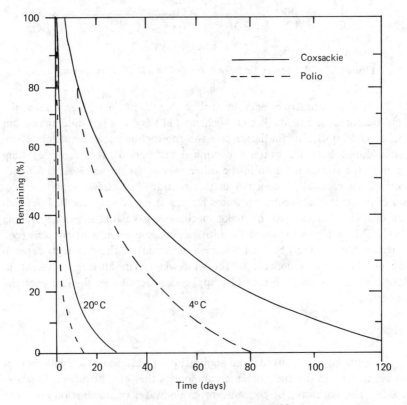

Figure 2. Virus survival in ponds [48].

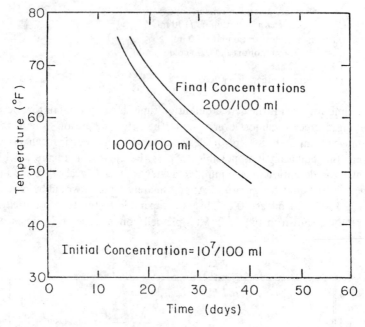

Figure 3. Fecal coliform die-away in ponds ("actual" time vs temperature).

Table XIII summarizes data on quality changes in the storage pond at the land treatment site in Muskegon, Michigan [11] and in a test holding pond in Haifa, Israel [50]. In the latter case the enterovirus were completely eliminated during both the winter and summer test periods. It is also interesting to note the significant drop in the other parameters in the Muskegon pond. Nitrogen is especially important in this regard. Work elsewhere with wastewater ponds indicates nitrogen losses ranging from 30% to over 90% depending on pH conditions, so the major mechanism is volatilization of ammonia [51]. This is a benefit where the nitrogen concentration is of concern (i.e., forested systems, etc.) but can also be a detriment to those systems depending on the nutrient content of the wastewater. The Muskegon system no longer disinfects prior to sprinkler application because of the low bacterial counts.

Organics

The principal concerns for land application of wastewaters are with respect to pesticides and the trace organic compounds that may be present in wastewaters. The pathways of concern are groundwater contamination and crop uptake. With respect to pesticides there is evidence that they are adsorbed in the soil profile and eventually biodegraded [52]. Table XIV summarizes

Table XIII. Quality Changes During Storage

Location	Input	Output
Muskegon, MI [11]		
Winter-spring storage BOD	81 mg/l	13 mg/l
SS	144 mg/l	20 mg/l
P	24 mg/l	1.4 mg/l
N	8.3 mg/l	5.6 mg/l
Fecal Coliform	$> 10^6$/100 ml	1000/100 ml
Haifa, Israel [50]		
73 day, winter season		
Total Coliform	2.3×10^7	1.84×10^4
Fecal Coliform	1.1×10^6	2.4×10^3
Fecal Strep.	1.1×10^6	5.0×10^2
Enteroviruses	1.1×10^3	0
35 day, summer season		
Total Coliform	1.4×10^7	2.3×10^4
Fecal Coliform	3.5×10^6	2.4×10^4
Fecal Strep.	6.0×10^5	3.7×10^3
Enterovirus	200	0
	no./100 ml	no./100 ml

Table XIV. Herbicide and Pesticide Removals

	Herbicides and Pesticides (ng/L)					
Location	Endrin	Lindane	Methoxychlor	Toxaphene	2,4-D	2,3,5 TP Silvex
Milton, WI, RI site [5]						
Wastewater	<0.03[a]	44	<0.01	<0.1	17.5	26.7
Groundwater A[b]	<0.03	157.6	<0.01	<0.1	92.4	41.5
Groundwater B	<0.03	3.9	<0.01	<0.1	23.6	38.6
Vineland, NH, RI site [8]						
Wastewater	<0.03	2830	<0.01	<0.1	9.5	72.0
Groundwater	<0.03	453	<0.01	<0.1	16.4	26.8
Roswell, NM, SR site [18]						
Wastewater	<0.03	560	<0.01	<0.1	29.0	28.0
Groundwater	<0.03	74.3	<0.01	<0.1	10.4	25.8
Dickinson, ND, SR site [53]						
Wastewater	<0.03	397	<0.01	<0.1	17.0	93
Groundwater	<0.03	53.6	<0.01	<0.1	6.2	47.1

[a]The "less than" symbol and associated number is the detection limit for the procedure used.

[b]At Milton, WI, groundwater A is beneath basins, groundwater B is approximately 150 ft down gradient.

results on herbicide and pesticide removals at four recently studied land application systems. Except for the Milton, Wisconsin, rapid infiltration site, which is continuously flooded and presumably continuously anaerobic beneath the basins, the systems showed very significant removals. Even the Milton site achieved comparable concentrations after another 160 ft of down gradient lateral travel. At the Milton, Vineland and Roswell sites the concentrations of lindane in the application site groundwater was higher than background. However, in every case the concentration of lindane, 2,4-D; and 2,35 TP Silvex were all well below the limitations proposed for drinking waters. Endrin, methoxychlor and toxaphene were never found in the wastewaters at any site above the detection limits specified in Table XIV. Unless there is a spill from an industrial plant producing these materials it is unlikely that excessive concentrations will appear in municipal wastewaters. Such pesticides and herbicides are more likely to be used during the agricultural management of slow rate sites producing crops for the market. In these cases they would be used at rates that do not exceed normal agronomic practice.

Concern over trace organics arose with detection of chlorinated hydrocarbons and similar material in drinking water supplies. For land treatment systems the concern is that such organic compounds will migrate through the soil profile and accumulate in drinking water aquifers. The principal concern is with respect to the chlorinated hydrocarbons which can occur as the inadvertent result of mixing industrial effluents or as a direct result of chlorinating municipal wastewaters. Work is currently underway to quantify the fate of specific trace organic compounds at operating land treatment sites in the U.S.

Some preliminary expectations are possible by considering the experience in West Germany with filtering contaminated river water through natural sand beds. It has been common practice for many decades on the lower Rhine and Ruhr rivers to draw the river water through natural sand beds prior to further treatment or distribution. The presence of organic compounds, especially chlorinated hydrocarbons, has become particularly severe. These are primarily due to industrial discharges into the river systems. A typical GC-MS chromatogram for the lower Rhine shows over 100 organic compounds in very significant amounts [54]. After passage through the sand bed the number is reduced to 40 or less and the concentration of these was significantly lower than the raw water [54]. The removal percentages of some specific chlorinated hydrocarbons in these same sand beds are listed in Table XV. It can be seen that removal efficiency decreases as the chlorine content of chloraromatic compounds increases. This would infer that chlorine disinfection of wastewater prior to land application will reduce the efficiency of organic removal.

The removal of chlorinated organics by natural sand and by activated carbon is compared in Table XVI. The source water was again the Rhine River

Table XV. Chlorinated Hydrocarbon Removals by a Sandy Soil [54]

Substance	% Removal
Chlorobenzene	96
Dichlorobenzene	62
Trichlorobenzene	18
Chlorotoluene	94
Dichlorotoluene	62

as described above [54]. For the categories listed the sand demonstrates generally comparable removal efficiencies to the activated carbon. An individual sand particle does not have anywhere near the adsorptive capacity of an activated carbon particle so the responsible factor must be related to detention time. The detention time in the sand system is measured in terms of days or weeks while it may be minutes or hours for the activated carbon system. The finer textured agricultural soils normally used for slow rate systems could be expected to demonstrate a greater efficiency than sand.

A study at the Muskegon, Michigan, land treatment site [11] identified 56 organic compounds in the raw sewage. This is higher than would be expected for municipal wastewater, but 60% of the flow in Muskegon comes from industrial sources. Only 17 of the compounds were present after aeration and detention in the storage ponds. Five of these were found in very low concentrations in the renovated water recovered from the site underdrains. Intake of these organics by plants grown on such sites is considered unlikely [55]. The large molecular weight organics do not tend to pass the semipermeable membrane of plant roots. Surface adherence to the crop is possible. However, as indicated previously, production of vegetables to be eaten raw is not encouraged and a lag time is recommended prior to access for grazing animals.

Table XVI. Natural Sand vs Activated Carbon for Chlorinated Organics Removal [54]

Substances	% Removal	
	Sand	Activated Carbon
Dissolved Organic Chlorine	38	72
Dissolved Organic Chlorine, Nonpolar	73	73
GC Detectable Organic Chlorine	49	37
Dissolved Organic Carbon	68	76
GC Detectable Organic Carbon	78	64

CONCLUSIONS

1. Proper design and operation of land treatment systems will eliminate or control adverse environmental or health impacts from the wastewater nutrients nitrogen and phosphorus.

2. The useful life of slow rate and rapid infiltration sites for removal of phosphorus and metals extends in the general case for at least several decades and exceeds the normal design life of the treatment system.

3. The accumulated phosphorus and metals in the soils after long term operation pose no significant environmental or health risk either during continued operation or after closure of the site.

4. Pathogens are the area of greatest concern but proper design and operation of the land treatment system provide for protection of public health. The potential impacts are transient in nature and should not increase with long term operation or prevail long after site closure.

5. Based on the limited data currently available land treatment systems seem to be more effective than conventional technologies in removing organics. For the general case the accumulated organics would not inhibit subsequent use of the site for other agricultural purposes.

6. Land treatment of wastewater is a viable treatment alternative. The long-term environmental and health aspects are generally of lesser concern than with the conventional mechanical technologies.

RECOMMENDATIONS

1. Location of land treatment sites above closed or "perched" groundwater systems should be carefully considered. If a high rate of groundwater withdrawal for public use is combined with little or no natural recharge or lateral flow a high level of nitrates and other dissolved minerals could result from the continued recycle and reuse of essentially the same water.

2. The management of soil pH in the near neutral range is recommended for agronomic use of sites after termination of land treatment when accumulated metal concentrations are high.

3. The cultivation of human food crops to be eaten raw is not recommended during operation or for a year or two after cessation of wastewater application.

4. Contact of dairy animals with either wastewaters or sludges should be avoided. Other grazing animals can be permitted on wastewater irrigated pasture after a week or two depending on the climate.

5. Chlorine disinfection of wastewaters prior to land application should be avoided if possible to maximize potential for removal of trace organic compounds in the soil system.

REFERENCES

1. *Process Design Manual for Land Treatment of Municipal Wastewaters,* U.S. EPA Manual 62519-81-006 (1981).
2. Pound, C. E., et al. "Long Term Effects of Land Application of Domestic Wastewater—Hollister, CA," U.S. EPA Report 600/2-78-084, ORD (1979).
3. Bouwer, H., and R. C. Rice. "The Flushing Meadows Project," *Proceedings, Land Treatment Symposium,* USACRREL. Hanover, NH (1978), pp. 213-220.
4. Satterwhite, M. B., et al. "Rapid Infiltration of Primary Sewage Effluent at Ft. Devens, MA," USACRREL Report 76-49 (1976).
5. Benham-Blair, et al. "Long-Term Effects of Land Application of Domestic Wastewater: Milton, WI, Rapid Infiltration Site," U.S. EPA Report 600/2-79-145, ORD (1979).
6. Smith, D. G., et al. "Treatment of Secondary Effluent by Infiltration-Percolation," U.S. EPA Report 600/2-79-174, ORD (1979).
7. Aulenbach, D. B. "Long Term Recharge of Trickling Filter Effluent into Sand—Lake George, NY," U.S. EPA Report 600/2-79-068, ORD (1979).
8. Koerner, E. L., and D. A. Haws. "Long-Term Effects of Land Application of Domestic Wastewater—Vineland, NJ," U.S. EPA Report 600/2-79-072, ORD (1979).
9. Thomas, R. E., et al. "Feasibility of Overland Flow for Treatment of Raw Domestic Wastewater," U.S. EPA Report 66/2-74-087, ORD (1974).
10. Iskandar, I. K., et al. "Wastewater Renovation by Slow Infiltration Land Treatment System," USACRREL Report 75-19 (1976).
11. U.S. EPA. "Is Muskegon County's Solution Your Solution?" U.S. EPA Region V (1976).
12. Overman, A. R. "Wastewater Irrigation at Tallahassee, FL," U.S. EPA Report 600/2-79-151, ORD (1979).
13. Sopper, W. E., and S. N. Kerr. "Utilization of Domestic Wastewater in Forest Ecosystems," *Proceedings of the Land Treatment Symposium,* USACRREL, Hanover, NH (1978), pp. 333-340.
14. Nutter, W. L., et al. "Renovation of Municipal Wastewater by Spray Irrigation on Steep Forest Slopes in the Southern Appalachians," *Proceedings, Recycling Treated Wastewater and Sludges Through Forest and Cropland,* Pennsylvania State University, June 1979.
15. Baillod, C. R., et al. "Evaluation of 88 Years Rapid Infiltration of Raw Municipal Sewage at Calumet, MI," in *Proceedings of the 1976 Cornell Agricultural Waste Management Conference* (Ann Arbor, MI: Ann Arbor Science Publishers, Inc., 1977), pp. 511-532.
16. "Application of Sludges and Wastewaters on Agricultural Land," MCD-35, OWPO, U.S. EPA (1978).
17. Leeper, G. W. "Reactions of Heavy Metals with Soils with Special Regard to Their Application in Sewage Wastes," U.S. Army Corps of Engineers (1972).
18. Koerner, E. L., and D. A. Haws. "Long-Term Effects of Land Application of Domestic Wastewater—Roswell, NM," U.S. EPA Report 600/2-79-ORT (1979).
19. Hinesly, T. D., et al. "Effects on Corn by Applications of Heated Anaerobically Digested Sludge," *Compost Sci.* 13(4):26-30 (1972).

20. Hinesly, T. D., et al. "Soybean Yield Responses and Assimilation of Zn and Cd from Sewage Sludge Amended Soil," *J. Water Poll. Control Fed.* 48:2137-2152 (1976).
21. Hyde, H. C., et al. "Effect of Heavy Metals in Sludge on Agricultural Crops," *J. Water Poll. Control Fed.* (SI) 10:2475-2486 (1979).
22. Reed, S. C., et al. "Pretreatment Requirements for Land Application of Wastewaters," presented at the National Conference of ASCE, Environmental Engineering Division (1975).
23. Anderson, N. Notice Paper No. 15, Victoria, Australia, Legislative assembly, June 2, 1976.
24. Anthony, R. G., et al. "Effects of Municipal Wastewater Irrigation on Selected Species of Animals," *Proceedings of the Land Treatment Symposium,* USACRREL, Hanover, NH (1978), pp. 281-288.
25. Page, A. L. "Fate and Effects of Trace Elements in Sewage Sludge When Applied to Agricultural Lands," U.S. EPA Report 600/2-74-005, ORD (1974).
26. Vaccaro, R. F., et al., "Wastewater Renovation and Retrieval on Cape Cod," U.S. EPA Report 600/2-79-176, ORD (1979).
27. Larkin, E. P., et al. "Land Application of Sewage Wastes: Potential for Contamination of Foodstuffs and Agricultural Soils by Viruses, Bacterial Pathogens and Parasites," *Proceedings of the Land Treatment Symposium,* USACRREL, Hanover, NH (1978), pp. 215-234.
28. Hays, B. D. "Potential for Parasitic Disease Transmission with Land Treatment of Sewage Plant Effluents and Sludges," Journal IAWPR, 11:583-595 (1977).
29. Dean, R. B. "The Sewage Farms of Paris," *Proceedings of the Land Treatment Symposium,* USACRREL, Hanover, NH (1978), pp. 253-256.
30. Larkin, E. P. "Persistence of Virus on Sewage Irrigated Vegetables," *J. Am. Soc. Civil Eng., Environ. Eng. Div.* 102:29-36 (1976).
31. Shuval, H. I. "Land Treatment of Wastewater in Israel," *Proceedings of the Land Treatment Symposium,* USACRREL, Hanover, NH (1978), pp. 429-436.
32. Popp, L. "Irrigation with Sewage from the Hygienic Point-of-View," International Commission on Irrigation and Drainage, presented at 5th Regional Meeting, Cambridge, England (1967).
33. Bell, R. G., and J. B. Bole. "Elimination of Fecal Coliform Bacteria from Soil Irrigated with Municipal Sewage Lagoon Effluent," *J. Environ. Qual.* 7:193-196 (1978).
34. Deemer, D. O. "Overland Flow," U.S. EPA Technology Transfer Seminar on Land Treatment, ERIC, June 1979.
35. Wellings, F. M., et al. "Demonstration of Virus in Groundwater after Effluent Discharge onto Soil," *Appl. Microbiol.* 29(6):751-757 (1975).
36. Wellings, F. M., et al. "Virus Considerations in Land Disposal of Sewage Effluents and Sludge," *Florida Sci.* 38(4):202-207 (1975).
37. Merrell, J. C., et al. "The Santee Recreation Project, Santee, CA," Report WP-20-7, Federal Water Pollution Control Administration (1967).
38. Vaughn, J. M., and E. F. Landry. "The Occurrence of Human Enteroviruses in a Long Island Groundwater Aquifer Recharged with Tertiary Wastewater Effluents," *Proceedings of the Land Treatment Symposium,* USACRREL, Hanover, NH (1978), pp. 233-244.

39. Lennette, E. H. "Problems Posed to Man by Viruses in Municipal Wastes," *Proceedings of the Symposium on Virus Aspects of Applying Municipal Waste to Land* (Gainesville, FL: University of Florida, 1978), pp. 1-7.

40. Adams, A. P., et al. "Bacterial Aerosols from Cooling Towers," *J. Water Poll. Control Fed.* 50(10):2362-2369 (1978).

41. Camann, D. E., et al. "Evaluating the Microbiological Hazard of Wastewater Aerosols," Contract Report DAMD 17-75-C-5072, USAMBRDL, Ft. Detrick, MD (1978).

42. Bergstron, et al. "Comparison of Bacterial Content in the Air of Some Sewage Treatment Plants," Forsvarets Forskningsanstalt, Sundbyberg, Sweden (1973).

43. Katzenelson, E., and B. Teltch. "Dispersion of Enteric Bacteria by Spray Irrigation," *J. Water Poll. Control Fed.* 48(4):710-716 (1976).

44. Sorber, C. A., et al. "A Study of Bacterial Aerosols at a Wastewater Irrigation Site," *J. Water Poll. Control Fed.* 48:2367-2379 (1976).

45. Johnson, D. E., et al. "Environmental Monitoring of a Wastewater Treatment Plant," U.S. EPA Report 600/1-79-027, Cincinnati, OH (1979).

46. Katzenelson, E. "Risk of Communicable Disease Infection Associated with Wastewater Irrigation in Agricultural Settlements," *Science* 194:944-946 (1976).

47. Shuval, H. I., and B. Teltsh. "Hygienic Aspects of the Dispersion of the Enteric Bacteria and Viruses by Sprinkled Irrigation of Wastewater," *Proceedings of the 1979 AWWA Water Reuse Symposium* (Denver, CO: American Water Works Association, 1979).

48. Sagik, B. P., et al. "The Survival of Human Enteric Virus In Holding Ponds," Contract Report DAMD 17-75-C-5062, U.S. Army Medical Research and Development Command (1978).

49. Bowles, D. S., et al. "Coliform Decay Rates in Waste Stabilization Ponds," *J. Water Poll. Control Fed.* 51(1):87-99 (1979).

50. Kott, Y., et al. "Lagooned Secondary Effluents as Water Source for Extended Agricultural Purposes," *Water Research* 12(12):1101-1106 (1978).

51. Reed, S. C. "Treatment-Storage Ponds for Land Application Systems," USACRREL Report, (in preparation).

52. Reed, S. C., et al. "Wastewater Management by Disposal on the Land," SR171, USACRREL (1972).

53. Benham-Blair, et al. "Long-Term Effects of Land Application of Domestic Wastewater: Dickinson, ND—Slow Rate," U.S. EPA Report 600/2-79-144, ORD (1979).

54. Kohn, W., et al. "Use of Ozone-Chlorine in Water Utilities in the Federal Republic of Germany," *J. Am. Water Works Assoc.* 70:326-331 (1978).

55. Kowal, N. E., and H. R. Pahren. "Health Effects Associated with Wastewater Treatment and Disposal," *J. Water Poll. Control Fed.* 51:1301 (1979).

CHAPTER 31

PERSISTENCE OF CHEMICAL POLLUTANTS
IN WATER REUSE

A. J. Englande, Jr. and Robert S. Reimers III
Department of Environmental Health Sciences
Tulane University School of Public Health
New Orleans, Louisiana

INTRODUCTION

Desired quality of water for reuse varies with anticipated usage. Reuse applications consequently require emphasis on the removal of selected physical, chemical and biological constituents to use-specific, nonadverse levels. Treatment plant operating experience has indicated several types of contaminants, particularly chemical parameters, to be difficult and expensive to remove adequately with currently employed waste treatment technology. These persistent constituents include: nitrogen (ammonia and NO_3), phenol, total dissolved solids, trace organics, boron and specific heavy metals [1-3].

This chapter focuses on chemical pollutants (and in particular on trace metals and trace organics) which tend to persist through biological and physical-chemical treatment systems for both municipal and industrial waste waters, and factors which affect leakage (complexation, redoxpotential, speciation, bioactivity, etc.). Removal efficiencies of specific unit operations for important parameters are also assessed based on operating data derived from pilot and full-scale waste treatment facilities. Average achievable effluent concentrations are compared to conventional criteria for potable, aquatic life propagation, groundwater injection and irrigation water purposes.

METHODS AND RESULTS

Allowable contaminant content of treated effluents depend on intended use. Table I indicates pertinent water quality criteria for various use classifications. Data are analyzed using Table I as a basis for comparison to "flag" specific parameters of concern for reuse application. Data analyzed herein include: (1) results of a survey comparing effluent quality from five AWT plants to potable water quality criteria; (2) long-term data evaluation from pilot and full-scale studies to determine the effectiveness of chemical contami-

Table I. Water Quality Criteria for Selected Chemical Parameters

Parameter	Drinking Water Permissible Levels [4,5]	Aquatic Water Minimum Risk to Hazard Level [5,6]	Groundwater Injection [7]	Irrigation
Ammonia Nitrogen (N)	–	0.01-0.04 mg/l	1.0 mg/l	–
Total Nitrogen (N)	–	–	10 mg/l	–
Nitrates (as N)	10 mg/l	–	–	–
Phenol	–	1 μg/l	1 μg/l	–
Total Dissolved Solids (TDS)	500 mg/l	–	–	–
COD	–	–	30 mg/l	–
CCE	70 μg/l[a]	–	–	–
Ag	50 μg/l	1-5.0 μg/l	50 μg/l	–
As	50 μg/l	–	50 μg/l	100 μg/l
B	–	–	0.5 mg/l	750 μg/l
Ba	1000 μg/l	–	1000 μg/l	–
Cd	10 μg/l	0.4-1.2 μg/l (soft to hard waters) 5 μg/l marine	10 μg/l	10 μg/l
Cr^{+6}	50 μg/l	100 μg/l	59 μg/l	100 μg/l
Cu	1000 μg/l	0.1 x 96 hours LC_{50}	1000 μg/l	200 μg/l
Fe	300 μg/l	1.0 mg/l	300 μg/l	5 mg/l
Hg	2 μg/l	0.05-0.10 μg/l (fresh to marine water)	5 μg/l	–
Mn	50 μg/l	100 μg/l (marine)	50 μg/l	200 μg/l
Pb	50 μg/l	30 μg/l	50 μg/l	5 mg/l
Se	10 μg/l	5-10 μg/l	10 μg/l	20 μg/l
Zn	5000 μg/l	20-100 μg/l	–	2.0 mg/l
Trihalo- methane	100 μg/l	–	–	–
Endrin	0.2 μg/l	0.004 μg/l	–	–
Lindane	4.0 μg/l	0.01 μg/l	–	–
Methoxychlor	100 μg/l	0.03 μg/l	–	–
Toxaphene	5 μg/l	0.005 μg/l	–	–
2,4,5-D	100 μg/l	0.02 μg/l	–	0.1 μg/l

[a]Omitted in final regulations.

nant removal by biological, tertiary (AWT) and physical-chemical processes; and (3) results of pilot and full-scale studies to derive relative removal efficiencies of selective unit operations for parameters of concern.

AWT Plant Survey

Extensive chemical, physical and biological analyses were conducted on periodic samples of effluent taken from five AWT demonstration facilities treating predominantly domestic wastewaters [8]. Results were compared to current U.S. drinking water regulations. Those pilot and full-scale plant sites evaluated included: Blue Plains, District of Columbia; Dallas, Texas; and Lake Tahoe, Pomona and Orange County, California. These systems were chosen primarily because of availability and because effluent quality exceeded that of secondary treatment systems. It should be noted that the Orange County facility was tested during a period of initial start-up. None of the systems were designed to produce a water of potable quality as each was part of an independent project with specific individual goals.

Spot samples taken over a 6- to 9-month period indicated that certain chemical parameters tended to persist through the various treatment schemes and were found to exceed drinking water standards in most of the treated effluents. These parameters included: nitrogen (ammonia and nitrate), phenol, carbon chloroform extract (CCE), total dissolved solids (TDS) and specific heavy metals. (Note that the CCE parameter was deleted from the 1975 Interim Drinking Water Standards.)

Survey data were evaluated by tabulating pertinent effluent constituent concentrations and their respective exceedance ratios (number of samples surpassing potable water quality/total number of spot samples evaluated). Nitrogen (ammonia and nitrate), total dissolved solids and phenol exhibited the highest exceedance ratios of the facilities evaluated. The Blue Plains facility, which employs biological nitrification-denitrification, was the only system to maintain residual ammonia concentration within acceptable limits. The treatment plants studied were not designed to remove TDS and hence in most cases the TDS content increased due to chemical treatment. Phenolic compounds in levels exceeding the 1 μg/l standard were found in all effluents. Except for Orange County, CCE concentrations also exceeded the originally proposed limit of 0.7 mg/l for all cases.

Selenium appeared to be the most persistent heavy metal characterized by high exceedance ratios for effluents from all facilities except Orange County. Chromium, arsenic and iron also were relatively stable and little removal was effected through the AWT plants evaluated. Cadmium, lead, mercury, manganese and silver occasionally exceeded potable criteria. AWT facilities employing lime treatment reported the highest and most consistent reduction in heavy metals concentrations due to precipitation, coprecipitation and adsorption by hydroxides and carbonates formed at high pH. Pomona, which did not employ lime treatment, experienced the highest diversity of specific heavy metal exceedance ratios. All systems, however, were characterized by high quality effluents and product water approaching potable quality. A detailed evaluation of these data can be found in Reference 8.

Removal of Chemical Contaminants by Conventional Waste Treatment Technology

The spot sampling approach employed during the preceding study underscored specific contaminants of potential concern including effluent variability considerations. Average long-term performance capabilities, however, were not defined. Two of the plants, Dallas and Orange County, have subsequently published data continuously collected over 18- and 6-month periods, respectively [9,10]. These data along with that from pilot and full-scale studies concerned with removal and pass through of pollutants in conventional biological, tertiary and physical-chemical waste treatment processes, are summarized as follows.

1. Chemical Pollutant Reduction by Biological Methods. A recent "state-of-the-art" review of trace metal removals from wastewaters by Cohen [11] indicated ranges of percent removals obtainable by biological treatment. These results, in parts, are presented in Table II. Also shown are projected secondary treatment reductions in metals content as experimentally determined by the Sanitation District of Los Angeles County for assessing industrial effluent concentration limitations [12]. Long-term chemical pollutant removal efficiencies are given for the Dallas activated sludge plant for the facility previously described. Results of a U.S. Environmental Protection Agency (EPA) survey [13] of secondary treatment effectiveness for chemical contaminant removal are also included, as are results from the EPA *Treatability Manual* [14] concerned solely with industrial wastewaters treatability experience.

Results tabulated in Table II show a wide range of demonstrated removal efficiencies which appear largely unpredictable. Inconsistencies are partly due to the variability of characteristics of the influent wastewaters, the complexities of the biological wastewater stabilization processes and overall process efficiencies. The Orange County AWT plant, for example, recently observed significant improvement in effluent quality following a change in secondary treatment from trickling filter to activated sludge and with the exclusion of a high nitrogen content industrial waste from discharging into the collection system [15]. This observation was substantiated by the EPA *Treatability*

Table II. Removal of Selected Contaminants by Biological Treatment

	Ranges of Removals Reported by Cohen [11]	LA Sanitary District Projected Removals [12] Residual Metal		Activated Sludge Removal Dallas [9] Residual Parameter		EPA Study Removal [13] Residual Parameter		EPA Treatability Manual Industrial Wastewater [14] [a]	Average Achievable
	(%)	(%)	(μg/l)	(%)	Concentration	(%)	Concentration	Removal Range (%)	Concentration (μg/l)
TOC	–	–	–	70.7	12 mg/l	60.0	25 mg/l	8->95	427 μg/l
COD	–	–	–	80.7	50 mg/l	73.0	110 mg/l	0-96	890 μg/l
NH3-N	–	–	–	67.3	5 mg/l	42.0	14 mg/l	–	–
TKN	–	–	–	61.3	9 mg/l	34.0	18 mg/l	26-63	174 μg/l
NO2 & NO3-N	–	–	–	Inc.[b]	5.1 mg/l	–	–	–	–
Phenol	–	–	–	–[b]	–	45.0	175 μg/l	82->99	79 μg/l
Ag	–	69	5.3	Inc.[b]	2.6 μg/l	–	–	31->96	32 μg/l
As	–	48	5.2	18.7	17.0 μg/l	–	–	43-96	35 μg/l
B	–	–	–	Inc.[b]	300.0 μg/l	–	–	–	–
Ba	–	–	–	33.3	120.0 μg/l	–	–	–	–
Cd	20-45	73	5.7	58.3	5 μg/l	18	30 μg/l	31-99	4 μg/l
Cr	40-80	77	240[c]	64.9	27 μg/l	42	218 μg/l	45-99	910 μg/l
Cu	0-70	76	–	79.4	29 μg/l	56	113 μg/l	98->99	95 μg/l
Fe	–	–	–	9.2	590 μg/l	57	1827 μg/l	–	–
Hg	20-75	84	0.19	44.8	0.16 μg/l	35	3.5 μg/l	30-87	<0.8 μg/l
Mn	–	–	–	42.3	41 μg/l	35	140 μg/l	–	–
Ni	–	–	–	–	–	21	182 μg/l	–	–
Pb	50-90	80	58	52.8	34 μg/l	38	92 μg/l	49-99	40 μg/l
Se	–	–	–	34.1	2.9 μg/l	–	–	0	41 μg/l
Zn	35-80	77	497[c]	44.2	63 μg/l	52	277 μg/l	35-92	200 μg/l

[a] Activated sludge treatment.
[b] Inc. = Increase.
[c] Source controls to meet ocean fallout criteria.

Manual for industrial wastewaters which indicated activated sludge to be the most effective biological waste treatment alternative for contaminants removal. The acceptability of variations in effluent quality will depend on the influent pollutant concentration, the intended use of the product water, and the dilution capacity at the point of discharge. This point is particularly illustrated by the Los Angeles Sanitary District results. Since significant industrial waste contributions are realized, source control will be required for chromium and zinc for compliance with ocean discharge limitations.

A comparison of nitrified Dallas effluent quality to water quality criteria for aquatic life (Table I) indicates exceedance with respect to ammonia nitrogen, silver, cadmium, mercury, lead and zinc. Results of an EPA survey of 163 biological treatment plants indicate aquatic life criteria exceedance with respect to ammonia, phenol, cadmium, chromium, iron, mercury, manganese, lead and zinc. Data from the EPA Industrial Pollutant Removal Survey indicated criteria exceedance for drinking water: chromium and selenium; for aquatic life: phenol, cadmium, chromium, mercury, lead and zinc; for ground water injection: COD, TKN, phenol, and chromium; and for irrigation: chromium. High residual concentrations were observed in many cases. The importance of adequate dilution capacity of the receiving water is again illustrated. Since the Dallas municipal plant is servicing approximately 12% industrial flow by volume and the plants surveyed by EPA an unknown industrial contribution, any further industrial increase would warrant added concern.

2. Chemical Pollutant Removal by AWT Methods. The treatment system which shows most promise in effectively achieving an effluent of consistently high quality is a combination of biological and physical-chemical processes. Table III illustrates overall tertiary, physical-chemical treatment removals of chemical contaminants from the Dallas and Orange County studies. The treatment sequence employed at Dallas following nitrifying activated sludge includes high lime coagulation (pH = 11) and clarification, recarbonation, mixed media filtration, carbon adsorption and chlorination, whereas that at Orange County following trickling filtration consists of high lime coagulation (pH > 11) and clarification, ammonia stripping, two-stage recarbonation, mixed media filtration, activated carbon adsorption and chlorination. Both plants treat wastewaters of low industrial content as demonstrated by the low incoming concentrations of metals.

As indicated in Table III, treated effluents were generally of excellent quality especially with respect to potable criteria. Some difficulties, however, are experienced in meeting stringent requirements for aquatic life, particularly if only minimum dilution is provided by the receiving water. Aquatic life criteria were exceeded at Dallas for ammonia, silver, cadmium, mercury, lead and zinc. Orange County experienced exceedance for aquatic quality (ammonia, phenol, silver, cadmium, mercury and zinc), groundwater injection

Table III. Removal of Selected Parameters by Tertiary Physical-Chemical Treatment

Parameter	Dallas [9]			Orange County [10]		
	Initial Concentration	Removal (%)	Residual Concentration	Initial Concentration	Removal (%)	Residual Concentration
TOC	12 mg/l	44.2	6.7 mg/l	–	–	6.7 mg/l
COD	50 mg/l	92.0	4.0 mg/l	142 mg/l	87.3	18 mg/l
NH$_3$-N	5 mg/l	28.0	3.6 mg/l	45 mg/l	93.0	3.1 mg/l
TKN	9 mg/l	50.0	4.5 mg/l	53 mg/l	91.0	4.8 mg/l
NO$_2$ & NO$_3$-N	5.1 mg/l	0	5.1 mg/l	–	–	–
TDS	479 mg/l	Inc.[a]	608 mg/l	1020 mg/l	–	–
Phenol	–	–	–	–	–	3.9 μg/l
Ag	2.6 μg/l	7.7	2.4 μg/l	5.5 μg/l	73.0	1.5 μg/l
As	17.0 μg/l	82.3	3.0 μg/l	3.3 μg/l	27.0	2.4 μg/l
B	300 μg/l	10.0	270 μg/l	1000 μg/l	16.0	840 μg/l
Ba	120 μg/l	Inc.[a]	140 μg/l	81 μg/l	62.0	31.0 μg/l
Cd	5 μg/l	60.0	2 μg/l	29 μg/l	94.0	1.7 μg/l
Cr	27 μg/l	25.9	20 μg/l	154 μg/l	83.0	26.0 μg/l
Cu	29 μg/l	Inc.[a]	46 μg/l	266 μg/l	88.0	32.0 μg/l
Fe	590 μg/l	83.0	100 μg/l	325 μg/l	80.0	66.0 μg/l
Hg	0.16 μg/l	Inc.[a]	0.51 μg/l	9 μg/l	26.0	6.7 μg/l
Mn	41 μg/l	73.1	11 μg/l	35 μg/l	86.0	4.9 μg/l
Pb	34 μg/l	Inc.[a]	35 μg/l	19 μg/l	72.0	5.3 μg/l
Se	2.9 μg/l	69.0	0.9 μg/l	1.8 μg/l	Inc.[a]	1.9 μg/l
Zn	63 μg/l	0	63 μg/l	412 μg/l	57.0	162 μg/l

[a]Inc. = Increase.

(ammonia, phenol and boron), drinking water (ammonia and mercury) and irrigation (boron).

Relatively poor overall removals (< 50%) were noted at Dallas for TOC (44.2%), nitrite and nitrate nitrogen (0%), TDS (increase), silver (7.7%), boron (10%), barium (increase), chromium (25.9%), copper (increase), mercury (increase), lead (increase) and zinc (0%). Orange County reported poor reductions for arsenic (12.1%), boron (16%), mercury (22.8%) and selenium (increase). Comparison of removal percentages for metals, however, may be misleading since with low incoming concentrations the numerical value of the residual metal content becomes increasingly sensitive to solubility product relationships, concentration gradients and the efficiency of suspended solids removal. In many cases residual concentrations approached limits of detection.

As will be subsequently discussed in more detail, the degree of metals reduction is plant specific and dependent on the specific metal, the presence of complexing constituents, oxidation-reduction potential, pH, carbon activity, coagulant employed, ionic makeup of carrier water and other operating and environmental factors.

3. Chemical Pollutant Removal by Physical-Chemical Treatment. Due to a lack of available data, only metals removal efficiency by physical-chemical treatment will be evaluated. Physical-chemical systems as demonstrated by Cohen [11] are generally more efficient than biological treatment in reducing variability and concentrations of metals in effluents. A 4-gpm pilot plant consisting of chemical coagulation and clarification, dual media filtration, and activated carbon was evaluated by spiking raw municipal waste water with elevated metals concentrations. The effect of various coagulants including lime, alum and ferric chloride on metals reduction was investigated and the results are summarized in Table IV.

Physical-chemical treatment is shown to be a most effective series of processes for reduction of metals from wastewaters. Coagulation with one or another of the coagulants followed by activated carbon obtained removal efficiencies > 95% for most metals.

Degree of removal was found to be dependent on the type of coagulant employed. Manganese, for example, was poorly removed by iron salts or alum, but readily precipitated by lime. The influence of influent concentrations was illustrated by selenium data. Removals decreased significantly as the initial Se level was reduced from 0.5 to 0.06 mg/l. It was concluded that although high removal efficiencies were reported, the stringent requirements for aquatic life would not be set for many metals particularly at high influent content. Alternative control measures, for example source control, would be required.

A 1978 review of metals removal by existing tertiary treatment was published by Sorg of U.S. EPA [16]. In this review, the influence of different

Table IV. Removal of Trace Metals by Physical-Chemical Processes [11]

Metal	Lime Precipitation-Activated Carbon			Alum-Activated Carbon			Ferric Chloride-Activated Carbon		
	Initial Concentration (mg/l)	Removal (%)	Residual Metal (µg/l)	Initial Concentration (mg/l)	Removal (%)	Residual Metal (µg/l)	Initial Concentration (mg/l)	Removal (%)	Residual Metal (µg/l)
Ag	0.5	98	10	0.6	99.2	5	0.5	99.1	5
As	5.0	84	800	–	–	–	5.0	97.1	145
Ba	5.0	81	950	0.5	92	40	5.0	95.6	220
Cd	5.0	99.6	20	0.7	55.5	312	5.0	98.6	70
Cr	5.0	98.2	90	$0.7\ Cr^{+3}$	99.1	6	5.0	99.3	35
				$0.7\ Cr^{+6}$	97.4	18			
Cu	5.0	90	500	0.7	98.3	12	5.0	96	200
Hg	0.5	92	40	0.06	98.3	1	0.05	99	1
Mn	5.0	98.2	90	0.7	33	469	5.0	17	4150
Pb	5.0	99.4	30	0.6	96.6	20	5.0	99.1	45
Se	0.5	95	25	0.5	56	220	0.1	80	20
Se	0.06	67	20	–	–	–	0.05	75	13
Zn	5.0	76	1200	2.5	28	1800	5.0	94	300

coagulants, pH and metal species were evaluated for arsenic, barium, cadmium, chromium, mercury, selenium and silver as shown in Table V. As indicated, the removal efficiency of arsenic can range from 95% to < 20% depending on the oxidation state and the coagulant employed. Similar results are observed for most other metals.

Removal of Chemical Pollutants by Selected Unit Operations

To better interpret overall chemical contaminant removal results from AWT and physical-chemical systems, it is necessary to evaluate reduction efficiencies of specific unit operations. The effects of precipitation, filtration and carbon adsorption are evaluated by summarizing data [9,10,14,16,17]. Results from the Hannah and Cohen study [17] were derived from the same system discussed by Cohen and described in the preceding section. Two carbon contactors were used: an "old carbon" having been in operation for about one year and in need of regeneration; the other was designed as "new carbon" characterized by minimal bioactivity.

Tables VI-VIII illustrate percent decreases with corresponding residual parameter concentrations for precipitation, filtration and carbon adsorption processes. Filtration is used as a polishing step to precipitation-clarification and is not effective in soluble reductions except where some organic sorption was observed. Both processes were evaluated as one at the Dallas facility. Great variation in removal efficiency was reported by the EPA *Treatability Manual* due to the great differences in industrial wastewater characteristics, concentrations of specific pollutants, etc. The effectiveness of these unit operations is discussed in detail in the following section.

DISCUSSION OF RESULTS

Results will be interpreted on an individual parameter basis. Reference should be made to Table IX which summarizes pertinent parameter removal efficiencies for selected unit operations and to Tables X-XII which indicate important properties of selected metals and trace organics, respectively.

Total Dissolved Solids

By nature of the treatment process, accumulation of mineral content is inherent in water reuse. Table VI indicates that treatment by the Dallas facility resulted in a net increase in TDS due primarily to lime addition during chemical precipitation. Influent waters sufficiently high in TDS will consequently require suitable blending water sources prior to reuse. Recent experience at Orange County has demonstrated reverse osmosis to be effective and

Table V. Removal of Heavy Metals by Physical-Chemical Treatment Process (% Removal) [16]

Constituent	Lime Coagulation	Alum Coagulation	Ferric Salt Coagulation	Ferrous Salt Coagulation	Ion Exchange or Reverse Osmosis	Carbon Absorption
Arsenic						
As^{+3}	70($>$10.8)[a] $<$20($<$10.8)	60	$<$20	—	55-100	—
As^{+5}	95($>$10.8) 30($<$10.8)	90	90	—	55-100	—
Boron	90(10-11)	20	35	—	—	—
Cadmium(+2)	98	10(7) 40(8)	20(7) 90(8)	—	—	—
Chromium						
Cr^{+3}	90($>$10.6) 70($<$ 9.2)	90(6.7-8.5) 78(9.2)	98	—	—	—
Cr^{+6}	$<$10	10	35	98(6.5-9.3)	—	—
Fluoride	—	—	—	—	—	80
Lead	98(8.5-11.3)	80-90	90	—	—	—
Mercury						
Inorganic	66-80(10.7-11) 30(9.2)	47(7) 38(8)	66(7) 97(8)	—	98	80
Organic	5(9-11)	0-4	0-40	—	98	80
Selenium						
Se^{+4}	45	20(5.5) 32(6.5)	85(5.5) 15(9.2)	—	97	—
Se^{+6}	$<$10	$<$10	$<$10	—	97	—
Silver	90(11.5) 70(9)	70(6-8)	70(6-8)	—	—	—

[a] Residual concentration, μg/l.

Table VI. Removal of Selected Parameters by Precipitation

Removal (%)	Dallas [9] High Lime	Orange County High Lime [10]	Hannah, et al. [17]			EPA Treatability Manual [14]			EPA Treatment Techniques [16]a
			High Lime	Alum	FeCl	Lime	Alum	Fe^{++} + Lime	
<10	TOC (10.9 mg/l) TDS (622 mg/l)b NO$_2$ & NO$_3$ (4.9 mg/l) Pb (34 µg/l)d Ag (4.0 µg/l)b Zn (68 µg/l)b B (290 µg/l) Ba (140 µg/l)b Cu (91 µg/l)b Fe (120 µg/l)	Se (1.8 µg/l)				Se (inc.)	Ag (120 µg/l) Phenol (inc.)c Trichloro-methane (inc.)		Lindane (>9000) Toxaphene (>9000) 2,4-D (>9000)
10-29	NH$_3$-N (4.0 mg/l) Hg (0.139 µg/l)	NH$_3$-N (37 mg/l) TKN (41 mg/l) As (2.5 µg/l) B (840 µg/l)		Zn (1900 µg/l)		Ag (<4 µg/l) Phenol (<10 µg/l)c Ttl Phenol (170 µg/l)	Pb (120 µg/l) As (32 µg/l) Ttl Phenol (56,000 µg/l)	Ag (12 µg/l) Cd (6 µg/l) Se (20 µg/l)	
30-49	TKN (5.9 mg/l) Cd (3 µg/l)	Zn (239 µg/l)	Hg (275 µg/l) Se (320 µg/l)e Se (32 µg/l)	Cd (400 µg/l)		Hg (<76 µg/l) Cd (>9 µg/l)	Zn (3800 µg/l) Cd (>9 µg/l) Cr (95 µg/l) Total Phenol (100 µg/l)c	Hg (<0.2 µg/l)	Endrin (6500)

								2,4,5-TP (3500)
50-69	COD (18 mg/l) Cr (12 µg/l)	COD (52 mg/l) Ba (36 µg/l)	Se (235 µg/l)[e] Cr+6 (273 µg/l)		Se (34 µg/l)[e] Se (16 µg/l)	Pb (51 µg/l) Cr (61 µg/l) As (<16 µg/l)	Hg (1.4 µg/l) Trichloromethane (140 µg/l)	Pb (<3 µg/l) Cr (<3.3 µg/l) As (<2 µg/l)
70-84	As (4.1 µg/l) Mn (11 µg/l) Se (0.5 µg/l)	Hg (2.6 µg/l) Cr (37 µg/l) Cu (73 µg/l) Pb (3.6 µg/l) Ag (0.8 µg/l)	Cu (210 µg/l)		Zn (640 µg/l) Cu (52 µg/l)		Cu (<37 µg/l) Dichloromethane (5600 µg/l)[c]	Zn (12 µg/l) Cu (21 µg/l)
85-100		Cd (2.4 µg/l) Fe (40 µg/l) Mn (4.4 µg/l)	Ag (30 µg/l) Hg (7 µg/l) Ba (65 µg/l) Pb (54 µg/l) Cr+3 (49 µg/l)	Ag (21 µg/l)	Ag (30 µg/l) Hg (4 µg/l)	Fe Mn	Phenol (<5 µg/l) Dichloromethane (<40 µg/l) Chloroform (1800 µg/l)[c]	

[a] Includes effect of filtration.
[b] Increase.
[c] With polymer addition.
[d] Values in parenthesis indicate residual concentration.
[e] Higher initial concentration.

Table VII. Removal of Selected Parameters by Filtration

Removal (%)	Orange County [10] High Lime[a]	Hannah, et al. [17]			EPA Treatability Study [14,16]
		High Lime	Alum	FeCl	
<10	Cr (41 μg/l)[b] Zn (412 μg/l)[b] Fe (207 μg/l)[b] Pb (4.7 μg/l)[b] Mn (6.2 μg/l)[b] Ag (1.3 μg/l)[b] Se (1.8 μg/l)[b] Hg (3.6 μg/l)[b]	Se (325 μg/l)[c] Se (37 μg/l)	Se (260 μg/l)[b] Zn (1900 μg/l)[b] Cu (231 μg/l)[b] Cd (385 μg/l)[b] Ba (105 μg/l)[b] Cr (252 μg/l)		
10-29	COD (45 mg/l) Ba (31 μg/l) Cd (1.8 μg/l) As (1.8 μg/l)			Se (25 μg/l)[c]	Lindane Isomer (55 μg/l)
30-49	Cu (49 μg/l)	Ag (12.5 μg/l) Hg (150 μg/l)	Ag (18.6 μg/l) Hg (3.6 μg/l)	Se (10 μg/l)	Chloroform (110 μg/l) Total Phenol (65,000 μg/l)
50-69			Pb (27 μg/l) Cr^{+3} (17 μg/l)		Dichloromethane (530 μg/l) Lindane Isomer (4 μg/l) Phenol (3400 μg/l)
70-84	NH$_3$-N (6.4 mg/l)			Ag (9 μg/l) Hg (1 μg/l)	Chlorodibromomethane (<10 μg/l)
85-100					Chloromethane (64 μg/l)

[a] Includes any removal or addition due to ammonia stripping and recarbonation.
[b] Increase.
[c] Higher initial concentration.

feasible in removing the dissolved salt content of a portion of the effluent with subsequent blending with the remaining portion following activated carbon treatment [10].

Nitrogen Removal

Even though high overall ammonia removal efficiencies were experienced at Dallas (67%) by the nitrifying activated sludge process and at Orange County (82%) by series operated ammonia stripping towers, residual concentrations remain above the criteria limits set for freshwater aquatic life and groundwater injection. Breakpoint chlorination practiced at Orange County resulted in reduction of ammonia nitrogen to the required 1 mg/l level for injection. Because of high operating cost and the formation of several chlorinated organics not effectively removed by activated carbon, the viability of this alternative is questionable. It should be noted that Orange County performance reflects operation with high influent NH_3-N concentrations of 43 mg/l.

The Blue Plains Treatment System 2 (low lime clarification, dispersed growth nitrification, denitrification, carbon adsorption and mixed media filtration) maintained residual ammonia concentrations within acceptable limits. Based on these data, biological nitrification-denitrification appears to offer a promising alternative for nitrogen removal.

A study by Idelovitch et al. [21] of the role of groundwater recharge in wastewater reuse in Israel indicates nitrification to NO_3 to be complete and reliable; however, denitrification was partial and fluctuating. Hence a concern over nitrate leakage into ground waters exists.

Specific Heavy Metals

1. Arsenic. Efficiency of physical-chemical methods for arsenic removal depends greatly on initial concentration. Precipitation with ferric chloride achieved $\sim$90% removal, and activated carbon increased reductions to 97%. Similar treatment with lime resulted in an 84% overall reduction of arsenic at an initial concentration of 5 mg/l (Table V). Precipitation efficiency was particularly low (24%) at the Orange County AWT plant, however, and carbon adsorption removal was negligible at both Orange County and Dallas. An effluent residual of $\sim$ 3 μg/l was reported by both facilities. Reduced removals are due in part to much lower concentration gradients. These low μg/l residual values are near the detection limits.

Maruyama et al. [22] reported arsenic to form slightly soluble compounds with a number of metals, including iron. Insoluble arsenic trisulfide was found to precipitate by reaction with hydrogen sulfide in acid solution,

Table VIII. Removal of Selected

| Removal (%) | Dallas [9] High Lime | Orange County [10] High Lime | Hannah, et al. [17] | | | | | |
| | | | High Lime | | Alum | | FeCl | |
			Old Carbon	New Carbon	Old Carbon	New Carbon	Old Carbon	New Carbon
<10	TDS (608 mg/l)	As (2.4 μg/l)[a]	Ag (13.5 μg/l)			Zn (1800 μg/l)	Hg (1 μg/l)	Hg (1 μg/l)
	NH$_3$-N (3.6 mg/l)	Ba (31 μg/l)					Se (12 μg/l)	Se (13 μg/l)
	NO$_2$ & NO$_3$	Hg (6.7 μg/l)[a]						
	(5.1 mg/l)[a]	Se (1.9 μg/l)[a]						
	Cr (20 μg/l)[a]	Cd (1.7 μg/l)						
	B (270 μg/l)	Ag (1.5 μg/l)[a]						
	Ba (140 μg/l)	Pb (5.3 μg/l)[a]						
	Hg (0.51 μg/l)[a]							
	Mn (11 μg/l)							
	Pb (35 μg/l)[a]							
	Se (0.9 μg/l)[a]							
	Zn (63 μg/l)							
10-29	TOC (6.7 mg/l)	Mn (4.9 μg/l)		Ag (10.0 μg/l)	Zn (1600 μg/l)	Se (220 μg/l)	Se 22 μg/l)	Se (20 μg/l)
	TKN (4.5 mg/l)				Pb (21 μg/l)	Cd (312 μg/l)		
	As (3.0 μg/l)					Pb (20 μg/l)		
	Fe (100 μg/l)							
30-49	Ag (2.4 μg/l)	NH$_3$-N (3.3 mg/l)	Se (21.2 μg/l)	Se (20 μg/l)			Ag (5 μg/l)	Ag (5 μg/l)
	Cd (2.0 μg/l)	Cr (26 μg/l)						
		Cu (32 μg/l)						
50-69	Cu (46 μg/l)	TOC (6.7 g/l)			Se (90 μg/l)	Ba (40 μg/l)		
		COD (18 mg/l)			Ba (35 μg/l)			
		Zn (162 μg/l)			Cr^{+3} (6 μg/l)			
		Fe (66 μg/l)						
70-84	COD (4 mg/l)		Hg (40 μg/l)	Hg (45 μg/l)	Ag (3 μg/l)	Ag (5 μg/l)		
					Hg (1 μg/l)	Hg (1 μg/l)		
						Cr^{+3} (5 μg/l)		
85-100			Se (20 μg/l)[d]	Se (25 μg/l)[d]	Cu (12 μg/l)	Cu (12 μg/l)		
					Cd (13 μg/l)	Cr^{+6} (18 μg/l)		
					Cr^{+6} (18 μg/l)			

[a] Increase.
[b] Laboratory.
[c] Low dosage: <10 μg/l of carbon.
[d] Higher initial concentration.
[e] High dosage: >50 μg/l of carbon.

Parameters by Carbon Adsorption

	EPA *Treatability Manual* [14]		EPA *Treatment Manual* [16]	
	Organic Removal		Coagulation, Filtration and Carbon Adsorption	
Metals Removal	Powdered	Granular	Powdered	Granular (5-7 min full bed contact time)
Cd (inc.)				
Ag (21 µg/l)				
Pb (46 µg/l) Se (19 µg/l)		Lindane (<1 µg/l)	Lindane (7000 µg/l)	
Zn (40 µg/l) Cu (66 µg/l) Hg (1.6 µg/l) Cr (60 µg/l) As (11 µg/l)		Dichloromethane (140 µg/l)		
	Dichloromethane (162 µg/l)[b] Phenol (95,000 µg/l) Total Phenol (1100 µg/l)	Chloroform (<10 µg/l) Phenol (2200 µg/l)[a] Phenol (0.7 µg/l)	2,4,5-TP (2000 µg/l)[c]	
	Endrin (0.6 µg/l)[b] Phenol (<50 µg/l)[b] Trichloromethane (56 µg/l)[b] Tribormomethane (13 µg/l)[b] Bromodichloromethane (11 µg/l)[b] Chloroform (30 µg/l)[b] Carbon Tetrachloride (17 µg/l)[b]	Total Phenol (23 µg/l) (with activated sludge)	Endrin (1500 µg/l)[c] (200 µg/l)[e] Lindane (100 µg/l)[e] Methoxychlor (<1000 µg/l) Toxaphene (700 µg/l)[e] 2,4-D (1000 µg/l)[c] (200 µg/l) 2,4,5-TP (500 µg/l)[e]	Endrin (<100 µg/l) Lindane (<100 µg/l) Methoxychlor (<100 µg/l) 2,4-F (<1000 µg/l) 2,4,5-TP (<100 µg/l) Toxaphene (<10 µg/l)

Table IX. Pollutant Removal by Wastewater Treatment Processes (Noted in Percent Removal)[a]

Constituent	Secondary Treatment (Biological)	Chemical Precipitation				Activated Carbon Adsorption	Comments on Activated Carbon	Residual Level (μg/l)	General Comments on Removal
		Lime	Ferrous Lime	Ferric Chloride	Alum				
Total Dissolved Solids	P	P	—	—	—	P	—	—	Generally increase TDS with treatment. Reverse osmosis effective in removal.
Ammonia Nitrogen	VG	P	—	—	—	P	—	—	Bionitrification most effective; breakpoint chlorination and stripping towers F to VG
Nitrate Nitrogen	VG	P	—	—	—	P to G	Depends on anaerobic bioactivity	—	Biodenitrification most feasible
Arsenic (As)	P to VG	P to G	G	G	F	P to G	Reacts with sulfides	2 μg/l	Depends on influent level, pH and redox potential
Barium (Ba)	F	P to G	—	—	G	P to G	Due to highly soluble nature	> 30 μg/l	Enhanced precipitation as sulfate concentration increases
Boron (B)	P	P	—	—	—	P	—	> 290 μg/l	Generally negligible
Cadmium (Cd)	P to VG	F to VG	P	—	F	P to VG	Old carbon better	2 μg/l	High removals due to precipitation of sulfide and hydroxide forms

Pollutant								Concentration	Remarks
Chromium (Cr)	F to VG	G	F	VG	G (Cr^{+6}) VG (Cr^{+3})	P to G (Cr^{+3}) VG (Cr^{+6})	Reduction with bioactivity Cr^{+3} less soluble than Cr^{+6}	20 μg/l	Depends on influent level and oxidation state
Copper (Cu)	P to VG	P to G	G	—	G	G to VG	Enhanced sorption with new carbon	> 20 μg/l	Influenced by influent concentration
Iron (Fe)	P to F	P to VG	—	—	—	P to G	Sulfide complexes ppt. but anaerobic bioactivity causes reduction to soluble Fe^{+2}	> 40 μg/l	Depends on influent level, pH and redox potential
Lead (Pb)	F to VG	F to G	G	—	P to VG	P to G		> 3 μg/l	Enhanced precipitation with higher sulfate levels
Manganese (Mn)	F	G to VG	—	—	P	P	Bioactivity on the carbon reduces Mn^{+4} to Mn^{+2} and release	5 μg/l	Depends on pH and redox potential
Mercury (Hg)	P to G	P to G	F	VG	G	P to G	Variability due to biological activity	> 2 μg/l	Removal is a function of pH, initial concentration and degree of complexation
Selenium (Se)	P to F	P to G	P	G	F	P to G	Variability due to highly soluble characteristics	2 μg/l	Depends on influent concentration
Silver (Ag)	P to G	P to VG	P	VG	P to VG	P to G	High affinity for sulfhydryl groups	2 μg/l	Depends on influent level
Zinc (Zn)	F to VG	P to G	G	—	P	P to G	Zinc sulfide ppt	> 12 μg/l	Depends on influent and sulfate levels

Table IX, continued

| Constituent | Secondary Treatment (Biological) | Chemical Precipitation | | | Activated Carbon Adsorption | Comments on Activated Carbon | Residual Level (μg/l) | General Comments on Removal |
		Lime	Ferrous Lime	Ferric Chloride	Alum				
Endrin	P	—			F	VG	Depends on the initial concentration of endrin and waste quality	<10 μg/l	Low affinity for oxidation by chemical means. High affinity for XAD-4 and clays. >99% removal by reverse osmosis and >99% destruction by incineration
Lindane	P	—			P	VG	Similar to endrin but with a lower affinity. Lindane has a high affinity for most solids	<10 μg/l	Oxidized by high ozone levels. High affinity for XAD-4 and clays. Can be incinerated but lower removal noted by reverse osmosis than for endrin
Methoxychlor	P	—			—	VG	Same as endrin, but there is apparently a slightly lower affinity for sorption	<10 μg/l	Has a high affinity for XAD-4 and clays and is destroyed by incineration. This pesticide is slowly reduced by hydrolytic destruction
Toxaphene	P	—			P	VG	Similar to endrin but with better affinity. Biodegration under anaerobic environment	<5 μg/l	Same as endrin, but is not removed by reverse osmosis

2,4-D	VG		—	P	VG	Has a high affinity for carbon and is also removed due to bioactivity	<5 µg/l	Similar to toxaphene except 2,4-D is destroyed rapidly in alkaline waters (95-98%) and is degradable in aerobic environments
2,4,5-TP (Silvex)	G		—	G	VG	Same as 2,4-D	<1 µg/l	Same as 2,4-D
Phenol	P	P	—	G	F	Limited by driving force to ~1 µg/l	<1 µg/l	Only with chemical oxidation are treatment methods effective to the 1 µg/l limit except for land treatment
Halomethane	F	F	—	F	VG (Powder) G (Granular)	In general needs high surface areas to be effective	<15 µg/l	Reverse osmosis is not effective due to low molecular weights. Filters appear to adsorb chloromethane derivatives

aP = poor, <30%; F = fair, 30-60%; G = good, 60-90%; VG = very good, >90%.

Table X. Chemical Properties of Selected Metals [18]

Heavy Metals	Oxidation States	Stable Inorganic Species		Stable Organic Species	Common Complexes	Volatility	Solubility
		Aerobic	Anaerobic				
Boron	+3	$B(OH)_3$ pH <9.2 & $B(OH)_4^-$ pH >9.2	$B(OH)_4^-$ except if BF_3 or boranes ($>B_6H_{10}$) present	$B(OR)_3$, $RB(OH)_2$, $R_2B(OH)$, R_3B (low alkyl boranes are unstable)	$B_3O_3(OH)_4^-$ at pH <7 and [B] >270 ppm	H_3BO_3 (volatile in steam)	H_3BO_3 (49 g/l @ 20°C, 379 g/l @ 100°C)
Mercury[a]	0,+1,+2	$HgCl_2$, pH <8.5 $HgO \cdot H_2O$ pH >8.5	Hg° & HgS_2^{-2} if [S] >32 ppm	$RHgX$ & R_2Hg; CH_3HgS^- $(CH_3Hg)_2S$, $(CH_3Hg)_3S^+$ CH_3HgCl (pH <8 and CH_3HgOH (pH >8)	$HgCl_4^{-2}$, $Hg(NH_3)^{+2}$ & $Hg(CN)^{-2}$	$(CH_3)_2Hg$, Hg°, & CH_3HgX plus most mercury compounds	Hg° (30 mg/l @ 25°C) $HgO \cdot H_2O$ (40 g/l @ 25°C)
Cadmium[a]	0,+2 (+1)[b]	Cd^{++} pH <8.5 $Cd(OH)^+$ pH >8.5	Cd° (very unstable)	R_2Cd; $RCdX$ & alkyl Cd (unstable); Cd complexes with organic acids, amines and sulfides)	$CdCl^+$, $Cd(NH_3)_4^{+2}$ & $Cd(CN)_4^{-2}$	Cd° & CH_3Cd X (decomposes in the presence of air)	$Cd(OH)^+$ (3 mg/l @ 20°C)
Zinc[a]	0,+2	Zn^{++}, $Zn(OH)^+$, $Zn(OH)_2$, $Zn(OH)_3^-$ & $Zn(OH)_4^{-2}$ (with → pH)	ZnS and Zn° (very unstable)	R_2Zn and $RZnX$; alkyl Zn (unstable)	$Zn(NH_3)_4^{+2}$ & $Zn(OH)_4^{-2}$	—	$Zn(OH)_2$ (3 mg/l @ 20°C)
Arsenic[c]	−3,0,+5	$H_3AsO_4 \to AsO_4^{-3}$ $H_3AsO_3 \to AsO_4^{-3}$ (with → pH)	As° and AsH_3	All oxidation states form stable organic compounds	Metals complex with arsenic	As°, AsH_3, & $(CH_3)_3As$	As (+5) → As (−3) (>100 g/l → μg/l)

Silver[a]	0,+1 (+2,+3)	Ag^+ pH <12 $Ag(OH)_2^-$ pH >12	$Ag°$	Ag organic olefins and aromatics (very stable and water soluble)	$AgCl_2^-$ & $Ag(NH_3)_2^+$	—	AgCl (0.9 mg/l) Ag_2O (1.2 mg/l)
Selenium[a]	$-2,0,+4,$ $+6$ $(+2)$	$H_3SeO_4 \rightarrow SeO_4^{-3}$ $H_3SeO_3 \rightarrow SeO_3^{-3}$ (with $\rightarrow$ pH)	H_2Se and $Se°$	Only Se(−2) forms organic compounds (generally decompose to Se°); stable organo selenium amino acids	Metals complex with selenium	$Se°$, H_2Se, & $(CH_3)_2$ Se	Se (+6)$\rightarrow$ Se (−2) (>100 g/l$\rightarrow$ μg/l)

[a]Pearson soft acid; [b]Parenthesis indicate unstable in water system; [c]Pearson hard acid.

Table XI. Physical and Chemical Properties of Selected Organics [14,19,20]

Properties/ Constituents	Physical Characteristics	Melting Point (°C)	Solubility (μg/l @ 25°C)	Volatility Vapor Pressure (mm of Hg @ 25°C)	Octanol Water Coefficient	Degradability	Comments on Chemical Reactivity
Endrin	Solid White crystalline	226-230 with decomposition	Insoluble in water. Soluble in alcohols, petroleum ether, acetone and benzene	2×10^{-7}	400,000	Degradation noted under low pH and in the presence of zinc	Stable to alkali and acids, but strong acids or heating > 200°C causes rearrangements
Lindane	Solid Brownish to white crystals	65-159 depending upon purity	10-32 in water Soluble in nonpolar solvents	9.4×10^{-6} Phosphene-like odor	5,250	10% within six weeks by soil microbes and same as endrin	Reactive with alkalies to form trichlorobenzenes from the B isomer
Methoxy-chlor	Solid Gray crystals	77-89 depending upon purity	0.62 in water. Soluble in ethanol, petroleum ether, kerosene and dichlorobenzene	—	—	Degraded in anoxic soils and same as endrin	Decomposes in the presence of light and alkaline solutions
Toxaphene	Solid Amber waxy	65-90 depending upon purity	0.4 to 0.3 in water Readily soluble in nonpolar solvents	0.4 φ 25°C = 1.64 mild terpene odor	825-3,300	Degradable in anaerobic environments but not aerobic. Same as endrin	Dehydrochlorinate upon prolonged exposure to sunlight, alkali, or temperature above 120°C
2,4-D	Solid White crystalline	135-141 depending upon purity	900 in water, $\sim 1 \times 10^{6}$ in acetone and ethanol	—	—	Degraded in soils under aerobic conditions	Rearranges in the presence of light

2,4,5-TP (Silvex)	Solid White crystalline	154-155	238 in water at 38°C, 590 in ethanol at 50°C. Soluble in petroleum ether	—	—	Degraded in soils under aerobic conditions	Rearranges in the presence of light
Phenol	Solid Clear, colorless, hydroscopic deliquescent	43	67,000 to complete solubility from 16-66°C. Soluble in nonpolar solvents	0.35 Specific gravity = 1.071	30	Degraded acclimatized microbes	pKa = 9.9 – 10.0. Forms phenoxides by hydrolytic oxidizing reactions. Reacts with many reduced compounds and is chlorinated

Table XII. Physical and Chemical Properties of Halomethanes[a] [14,19]

Properties/ Constituents	Physical Characteristics (all colorless)	Melting Point (°C)	Boiling Point (°C)	Solubility in water (µg/l@25°C)	Solubility for Organic Solvents	Octanol Water Coefficient	Specific Gravity at 20°C	Vapor Pressure (mm Hg @ 20°C)
Chloromethane	Gas	-97.7	-24.2	5,380	alcohol, ether, acetone, benzene, chloroform, acetic acid	8	0.973 @ 10°C	—
Bromomethane	Gas	-93.6	3.6	1,000	alcohol, ether, acetic acid	13	1.737 @ 10°C	—
Dichloro-methane	Liquid	-95.0	49.8	16,700	alcohol, ether, acetic acid	18	1.327	362
Trichlorofluoro-methane	Liquid	-111.0	23.8	1,100	alcohol, ether, acetic acid	340	1.467	667
Dichlorodi-fluoromethane	Gas	-158.0	-30.0	280	alcohol, ether, acetic acid	140	1.75 @ 115°C	4,306
Tribromo-methane	Liquid	8.3	149.5	slightly soluble	alcohol, ether, chloroform ligroin	200	2,890	10 @ 34°C
Bromodichloro-methane	Liquid	-57.1	90.0	Insoluble	alcohol, ether, acetone, benzene, chloroform	76	1,980	50

Trichloro-methane	Liquid	−63.5	61.7	9,600	alcohol, ether, acetone, benzene, chloroform	93	—	150
Tetrachloro-methane	Liquid	−22.9	76.5	800	alcohol, ether, acetone, benzene, chloroform	440	—	90
Chlorodi-bromomethane	Liquid	−20.0	120.0	—	alcohol, ether, acetone, benzene, chloroform	120	—	15 @ 10.5°C

[a]Chemical characteristics of halomethanes are: 1. slowly hydrolyzed in neutral waters forming methanol and hydrogen halide; 2. monohalomethanes are not oxidized readily under ambient conditions (25°C and 1 atmosphere of pressure); 3. photodecomposition occurs more readily in the following order: I > Br > Cl > F.

but readily dissolve in basic solutions. Iron precipitation was recommended for arsenic reduction.

An important phenomenon affecting arsenic removal is its oxidation state. Arsenic (+5) tends to precipitate more readily in the presence of polyelectrolytes than does the less soluble arsenic (+3) as noted by Sorg [16] in Table V. With respect to adsorption, arsenic (+3) has a higher affinity (by 2 orders of magnitude) to adsorb onto clays and minerals [22]. Arsenic (+3) only appears to react with organosulfhydryl groups, and arsenic (+5) reacts only with amine groups [23]. The oxidation state of arsenic therefore influences its precipitation, solubility, adsorption and reactivity with organics.

2. Barium. The effectiveness of physical-chemical treatment processes for barium removal increases with greater influent barium and sulfate levels. Barium forms slightly soluble barium carbonate and barium sulfate complexes. The hydroxide is relatively soluble and not thought involved in the removal process. Low lime, iron and alum precipitation was found efficient in removing barium by 99.5, 95 and 92%, respectively. High lime addition resulted in lower reductions (81% at an initial concentration of 5 mg/l) due to the formation of relatively soluble barium hydroxide.

Removal of barium by AWT processes was not as promising as indicated by the above results. Reductions varied from < 10% at Dallas to 56% at Orange County by high lime precipitation. Activated carbon treatment resulted in negligible removals (< 10%) as predicted by Sigworth and Smith [24] due to the highly soluble nature of barium. Enhanced precipiation reductions at Orange County are attributed to the relatively high influent sulfate concentration of 334 mg/l compared to 98 mg/l at Dallas.

3. Boron. The highly soluble and complexing properties of boron characterize it as a persistent element through both biological and chemical processes. Data indicate negligible removals through both the Dallas and Orange County AWT facilities. Residual effluent concentrations were 290 μg/l (< 10% removed) and 840 μg/l (16% reduction), respectively.

Idelovitch [21] has observed boron to be effectively adsorbed and coprecipitated with $Mg(OH)_2$ during high lime-magnesium treatment. The boron adsorption capacity of the soil-aquifer system was found very low (exceeded when effluent boron was 0.3 mg/l).

4. Cadmium. Physical-chemical processes employing lime and ferric chloride as coagulants experienced high removals of cadmium (99.6 and 98.6% respectively). When alum was employed the removal efficiency decreased to 55%. Best removals were observed for the high lime system with aged carbon. This result was not unexpected, since cadmium forms an insoluble hydroxide at high pH and is precipitated by sulfide over a broad pH range. Cadmium may form soluble complexes with ammonia and cyanide which can reduce precipitation removal efficiency. High lime precipitation followed by carbon

adsorption at the AWT facilities resulted in what appears to be a minimum cadmium solubility for these processes at approximately 2 μg/l.

5. Chromium. Removal of chromium depends primarily on its oxidation state and initial concentration. Trivalent chromium behaves much differently than the hexavalent form and is much less soluble. Hydroxide ions readily react with Cr^{+3} to precipitate hydrous oxide which is amphoteric dissolving in excess base. As expected, good removals of chromium ($>$ 50%) were observed by precipitation in physical-chemical and AWT systems. Filtration effectively removed trivalent chromium, but the soluble Cr^{+6} was relatively unaffected. Chromium removal efficiencies decreased with lower influent concentrations as illustrated by comparing AWT and physical-chemical results (except for alum coagulation). Carbon adsorption was effective in reducing Cr^{+6} content by reduction to the less soluble trivalent state. Reduction is probably due to chemical and biological activity within the carbon column. Data appear to indicate that below a residual of 20 μg/l, the sorption of chromium becomes negligible.

6. Copper. Good removals were experienced in removing copper from wastewaters by physical-chemical processes. Chemical precipitation using iron and low lime resulted in 96 and 93% reductions, respectively, with influent concentrations of 5 mg/l. High lime removals were somewhat erratic, averaging $\sim$ 90%. Alum was the least effective of the coagulants employed, resulting in only a 70% reduction at an influent concentration of 0.7 mg/l. Activated carbon appeared more efficient with alum addition increasing removals to 98%. New carbon exhibited better removal capacity than aged carbon.

The AWT facilities while characterized by lower removal efficiencies also produced lower residual metals content which again indicates the significance of influent concentration and minimum solubility levels. High lime precipitation indicated copper to have a limiting solubility of approximately 70 μg/l at a pH of 11.5. Data also indicate some affinity of copper for carbon in AWT systems ($>$ 35% removal) contrary to other findings [17]. Enhanced sorption could be due to the formation of complexes which may be sorbed to the carbon or biological growth within the column. Efficient residual levels of as low as 20 μg/l have been reported for specific industrial waste treatment facilities [14].

7. Iron. High levels of residual iron in treated effluents are primarily the result of $FeCl_3$ or $FeSO_4$ addition to enhance sedimentation. Chemical precipitation can be effective in reducing iron content depending on pH, oxidation-reduction potential and initial concentration. Iron can easily be precipitated as ferric hydroxide at high pH under aerobic conditions. High lime coagulation at the AWT facilities resulted in residual iron levels of 120 μg/l and 40 μg/l for Dallas and Orange County, respectively. Bioactivity within carbon columns can reduce insoluble Fe^{+3} to soluble Fe^{+2} causing release. Added complexity results, however, since iron can react within the

column to form insoluble iron sulfides. More than 68% of iron was removed by carbon adsorption at Orange County, whereas less than half that efficiency was observed at the Dallas facility.

8. Lead. Physical-chemical treatment gave excellent removals of lead yielding $\sim$99% reduction at an initial level of 5.0 mg/l. At reduced concentrations, however, variable efficiencies were observed. Dallas and Orange County reported overall removals of < 10% and 72%, respectively. Fluctuations may be due to organic complexation, chemical speciation and solubility limitations. Lead also forms insoluble salts with sulfate, carbonate and sulfides which may account for the enhanced Orange County removals by precipitation (81%) since this wastewater was characterized by a higher initial sulfate content. Negligible removals were observed by activated carbon treatment. Lead solubility is controlled by $PbSO_4$ at low pH and by $PbCO_3$ at high pH.

9. Manganese. The effect of coagulant selection on manganese removal was demonstrated by Cohen [11] where high removals with lime (> 90%) but poorer reductions with alum and $FeCl_3$ (< 3%) were observed during precipitation. Inorganic manganese salts are soluble in acid and neutral solution but form insoluble manganous hydroxide precipitates with increasing pH. Under alkaline and aerobic conditions Mn^{+2} will be oxidized and converted to insoluble manganese oxide and hydroxide. Precipitation results are therefore as expected. A residual concentration of $\sim$5 μg/l appears to represent the lowest concentration effectively removed by adsorption. The reducing conditions within the carbon column may cause reduction to the soluble divalent manganese ion with subsequent release.

10. Mercury. Physical-chemical removal of mercury is influenced by pH, initial mercury content and degree of complexation. Precipitation in physical-chemical plants was found very effective (> 90% removal) when alum and ferric chloride were used as coagulants. When lime was employed, however, removal decreased to 45% at an initial mercury level of 500 μg/l. AWT plants indicated low and inconsistent reductions through each unit operation at the low residual concentrations present (< 5 μg/l). At pH values > 8.5 the formation of relatively soluble mercuric oxide may result in poorer removals. Filtration can be used to enhance recovery since most mercury precipitates are not easily removed by clarification. Sorption efficiency of activated carbon drops markedly as the influent level decreases to < 5 μg/l. Due to its nonionic properties and its affinity for sulfhydryl groups carbon adsorption may be effective in mercury removal. Bioactivity resulting in methylation or demethylation will reduce adsorption capacity however. As indicated by Table IV, mercury was not removed to any significant degree at either the Dallas or Orange County facilities. Residue Hg concentrations in specific industrial waste were reported as low as 2 μg/l (14).

11. Selenium. None of the coagulants tested were highly effective in removing selenium by coagulation-filtration. Activated carbon performance was unpredictable. Good removals (except for alum) were attained with sorption until influent carbon levels of selenium decreased to < 20 $\mu g/l$. Higher effluent concentrations were observed when alum was employed due to the formation of selenate complexes or perhaps competitive sorption of the chemically similar sulfur compounds present.

The oxidation state of selenium plays a major role in determining the efficiency of unit operation removal efficiency. This is especially in evidence where coagulation is employed. From Tables VI and VIII, the more reduced form of selenium is 10-30% less soluble depending upon the coagulant [15] employed.

Idelovitch [21] found the unsaturated soil mantle and aquifer effective in reducing selenium levels from 6 to 9 $\mu g/l$ to < 1-2 $\mu g/l$ presumably due to chemical precipitation and adsorption.

12. Silver. Physical-chemical treatment of elevated silver concentrations (0.5 mg/l) produced very high removals of silver ($> 90\%$) independent of the coagulant used. At the lower influent levels experienced during AWT systems operation removals were variable. Residual concentrations, however, were < 4 $\mu g/l$. Carbon adsorption proved fairly effective in all systems studied (except for Orange County where low residuals of 1.5 $\mu g/l$ existed) due to the relatively high affinity of silver for reduced sulfhydryl groups. Data indicate that for the processes and conditions evaluated herein an apparent limiting silver solubility of ~ 2 $\mu g/l$ exists.

13. Zinc. The amphoteric properties of zinc affect precipitation results. Using alum as a coagulant resulting in a pH reduction (6-7) with only 25% of the zinc being precipitated. $FeCl_3$ application resulted in a 63% removal, whereas with low lime addition better removals were observed in the physical-chemical system (94%) but more variability in effluent zinc content was noted due to the formation of soluble zinc complexes [12]. At lower influent concentrations experienced by the AWT facilities (63 $\mu g/l$ and 300 $\mu g/l$ for Dallas and Orange County, respectively) overall plant removals ranged from 0 to 59%. None of the unit operations were significantly effective in reducing zinc content except at the Orange County plant where carbon adsorption provided a 61% reduction. This may be due to zinc sulfide precipitation in the carbon columns. Specific industrial waste treatment effluents contained zinc at levels as low as 12 $\mu g/l$ [14].

Trace Organics

Residual COD and TOC concentrations often are used as indicators of trace organic effluent content. As shown in Tables VI and VIII, high removals (> 50%) of COD were experienced at both Dallas and Orange County by precipitation and carbon adsorption methods. While TOC removal efficiency varied through each facility, residual concentrations of 6.7 mg/l TOC were identical. The CCE content is also sometimes employed as a measure of trace organics. This too does not represent any indication of individual organics which may or may not be of toxicological significance. Due to the inaccuracy inherent in the CCE procedure and the questionable significance of the values obtained, this parameter was dropped as a potable water quality parameter.

Besides clarification and activated-carbon adsorption, recent work by McCarty et al. [10] at the Orange County facility demonstrates that air stripping for ammonia removal is also highly effective in removing a wide range of highly volatile, low molecular weight refractory organics. These organics are relatively nonpolar and are not easily removed by activated carbon or reverse osmosis. Recent work by McCarty et al. [25] has noted poor removals of trace organic compounds having molecular weights < 200 atomic mass units and nonionic by reverse osmosis; higher molecular weight organics are effectively removed. It should be emphasized that the specific membrane employed will have a major influence on the effectiveness of the reverse osmosis process.

1. Phenol. Phenolic compounds in levels exceeding the 1 μg/l standard were found in all effluents studied. Biological-AWT methods as currently employed appear to be inadequate in removing phenol to the 1 μg/l requirement. A recent EPA study evaluating the fate of priority organic pollutants observed only phenol to pass through the two wastewater plants tested [26]. For industrial waste treatment systems, alum precipitation removed 80 to > 90% of phenol with residues < 5 μg/l. Carbon adsorption was effective depending upon the industry due to the great variability of raw wastewater quality [27]. In general, both chemical oxidation and land treatment of tertiary treated wastewater has been found reliable for phenol removals below the 1 μg/l limit [21,27].

2. Endrin. The removal of endrin wastewaters was found to be only effective by adsorptive processes with high sorptive capacities (> 99%) being demonstrated by carbon, clays and XAD-4 resin. The efficiency of carbon adsorption treatment was found to decrease with lower influent levels. The resulting effluent quality, however, increases with lower influent concentrations of endrin. Coagulation, chemical oxidation and biological waste treatment were all ineffective in removing endrin from wastewaters [14]. Reverse osmosis has been found effective in reducing endrin concentrations

by > 99% [16,24,28]. Lowest achievable residual levels are estimated at < 10 μg/l. More data, however, are needed to delineate this value with respect to influent water quality.

3. Lindane. As with endrin, lindane is ineffectively removed by coagulation and biological waste treatment but has a high affinity for activated carbon, XAD-4 resin, filter media and clays. Granular activated carbon capacities are similar as for endrin; however, lindane exhibited a lower affinity for powdered carbon than endrin at low dosages (30-85% removal respectively). Oxidation by ozone (O_3-) at a dose of 38 mg/l reduced lindane by 55%. Due to its lower molecular weight, reverse osmosis was not as effective as for endrin (73% vs 99% removal). Effluent lindane residue levels are estimated to be < 10 μg/l. More data, however, are required under various conditions.

4. Methoxychlor. There is little information concerning the removal of methoxychlor by wastewater treatment processes. As expected, methoxychlor displayed a high affinity (> 99%) for activated carbon, clays and XAD-4 resin. Biological processes were found to have little effect on the removal or degradation of methoxychlor except under anaerobic conditions. Methoxychlor has potential for degradation and destruction in the presence of ultraviolet light and alkaline solutions [19,20]. Since methoxychlor has similar properties to the previous two pesticides, its residual levels are estimated to < 10 μg/l.

5. Toxaphene. The removal of toxaphene has been observed by adsorptive processes. Toxaphene has demonstrated a high affinity (> 99%) for activated carbon (93% for powdered carbon at low dosages and > 99% for granular carbon), for clays and for synthetic resins such as XAD-4 as noted in the literature [20]. Recent studies have shown a low efficiency of toxaphene removals by coagulation, filtration, chemical oxidation and biological waste treatment [19]. Even though toxaphene is of higher molecular weight than endrin, the reverse osmosis process was found to have no noticeable ability to remove toxaphene [20]. This low affinity could be due to a different waste and RO membrane. Since toxaphene appears to have a very high sorptive affinity, its residual concentration is estimated at less than 5 μg/l. As with the other pesticides, this value should be ascertained from laboratory and field verification.

6. 2,4-D. Due to its chemical structure, 2,4-D is one of the least stable of the six pesticides elucidated. Recent studies have noted considerable reductions of 2,4-D by activated carbon sorption, biological waste treatment processes and alkaline digestion processes [28]. In these processes, removals were found to be > 95% but subject to fluctuations in waste characteristics. Filtration, coagulation and chemical oxidation have been observed ineffective. (< 10%) for 2,4-D removal [16]. Since 2,4-D is readily reduced by several unit operations, the residual level is estimated at < 5 μg/l.

7. 2,4,5-TP (Silvex). As with 2,4-D, 2,4,5-TP being of similar structure has similar characteristics. For example, 2,4,5-TP can be readily reduced (> 95%) by activated carbon adsorption, biological waste treatment processes and alkaline digestion but is resistent to chemical oxidation [28]. The major difference was noted for coagulation and filtration processes which had > 60% reductions of 2,4,5-TP (probably due to sorption on the filter media) [16]. Since 2,4,5-TP is readily reduced by various processes, the residual concentration is predicted to be ~ 5 μg/l, but this value should vary depending on a number of variables.

8. Halomethanes. With recent health concern over the sublethal effects due to halogenated methanes, removal potential of various unit operations becomes increasingly important. In general, halomethanes have been found to be removed by biological processes anywhere from 0 to 100%, generally in the range of ~ 40-55% reduction. Removal by filtration was from 0 to 94% depending upon the initial level and type of halomethanes and specific chemical additive. The most promising removal techniques were found to be ammonia stripping (67-94% removal) and powdered activated carbon adsorption (84-99% removal). The granular carbon was not as effective because there was less adsorbing surface area and these compounds were of lower molecular weights (< 200 atomic mass units). These low weights make the potential of removal by reverse osmosis ineffective [14]. Data appear to indicate an effluent residual concentration of ~ 15 μg/l. As with the other trace organics, this value requires verification.

CONCLUSIONS

Data from biological, tertiary and physical-chemical treatment processes illustrate the necessity of a systems approach in design of wastewater renovation facilities. To attain maximum effectiveness and reliability, the treatment system must consist of several unit operations in series with overlapping functions that rely on different and unrelated removal mechanisms. Biological treatment, for example, while not particularly effective in reducing pollutants to trace levels, can decrease the amount of pollutant to be handled by subsequent tertiary treatment, and very importantly, minimize influent variability by providing buffer capacity. The coagulant selected in the treatment sequences can greatly affect removal efficiencies by coagulation and subsequent treatment processes. High lime was found to be most effective for removal of various metals studied. The recommended coagulant, however, should be metal and system specific.

Chemical parameters found to consistently exceed water quality criteria for intended use included: total dissolved solids, nitrogen (ammonia and nitrate), phenol, trace organics and specific heavy metals. Total dissolved solids were not removed to any appreciable extent by the treatment sequences evaluated except for the reverse osmosis facility operated at Orange County. Ammonia nitrogen was removed to a level ≤ 1 mg/l only when breakpoint chlorination polished stripping tower effluent or when biological nitrification-denitrification was employed. Concern, however, exists over health effects of chlorinated organics generated by the breakpoint chlorination process. Significant organic removal was effected by chemical precipitation, carbon adsorption and ammonia stripping. The air stripping process was found highly effective in removing a wide range of highly volatile, low molecular weight refractory organics not easily removed by activated carbon or reverse osmosis. None of the treatment sequences evaluated appeared capable of reducing phenol concentrations to a 1 µg/l limit. The preceding results underscore the importance of suitable blending water sources prior to reuse.

In general AWT and physical-chemical facilities produced effluents of excellent metals quality, especially with respect to potable water criteria. Metals found to persist through the treatment scheme evaluated included boron, mercury, selenium and zinc. Removal of many metals is inconsistent due to metal concentration, nature of carrier water, etc. Residual concentrations of boron and zinc exceeded recommended criteria for irrigation and aquatic life respectively. Data appeared to indicate minimum metals residual concentrations obtainable by the processes studied. Residuals of arsenic, cadmium, chromium, mercury, selenium and silver were found to be 2, 2, 20, > 2, 2-20, and 2 µg/l, respectively. It should be noted that most of these values are near their respective limits of detection. These values exceed potable water standards for selenium and mercury; ground water injection recommendations for selenium; and aquatic life criteria for cadmium, mercury, selenium and silver. The importance of adequate dilution capacity of receiving water is therefore self-evident.

A significant decrease in removal efficiency was observed with reduced influent metals concentration. It should be noted, however, that comparison of removal percentages may be misleading since with low incoming concentrations of numerical value of the residual metals content becomes increasingly sensitive to solubility product relationships, concentration gradients and the efficiency of suspended solids removal. High levels of metals could result in unacceptable effluent concentrations of specific metals, particularly in plants experiencing significant industrial input. Alternative methods of reduction such as source control will be required for such cases.

It is noteworthy that industrial waste treatment facilities oftentimes yielded better percent removals and lower effluent residual concentrations that combined treatment systems. Increased biological removals (percent) were often observed for arsenic, cadmium, chromium, copper, lead and zinc. This is probably attributed to the fact that for most cases higher influent concentrations were experienced. Lower residual concentrations for pure industrial wastewater effluents of arsenic (2 vs 3 μg/l), copper (20 vs 70 μg/l), lead (> 3 vs > 5 μg/l), mercury (> 2 vs 5 μg/l) and zinc (> 12 vs > 60 μg/l) were indicated by data comparison. This is not surprising since many industrial wastes require specific treatment methods which may be subject to fewer variables than mixed waste combinations and hence fewer interferences due to complexation, competing reactions, etc., may be achieved.

Selected pesticides, phenol and halomethanes were removed fairly readily by activated carbon adsorption (generally > 90% reduction). Other unit processes (biological secondary treatment, chemical precipitation and flotation), however, were in general ineffective (< 30%) in terms of removal. Exceptions included reverse osmosis (removed organics > 200 atomic mass units) and ammonia stripping (removed organics < 200 atomic mass units). Effluent residual levels for organics are not well defined. It is therefore recommended that a study be initiated to monitor various industrial, combined and municipal plants in order to ascertain residual levels for trace organic constituents.

REFERENCES

1. Englande, A. J., J. K. Smith and J. N. English. "Potable Water Quality of Advanced Wastewater Treatment Plant Effluents," paper presented at the IAWPR Workshop in Advanced Treatment and Reclamation of Wastewater, South Africa, 1977.
2. Idelovitch, E. "Wastewater Reuse by Biological-Chemical Treatment and Groundwater Recharge," paper presented at the Water Pollution Control Federation Meeting, Philadelphia, PA, October 2-7, 1977.
3. Hahingh, W. H. J. "Reclaimed Water: A Health Hazard?," *Water South Africa* 3(104) (1977).
4. "National Interim Primary Drinking Water Regulations," *Federal Register,* 40(248):59565 (December 24, 1975) with amendments and corrections through August 27, 1980.
5. "Quality Criteria for Water," U.S. EPA Report-440/9-76-023 (1976).
6. "Water Quality Criteria 1972," in *Ecological Research Series,* a report of the Committee for Water Quality Criteria, Washington, DC, 1972.
7. Argo, D. G. "The Cost of Water Reclamation by Advanced Wastewater Treatment," paper presented at the Water Pollution Control Federation 51st Annual Conference, Anaheim, CA, October 1-6, 1978.
8. Smith, J. K., A. J. Englande, Jr., M. M. McKown and S. C. Lynch. "Characterization of Reuseable Municipal Wastewater Effluents and Concentration of Organic Constituents," U.S. EPA Report-600/2-78-016)(1978).
9. *Municipal Wastewater Reuse News.* American Water Works Research Foundation 6:6, Denver, CO (March 1978).
10. McCarty, P. L., M. Reinhard, C. Dolu, H. Nguyen and D. G. Argo. "Water Factory 21: Inorganic, Organic and Biological Quality of Reclaimed Wastewater and Plant Performance," Department of Civil Engineering, Stanford University, Technical Report No. 226 (January 1978).

11. Cohen, J. M. "Trace Metal Removal by Wastewater Treatment," U.S. EPA Technology Transfer (January 1977).
12. Kremer, J. G., and J. D. Thomas. "The Industrial Waste Program of the Los Angeles County Sanitation Districts," *Water Poll. Control Fed. Highlights* 14 (August 1977).
13. "State and Local Pretreatment Programs," U.S. EPA Report-430/19-76-0176 (1977).
14. *Treatability Manual,* U.S. EPA-60018-80-042 a,b,c,d,e), Office of Research and Development (July 1980).
15. Argo, D. G. Orange County Water District, Personal communication (January 1979).
16. Sorg, T. J. "Treatment Techniques for the Removal of Inorganic Contaminants from Drinking Water," U.S. EPA Report-600/8-77-005 (1978).
17. Hannah, S. A., and J. M. Cohen. "Removal of Uncommon Trace Metals by Physical and Chemical Treatment Processes," *J. Water Poll. Control Fed.* 49:2297 (1977).
18. Reimers, R. S., and A. J. Englande. "Trace Metals Removal by Selected Wastewater Treatment Unit Operations," *The National Conference in Hazardous and Toxic Wastes,* Newark, NJ (June 1980).
19. Sittig, M. "Priority Toxic Pollutants-Health Impacts and Allowable Limits," Noyes Data Corporation, Park Ridge, NJ (1980), p. 370.
20. Sanborn, J. R., R. L. Metcalf and B. M. Francis. "The Degradation of Selected Pesticides in Soil: A Review of the Published Literature," U.S. Department of Commerce National Technical Information Service, PB-272-353 (August 1977).
21. Idelovitch, E., R. Terkeltoub and M. Michael. "The Role of Groundwater Recharge in Wastewater Reuse: Israel's Dan Region Project," *J. Am. Water Works Assoc.* 72:7:391-400 (1980).
22. Maruyama, T., S. A. Hannah and J. M. Cohen. "Metal Removal by Physical and Chemical Treatment Processes," *J. Water Poll. Control Fed.* 47:5: 962-975 (1975).
23. Reimers, R. S., et al. "The Effectiveness of Land Treatment for Removal of Heavy Metals," 31st Purdue Industrial Waste Conference, West Lafayette, IN (May 1976).
24. Sigworth, E. Z., and S. B. Smith. "Adsorption of Inorganic Compounds by Activated Carbon," *Taste Odor J.* 38 (3rd quarter):1-7 (1972).
25. McCarty, P. L., et al. "Advanced Treatment of Wastewater Reclamation at Water Factory 21," Department of Civil Engineering, Stanford University, Technical Report No. 236 (January 1980).
26. Feiler, H. "Fate of Priority Pollutants in Publicly Owned Treatment Works-Pilot Study," Effluent Guidelines Division, Water Planning and Standards, Office of Water and Waste Management, U.S. EPA Report-1-79-300 (1979).
27. Patterson, J. W. *Wastewater Treatment Technology* (Ann Arbor, MI: Ann Arbor Science Publishers, Inc., 1975), p. 265.
28. Hackman, E. E. *Toxic Organic Chemicals-Destruction and Waste Treatment,* Noyes Data Corporation, Park Ridge, NJ (1978), p. 356.
29. Liptak, B. G. *Environmental Engineers' Handbook, Vol. 1, Water Pollution* (Radner, PA: Chilton Book Company, 1974), p. 178.